Testing Procedures are provided at the end of some chapters to outline potential steps and procedures for conducting many of the tests mandated by NFPA 25. It is important to note that these processes and procedures are not part of NFPA 25. The actions identified and the steps provided in these procedures represent one approach to conducting a test required by this standard.

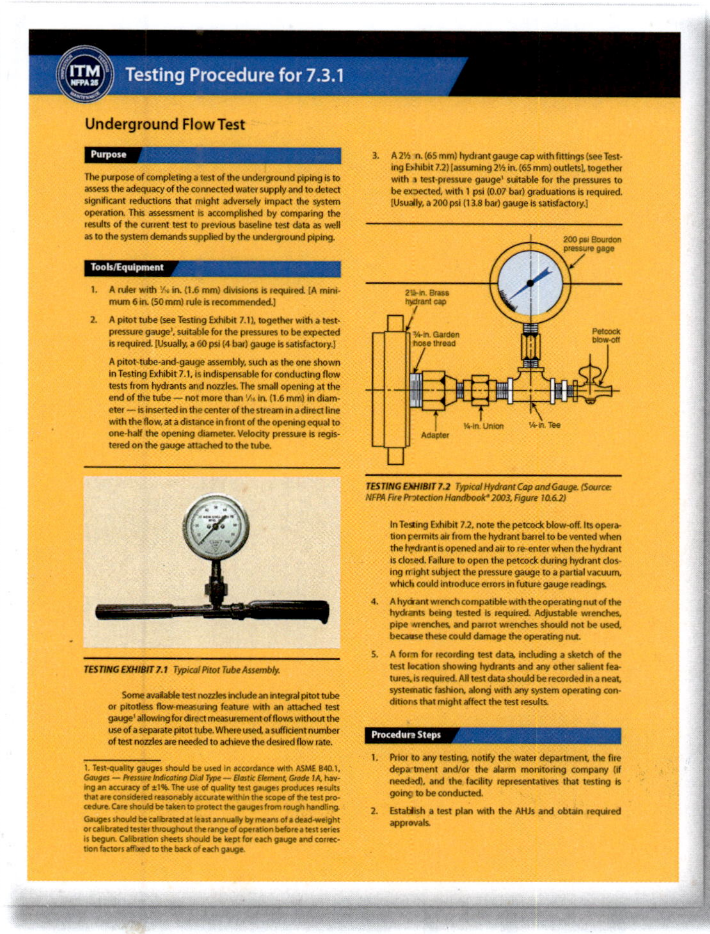

Testing Procedure for 7.3.1

Underground Flow Test

Purpose

The purpose of completing a test of the underground piping is to assess the adequacy of the connected water supply and to detect significant reductions that might adversely impact the system operation. This assessment is accomplished by comparing the results of the current test to previous baseline test data as well as to the system demands supplied by the underground piping.

Tools/Equipment

1. A ruler with 1/16 in. (1.6 mm) divisions is required. [A minimum 6 in. (50 mm) rule is recommended.]
2. A pitot tube (see Testing Exhibit 7.1), together with a test-pressure gauge[1], suitable for the pressures to be expected is required. [Usually, a 60 psi (4 bar) gauge is satisfactory.]

   A pitot-tube-and-gauge assembly, such as the one shown in Testing Exhibit 7.1, is indispensable for conducting flow tests from hydrants and nozzles. The small opening at the end of the tube — not more than 1/16 in. (1.6 mm) in diameter — is inserted in the center of the stream in a direct line with the flow, at a distance in front of the opening equal to one-half the opening diameter. Velocity pressure is registered on the gauge attached to the tube.

TESTING EXHIBIT 7.1 Typical Pitot Tube Assembly.

Some available test nozzles include an integral pitot tube or pitotless flow-measuring feature with an attached test gauge[1] allowing for direct measurement of flows without the use of a separate pitot tube. Where used, a sufficient number of test nozzles are needed to achieve the desired flow rate.

3. A 2½ in. (65 mm) hydrant gauge cap with fittings (see Testing Exhibit 7.2) [assuming 2½ in. (65 mm) outlets], together with a test-pressure gauge[1] suitable for the pressures to be expected, with 1 psi (0.07 bar) graduations is required. [Usually, a 200 psi (13.8 bar) gauge is satisfactory.]

TESTING EXHIBIT 7.2 Typical Hydrant Cap and Gauge. (Source: NFPA Fire Protection Handbook® 2003, Figure 10.6.2)

In Testing Exhibit 7.2, note the petcock blow-off. Its operation permits air from the hydrant barrel to be vented when the hydrant is opened and air to re-enter when the hydrant is closed. Failure to open the petcock during hydrant closing might subject the pressure gauge to a partial vacuum, which could introduce errors in future gauge readings.

4. A hydrant wrench compatible with the operating nut of the hydrants being tested is required. Adjustable wrenches, pipe wrenches, and parrot wrenches should not be used, because these could damage the operating nut.
5. A form for recording test data, including a sketch of the test location showing hydrants and any other salient features, is required. All test data should be recorded in a neat, systematic fashion, along with any system operating conditions that might affect the test results.

Procedure Steps

1. Prior to any testing, notify the water department, the fire department and/or the alarm monitoring company (if needed), and the facility representatives that testing is going to be conducted.
2. Establish a test plan with the AHJs and obtain required approvals.

1. Test-quality gauges should be used in accordance with ASME B40.1, *Gauges — Pressure Indicating Dial Type — Elastic Element, Grade 1A*, having an accuracy of ±1%. The use of quality test gauges produces results that are considered reasonably accurate within the scope of the test procedure. Care should be taken to protect the gauges from rough handling. Gauges should be calibrated at least annually by means of a dead-weight or calibrated tester throughout the range of operation before a test series is begun. Calibration sheets should be kept for each gauge and correction factors affixed to the back of each gauge.

Topics covered in the testing procedures include the purpose for the test, a list of tools or equipment needed, detailed steps required to test the system, and an explanation of how to evaluate the results.

Much of the artwork throughout the book has been updated and improved.

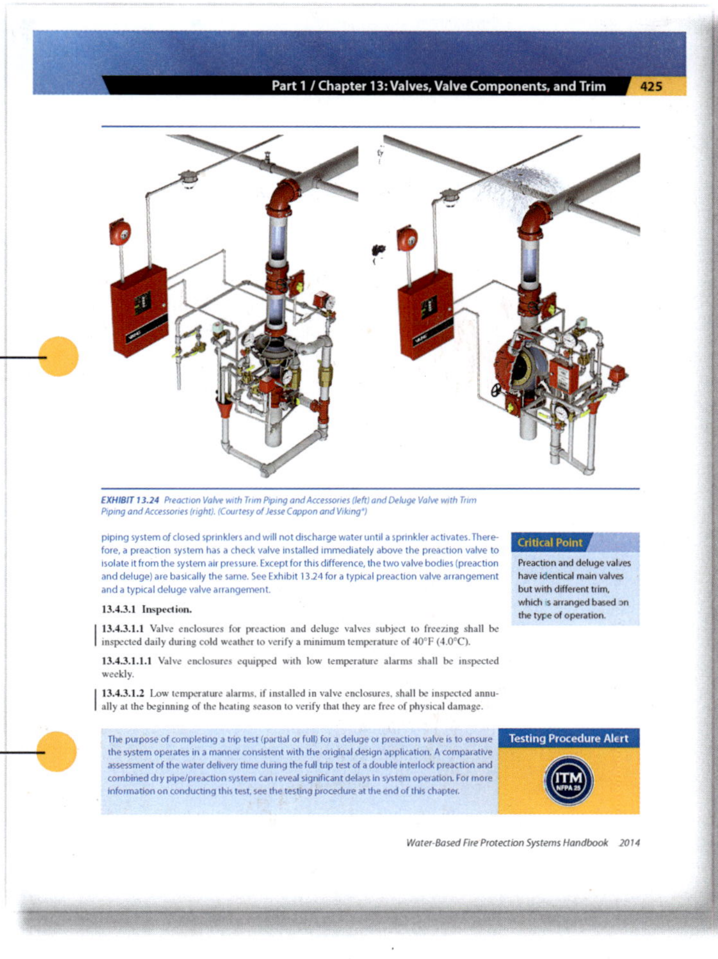

Part 1 / Chapter 13: Valves, Valve Components, and Trim 425

EXHIBIT 13.24 Preaction Valve with Trim Piping and Accessories (left) and Deluge Valve with Trim Piping and Accessories (right). (Courtesy of Jesse Cappon and Viking®)

piping system of closed sprinklers and will not discharge water until a sprinkler activates. Therefore, a preaction system has a check valve installed immediately above the preaction valve to isolate it from the system air pressure. Except for this difference, the two valve bodies (preaction and deluge) are basically the same. See Exhibit 13.24 for a typical preaction valve arrangement and a typical deluge valve arrangement.

Critical Point
Preaction and deluge valves have identical main valves but with different trim, which is arranged based on the type of operation.

13.4.3.1 Inspection.

13.4.3.1.1 Valve enclosures for preaction and deluge valves subject to freezing shall be inspected daily during cold weather to verify a minimum temperature of 40°F (4.0°C).

13.4.3.1.1.1 Valve enclosures equipped with low temperature alarms shall be inspected weekly.

13.4.3.1.2 Low temperature alarms, if installed in valve enclosures, shall be inspected annually at the beginning of the heating season to verify that they are free of physical damage.

Testing Procedure Alert
The purpose of completing a trip test (partial or full) for a deluge or preaction valve is to ensure the system operates in a manner consistent with the original design application. A comparative assessment of the water delivery time during the full trip test of a double interlock preaction and combined dry pipe/preaction system can reveal significant delays in system operation. For more information on conducting this test, see the testing procedure at the end of this chapter.

Water-Based Fire Protection Systems Handbook 2014

Testing Procedure Alert boxes are located throughout the handbook to let the reader know that a more complete list of suggested procedures can be found at the end of the chapter.

Water-Based Fire Protection Systems Handbook

FOURTH EDITION

Edited by

Matthew J. Klaus
Principal Fire Protection Engineer

With the complete text of the 2014 edition of NFPA® 25, *Standard for the Inspection, Testing, and Maintenance of Water-Based Fire Protection Systems*

NFPA

National Fire Protection Association®
Quincy, Massachusetts

Product Management: Michael S. Barresi, Jr.
Development and Production: Khela Thorne
Copyediting: Kenneth Ritchie
Art Direction: Cheryl Langway
Permissions: Josiane Domenici
Illustrations: George Nichols,
Stephan Laforest

Cover Design: Greenwood Associates Design
Composition: Shepherd, Inc.
Manufacturing: Ellen Glisker
Printing: King Printing/Lowell, MA.

Important Notices and Disclaimers: Publication of this handbook is for the purpose of circulating information and opinion among those concerned for fire and electrical safety and related subjects. While every effort has been made to achieve a work of high quality, neither the NFPA® nor the contributors to this handbook guarantee the accuracy or completeness of or assume any liability in connection with the information and opinions contained in this handbook. The NFPA and the contributors shall in no event be liable for any personal injury, property, or other damages of any nature whatsoever, whether special, indirect, consequential, or compensatory, directly or indirectly resulting from the publication, use of, or reliance upon this handbook.

This handbook is published with the understanding that the NFPA and the contributors to this handbook are supplying information and opinion but are not attempting to render engineering or other professional services. If such services are required, the assistance of an appropriate professional should be sought.

NFPA codes, standards, recommended practices, and guides ("NFPA Documents"), including the NFPA Document that is the subject of this handbook, are made available for use subject to Important Notices and Legal Disclaimers, which appear at the end of this handbook and can also be viewed at *www.nfpa.org/disclaimers.*

Notice Concerning Code Interpretations: This fourth edition of the *Water-Based Fire Protection Systems Handbook* is based on the 2014 edition of NFPA 25, *Standard for the Inspection, Testing, and Maintenance of Water-Based Fire Protection Systems.* All NFPA codes, standards, recommended practices, and guides ("NFPA Documents") are developed in accordance with the published procedures of the NFPA by technical committees comprised of volunteers drawn from a broad array of relevant interests. The handbook contains the complete text of NFPA 25 and any applicable Formal Interpretations issued by the NFPA. This NFPA Document is accompanied by explanatory commentary and other supplementary materials.

The commentary and supplementary materials in this handbook are not a part of the NFPA Document and do not constitute Formal Interpretations of the NFPA (which can be obtained only through requests processed by the responsible technical committees in accordance with the published procedures of the NFPA). The commentary and supplementary materials, therefore, solely reflect the personal opinions of the editor or other contributors and do not necessarily represent the official position of the NFPA or its technical committees.

REMINDER: UPDATING OF NFPA DOCUMENTS

NFPA 25, *Standard for the Inspection, Testing, and Maintenance of Water-Based Fire Protection Systems,* like all NFPA codes, standards, recommended practices, and guides ("NFPA Documents"), may be amended from time to time through the issuance of Tentative Interim Amendments or corrected by Errata. An official NFPA Document at any point in time consists of the current edition of the document together with any Tentative Interim Amendment and any Errata then in effect. In order to determine whether an NFPA Document has been amended through the issuance of Tentative Interim Amendments or corrected by Errata, visit the Document Information Pages on NFPA's website. The Document Information Pages provide up-to-date, document specific information including any issued Tentative Interim Amendments and Errata. To access the Document Information Page for a specific NFPA Document go to http://www.nfpa.org/document for a list of NFPA Documents, and click on the appropriate Document number (e.g., NFPA 25). In addition to posting all existing Tentative Interim Amendments and Errata, the Document Information Page also includes the option to sign-up for an "Alert" feature to receive an email notification when new updates and other information are posted regarding the document.

NFPA No.: 25HB14
ISBN (book): 978-1-4559-0664-2
ISBN (e-book): 978-1-4559-0835-6

Library of Congress Control No.: 2013939754

Printed in the United States of America
17 18 7 6

CONTENTS

PREFACE

The amount of data and the number of resources available to sprinkler system designers has never been greater. With the proper application of this information, along with the use of state-of-the-art equipment and system components, water-based fire protection system effectiveness and reliability continues to improve upon what is already considered to be a stellar track record. Even with this data and the latest products on the market, however, a system is only as effective as the inspection, testing, and maintenance (ITM) program that is used to keep it running once a building is occupied. The standard of care for proper water-based fire protection system ITM execution is NFPA 25, *Standard for the Inspection, Testing, and Maintenance of Water-Based Fire Protection Systems*. This standard is the foundation of the *Water-Based Fire Protection Systems Handbook.*

In addition to the requirements of NFPA 25, this handbook contains commentary on why the requirements have been included in the standard and how they can be executed. Dispersed throughout the handbook are several features intended to give the user additional information that will help them properly apply the standard. These recurring features include:

- FAQs
- Criti
- ITM Deficiency, Impairment, or Hazard Evaluation examples
- Testing Procedures
- System Tagging case studies

The frequently asked questions, or FAQ, feature is based on the questions most commonly asked of the NFPA 25 staff. Each Critical Point feature highlights important information found in the commentary text. Several Case In Point boxes are also included throughout, to expand on material presented in the commentary.

This edition also includes a new feature — ITM Deficiency, Impairment, or Hazard Evaluation. This feature provides examples to illustrate the differences between ITM-related field conditions that fall within the scope of NFPA 25 and items that are not considered to be "wear-and-tear" issues. Conditions that require a design evaluation are outside of the scope of the NFPA 25 standard and are not considered part of the NFPA 25 inspector's work. Knowing when something should or should not be covered on an NFPA 25 inspection report can be critical, especially when it comes to liability.

Another new feature in this handbook is the addition of testing procedures, which are provided at the end of most chapters, to discuss in detail what steps should be taken in order to properly test systems. These procedures are identified with a brief Testing Procedure Alert box within the chapter, alerting the reader that a more complete list of procedures can be found at the end of the chapter.

While NFPA 25 identifies a myriad of tests that must be conducted in order to gain some assurance of system functionality, the standard does not provide processes and procedures for conducting these tests. This is largely done because in a standard it is not possible to identify a singular approach to testing a system that could have been designed dozens of different ways depending on when it was built. It is also difficult to write a procedure where the resources available to the owner, inspector, and AHJ will vary greatly from facility to facility. As such, NFPA 25 simply identifies the activity that must be performed and the frequency at which it must be performed. This handbook contains a new feature to fill in some of the gaps that cannot be addressed within a standard.

These testing procedures outline potential processes and procedures for conducting many of the tests that are mandated by NFPA 25. It is important to note that these processes and procedures are not part of NFPA 25. The actions identified and the steps provided in these procedures represent one approach to conducting a test required by this standard. In some instances, the procedures outlined in this handbook cannot be performed as described or it might not be appropriate to perform them based on the specific equipment that has been installed or the arrangement of the equipment.

A System Tagging feature has also been added to this edition of the handbook to help readers identify field conditions that warrant action and determine the criticality of the situation. Assigning criticality to the field conditions that are identified during inspections is a critical part of the ITM process. This new feature is intended to help readers understand what goes into the process of classifying deficiencies and impairments. The feature also examines various specific field conditions and explains the process of determining the severity of the condition and whether the condition should be listed as an impairment, a critical deficiency, or a noncritical deficiency.

In addition to the commentary and the features that enhance it are eight supplements that take a deeper look at specific subjects. The subjects include microbiologically induced corrosion (Supplement 1), fire pump testing and analysis (Supplement 2), NFPA 25 roles and responsibilities (Supplement 3), and in-depth summaries on the role of the inspector (Supplement 4), the role of the owner (Supplement 5), and the role of the authority having jurisdiction (Supplement 6). These supplements are prepared by industry experts and technical committee members who use their experience and firsthand knowledge of these issues to better explain topics covered in NFPA 25. The handbook also includes excerpts from a recent Fire Protection Research Foundation report, "Fire Pump Field Data Collection and Analysis," (Supplement 7) as well as a detailed list of the technical/substantive changes from the 2011 edition to the 2014 edition of NFPA 25 (Supplement 8).

Acknowledgments

Producing this handbook has taken a tremendous amount of time and effort on the part of a number of committed individuals outside of NFPA. For their work in contributing new commentary, revising old commentary, and providing new features and supplements, I would like to acknowledge and thank Bob Caputo, Russ Leavitt, Bruce Clarke, Tracey Bellamy, Terry Victor, Gayle Pennel, Rich Ray, and John Lake. I would also like to thank Byron Blake, Bill Koffel, John Munno, George Stanley, Tom Multer, Peter Petrus, Damon Pietraz, Josh Elvove, Cecil Bilbo, David Martinez, and Peter Schwab for sending in their photographs that were used throughout this book.

Producing this handbook has taken a tremendous amount of effort on the part of a number of people on the NFPA staff as well. Specifically, Mike Barresi, product manager, who has the unenviable task of keeping our team moving forward even when it looks like we might discuss the merits of a single piece of artwork for 4 hours, as well as Khela Thorne, senior developmental editor, who helped enhance the content in this handbook by challenging the concept of a what an NFPA handbook is (and by adding a new feature after we had finished all of our other work!). Also a special thanks to Kim Cervantes, senior project editor; Cheryl Langway, art coordinator; Josiane Domenici, permissions editor; and Ken Ritchie, copy editor, for their professionalism and dedication to the project.

Last, but certainly not least, I would like to thank my family — my wife, Lisa, and my children, Olivia, Mackenzie, Riley, and Elena — for giving me the drive and the opportunity to take on projects like this one.

Matt Klaus

ABOUT THE CONTRIBUTORS

Anthony M. Aguilera, CFPS (Supplement 1)

Anthony Aguilera is director of loss prevention and risk management for Honeywell Aerospace's Health, Safety & Environmental department and holds a Honeywell Fellos designation for technical expertise. He has loss prevention oversight responsibilities for over 100 global manufacturing facilities, including those for business continuity management and process safety management. Anthony has a Bachelor of Science in Management from the University of Phoenix and is a certified fire protection specialist. He is a 26-year member of NFPA and served on the Industrial Fire Protection Section (IFPS) Board of Directors from 1999 to 2003. Anthony is also a patented inventor of a microbial fumigation process for fire protection sprinkler system piping.

Tracey D. Bellamy, PE, CFPS (Testing Procedures Feature)

Tracey Bellamy is the chief engineering officer with Telgian Corporation. He is a graduate of the University of South Carolina holding both a Bachelor of Science and Master of Engineering in Civil Engineering as well as an Advanced Graduate Certificate in Fire Protection Engineering from Worchester Polytechnic Institute. Tracey has more than 25 years of experience in fire protection engineering, with a diverse range of job assignments that includes working as an AHJ with the South Carolina State Fire Marshal, supporting facility operations and construction at a Department of Energy facility, and overseeing fire protection engineering operations for Telgian Corporation. He holds professional registration as both a fire protection and civil engineer with licenses in 49 states, Guam, Puerto Rico, and the District of Columbia and is also a Certified Fire Protection Specialist. Tracey is an NFPA member and currently serves on multiple NFPA technical committees, including those responsible for NFPA 13, NFPA 15, NFPA 16, NFPA 25, NFPA 30, NFPA 30B, NFPA *101*®, and *NFPA 5000*®.

Bob Caputo, CFPS, CET (Chapters 5–7, 9, 10, and 16)

Bob Caputo has been working in the fire protection industry for more than 35 years. He is a member of the NFPA 13 Sprinkler System Installation, Discharge, and Hanging & Bracing committees as well as the NFPA 4 (Commissioning and Integrated Testing) and NFPA 25 (Inspection, Testing, and Maintenance of Water-Based Systems) committees. Bob is the Chairman of the NFPA 16 Technical Committee for Foam-Water Sprinklers. As a senior member of both NFPA and American Fire Sprinkler Association faculties, he has written and presented seminars and webinars throughout the world on fire protection and life safety systems design and installation, hydraulic calculations, plan review, fire pumps, standpipes, special hazards, and system applications. His consulting practice provides training, plan review, and inspection services; technical opinion reports; and expert witnessing in legal cases related to fire protection and life safety systems. Bob attended the University of Albuquerque, New Mexico, and is a U.S. Navy veteran and former volunteer fire fighter.

Bruce H. Clarke, CFPS (Chapters 8, 11–13, Supplements 1 and 2)

Bruce Clarke is the global property engineering training coordinator for AIG. Bruce has a Bachelor of Science from California State University, Chico, and more than 20 years of direct experience in fire protection systems code consulting, property loss prevention inspections and consulting, design review, construction management, and fire systems commissioning, having worked throughout the United States, Canada, Mexico, and the APAC countries. Bruce is also Six Sigma Green Belt Certified and a past green building construction LEED AP. In addition to contributing to multiple technical publications and task groups related to industry fire protection, he has served as a member of NFPA 3 (Commissioning Fire Protection Systems), NFPA 4 (Commissioning and Integrated Testing), NFPA 25 (Inspection, Testing, and Maintenance of Water-Based Systems), NFPA 318 (Semiconductor and Related Facilities), and NFPA 750 (Water Mist Fire Suppression Systems) committees. He has been on the Certified Fire Protection Specialist Board of Directors since 2004 and is also past president of both the Arizona and Carolinas chapters of the Society of Fire Protection Engineers (SFPE).

John A.M. Lake, CFPS (Supplements 3 and 6)

John Lake is a fire protection specialist for the City of Gainesville, Florida. He is a past fire marshal for Marion County Fire Rescue for 25 years, the 2002 Florida Fire Marshal of the Year, the chairman of the NFPA 501 and NFPA 501A (Manufactured Housing) committee, and a member of the NFPA 25 (Inspection, Testing, and Maintenance of Water-Based Systems), NFPA 24 and NFPA 291 (Private Water Supply Piping Systems), NFPA *101*® (Assembly Occupancies), and NFPA 1 (Fire Code) committees. He has 32 years of experience in management of fire and rescue services, including preparation of budgets, management of life safety programs, and development of strategic plans, while developing partnerships to accomplish goals of the department.

Russell Leavitt, SET, CFPS (Chapters 1–4, 14, and 15, Supplements 3 and 5)

Russ Leavitt is the executive chairman of Telgian Corporation, which consists of TVA Fire and Life Safety, Inc. and Fire Materials Group, LLC. He is a graduate of the University of Nevada, Las Vegas; holds a Level IV certification from NICET in Fire Sprinkler Layout; and is a certified fire protection specialist. Russ is a licensed contractor in 40 states and has 30 years of experience in the design, installation, and testing of fire protection systems. He currently serves as the chairman of the Technical Committee for Sprinkler System Discharge Criteria, a principal member of the NFPA 25 Technical Committee for Inspection, Testing, and Maintenance of Water-Based Systems, the Technical Correlating Committee on Automatic Sprinkler Systems, and the Technical Correlating Committee on Building Code. Russ is a senior instructor for NFPA's Professional Development Series and has authored a number of articles and training materials, including the AFSA Study Guide for NICET Certification on Inspections and Testing of Water-Based Systems, the AFSA online hydraulics course, and the Fire Smarts Residential Sprinkler training guide. He is a member of NFPA and the Society of Fire Protection Engineers (SFPE).

T. Gayle Pennel, PE
(Supplement 7)

Gayle Pennel is a project manager at Aon Fire Protection Engineering Corporation, specializing in all aspects of water supplies and fire protection systems. His experience includes designing fire protection for exhibition halls, high-rise and ultra-high buildings, large box retail stores, refineries, distribution centers, and factories. Gayle has provided investigative and expert testimony for fire protection failures. He has also designed and supervised large-scale fire tests for exposed expanded plastics, water miscible flammable liquids in plastic containers, and other hazards where existing standards did not provide protection criteria. Gayle has been active in code development, including incorporating large-scale fire test results into existing standards. He is currently chairman of the NFPA 20 (Fire Pumps) committee and a member of the NFPA 25 (Inspection, Testing, and Maintenance of Water-Based Systems) committee.

Rich Ray
(Supplement 3)

Rich Ray is president of Cybor Fire Protection Company & Global Fire Protection Company in Downers Grove, Illinois. He is the vice chairman of the Illinois State Board of Professional Engineers and a member of several organizations, including the Illinois Fire Prevention Association and the National Fire Sprinkler Association, where he is on the board of directors. Rich is also on the technical committee for NFPA 25 (Inspection, Testing, and Maintenance of Water-Based Systems). He also serves on several of Underwriters Laboratories' Standards Technical Panels. Rich graduated from the Illinois Institute of Technology with a Bachelor of Science in Fire Protection and Safety Engineering, and he has a Master in Business Administration from Illinois Benedictine College.

J. William Sheppard, FSFPE
(Supplement 5)

Bill Sheppard recently retired as manager of fire protection and emergency response, global security operations, for General Motors, where he was responsible for the implementation and administration of fire protection and emergency response policy throughout GM facilities worldwide. He is a member of NFPA, serving as a principal member of the technical committees for Private Water Supply Piping Systems (past chairman); Inspection, Maintenance, and Testing of Water-Based Systems; and Ovens and Furnaces. Bill also serves as a principal on the Technical Correlating Committee on Automatic Sprinkler Systems and is a member and fellow of the Society of Fire Protection Engineers (SFPE).

Terry L. Victor
(Supplement 4 and ITM Deficiency, Impairment, or Hazard Evaluation Feature)

Terry Victor is national manager of sprinkler business processes at SimplexGrinnell and has more than 40 years of technical experience in the sprinkler industry. He serves on various industry boards and committees, including the NFSA Contractors Council; the NFSA Engineering and Standards Committee, where he has served as chairman; the NFSA ITM Committee, where he has served as chairman; and the NICET Fire Protection Advisory Group. Terry is NICET Level IV–certified in both Sprinkler System Layout and Special Hazards System Layout. He is a member of several NFPA technical committees, including those for NFPA 3, NFPA 4, NFPA 13, NFPA 13D, NFPA 13R, NFPA 20, and NFPA 25. He has been on the NFPA 25 (Inspection, Testing, and Maintenance of Water-Based Systems) technical committee since it was first formed for the 1992 edition.

ABOUT THE EDITOR

Matthew J. Klaus

Matt Klaus is a principal fire protection engineer at the National Fire Protection Association, where he is responsible for NFPA documents addressing commissioning, integrated system testing, and automatic sprinkler systems. He holds a Bachelor of Science in Civil Engineering and a Master of Science in Fire Protection Engineering from Worcester Polytechnic Institute. He is a member of the Salamander Honorary Fire Protection Engineering Society. Matt has extensive fire protection engineering consulting experience as a project manager for projects in Dubai, Abu Dhabi, Qatar, and the Kingdom of Bahrain, as well as for projects across the United States. He has also designed and commissioned various fire protection systems including smoke control systems, suppression systems, and fire alarm systems. His project work includes the use of fire and egress modeling software for engineering analyses of roadway tunnels, rail systems, football stadiums, high-rise buildings, shopping malls, and transportation hubs.

PART 1

ITM
NFPA 25
INSPECTION TESTING MAINTENANCE

NFPA® 25, *Standard for the Inspection, Testing, and Maintenance of Water-Based Fire Protection Systems*, 2014 Edition, with Commentary

Part 1 of this handbook includes the complete text of the 2014 edition of NFPA 25, *Standard for the Inspection, Testing, and Maintenance of Water-Based Fire Protection Systems,* which consists of 16 mandatory chapters and 6 nonmandatory annexes. Working within the framework of NFPA's consensus-based codes- and standards-making process, the Technical Committee on Inspection, Testing, and Maintenance of Water-Based Fire Protection Systems prepared both the mandatory provisions found in Chapter 1 through Chapter 16 and the nonmandatory material found in the annexes.

The annex material is advisory or informational in nature and is provided to assist users in interpreting the mandatory standard provisions. It is not considered part of the requirements of the standard. An asterisk (*) following a standard paragraph number indicates that advisory material pertaining to that paragraph appears in Annex A. For the reader's convenience, in this handbook Annex A material appears immediately following the mandatory paragraph to which it relates.

The explanatory commentary in this handbook was prepared by the handbook editor, with the assistance of those persons mentioned in the acknowledgments, and is intended to provide the reader with an understanding of the provisions of the standard and to serve as a resource and reference for implementing the provisions of or enforcing the standard. It is not a substitute for the actual wording of the standard or the text of the many codes and standards that are incorporated by reference. The commentary immediately follows the standard text it discusses and is set in blue type for easy identification.

The handbook also features a tool designed to help readers easily identify new or revised elements in the 2014 edition of the standard. Changes other than editorial are indicated by a vertical rule beside the paragraph, table, or figure in which the change occurred. Where one or more complete paragraphs have been deleted, the deletion is indicated by a bullet (•) between the paragraphs that remain.

This edition of the handbook includes several new features to assist the reader in using this handbook. These include the following:

- Frequently Asked Questions (FAQs)
- Critical Points
- Case in Points
- ITM vs Hazard Evaluation Examples
- Testing Procedures
- System Tagging Case Studies

See the Preface of this handbook and the front and back inside covers for a more detailed explanation of these features.

1 ADMINISTRATION

Chapter 1 of NFPA 25, *Standard for the Inspection, Testing, and Maintenance of Water-Based Fire Protection Systems*, covers the administrative requirements for the periodic inspection, testing, and maintenance (ITM) of water-based fire protection systems. In addition to the scope and purpose of the standard, Chapter 1 provides guidance on the application of the standard and explains how units of measurement are expressed throughout the document. Note that the most important point when using NFPA 25 is that the purpose of the standard is to verify the operational status of a system and to provide a reasonable degree of certainty that the system will perform when needed.

1.1 Scope

This document establishes the minimum requirements for the periodic inspection, testing, and maintenance of water-based fire protection systems and the actions to undertake when changes in occupancy, use, process, materials, hazard, or water supply that potentially impact the performance of the water-based system are planned or identified.

The scope of NFPA 25 is intended to help users determine if they are using the correct standard. NFPA 25 contains the minimum requirements for the inspection, testing, and maintenance of a water-based system to ensure the system performs properly. It also contains requirements for investigating and addressing changes that occur to a building, its use, or water supply and the potential impact on the water-based systems. The minimum requirements specified in NFPA 25 must be met in order for a system to comply with this standard. However, nothing in the standard is intended to prevent more frequent or more extensive inspection, testing, and maintenance activities if the level of safety or performance of the system is at stake, or if the owner would like a more comprehensive assessment of the fire protection systems in a particular building. Also, Section 4.7 permits alternative means of compliance using a performance-based program that could result in less frequent inspection, testing, and maintenance activities.

Critical Point

Nothing in the standard is intended to prevent more frequent or more extensive inspection, testing, and maintenance activities if the level of safety or performance of the system is at stake, or if the owner would like a more comprehensive assessment of the fire protection systems in a particular building.

1.1.1 Coordination with *NFPA 72* Testing Requirements. This standard does not address all of the inspection, testing, and maintenance of the electrical components of the automatic fire detection equipment used to activate preaction and deluge systems that are addressed by *NFPA 72, National Fire Alarm and Signaling Code*.

1.1.1.1 The inspection, testing, and maintenance required by this standard and *NFPA 72, National Fire Alarm and Signaling Code*, shall be coordinated so that the system operates as intended.

The requirements in 1.1.1 and 1.1.1.1 clarify that not all of the inspection, testing, and maintenance requirements for electrical devices that are used with preaction, deluge, and other specialty systems are addressed by NFPA 25. For the complete requirements of those systems, refer to *NFPA 72®, National Fire Alarm and Signaling Code*. In some cases, it might be necessary to coordinate the inspection, testing, and maintenance of electrical initiating devices and other components with a qualified electrician or fire alarm technician.

1.1.1.2* All inspections, testing, and maintenance required by *NFPA 72* shall conform to *NFPA 72*, and all inspections, testing, and maintenance required by this standard shall conform to this standard.

A.1.1.1.2 There are times when a single inspection or test can meet the requirements of both NFPA 25 and *NFPA 72* (e.g., operation of a tamper switch). This standard does not necessarily require that two separate inspections or tests be conducted on the same component, provided the inspection or test meets the requirements of both standards and the individual performing the inspection or test is qualified to perform the inspection or test required by both standards.

NFPA 25 and *NFPA 72* address common components such as waterflow indicators. Although the technical committees for each document have worked to coordinate the requirements of the two standards, the frequency and methods for inspecting, testing, and maintaining these components might differ. Paragraph 1.1.1.2 and A.1.1.1.2 state that devices must be inspected, tested, and maintained in accordance with the requirements of the applicable standard. When contemplating the use of a single test to satisfy both NFPA 25 and *NFPA 72*, it is important to investigate the jurisdictional requirements regarding licensing and certification. Some jurisdictions prohibit a single individual from performing tests for both suppression and alarm systems due to licensing restrictions. However, in many jurisdictions, a single individual could be licensed or certified for multiple competencies.

1.1.2 Types of Systems.

1.1.2.1 The types of systems addressed by this standard include, but are not limited to, sprinkler, standpipe and hose, fixed water spray, private fire hydrants, water mist, and foam water.

1.1.2.2 Water supplies that are part of these systems, such as private fire service mains and appurtenances, fire pumps and water storage tanks, and valves that control system flow, are also included in this standard.

1.1.3* This standard addresses the operating condition of fire protection systems as well as impairment handling and reporting and applies to fire protection systems that have been properly installed in accordance with generally accepted practice.

FAQ Do the requirements of NFPA 25, including a service program, inspection, maintenance, tests, and record or document retention, not apply if the system was not originally installed properly? If not, what is the meaning of 1.1.3?

It is the intent of the standard to require inspection, testing, and maintenance of all water-based fire protection systems, regardless of the quality of the design and installation. The phrase "installed in accordance with generally accepted practice" is used so that implementation of the standard is not compromised because of design or installation variances from the applicable installation standard.

A.1.1.3 Generally accepted NFPA installation practices for water-based fire protection systems relevant to this standard are found in the following:

NFPA 13, *Standard for the Installation of Sprinkler Systems*
NFPA 13R, *Standard for the Installation of Sprinkler Systems in Low-Rise Residential Occupancies*
NFPA 14, *Standard for the Installation of Standpipe and Hose Systems*

NFPA 15, *Standard for Water Spray Fixed Systems for Fire Protection*
NFPA 16, *Standard for the Installation of Foam-Water Sprinkler and Foam-Water Spray Systems*
NFPA 20, *Standard for the Installation of Stationary Pumps for Fire Protection*
NFPA 22, *Standard for Water Tanks for Private Fire Protection*
NFPA 24, *Standard for the Installation of Private Fire Service Mains and Their Appurtenances*
NFPA 750, *Standard on Water Mist Fire Protection Systems*

1.1.3.1* This standard does not require the inspector to verify the adequacy of the design of the system.

A.1.1.3.1 The requirement to evaluate the adequacy of the design of the installed system is not a part of the periodic inspection, testing, and maintenance requirements of this standard. However, such evaluation is the responsibility of the property owner or designated representative as indicated in 4.1.6 and 4.1.7.

The inspections required by NFPA 25 are not intended to reveal installation flaws or code compliance violations. As a result of practical limitations and an effort to keep the cost-benefit ratio of implementing the standard reasonable, the inspections required by NFPA 25 are specifically intended to assess the operating condition of the system, with the goal that the system will function as intended.

The installation standards are typically not retroactive; therefore, inspecting a system for compliance to the original installation or design criteria would require extensive research and knowledge. For example, inspecting sprinklers to ensure that they are positioned correctly to meet the requirements related to obstructions would require the inspector to determine which edition of the appropriate installation standard and which associated obstruction rules were in effect and enforced at the time of the original installation. Local authority having jurisdiction (AHJ) amendments and variances would also need to be researched. This research would be complicated by the fact that the obstruction rules for sprinklers have greatly expanded over the last two decades with the continuous introduction of new sprinkler listings and applications. What complicates the task even further is that it is virtually impossible to identify many obstruction deficiencies from the floor, which is where NFPA 25 expects the inspection to be conducted.

Therefore, to manage the cost of inspections while maintaining a reasonable degree of protection, NFPA 25 assumes that the original installation was approved by all applicable AHJs and is still considered acceptable.

Critical Point

To manage the cost of inspections while maintaining a reasonable degree of protection, NFPA 25 assumes that the original installation was approved by all applicable AHJs and is still considered acceptable.

However, where modifications to the building or its use are contemplated or have been made, such as new or relocated walls or ceilings, changes in occupancy, or the construction of an addition, the owner must comply with the requirements in Chapter 4.

FAQ Is an NFPA 25 inspection intended to address violations of NFPA 13 and other issues of improper or inadequate installation?

NFPA 25 inspections are intended to address the operating condition of a system or system components, not the adequacy of the system design when compared to the applicable design and installation standards. This is why NFPA 25 is commonly referred to as a "wear and tear" document.

The types of deficiencies that are intended to be identified during an NFPA 25 inspection include system components that are leaking, damaged, corroded, or in disrepair. Exhibit 1.1 shows a section of FDC piping and a coupling that are heavily corroded and need to be replaced. This is not to say that all system components that show signs of wear and tear will need to be immediately replaced. In some instances, there might be some exterior wear and tear or corrosion on a system component, but replacement of the component is not necessary. Exhibit 1.2 shows a flanged elbow displaying exterior corrosion and rusted bolts. In this instance, the wear

EXHIBIT 1.1 *Corroded FDC Piping. (Courtesy of Byron Blake and SimplexGrinnell)*

EXHIBIT 1.2 *Lightly Corroded Fitting. (Courtesy of Wayne Automatic Fire Sprinklers Inc.)*

Critical Point

NFPA 25 inspections are intended to address the operating condition of a system or system components, not the adequacy of the system design when compared to the applicable design and installation standards.

and tear on the component might not warrant the replacement of the components, but possibly just some new bolts and some scouring of the rusted exterior surface. The determination of the necessary corrective action is somewhat subjective and might require the input of the municipal AHJ or insurance representative.

It is not the intent of NFPA 25 to place the burden of a complete system evaluation on the inspector. However, apparent deficiencies related to design or installation should be reported to the building owner. In reporting alleged design or installation deficiencies to the owner, the inspector must keep in mind that variations from current installation standards are not necessarily deficiencies and could be compliant with the installation standard or equivalency used for the original installation or variances permitted by the AHJ.

Case In Point

It is not practical for an inspector to perform an evaluation of the design and installation aspects of a fire protection system because the cost of doing so would be prohibitive. This photo shows a typical warehouse space. In order for an inspector to properly evaluate the adequacy of the system in this space, the inspector would need to determine the ceiling height, storage height, ceiling slope, commodity classification, clearance to ceiling, aisle widths, sprinkler design approach, and many other factors that went into the sprinkler system design. It would take an intimate knowledge of the building, its history, and operation for an inspector to be able to properly evaluate the system. For this reason, it is simply not reasonable to include this level of evaluation as part of an annual inspection of the system.

Typical Large Warehouse Space. (Courtesy of Indonesian Fire & Rescue Foundation)

Such an evaluation should be conducted only by a qualified contractor, engineer, or design professional, and it is required only when changes are made to the building, its use, or its water supplies. As with most building codes, the requirements in 1.3.4 of *NFPA 5000®, Building Construction and Safety Code®*, apply to existing structures when the following situations occur:

- A change of occupancy classification
- A repair, renovation, alteration, reconstruction, or addition is made

Similar provisions for existing buildings are found in 10.3.2 of NFPA 1, *Fire Code*:

> Existing buildings that are occupied at the time of adoption of this *Code* shall remain in use provided that the following conditions are met:
> (1) The occupancy classification remains the same.
> (2) No condition deemed hazardous to life or property exists that would constitute an imminent danger.

Note that neither code requires a periodic evaluation of the system or building, but that an evaluation is needed when certain changes occur. This evaluation is covered in detail in Chapter 4 of NFPA 25.

Many owners, insurance providers, and government entities require a periodic evaluation or assessment of their existing systems. In these instances, NFPA 3, *Recommended Practice on Commissioning and Integrated Testing of Fire Protection and Life Safety Systems*, can be used as a basis for setting up a re-commissioning or retro-commissioning plan for a system or series of systems within a building.

1.1.4* Corrective action needed to ensure that a system operates in a satisfactory manner shall be in accordance with this standard unless this standard specifically refers to an appropriate installation standard.

While it is not within the scope of NFPA 25 to require the evaluation of a system installation, there are times when an installation standard is used. Such is the case when a corrective action is necessary. For example, if the testing of sprinklers in accordance with 5.3.1 results in the replacement of more than 20 sprinklers in the system, the provisions of 5.5.1 and Table 5.5.1 require a hydrostatic test on the piping. Table 5.5.1 refers to NFPA 13, *Standard for the Installation of Sprinkler Systems*, for this test.

When corrective action is required for an impaired component, the corrective action must be executed in accordance with the appropriate design and installation standard. For example, when a sprinkler is leaking or the glass bulb has been damaged, stopping the leak or replacing the bulb with a nail, as shown in Exhibit 1.3, is not considered making a corrective action in accordance with the design standard. This sprinkler would need to be replaced, as opposed to the attempted repair work shown in this photo.

EXHIBIT 1.3 *Improper Corrective Action. (Courtesy of Academy of Sprinkler Technology)*

A.1.1.4 For systems originally installed in accordance with one of these standards, the repair, replacement, alteration, or extension of such systems should also be performed in accordance with that same standard. When original installations are based on other applicable codes or standards, repair, replacement, alteration, or extension practices should be conducted in accordance with those other applicable codes or standards.

1.1.5 This standard shall not apply to sprinkler systems designed, installed, and maintained in accordance with NFPA 13D, *Standard for the Installation of Sprinkler Systems in One- and Two-Family Dwellings and Manufactured Homes*.

Critical Point

Systems installed in accordance with NFPA 13D are not covered by this standard due to their simplicity, although some of the procedures contained herein can be applied as good practice.

FAQ Does NFPA 25 apply to sprinkler systems in one- and two-family homes?

Systems installed in accordance with NFPA 13D, *Standard for the Installation of Sprinkler Systems in One- and Two-Family Dwellings and Manufactured Homes*, are not covered by this standard due to their simplicity, although some of the procedures contained herein can be applied as good practice. Chapter 12 of NFPA 13D provides the requirements for the inspection, testing, and maintenance for these systems. See A.12.2 of NFPA 13D for a recommended inspection and testing program.

While the scope of NFPA 25 specifically excludes NFPA 13D systems that are installed in one- and two-family homes, NFPA 13D systems are now being installed in some occupancies that the systems were not originally intended for. A primary example of this is in assisted living facilities. NFPA *101*®, *Life Safety Code*®, allows assisted living facilities to employ sprinkler systems designed to NFPA 13D instead of NFPA 13 or NFPA 13R, *Standard for the Installation of Sprinkler Systems in Low-Rise Residential Occupancies*. When this design option is exercised, NFPA *101* requires that certain ITM activities identified in NFPA 25 must be followed even though NFPA 25 specifically excludes NFPA 13D systems from its scope.

In order to address this situation, the Technical Committee on Inspection, Testing, and Maintenance of Water-Based Systems included a new chapter in the 2014 edition of NFPA 25 (Chapter 16) to formally extract these requirements from NFPA *101*.

Critical Point

The purpose of NFPA 25 is to ensure that the operational status of a system is maintained.

1.2* Purpose

The purpose of NFPA 25 is to ensure that the operational status of a system is maintained. This section also provides the AHJ with the flexibility to deal with extenuating circumstances such as a product recall or other situations specific to a particular area or project.

A.1.2 History has shown that the performance reliability of a water-based fire protection system under fire-related conditions increases where comprehensive inspection, testing, and maintenance procedures are enforced. Diligence during an inspection is important. The inspection, testing, and maintenance of some items in the standard might not be practical or possible, depending on existing conditions. The inspector should use good judgment when making inspections.

Sprinkler systems have an excellent success record. In those instances when they do fail, the majority of failures can be attributed to closure of control valves or lack of proper maintenance. Commentary Table 1.1 lists some types of failures related to maintaining the operational status of the fire protection system. The failures attributed to manual intervention defeating the equipment relates to equipment being manually shut down after a fire occurred but before sprinklers operated. NFPA 25 addresses wholly or in part all of the issues listed in Commentary Table 1.1.

COMMENTARY TABLE 1.1 *Reasons for Failure to Operate When Fire Was Large Enough to Activate Equipment and Equipment Was Present in Area of Fire*

Cause of Failure	*Percentage of Cases*
Equipment shut off before the fire	63
Manual intervention defeated the equipment	18
Lack of maintenance	6
Damaged component	8

Source: Adapted from John R. Hall, Jr., "U.S. Experience with Sprinklers," NFPA, March 2012, Table 3.2.

1.2.1 The purpose of this document is to provide requirements that ensure a reasonable degree of protection for life and property from fire through minimum inspection, testing, and maintenance methods for water-based fire protection systems.

Case In Point

NFPA 25 was first published as a standard in 1992. The concept of inspection, testing, and maintenance of water-based fire protection systems has a considerably longer history. In 1939, a preliminary progress report of the sprinkler committee was presented during NFPA's 43rd annual meeting. A year later, in 1940, the Association membership officially adopted the first edition of NFPA 13A, *Recommended Practice for the Inspection, Testing, and Maintenance of Sprinkler Systems,* at its annual meeting. Subsequently, NFPA 13A went through 11 revisions, the last of which was published in 1987. The recommended practice was officially withdrawn during the Fall 1993 Association meeting to be replaced by the new NFPA 25, which had been adopted the previous year.

During the Fall 1988 meeting, the NFPA membership adopted the first edition of NFPA 14A, *Recommended Practice for the Inspection, Testing, and Maintenance of Standpipe and Hose Systems*, presented by the Technical Committee on Standpipes. NFPA 14A was developed from existing industry standards and model building codes. It too was withdrawn when NFPA 25 was first introduced.

Both NFPA 13A and 14A were recommended practices, so they contained no mandatory language but offered suggestions for the proper care of sprinkler and standpipe systems. During the development of the final revision of NFPA 13A, several areas needed to be addressed that were outside of the scope of the Technical Committee on Automatic Sprinklers. Discussions on equipment such as underground mains, tanks, and pumps were necessary to provide guidance on the care of these systems. These issues surfaced once again during development of NFPA 14A. As a result, the Correlating Committee on Water Extinguishing Systems recommended that a new document be developed as a standard that contained mandatory language and that included a chapter on each type of basic system under their committee scope. An ad hoc committee was formed in 1989 to develop a committee scope, document scope and purpose, and an outline draft of each chapter. The NFPA Standards Council approved the proposed project in 1990, a technical committee was appointed, and NFPA 25 was born. Although the standard was met with some resistance initially because some felt that maintenance of systems should be covered in a recommended practice and not mandated, the standard now has wide acceptance.

Prior to the development of NFPA 25, maintenance of systems had been well documented as a significant contributing factor in the successful operation of a sprinkler system. First published in the *Fire Journal* in 1965 and again in 1970, "Automatic Sprinkler Performance Tables, 1970 Edition," indicated that sprinklers perform exceptionally well. However, when systems do fail, 63 percent of those failures can be directly attributed to lack of proper maintenance. NFPA 25 addresses each of the leading causes of failures, from water control valves being shut off to inadequate water supplies (addressed in Chapter 7), obstruction to water distribution (addressed in Chapters 5 and 14), defective dry pipe valves (addressed in Chapter 13), and frozen systems (addressed in all "systems" chapters).

1.2.2 In those cases where it is determined that an existing situation involves a distinct hazard to life or property, the authority having jurisdiction shall be permitted to require inspection, testing, and maintenance methods in excess of those required by the standard.

1.3* Application

A.1.3 An entire program of quality control includes, but is not limited to, maintenance of equipment, inspection frequency, testing of equipment, on-site fire brigades, loss control provisions, and personnel training. Personnel training can be used as an alternative even if a specific frequency differs from that specified in this standard.

The increasing availability of new products, materials, and technologies, such as ultrasound examination of piping systems, makes the language in Sections 1.3 and A.1.3 necessary. Products or methods not specifically addressed in the standard are recognized in Section 1.3, provided that the level of system integrity and performance is not lowered.

FAQ Can the specified frequencies in NFPA 25 be altered?

Increasing frequencies for inspections, tests, and maintenance is generally permitted without restriction since NFPA 25 provides the minimum requirements. However, decreasing frequencies from those specified in NFPA 25 cannot be done unconditionally. The standard allows alternative programs for keeping systems operational, but it must be demonstrated that the programs developed deliver equal system integrity and performance. These alternative programs cannot simply be indoctrinated by the entities that develop them, but rather, they must be reviewed and approved by the AHJ (see 1.3.2). A performance-based option for inspection, testing, and maintenance is provided in Section 4.7. This performance-based option is permitted for establishing alternative methods or frequencies for inspection, testing, and maintenance but must be approved by the AHJ.

1.3.1 It is not the intent of this document to limit or restrict the use of other inspection, testing, or maintenance programs that provide an equivalent level of system integrity and performance to that detailed in this document.

1.3.2 The authority having jurisdiction shall be consulted and approval obtained for such alternative programs.

1.4* Units

Metric units of measurement in this standard are in accordance with the modernized metric system known as the International System of Units (SI).

A.1.4 The liter and bar units, which are not part of but are recognized by SI, commonly are used in international fire protection. These units are provided in Table A.1.4 with their conversion factors.

***TABLE A.1.4** Metric Conversions*

Name of Unit	*Unit Symbol*	*Conversion Factor*
liter	L	1 gal = 3.785 L
liter per minute per square meter	$L/min \cdot m^2$	1 gpm/ft^2 = 40.746 $L/min \cdot m^2$
cubic decimeter	dm^3	1 gal = 3.785 dm^3
pascal	Pa	1 psi = 6894.757 Pa
bar	bar	1 psi = 0.0689 bar
bar	bar	1 bar = 10^5 Pa

Note: For additional conversions and information, see IEEE/ASTM-SI-10, *American National Standard for Use of the International System of Units (SI): The Modern Metric System.*

1.4.1 If a value for measurement as given in this standard is followed by an equivalent value in other units, the first stated shall be regarded as the requirement. A given equivalent value shall be considered to be approximate.

1.4.2 SI units have been converted by multiplying the quantity by the conversion factor and then rounding the result to the appropriate number of significant digits. Where nominal or trade sizes exist, the nominal dimension has been recognized in each unit.

All units of measure in this standard are direct mathematical conversions, except pipe diameters, in which case a "soft conversion" has been used to recognize trade sizes for metric pipe. Direct mathematical conversions can be problematic because direct conversions are still subject to rounding. It is important to note that the first stated unit is the requirement and the equivalent (second) unit is considered approximate. In NFPA 25, U.S. customary (English) units are listed first, followed by the SI (metric) equivalent.

References Cited in Commentary

National Fire Protection Association, 1 Batterymarch Park, Quincy, MA 02169-7471.

NFPA 1, *Fire Code,* 2012 edition.

NFPA 3, *Recommended Practice on Commissioning and Integrated Testing of Fire Protection and Life Safety Systems,* 2012 edition.

NFPA 13, *Standard for the Installation of Sprinkler Systems,* 2013 edition.

NFPA 13A, *Recommended Practice for the Inspection, Testing, and Maintenance of Sprinkler Systems,* withdrawn in 1993.

NFPA 13D, *Standard for the Installation of Sprinkler Systems in One- and Two-Family Dwellings and Manufactured Homes,* 2013 edition.

NFPA 13R, *Standard for the Installation of Sprinkler Systems in Low-Rise Residential Occupancies,* 2013 edition.

NFPA 14A, *Recommended Practice for the Inspection, Testing, and Maintenance of Standpipe and Hose Systems,* withdrawn in 1992.

NFPA 72®, National Fire Alarm and Signaling Code, 2013 edition.

NFPA 101®, Life Safety Code®, 2012 edition.

NFPA 5000®, Building Construction and Safety Code®, 2012 edition.

"Automatic Sprinkler Performance Tables, 1970 edition," *Fire Journal,* July 1970, pp. 35–39.

Hall, John R., Jr., "U.S. Experience with Sprinklers," March 2012.

ITM NFPA 25 — INSPECTION TESTING MAINTENANCE

2

REFERENCED PUBLICATIONS

Chapter 2 lists publications that contain mandatory references for use with NFPA 25. Annex G lists publications having nonmandatory informational references. By locating the mandatory referenced publications immediately after Chapter 1, Administration, the user is presented with the complete list of publications needed for effective use of the standard before reading the specific requirements.

The applicable provisions of the publications that are listed in Chapter 2 are considered requirements of NFPA 25. Therefore, whether a requirement resides within NFPA 25 or it is mandatorily referenced in a document from the referenced publications list, the requirement must be met in order to achieve compliance with NFPA 25. However, the requirements contained in the referenced publications are only applicable to the extent called for in NFPA 25.

2.1 General

The documents or portions thereof listed in this chapter are referenced within this standard and shall be considered part of the requirements of this document.

2.2 NFPA Publications

National Fire Protection Association, 1 Batterymarch Park, Quincy, MA 02169-7471.

NFPA 11, *Standard for Low-, Medium-, and High-Expansion Foam,* 2010 edition.

NFPA 13, *Standard for the Installation of Sprinkler Systems,* 2013 edition.

NFPA 13D, *Standard for the Installation of Sprinkler Systems in One- and Two-Family Dwellings and Manufactured Homes,* 2013 edition.

NFPA 14, *Standard for the Installation of Standpipe and Hose Systems,* 2013 edition.

NFPA 15, *Standard for Water Spray Fixed Systems for Fire Protection,* 2012 edition.

NFPA 16, *Standard for the Installation of Foam-Water Sprinkler and Foam-Water Spray Systems,* 2011 edition.

NFPA 20, *Standard for the Installation of Stationary Pumps for Fire Protection,* 2013 edition.

NFPA 22, *Standard for Water Tanks for Private Fire Protection,* 2013 edition.

NFPA 24, *Standard for the Installation of Private Fire Service Mains and Their Appurtenances,* 2013 edition.

NFPA 72®, *National Fire Alarm and Signaling Code,* 2013 edition.

NFPA *101*®, *Life Safety Code*®, 2012 edition.

NFPA 110, *Standard for Emergency and Standby Power Systems,* 2013 edition.

NFPA 307, *Standard for the Construction and Fire Protection of Marine Terminals, Piers, and Wharves,* 2011 edition.

NFPA 409, *Standard on Aircraft Hangars,* 2011 edition.

NFPA 1962, *Standard for the Care, Use, Inspection, Service Testing, and Replacement of Fire Hose, Couplings, Nozzles, and Fire Hose Appliances,* 2013 edition.

2.3 Other Publications

2.3.1 ASTM Publications. ASTM International, 100 Barr Harbor Drive, P.O. Box C700, West Conshohocken, PA 19428-2959.

ASTM D 975-11b, *Standard Specification for Diesel Fuel Oils*, 2011.

ASTM D 3359, *Standard Test Methods for Measuring Adhesion by Tape Test*, 2008.

ASTM D 6751-11b, *Standard Specification for Biodiesel Fuel Blend Stock (B100) for Middle Distillate Fuels*, 2011.

ASTM D 7462-11, *Standard Test Method for Oxidation Stability of Biodiesel (B100) and Blends of Biodiesel with Middle Distillate Petroleum Fuel (accelerated Method)*, 2011.

2.3.2 HI Publications. Hydraulic Institute, 6 Campus Drive, First Floor North, Parsippany, NJ 07054-4406.

HI 3.6, *Rotary Pump Tests*, 2010.

2.3.3 Other Publications.

Merriam-Webster's Collegiate Dictionary, 11th edition, Merriam-Webster, Inc., Springfield, MA, 2003.

2.4 References for Extracts in Mandatory Sections

NFPA 11, *Standard for Low-, Medium-, and High-Expansion Foam,* 2010 edition.

NFPA 13, *Standard for the Installation of Sprinkler Systems,* 2013 edition.

NFPA 14, *Standard for the Installation of Standpipe and Hose Systems,* 2013 edition.

NFPA 15, *Standard for Water Spray Fixed Systems for Fire Protection,* 2012 edition.

NFPA 16, *Standard for the Installation of Foam-Water Sprinkler and Foam-Water Spray Systems,* 2011 edition.

NFPA 20, *Standard for the Installation of Stationary Pumps for Fire Protection,* 2013 edition.

NFPA 24, *Standard for the Installation of Private Fire Service Mains and Their Appurtenances,* 2013 edition.

NFPA 96, *Standard for Ventilation Control and Fire Protection of Commercial Cooking Operations,* 2014 edition.

NFPA *101*®, *Life Safety Code*®, 2012 edition.

NFPA 110, *Standard for Emergency and Standby Power Systems,* 2013 edition.

NFPA 750, *Standard on Water Mist Fire Protection Systems,* 2010 edition.

NFPA 820, *Standard for Fire Protection in Wastewater Treatment and Collection Facilities,* 2012 edition.

NFPA 1141, *Standard for Fire Protection Infrastructure for Land Development in Wildland, Rural, and Suburban Areas,* 2012 edition.

NFPA 1911, *Standard for the Inspection, Maintenance, Testing, and Retirement of In-Service Automotive Fire Apparatus,* 2012 edition.

ITM NFPA 25 — INSPECTION TESTING MAINTENANCE

3

DEFINITIONS

Chapter 3 of NFPA 25 contains definitions of specialized terms used in the standard and systems descriptions to assist the user. The NFPA official definitions listed in Section 3.2 are used throughout NFPA codes and standards for consistency. Section 3.3 lists general fire protection terms. Some of the definitions in this chapter are extracted from other NFPA standards, and they are identified with the standard and edition year in brackets following the definition.

3.1 General

The definitions contained in this chapter shall apply to the terms used in this standard. Where terms are not defined in this chapter or within another chapter, they shall be defined using their ordinarily accepted meanings within the context in which they are used. *Merriam-Webster's Collegiate Dictionary*, 11th edition, shall be the source for the ordinarily accepted meaning.

3.2 NFPA Official Definitions

3.2.1* Approved. Acceptable to the authority having jurisdiction.

A.3.2.1 Approved. The National Fire Protection Association does not approve, inspect, or certify any installations, procedures, equipment, or materials; nor does it approve or evaluate testing laboratories. In determining the acceptability of installations, procedures, equipment, or materials, the authority having jurisdiction may base acceptance on compliance with NFPA or other appropriate standards. In the absence of such standards, said authority may require evidence of proper installation, procedure, or use. The authority having jurisdiction may also refer to the listings or labeling practices of an organization that is concerned with product evaluations and is thus in a position to determine compliance with appropriate standards for the current production of listed items.

The term *approved* is not used often in this standard, since during the course of an inspection or test, the inspector is not required to verify the use of approved components or installation methods. However, there are a number of instances where the standard calls for procedures or programs to be approved.

3.2.2* Authority Having Jurisdiction (AHJ). An organization, office, or individual responsible for enforcing the requirements of a code or standard, or for approving equipment, materials, an installation, or a procedure.

A.3.2.2 Authority Having Jurisdiction (AHJ). The phrase "authority having jurisdiction," or its acronym AHJ, is used in NFPA documents in a broad manner, since jurisdictions and approval agencies vary, as do their responsibilities. Where public safety is primary, the authority having jurisdiction may be a federal, state, local, or other regional department or individual such as a fire chief; fire marshal; chief of a fire prevention bureau, labor department, or health department; building official; electrical inspector; or others having statutory authority. For insurance purposes, an insurance inspection department, rating bureau, or other insurance company representative may be the authority having jurisdiction. In many circumstances, the property owner or his or her designated agent assumes the role of the authority having jurisdiction; at government installations, the commanding officer or departmental official may be the authority having jurisdiction.

The *authority having jurisdiction* (AHJ) is the person or entity enforcing the standard. In cases where the standard is to be legally enforced, the AHJ is usually a representative of a governmental agency such as a fire marshal or building official. The AHJ can also be an insurance company representative or the property owner. It is common for multiple AHJs to have an interest in the same property.

To comply with NFPA 25, a program for inspection, testing, and maintenance (ITM) of systems can exceed the NFPA 25 requirements but cannot be less stringent than those specified unless an alternative program provides an equal level of system integrity and performance. Refer to Section 1.3 and Section 4.6 for more information.

Critical Point

To comply with NFPA 25, a program for inspection, testing, and maintenance of systems can exceed the NFPA 25 requirements but cannot be less stringent than those specified unless an alternative program provides an equal level of system integrity and performance.

3.2.3* Listed. Equipment, materials, or services included in a list published by an organization that is acceptable to the authority having jurisdiction and concerned with evaluation of products or services, that maintains periodic inspection of production of listed equipment or materials or periodic evaluation of services, and whose listing states that either the equipment, material, or service meets appropriate designated standards or has been tested and found suitable for a specified purpose.

A.3.2.3 Listed. The means for identifying listed equipment may vary for each organization concerned with product evaluation; some organizations do not recognize equipment as listed unless it is also labeled. The authority having jurisdiction should utilize the system employed by the listing organization to identify a listed product.

The use of the term *listed* is most often used with installation standards. The term is typically used in NFPA 25 when referring to corrective actions or inspection, testing, or maintenance procedures that are specified with a product's listing. However, it should be kept in mind that listed devices are required or intended for use in conducting certain tests. For example, NFPA 20, *Standard for the Installation of Stationary Pumps for Fire Protection*, requires that listed fixed nozzles, hose valves, or flow meters must be installed for fire pump testing. Where listed devices are not installed, steps should be taken to obtain approval from the AHJ for the use of non-listed test devices as specified in 8.3.3.1, which states that the annual test must be conducted through *approved* test devices.

FAQ Do the terms *classified* and *approved* mean the same as the term *listed*?

When system components or parts are replaced, the related installation standard to be followed identifies the components that are required to be listed. As defined, a listing means that the component has been evaluated under the conditions in which it is expected to perform and that it meets certain product standards. Some testing laboratories refer to "classified" or "approved" components, each of which meets the definition of *listed* in this standard.

3.2.4 Shall. Indicates a mandatory requirement.

The term *shall* indicates a requirement of this standard and mandates that a specific provision of NFPA 25 be followed. Mandatory requirements of NFPA 25 are found in the main body of the standard (Chapter 1 through Chapter 16).

3.2.5 Should. Indicates a recommendation or that which is advised but not required.

The term *should* indicates a recommendation of this standard. It identifies a good idea or a best practice. When the term *should* is used with a provision of the standard, the provision is not meant to be a requirement. Recommended provisions are limited to Annex A through Annex F of NFPA 25. However, some AHJs could mandate that the annex material be enforceable. In such cases, the annex material should be treated as mandatory, but it is not the intent of any NFPA standard for the annex material to be mandatory.

The user of this handbook is reminded that any section number preceded by a letter (e.g., A.3.4) is an annex item. In this handbook, the annex material is placed immediately following the code section it references for the reader's convenience.

3.2.6 Standard. A document, the main text of which contains only mandatory provisions using the word "shall" to indicate requirements and which is in a form generally suitable for mandatory reference by another standard or code or for adoption into law. Nonmandatory provisions are not to be considered a part of the requirements of a standard and shall be located in an appendix, annex, footnote, informational note, or other means as permitted in the *Manual of Style for NFPA Technical Committee Documents*.

3.3 General Definitions

3.3.1 Adjust. To maintain or regulate, within prescribed limits, by setting the operating characteristics to specified parameters. [**1911,** 2012]

3.3.2* Alarm Receiving Facility. The place where alarm or supervisory signals are received.

A.3.3.2 Alarm Receiving Facility. This can include proprietary supervising stations, central supervising stations, remote supervising stations, or public fire service communications centers.

3.3.3* Automatic Detection Equipment. Equipment that automatically detects heat, flame, products of combustion, flammable gases, or other conditions likely to produce fire or explosion and cause other automatic actuation of alarm and protection equipment.

A.3.3.3 Automatic Detection Equipment. Water spray systems can use fixed temperature, rate-of-rise, rate-compensation fixed temperature, optical devices, flammable gas detectors, or products of combustion detectors.

The term *automatic detection equipment* is normally used to describe an electronic detection system and should not be confused with a pilot sprinkler system where the sprinkler is used as a fixed temperature heat detector to hydraulically or pneumatically release the system actuation valve.

3.3.4* Automatic Operation. Operation without human intervention.

A.3.3.4 Automatic Operation. This operation includes, but is not limited to, heat, rate of heat rise, smoke, or pressure change.

3.3.5 Automatic Transfer Switch. Self-acting equipment for transferring the connected load from one power source to another power source. [**110,** 2013]

3.3.6 Clean. To remove dirt, scale, and debris.

3.3.7* Deficiency. For the purposes of inspection, testing, and maintenance of water-based fire protection systems, a condition that will or has the potential to adversely impact the performance of a system or portion thereof but does not rise to the level of an impairment.

A.3.3.7 Deficiency. Depending on the nature and significance of the deficiency, it can result in a system impairment. Critical deficiencies will adversely impact performance but without the need for the implementing impairment procedures. Noncritical deficiencies have the potential to impact performance.

Table A.3.3.7 provides examples for classifying conditions needing repair or correction that are identified during the inspection, testing, and maintenance of water-based suppression systems. The conditions are classified as an impairment, critical deficiency, or noncritical deficiency. The table is not all-inclusive but is included to provide guidance in responding to these conditions. For example, an impairment should be addressed promptly by either immediately correcting the condition or implementing the impairment procedures found in Chapter 15. Critical and noncritical deficiencies should be corrected as soon as practical after considering the nature and severity of the risk. It should be noted that many jurisdictions have requirements for the timely correction of impairments and/or deficiencies.

TABLE A.3.3.7 *Water-Based Fire Protection System Inspection and Testing Findings*

Item	*Finding*	*Reference*	*Impairment*	*Critical Deficiency*	*Noncritical Deficiency*
Chapter 5: Sprinkler Systems — Inspection					
All sprinklers	Leaking — spraying or running water	5.2.1.1.1	X		
All sprinklers	Leaking — dripping water	5.2.1.1.1		X	
All sprinklers	Foreign material attached or suspended from	5.2.1.1.1	X		
All sprinklers	Spray pattern obstructed — less than 18 in. (457 mm) or 36 in. 915 mm) below deflector (stock, furnishings, and equipment), temporary or nonpermanent (signs, banners, decorations, etc.)	5.2.1.1.1		X	
All sprinklers	Lightly loaded	5.2.1.1.1			X
Standard-response sprinklers in nonresidential occupancies	One sprinkler and less than 50% of sprinklers in compartment is heavily loaded or corroded; painted operating element, bulb, deflector, or coverplate; improper orientation; glass bulb has lost fluid	5.2.1.1.1		X	
Standard-response sprinklers in nonresidential occupancies	Two or more sprinklers in compartment are heavily loaded or corroded; painted operating element, bulb, deflector, or coverplate; improper orientation; glass bulb has lost fluid	5.2.1.1.1	X		
Fast-response element, quick-response, residential sprinklers and standard-response in residential occupancies	Any sprinklers, heavily loaded or corroded; painted operating element, bulb, deflector, or coverplate; improper orientation; glass bulb has lost fluid	5.2.1.1.1	X		
Coverplates	Concealed sprinkler coverplates caulked or glued to ceiling	5.2.1.1.1	X		
Escutcheons and coverplates	Missing recessed or flush escutcheons, concealed coverplate with deflector and operating element in correct position	5.2.1.1.6			X

TABLE A.3.3.7 Continued

Item	*Finding*	*Reference*	*Impairment*	*Critical Deficiency*	*Noncritical Deficiency*
Escutcheons and coverplates	Missing recessed or flush escutcheons, concealed coverplate with deflector and operating element not in correct position	5.2.1.1.6	X		
Escutcheons	Recessed or flush escutcheons caulked or glued to ceiling	5.2.1.1.1		X	
Spare sprinkler cabinet	Cabinet missing, temperature over 100°F, not proper number and type, missing wrench for each type	new 5.2.1.3(1), 5.2.1.3(2)			X
Pipe and fittings	Leaking — slowly dripping and/or moisture on surface	5.2.2.1		X	
Pipe and fittings	Leaking — spraying or running water	5.2.2.1	X		
Pipe and fittings	Critical mechanical damage			X	
Hangers and seismic braces	Damaged or loose	5.2.3.2			X
Hangers and seismic braces	Unattached	5.2.3.2		X	
Gauges	Poor condition	5.2.4.1			X
Gauges	Not showing normal water/air pressure	5.2.4.1, 5.2.4.2		X	
Gauges	Freezer — system pressure lower than compressor	5.2.4.4	X		
Alarm devices	Physical damage apparent	5.2.5			X
Hydraulic design information sign	Not attached properly, illegible or missing	5.2.6			X
Information sign	Not attached, illegible or missing	new			X
Heat tape	Not in accordance with manufacturer's instructions	5.2.7		X	
Chapter 5: Sprinkler Systems — Testing					
Gauges	Not replaced or calibrated in 5 years, not accurate within 3% of scale	5.3.2			X
Alarm devices	Water motor and gong not functioning	5.3.3			X
Alarm devices	Pressure switch– or vane-type switch not functioning or no alarm	5.3.3		X	
Antifreeze systems	Mixture and concentration does not meet requirements of 5.3.4.2.1	5.3.4		X	
Antifreeze systems	Concentration is inadequate to prevent freezing	Table A.5.3.4.2.1(1)	X		
Main drain	More than 10% drop in full flow pressure	13.2.5.2		X	
Assessment of internal condition	Inspection revealed presence of MIC, zebra mussels, rust, and scale	14.2.1		X	
Chapter 6: Standpipe and Hose Systems — Inspection					
Pipe and fittings	Leaking — slowly dripping and/or moisture on surface	6.2.1		X	
Pipe and fittings	Leaking — spraying or running water	6.2.1	X		
Pipe and fittings	Critical mechanical damage	6.2.1		X	
Hose	Cuts, couplings not of compatible threads	6.2.1, NFPA 1962		X	
Hose	Deterioration, no gasket or damaged gaskets	6.2.1, NFPA 1962		X	

(continues)

TABLE A.3.3.7 Continued

Item	Finding	Reference	Impairment	Critical Deficiency	Noncritical Deficiency
Hose	Mildew present, corrosion present, hose not connected	6.2.1, NFPA 1962			X
Hose nozzle	Missing, broken parts or thread gasket damaged	6.2.1, NFPA 1962		X	
Hose storage	Hose not properly racked or rolled, nozzle clip missing, nozzle not contained, damaged, obstructed	6.2.1, NFPA 1962		X	
Cabinet	Corroded or damaged parts, not easy to open, not accessible, not identified, door glazing in poor condition, lock not functioning in break glass type, valve, hose nozzle, fire extinguisher, etc. not readily accessible	6.2.1, NFPA 1962		X	
Hydraulic design information sign	Missing	6.2.3			X
Chapter 6: Standpipe and Hose Systems — Testing					
Hose storage device	Rack will not swing out of cabinet at least 90 degrees	6.2.1, NFPA 1962			X
Standpipe system	Test results did not provide design pressure at required flow	6.3.1.1		X	
Hydrostatic test of manual and semiautomatic dry standpipe systems	Leakage in inside piping	6.3.2			X
Main drain	More than 10% drop in full flow pressure	6.3.1.5		X	
Assessment of internal condition	Inspection revealed presence of MIC, zebra mussels, rust, and scale	14.2.1		X	
Chapter 7: Private Fire Service Mains — Inspection					
Exposed piping	Leaking — slowly dripping, and/or moisture on surface	7.2.2.1.2		X	
Exposed piping	Leaking — spraying or running water	7.2.2.1.2	X		
Exposed piping	Mechanical damage, corroded, not properly restrained	7.2.2.1.2		X	
Mainline strainers	Plugged, fouled	7.2.2.3	X		
Mainline strainers	Corroded	7.2.2.3		X	
Dry barrel, wet barrel, and wall hydrant	Inaccessible, barrel contains ice, cracks in barrel	7.2.2.4	X		
Dry barrel, wet barrel, and wall hydrant	Barrel contains water, improper drainage from barrel, leaks at outlets or top of hydrant	7.2.2.4		X	
Dry barrel, wet barrel, and wall hydrant	Tightness of outlets, worn nozzle threads, worn operating nut, missing wrench	7.2.2.4			X
Monitor nozzles	Damaged, corroded, leaking	7.2.2.6		X	
Hose/hydrant houses	Inaccessible	7.2.2.7	X		
Hose/hydrant houses	Damaged	7.2.2.7		X	
Hose/hydrant houses	Not fully equipped	7.2.2.7			X

TABLE A.3.3.7 Continued

Item	Finding	Reference	Impairment	Critical Deficiency	Noncritical Deficiency
Chapter 7: Private Fire Service Mains — Testing					
Underground and exposed piping	Test results not comparable to previous results	7.3.1		X	
Dry barrel and wall hydrant	Hydrant did not flow clear or did not drain within 60 minutes	7.3.2.1, 7.3.2.4			X
Monitor nozzles	Did not flow acceptable amount of water, did not operate throughout their full range	7.3.3.1, 7.3.3.2		X	
Chapter 8: Fire Pumps — Inspection					
Pump house/room	Ventilating louvers not free to operate	8.2.2		X	
Pump house/room	Heat not adequate, temperature less than 40°F	8.2.2(1)	X		
Pump house/room	Heat not adequate, temperature less than 70°F for diesel pumps without engine heaters	8.2.2(1)	X		
Pump house/room	Heat not adequate, temperature less than 40°F, not as recommended by the engine manufacturer, for diesel pumps with engine heaters	8.2.2(1)	X		
Pump system	Suction, discharge, or bypass valves not fully open, pipe leaking, suction line and system line pressure not normal, wet pit suction screens obstructed	8.2.2	X		
Pump system suction	Reservoir empty	8.2.2	X		
Pump system	Suction reservoir does not have required water level, wet pit suction screens missing	8.2.2		X	
Pump system	Minor leaking or drips on floor	8.2.2(2)			X
Pump system	Suction, discharge, or bypass valves not fully open, major leaking such as spraying or leaking to extent that pump performance might be questioned	8.2.2(2)	X		
Electrical power to pump system	No electrical power — controller pilot light not illuminated, transfer switch pilot light not illuminated, isolating switch not closed, reverse phase alarm pilot light on or normal phase light is off, oil level in vertical motor sight glass not normal	8.2.2(3)	X		
Electrical power to pump system	Electrical power is provided — controller pilot light not illuminated, transfer switch pilot light not illuminated, reverse phase alarm pilot light on, normal phase light is not illuminated	8.2.2(3)			X
Electrical power to pump system	Circuit breakers and fuses tripped/open	8.2.2(3)	X		
Diesel engine system	Fuel tank empty	8.2.2	X		

(continues)

TABLE A.3.3.7 Continued

Item	Finding	Reference	Impairment	Critical Deficiency	Noncritical Deficiency
Diesel engine system	Alarm pilot lights are on	8.2.2(4)		X	
Diesel engine system	Battery charging current not normal	8.2.2(4)		X	
Diesel engine system	Battery failure pilot lights on	8.2.2(4)		X	
Diesel engine system	Battery pilot lights off	8.2.2(4)		X	
Diesel engine system	Battery terminals corroded	8.2.2(4)		X	
Diesel engine system	Battery voltage readings not normal	8.2.2(4)		X	
Diesel engine system	Controller selector switch not in auto position	8.2.2(4)	X		
Diesel engine system	Cooling water level not normal	8.2.2(4)			X
Diesel engine system	Cooling water level not visible	8.2.2(4)		X	
Diesel engine system	Crankcase oil level not normal	8.2.2(4)			X
Diesel engine system	Crankcase oil level below low level	8.2.2(4)	X		
Diesel engine system	Electrolyte level in batteries not normal	8.2.2(4)			X
Diesel engine system	Electrolyte level in batteries below top of battery plates	8.2.2(4)		X	
Diesel engine system	Engine running time meter not reading	8.2.2(4)			X
Diesel engine system	Fuel tank less than two-thirds full	8.2.2(4)		X	
Diesel engine system	Water-jacket heater not operating	8.2.2(4)		X	
Diesel engine system	Oil level in right angle gear drive not normal (not at level mark but visible in sight glass)	8.2.2(4)			X
Diesel engine system	Oil level in right angle gear drive below low level (not visible in sight glass or below one finger knuckle for inspection hole)	8.2.2(4)		X	
Steam system	Steam pressure gauge reading not normal	8.2.2		X	
Chapter 8: Fire Pumps — Testing					
Fire pump test	Pump did not start automatically	8.3.2.2	X		
	Pump failed to run for 10 minutes	8.3.2.3	X		
	Pump failed to run for 30 minutes	8.3.2.4	X		
Fire pump test — pump system	System suction and discharge gauge reading, or pump starting pressure not acceptable	8.3.2.8(1)		X	
Fire pump test — pump system	Pump packing gland discharge not acceptable, unusual noise or vibration, packing boxes, bearings, or pump casing overheating	8.3.2.8(1)		X	

TABLE A.3.3.7 Continued

Item	Finding	Reference	Impairment	Critical Deficiency	Noncritical Deficiency
Fire pump test — electrical motor–driven system	Time for motor to accelerate to full speed, time controller is on first step, or time pump runs after starting not acceptable	8.3.2.8(2)	X		
Fire pump test — diesel engine–driven system	Time for engine to crank and time for engine to reach running speed not acceptable (engine to reach rated speed within 20 seconds per 11.2.7.1 of NFPA 20, 2013 edition)	8.3.2.8(3)		X	
Fire pump test — diesel engine–driven system	Low rpm	8.3.2.8(3)	X		
Fire pump test — diesel engine–driven system	Low oil pressure, high temperature, high cooling water pressure	8.3.2.8(3)		X	
Fire pump test — diesel engine–driven system	Time for engine to crank and time for engine to reach running speed not acceptable, low rpm, low oil pressure, high temperature, high cooling water pressure	8.3.2		X	
Fire pump test — steam system	Gauge reading and time for turbine to reach running speed not acceptable	8.3.2		X	
Fire pump test — steam system	Gauge reading and time for turbine to reach running speed not acceptable	8.3.2.8(4)	X		
Fire pump annual test	Circulation relief valve and/or pressure relief valve did not work properly at churn condition	8.3.3.2(1)		X	
Fire pump annual test	Pressure relief valve did not work properly at each flow condition	8.3.3.3		X	
Fire pump annual test (with transfer switch)	Overcurrent protective devices opened when simulating a power failure condition at peak load, power not transferred to alternate source, pump did not continue to perform at peak load, pump did not reconnect to normal power after removing power failure condition	8.3.3.4	X		
Fire pump annual test	Alarms did not properly operate	8.3.3.5		X	
Pump house/room	Heating, lighting, ventilating systems did not pass test	8.3.4.3		X	
Fire pump annual test	Parallel or angular alignment not correct	8.3.4.4		X	
Fire pump annual test	Flow test results not within 5% of acceptance test or nameplate	8.3.5.4		X	
Fire pump annual test	Voltage readings at motor not within 5% below or 10% above rated (nameplate)	8.3.5.6		X	
Fire pump annual test	Flow test results not within 5% of initial unadjusted acceptance test or nameplate	8.3.5.4		X	
Chapter 9: Water Storage Tanks — Inspection					
Water level	Water level and/or condition not correct	9.2.1		X	
Water level	Tank is empty	9.2.1	X		

(continues)

TABLE A.3.3.7 Continued

Item	*Finding*	*Reference*	*Impairment*	*Critical Deficiency*	*Noncritical Deficiency*
Air pressure	Air pressure in pressure tanks not correct	9.2.2	X		
Heating system	Heating system not operational, water temperature below 40°F	9.2.3		X	
Heating system	Water temperature at or below 32°F	9.2.3	X		
Exterior	Tank exterior, supporting structure, vents, foundation, catwalks, or ladders where provided damaged	9.2.5.1			X
Exterior	Area around tank has fire exposure hazard in form of combustible storage, trash, debris, brush, or material	9.2.5.2			X
Exterior	Accumulation of material on or near parts that could result in accelerated corrosion or rot	9.2.5.2			X
Exterior	Ice buildup on tank and support	9.2.5.2		X	
Exterior	Erosion exists on exterior sides or top of embankments supporting coated fabric tanks	9.2.5.2			X
Exterior	Expansion joints leaking or cracking	9.2.5.3		X	
Exterior	Hoops and grilles of wooden tanks in poor condition	9.2.5.4			X
Exterior	Exterior painted, coated, or insulated surfaces of tanks or supporting structure degraded	9.2.5.5			X
Interior (pressure tanks or steel tanks w/o corrosion protection every 3 years, all others every 5 years)	Pitting, corrosion, spalling, rot, other forms of deterioration, waste materials exist, aquatic growth, local or general failure of interior coating	9.2.6.3			X
Interior (pressure tanks or steel tanks w/o corrosion protection every 3 years, all others every 5 years)	Voids beneath floor, with sand in middle of tanks on ring-type foundations	9.2.6.5			X
Interior (pressure tanks or steel tanks w/o corrosion protection every 3 years, all others every 5 years)	Heating system components or piping in poor condition but working	9.2.6.6			X
Interior (pressure tanks or steel tanks w/o corrosion protection every 3 years, all others every 5 years)	Heating system components or heating system piping in poor condition and not working	9.2.6.6	X		

TABLE A.3.3.7 Continued

Item	*Finding*	*Reference*	*Impairment*	*Critical Deficiency*	*Noncritical Deficiency*
Interior (pressure tanks or steel tanks w/o corrosion protection every 3 years, all others every 5 years)	Blockage of antivortex plate	9.2.6.7	X		
Interior (pressure tanks or steel tanks w/o corrosion protection every 3 years, all others every 5 years)	Deterioration of antivortex plate	9.2.6.7		X	
Chapter 9: Water Storage Tanks — Testing					
Interior testing	Tank coating did not pass adhesion, coating thickness, or wet sponge test	9.2.7			X
Interior testing	Tank walls and bottoms did not pass ultrasonic test	9.2.7			X
Interior testing	Tank bottom seams did not pass vacuum-box test	9.2.7			X
Testing	Level indicator not tested after 5 years, lacked freedom of movement, or not accurate	9.3.1		X	
Testing	Low water temperature alarm did not pass test	9.3.3		X	
Testing	High water temperature limit switch did not pass test	9.3.4			X
Testing	High and low water level alarms did not pass test	9.3.5		X	
Gauges	Not tested in 5 years, not accurate within 3% of scale	9.3.6			X
Chapter 10: Water Spray Fixed Systems — Inspection					
Pipe and fittings	Mechanical damage, missing or damaged paint or coating, rusted or corroded, not properly aligned or trapped sections, low point drains not functioning, improper location of rubber-gasketed fittings	10.2.4.1		X	
Hangers and seismic braces	Damaged or missing, not securely attached to structural or piping, missing or damaged paint or coating, rusted or corroded	10.2.4.2		X	
Water spray nozzles	Discharge devices missing, not properly positioned or pointed in design direction, loaded or corroded	10.2.5.1		X	
Water spray nozzles	Missing caps or plugs if required, or not free to operate as intended	10.2.5.2		X	
Strainers	Strainer plugged or fouled	10.2.7	X		

(continues)

TABLE A.3.3.7 Continued

Item	*Finding*	*Reference*	*Impairment*	*Critical Deficiency*	*Noncritical Deficiency*
Strainers	Strainer damaged or corroded	10.2.7			X
Drainage	Trap sumps and drainage trenches blocked, retention embankments or dikes in disrepair	10.2.8			X
Ultra-high-speed	Detectors have physical damage or deposits on lenses of optical detectors	10.4.2		X	
Ultra-high-speed	Controllers found to have faults	10.4.3		X	
Chapter 10: Water Spray Fixed Systems — Testing					
Operational test	Heat detection system did not operate within 40 seconds, flammable gas detection system did not operate within 20 seconds	10.3.4.1.1	X		
Operational test	Nozzles plugged	10.3.4.3.1	X		
Operational test	Nozzles not correctly positioned	10.3.4.3.1		X	
Operational test	Pressure readings not comparable to original design requirements	10.3.4.4		X	
Operational test	Manual actuation devices did not work properly	10.3.6	X		
Main drain	More than 10% drop in full flow pressure	10.3.7.1		X	
Ultra-high-speed operational test	Response time was more than 100 milliseconds	10.4.5	X		
Assessment of the internal condition	Inspection revealed presence of MIC, zebra mussels, rust, and scale	14.2.1		X	
Chapter 11: Foam-Water Sprinkler Systems — Inspection					
Alarm devices	Physical damage apparent	11.1.3.1.3			X
Pipe and fittings	Mechanical damage, missing or damaged paint or coating, rusted or corroded, not properly aligned or trapped sections, low point drains not functioning, improper location or poor condition of rubber-gasketed fittings	11.2.3		X	
Hangers and seismic braces	Damaged or missing, not securely attached to structural or piping, missing or damaged paint or coating, rusted or corroded	11.2.4		X	
Foam-water discharge devices	Discharge devices missing	11.2.5.1	X		
Foam-water discharge devices	Discharge devices not properly positioned or pointed in design direction, loaded or corroded	11.2.5.1		X	
Foam-water discharge devices	Not free to operate as intended	11.2.5.2		X	
Foam-water discharge devices	Missing caps or plugs if required	11.2.5.2		X	
Foam-water discharge devices	Incorrect foam concentrate for application and devices	11.2.5.4		X	
Foam concentrate strainers	Blow-down valve open or not plugged	11.2.7.2		X	
Drainage	Trap sumps and drainage trenches blocked, retention embankments or dikes in disrepair	11.2.8			X

TABLE A.3.3.7 Continued

Item	*Finding*	*Reference*	*Impairment*	*Critical Deficiency*	*Noncritical Deficiency*
Proportioning systems (all)	Proportioning system valves not in correct open/closed position in accordance with specified operating conditions	11.2.9.3	X		
Proportioning systems (all)	Concentrate tank does not have correct quantity required by original design	11.2.9.4		X	
Proportioning systems (all)	Concentrate tank empty	11.2.9.4	X		
Standard pressure proportioner	Automatic drains (ball drip valves) not free or open, external corrosion on foam concentrate tanks	11.2.9.5.1			X
Bladder tank proportioner	Water control valve to foam concentrate in "closed" position	11.2.9.5.2	X		
Bladder tank proportioner	Foam in water surrounding bladder	11.2.9.5.2	X		
Bladder tank proportioner	External corrosion on foam concentrate tank	11.2.9.5.2			X
Line proportioner	Strainer damaged, corroded, pressure vacuum vent not operating freely	11.2.9.5.3		X	
Line proportioner	Strainer plugged or fouled	11.2.9.5.3	X		
Line proportioner	External corrosion on foam concentrate tank	11.2.9.5.3			X
Standard balanced pressure proportioner	Sensing line valves not open, no power to foam liquid pump	11.2.9.5.4	X		
Standard balanced pressure proportioner	Strainer damaged, corroded, plugged, or fouled, pressure vacuum vent not operating freely, gauges damaged or not showing proper pressures	11.2.9.5.4		X	
In-line balanced pressure proportioner	Sensing line valves at pump unit or individual proportioner stations not open, no power to foam liquid pump	11.2.9.5.5	X		
In-line balanced pressure proportioner	Strainer damaged, corroded, pressure vacuum vent not operating freely, gauges damaged or not showing proper pressures	11.2.9.5.5		X	
In-line balanced pressure proportioner	Strainer plugged or fouled	11.2.9.5.5	X		
Orifice plate proportioner	No power to foam liquid pump	11.2.9.5.6	X		
Orifice plate proportioner	Strainer damaged, corroded, pressure vacuum vent not operating freely, gauges damaged or not showing proper pressures	11.2.9.5.6		X	
Orifice plate proportioner	Strainer plugged or fouled	11.2.9.5.6	X		
Chapter 11: Foam-Water Sprinkler Systems — Testing					
Alarm devices	Water motor and gong not functioning	11.1.3.1.1, 11.3.1.1		X	
Alarm devices	Pressure switch or vane-type switch not functioning or no alarm	11.1.3.1.2, 11.3.1.2		X	

(continues)

TABLE A.3.3.7 Continued

Item	*Finding*	*Reference*	*Impairment*	*Critical Deficiency*	*Noncritical Deficiency*
Operational test	Fire detection system did not operate within requirements of *NFPA 72*	11.3.2.4		X	
Operational test	Nozzles plugged	11.3.2.6.1	X		
Operational test	Nozzles not correctly positioned	11.3.2.6.1		X	
Operational test	Pressure readings not comparable to original design requirements	11.3.2.7.3		X	
Operational test	Manual actuation devices not working properly	11.3.4	X		
Operational test	Foam sample failed concentration test	11.3.5	X		
Main drain	More than 10% drop in full flow pressure	13.2.5.2		X	
Assessment of internal condition	Inspection revealed presence of MIC, zebra mussels, rust, and scale	14.2.1		X	
Chapter 13: Valves, Valve Components, and Trim — Inspection					
Gauges	Poor condition	13.2.7.1			X
Gauges	Not showing normal water/air pressure	13.2.7.1		X	
Control valve	Improper closed position	13.3.2.2	X		
Control valve	Improper open position, leaking	13.3.2.2		X	
Control valve	Not accessible, no appropriate wrench if required, no identification	13.3.2.2			X
Control valve	Not sealed, locked, or supervised	13.3.2.2		X	
Alarm valve	External physical damage, trim valves not in appropriate open or closed position, retard chamber or alarm drain leaking	13.4.1.1		X	
Valve enclosure	Upon visual observation, enclosure not maintaining minimum 40°F (4°C) temperature	13.4.3.1.1, 13.4.4.1.1		X	
Valve enclosure	Low temperature alarms (if installed) are physically damaged	13.4.3.1.1, 13.4.4.1.1		X	
Preaction valve and deluge valve	External physical damage, trim valves not in appropriate open or closed position, valve seat leaking	13.4.3.1.6		X	
Preaction valve and deluge valve	Electrical components not in service	13.4.3.1.6	X		
Dry pipe valve/ quick-opening device	External physical damage, trim valves not in appropriate open or closed position, intermediate chamber leaking	13.4.4.1.4		X	
Sprinkler pressure-reducing control valves	Not in open position	13.5.1.1	X		
Sprinkler pressure-reducing control valves	Not maintaining downstream pressures in accordance with design criteria	13.5.1.1		X	
Sprinkler pressure-reducing control valves	Leaking, valve damaged, hand wheel missing or broken	13.5.1.1		X	
Hose connection pressure-reducing valves	Hand wheel broken or missing, hose threads damaged, leaking, reducer missing	13.5.2.1		X	

TABLE A.3.3.7 Continued

Item	Finding	Reference	Impairment	Critical Deficiency	Noncritical Deficiency
Hose connection pressure-reducing valves	Cap missing	13.5.2.1			X
Hose rack assembly pressure-reducing valve	Hand wheel broken or missing, leaking	13.5.3.1		X	
Hose valves	Leaking, visible obstructions, caps, hose threads, valve handle, cap gasket, no restricting device, damaged, or in poor condition	13.5.6.1		X	
Hose valves	Hose threads not compatible	13.5.6.1	X		
Backflow prevention assemblies	Reduced-pressure assemblies, differential-sensing valve relief port continuously discharging	13.6.1.2		X	
Fire department connection	Not accessible, damaged couplings, or clapper not operating properly or missing	13.7.1	X		
Fire department connection	Couplings and swivels damaged, do not rotate smoothly, check valve leaking, automatic drain not operating properly or missing	13.7.1		X	
Fire department connection	Missing identification sign	13.7.1			X
Chapter 13: Valves, Valve Components, and Trim — Testing					
Main drain	More than 10% drop in full flow pressure	13.2.5.2		X	
Alarm devices	Water motor and gong not functioning	13.2.6.1		X	
Alarm devices	Pressure switch or vane-type switch not functioning, no alarm	13.2.6.2		X	
Gauges	Not replaced or calibrated in 5 years, not accurate within 3% of scale	13.2.7.2, 13.2.7.3			X
Control valve	Valve not operating through its full range	13.3.3.1		X	
Control valve	No spring or torsion felt in rod when opening post indicator valve	13.3.3.2	X		
Supervisory switches	No signal from two revolutions of hand wheel from normal position or when stem has moved one-fifth of distance from normal position, signal restored in position other than normal	13.3.3.5.2		X	
Preaction valve	Priming water level not correct	13.4.3.2.1		X	
Preaction valve	Pressure reading at hydraulically most remote nozzle and/or at valve not comparable to original design values	13.4.4.2.2.2		X	
Preaction valve	Three-year leakage test failed	13.4.3.2.6		X	
Deluge valve	Annual full flow trip test revealed plugged nozzles, manual actuation devices did not operate properly	13.4.3.2.2.3	X		
Deluge valve	Pressure reading at hydraulically most remote nozzle and/or at valve not compatible with original design values	13.4.3.2.2.3		X	

(continues)

TABLE A.3.3.7 Continued

Item	Finding	Reference	Impairment	Critical Deficiency	Noncritical Deficiency
Preaction valve	Low air pressure switch did not send signal, no alarm	13.4.3.2.12		X	
Preaction and deluge valve	Low temperature switch did not send signal, no alarm	13.4.3.2.13		X	
Preaction valve	Automatic air maintenance device did not pass test	13.4.3.2.14			X
Dry pipe valve	Priming water level not correct	13.4.4.2.1		X	
Dry pipe valve	Test results not comparable with previous results	13.4.4.2.2		X	
Quick-opening device	Quick-opening device did not pass test	13.4.4.2.4		X	
Dry pipe valve	Low air pressure switch did not send signal, no alarm	13.4.4.2.6		X	
Dry pipe valve	Low temperature switch did not send signal, no alarm	13.4.4.2.7		X	
Dry pipe valve	Automatic air maintenance device did not pass test	13.4.4.2.8		X	
Dry pipe system	Three-year leakage test failed	13.4.4.2.9		X	
Sprinkler pressure-reducing control valves	Test results not comparable to previous results	13.5.1.2		X	
Hose connection pressure-reducing valves	Test results not comparable to previous results	13.5.2.2		X	
Hose rack assembly pressure-reducing valve	Test results not comparable to previous results	13.5.3.2		X	
Hose valves (Class I and Class III standpipe system)	Annual test revealed valve leaking or difficult to operate	13.5.6.2.1.1		X	
Hose valves (Class II standpipe system)	Test revealed valve leaking or difficult to operate	13.5.6.2.2, 13.5.6.2.2.1		X	
Backflow prevention assemblies	Did not pass forward flow test	13.6.2.1	X		

The table does not take into account every variation of the conditions needing repair or correction. For example, a single lightly painted sprinkler in a large warehouse might be noncritical in its risk while a single painted sprinkler in a battery-charging station might be considered a critical deficiency or perhaps an impairment. In addition, the nature of the hazard or the life safety exposure of the occupancy should be considered when assigning a classification. The table should be used with good judgment and could require input from the authority having jurisdiction.

In previous editions of NFPA 25, the inspection, testing, and maintenance requirements were presented as pass/fail requirements, because no other options for identifying system status were presented in the standard. In other words, the standard considered problems with a system only in the context of passing or failing a requirement within the standard. In the 2008 edition, tables were added to each chapter to provide guidance on component replacement action, but the standard still did not address when or if a certain violation of a requirement within the standard would result in the system being impaired. As a result, some AHJs treated any and all deficiencies as "impairments" and applied the requirements of Chapter 15, Impairments. For example, some AHJs considered a system "impaired" when minor problems were noted, such as a missing hydraulic nameplate or a missing sprinkler in the spare sprinkler cabinet.

Guidance was needed to determine which deficiency would result in what level of operational system status tagging for consistency of interpretation and enforcement. The solution was to develop sufficient guidance within the standard to clarify which deficiencies result in systems that need to be fixed but are otherwise anticipated to function properly in an emergency, and which deficiencies affect system operation or even rise to the level of an impairment.

As a result, in the 2011 edition of NFPA 25, the definition of the term *deficiency* was revised and two new definitions were added — *critical deficiency* and *noncritical deficiency* — with the intent of providing clearer guidance for inspectors and AHJs in determining the severity of violations and the subsequent corrective actions that are necessary. Table A.3.3.7 was developed to provide examples of classifications of some of the needed corrections and repairs that are identified in the system chapters of NFPA 25. While Table A.3.3.7 is not all-inclusive, it can help users determine the potential categories of deficiencies. These categories are strictly advisory, but should add some degree of consistency in interpreting when a system should be tagged as deficient or as impaired.

In the 2014 edition, Annex E was relocated to Annex A because it was often overlooked due to its location in the back of the book. Moving the table into what is considered by many users to be the "main annex" should make it more accessible, and hopefully it will be used more frequently by those trying to classify deficiencies. Since the table addresses potential classifications of deficiencies, it made sense to attach it to the definition of the term *deficiency*. While the table is moving forward in the standard, its presence in Annex A does not make it any more or less enforceable than when it was in Annex E, just more accessible.

Assigning the appropriate deficiency or impairment level to field conditions that are identified while conducting inspections is a critical part of the ITM process. It is also what many believe to be the most difficult part of the process, since it is so subjective. While NFPA 25 does not specifically define a "tagging structure" for identifying these levels, many states have developed a series of different color tags that are placed on a system or component to identify the significance of the condition identified.

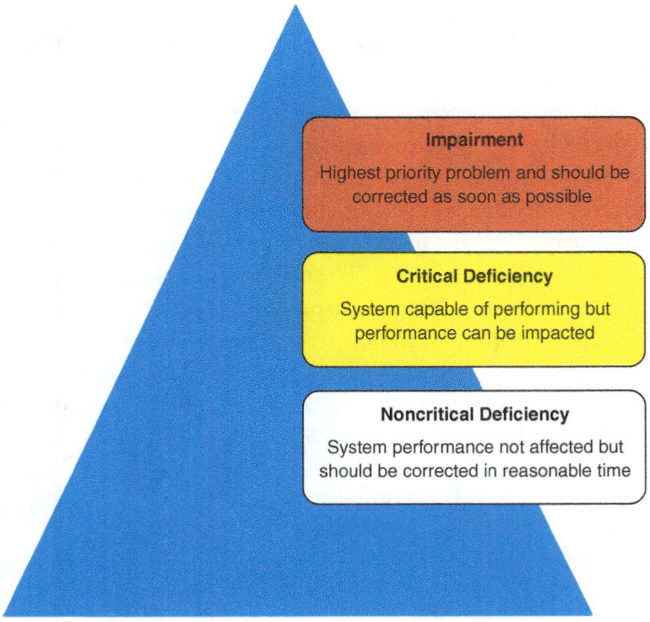

Deficiency Levels.

This handbook contains a new feature that is intended to help readers learn about what goes into the process of classifying deficiencies and impairments. Using Table A.3.3.7 as a guide, this recurring feature examines dozens of field conditions and provides a glimpse at the thought process behind determining the severity of the condition and, ultimately, if the condition should be listed as an impairment, critical deficiency, or noncritical deficiency.

In order to properly classify a deficient condition, the definitions in Chapter 3 become critically important. An impairment, as defined in 3.3.21, is a condition that renders a system inoperable or ineffective during a fire event. Many people believe that an impaired system is a system that has been shut off; however, that is only one application of the term. When a sprinkler is leaking, painted, or damaged, that device, and by extension a portion of the system, is impaired. Simply put, if a fire were to occur and a system or a portion of the system would not function as designed, it would qualify as an impairment.

A deficiency that does not leave a system impaired but, if it is not corrected, could lead to an impairment would qualify as a critical deficiency. A pipe that is showing corrosion or a sprinkler that has loading on the frame arm will likely still function during a fire event; however, if the condition were to get any worse it could lead to an impairment. Finally, noncritical deficiencies, as defined by 3.3.7.2, refer to noncompliant conditions that will not impact system functionality, such as missing signs and broken gauges, or that have an unlikely chance of causing a given system to be ineffective during a fire.

While these definitions help the user to classify deficiencies, in many cases, the inspector will need to make some judgments. Judgments must be made since these classifications are looking for some sort of prognostication on what the future holds for a system based on, in many cases, what a single component looks like. In a perfect world, classifying deficiencies would be black and white; however, in most cases, the answer is some shade of gray.

When looking at these case studies of deficient conditions, there are a few things to keep in mind. First and foremost, there is no "correct" classification for most deficiencies. That is why Table A.3.3.7 resides in the annex as opposed to the body of NFPA 25. Even the technical committee that writes the standard, which is composed of experts on the ITM process, is not in a position to mandate specific levels of criticality for common deficiencies.

The purpose of this feature is not to tell the user exactly how to classify a deficiency if they run into something that looks like a case study shown in this handbook, but rather to study the thought process associated with applying the appropriate tag. It is important to remember that the photographs that are used in this feature only provide a snapshot of the field condition found, without all of the surrounding facts that might play into the proper classification of a condition. In some instances, the inspector needs to be aware of the cumulative effects of the observed combination of conditions for all of the system components and the protected area that might lead to a "likely" adverse impact on the system operation in order to make good judgments. The dynamic nature of deficient mechanical systems makes it difficult to show the layered impact of multiple conditions in a single photograph; therefore, this feature takes a more static approach and focuses on the likelihood of a condition adversely affecting system operation.

To that end, the snapshot does not necessarily tell the whole story. Therefore, the way that the condition is classified in this handbook might not always be the appropriate classification for a similar, but not identical, condition identified during an inspection. Another critical concept to keep in mind is that these images assume that the inspector could see the condition being addressed from the floor level. If a condition similar to what is being shown in this feature exists in a building, but it is not something the inspector could see as part of the inspection from floor level, it would not need to be identified in the NFPA 25 inspection report.

This feature uses a series of tag colors to represent the level of deficiency being depicted. For the purpose of this handbook, the following color scheme will be used:

Red tag = impairment

Yellow tag = critical deficiency

White tag = noncritical deficiency

System Tagging

These photos represent the types of case studies that will be reviewed throughout this book. One has been provided for each of the three tag conditions that will be considered.

In the photo below on the left, the presence of corrosion on the flange bolts in and of itself is not a major concern, however note that the bolts on the upper flange have already been replaced. Also note that one of the bolts on the lower flange has sheared off. Because some of the bolts on this elbow already needed to be replaced and other bolts are brittle enough to break, these bolts, although not in sufficiently bad shape to cause leaking, should be replaced.

This image in the center below shows a single sprinkler with storage located less than 1 ft (0.3 m) immediately below the deflector and more storage located behind the sprinkler that is almost level with the deflector. In many cases, the operations manager is unaware that storing goods within 18 in. (0.46 m) of the deflector is not permitted, and this condition exists over hundreds if not thousands of feet of floor area. These boxes will interrupt spray pattern development and can lead to large unprotected areas of stored goods. This condition can lead to system ineffectiveness, and therefore, it must be considered a critical deficiency. In situations where there are one or two boxes obstructing a single sprinkler in a large room, the inspector might consider looking at it as a noncritical deficiency and providing recommendations for the facility operations and maintenance manual.

A single residential sprinkler that is painted, like the one in the photo below on the right, or one that is heavily loaded would need to be considered an impairment. Table A.3.3.7 makes a distinction between classifications in residential and nonresidential occupancies, highlighting the importance of not just looking at a single component, but getting the complete picture of the facility and the impact of the observed condition in that facility.

(Courtesy of Byron Blake and SimplexGrinnell)

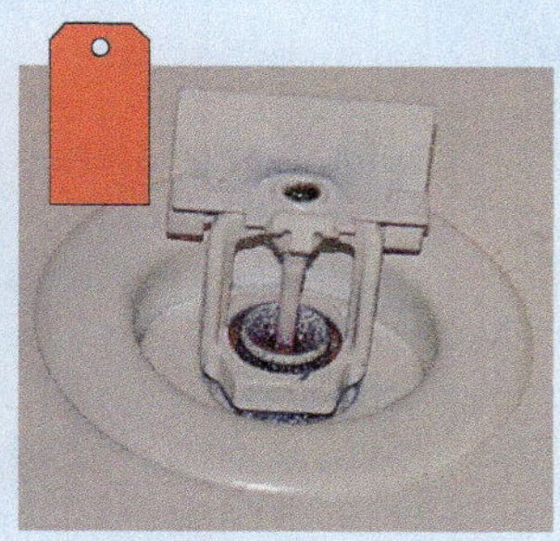

(Courtesy of Byron Blake and SimplexGrinnell)

Noncritical Deficiency — Critical Deficiency — Impairment

At present, there are several different tagging programs that are observed by various state, local, and private jurisdictions. These tagging programs use anywhere from two to five tags, with varying color schemes. For example, in the state of Florida, green tags are used to represent systems that are in compliance with the applicable provisions, and red tags represent systems that contain deficiencies or impairments. The system used in this handbook is not intended to replicate one of these systems, but rather to be used in the absence of a universally accepted tagging program.

3.3.7.1 Critical Deficiency. A deficiency that, if not corrected, can have a material effect on the ability of the fire protection system or unit to function as intended in a fire event.

3.3.7.2 Noncritical Deficiency. A deficiency that does not have a material effect on the ability of the fire protection system or unit to function in a fire event, but correction is needed to meet the requirements of this standard or for the proper inspection, testing, and maintenance of the system or unit.

Critical Point

While it is clear in the standard that all deficiencies are required to be corrected, NFPA 25 recognizes that all deficiencies do not have the same impact on a system's ability to provide a reasonable degree of protection.

One of the more difficult tasks associated with NFPA 25 is determining how to treat the various deficiencies that are found while conducting inspections and tests. While it is clear in the standard that all deficiencies are required to be corrected, NFPA 25 recognizes that all deficiencies do not have the same impact on a system's ability to provide a reasonable degree of protection. So although it might be desirable to treat the correction of all deficiencies the same, it is not practical.

In deference to the concept of reasonableness, the standard recognizes that judgment should be used when determining how to address the corrective actions for deficiencies. To assist with this, the standard separates deficiencies into three basic categories: impairment, critical deficiency, and noncritical deficiency. An impairment, which is defined as a condition where a system or a portion of a system might not function in a fire event, requires immediate action by either correcting the deficiency or implementing the impairment procedures found in Chapter 15.

A system with critical and noncritical deficiencies will still operate, but its performance might be impacted negatively in some manner. The term *material effect* is used in both definitions to provide separation between these deficiencies. There is no way for judgment to be excluded with the classification of deficiencies. Table A.3.3.7 provides guidance, but even with the inclusion of this information, it is impossible to classify every situation that might be encountered. Some inspectors classify deficiencies, but the most common industry practice is to have the inspector record all observed deficiencies, followed by a subsequent classification conducted by a qualified individual or team. However, if the inspector identifies any severe impairments, such as a shut water control valve or a inoperable fire pump, then those instances would need to be reported immediately.

FAQ Must all deficiencies be treated equally with regard to reporting and corrective action?

Many jurisdictions have enacted differing reporting procedures, depending on the type or severity of the deficiency. Other jurisdictions treat all deficiencies with equal measure without regard to the type or severity. It is the intent of the Technical Committee on Inspection, Testing, and Maintenance of Water-Based Systems to define the differing levels of deficiencies in order to provide better consistency in interpretation and enforcement. Examples of critical deficiencies include a painted or damaged sprinkler, a faulty pressure-reducing hose valve, or an inaccessible fire hydrant. Noncritical deficiencies include missing signs, a broken handle on an inspector's test valve, or an inadequate number of spare sprinklers. Thus, a large number of painted sprinklers in a warehouse typically would be of more concern than a single painted sprinkler, and damaged threads on a hose connection might require a more immediate response than a missing sign on a control valve. While all deficiencies must be corrected, it is expected that the type of deficiency will dictate the appropriate level of response.

The correction of some deficiencies might require the system or a portion of it to be taken out of service. Such an action would be considered a preplanned impairment. Chapter 15 of the standard addresses the actions needed for preplanned impairments. Information in Annex A provides guidance for determining what level of corrective action should take place to bring the system back on line. It is not the intent of the standard that noncritical and critical deficiencies require impairment procedures such as fire watches, building evacuation, or standby fire protection during the time that the deficiency is being corrected, unless the corrective action requires impairment of the system, such as shutting off the water supply to replace a sprinkler.

3.3.8 Discharge Device. A device designed to discharge water or foam-water solution in a predetermined, fixed, or adjustable pattern. Examples include, but are not limited to, sprinklers, spray nozzles, and hose nozzles. [**16,** 2011]

3.3.9 Double Check Valve Assembly (DCVA). This assembly consists of two internally loaded check valves, either spring-loaded or internally weighted, installed as a unit between two tightly closing resilient-seated shutoff valves as an assembly, and fittings with properly located resilient-seated test cocks.

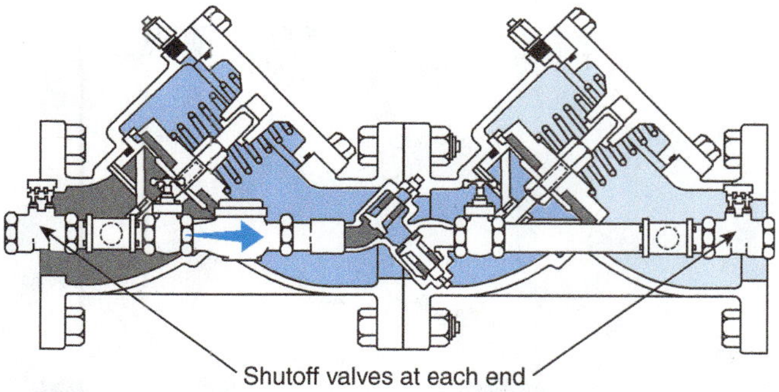

EXHIBIT 3.1 Double Check Valve Assembly (top) and Interior of a Double Check Detector Assembly (bottom). (Courtesy of CRD Water Services, Canada)

The double check valve assembly (DCVA) is also referred to as a double check detector assembly (DCDA). The two differ slightly, but the purpose of both is to protect the potable water supply line from possible contamination or pollution from the fire system, back-pressure from fire line booster pumps, or stagnant "black water" that sits in fire lines over extended periods of time. The DCDA contains a bypass assembly with a water meter along with the double check valve and two tightly closing control valves or gate valves. Exhibit 3.1 shows both a DCVA and the interior of a DCDA.

3.3.10 Drain.

| **3.3.10.1 Main Drain.** The primary drain connection located on the system riser.

Main drains are used as a test connection (see 13.2.5). Although flow can be estimated from a main drain connection, the main drain test is not intended to evaluate a water supply for hydraulic purposes. The purpose of the main drain test is to verify that water supply valves are open and to reveal any changes in the condition of the water supply by comparing results to those of previous tests. A main drain test on a system with large supply piping might not indicate a drop in pressure, even though the water supply might have deteriorated. For systems with large supply piping, additional flow testing outside of what is required by NFPA 25 might need to be conducted to ascertain a more thorough understanding of the condition of the water supply.

3.3.10.2* Sectional Drain. A drain located beyond a sectional control valve that drains only a portion of the system.

A.3.3.10.2 Sectional Drain. An example of a sectional drain is a drain located beyond a floor control valve on a multistory building.

Sectional drains are not same as "auxiliary drains." Auxiliary drains are required for trapped sections of piping that cannot be drained using the main or sectional drain valves. Auxiliary drains for a wet sprinkler system can consist of a fitting with a plug or they can be complete with a valve, nipple, and cap. The size and configuration of the auxiliary drain depend on the type of system and the capacity of the trapped section of the system. Exhibit 3.2 illustrates an auxiliary drain for a dry pipe system where the volume of trapped water is 5 gal (18.9 L) or more.

EXHIBIT 3.2 Dry System Auxiliary Drain.

3.3.11 Fire Department Connection. A connection through which the fire department can pump supplemental water into the sprinkler system, standpipe, or other system furnishing water for fire extinguishment to supplement existing water supplies.

Fire department connections (FDCs) are an integral part of a fire protection system, because they provide supplemental flow and pressure to the sprinkler and standpipe systems. Because of their location on the outside of the building, FDCs are prone to failure due to tampering or the exterior environment. Exhibit 3.3 through Exhibit 3.5 show FDCs that have been compromised for a multitude of reasons.

EXHIBIT 3.3 Birds Nest in FDC. (Courtesy of Wayne Automatic Fire Sprinklers Inc.)

EXHIBIT 3.4 Inaccessible FDC. (Courtesy of Wayne Automatic Fire Sprinklers Inc.)

EXHIBIT 3.5 Trash-Filled FDC.

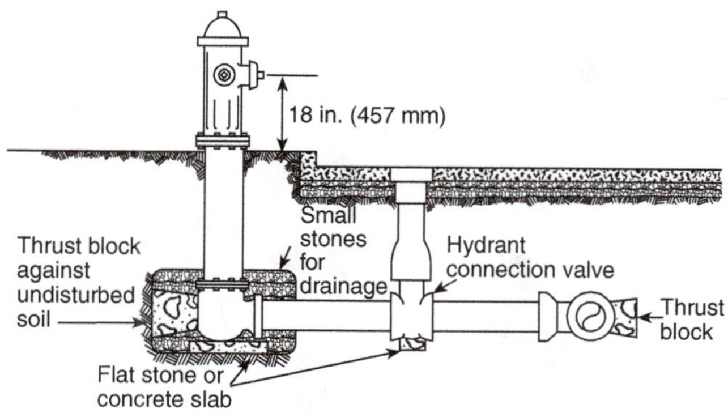

FIGURE A.3.3.12(a) Typical Fire Hydrant Connection.

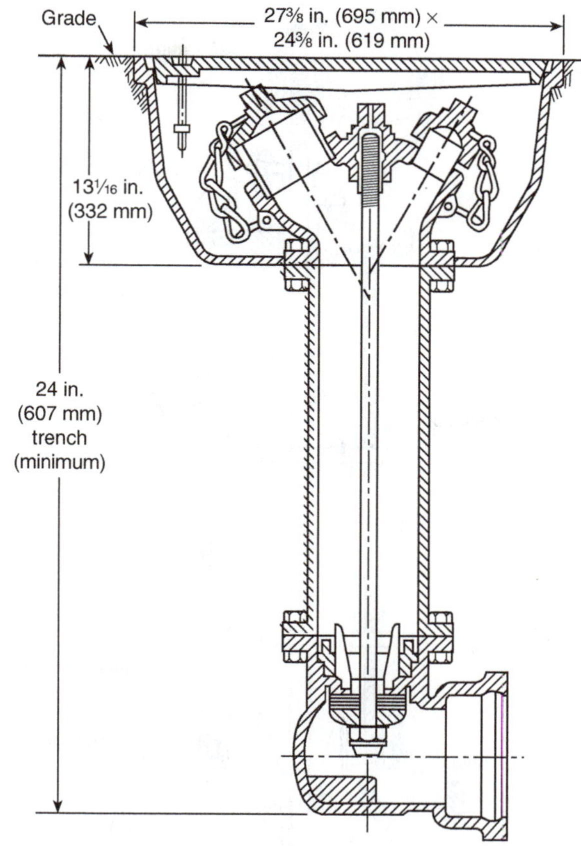

FIGURE A.3.3.12(b) Flush-Type Hydrant.

3.3.12* Fire Hydrant. A valved connection on a water supply system having one or more outlets and that is used to supply hose and fire department pumpers with water. [**1141,** 2012]

A.3.3.12 Fire Hydrant. See Figure A.3.3.12(a) and Figure A.3.3.12(b).

3.3.12.1* Dry Barrel Hydrant (Frostproof Hydrant). A type of hydrant with the main control valve below the frost line between the footpiece and the barrel.

A.3.3.12.1 Dry Barrel Hydrant (Frostproof Hydrant). A drain is located at the bottom of the barrel above the control valve seat for proper drainage after operation to prevent freezing. See Figure A.3.3.12.1.

3.3.12.2* Monitor Nozzle Hydrant. A hydrant equipped with a monitor nozzle capable of delivering more than 250 gpm (946 L/min).

A.3.3.12.2 Monitor Nozzle Hydrant. See Figure A.3.3.12.2.

3.3.12.3* Wall Hydrant. A hydrant mounted on the outside of a wall of a building, fed from interior piping, and equipped with control valves located inside the building that normally are key-operated from the building's exterior.

A.3.3.12.3 Wall Hydrant. See Figure A.3.3.12.3.

3.3.12.4* Wet Barrel Hydrant. A type of hydrant that sometimes is used where there is no danger of freezing weather. Each outlet on a wet barrel hydrant is provided with a valved outlet threaded for fire hose. [**24,** 2013]

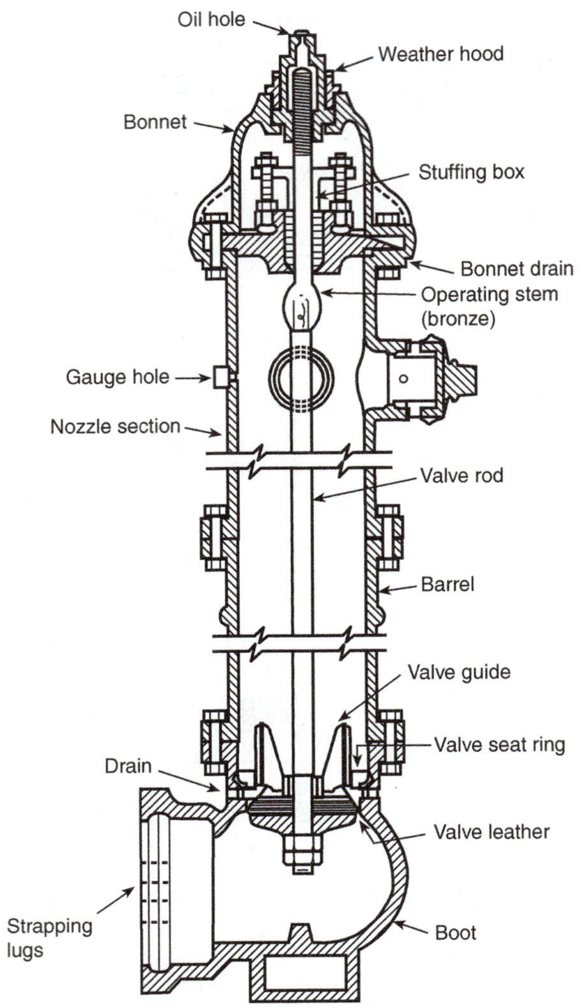

FIGURE A.3.3.12.1 *Dry Barrel Hydrant.*

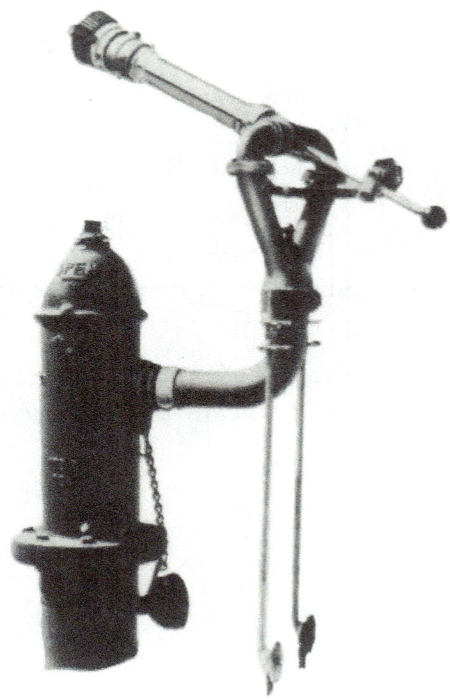

FIGURE A.3.3.12.2 *Hydrant with Monitor Nozzle.*

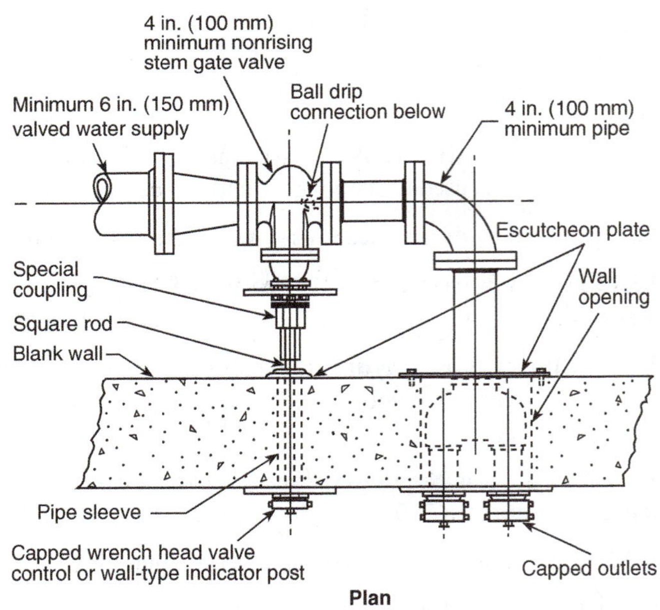

FIGURE A.3.3.12.3 *Wall Hydrant.*

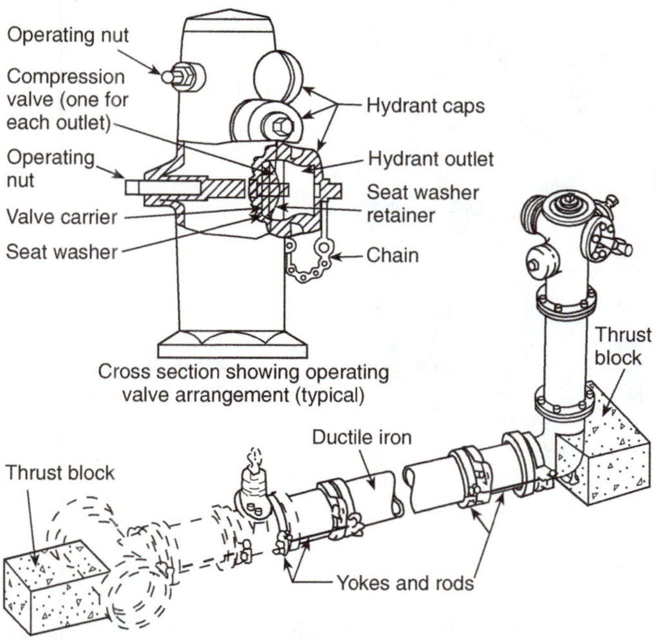

FIGURE A.3.3.12.4 Wet Barrel Hydrant. (Courtesy of the Los Angeles Department of Water and Power.)

A.3.3.12.4 Wet Barrel Hydrant. See Figure A.3.3.12.4.

3.3.13* Foam Concentrate. A concentrated liquid foaming agent as received from the manufacturer. [**11,** 2010]

A.3.3.13 Foam Concentrate. For the purpose of this document, *foam concentrate* and *concentrate* are used interchangeably.

3.3.14 [Foam] Discharge Device. A device designed to discharge water or foam-water solution in a predetermined, fixed, or adjustable pattern. Examples include, but are not limited to, sprinklers, spray nozzles, and hose nozzles. [**16,** 2011]

3.3.15 Hose Connection. A combination of equipment provided for connection of a hose to the standpipe system that includes a hose valve with a threaded outlet. [**14,** 2013]

An example of a hose connection is shown in Exhibit 3.6.

EXHIBIT 3.6 Hose Connection. (Courtesy of Byron Blake and SimplexGrinnell)

3.3.16* Hose House. An enclosure located over or adjacent to a hydrant or other water supply designed to contain the necessary hose nozzles, hose wrenches, gaskets, and spanners to be used in fire fighting in conjunction with and to provide aid to the local fire department.

A.3.3.16 Hose House. See Figure A.3.3.16(a) through Figure A.3.3.16(c).

3.3.17 Hose Nozzle. A device intended for discharging water for manual suppression or extinguishment of a fire.

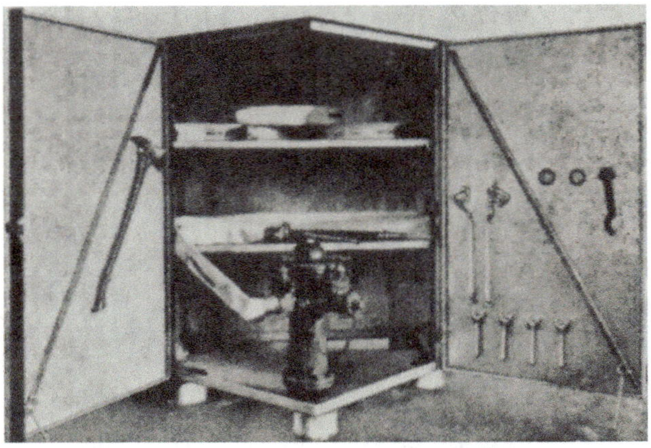

FIGURE A.3.3.16(a) Hose House of Five-Sided Design for Installation over Private Hydrant.

FIGURE A.3.3.16(b) Steel Hose House of Compact Dimensions for Installation over Private Hydrant. House is shown closed; top lifts up, and doors on front side open for complete accessibility.

FIGURE A.3.3.16(c) Hose House That Can Be Installed on Legs, As Pictured, or on Wall Near, but Not Directly over, Private Hydrant.

3.3.18 Hose Station. A combination of a hose rack, hose nozzle, hose, and hose connection. [**14,** 2013]

An example of a hose station is shown in Exhibit 3.7.

3.3.19 Hose Storage Devices.

3.3.19.1* Conventional Pin Rack. A hose rack where the hose is folded vertically and attached over the pins.

A.3.3.19.1 Conventional Pin Rack. See Figure A.3.3.19.1.

3.3.19.2* Horizontal Rack. A hose rack where the hose is connected to the valve, then stack-folded horizontally to the top of the rack.

A.3.3.19.2 Horizontal Rack. See Figure A.3.3.19.2.

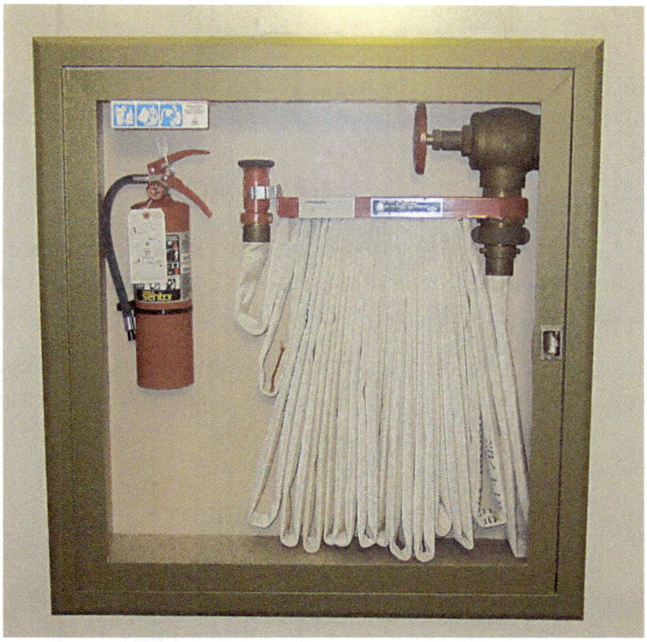

EXHIBIT 3.7 *Hose Station. (Courtesy of Byron Blake and SimplexGrinnell)*

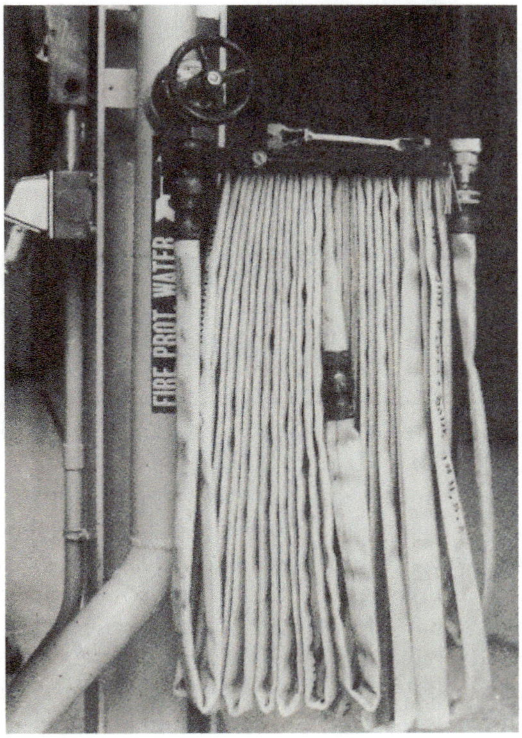

FIGURE A.3.3.19.1 *Conventional Pin Rack.*

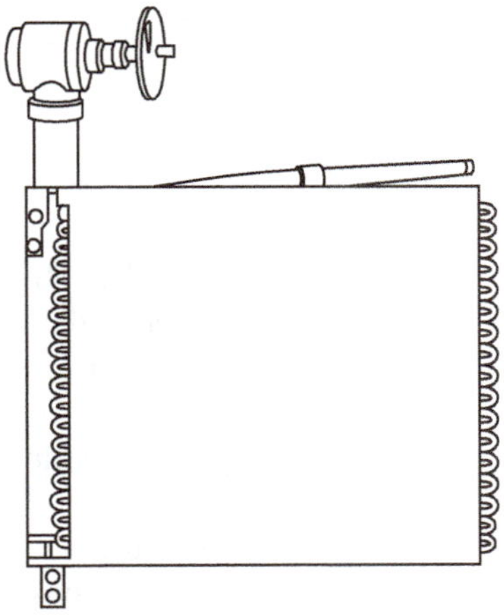

FIGURE A.3.3.19.2 *Horizontal Rack.*

3.3.19.3* Hose Reel. A circular device used to store hose.

A.3.3.19.3 Hose Reel. See Figure A.3.3.19.3.

3.3.19.4* Semiautomatic Hose Rack Assembly. The same as a "conventional" pin rack or hose reel except that, after the valve is opened, a retaining device holds the hose and water until the last few feet are removed.

A.3.3.19.4 Semiautomatic Hose Rack Assembly. See Figure A.3.3.19.4.

3.3.20 Hydrostatic Test. A test of a closed piping system and its attached appurtenances consisting of subjecting the piping to an increased internal pressure for a specified period of duration to verify system integrity and leak rates. [**24,** 2013]

3.3.21* Impairment. A condition where a fire protection system or unit or portion thereof is out of order, and the condition can result in the fire protection system or unit not functioning in a fire event.

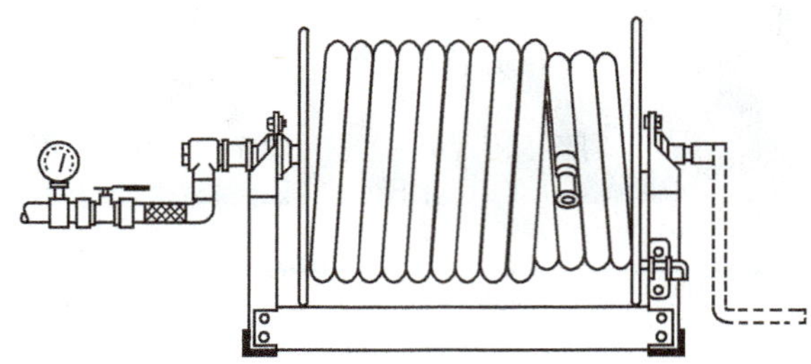

FIGURE A.3.3.19.3 Constant Flow Hose Reel.

FIGURE A.3.3.19.4 Semiautomatic Hose Rack Assembly.

A.3.3.21 Impairment. The use of the phrase *fire protection system or unit* is a broad reference to those terms used in titles of Chapters 5 through 12. Some fire protection features are referred to as systems in the installation standards (e.g., sprinkler, standpipe, water spray, foam-water, and water mist), or are referred to as units (e.g., fire pumps), and others use neither term (e.g., private service fire mains and water tanks). For the purpose of this standard, the term *unit* refers to a fire pump and its connections required by NFPA 20, or a water storage tank and its connections required by NFPA 22, or a private service fire main and its connections required by NFPA 24. The use of the term *unit* in the definitions of impairment, deficiency, critical deficiency, and noncritical deficiency is not referring to an individual component such as a sprinkler, valve, fitting, switch, piece of pipe, and so forth.

Temporarily shutting down a system as part of performing the routine inspection, testing, and maintenance on that system while under constant attendance by qualified personnel, and where the system can be restored to service quickly, should not be considered an impairment. Good judgment should be considered for the hazards presented.

3.3.21.1* Emergency Impairment. A condition where a water-based fire protection system or portion thereof is out of order due to an unplanned occurrence, or the impairment is found while performing inspection testing or maintenance activities.

A.3.3.21.1 Emergency Impairment. Examples of emergency impairments might include a ruptured pipe, an operated sprinkler, or an interruption of the water supply to the system.

FAQ What precautions can be taken to prepare for an emergency impairment?

An emergency impairment can be difficult to plan for or predict. However, some precautions can be taken to minimize system "downtime." These precautions include providing ready access to as-built plans, original test results, manufacturers' data sheets, and other "close-out" system documents, which can greatly aid in the timely correction of impairments (see 4.3.4). In addition, following the requirements for the on-site maintenance of spare sprinklers or nozzles, including pilot sprinklers, and repair kits for all valves, including alarm, dry pipe, deluge, preaction, and backflow prevention devices, can eliminate the need for ordering replacement equipment or special materials. (See the commentary in Chapter 15 for more information on impairments.) For example, the practice of maintaining a reserve supply of foam concentrate is essential to the timely re-commissioning of a foam-water sprinkler or foam-water spray system.

Critical Point

An emergency impairment can be difficult to predict or plan for. However, some precautions can be taken to minimize system "downtime."

3.3.21.2 Preplanned Impairment. A condition where a water-based fire protection system or a portion thereof is out of service due to work planned in advance, such as revisions to the water supply or sprinkler system piping.

Any system impairment can lead to the danger of a shut valve. Of system failures noted in Commentary Table 1.1 that were attributed to shut valves, it was not apparent whether the valve was shut due to maintenance (preplanned or emergency impairment) or from willful or unintentional neglect. Either way, during an impairment, it is essential to ensure that impairment procedures are in place and that valves are left opened completely after the system corrections are completed. A programmed approach to monitoring system valves through a lockout/tagout program can ensure that no valve is left completely or partially closed. See the commentary in Chapter 15 for more information on impairments. Also see Article 120 of *NFPA 70E®, Standard for Electrical Safety in the Workplace®*, for further information regarding lockout/tagout procedures.

3.3.22 Inspect. See 3.3.23, Inspection.

3.3.23 Inspection. A visual examination of a system or portion thereof to verify that it appears to be in operating condition and is free of physical damage. [**820,** 2012]

FAQ Is a sprinkler system *inspection* different from a sprinkler system *evaluation* in the context of NFPA 25? If so, is *evaluation* defined by NFPA?

An inspection is a visual examination of a system or component to verify that the system or component is in apparent working order. The intent of the inspections described in NFPA 25 is to look for obvious signs of damage or wear and, most importantly, to verify that water control valves are open. Inspections are the most frequently required activity in the standard and provide a minimum assurance that the system and components are in operating condition. The testing and maintenance activities verify that the system and its components are functioning.

The term *evaluation,* or *evaluate,* is not defined in any NFPA code or standard, and thus, the user should refer to *Merriam-Webster's Collegiate Dictionary,* which defines the word *evaluate* as follows: "to determine the significance, worth, or condition of by careful appraisal and study."

An evaluation of a system is more thorough than a visual examination. An evaluation might examine the design criteria or the system's compliance to installation standards and should be conducted by a registered design professional, a licensed contractor, or other qualified individual. (See 3.3.34 for the definition of the term *qualified* and related commentary.)

3.3.24* Inspection, Testing, and Maintenance Service. A service program provided by a qualified contractor or qualified property owner's representative in which all components unique to the property's systems are inspected and tested at the required times and necessary maintenance is provided.

A.3.3.24 Inspection, Testing, and Maintenance Service. This program includes logging and retention of relevant records.

3.3.25 Maintenance. In water-based fire protection systems, work performed to keep equipment operable or to make repairs.

Maintenance includes not only the required functions in the standard, but also practices and procedures recommended by the manufacturer. When contracting inspection, testing, and maintenance services, it is important for owners and contractors to have a clear understanding of the maintenance services that are included in the agreement. For example, the ongoing maintenance to keep dry system or preaction piping free of water is not the responsibility of the service provider unless it is specifically identified in the agreement. However, many owners are not aware of the need to monitor the condition of the piping by the periodic operation of any auxiliary drains (see A.13.4.3.3.3 and A.13.4.4.3.2). Good communication between owners and service providers regarding system maintenance can significantly enhance system reliability and performance.

3.3.26 Manual Operation. Operation of a system or its components through human action.

3.3.27 Nozzle.

3.3.27.1* Monitor Nozzle. A permanently mounted device specifically designed with a high flow rate to provide a far-reaching stream for locations where large amounts of water need to be available without the delay of laying hose lines.

A.3.3.27.1 Monitor Nozzle. Monitor nozzles can be used to protect large amounts of combustible materials, aircraft, tank farms, and any other special hazard. See Figure A.3.3.27.1(a) and Figure A.3.3.27.1(b).

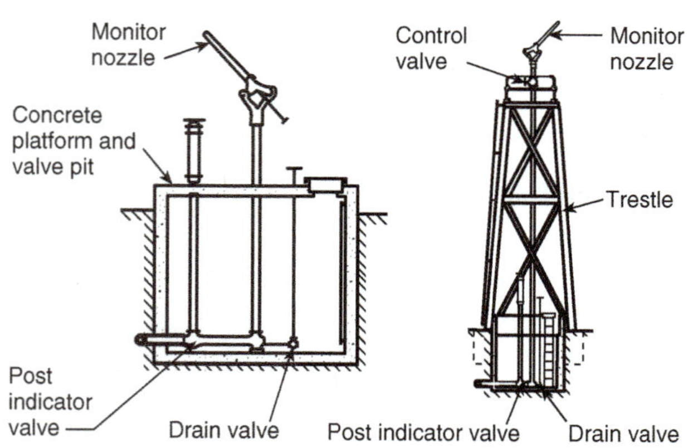

FIGURE A.3.3.27.1(a) Standard Monitor Nozzles; Gear Control Nozzles Also Are Permitted.

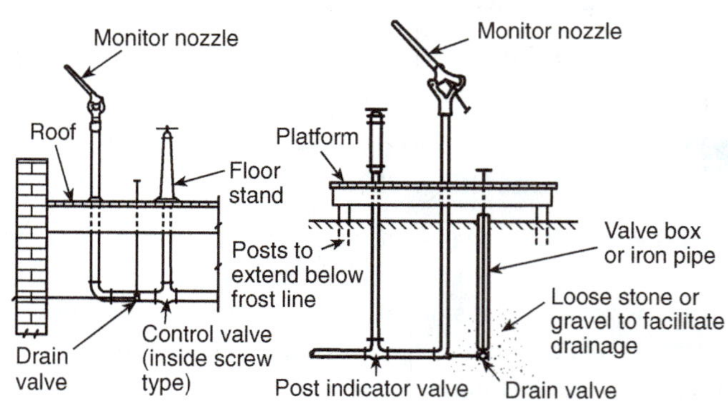

FIGURE A.3.3.27.1(b) Alternative Arrangement of Standard Monitor Nozzles.

3.3.27.2* Water Spray Nozzle. An open or automatic water discharge device that, when discharging water under pressure, will distribute the water in a specific, directional pattern.

A.3.3.27.2 Water Spray Nozzle. The selection of the type and size of spray nozzles should have been made with proper consideration given to factors such as physical character of the hazard involved, draft or wind conditions, material likely to be burning, and the general purpose of the system.

High velocity spray nozzles, generally used in piped installations, discharge in the form of a spray-filled cone. Low velocity spray nozzles usually deliver a much finer spray in the form of either a spray-filled spheroid or cone. Due to differences in the size of orifices or waterways in the various nozzles and the range of water particle sizes produced by each type, nozzles of one type cannot ordinarily be substituted for those of another type in an individual installation without seriously affecting fire extinguishment. In general, the higher the velocity and the coarser the size of the water droplets, the greater the effective "reach" or range of the spray.

Another type of water spray nozzle uses the deflector principle of the standard sprinkler. The angle of the spray discharge cones is governed by the design of the deflector. Some manufacturers make spray nozzles of this type individually automatic by constructing them with heat-responsive elements as used in standard automatic sprinklers.

Exhibit 3.8 illustrates directional, non-automatic water spray nozzles. These nozzles are available with a variety of spray patterns ranging from 30°F to 180°F (1.1°C to 82.2°C). They can also be used in an automatic configuration with a fusible link, with temperature ratings comparable to those of a standard spray sprinkler.

EXHIBIT 3.8 *Directional, Non-Automatic Water Spray Nozzles.*

3.3.28 Orifice Plate Proportioning. This system utilizes an orifice plate(s) through which passes a specific amount of foam concentrate at a specific pressure drop across the orifice plate(s).

This type of system depends on a properly sized orifice plate to control the amount of foam concentrate from a foam concentrate pump.

3.3.29 Performance-Based Program. Methods and frequencies that have been demonstrated to deliver equivalent or superior levels of performance through quantitative performance-based analysis.

3.3.30* Pressure-Regulating Device. A device designed for the purpose of reducing, regulating, controlling, or restricting water pressure. [**14,** 2013]

A.3.3.30 Pressure-Regulating Device. Examples include pressure-reducing valves, pressure control valves, and pressure-restricting devices.

3.3.31 Pressure-Restricting Device. A valve or device designed for the purpose of reducing the downstream water pressure under flowing (residual) conditions only. [**14,** 2013]

A pressure-restricting device (see Exhibit 3.9) is typically found on standpipe systems where flowing pressures exceed 100 psi (6.9 bar) for 1½ in. (40 mm) hose connections or 175 psi (12.1 bar) for 2½ in. (65 mm) hose connections.

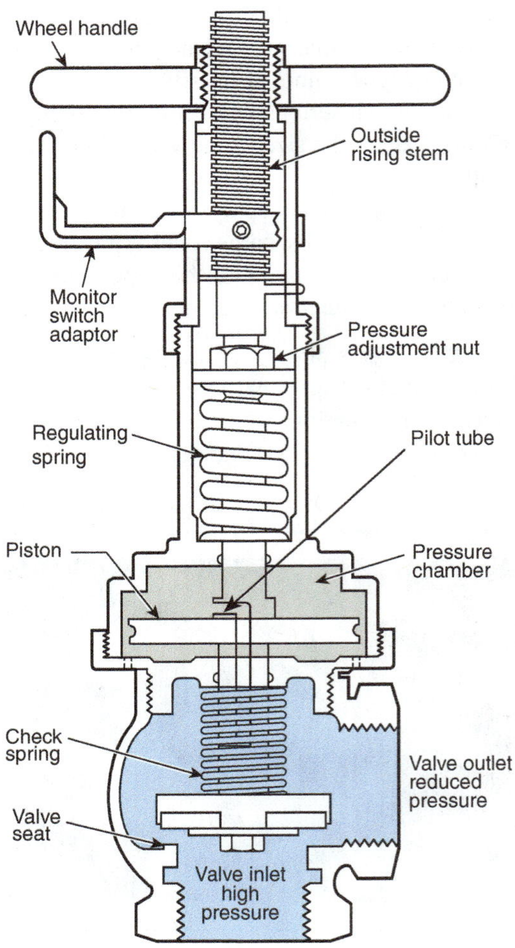

EXHIBIT 3.9 Pressure-Restricting Device. (Source: Fire Protection Handbook, 2008, Figure 16.10.9)

3.3.32* Pressure Vacuum Vent. A venting device mounted on atmospheric foam concentrate storage vessels to allow for concentrate expansion and contraction and for tank breathing during concentrate discharge or filling.

A.3.3.32 Pressure Vacuum Vent. At rest (static condition), this device is closed to prevent free breathing of the foam concentrate storage tank. See Figure A.3.3.32.

Exhibit 3.10 shows a pressure vacuum vent on a foam concentrate storage tank dome.

3.3.33* Proportioner.

A.3.3.33 Proportioner. See Figure A.3.3.33.

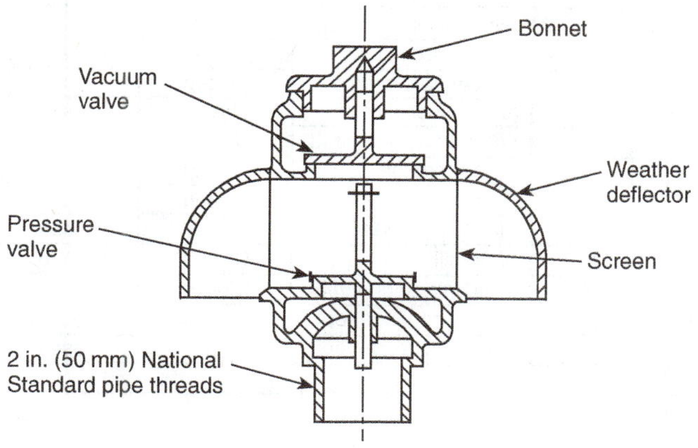

FIGURE A.3.3.32 Pressure Vacuum Vent.

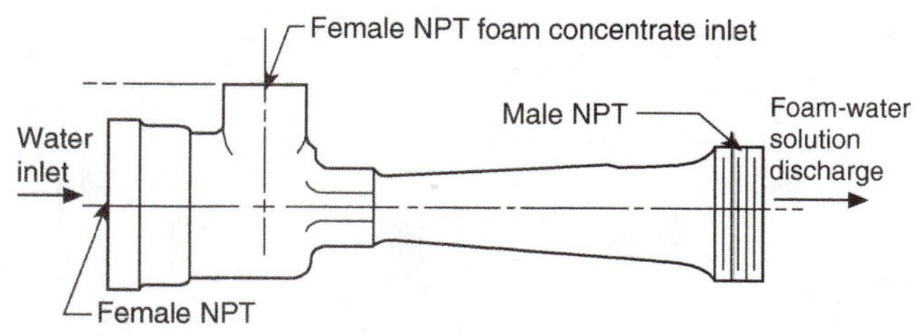

FIGURE A.3.3.33 Proportioner.

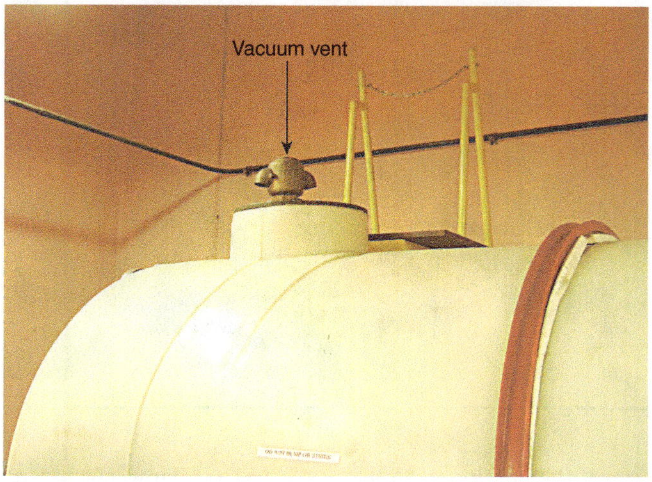

EXHIBIT 3.10 *Pressure Vacuum Vent.*

3.3.33.1* Bladder Tank Proportioner. A system that is similar to a standard pressure proportioner, except the foam concentrate is contained inside a diaphragm bag that is contained inside a pressure vessel.

A.3.3.33.1 Bladder Tank Proportioner. Operation is the same as a standard pressure proportioner, except that, because of the separation of the foam concentrate and water, this system can be used with all foam concentrates, regardless of specific gravity. See Figure A.3.3.33.1.

The bladder tank proportioner relies on water pressure externally applied to a bladder installed inside the tank as the means of expelling the foam concentrate. This arrangement allows relatively accurate proportioning over a wide range of flow. It is also economical, since external power is not required; however, there are limitations on the maximum size of the tank. Exhibit 3.11 shows a bladder tank proportioning system.

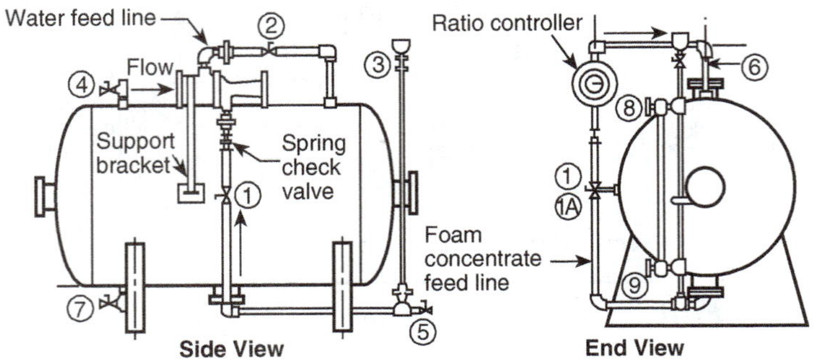

Valve description		Normal position	
Valve no.	Description	Manual system	Auto system
1	Concentrate shutoff	Closed	Closed
1A	Auto. conc. shutoff	N/A	Closed
2	Water pres. shutoff	Open	Open
3	Fill cup shutoff	Closed	Closed
4	Tank water vent	Closed	Closed
5	Diaph. conc. vent	Closed	Closed
6	Water fill	Closed	Closed
7	Concentrate drain/fill	Closed	Closed
8	Upr. sight gauge (opt.)	Closed	Closed
9	Lwr. sight gauge (opt.)	Closed	Closed

FIGURE A.3.3.33.1 Bladder Tank Proportioner.

EXHIBIT 3.11 *Bladder Tank Proportioning System. (Courtesy of Tyco Fire Suppression & Building Products)*

3.3.33.2* In-Line Balanced Pressure Proportioner. A system that is similar to a standard balanced pressure system, except the pumped concentrate pressure is maintained at a fixed preset value.

A.3.3.33.2 In-Line Balanced Pressure Proportioner. Balancing of water and liquid takes place at individual proportioners located in the system riser or in segments of multiple systems. See Figure A.3.3.33.2.

3.3.33.3* Line Proportioner. A system that uses a venturi pickup-type device where water passing through the unit creates a vacuum, thereby allowing foam concentrate to be picked up from an atmospheric storage container.

A.3.3.33.3 Line Proportioner. See Figure A.3.3.33.3.

The line proportioner relies on venturi action. In other words, foam concentrate is drawn into the water stream as water flows through the pipe, drafting the foam concentrate from a tank. Foam proportioning is sensitive to fluctuations in flow and pressure, and variations in flow and pressure will result in incorrect concentrations of foam concentrate.

3.3.33.4* Standard Balanced Pressure Proportioner. A system that utilizes a foam concentrate pump where foam concentrate is drawn from an atmospheric storage tank, is pressurized by the pump, and passes back through a diaphragm balancing valve to the storage tank.

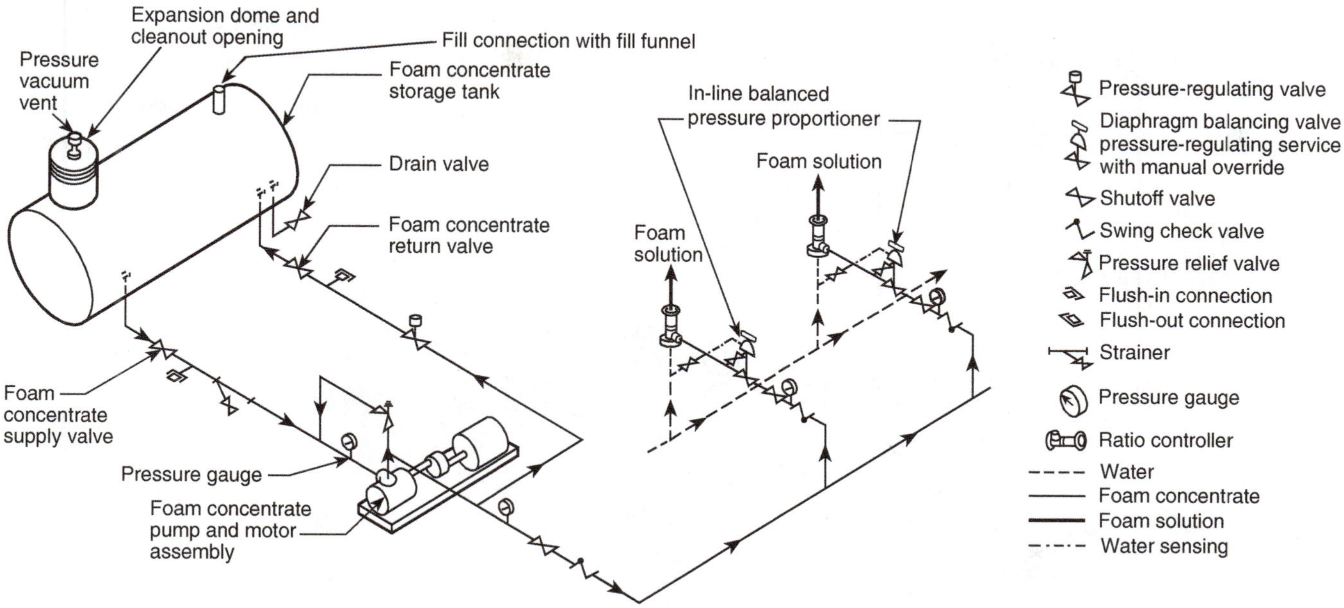

FIGURE A.3.3.33.2 In-Line Balanced Pressure Proportioner.

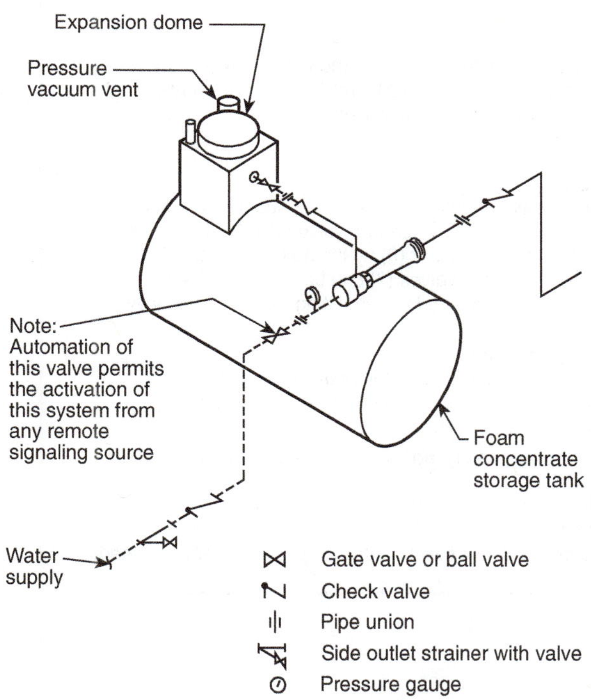

FIGURE A.3.3.33.3 Line Proportioner.

A.3.3.33.4 Standard Balanced Pressure Proportioner. Water and foam concentrate-sensing lines are directed to the balancing valve and maintain the foam liquid at a pressure equal to that of the water pressure. The two equal pressures are fed to the proportioner proper and are mixed at a predetermined rate. See Figure A.3.3.33.4.

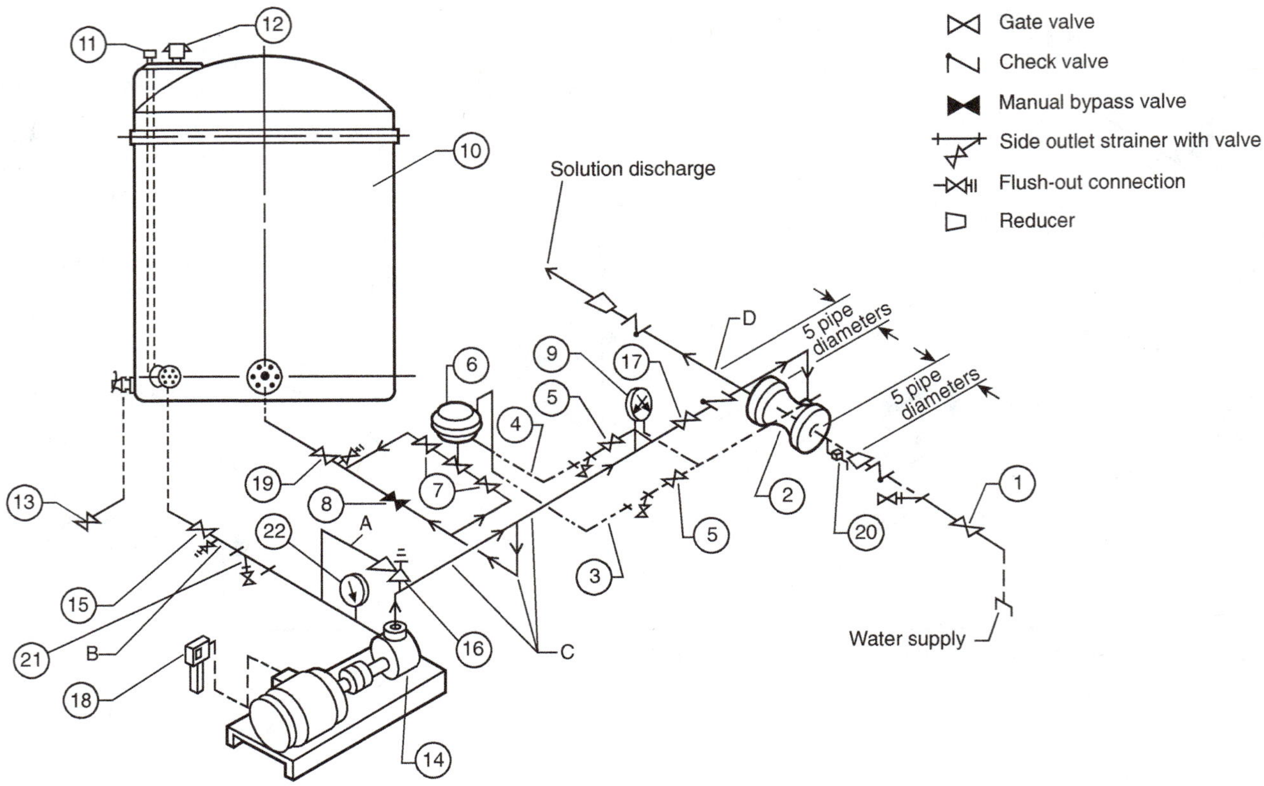

Legend:

1 Water supply valve (normally closed)
2 Ratio controller
3 Water balance line — minimum 3/16 in. (5 mm) I.D. pipe or tubing recommended
4 Concentrate balance line — minimum 3/16 in. (5 mm) I.D. pipe or tubing recommended
5 Sensing line valves (normally open)
6 Diaphragm control valve — automatic pressure balance — must be in vertical position
7 Block valves (normally open)
8 Manual bypass valve (normally open)
9 Water and concentrate pressure gauge (duplex)
10 Foam concentrate storage tank
11 Concentrate storage tank fill connection
12 Pressure vacuum vent
13 Concentrate storage tank drain valve (normally closed)
14 Foam concentrate pump and motor
15 Concentrate pump supply valve (normally open)
16 Pressure relief valve (setting as required by system)
17 Concentrate pump discharge valve (normally open)
18 Electric motor starter and switch
19 Concentrate return line valve (normally open)
20 Ball drip valve — 3/4 in. (20 mm) (install in horizontal position)
21 Strainer with valved side outlet
22 Compound gauge

Operation:
Start concentrate pump (18). Open water supply valve (1). Open concentrate pump discharge valve (17). Equal gauge readings then maintained at (9) by the automatic valve (6). For manual operation, valves (7) can be closed and equal gauge readings maintained by regulating valve (8) manually.

System Automation:
By automating certain valves, the balanced pressure proportioning system can be activated from any remote signaling source.

- Water supply valve (1), normally closed, to be automatically operated;
- Concentrate pump discharge valve (17), normally closed, to be automatically operated;
- Electric motor starter switch (18) to be automatically operated.

FIGURE A.3.3.33.4 Standard Balanced Pressure Proportioner.

Balanced pressure proportioning relies on the use of a separate pump for pressurizing foam concentrate. This system can be used across a wide range of flow rates.

3.3.33.5* Standard Pressure Proportioner. A system that uses a pressure vessel containing foam concentrate where water is supplied to the proportioner, which directs an amount of the supply downward onto the contained concentrate, thereby pressurizing the tank.

A.3.3.33.5 Standard Pressure Proportioner. Pressurized concentrate then is forced through an orifice back into the flowing water stream. This type of system is applicable for

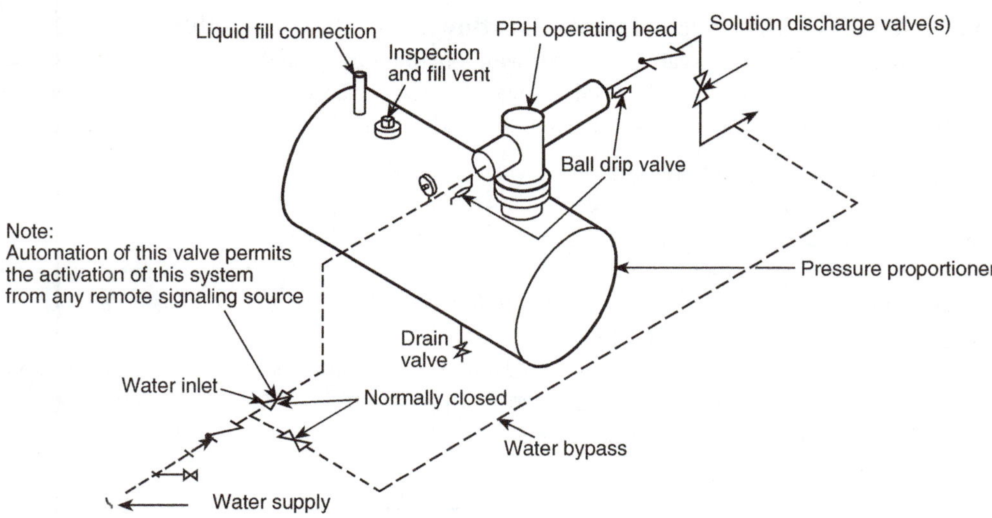

FIGURE A.3.3.33.5 Standard Pressure Proportioner.

use with foam concentrates having a specific gravity substantially higher than water. It is not applicable for use with foam concentrates with a specific gravity at or near that of water. See Figure A.3.3.33.5.

Similar to the bladder tank proportioner, the standard pressure proportioner relies on a viscosity of the foam concentrate that is higher than water. Water pressure is used to move the foam concentrate through a proportioner and mix it with the water stream.

3.3.34 Qualified. A competent and capable person or company that has met the requirements and training for a given field acceptable to the AHJ. [**96,** 2014]

The use of the term *qualified* in the standard is intended to be general in nature. The intent of NFPA 25 is that the owner or the owner's designated representative(s) perform many of the required inspections. For example, the owner or the owner's representative could be trained to a level acceptable to the AHJ, and therefore, be *qualified* to perform the monthly inspection to verify that the system control valve(s) are open. A facility maintenance person could be trained to inspect pressure gauges, exterior conditions of water storage tanks, accessibility of fire hydrants, and so forth.

Critical Point

The intent of NFPA 25 is that the owner or the owner's designated representative(s) perform many of the required inspections.

FAQ Does NFPA 25 require an inspector or service company to be licensed?

NFPA 25 does not require licensing of the individual or company performing the work, because many jurisdictions do not require licensing and because licensing laws vary considerably from one jurisdiction to the next. As a result, NFPA 25 requires an inspector or service provider simply to possess the appropriate knowledge and experience needed to perform the work specified therein. It is important to note that many licensing laws specifically require that an individual who is *"engaged in the business of providing this service for a fee"* must be licensed. This stipulation means that for properties that have a *qualified* maintenance staff, licensing is not necessarily needed for work that is done "in house." Before performing any inspection, testing, or maintenance work, any individual or business entity should contact the appropriate jurisdiction, such as the local or state fire marshal's office or the state contractor's licensing board, for any licensing requirements.

3.3.35 Rebuild. To restore working condition by replacement or repair of worn or damaged parts.

3.3.36 Reduced-Pressure Principle Backflow Prevention Assembly (RPBA). Two independently acting check valves together with a hydraulically operating, mechanically independent pressure differential relief valve located between the check valves, along with two resilient-seated shutoff valves, all as an assembly, and equipped with properly located test cocks.

In a normal flowing condition, both check valves of the reduced-pressure principle backflow prevention assembly (RPBA) will be open to meet the demand, and the relief valve will remain closed. Should a back-pressure condition occur, both check valves will close. The section between the two check valves provides additional security should the second check valve fail or leak. In such a case the back-pressure would flow into the middle zone, causing an increase in pressure to the point that the differential relief valve will open, discharging water to atmosphere. This could result in a considerable amount of water being discharged in the valve room. It is important to ensure that the drainage in rooms containing RPBAs is maintained free and clear of any obstructions. Exhibit 3.12 shows an example of a typical RPBA.

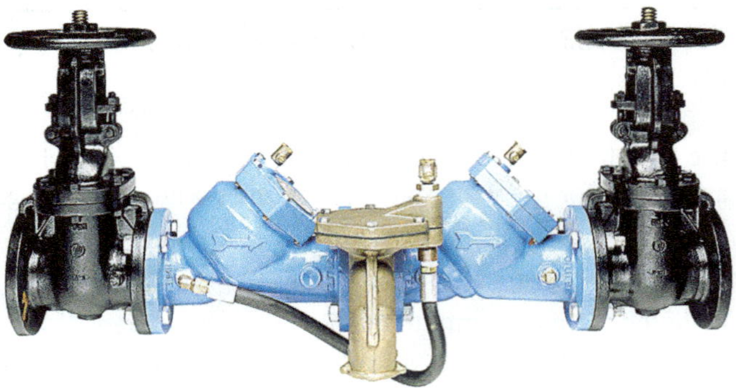

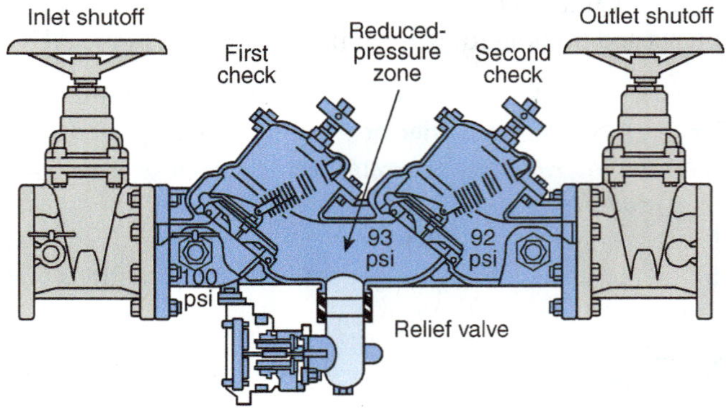

EXHIBIT 3.12 *Reduced-Pressure Principle Backflow Prevention Assembly (top) (Courtesy of Watts Regulator Co.) and Interior View of the Valve (bottom).*

3.3.37 Remove. To physically take away or eliminate.

3.3.38 Repair. Restore to sound working condition or to fix damage.

3.3.39 Replace. To remove a component and install a new or equivalent component.

3.3.40 Sprinkler.

3.3.40.1 Installation Orientation. The following sprinklers are defined according to orientation.

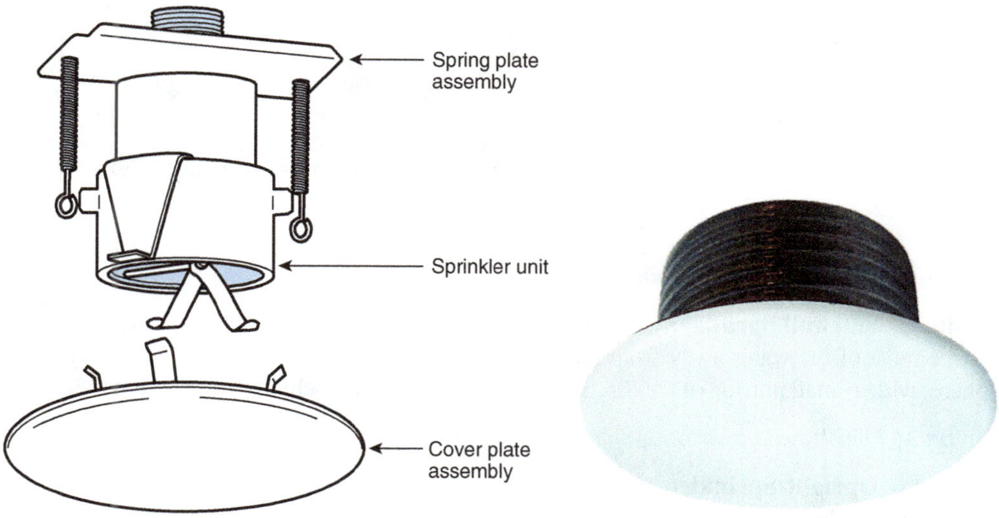

EXHIBIT 3.13 Standard Model G Concealed Ceiling Sprinkler: (left) Diagram of Unit Components, (right) Photo of Typical Unit. (Courtesy of Reliable Automatic Sprinkler Company, Inc.)

EXHIBIT 3.14 Viking Model H Standard Spray Pendent Flush-Mount Sprinkler. (Courtesy of Viking®)

EXHIBIT 3.15 Standard Spray Pendent Sprinkler. (Courtesy of Reliable Automatic Sprinkler Company, Inc.)

The sprinklers described in 3.3.40.1 are based on the geometry of the sprinkler installation. Representative discharge patterns for these devices allow them to be categorized as concealed, flush, pendent, and recessed sprinklers, all of which have similar discharge patterns and are mounted to the bottom of a branch line or pipe drop. Sidewall sprinklers are typically installed along a wall or lintel and discharge water away from the wall into the room or space. Sidewall sprinklers can be mounted on the side, bottom, or top of a branch line, as specified in their listings. Upright sprinklers have a spray pattern that appears similar to that of a pendent sprinkler. The difference is that upright sprinklers are mounted to the top of branch lines or sprigs.

3.3.40.1.1 Concealed Sprinkler. A recessed sprinkler with cover plate. [**13,** 2013]

Exhibit 3.13 shows a photo of an assembled concealed sprinkler as well as a diagram showing the various components of the device. As shown in the diagram, the cover plate drops away when exposed to a certain amount of heat. The fusible elements holding the cover plate are designed to operate prior to the activation of the sprinkler's thermal element. The cover plate is included as part of the listed sprinkler assembly.

3.3.40.1.2 Flush Sprinkler. A sprinkler in which all or part of the body, including the shank thread, is mounted above the lower plane of the ceiling. [**13,** 2013]

Exhibit 3.14 illustrates a flush sprinkler.

3.3.40.1.3 Pendent Sprinkler. A sprinkler designed to be installed in such a way that the water stream is directed downward against the deflector. [**13,** 2013]

Exhibit 3.15 shows a pendent sprinkler. The flat profile of the deflector is the most obvious characteristic of this type of sprinkler and is usually noticeable even when viewing from floor level.

3.3.40.1.4 Recessed Sprinkler. A sprinkler in which all or part of the body, other than the shank thread, is mounted within a recessed housing. [**13,** 2013]

EXHIBIT 3.16 Recessed Sprinkler. (Courtesy of Reliable Automatic Sprinkler Company, Inc.)

EXHIBIT 3.17 Horizontal Sidewall Sprinkler. (Courtesy of Reliable Automatic Sprinkler Company, Inc.)

Exhibit 3.16 shows a recessed sprinkler.

3.3.40.1.5 Sidewall Sprinkler. A sprinkler having special deflectors that are designed to discharge most of the water away from the nearby wall in a pattern resembling one-quarter of a sphere, with a small portion of the discharge directed at the wall behind the sprinkler. [**13,** 2013]

Exhibit 3.17 illustrates a horizontal sidewall sprinkler.

3.3.40.1.6 Upright Sprinkler. A sprinkler designed to be installed in such a way that the water spray is directed upwards against the deflector. [**13,** 2013]

Exhibit 3.18 shows several standard spray upright sprinklers. The most noticeable characteristics of upright sprinklers are a larger deflector than the pendent sprinkler and the downturned prongs at the deflector edge. As with the pendent sprinkler, upright sprinklers typically can be identified from floor level. Also, the upright sprinkler has a larger diameter glass bulb than the quick-response sprinkler shown in Exhibit 3.26.

3.3.40.2* Control Mode Specific Application (CMSA) Sprinkler. A type of spray sprinkler that is capable of producing characteristic large water droplets and that is listed for its capability to provide fire control of specific high-challenge fire hazards. [**13,** 2013]

A.3.3.40.2 Control Mode Specific Application (CMSA) Sprinkler. A large drop sprinkler is a type of CMSA sprinkler that is capable of producing characteristic large water droplets and that is listed for its capability to provide fire control of specific high-challenge fire hazards.

Control mode specific application (CMSA) sprinklers incorporate a wide variety of sprinklers that are capable of fire control in high-challenge fire scenarios. The large-drop sprinkler, which was defined in previous editions of the standard, is now included within this category. Exhibit 3.19

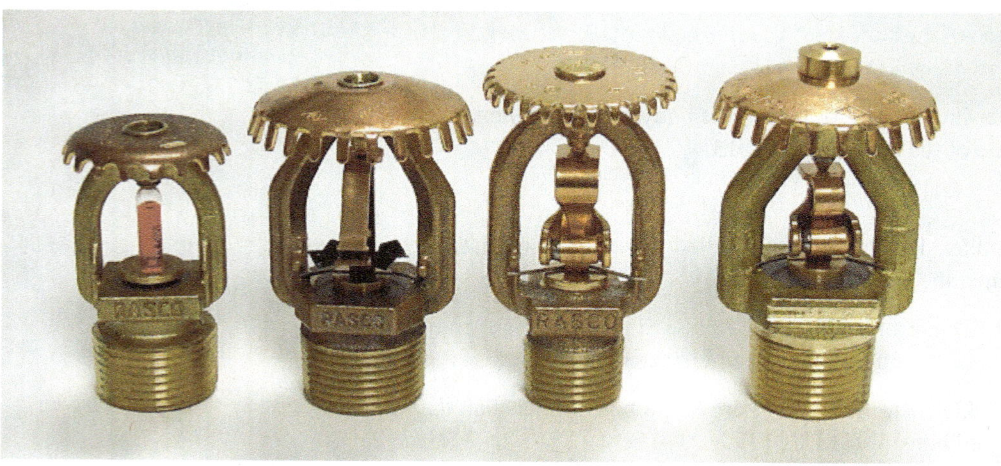

EXHIBIT 3.18 Upright Sprinklers. (Courtesy of Reliable Automatic Sprinkler Company, Inc.)

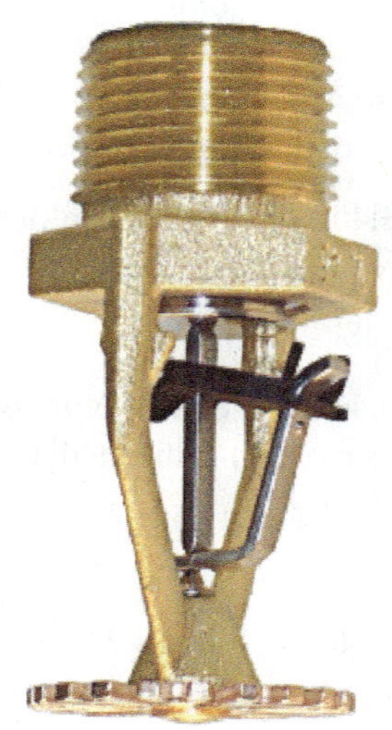

EXHIBIT 3.19 Control Mode Specific Application (CMSA) Sprinkler. (Courtesy of Victaulic®)

shows a CMSA sprinkler. Note that it has a similar appearance to the early suppression fast-response (ESFR) sprinkler, which is covered in 3.3.40.5, but it has a control mode objective rather than a suppression objective.

3.3.40.3 Corrosion-Resistant Sprinkler. A sprinkler fabricated with corrosion-resistant material, or with special coatings or platings, to be used in an atmosphere that would normally corrode sprinklers. [**13,** 2013]

A corrosion-resistant sprinkler is usually a standard sprinkler to which a corrosion-resistant coating, such as wax or lead, has been applied. Any coating of this type can be applied only by the manufacturer. Exhibit 3.20 shows a typical corrosion-resistant sprinkler.

EXHIBIT 3.20 Corrosion-Resistant Sprinkler. (Courtesy of American Fire Sprinkler Association)

3.3.40.4 Dry Sprinkler. A sprinkler secured in an extension nipple that has a seal at the inlet end to prevent water from entering the nipple until the sprinkler operates. [**13,** 2013]

Exhibit 3.21 shows an example of one type of dry sprinkler. Dry sprinklers are available as upright, pendent, or sidewall types.

3.3.40.5 Early Suppression Fast-Response (ESFR) Sprinkler. A type of fast-response sprinkler that has a thermal element with an RTI of 50 (meters-seconds)$^{1/2}$ or less and is listed for its capability to provide fire suppression of specific high-challenge fire hazards. [**13,** 2013]

The early suppression fast-response (ESFR) sprinkler evolved by examining the combined effects of sprinkler sensitivity and water distribution characteristics to achieve early fire suppression. The ESFR concept is to apply a sufficient amount of water to the burning fuel during the early phases of a fire and penetrate the developing fire plume, then fire suppression can be achieved. If the fire plume reaches velocities that prevent the water droplets from reaching the burning fuel, the likelihood of suppression is greatly reduced. ESFR sprinklers, in contrast to standard sprinklers, operate earlier in the fire and provide adequate discharge to suppress the fire before a severe fire plume develops. Exhibit 3.22 and Exhibit 3.23 show examples of ESFR sprinklers.

3.3.40.6 Extended Coverage Sprinkler. A type of spray sprinkler with maximum coverage areas as specified in Sections 8.8 and 8.9 of NFPA 13, *Standard for the Installation of Sprinkler Systems*. [**13,** 2013]

EXHIBIT 3.21 Dry Sprinkler. (Courtesy of Viking®)

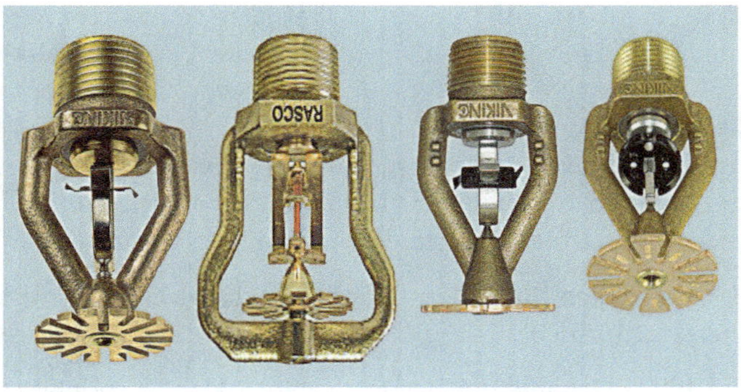

EXHIBIT 3.22 Examples of Various ESFR Pendent Sprinklers. (Courtesy of Viking® and Reliable Automatic Sprinkler Company, Inc.)

EXHIBIT 3.23 Upright-Type ESFR Sprinkler. (Courtesy of Viking®)

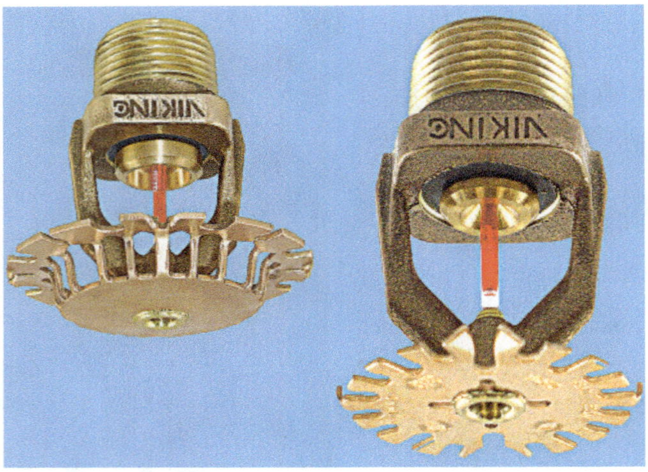

EXHIBIT 3.24 *Examples EC Sprinklers. (Courtesy of Viking®)*

EXHIBIT 3.25 *Example of Sidewall-Type EC Sprinkler. (Courtesy of Viking®)*

The extended coverage (EC) sprinkler is available in pendent, upright, and sidewall configurations. Some examples are shown in Exhibit 3.24 and Exhibit 3.25. The advantage of EC sprinklers is that their areas of coverage are greater than those established for standard spray and sidewall sprinklers.

3.3.40.7 Nozzles. A device for use in applications requiring special water discharge patterns, directional spray, or other unusual discharge characteristics. [**13,** 2013]

Nozzles, such as the one shown in Exhibit 3.8, are special application devices that have unique characteristics designed to meet specific needs. These devices might be equipped with heat-responsive elements or might be open to the atmosphere, and they are designed to spray water in a specific direction. NFPA 15, *Standard for Water Spray Fixed Systems for Fire Protection,* provides more information on nozzles.

3.3.40.8 Old-Style/Conventional Sprinkler. A sprinkler that directs from 40 percent to 60 percent of the total water initially in a downward direction and that is designed to be installed with the deflector either upright or pendent. [**13,** 2013]

Old-style/conventional sprinklers were predominantly used in North America prior to the development of spray sprinklers. The major difference between old-style/conventional and standard spray sprinklers is the design of their deflectors. Old-style/conventional sprinklers, whether upright or pendent, direct at least 40 percent of their water discharge up against the ceiling, with the remaining discharge directed downward. Spray sprinklers direct 100 percent of their water discharge downward.

Old-style/conventional sprinklers are still available and are used in Europe. These sprinklers are also used in North America for special applications in which an upward sprinkler discharge is desired, such as for fur storage vaults and for pier and wharf protection.

3.3.40.9 Open Sprinkler. A sprinkler that does not have actuators or heat-responsive elements. [**13,** 2013]

3.3.40.10 Ornamental/Decorative Sprinkler. A sprinkler that has been painted or plated by the manufacturer. [**13,** 2013]

Sometimes an ornamental sprinkler is nothing more than a corrosion-resistant sprinkler with a special coating similar in color to that of the interior building finish. The same holds true for the covers of concealed sprinklers.

It is important to remember that any coating for decorative or protective purposes must be applied only by the sprinkler manufacturer.

3.3.40.11 Quick-Response Early Suppression (QRES) Sprinkler. A type of quick-response sprinkler that has a thermal element with an RTI of 50 (meter-seconds)$^{1/2}$ or less and is listed for its capability to provide fire suppression of specific fire hazards. [**13,** 2013]

3.3.40.12 Quick-Response Extended Coverage Sprinkler. A type of quick-response sprinkler that has a thermal element with an RTI of 50 (meter-seconds)$^{1/2}$ or less and complies with the extended protection areas defined in Chapter 8 of NFPA 13, *Standard for the Installation of Sprinkler Systems*. [**13,** 2013]

3.3.40.13 Quick-Response (QR) Sprinkler. A type of spray sprinkler that has a thermal element with an RTI of 50 (meter-seconds)$^{1/2}$ or less and is listed as a quick-response sprinkler for its intended use. [**13,** 2013]

Exhibit 3.26 shows a quick-response (QR) sprinkler. Note the smaller diameter of the glass bulb as compared to that of the standard-response upright sprinkler shown in Exhibit 3.18.

EXHIBIT 3.26 *Quick-Response Sprinkler. (Courtesy of Viking®)*

3.3.40.14 Residential Sprinkler. A type of fast-response sprinkler having a thermal element with an RTI of 50 (meters-seconds)$^{1/2}$ or less, that has been specifically investigated for its ability to enhance survivability in the room of fire origin, and that is listed for use in the protection of dwelling units. [**13,** 2013]

3.3.40.15 Special Sprinkler. A sprinkler that has been tested and listed as prescribed in 8.4.8 of NFPA 13, *Standard for the Installation of Sprinkler Systems*. [**13,** 2013]

NFPA 13, *Standard for the Installation of Sprinkler Systems,* provides detailed information for special sprinklers. This type of sprinkler does not fall into the other categories, as it is usually intended to protect specific hazards or construction features. Special sprinklers must be evaluated and listed for their intended purpose.

3.3.40.16 Spray Sprinkler. A type of sprinkler listed for its capability to provide fire control for a wide range of fire hazards. [**13,** 2013]

3.3.40.17 Standard Spray Sprinkler. A spray sprinkler with maximum coverage areas as specified in Sections 8.6 and 8.7 of NFPA 13, *Standard for the Installation of Sprinkler Systems*. [**13,** 2013]

The standard spray sprinkler is installed in accordance with the coverage area limitations in Sections 8.6 and 8.7 of NFPA 13 and is available in pendent, upright, and sidewall configurations. The standard spray sprinkler has proven to be effective for a broad range of hazards and applications by adjusting the water discharge density; it is very popular and, to a certain degree, serves as the benchmark for sprinkler criteria and performance. Exhibit 3.27 illustrates a standard spray pendent sprinkler and a standard spray upright sprinkler.

3.3.41* Standpipe System. An arrangement of piping, valves, hose connections, and allied equipment installed in a building or structure, with the hose connections located in such a manner that water can be discharged in streams or spray patterns through attached hose and nozzles, for the purpose of extinguishing a fire, thereby protecting a building or structure and its contents in addition to protecting the occupants. [**14,** 2013]

Manual and automatic standpipe systems can be dry or wet. A manual standpipe requires the fire department to supply the volume and pressure required to meet the system design criteria.

EXHIBIT 3.27 *Examples of Standard Spray Sprinklers: Pendent Sprinkler (left), and Upright Sprinkler (right). (Courtesy of Reliable Automatic Sprinkler Company, Inc.)*

An automatic standpipe supplies both the required volume and pressure automatically. Manual standpipes are traditionally dry, but there are many situations where the standpipe is connected to a water supply and filled with water. For example, where buildings are retrofitted with fire sprinklers, it is common to have the existing manual dry standpipe connected to the water supply and used to supply the water for the sprinkler system. The standpipe is still manual by design criteria but is designated as a wet standpipe because of its connection to the water supply.

Automatic standpipe systems are typically wet, but where they are subject to freezing, a dry pipe valve can be added to the system and the piping filled with air pressure much like a dry sprinkler system. An automatic dry system is installed only where there is a need for freeze protection.

A.3.3.41 Standpipe System. This is accomplished by means of connections to water supply systems or by means of pumps, tanks, and other equipment necessary to provide an adequate supply of water to the hose connections.

Standpipe systems differ substantially from the hose connections required by NFPA 13 for sprinkler systems. Standpipe systems are fire extinguishing systems and, as a result, require much higher flows and pressures for trained fire professionals to use them effectively. Therefore, standpipe systems require specific inspection, testing, and maintenance procedures, as referenced in Chapter 6.

Sometimes it is expedient to combine the standpipe piping and sprinkler piping into one system. Exhibit 3.28 illustrates a combined sprinkler/standpipe system. The standpipe system shown is a single-zone type. The detail views illustrate arrangements for the piping on each floor of a building protected by a combined sprinkler/standpipe system.

3.3.41.1 Automatic Standpipe System. A standpipe system that is attached to a water supply capable of supplying the system demand and that requires no action other than opening a hose valve to provide water at hose connections.

3.3.41.2 Dry Standpipe. A standpipe system designed to have piping contain water only when the system is being utilized.

3.3.41.3 Manual Standpipe. Standpipe system that relies exclusively on the fire department connection to supply the system demand.

3.3.41.4 Wet Standpipe System. A standpipe system having piping containing water at all times. [**14,** 2013]

3.3.42 Semiautomatic Dry Standpipe System. A standpipe system permanently attached to a water supply that is capable of supplying the system demand at all times arranged through the use of a device such as a deluge valve and that requires activation of a remote control device to provide water at hose connections. [**14,** 2013]

3.3.43 Standpipe System Classes.

3.3.43.1 Class I System. A system that provides 2½ in. (65 mm) hose connections to supply water for use by fire departments. [**14,** 2013]

3.3.43.2 Class II System. A system that provides 1½ in. (40 mm) hose stations to supply water for use primarily by trained personnel or by the fire department during initial response. [**14,** 2013]

3.3.43.3 Class III System. A system that provides 1½ in. (40 mm) hose stations to supply water for use by trained personnel and 2½ in. (65 mm) hose connections to supply a larger volume of water for use by fire departments. [**14,** 2013]

FAQ Is hose required to be installed in a Class III standpipe system?

Hose is permitted but not required to be installed in a Class III standpipe system. NFPA 14, *Standard for the Installation of Standpipe and Hose Systems,* allows Class III systems in a fully sprinklered building to be equipped with a 2½ in. (65 mm) hose valve and a 2½ in. × 1½ in. (65 mm × 40 mm) reducer, cap, and chain, and no hose. Hose is being installed in fewer and fewer sprinklered buildings. Fire departments will not use fire hose that they do not maintain, and the consensus of most industry experts is that building occupants should not use fire hose unless they are properly trained and equipped with the required protective gear and breathing apparatus. NFPA 25 allows existing hose to be removed where approved by the AHJ (see 6.1.7). However, where hose is provided, it must be inspected, tested, and maintained in accordance with the requirements of Chapter 6.

3.3.44* Strainer. A device capable of removing from the water all solids of sufficient size that are obstructing water spray nozzles.

A.3.3.44 Strainer. There are two types of strainers. Pipeline strainers are used in water supply connections. These are capable of removing from the water all solids of sufficient size to obstruct the spray nozzles [⅛ in. (3.2 mm) perforations usually are suitable]. Pipeline strainer designs should incorporate a flushout connection or should be capable of flushing through the main drain.

Individual strainers for spray nozzles, where needed, are capable of removing from the water all solids of sufficient size to obstruct the spray nozzle that they serve.

3.3.45 Supervision. In water-based fire protection systems, a means of monitoring system status and indicating abnormal conditions.

3.3.46 Test. The operation of a device to verify that it is functioning correctly, or the measurement of a system characteristic to determine if it meets requirements.

3.3.47* Testing. A procedure used to determine the operational status of a component or system by conducting periodic physical checks, such as waterflow tests, fire pump tests, alarm tests, and trip tests of dry pipe, deluge, or preaction valves.

A.3.3.47 Testing. These tests follow up on the original acceptance test at intervals specified in the appropriate chapter of this standard.

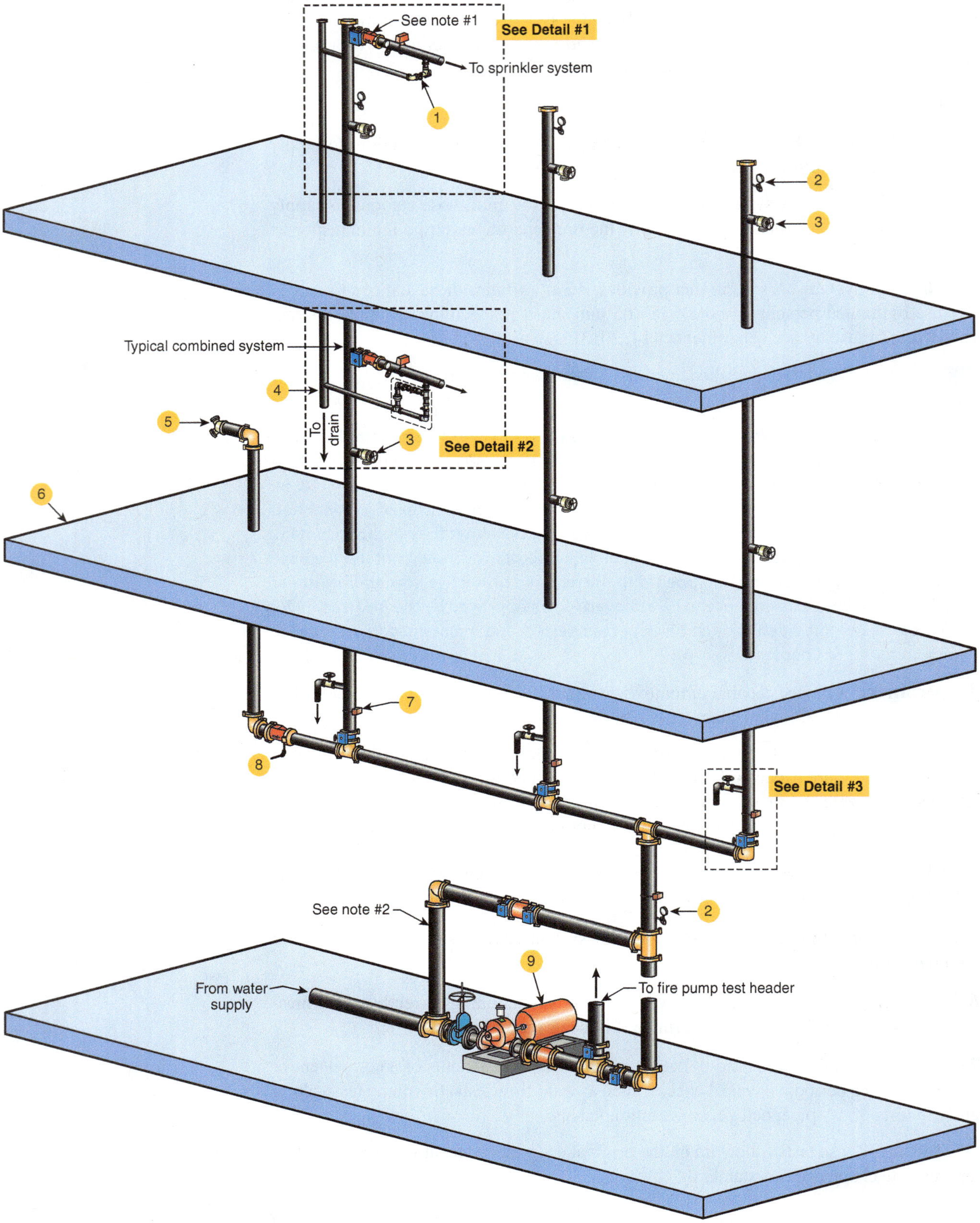

EXHIBIT 3.28 *Combination Sprinkler/Standpipe System with Acceptable Piping Arrangements. (Courtesy of Stephan Laforest)*

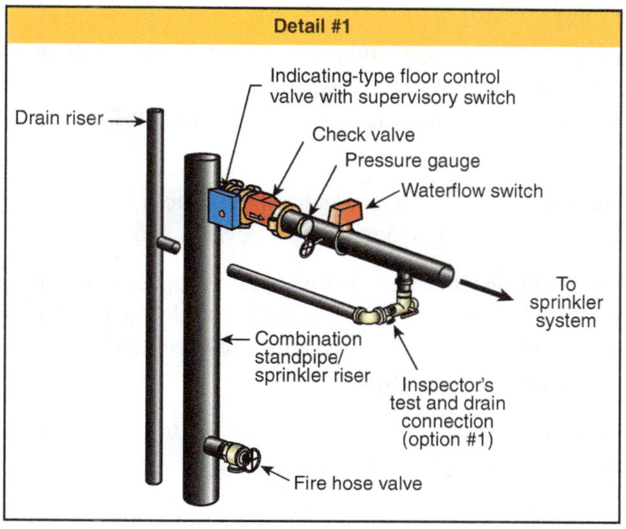

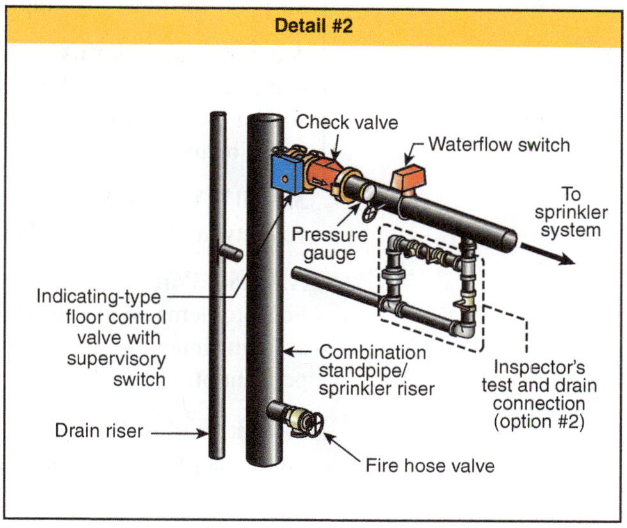

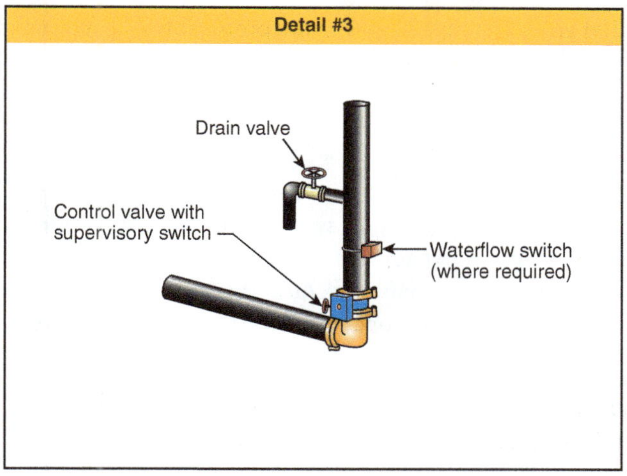

Legend

1. Test and drain
2. Pressure gauge
3. Fire hose valve
4. Drain riser
5. Fire department connection
6. Grade level
7. Waterflow switch (where required)
8. Check valve with ball drip
9. Fire pump

Notes:
1. Sprinkler floor assembly in accordance with NFPA 13, *Standard for the Installation of Sprinkler Systems.*
2. Bypass in accordance with NFPA 20, *Standard for the Installation of Stationary Pumps for Fire Protection.*

EXHIBIT 3.28 *Continued*

Critical Point

An important function of the standard is to verify that water supplies are in operating condition.

3.3.48 Valve Status Test. Flowing water to verify that valves for a portion of the system are not closed.

The concept of a valve status test is new to the 2014 edition, and it is simply to assure that the valve is in the proper position based on its desired function. One of the leading causes of fires overwhelming sprinklered buildings is that the water could not reach the sprinklers or hose outlets due to partially or fully closed valves. This test is intended to cut down on the number of times a valve is inadvertently left closed or partially closed.

3.3.49* Valve Status Test Connection. A point in the system where water is discharged for purposes of performing a valve status test.

A.3.3.49 Valve Status Test Connection. These connections can include the main drain, fire pump test header, backflow preventer forward flow test connection, fire hydrant, and other similar locations. In the absence of the aforementioned devices, an inspector's test connection might be used.

3.3.50* Water Spray. Water in a form having a predetermined pattern, particle size, velocity, and density discharge from specially designed nozzles or devices. [**15,** 2012]

A.3.3.50 Water Spray. Water spray fixed systems are usually applied to special fire protection problems, since the protection can be specifically designed to provide for fire control, extinguishment, or exposure protection. Water spray fixed systems are permitted to be independent of, or supplementary to, other forms of protection.

3.3.51 Water Supply. A source of water that provides the flows [gal/min (L/min)] and pressures [psi (bar)] required by the water-based fire protection system.

FAQ What are examples of acceptable water supplies for use in fire protection systems?

Some examples of water supplies that are acceptable for use in fire protection systems include the following:

- Connections to water works systems
- Fire pumps installed in accordance with NFPA 20
- Pressure and gravity tanks

Other sources of water can be used for fire protection systems, subject to the approval of the AHJ. Ponds, reservoirs, lakes, rivers, and other sources can be acceptable provided the year-round availability of a sufficient volume can be confirmed. All of these sources will be used typically in conjunction with a fire pump. An important function of the standard is to verify that water supplies are in operating condition.

3.3.52 Waterflow Alarm Device. An attachment to a water-based fire protection system that detects and signals a predetermined waterflow.

3.4 Deluge Foam-Water Sprinkler and Foam-Water Spray Systems Definitions

3.4.1 Foam-Water Spray System. A special system that is pipe-connected to a source of foam concentrate and to a water supply. The system is equipped with foam-water spray nozzles for extinguishing agent discharge (foam followed by water or in reverse order) and for distribution over the area to be protected. System operation arrangements parallel those for foam-water sprinkler systems as described in the definition of Foam-Water Sprinkler System. [**16,** 2011]

Foam-water spray systems are almost identical to water spray fixed systems; however, foam systems have foam-generating equipment attached to the feed main piping. The discharge device or nozzle must be listed for use with foam concentrate.

3.4.2 Foam-Water Sprinkler System. A special system that is pipe-connected to a source of foam concentrate and to a water supply. The system is equipped with appropriate discharge devices for extinguishing agent discharge and for distribution over the area to be protected. The piping system is connected to the water supply through a control valve that usually is actuated by operation of automatic detection equipment that is installed in the same areas as the sprinklers. When this valve opens, water flows into the piping system, foam concentrate is injected into the water, and the resulting foam solution discharging through the discharge devices generates and distributes foam. Upon exhaustion of the foam concentrate supply, water discharge follows and continues until shut off manually. Systems can be used for discharge of water first, followed by discharge of foam for a specified period, and then followed by water until manually shut off. Existing deluge sprinkler systems that have been converted to the use of aqueous film-forming foam or film-forming fluoroprotein foam are classified as foam-water sprinkler systems. [**16,** 2011]

Foam-water sprinkler systems and deluge systems are almost identical; they differ in that foam-water sprinkler systems have foam-generating equipment attached to the feed main piping. Like foam-water spray systems, foam-water sprinkler systems must employ discharge devices or sprinklers that are listed for use with foam concentrate. See the commentary in Chapter 11 for more information on foam-water sprinkler systems.

3.5 Valve Definitions

3.5.1* Control Valve. A valve controlling flow to water-based fire protection systems.

A.3.5.1 Control Valve. Experience has shown that closed valves are the primary cause of failure of water-based fire protection systems in protected occupancies. Control valves do not include hose valves, inspector's test valves, drain valves, trim valves for dry pipe, preaction and deluge valves, check valves, or relief valves.

Shut or partially shut control valves account for the majority of all sprinkler system failures. Therefore, the control valve requires vigorous inspection and monitoring. Control valves require close scrutiny, particularly when there are no mechanical monitoring systems (e.g., chains with locks) or electronic monitoring systems (e.g., tamper switches). When valves are provided with only a seal, the inspection frequency is weekly. With the aforementioned mechanical or electronic monitoring system in place, the inspection frequency is relaxed to monthly.

Further, whenever a control valve at the system riser is operated, a main drain flow test must be performed to verify that the valve has been opened fully. See the commentary in Chapter 13 for more information on valves and valve components.

3.5.2* Deluge Valve. A water supply control valve intended to be operated by actuation of an automatic detection system that is installed in the same area as the discharge devices.

A.3.5.2 Deluge Valve. Each deluge valve is intended to be capable of automatic and manual operation.

3.5.3 Hose Valve. The valve to an individual hose connection. [**14,** 2013]

3.5.4 Pressure Control Valve. A pilot-operated pressure-reducing valve designed for the purpose of reducing the downstream water pressure to a specific value under both flowing (residual) and nonflowing (static) conditions. [**14,** 2013]

3.5.5 Pressure-Reducing Valve. A valve designed for the purpose of reducing the downstream water pressure under both flowing (residual) and nonflowing (static) conditions. [**14,** 2013]

Typically, pressure control and pressure-reducing valves are found on systems that are subjected to high pressures, usually in excess of 175 psi (12.1 bar). These valves are intended to protect system components from failure due to high pressure. They must be tested periodically to verify that they are functioning correctly, since failure of the valve can damage system components and could present a danger to personnel.

A typical application for these valves is in a high-rise building where high pressures are needed to move water to the upper floors. They can be used to control pressures for entire sprinkler or standpipe systems or individual fire department hose connections. In the case of valves controlling pressure to individual fire department valves, pressures might be either preset at the factory or adjustable in the field.

Exhibit 3.29 shows an example of both types of pressure-reducing valves.

EXHIBIT 3.29 *Pressure-Reducing Valves: Factory Set (left) and Field Set (right). (Courtesy of Fire-End & Croker Corp.)*

3.5.5.1* Master Pressure-Reducing Valve. A pressure-reducing valve installed to regulate pressures in an entire fire protection system and/or standpipe system zone.

A.3.5.5.1 Master Pressure-Reducing Valve. Master pressure-reducing valves are typically found downstream of a fire pump's discharge.

A master pressure-reducing valve is used to control pressure to an entire sprinkler or standpipe system. This valve controls water pressure to the entire system, so it must be tested more frequently than other valves in the system to verify that it is functioning properly. For more information regarding the inspection, testing, and maintenance of this valve, see Chapter 13. See Exhibit 3.30 for a typical master pressure-reducing valve.

3.5.6 [Pressure] Relief Valve. A device that allows the diversion of liquid to limit excess pressure in a system. [**20,** 2013]

3.5.6.1 Circulation Relief Valve. A valve used to cool a pump by discharging a small quantity of water. This valve is separate from and independent of the main relief valve. [**20,** 2013]

EXHIBIT 3.30 Master Pressure-Reducing Valve. (Courtesy of CLA-VAL, Canada)

3.6 Water-Based Fire Protection System Definitions

3.6.1 Combined Standpipe and Sprinkler System. A system where the water piping services both 2½ in. (65 mm) outlets for fire department use and outlets for automatic sprinklers.

A combined standpipe and sprinkler system, such as the one shown in Exhibit 3.28, has common piping through the feed main and riser portion of the system. A sprinkler system and a standpipe system that share a common supply header at the water service entrance to the building do not constitute a "combined system." With combined systems, the sprinkler system is designed and installed in such a way that it is isolated from the standpipe system, usually by means of floor control assemblies. Isolating the sprinkler system(s) for the standpipe system allows manual fire fighting to continue even after a sprinkler system is shut off.

3.6.2 Fire Pump. A pump that is a provider of liquid flow and pressure dedicated to fire protection. [**20,** 2013]

Whether taking suction from a natural body of water, a storage tank, or a reliable water works, the presence of a fire pump, such as the one shown in Exhibit 3.31, indicates that the primary water source lacks sufficient pressure to meet the fire protection system demand. The fire pump must be properly maintained to be able to function in the event of a fire. Failure of the fire pump will result in inadequate pressure to meet the demands of the fire protection system. Therefore, frequent operation of the pump and an annual evaluation of performance are necessary. See the commentary in Chapter 8 for more information on fire pumps.

3.6.3* Private Fire Service Main. Private fire service main, as used in this standard, is that pipe and its appurtenances on private property (1) between a source of water and the base of the system riser for water-based fire protection systems, (2) between a source of water and inlets to foam-making systems, (3) between a source of water and the base elbow of private hydrants or monitor nozzles, and (4) used as fire pump suction and discharge piping, (5) beginning at the inlet side of the check valve on a gravity or pressure tank. [**24,** 2013]

See the commentary in Chapter 7 for more information on private fire service mains.

EXHIBIT 3.31 Horizontal Fire Pump Installation.

A.3.6.3 Private Fire Service Main. See Figure A.3.6.3.

3.6.4* Sprinkler System. A system that consists of an integrated network of piping designed in accordance with fire protection engineering standards that includes a water supply source, a water control valve, a waterflow alarm, and a drain and is commonly activated by heat from a fire, discharging water over the fire area. The portion of the sprinkler system above ground is a network of specifically sized or hydraulically designed piping installed in a building, structure, or area, generally overhead, and to which sprinklers are attached in a systematic pattern. The system is commonly activated by heat from a fire and discharges water over the fire area. [**13,** 2013]

A sprinkler system can be identified by the presence of a water supply control valve, a drain, a pressure gauge, a waterflow alarm, a network of piping, and sprinklers. Most systems also have a fire department connection. See the commentary in Chapter 5 for more information on sprinkler systems.

A.3.6.4 Sprinkler System. A sprinkler system is considered to have a single system riser control valve. The design and installation of water supply facilities such as gravity tanks, fire pumps, reservoirs, or pressure tanks are covered by NFPA 20, *Standard for the Installation of Stationary Pumps for Fire Protection,* and NFPA 22, *Standard for Water Tanks for Private Fire Protection.*

3.6.4.1 Antifreeze Sprinkler System. A wet pipe system using automatic sprinklers that contains a liquid solution to prevent freezing of the system, intended to discharge the solution upon sprinkler operation, followed immediately by water from a water supply. [**13,** 2013]

The use of new antifreeze systems and the inspection, testing, and maintenance of existing antifreeze systems have undergone significant changes. For example, antifreeze solutions used in new systems must be listed, and existing systems must have testing conducted to determine the type and concentration of antifreeze solution. Such maintenance requirements vary depending on a number of conditions; see Chapter 5 for the requirements. Exhibit 3.32 shows one of three arrangements of supply piping and valves for an antifreeze system that are permitted by NFPA 13.

3.6.4.1.1 Premixed Antifreeze Solution. A mixture of an antifreeze material with water that is prepared and factory-mixed by the manufacturer with a quality control procedure in place that ensures that the antifreeze solution remains homogeneous and that the concentration is as specified. [**13,** 2013]

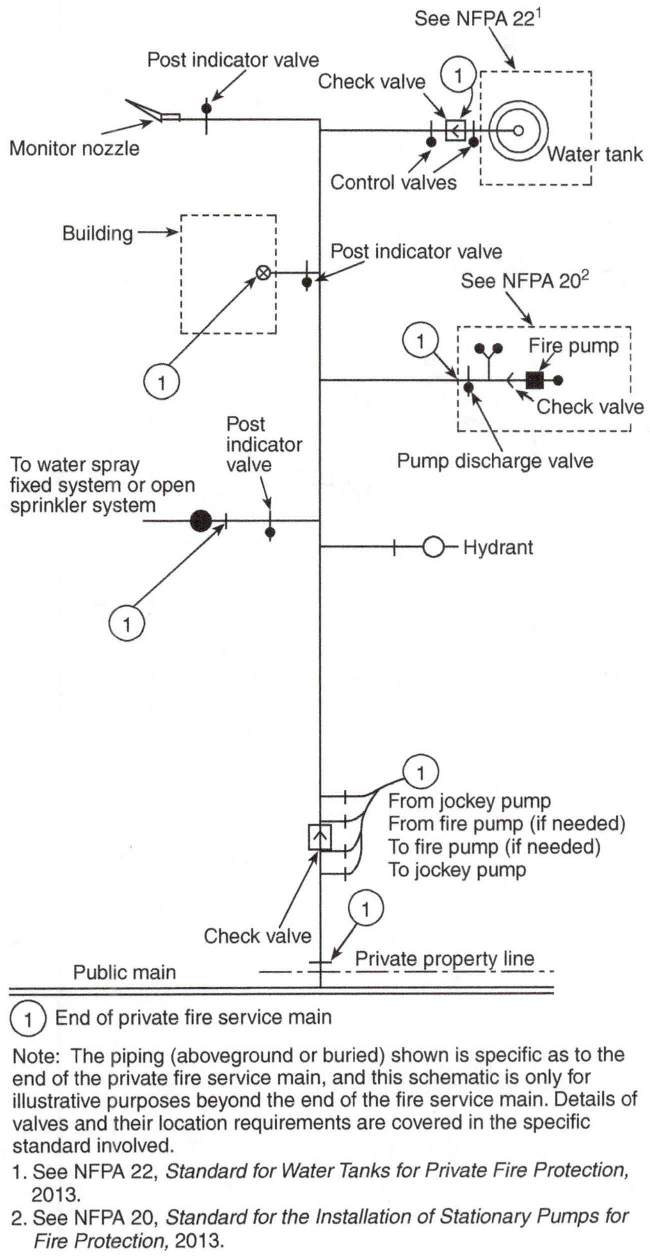

FIGURE A.3.6.3 *Typical Private Fire Service Main. [24:Figure A.3.3.11]*

3.6.4.2 Deluge Sprinkler System. A sprinkler system employing open sprinklers or nozzles that are attached to a piping system that is connected to a water supply through a valve that is opened by the operation of a detection system installed in the same areas as the sprinklers or the nozzles. When this valve opens, water flows into the piping system and discharges from all sprinklers or nozzles attached thereto. [**13,** 2013]

Deluge sprinkler systems are typically used in high-challenge fire areas where water is required for the entire area. Deluge systems are used to extinguish or control a fire, and they are also installed to provide exposure protection to areas adjacent to a fire. Since the deluge piping

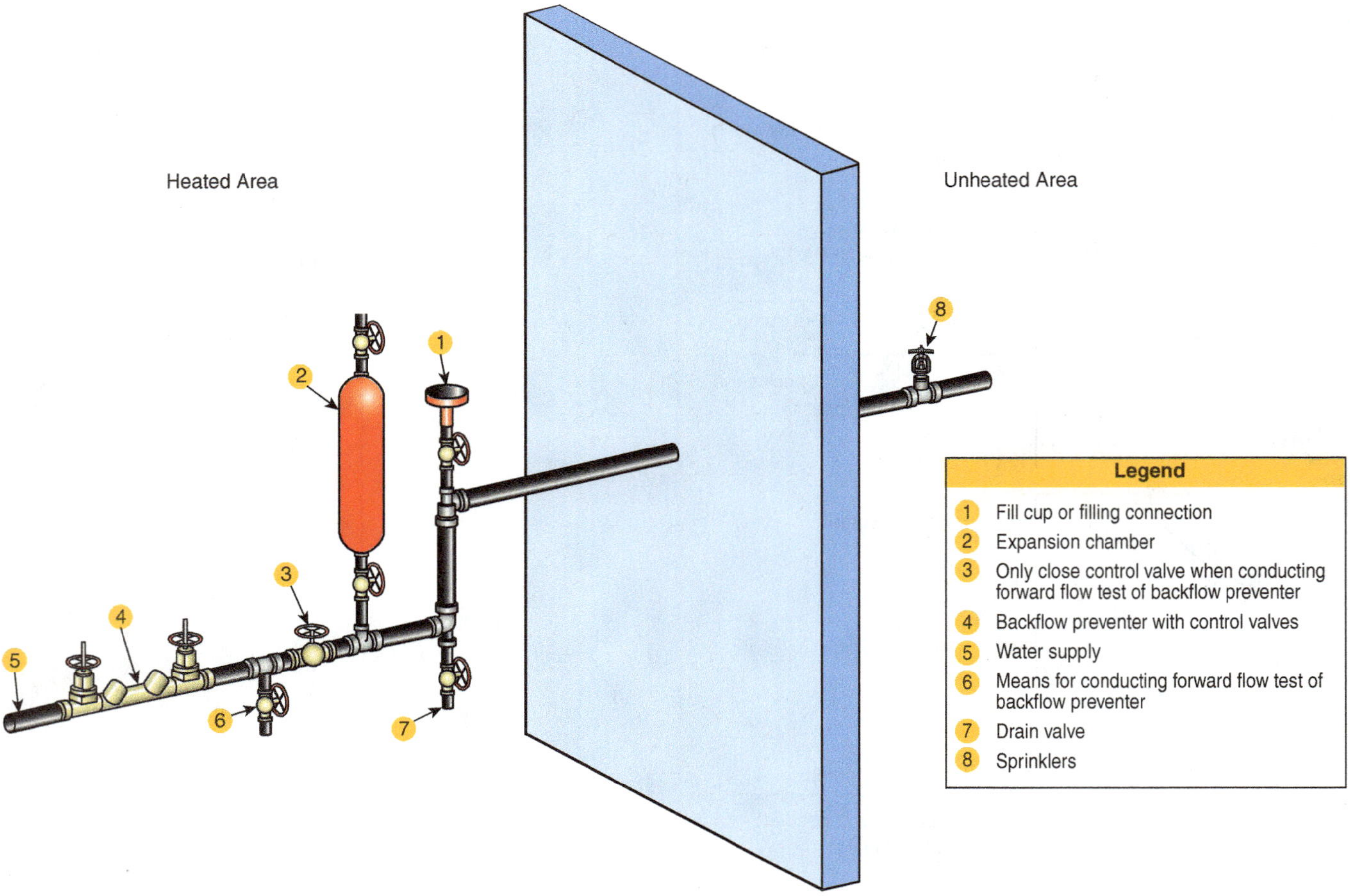

EXHIBIT 3.32 *Valve Arrangement for an Antifreeze System Loop. (Courtesy of Stephan Laforest)*

contains no water and is exposed to the atmosphere, the pipe is subject to the effects of corrosion resulting from condensation or improper draining of the system.

Deluge systems, such as the one shown in Exhibit 3.33, are more complicated and might require more attention than wet pipe systems, due to the actuation systems used. These actuation systems can be in the form of electrical detection systems or wet or dry pilot sprinklers.

3.6.4.3 Dry Pipe Sprinkler System. A sprinkler system employing automatic sprinklers that are attached to a piping system containing air or nitrogen under pressure, the release of which (as from the opening of a sprinkler) permits the water pressure to open a valve known as a dry pipe valve, and the water then flows into the piping system and out the opened sprinklers. [**13,** 2013]

A dry pipe sprinkler system is typically installed where water-filled piping is subject to freezing. (See Exhibit 3.34.) A dry pipe system requires an extensive inspection, testing, and maintenance process because of the need to keep the piping free of water, the requirement for constant pressurized air, and a complicated valve arrangement.

Exposure to fluctuating temperatures and corrosive environments in a dry pipe system can accelerate the corrosion of pipe, fittings, hangers, and sprinklers. The corrosion of the inner surfaces of the piping system is of particular concern because it can lead to the presence of obstructions in the form of rust or scale. Careful observation of water discharge during testing of dry pipe systems is essential to detect this obstructing material.

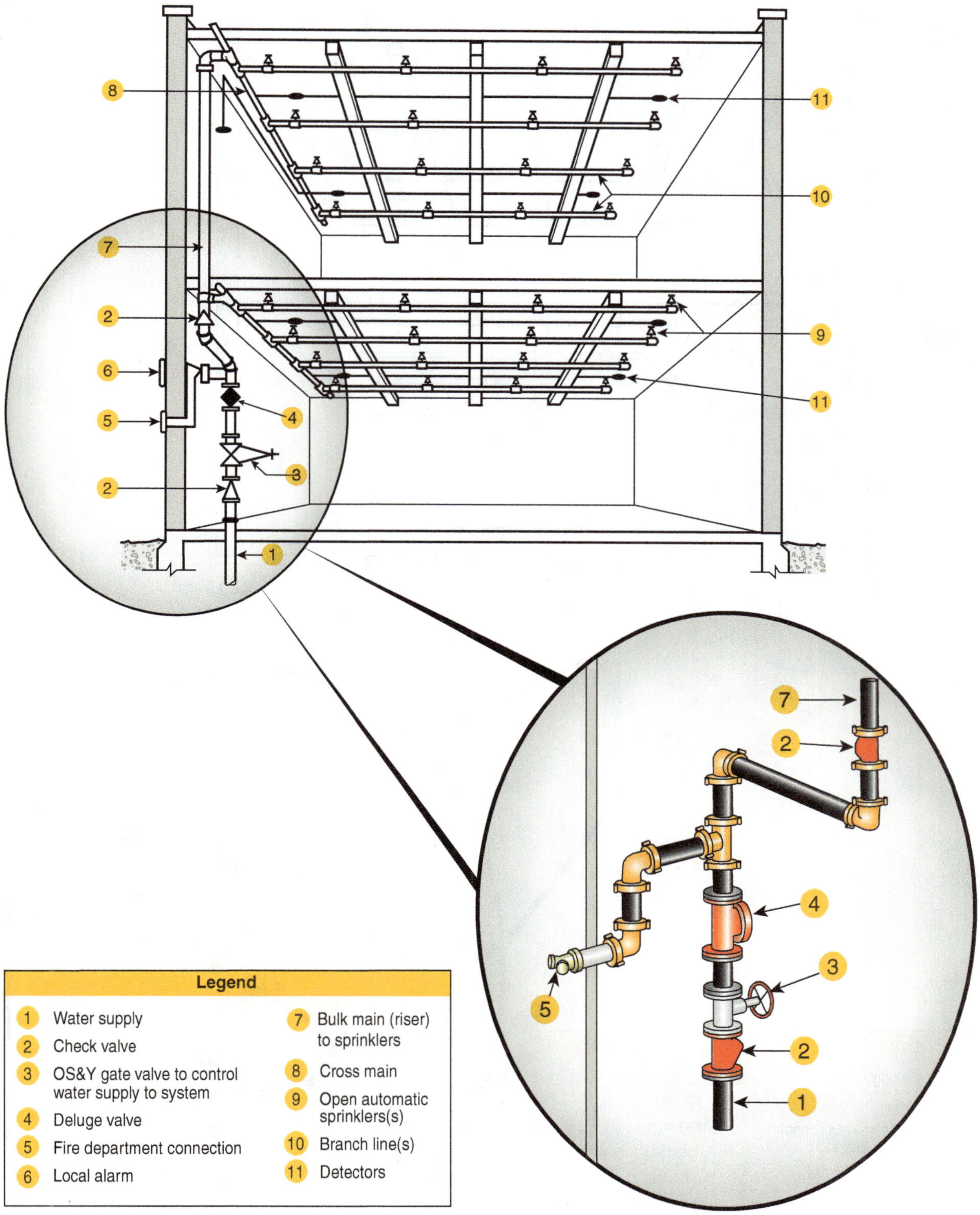

EXHIBIT 3.33 *Deluge System. (Courtesy of Stephan Laforest)*

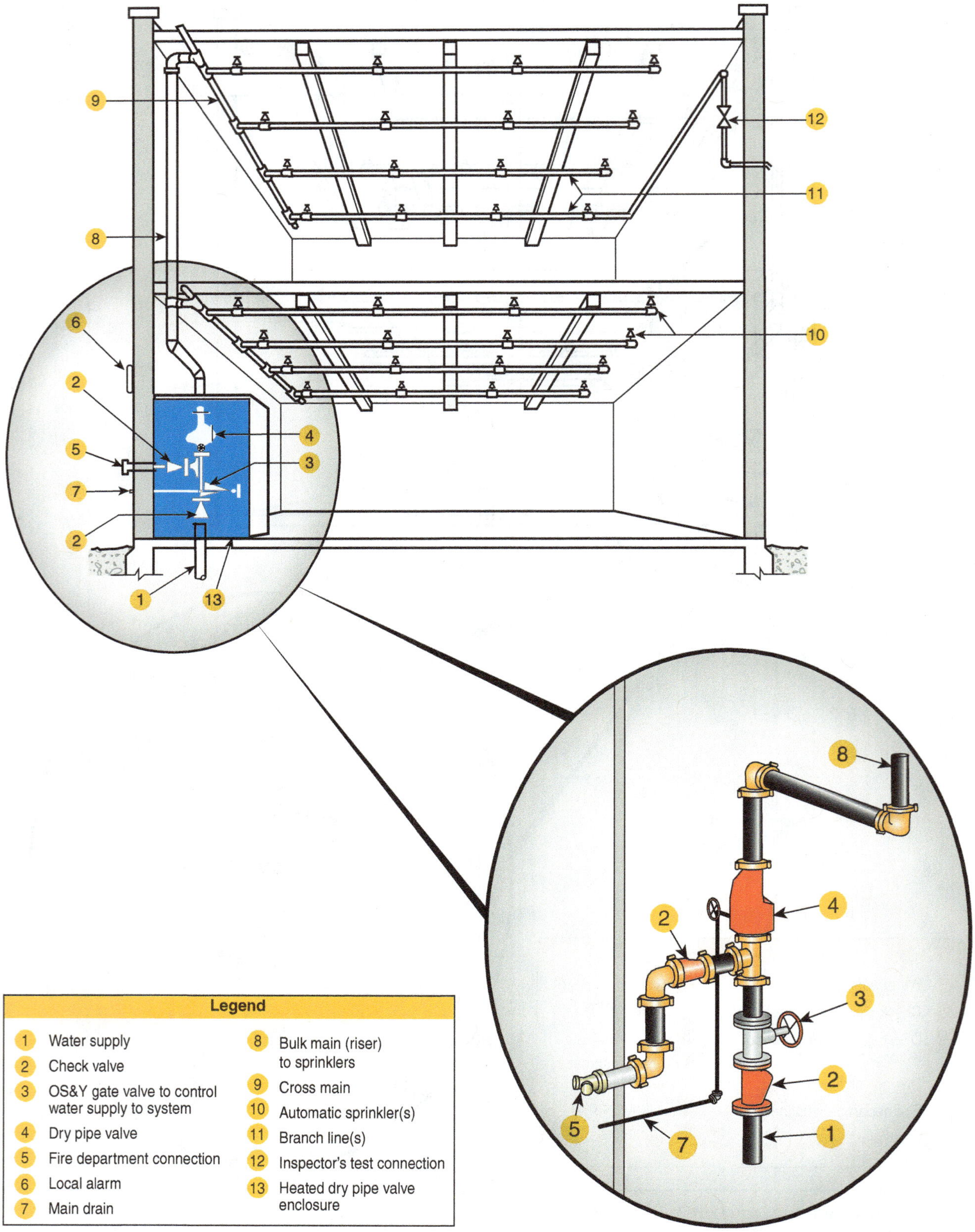

EXHIBIT 3.34 *Dry-Pipe System for Unheated Properties. (Courtesy of Stephan Laforest)*

The dry pipe valve is a differential type valve, meaning that very little air pressure can hold back significantly greater water pressure, usually around 1 psi (0.07 bar) of air pressure to 5 psi or 6 psi (0.34 bar or 0.41 bar) of water pressure. Since a dry pipe system is prone to the formation of internal rust and scale, particular attention to the dry pipe valve is necessary to prevent the accumulation of this material in the valve seat. Accumulated rust and scale can prevent the clapper from sealing properly and result in malfunction of the dry pipe valve, which would necessitate an internal inspection.

The principle of a dry pipe valve is illustrated in Exhibit 3.35. Compressed air in the sprinkler system holds the dry pipe valve closed, preventing water from entering the sprinkler piping until a sprinkler opens and releases air and the air pressure has dropped below a predetermined point.

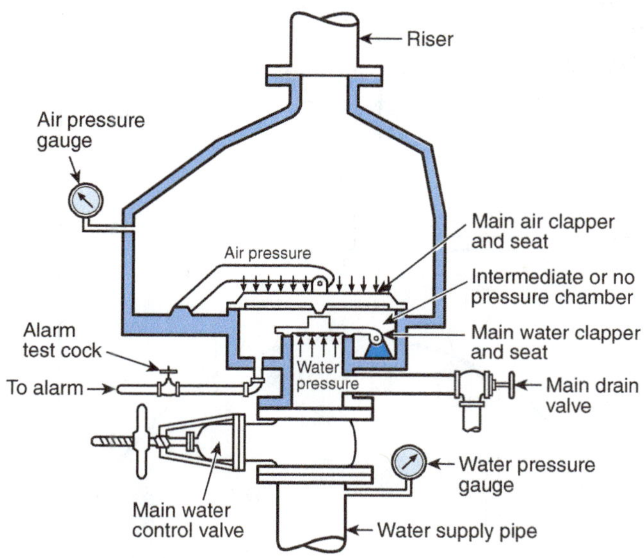

EXHIBIT 3.35 *Dry Pipe Valve. (Source: Fire Protection Handbook, 2008, Figure 16.3.9)*

3.6.4.4 Marine System. A sprinkler system installed on a ship, boat, or other floating structure that takes its supply from the water on which the vessel floats.

3.6.4.5 Preaction Sprinkler System. A sprinkler system employing automatic sprinklers that are attached to a piping system that contains air that might or might not be under pressure, with a supplemental detection system installed in the same areas as the sprinklers. [**13,** 2013]

Several types of preaction systems can be found in use, including the following:

- *Single interlock systems,* which allow water to enter the piping only upon the operation of detection devices
- *Non-interlock systems,* which allow water to enter the piping system upon activation of either detection devices or automatic sprinklers
- *Double interlock systems,* which allow water to enter the piping system only upon operation of both detection devices and automatic sprinklers

A typical preaction sprinkler system is shown in Exhibit 3.36.

3.6.4.6* Wet Pipe Sprinkler System. A sprinkler system employing automatic sprinklers attached to a piping system containing water and connected to a water supply so that water discharges immediately from sprinklers opened by heat from a fire. [**13,** 2013]

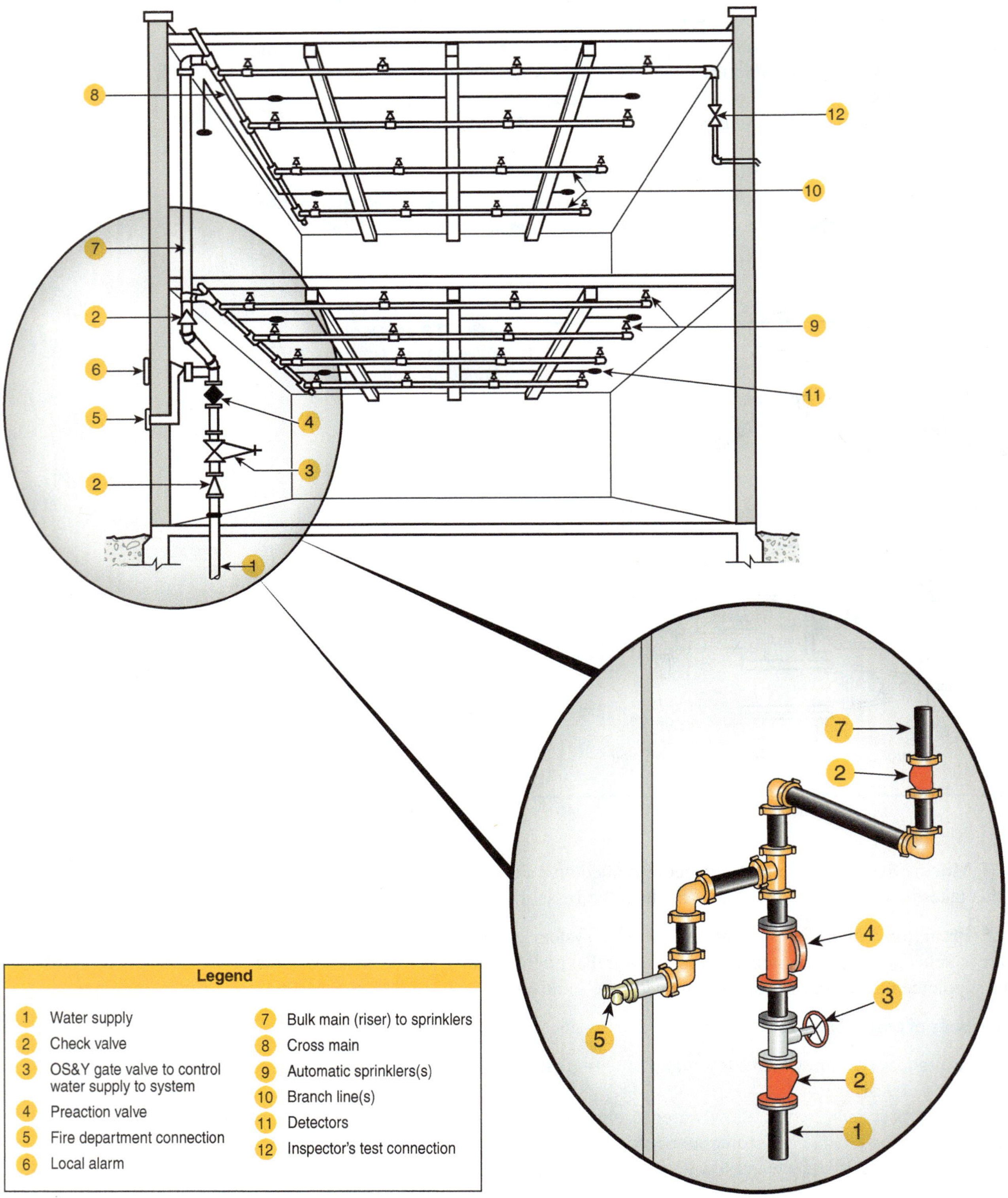

EXHIBIT 3.36 *Typical Preaction Sprinkler System. (Courtesy of Stephan Laforest)*

A.3.6.4.6 Wet Pipe Sprinkler System. Hose connections [1½ in. (40 mm) hose, valves, and nozzles] supplied by sprinkler system piping are considered components of the sprinkler system.

The wet pipe system (see Exhibit 3.37) is the most common of all sprinkler systems and the simplest to maintain. This system is filled with water under pressure at all times so that water is discharged immediately upon activation of a sprinkler. The alarm valve shown in Exhibit 3.37 causes a warning signal to sound when water flows through the system. However, it must be noted here that sprinklers operate independently of each other; in many cases, fire is controlled by a single sprinkler. One of the few limitations of a wet pipe system is that it must be protected from freezing.

3.6.5 Water Mist System. A distribution system connected to a water supply or water and atomizing media supplies that is equipped with one or more nozzles capable of delivering water mist intended to control, suppress, or extinguish fires and that has been demonstrated to meet the performance requirements of its listing and [the applicable] standard. [**750,** 2010]

Water spray fixed systems are usually controlled by a deluge valve supplying water to open or non-automatic discharge devices. For automatic systems, the actuating mechanism might be in the form of an electrical detection system or a system of hydraulic or pneumatic pilot sprinklers. These systems are usually more complex than either wet or dry pipe sprinkler systems. See the commentary in Chapter 10 for more information on water spray fixed systems.

3.6.6* Water Spray System. An automatic or manually actuated fixed pipe system connected to a water supply and equipped with water spray nozzles designed to provide a specific water discharge and distribution over the protected surfaces or area. [**15,** 2012]

A.3.6.6 Water Spray System. Automatic systems can be actuated by separate detection equipment installed in the same area as the water spray nozzles or by the water spray nozzles using an operating element. In some cases, the automatic detector can also be located in another area. [**15:** A.3.3.21]

3.6.6.1 Ultra High-Speed Water Spray System. A type of automatic water spray system where water spray is rapidly applied to protect specific hazards where deflagrations are anticipated. [**15,** 2012]

3.6.7 Water Tank. A tank supplying water for water-based fire protection systems.

A water tank is used primarily when there is an insufficient water supply. Ground-mounted suction tanks and embankment-supported coated fabric suction tanks are always used in conjunction with a fire pump. Pressure tanks and elevated tanks provide pressure in addition to the water. Water tanks are also utilized with high-value or high-risk properties as a redundant or back-up water supply, but in most cases, the tanks are the only source of water for a fire protection system and must be cared for to maintain system reliability. Water storage tanks are usually designed and installed in accordance with NFPA 22, *Standard for Water Tanks for Private Fire Protection*. See the commentary in Chapter 9 for more information on water tanks.

3.7 ITM Task Frequencies

3.7.1* Frequency. Minimum and maximum time between events.

A.3.7.1 Frequency. The frequencies in NFPA 25 are intended to establish an optimal time between tasks that are required by this document. When scheduling conflicts or other conditions do not allow the tasks to be performed on a strict calendar schedule, it is important that the required task frequencies be identified and complied with according to the variances described in the frequency definitions. When the required task frequencies have not been followed, it should be noted on the inspection report, the task should be performed, and the task

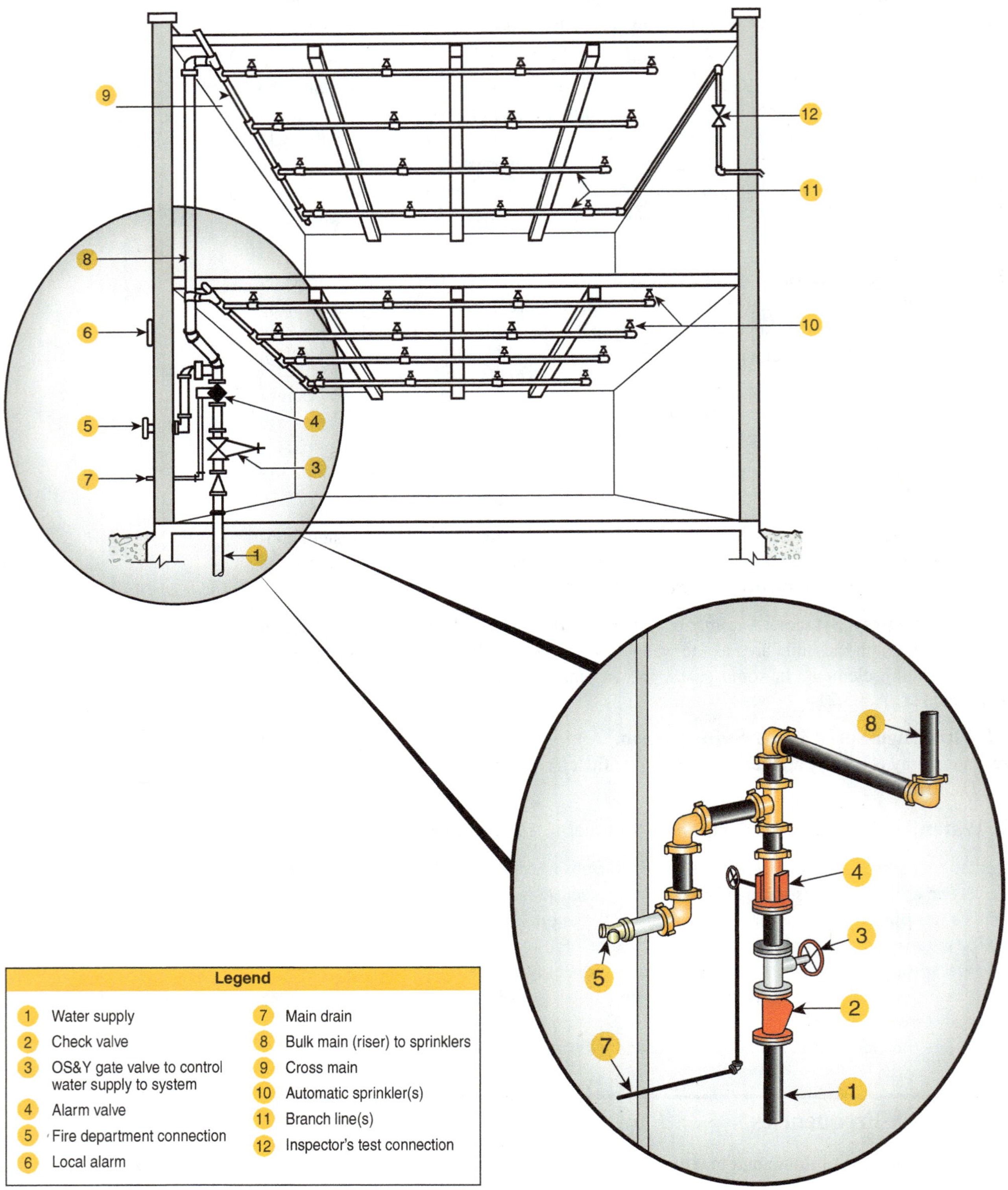

EXHIBIT 3.37 *Wet Pipe Sprinkler System. (Courtesy of Stephan Laforest)*

frequencies should be followed for all future tasks. The variances should not be used to "skip" tasks or to perform fewer tasks than called for in this document.

3.7.1.1 Daily Frequency. Occurring every day.

3.7.1.2 Weekly Frequency. Occurring once per calendar week.

3.7.1.3 Monthly Frequency. Occurring once per calendar month.

3.7.1.4 Quarterly Frequency. Occurring four times per year with a minimum of 2 months and a maximum of 4 months.

3.7.1.5 Semiannual Frequency. Occurring twice per year with a minimum of 4 months and a maximum of 8 months.

3.7.1.6 Annual Frequency. Occurring once per year with a minimum of 9 months and a maximum of 15 months.

3.7.1.7 Three Years Frequency. Occurring once every 36 months with a minimum of 30 months and a maximum of 40 months.

3.7.1.8 Five Years Frequency. Occurring once every 60 months with a minimum of 54 months and a maximum of 66 months.

References Cited in Commentary

Merriam-Webster, Inc., 47 Federal Street, P.O. Box 281, Springfield, MA 01102.

Merriam-Webster's Collegiate Dictionary, 11th edition, 2003.

National Fire Protection Association, 1 Batterymarch Park, Quincy, MA 02169-7471.

Cote, A. E., ed., *Fire Protection Handbook®*, 20th edition, 2008.

NFPA 13, *Standard for the Installation of Sprinkler Systems,* 2013 edition.

NFPA 14, *Standard for the Installation of Standpipe, Private Hydrant, and Hose Systems,* 2013 edition.

NFPA 15, *Standard for Water Spray Fixed Systems for Fire Protection,* 2012 edition.

NFPA 20, *Standard for the Installation of Stationary Pumps for Fire Protection,* 2013 edition.

NFPA 22, *Standard for Water Tanks for Private Fire Protection,* 2013 edition.

NFPA 70E®, Standard for Electrical Safety in the Workplace®, 2012 edition.

4

GENERAL REQUIREMENTS

Chapter 4 of NFPA 25 contains general requirements for the different types of systems covered under the scope of the standard. This chapter outlines the administrative guidelines for compliance with the standard, which include such topics as changes in building use or occupancy, impairments, corrective action, record keeping, and essential safety requirements for those performing the work specified in Chapter 5 through Chapter 16. The primary focus of this chapter is on the property owner or designated representative. The standard aims to clearly identify that the owner or the owner's designated representative is responsible for all aspects of maintaining the systems addressed by the standard. These responsibilities include changes to building contents, changes to building use, or renovations to interior building arrangement that can hinder the efficacy of the fire protection systems operation.

4.1 Responsibility of Property Owner or Designated Representative

4.1.1* Responsibility for Inspection, Testing, Maintenance, and Impairment. The property owner or designated representative shall be responsible for properly maintaining a water-based fire protection system.

In many jurisdictions, NFPA 25 is adopted through a model code or state law. However, due to a lack of manpower, authorities having jurisdiction (AHJs) often cannot actively enforce the standard in those jurisdictions. Yet just because the AHJ is not banging on the door looking for an inspection, testing, and maintenance (ITM) inspection report, the owner still must follow the ITM program outlined by NFPA 25.

FAQ Who is responsible for the inspection, testing, and maintenance of the water-based fire protection system?

Paragraph 4.1.1 sets the tone for the enforcement of the standard by stating that the property owner or designated representative (see A.4.1.1.3) is responsible for maintaining a water-based fire protection system. Maintaining a system includes the inspection, testing, and maintenance tasks, as mandated by the standard.

A.4.1.1 Any portion or all of the inspection, testing, and maintenance can be permitted to be contracted with an inspection, testing, and maintenance service.

Most owners contract a service for at least some of the inspection, testing, and maintenance requirements of NFPA 25. Many misunderstandings concerning the expectations of the owner and the contracted service provider can be eliminated with well-written agreements that include detailed scopes of work, any exclusions, schedule constraints, and terms and conditions.

Critical Point

When contracting services, owners should require verification that the service provider and personnel are qualified and, where required, properly licensed to perform the contracted tasks.

When contracting services, owners should require verification that the service provider and personnel are qualified and, where required, properly licensed to perform the contracted tasks.

•

4.1.1.1* Inspection, testing, maintenance, and impairment procedures shall be implemented in accordance with those established in this document and in accordance with the manufacturer's instructions.

As stated in Chapter 1, the requirements found in NFPA 25 are minimum requirements. As such, any program meeting or exceeding these requirements is considered to be in compliance with this standard.

A.4.1.1.1 In order to ensure compliance, the owner should verify that windows, skylights, doors, ventilators, other openings and closures, concealed spaces, unused attics, stair towers, roof houses, and low spaces under buildings do not expose water-filled piping to freezing. This should occur prior to the onset of cold weather and periodically thereafter.

4.1.1.2 Inspection, testing, and maintenance shall be performed by qualified personnel.

FAQ Is the property owner required to hire a contractor to perform the activities required by this standard, or can the in-house maintenance staff perform these functions?

Although extensive training and experience are necessary for many of the prescribed tasks, property owners or their designated representatives can perform some of the required inspection activities, such as basic inspections of valve position or pressure gauge readings, which require a minimal amount of training and instruction. See 3.3.34 for the definition of the term *qualified*.

It should be noted that many state and local jurisdictions have specific mandates regarding the qualifications for performing the inspection, testing, and maintenance of water-based suppression systems. The requirements should be researched to ensure that all work is performed in accordance with jurisdictional requirements. However, it is important to emphasize that even in the absence of jurisdictional oversight, no one should perform any of the activities prescribed by this standard without sufficient training.

4.1.1.3* Where the property owner or designated representative is not the occupant, the property owner or designated representative shall be permitted to delegate the authority for inspecting, testing, maintenance, and the managing of impairments of the fire protection system to a designated representative.

A.4.1.1.3 Examples of designated representatives can include the occupant, management firm, or managing individual through specific provisions in the lease, written use agreement, or management contract.

4.1.1.4 Where a designated representative has received the authority for inspecting, testing, maintenance, and the managing of impairments, the designated representative shall comply with the requirements identified for the property owner or designated representative throughout this standard.

4.1.2* Freeze Protection. The property owner or designated representative shall ensure that water-filled piping is maintained at a minimum temperature of 40°F (4°C) unless an approved antifreeze solution is utilized.

When a sprinkler system freezes, it often requires expensive repairs to the system and to the building if water is discharged through broken pipe, fittings, or sprinklers. In addition, the sprinkler system is impaired, which creates a heightened risk of loss from a fire event. This standard places the responsibility with the owner for maintaining water-filled piping at a minimum temperature of 40°F (4°C) or, in cases where the piping itself cannot be protected, maintaining a minimum temperature of 40°F (4°C) within any areas of the building containing unprotected water-filled pipe.

This subsection specifically refers to an "approved" antifreeze solution. The requirements for antifreeze systems have changed drastically since the 2011 edition of NFPA 25 was published. As outlined in Chapter 5, an acceptable antifreeze solution would be a solution with 30 percent propylene glycol or less, or a solution with 38 percent glycerine or less. These solutions have been deemed acceptable for existing systems based on test data that have shown that solutions with this concentration of antifreeze did not increase the heat release rate of a fire when the solution was discharged.

Another design approach that might be acceptable would be an antifreeze solution that has been approved by the AHJ as described in a deterministic risk assessment. Refer to 5.3.4.2.1(3) for more information on deterministic risk assessments for antifreeze systems.

A.4.1.2 Other means of freeze protection for water-filled piping including heated valve enclosures, heat tracing, insulation, antifreeze solutions, or other methods are allowed by the applicable installation standard. Installation standards require heat tracing protecting fire protection piping against freezing to be supervised.

4.1.2.1 All areas of the building containing water-filled piping that does not have another means of freeze protection shall be maintained at a minimum temperature of 40°F (4°C).

Critical Point

NFPA 25 places the responsibility with the owner for maintaining water-filled piping at a minimum temperature of 40°F (4°C) or, in cases where the piping itself cannot be protected, maintaining a minimum temperature of 40°F (4°C) within any areas of the building containing unprotected water-filled pipe.

Case In Point

When owners find a frozen pipe or frozen system component in a building, they often look to their NFPA 25 inspection provider to find out why the potential freezing condition was not noted in the previous annual system inspection report. However, instances in which the ambient condition has not been properly maintained above a freezing temperature are not considered by NFPA 25 as information that should be recorded during an inspection. This is largely because the ambient temperature in a given room or space varies from day to day. If an inspector recorded a temperature of 75°F (21.1°C) during an annual inspection of a building in July, it probably would not be reflective of the temperature found during an inspection of the building in January. For this reason, frozen system components, such as the valve and the pipe shown in the accompanying photos, cannot be predicted by an inspector during an inspection.

Valve Rupture Due to Freezing. (Courtesy of Byron Blake and SimplexGrinnell)

Pipe Split Due to Freezing. (Courtesy of Byron Blake and SimplexGrinnell)

4.1.2.2 Aboveground water-filled pipes that pass through open areas, cold rooms, passageways, or other areas exposed to temperatures below 40°F (4°C), protected against freezing by insulating coverings, frostproof casings, listed heat tracing systems, or other reliable means, shall be maintained at temperatures between 40°F (4°C) and 120°F (48.9°C).

4.1.2.3 Where other approved means of freeze protection for water-filled piping as described in 4.1.2.2 are utilized, they shall be inspected, tested, and maintained in accordance with this standard.

FAQ Who is responsible for keeping the building heated and the heating system operational?

The requirements in 4.1.2 and 4.1.2.1 place the responsibility for keeping the building and all heating systems operational on the owner. In addition, all means that are used to protect water-filled piping itself must be inspected, tested, and maintained in accordance with NFPA 25. The intent behind the requirements in 4.1.2 and the related subsections is that wet pipe sprinkler systems require an ambient temperature of 40°F (4.4°C) to prevent freezing, which can lead to broken pipes or fittings and subsequently an inoperable system.

During warm weather, freezing is seldom an issue, so problems with the building enclosure often go unnoticed until the weather gets colder. Prior to the onset of cold weather, it is important for the owner to check for areas in the building where cold air could enter. Particular attention should be paid to unoccupied areas such as openings and closures, blind spaces, unused attics, and so on, because problems in unattended areas are most likely to go unnoticed. Heating equipment should also be inspected and tested to ensure it operates as needed when cold weather arrives.

4.1.3* Accessibility. The property owner or designated representative shall provide ready accessibility to components of water-based fire protection systems that require inspection, testing, and maintenance.

A.4.1.3 The components are not required to be open or exposed. Doors, removable panels, or valve pits can be permitted to satisfy the need for accessibility. Such equipment should not be obstructed by features such as walls, ducts, columns, direct burial, or stock storage.

The requirement in 4.1.3 for accessibility is intended to address the all-too-common practice of placing objects such as file cabinets or stock in front of sprinkler risers and other control equipment. This requirement is not limited to the interior of the building. Exterior components of the system must be kept accessible as well. Exhibit 4.1 shows a fire department connection (FDC) that has been obstructed by shrubbery. Exterior components such as post indicator valves and

Case In Point

While A.4.1.3 states that the components are not required to be exposed and that access panels are permitted for aesthetic reasons, the access panel and signage cannot be intentionally obscured. The accompanying photos show an access panel and inspector test connection (ITC) sign that are blocked by a clock. An inspector will not necessarily remember year to year that the clock needs to be taken down in order to access the valve. Furthermore, if there is turnover within the facilities management personnel, it is possible for the owner and the in-house team to be unaware of system components that are hidden behind decorative items.

ITC Hidden by Clock. (Photo courtesy of Byron Blake and SimplexGrinnell)

ITC Sign and Access Panel. (Photo courtesy of Byron Blake and SimplexGrinnell)

ITC Accessible with Panel Removed. (Photo courtesy of Byron Blake and SimplexGrinnell)

FDCs might also be blocked by dumpsters and outdoor storage as shown in Exhibit 4.2, and even snow piles as a result of plowing and shoveling, as shown in Exhibit 4.3. The property owner or the designated representative must maintain access to equipment for inspection, testing, and maintenance purposes. The standard refers to the *property owner or designated representative* throughout the standard to distinguish between the property owner and the owner of the contracting firm.

In addition, when conducting inspections of sprinklered residential-type buildings such as hotels, motels, apartments, and condominiums, gaining access to individual dwelling units can be challenging. NFPA 25 provides no exceptions to the requirement to perform inspections, tests, and maintenance due to limited accessibility. For example, when conducting the annual inspection of sprinklers, performing a "sample" inspection of a few dwelling units does not meet the intent of the standard. The inspection of systems housed in dwelling units might require extensive coordination between the owner, the service provider (if one is used), and dwelling unit occupants.

Critical Point

The requirement in 4.1.3 for accessibility is intended to address the all-too-common practice of placing objects such as file cabinets or stock in front of sprinkler risers and other control equipment.

EXHIBIT 4.1 *FDC Obstructed by Shrubbery.*

EXHIBIT 4.2 *FDC Obstructed and Inaccessible Due to Exterior Storage. (Courtesy of Byron Blake and SimplexGrinnell)*

EXHIBIT 4.3 *FDC Obstructed by Snow. (Courtesy of Byron Blake and SimplexGrinnell)*

4.1.4 Notification of System Shutdown or Testing. The property owner or designated representative shall notify the authority having jurisdiction, the fire department, if required, and the alarm-receiving facility before testing or shutting down a system or its supply.

Paragraph 4.1.4 establishes the requirement for notification when systems are removed from service. Testing a system without proper notification might — and often does — result in a false alarm. False alarms must be avoided, since they remove emergency services personnel from service at a time when their services might be needed for an actual emergency. In many jurisdictions, repeated false alarms can result in a fine or other penalty for property owners. When flowing an inspector's test connection, it may be obvious that an alarm is intended to sound upon waterflow and that the supervisory service must be notified as prescribed by 4.1.4. Perhaps not so obvious is the fact that other tests, such as a fire pump test, hydrant test, or main drain test, can also lead to the tripping of a waterflow switch, which will sound an alarm and, potentially, send a signal to the fire department. At the conclusion of such tests, the alarm-receiving facility should be notified that any alarms received from that point on are not test-generated and should be responded to appropriately.

4.1.4.1 The notification of system shutdown or test shall include the purpose for the shutdown or test, the system or component involved, the estimated time of shutdown or test, and the expected duration of the shutdown or test.

4.1.4.2 The authority having jurisdiction, the fire department, and the alarm-receiving facility shall be notified when the system, supply, or component is returned to service or when the test is complete.

The requirement in 4.1.4.2 assists the AHJ or alarm company in establishing a follow-up program to ensure that systems are properly returned to service.

4.1.5* Corrections and Repairs.

A.4.1.5 Recalled products should be replaced or remedied. Remedies include entrance into a program for scheduled replacement. Such replacement or remedial product should be installed in accordance with the manufacturer's instructions and the appropriate NFPA installation standards. A recalled product is a product subject to a statute or administrative regulation specifically requiring the manufacturer, importer, distributor, wholesaler, or retailer of a product, or any combination of such entities, to recall the product, or a product voluntarily recalled by a combination of such entities.

Needed corrections and repairs should be classified as an impairment, critical deficiency, or noncritical deficiency according to the effect on the fire protection system and the nature of the hazard protected.

Impairments are the highest priority problem found during inspection, testing, and maintenance and should be corrected as soon as possible. The fire protection system cannot provide an adequate response to a fire, and implementation of impairment procedures outlined in Chapter 15 is required until the impairment is corrected.

Critical deficiencies need to be corrected in a timely fashion. The fire protection system is still capable of performing, but its performance can be impacted and the implementation of impairment procedures might not be needed. However, special consideration must be given to the hazard in the determination of the classification. A deficiency that is critical for one hazard might be an impairment in another.

Noncritical deficiencies do not affect the performance of the fire protection system but should be corrected in a reasonable time period so that the system can be properly inspected, tested, and maintained.

Assembly occupancies, health care facilities, prisons, high-rise buildings, other occupancies where the life safety exposure is significant, or facilities that cannot be evacuated in a timely manner require special consideration. As an example, a nonfunctioning waterflow alarm might be considered a critical deficiency in a storage warehouse but an impairment in a hospital.

High hazard occupancies where early response to a fire is critical also require special consideration. A small number of painted sprinklers could be considered an impairment for a system protecting a high hazard occupancy but might be considered a critical deficiency in a metal working shop.

Classifications of needed corrections and repairs are shown in Table A.3.3.7.

One of the most common misapplications of NFPA 25 comes from not understanding the document scope completely. NFPA 25 is intended to confirm the functionality of the system components that are installed, not to make sure that the designer has designed the system correctly and observed the rules of the design and installation standard.

This misapplication of scope can often lead to arguments between owners, inspectors, and AHJs when certain conditions are present in a building. In some cases, the owner has the expectation that the hired inspector is not only looking for "wear and tear" items associated with the water-based systems but also giving the system a clean bill of health from a design perspective as well. The level of effort to produce this clean bill of health is significantly more than what the standard intends and requires an inspector to deliver.

Simply put, if a condition that is found in the field does not comply with the design standard, but the components observed are in good working condition, it would not be considered a deficiency or impairment per NFPA 25.

In an attempt to provide some examples of where the difference between a hazard evaluation and an ITM-related deficiency exists, this handbook contains a recurring feature (see below) that provides examples of the types of situations that are within the scope of the standard and those that are not.

ITM Deficiency, Impairment, or Hazard Evaluation?

ANSWER: Hazard Evaluation

Recalled sprinklers or other recalled components should be addressed by having them replaced or by other remedies as described by the recall. However, the specific knowledge needed to identify recalled products makes it next to impossible to include a requirement in NFPA 25 that could be reasonably complied with and enforced. If an inspector does have specific knowledge about a recalled product, such as this Omega sprinkler, it should be reported to the owner as a recommendation to have the entire building inspected for other recalled sprinklers, including those in concealed spaces.

The first mention of *recalled products* appeared in Annex A of the 2002 edition of NFPA 25, as guidance to the general requirements section on corrections and repairs, where it still resides in the current edition. A requirement was added to Chapter 5 in the 2008 edition that made the general statement, "Sprinklers that are subject to recall shall be replaced per the manufacturer's requirements." This statement was deleted in the 2011 edition, and proposals to bring it back into the 2014 edition failed.

(Courtesy of Byron Blake and SimplexGrinnell)

4.1.5.1* The property owner or designated representative shall correct or repair deficiencies or impairments that are found during the inspection, test, and maintenance required by this standard.

As noted in Chapter 1, corrective action to remediate deficiencies and impairments must be done in accordance with the applicable design and installation standard. Exhibit 4.4 shows a leaking fitting that was simply wrapped with a cloth. Wrapping any absorbent material around a leaking fitting can actually create a worse condition as the absorbent material becomes drenched and allows water to be in constant contact with a larger surface area of the pipe and/or fitting. This could then lead to more rapid corrosion and system impairment.

EXHIBIT 4.4 Corrective Action Not Executed per Installation Standard. (Courtesy of Byron Blake and SimplexGrinnell)

A.4.1.5.1 System deficiencies not explained by normal wear and tear, such as hydraulic shock, can often be indicators of system problems and should be investigated and evaluated by a qualified person or engineer. Failure to address these issues could lead to catastrophic failure. Examples of deficiencies that can be caused by issues beyond normal wear and tear are as follows:

(1) Pressure gauge deficiencies as follows:
   (a) Gauge not returning to zero
   (b) Gauge off scale
   (c) Gauge with bent needle
(2) Support devices deficiencies as follows:
   (a) Bent hangers and/or rods
   (b) Hangers pulled out/off structure
   (c) Indication of pipe or hanger movement such as the following:
      i. Hanger scrape marks on pipe, exposed pipe surface where pipe and hangers are painted
      ii. Firestop material damaged at pipe penetration of fire-rated assembly
(3) Unexplained system damage as follows:
   (a) Unexplained system damage beyond normal wear and tear
   (b) Bent or broken shafts on valves
   (c) Bent or broken valve clappers
   (d) Unexplained leakage at branch lines, cross main, or feed main piping
   (e) Unexplained leakage at closed nipples
   (f) Loose bolts on flanges and couplings

(4) Fire pump deficiencies as follows:
   (a) Fire pump driver out of alignment
   (b) Vibration of fire pump and/or driver
   (c) Unusual sprinkler system piping noises (sharp report, loud bang)

While NFPA 25 focuses on the specific requirements for inspection, testing, and maintenance of water-based fire protection systems, the ultimate goal of such a program is to maintain the operational status of a system. Paragraph 4.1.5.1 provides the enforcement tool to initiate corrective action to ensure that systems are functioning at all times. However, the standard does not provide specific timelines on how quickly corrections or repairs must be completed.

4.1.5.2 Corrections and repairs shall be performed by qualified maintenance personnel or a qualified contractor.

It should be noted that many jurisdictions regulate who can perform corrections and repairs, along with regulating specific licensing or certification requirements. See the commentary following 3.3.34 and 4.1.1.2.

4.1.6* Changes in Occupancy, Use, Process, or Materials. The property owner or designated representative shall not make changes in the occupancy, the use or process, or the materials used or stored in the building without evaluation of the fire protection systems for their capability to protect the new occupancy, use, or materials.

A.4.1.6 The inspections and tests specified in this standard do not address the adequacy of design criteria or the capability of the fire protection system to protect the building or its contents. It is assumed that the original system design and installation were appropriate for the occupancy and use of the building and were approved by all applicable authorities having jurisdiction. If no changes to the water supply or to the building or its use have transpired since it was originally occupied, no evaluation is required. If changes are contemplated, it is the owner's responsibility to arrange for the evaluation of the fire protection system(s). Where the inspections and tests specified in the standard have been contracted to a qualified inspection provider or contractor, it is not the role of the inspector or contractor to determine if any changes have been made or the subsequent evaluation of the fire protection system. The evaluation of any building changes should be conducted before any proposed change is incorporated and should utilize the appropriate installation standard and input from applicable authorities having jurisdiction.

Fire protection systems should not be removed from service when the building is not in use; however, where a system that has been out of service for a prolonged period (such as in the case of idle or vacant properties) is returned to service, it is recommended that a responsible and experienced contractor be retained to perform all inspections and tests.

4.1.6.1 The evaluation required by 4.1.6 shall not be considered part of the normal inspection, testing, and maintenance required by this standard.

As outlined in Chapter 1, the scope of NFPA 25 does not require the inspector to evaluate the design and layout of the systems covered in the standard. The standard presupposes that the system, as designed and installed, complies with the applicable standards at the time of construction and at the points where changes were made to the building that would necessitate system modifications. The requirements in 4.1.6 address this concept by identifying that is it the owner's responsibility to understand the impact of changes in occupancy, use, process, or material. The requirements in 4.1.6 also state that at the time these changes are made, a hazard evaluation must be conducted that will either confirm that the system(s) meets the design requirements of the applicable standard as currently installed, or that modifications to the system(s) must be made. In the event that changes are made to the building and this evaluation does not take place, any noncompliant conditions created by the changes are not the responsibility of the inspector to identify.

Case In Point

In some cases, a condition that does not comply with the design standard is relatively easy to identify. It can be identified from the floor level without the need of a lift, ladder, tape measure, or any other equipment. The first photo shows a sprinkler that is several feet below the ceiling. In this instance, a drop ceiling was removed and the sprinkler was not relocated to within 12 in. (305 mm) of the ceiling as required by NFPA 13, *Standard for the Installation of Sprinkler Systems*. A similar case would be a sprinkler located too close to a hot air diffuser, as shown in the second photo. Although the sprinkler shown is too close to the hot air diffuser per the spacing rules of NFPA 13, this is not a deficiency per NFPA 25 and should not be called out in an NFPA 25 inspection report.

That is not to say that inspectors cannot inform building owners that there might be an issue with a system as it relates to a design standard such as NFPA 13. The problem with providing a client with more information, though, is that it can open the inspector to liability. If an inspector is identifying some design deficiencies on an NFPA 25 inspection report, but not all of them, the line can become blurred as to the depth of the review and the inspector's contractual obligation to the owner. Many owners believe that NFPA 25 requires the inspector to conduct a hazard evaluation, so when an inspector includes a few design deficiencies mixed in with proper NFPA 25 deficiencies, it can further muddy the waters.

Many inspectors feel that they have a moral and ethical obligation, as members of the fire protection and life safety community, to inform their clients of all potential concerns. One way to comply with the requirements of NFPA 25 while still satisfying that moral obligation is to keep two deficiency reports: one dealing with NFPA 25 issues and another that could be called the "good Samaritan" list. The NFPA 25 report would catalog wear-and-tear issues while clearly identifying the scope of NFPA 25. The "good Samaritan" report would list all other items that are beyond the scope of NFPA 25, along with a disclaimer stating that these items are not considered part of an NFPA 25 inspection program. This way, the scope of NFPA 25 is acknowledged, as are any and all problems observed during the inspection.

No Drop Ceiling. (Courtesy of Byron Blake and SimplexGrinnell)

Heat Source. (Courtesy of Byron Blake and SimplexGrinnell)

The requirement in 4.1.6.1 states that the inspections and tests found in NFPA 25 do not include an evaluation of the capability of the system to adequately protect the property or its contents. The owner or designated representative is responsible for an evaluation of the adequacy (i.e., design criteria) of the fire protection systems when changes are proposed to the building or its use. However, most owners do not understand the significance of this requirement, nor do they know how to have the evaluation conducted. In addition, many users of the standard assume that the inspections and tests specified in the standard include an evaluation of the adequacy of the system to protect the hazard presented by the building, its use, and contents. This assumption is incorrect. For example, an evaluation of the system design criteria would include an examination of the area of coverage and spacing of the sprinklers, and these tasks are not part of the annual inspection required in the standard.

Critical Point

Many users of the standard assume that the inspections and tests specified in the standard include an evaluation of the adequacy of the system to protect the hazard presented by the building, its use, and contents. This assumption is incorrect.

To further clarify this point, Exhibit 4.5 shows a retrofitted cross-corridor door. Prior to the retrofit, the sprinklers in the corridor were spaced from each other according to the requirements of NFPA 13. However, after the retrofit, the sprinkler in the foreground must be evaluated against different criteria for spacing from a wall. In this case, the sprinkler is too far from the wall. This evaluation is required by 4.1.6.2, but the criteria for the evaluation are not within the scope of the inspections required by NFPA 25. The criteria for the evaluation of a sprinkler system are found in installation requirements specified by NFPA 13. This is an example of how the scopes of the two standards are interrelated, and this is clarified further in 4.1.7.

EXHIBIT 4.5 *Retrofitted Cross-Corridor Door Impacts Sprinkler Spacing.*

As stated in Chapter 1, the purpose of the inspections and tests required by this standard is to maintain the operational status of the system(s). Most inspectors are not trained to evaluate the design criteria of a system, and they should not attempt to do so. It is important that contractors or inspection providers are clear in explaining to the owner the scope of the work that they have been contracted to do. An owner might want to have the system evaluated for its adequacy — and it is certainly appropriate for the contractor or inspection provider to conduct the evaluation if qualified to do so — but this evaluation is separate from the inspection and testing activities specified by this standard.

Critical Point

The owner or designated representative is responsible for an evaluation of the adequacy (i.e., design criteria) of the fire protection systems when changes are proposed to the building or its use.

4.1.6.2 The evaluation shall consider factors that include, but are not limited to, the following:

(1) Occupancy changes such as converting office or production space into warehousing
(2) Process or material changes such as metal stamping to molded plastics
(3) Building revisions such as relocated walls, added mezzanines, and ceilings added below sprinklers
(4) Removal of heating systems in spaces with piping subject to freezing

Critical Point

By monitoring building alterations and performing inspections, costly damage to buildings and contents can be avoided.

Occupancy changes usually involve a building permit and the subsequent review process by the AHJ. Yet, the more subtle changes listed in 4.1.6.2, items 3 and 4, might not always be reported. Moving walls and adding mezzanines and ceilings can severely affect the spray pattern of sprinklers. Sprinkler protection in unheated spaces frequently causes damage to systems and water damage to buildings and their contents. By monitoring building alterations and performing inspections, costly damage to buildings and contents can be avoided.

In recent years, a growing number of installations have included sprinklers attached to flexible sprinkler hose fittings. As a result, the relocation of sprinklers attached to flexible hose fittings, and who should perform the work, is an area of concern.

Flexible sprinkler hose fittings are recognized in NFPA 13. However, due to growing concern regarding the ability to move sprinklers without proper evaluation, NFPA 13 contains the following language in Chapter 9:

> **9.2.1.3.3.4*** Where flexible sprinkler hose fittings are used to connect sprinklers to branch lines in suspended ceilings, a label limiting relocation of the sprinkler shall be provided on the anchoring component.

The annex material for this requirement in NFPA 13 provides suggested language for the label as follows:

CAUTION: DO NOT REMOVE THIS LABEL.

> Relocation of this device should only be performed by qualified and/or licensed individuals that are aware of the original system design criteria, hydraulic criteria, sprinkler head listing parameters, and knowledge of the state and local codes including NFPA 13 installation standards. Relocation of the device without this knowledge could adversely affect the performance of this fire protection and life safety system.

Where flexible sprinkler hose fittings allow sprinklers to be adjusted and relocated, adjustments and relocations made by unqualified individuals could violate requirements, such as those for maximum or minimum sprinkler spacing or obstruction avoidance. For this reason, the committee has added a requirement for a label warning against relocation. To further discourage relocation, the length of flexible drops should be kept to the minimum required for adequate installation and proper performance.

Exhibit 4.6 shows a flexible sprinkler hose fitting along with its mounting and a warning label.

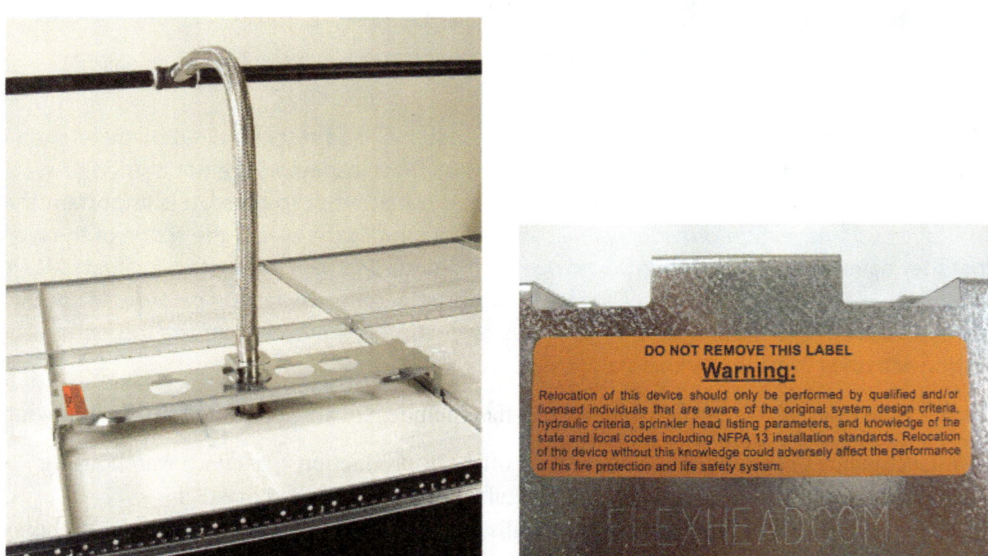

EXHIBIT 4.6 *Flexible Sprinkler Hose Fitting (left) and Warning Label (right). (Courtesy of FlexHead Industries)*

ITM Deficiency, Impairment, or Hazard Evaluation?

ANSWER: Hazard Evaluation

This combustible concealed space was discovered by an inspector. Perhaps a new room or area was added in the building, or perhaps a ceiling was added and the sprinklers that previously protected the combustible construction were turned pendent and extended through the ceiling. It is even possible that the system was installed this way and a variance was provided by the AHJ that allowed sprinklers to be omitted from this space. Under any of these circumstances, this finding warrants a hazard evaluation as required by 4.1.7.1 whenever ". . . changes in the occupancy, hazard, water supply, storage commodity, storage arrangement, building modification, or other condition that affects the installation criteria of the system are identified . . ."

The NFPA 25 inspector is not required to know the installation standard requirements. However, it is generally known that combustible concealed spaces in most occupancies require sprinkler protection. While NFPA 25 does not require the inspector to look or inspect above a ceiling when performing an annual visual inspection, this condition could have been discovered while investigating a visible leak or while checking for adequate heat as required by 4.1.2.1.

(Courtesy of Byron Blake and SimplexGrinnell)

4.1.7* Addressing Changes in Hazard.

A.4.1.7 See Annex E for an example of a hazard evaluation form. A hazard evaluation is not part of a system inspection.

In addition to Annex E, another tool for evaluating an existing system is NFPA 3, *Recommended Practice for Commissioning and Integrated Testing of Fire Protection and Life Safety Systems*. NFPA 3 provides guidance on the retro-commissioning of individual systems or integrated systems. The purpose of NFPA 3 is to provide a structure and reporting mechanism for the design and operation of a system. Properly documenting a system and its design at the same time as design decisions are being made and put on paper, which is what commissioning is intended to do, makes future assessments and evaluations easier and more cost effective to conduct.

4.1.7.1 Where changes in the occupancy, hazard, water supply, storage commodity, storage arrangement, building modification, or other condition that affects the installation criteria of the system are identified, the property owner or designated representative shall promptly take steps to evaluate the adequacy of the installed system in order to protect the building or hazard in question.

Management of change has become a critical concept for property owners, manufacturers, and just about any other business trying to keep its head above water. In these tough economic times, some companies are shying away from constructing new, customized storage facilities, instead choosing to "recycle" old warehouse spaces to fit their needs. Other companies are manufacturing, storing, or selling different products to remain viable, which can create a change in hazard within the building. If these changes in business structure, philosophy, and operation that result in a change in hazard are not properly evaluated, they can have a major impact on the effectiveness of fire protection systems.

4.1.7.2 Where the evaluation reveals that the installed system is inadequate to protect the building or hazard in question, the property owner or designated representative shall make the required corrections.

Case In Point

Storage facilities are a great example as to why management of change is so critical. If a company changes a product that they store from a Class IV commodity to a Group A plastic, the sprinkler system could be grossly inadequate for the fire that would result from the new commodity. This is why a review of the sprinkler system for effectiveness must be conducted wherever there is a change in one or more of the critical characteristics that drive sprinkler system design for storage areas. These changes include, but are not limited to, change in commodity classification; change in storage height; change in clearance to ceiling; change in packaging, such as encapsulated to exposed; the addition of solid shelving; change in storage type, such as from shelf storage to rack storage; and change in pallet type.

4.1.7.3 Corrections shall be approved.

The requirement in 4.1.7.3 clarifies the role of the AHJ in the evaluation of the adequacy of the system(s) when changes in the hazard are identified. The owner is not required to include the AHJ in evaluation of the buildings and its systems, but where a system is determined to be inadequate, the corrections must be acceptable to the AHJ. Annex E includes a sample evaluation form. It is critical that whoever conducts the evaluation is qualified to do so.

Critical Point

The owner is not required to include the AHJ in evaluation of the buildings and its systems, but where a system is determined to be inadequate, the corrections must be acceptable to the AHJ.

4.1.8 Valve Location. The location of shutoff valves shall be identified at the system riser or other approved locations.

4.1.9 Information Sign.

The requirements in 4.1.8 and 4.1.9 state the importance of clearly identifying and marking the location of shutoff valves. In many cases, time is wasted after a system begins to discharge water while the building occupants search for the location of the shutoff valve. This search results in additional water damage that could have been avoided.

4.1.9.1 A permanently marked metal or rigid plastic information sign shall be placed at the system control riser supplying an antifreeze loop, dry system, preaction system, or auxiliary system control valve.

4.1.9.2 Each sign shall be secured with a corrosion-resistant wire, chain, or other approved means and shall indicate at least the following information:

(1) Location of the area served by the system
(2) Location of auxiliary drains and low-point drains for dry pipe and preaction systems
(3) The presence and location of antifreeze or other auxiliary systems
(4) The presence and location(s) of heat tape

Critical Point

It is important to recognize that the presence of the system information sign is a retroactive requirement. These signs were not required by NFPA 13 prior to the 2007 edition, so many systems might not have them. This information is critical to the assessment of a building while conducting certain ITM tasks, as evidenced by NFPA 13 requiring it in the last three editions of the standard.

The sign required in 4.1.9 is needed to provide anyone maintaining a system with vital information that might not otherwise be readily known. The sign is particularly useful in addressing freeze threats by identifying any locations with heat tape, and acknowledging that auxiliary and low-point drains are installed on the system along with their number and location. If this information is not known, the system can be damaged due to freezing of water in the trapped sections of pipe. It is important to recognize that 4.1.9.2 is a retroactive requirement. For example, system information signs were not required by NFPA 13 prior to the 2007 edition, so many systems might not have them. This information is critical to the assessment of a building while conducting certain ITM tasks, as evidenced by NFPA 13 requiring it in the last three editions of the standard. If this sign is not present, contractors and inspection providers should alert the owner to the need for this sign and the importance of the information. The required information could come from the building records, which include the as-built sprinkler system drawing(s),

System Tagging

A coupling that has been corroded to the point where the head of the bolt has separated from its body would typically cause significant leaking and would render a system impaired. It is important to keep in mind that if the coupling were somehow not leaking and the corroded side of the coupling were up against a wall or in a location that could not be seen from the floor level, it might not be noticed during an inspection.

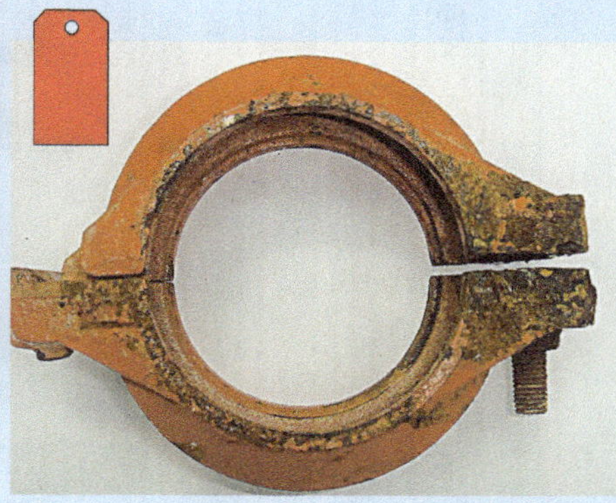

(Courtesy of Byron Blake and SimplexGrinnell)

Noncritical Deficiency Critical Deficiency Impairment

the Contractor's Material and Test Certificate, or other documents. If these are not available, the owner might need to ask the contractor or other qualified entity or individual to survey the system and provide the information needed for the sign.

4.1.10 Impairments.

4.1.10.1 Where an impairment to a water-based fire protection system occurs or is identified during inspection, testing, or maintenance activities, the procedures outlined in Chapter 15 shall be followed, including the attachment of a tag to the impaired system.

Exhibit 4.7 shows an impaired system that has been identified with an impairment tag by the AHJ. NFPA 25 does not mandate a certain style or color of impairment tag; however, many states have their own tagging system that should be consulted. Where a system impairment occurs in a location that is remote from the system control valve, the inspector will often provide a tag both at the system riser and at the impaired component. For instance, if a system has to be shut down due to a failed coupling in a section of feed main piping outside of the sprinkler control room, it makes sense to tag both the coupling and the closed control valve. The tag on the control valve should note where the impairment is, leading the AHJ, owner, or inspector to the other tag.

4.1.10.2 Where a water-based fire protection system is returned to service following an impairment, the system shall be verified to be working properly by means of an appropriate inspection or test as described in the table "Summary of Component Replacement [Action] Requirements" in the applicable chapters of this document.

Component replacement and testing tables are included in NFPA 25 for each system type to assist the user in determining the type and extent of actions that are needed following a repair

EXHIBIT 4.7 *Example of an Impaired System Tag. (Courtesy of Byron Blake and SimplexGrinnell)*

Critical Point

Component replacement and testing tables are included in NFPA 25 for each system type to assist the user in determining the type and extent of actions that are needed following a repair or adjustment.

or adjustment. See the appropriate systems chapter for the pertinent component replacement and testing table. There are times when these actions are in accordance with the requirements of the appropriate installation standard. For example, the tests needed when replacing an air maintenance device in a dry pipe sprinkler system would be performed in accordance with NFPA 13.

4.2 Manufacturer's Corrective Action

Manufacturers shall be permitted to make modifications to their own listed product in the field with listed devices that restore the original performance as intended by the listing, where acceptable to the authority having jurisdiction.

4.3 Records

4.3.1* Records shall be made for all inspections, tests, and maintenance of the system and its components and shall be made available to the authority having jurisdiction upon request.

A.4.3.1 Inspection reports used for system inspections should contain an "Owner's Section" as shown in Figure A.4.3.1 that the property owner or designated representative should complete. Typical records include, but are not limited to, valve inspections; flow, drain, and pump tests; and trip tests of dry pipe, deluge, and preaction valves.

Acceptance test records should be retained for the life of the system or its special components. Subsequent test records should be retained for a period of 1 year after the next test. The comparison determines deterioration of system performance or condition and the need for further testing or maintenance.

4.3.1.1* Records shall be permitted to be stored and accessed electronically.

A.4.3.1.1 Computer programs that file inspection and test results should provide a means of comparing current and past results and should indicate the need for corrective maintenance or further testing.

The inspection, testing, and maintenance records referred to in Section 4.3 provide written documentation of compliance with NFPA 25. These records also offer a service history of the installed systems by indicating their performance during testing and/or maintenance. The

Owner's Section

A. Is the building occupied?	❏ Yes	❏ No
B. Has the occupancy and hazard of contents remained the same since the last inspection?	❏ Yes	❏ No
C. Are all fire protection systems in service?	❏ Yes	❏ No
D. Has the system remained in service without modification since the last inspection?	❏ Yes	❏ No
E. Was the system free of actuation of devices or alarms since the last inspection?	❏ Yes	❏ No

Explain any "no" answers:

____

____

____

____

____

____________________	____________________
Owner or Designated Representative (print)	Signature and Date

 NFPA 25

FIGURE A.4.3.1 Owner's Section on Inspection Report.

advances in electronic database software allow for complete and highly detailed histories of system maintenance. Many AHJs allow owners to use their own central database or that of a service provider to electronically file and maintain records, rather than to store traditional paper records on site.

4.3.2 Records shall indicate the following:

(1) The procedure/activity performed (e.g., inspection, test, or maintenance)
(2) The organization that performed the activity
(3) The required frequency of the activity
(4) The results and date of the activity
(5) The name and contact information of the qualified contractor or owner, including lead person for activity

Any form, whether produced commercially or developed independently, can be used to document inspection, testing, and maintenance activities, provided it details the activity sufficiently to verify compliance with NFPA 25. Many people assume that NFPA 25 requires a specific series of forms that need to be filled out; however in reviewing the standard, it is completely devoid of forms. This absence of forms in the standard is not intended to lessen the importance of recording ITM data, but rather it allows the owners and inspectors the flexibility to use their own project management programs, software, and ITM forms to track their ITM work. This handbook contains several examples of sample forms that can be used for recording system information or documenting an ITM task that has been completed. Exhibit 4.8 is a form that can be used to record general system information for a given project. This form establishes a baseline for conducting ITM work and can be used as a point of reference by the inspector each time a subsequent ITM task is conducted.

AUTOMATIC SPRINKLER SYSTEMS: GENERAL INFORMATION

Date: ______________ **Inspector:** ______________ **System:** ______________

Location: ______________________________

General

System designation ______________________

Building ______________________

Location of sprinkler valve ______________________

Type of sprinkler system ❑ Wet ❑ Dry ❑ Deluge ❑ Preaction

Make and model of sprinkler valve ______________________

Is building fully sprinklered? ❑ Yes ❑ No

Is entire sprinkler system in service? ❑ Yes ❑ No

Has sprinkler system been modified since last inspection? ❑ Yes ❑ No

Valves

How are valves supervised? ❑ Seated ❑ Locked ❑ Tamper switch

Are valves identified with signs? ❑ Yes ❑ No

Water Supply (See Chapters 7 and 9 of NFPA 25)

When was last water supply made? ______________________

Are reservoirs, tanks, or pressure tanks in good condition? ______________________

Pumps (See Chapter 8 of NFPA 25)

Is fire pump ❑ Diesel ❑ Electric ❑ Gasoline ❑ None?

When was pump last inspected? ______________________

Is pump in good condition? ❑ Yes ❑ No

Fire Department Connections

Location ______________________

Are identification signs provided? ❑ Yes ❑ No

Wet Systems

Is building adequately heated? ❑ Yes ❑ No

Is system hydraulically calculated? ❑ Yes ❑ No

If yes, is hydraulic information sign provided at valve? ❑ Yes ❑ No

Dry Systems

Is dry pipe valve in heated room? ❑ Yes ❑ No

Does heated room have low-temperature alarm? ❑ Yes ❑ No

Notes

NFPA® NFPA Water-Based Fire Protection Systems Forms

EXHIBIT 4.8 *Sample Automatic Sprinkler Systems Form.*

4.3.3* Records shall be maintained by the property owner.

The requirement in 4.3.3 emphasizes that it is the owner's responsibility to maintain the required records; however, it is a good idea for the inspector to keep records as well. It allows the inspector to identify commonalities in deficiencies in a specific property or even a certain type of system. See the commentary following 4.3.1.1 on how the records are permitted to be stored.

A.4.3.3 See Section B.2 for information regarding sample forms.

4.3.4 As-built system installation drawings, hydraulic calculations, original acceptance test records, and device manufacturer's data sheets shall be retained for the life of the system.

4.3.5 Subsequent records shall be retained for a period of 1 year after the next inspection, test, or maintenance of that type required by the standard.

Compliance with 4.3.5 results in maintaining records for at least 1 year following each successive inspection, testing, or maintenance activity. This requirement for records retention is intended to provide evidence of trending. For example, a main drain test performed on an annual basis should result in a total of three test reports: the original test, results for the current year, and results from the prior year. The record from 2 years ago could be discarded at this time. With this information, fluctuations in the test results are shown and used for evaluating the operational condition of the system.

4.4 Water Supply Status

During inspection, testing, and maintenance, water supplies, including fire pumps, shall remain in service unless under constant attendance by qualified personnel or unless impairment procedures in Chapter 15 are followed.

The requirement in Section 4.4 is intended to avoid unnecessary system impairment procedures for any water supply that is turned off for a length of time. Provided that water supplies, including fire pumps, are constantly attended, impairment procedures are not needed for valves that are closed for short periods for testing. An example of such a situation would be closing the discharge isolation valve of a fire pump for the annual flow test.

4.5* Inspection

System components shall be inspected at intervals specified in the appropriate chapters.

A.4.5 Inspection and periodic testing determine what, if any, maintenance actions are required to maintain the operability of a water-based fire protection system. The standard establishes minimum inspection/testing frequencies, responsibilities, test routines, and reporting procedures but does not define precise limits of anomalies where maintenance actions are required.

Substandard conditions, such as a closed valve, subnormal water pressure, loss of building heat or power, or obstruction of sprinklers, nozzles, detectors, or hose stations, can delay or prevent system actuation and impede manual fire-fighting operations.

Section 4.5 refers to inspection intervals for specific system components that can be found in subsequent chapters of NFPA 25. Those chapters are as follows:

Chapter 5 — Sprinkler Systems
Chapter 6 — Standpipe and Hose Systems
Chapter 7 — Private Fire Service Mains
Chapter 8 — Fire Pumps
Chapter 9 — Water Storage Tanks

Chapter 10 — Water Spray Fixed Systems

Chapter 11 — Foam-Water Sprinkler Systems

Chapter 12 — Water Mist Systems

Chapter 13 — Valves, Valve Components, and Trim

The inspection, testing, and maintenance frequencies established in the standard vary from daily, weekly, monthly, quarterly, semiannually, annually, to every 5 years. Each chapter provides a summary table at the beginning that lays out all of the inspection tasks for each system and system component, along with the required frequency for each inspection. In the 2014 edition, Chapter 3 now provides ranges for the amount of time that can be permitted for a task to be completed.

4.6 Testing

Critical Point

Section 4.6 provides a detailed procedure for determining an alternative inspection/testing frequency for fire protection systems and equipment.

Section 4.6 provides a detailed procedure for determining an alternative inspection/testing frequency for fire protection systems and equipment. Since its inception, NFPA 25 has included a provision allowing an alternative method of performing inspection, testing, and maintenance (see Section 1.3). But this provision does not detail exactly how such an alternative method should be implemented.

Section 4.6 explains in detail how to implement such a program. It is important to note that the performance-based option requires due diligence in conducting inspections and testing and demands a consistent approach to reporting and record keeping. The challenge in developing a performance-based program is determining an acceptable failure rate.

There is a wide difference of opinions on establishing an acceptable failure rate, according to a recent study on fire pump performance by the NFPA Research Foundation, "Fire Pump Field Data Collection and Analysis." The challenge entails pinning down the break point of a cost benefit analysis. Is it worth conducting a weekly test if it makes a system or system component 1 percent or 2 percent more reliable than a system or component that is tested monthly? The cost of running 52 tests instead of 12 is significant to the owner, and it is difficult to quantify the amount of additional reliability that is necessary to justify spending that money. It ends up being determined on a case-by-case basis, based on the liability that the owner carries and the specifics of the building and fire protection system.

Performance-based options cannot be implemented without approval by the AHJ. The documentation submitted to the AHJ for review must be complete and sufficiently detailed to clearly justify the deviation in inspection and test frequencies from those specified in NFPA 25.

4.6.1 All components and systems shall be tested to verify that they function as intended.

4.6.2 The frequency of tests shall be in accordance with this standard.

FAQ Must a monthly test be conducted on the exact date each month, or can frequency be varied as long as the test is on or about the same day each month?

It is impractical to require an inspection or test to be performed on the exact date for each subsequent interval. See the commentary following Section 4.5 and the definitions in Chapter 3 for more information on the acceptable ranges to conduct an ITM task. Allowances might be needed for unexpected weather conditions or other uncontrolled factors that impact the scheduled activity.

4.6.3 Fire protection system components shall be restored to full operational condition following testing, including reinstallation of plugs and caps for auxiliary drains and test valves.

•

4.6.4* Test results shall be compared with those of the original acceptance test (if available) and with the most recent test results.

This requirement highlights the reason why inspectors should keep their own copies of test and inspection reports. While owners are required to maintain these records, if inspectors keep their own records, they can make better comparisons to past testing and essentially create a building or system history.

A.4.6.4 The types of tests required for each protection system and its components, and the specialized equipment required for testing, are detailed in the appropriate chapter.

As referred to in 4.3.4, original records should include, at a minimum, the contractor's material and test certificate, "as-built" drawings and calculations, and any other required or pertinent test reports. These documents establish the conditions under which the systems were first installed and offer some insight to the design intent, installation standards used, and water supply present at the time of installation. Original records are instrumental in determining any subsequent changes or modifications to the buildings or system.

4.6.5* When a component or subsystem is adjusted, repaired, reconditioned, or replaced, it shall be tested in accordance with the original acceptance test required for that subsystem or the requirements where specified by the standard.

The individual system chapters of NFPA 25 contain corrective action tables at the end of the chapter that provide references to many of these required tests. These tests are not intended to create an additional burden; they are simply to assure that the functionality of the repaired or replaced components can be confirmed as it would be for a component in a new system.

A.4.6.5 Examples of subsystems or components include fire pumps, drivers or controllers, pressure-regulating devices, detection systems and controls, alarm check, and dry pipe, deluge, and preaction valves. The required tests for components are contained in the corresponding chapter in tables titled *Summary of [Component] Inspection, Testing, and Maintenance.*

4.6.6* Automated Testing. (Reserved)

A.4.6.6 Some devices, such as waterflow alarms, can be tested automatically. Some things to consider include the following:

(1) Automatic tests should be arranged to test the same functions as the required tests.
(2) Components and arrangements of automated tests should be listed for the purpose and designed so that failure of the testing equipment should not impair the operation of the system unless indicated by a trouble or supervisory signal in accordance with *NFPA 72.*
(3) Failure of a component or system to pass an automatic test should result in an audible trouble signal in accordance with *NFPA 72.*
(4) Not all tests required by NFPA 25 can be suitable for automatic testing.
(5) Periodic visual inspection including the use of video should be performed.

4.7* Performance-Based Programs

As an alternative means of compliance and where approved by the authority having jurisdiction, components and systems shall be permitted to be inspected, tested, and maintained under a performance-based program.

A.4.7 Section 4.7 provides the option to adopt a performance-based test and inspection method as an alternative means of compliance with 4.6.2. The prescriptive test and requirements contained in this standard are essentially qualitative. Equivalent or superior levels of performance can be demonstrated through quantitative performance-based analyses. This section provides a basis for implementing and monitoring a performance-based program acceptable under this option (providing approval is obtained by the authority having jurisdiction).

The concept of a performance-based testing and inspection program is to establish the requirements and frequencies at which inspection must be performed to demonstrate an acceptable level of operational reliability. The goal is to balance the inspection/test frequency with proven reliability of the system or component. The goal of a performance-based inspection program is also to adjust test/inspection frequencies commensurate with historical documented equipment performance and desired reliability. Frequencies of test/inspection under a performance-based program can be extended or reduced from the prescriptive test requirements contained in this standard when continued testing has been documented indicating a higher or lower degree of reliability compared to the authority having jurisdiction's expectations of performance. Additional program attributes that should be considered when adjusting test/inspection frequencies include the following:

(1) System/component preventive maintenance programs
(2) Consequences of system maloperation
(3) System/component repair history
(4) Building/service conditions

Fundamental to implementing a performance-based program is that adjusted test and inspection frequencies must be technically defensible to the authority having jurisdiction and supported by evidence of higher or lower reliability. Data collection and retention must be established so that the data utilized to alter frequencies are representative, statistically valid, and evaluated against firm criteria. Frequencies should not arbitrarily be extended or reduced without a suitable basis and rationale. It must be noted that transitioning to a performance-based program might require additional expenditures of resources in order to collect and analyze failure data, coordinate review efforts, replace program documents, and seek approval from the authority having jurisdiction. The following factors should be considered in determining whether a transition to a performance-based test program as permitted in Section 4.7 is appropriate:

(1) Past system/component reliability — have problems routinely been identified during the performance of the prescriptive test requirements of 4.6.2, or have systems consistently performed with minimal discrepancies noted?
(2) Do the recurring resource expenditures necessary to implement the prescriptive test requirements in 4.6.2 justify conducting the detailed analysis needed to support a performance-based testing program?
(3) Is the increased administrative burden of implementing, documenting, and monitoring a performance-based program worthwhile?

Failure Rate Calculation

A performance-based program requires that a maximum allowable failure rate be established and approved by the authority having jurisdiction in advance of implementation. The use of historical system/component fire system inspection records can be utilized to determine failure rates. One method of calculating the failure rate of a fire system is based on the following equation:

$$FSFR(t) = \frac{NF}{(NC)(t)}$$

where:

$FSFR(t)$ = fire system failure rate (failures per year)
t = time interval of review in years
NF = number of failures
NC = total number of fire systems inspected or tested

Example

Data are collected for 50 fire pump weekly tests over a 5-year period. The testing is conducted, as described in 8.3.1. A review of the data has identified five failures:

Total components: 280

Data collection period: 5 years

Total failures: 5

$$FSFR = \frac{5}{280 \times 5} = 0.003/\text{year}$$

A fundamental requirement of a performance-based program is the continual monitoring of fire system/component failure rates and determining whether they exceed the maximum allowable failure rates as agreed upon with the authority having jurisdiction. The process used to complete this review should be documented and repeatable.

Coupled with this ongoing review is a requirement for a formalized method of increasing or decreasing the frequency of testing/inspection when systems exhibit either a higher than expected failure rate or an increase in reliability as a result of a decrease in failures, or both.

A formal process for reviewing the failure rates and increasing or decreasing the frequency of testing must be well documented. Concurrence of the authority having jurisdiction on the process used to determine test frequencies should be obtained in advance of any alterations to the test program. The frequency required for future tests might be reduced to the next inspection frequency and maintained there for a period equaling the initial data review or until the ongoing review indicates that the failure rate is no longer being exceeded — for example, going from annual to semiannual testing when the failure rate exceeds the authority having jurisdiction's expectations or from annual to every 18 months when the failure trend indicates an increase in reliability.

References

Edward K. Budnick, P.E., "Automatic Sprinkler System Reliability," *Fire Protection Engineering*, Society of Fire Protection Engineers, Winter 2001.

Fire Protection Equipment Surveillance Optimization and Maintenance Guide, Electric Power Research Institute, July 2003.

William E. Koffel, P.E., *Reliability of Automatic Sprinkler Systems, Alliance for Fire Safety*.

NFPA's Future in Performance Based Codes and Standards, July 1995.

NFPA Performance Based Codes and Standards Primer, December 1999.

4.8* Maintenance

Maintenance shall be performed to keep the system equipment operable or to make repairs.

A.4.8 Preventive maintenance includes, but is not limited to, lubricating control valve stems; adjusting packing glands on valves and pumps; bleeding moisture and condensation from air compressors, air lines, and dry pipe system auxiliary drains; and cleaning strainers. Frequency of maintenance is indicated in the appropriate chapter.

Corrective maintenance includes, but is not limited to, replacing loaded, corroded, or painted sprinklers; replacing missing or loose pipe hangers; cleaning clogged fire pump impellers; replacing valve seats and gaskets; restoring heat in areas subject to freezing temperatures where water-filled piping is installed; and replacing worn or missing fire hose or nozzles.

Emergency maintenance includes, but is not limited to, repairs due to piping failures caused by freezing or impact damage; repairs to broken underground fire mains; and replacement of frozen or fused sprinklers, defective electric power, or alarm and detection system wiring.

In addition to the maintenance requirements of NFPA 25, it is important to review the preventative maintenance information provided by the manufacturer. The requirements in the manufacturer's literature are often tied to the warranty, and the argument that the NFPA 25 maintenance schedule was followed will not hold water if the frequencies do not match up.

4.9 Safety

4.9.1 General. Inspection, testing, and maintenance activities shall be conducted in accordance with applicable safety regulations.

4.9.2 Confined Spaces. Legally required precautions shall be taken prior to entering confined spaces such as tanks, valve pits, or trenches.

Paragraph 4.9.2 addresses precautions to be taken when working in confined spaces. A confined space is any space that is large enough for a person to enter and perform work and that has limited or restricted access, such as a water storage tank. At times, an inspector will find it necessary to enter a confined space, such as a valve pit, water storage tank, or trench.

Confined space entry is regulated by the U.S. Department of Labor, Occupational Safety and Health Administration (OSHA), in 29 CFR 1910.146, "Permit-Required Confined Spaces." That standard establishes the requirement for a permit-required confined space entry program to protect employees from confined space hazards and regulate entry into these spaces. In areas not subject to OSHA regulations, such as those outside the United States, other government regulations might apply. Exhibit 4.9 illustrates a warning sign identifying a confined space as part of a permit-required confined space entry program.

Before a confined space can be entered, the internal atmosphere must be tested for oxygen content, flammable gases and vapors, and potential toxic air contaminants. The space must also be ventilated to eliminate or control any hazardous atmosphere.

Access to any confined space must be coordinated with the property owner, contractor, or host employer. (A host employer is the customer of the contractor or inspector.) Controlled access to a confined space might be managed through the use of a permit. The permit should indicate that atmospheric testing has been completed and that proper safety equipment (such as ventilation, communications, personal protective equipment, lighting, ladders, and rescue and emergency equipment) is in place. The permit should also address any other requirements as established in the host employer's confined space entry program. Exhibit 4.10 illustrates a sample confined space entry permit.

EXHIBIT 4.9 Sample Confined Space Entry Sign.

Confined Space Entry Permit

Date and Time Issued: ____________ Date and Time Expires: ____________
Job Site/Space I.D.: ____________ Job Supervisor: ____________
Equipment to be worked on: ____________ Work to be performed: ____________

Stand-by personnel: ____________

1. Atmospheric checks: Time ________
   Oxygen ________ %
   Explosive ________ % L • F • L •
   Toxic ________ PPM

2. Tester's signature: ____________

3. Source isolation (No Entry):

	N/A	Yes	No
Pumps or lines blinded,	( )	( )	( )
disconnected, or blocked	( )	( )	( )

4. Ventilation modification:

	N/A	Yes	No
Mechanical	( )	( )	( )
Natural ventilation only	( )	( )	( )

5. Atmospheric check after isolation and ventilation:

Oxygen ________ %		>	19.5	%
Explosive ________ % L • F • L •		<	10	%
Toxic ________ PPM		<	10	PPM H(2)S

   Time ________
   Testers signature: ____________

6. Communication procedures: ____________

7. Rescue procedures: ____________

8. Entry, standby, and backup persons:

	Yes	No
Successfully completed required training? Is it current?	( )	( )

9. Equipment:

	N/A	Yes	No
Direct reading gas monitor - tested	( )	( )	( )
Safety harnesses and lifelines for entry and standby persons	( )	( )	( )
Hoisting equipment	( )	( )	( )
Powered communications	( )	( )	( )
SCBA's for entry and standby persons	( )	( )	( )
Protective Clothing	( )	( )	( )
All electric equipment listed Class I, Division I, Group D and Non-sparking tools	( )	( )	( )

10. Periodic atmospheric tests:

Oxygen	______%	Time ______	Oxygen	______%	Time ______
Oxygen	______%	Time ______	Oxygen	______%	Time ______
Explosive	______%	Time ______	Explosive	______%	Time ______
Explosive	______%	Time ______	Explosive	______%	Time ______
Toxic	______%	Time ______	Toxic	______%	Time ______
Toxic	______%	Time ______	Toxic	______%	Time ______

We have reviewed the work authorized by this permit and the information contained here-in. Written instructions and safety procedures have been received and are understood. Entry cannot be approved if any squares are marked in the "No" column. This permit is not valid unless all appropriate items are completed.

Permit Prepared By: (Supervisor) ____________
Approved By: (Unit Supervisor) ____________
Reviewed By (Cs Operations Personnel):

____________ ____________
(printed name) (signature)

This permit to be kept at job site. Return job site copy to Safety Office following job completion.

EXHIBIT 4.10 *Sample Confined Space Entry Permit. (Source: OSHA, 29 CFR 1910.146, Appendix D)*

Ongoing testing or monitoring of the confined space atmosphere is necessary to ensure that safe conditions are maintained for the duration of entry operations.

During confined space entry, at least one attendant trained in summoning rescue and emergency services, rescuing entrants from the space, and providing emergency services to rescued employees must remain outside of the permit space.

It is important for anyone involved in the inspection, testing, and maintenance of systems to recognize a confined space and be prepared to follow the host employer's confined space entry program. Paragraph 4.9.2 is not intended to outline a comprehensive confined space entry program.

4.9.3 Fall Protection. Legally required equipment shall be worn or used to prevent injury from falls to personnel.

Fall protection, as referred to in 4.9.3, can include wearing safety harnesses, as well as the proper use of ladders, staging, and powered lifts. The inspector should be familiar with any regulations in effect governing this type of equipment.

4.9.4 Hazards. Precautions shall be taken to address any hazards, such as protection against drowning where working on the top of a filled embankment or a supported, rubberized fabric tank, or over open water or other liquids.

4.9.5* Hazardous Materials.

The hazardous materials referenced in 4.9.5 can be any material capable of causing harm or bodily injury. These materials can include cleaning agents, solvents, paint, adhesives, or other chemicals used for a variety of purposes. It is important that inspectors and service personnel be able to both recognize the presence of a hazardous material and identify and classify the material to prevent accident, injury, or fire.

Critical Point

It is important that inspectors and service personnel be able to both recognize the presence of a hazardous material and identify and classify the material to prevent accident, injury, or fire.

In 1983, OSHA established the Hazard Communication Standard, 29 CFR 1910, Subpart Z, "Toxic and Hazardous Substances." That standard requires manufacturers to provide information about hazardous materials to their customers in the form of Material Safety Data Sheets (MSDS), such as the one shown in Exhibit 4.11.

A completed MSDS provides chemical and toxicological data regarding hazardous materials and the need for any personal protective equipment. Prior to entering an area containing hazardous materials, an inspector or service personnel should request a copy of the MSDS for all substances in the area. The owner is required per 4.9.5.2 to notify the inspector of the hazardous materials stored on the property. The owner should have a catalog of MSDSs on site that can be used by the inspector as a quick reference guide while the inspection is conducted.

A.4.9.5 Most places using or storing hazardous materials have stations set up for employees where material safety data sheets (MSDSs) are stored. The inspector should be familiar with the types of materials present and the appropriate actions to take in an emergency.

4.9.5.1 Legally required equipment shall be used where working in an environment with hazardous materials present.

4.9.5.2 The property owner or designated representative shall advise anyone performing inspection, testing, and maintenance on any system under the scope of this document, with regard to hazardous materials stored on the premises.

4.9.6* Electrical Safety. Legally required precautions shall be taken when testing or maintaining electric controllers for motor-driven fire pumps.

A.4.9.6 WARNING: NFPA 20, *Standard for the Installation of Stationary Pumps for Fire Protection*, includes electrical requirements that discourage the installation of a disconnect means in the power supply to electric motor-driven fire pumps. This is intended to ensure the availability of power to the fire pumps. Where equipment connected to those circuits is serviced or maintained, the service person could be subject to unusual exposure to electrical and other hazards. It could be necessary to establish special safe work practices and to use safeguards or personal protective clothing, or both. See also *NFPA 70E* for additional safety guidance.

Safety-related work practices and procedures for persons who work on energized electrical equipment should be in place prior to engaging in such activity. As a minimum, procedures should be in accordance with 29 CFR 1926, "Safety and Health Regulations for Construction," and *NFPA 70E®, Standard for Electrical Safety in the Workplace®*. A fire pump controller is classified by *NFPA 70E* as a 600 V Class Motor Control Center. Commentary Table 4.1, excerpted from

Material Safety Data Sheet
May be used to comply with
OSHA's Hazard Communication Standard,
29 CFR 1910. 1200. Standard must be
consulted for specific requirements.

U.S. Department of Labor
Occupational Safety and Health Administration
(Non-Mandatory Form)
Form Approved
OMB No. 1218-0072

IDENTITY (*As Used on Label and List*)	Note: *Blank spaces are not permitted. If any item is not applicable, or no information is available, the space must be marked to indicate that.*

Section I

Manufacturer's Name	Emergency Telephone Number
Address (*Number, Street, City, State, and ZIP Code*)	Telephone Number for Information
	Date Prepared
	Signature of Preparer (*optional*)

Section II—Hazardous Ingredients/Identity Information

Hazardous Components (*Specific Chemical Identity; Common Name(s)*)	OSHA PEL	ACGIH TLV	Other Limits Recommended	% (*optional*)

Section III—Physical/Chemical Characteristics

Boiling Point		Specific Gravity (H_2O = 1)	
Vapor Pressure (mm Hg)		Melting Point	
Vapor Density (Air = 1)		Evaporation Rate (Butyl Acetate = 1)	
Solubility in Water			
Appearance and Odor			

Section IV—Fire and Explosion Hazard Data

Flash Point (*Method Used*)	Flammable Limits	LEL	UEL
Extinguishing Media			
Special Fire-Fighting Procedures			
Unusual Fire and Explosion Hazards			

EXHIBIT 4.11 *OSHA Form 174 — Material Safety Data Sheet. (Source: OSHA, 29 CFR 1910.1200, Subpart Z)*

(continues)

Section V—Reactivity Data

Stability	Unstable		Conditions to Avoid
	Stable		
Incompatibility (*Materials to Avoid*)			
Hazardous Decomposition or By-products			
Hazardous Polymerization	May Occur		Conditions to Avoid
	Will Not Occur		

Section VI—Health Hazard Data

Route(s) of Entry:	Inhalation?	Skin?	Ingestion?
Health Hazards (*Acute and Chronic*)			
Carcinogenicity:	NTP?	IARC Monographs?	OSHA Regulated?
Signs and Symptoms of Exposure			
Medical Conditions Generally Aggravated by Exposure			
Emergency and First Aid Procedures			

Section VII—Precautions for Safe Handling and Use

Steps to Be Taken in Case Material Is Released or Spilled
Waste Disposal Method
Precautions to Be Taken in Handling and Storing
Other Precautions

Section VIII—Control Measures

Respiratory Protection (*Specify Type*)		
Ventilation	Local Exhaust	Special
	Mechanical (*General*)	Other
Protective Gloves		Eye Protection
Other Protective Clothing or Equipment		
Work/Hygienic Practices		

*U.S.G.P.O.: 1986-491-529/45775

EXHIBIT 4.11 *Continued*

Critical Point

A fire pump controller is classified by *NFPA 70E* as a 600 V Class Motor Control Center. Commentary Table 4.1, excerpted from *NFPA 70E*, defines the hazard/risk category and voltage-rated gloves and tools necessary for work specifically involving fire pump controllers.

NFPA 70E, defines the hazard/risk category and voltage-rated gloves and tools necessary for work specifically involving fire pump controllers.

Once the hazard/risk categories for testing and maintenance of fire pump controllers have been determined from Commentary Table 4.1, the need for and type of protective clothing and personal protective equipment can be determined from Commentary Table 4.2, which is also excerpted from *NFPA 70E*.

Exhibit 4.12 illustrates the necessary personal protective equipment needed for service and maintenance of fire pump controllers.

COMMENTARY TABLE 4.1 *Hazard/Risk Category Classifications and Use of Rubber Insulating Gloves and Insulated and Insulating Hand Tools Alternating Current Equipment*

Tasks Performed on Energized Equipment	*Hazard/Risk Category*	*Rubber Insulating Gloves*	*Insulated and Insulating Hand Tools*
Panelboards or Other Equipment Rated 240 V and Below			
Parameters:			
Maximum of 25 kA short circuit current available; maximum of 0.03 sec (2 cycle) fault clearing time; minimum 18 in. working distance			
Potential arc flash boundary with exposed energized conductors or circuit parts using above parameters: 19 in.			
Perform infrared thermography and other non-contact inspections outside the restricted approach boundary	0	N	N
Circuit breaker (CB) or fused switch operation with covers on	0	N	N
CB or fused switch operation with covers off	0	N	N
Work on energized electrical conductors and circuit parts, including voltage testing	1	Y	Y
Remove/install CBs or fused switches	1	Y	Y
Removal of bolted covers (to expose bare, energized electrical conductors and circuit parts)	1	N	N
Opening hinged covers (to expose bare, energized electrical conductors and circuit parts)	0	N	N
Work on energized electrical conductors and circuit parts of utilization equipment fed directly by a branch circuit of the panelboard	1	Y	Y
Panelboards or Other Equipment Rated > 240 V and up to 600 V			
Parameters:			
Maximum of 25 kA short circuit current available; maximum of 0.03 sec (2 cycle) fault clearing time; minimum 18 in. working distance			
Potential arc flash boundary with exposed energized conductors or circuit parts using above parameters: 30 in.			
Perform infrared thermography and other non-contact inspections outside the restricted approach boundary	1	N	N
Circuit breaker (CB) or fused switch operation with covers on	0	N	N
CB or fused switch operation with covers off	1	Y	N
Work on energized electrical conductors and circuit parts, including voltage testing	2	Y	Y
Remove/install CBs or fused switches	2	Y	Y
Removal of bolted covers (to expose bare, energized electrical conductors and circuit parts)	1	N	N
Opening hinged covers (to expose bare, energized electrical conductors and circuit parts)	0	N	N
Work on energized electrical conductors and circuit parts of utilization equipment fed directly by a branch circuit of the panelboard	2	Y	Y
600 V Class Motor Control Centers (MCCs)			
Parameters:			
Maximum of 65 kA short circuit current available; maximum of 0.03 sec (2 cycle) fault clearing time; minimum 18 in. working distance			
Potential arc flash boundary with exposed energized conductors or circuit parts using above parameters: 53 in.			
Perform infrared thermography and other non-contact inspections outside the restricted approach boundary	1	N	N
CB or fused switch or starter operation with enclosure doors closed	0	N	N
Reading a panel meter while operating a meter switch	0	N	N
CB or fused switch or starter operation with enclosure doors open	1	N	N
Work on energized electrical conductors and circuit parts, including voltage testing	2	Y	Y

(continues)

COMMENTARY TABLE 4.1 *Continued*

Tasks Performed on Energized Equipment	*Hazard/Risk Category*	*Rubber Insulating Gloves*	*Insulated and Insulating Hand Tools*
Work on control circuits with energized electrical conductors and circuit parts 120 V or below, exposed	0	Y	Y
Work on control circuits with energized electrical conductors and circuit parts >120 V, exposed	2	Y	Y
Application of temporary protective grounding equipment, after voltage test	2	Y	N
Work on energized electrical conductors and circuit parts of utilization equipment fed directly by a branch circuit of the motor control center	2	Y	Y
600 V Class Motor Control Centers (MCCs)			
Parameters: Maximum of 42 kA short circuit current available; maximum of 0.33 sec (20 cycle) fault clearing time; minimum 18 in. working distance			
Potential arc flash boundary with exposed energized conductors or circuit parts using above parameters: 165 in.			
Insertion or removal of individual starter "buckets" from MCC	4	Y	N
Removal of bolted covers (to expose bare, energized electrical conductors and circuit parts)	4	N	N
Opening hinged covers (to expose bare, energized electrical conductors and circuit parts)	1	N	N
600 V Class Switchgear (with power circuit breakers or fused switches) and 600 V Class Switchboards			
Parameters: Maximum of 35 kA short circuit current available; maximum of up to 0.5 sec (30 cycle) fault clearing time; minimum 18 in. working distance			
Potential arc flash boundary with exposed energized conductors or circuit parts using above parameters: 233 in.			
Perform infrared thermography and other non-contact inspections outside the restricted approach boundary	2	N	N
CB or fused switch operation with enclosure doors closed	0	N	N
Reading a panel meter while operating a meter switch	0	N	N
CB or fused switch operation with enclosure doors open	1	N	N
Work on energized electrical conductors and circuit parts, including voltage testing	2	Y	Y
Work on control circuits with energized electrical conductors and circuit parts 120 V or below, exposed	0	Y	Y
Work on control circuits with energized electrical conductors and circuit parts >120 V, exposed	2	Y	Y
Insertion or removal (racking) of CBs from cubicles, doors open or closed	4	N	N
Application of temporary protective grounding equipment after voltage test	2	Y	N
Removal of bolted covers (to expose bare, energized electrical conductors and circuit parts)	4	N	N
Opening hinged covers (to expose bare, energized electrical conductors and circuit parts)	2	N	N
Other 600 V Class (277 V through 600 V, nominal) Equipment			
Parameters: Maximum of 65 kA short circuit current available; maximum of 0.03 sec (2 cycle) fault clearing time; minimum 18 in. working distance (except as indicated)			
Potential arc flash boundary with exposed energized conductors or circuit parts using above parameters: 53 in.			

COMMENTARY TABLE 4.1 *Continued*

Tasks Performed on Energized Equipment	*Hazard/Risk Category*	*Rubber Insulating Gloves*	*Insulated and Insulating Hand Tools*
Lighting or small power transformers (600 V, maximum)			
Removal of bolted covers (to expose bare, energized electrical conductors and circuit parts)	2	N	N
Opening hinged covers (to expose bare, energized electrical conductors and circuit parts)	1	N	N
Work on energized electrical conductors and circuit parts, including voltage testing	2	Y	Y
Application of temporary protective grounding equipment, after voltage test	2	Y	N
Revenue meters (kW-hour, at primary voltage and current)—insertion or removal	2	Y	N
Cable trough or tray cover removal or installation	1	N	N
Miscellaneous equipment cover removal or installation	1	N	N
Work on energized electrical conductors and circuit parts, including voltage testing	2	Y	Y
Application of temporary protective grounding equipment, after voltage test	2	Y	N
Insertion or removal of plug-in devices into or from busways	2	Y	N
NEMA E2 (fused contactor) Motor Starters, 2.3 kV Through 7.2 kV			
Parameters:			
Maximum of 35 kA short circuit current available; maximum of up to 0.2 sec (12 cycle) fault clearing time; minimum 36 in. working distance			
Potential arc flash boundary with exposed energized conductors or circuit parts using above parameters: 422 in.			
Perform infrared thermography and other non-contact inspections outside the restricted approach boundary	3	N	N
Contactor operation with enclosure doors closed	0	N	N
Reading a panel meter while operating a meter switch	0	N	N
Contactor operation with enclosure doors open	2	N	N
Work on energized electrical conductors and circuit parts, including voltage testing	4	Y	Y
Work on control circuits with energized electrical conductors and circuit parts 120 V or below, exposed	0	Y	Y
Work on control circuits with energized electrical conductors and circuit parts >120 V, exposed	3	Y	Y
Insertion or removal (racking) of starters from cubicles, doors open or closed	4	N	N
Application of temporary protective grounding equipment, after voltage test	3	Y	N
Removal of bolted covers (to expose bare, energized electrical conductors and circuit parts)	4	N	N
Opening hinged covers (to expose bare, energized electrical conductors and circuit parts)	3	N	N
Insertion or removal (racking) of starters from cubicles of arc-resistant construction, tested in accordance with IEEE C37.20.7, doors closed only	0	N	N
Metal Clad Switchgear, 1 kV Through 38 kV			
Parameters:			
Maximum of 35 kA short circuit current available; maximum of up to 0.2 sec (12 cycle) fault clearing time; minimum 36 in. working distance			
Potential arc flash boundary with exposed energized conductors or circuit parts using above parameters: 422 in.			
Perform infrared thermography and other non-contact inspections outside the restricted approach boundary	3	N	N

(continues)

COMMENTARY TABLE 4.1 *Continued*

Tasks Performed on Energized Equipment	*Hazard/Risk Category*	*Rubber Insulating Gloves*	*Insulated and Insulating Hand Tools*
CB operation with enclosure doors closed	2	N	N
Reading a panel meter while operating a meter switch	0	N	N
CB operation with enclosure doors open	4	N	N
Work on energized electrical conductors and circuit parts, including voltage testing	4	Y	Y
Work on control circuits with energized electrical conductors and circuit parts 120 V or below, exposed	2	Y	Y
Work on control circuits with energized electrical conductors and circuit parts > 120 V, exposed	4	Y	Y
Insertion or removal (racking) of CBs from cubicles, doors open or closed	4	N	N
Application of temporary protective grounding equipment, after voltage test	4	Y	N
Removal of bolted covers (to expose bare, energized electrical conductors and circuit parts)	4	N	N
Opening hinged covers (to expose bare, energized electrical conductors and circuit parts)	3	N	N
Opening voltage transformer or control power transformer compartments	4	N	N
Arc-Resistant Switchgear Type 1 or 2 (for clearing times of < 0.5 sec with a perspective fault current not to exceed the arc-resistant rating of the equipment)			
Parameters: Maximum of 35 kA short circuit current available; maximum of up to 0.2 sec (12 cycle) fault clearing time; minimum 36 in. working distance			
Potential arc flash boundary with exposed energized conductors or circuit parts using above parameters: 422 in.			
CB operation with enclosure door closed	0	N	N
Insertion or removal (racking) of CBs from cubicles, doors closed	0	N	N
Insertion or removal of CBs from cubicles with door open	4	N	N
Work on control circuits with energized electrical conductors and circuit parts 120 V or below, exposed	2	Y	Y
Insertion or removal (racking) of ground and test device with door closed	0	N	N
Insertion or removal (racking) of voltage transformers on or off the bus door closed	0	N	N
Other Equipment 1 kV Through 38 kV			
Parameters: Maximum of 35 kA short circuit current available; maximum of up to 0.2 sec (12 cycle) fault clearing time; minimum 36 in. working distance			
Potential arc flash boundary with exposed energized conductors or circuit parts using above parameters: 422 in.			
Metal-enclosed interrupter switchgear, fused or unfused			
Switch operation of arc-resistant-type construction, tested in accordance with IEEE C37.20.7, doors closed only	0	N	N
Switch operation, doors closed	2	N	N
Work on energized electrical conductors and circuit parts, including voltage testing	4	Y	Y
Removal of bolted covers (to expose bare, energized electrical conductors and circuit parts)	4	N	N
Opening hinged covers (to expose bare, energized electrical conductors and circuit parts)	3	N	N
Outdoor disconnect switch operation (hookstick operated)	3	Y	Y
Outdoor disconnect switch operation (gang-operated, from grade)	2	Y	N

COMMENTARY TABLE 4.1 *Continued*

Tasks Performed on Energized Equipment	Hazard/Risk Category	Rubber Insulating Gloves	Insulated and Insulating Hand Tools
Insulated cable examination, in manhole or other confined space	4	Y	N
Insulated cable examination, in open area	2	Y	N

Y = Yes (required). N: No (not required).

Notes:

(1) Rubber insulating gloves are gloves rated for the maximum line-to-line voltage upon which work will be done.

(2) Insulated and insulating hand tools are tools rated and tested for the maximum line-to-line voltage upon which work will be done, and are manufactured and tested in accordance with ASTM F 1505, *Standard Specification for Insulated and Insulating Hand Tools*.

(3) The use of "N" does not indicate that rubber insulating gloves and insulated and insulating hand tools are not required in all cases. Rubber insulating gloves and insulated and insulating hand tools may be required by 130.4, 130.7 (C) (7)(a), and 130.7(D)(1).

(4) For equipment protected by upstream current limiting fuses with arcing fault current in their current limiting range (½ cycle fault clearing time or less), the hazard/risk category required may be reduced by one number.

(5) For power systems up to 600 V the arc flash boundary was determined by using the following information: When 0.03 second trip time was used, that indicated MCC or panelboard equipment protected by a molded-case circuit breaker. Working distance used was 18 in. (455 mm). Arc gap used was 32 mm for switchgear and 25 mm for MCC and protective device type 0 for all. When 0.33 or 0.5 second trip time was used, that indicated a LVPCB (drawout circuit breaker) in switchgear. Working distance was 18 in. (455 mm). Arc gap used was 32 mm and protective device type 0 for all. All numbers were rounded up or down depending on closest multiple of 5.

(6) For power systems from 1 kV to 38 kV the arc flash boundary was determined by using the following information: No maximum values were given in the 2009 edition of NFPA 70E for short-circuit current or operating time. Two sets of equations were performed: 35 kA AIC and 0.2 second operating time and 26 kA AIC and 0.2 second operating time. 0.2 seconds was used by adding the typical maximum total clearing time of the circuit breaker to an estimated value for relay operation. This coincides with the IEEE 1584 values of 0.18 second operating time and 0.08 tripping time rounded off. A short-circuit current of 35 kA was used as a maximum (HRC-4 @ ~ 40 cal/cm²) and 26 kA was used to compare the effects of lowering the short circuit current (HRC-4 @ ~ 30 cal/cm²). Working distance used was 36 in. (909 mm), arc gap was 6 in. (455 mm), and protective device type 0 for all.

Source: NFPA 70E, 2012, Table 130.7(C)(15)(a).

EXHIBIT 4.12 *Personal Protective Equipment.*

COMMENTARY TABLE 4.2 Protective Clothing and Personal Protective Equipment (PPE)

Hazard/Risk Category	Protective Clothing and PPE
0	**Protective Clothing, Nonmelting or Untreated Natural Fiber (i.e., untreated cotton, wool, rayon, or silk, or blends of these materials) with a Fabric Weight of at Least 4.5 oz/yd²** Shirt (long sleeve) Pants (long) **Protective Equipment** Safety glasses or safety goggles (SR) Hearing protection (ear canal inserts) Heavy duty leather gloves (AN) *(See Note 1.)*
1	**Arc-Rated Clothing, Minimum Arc Rating of 4 cal/cm²** *(See Note 3.)* Arc-rated long-sleeve shirt and pants or arc-rated coverall Arc-rated face shield *(see Note 2)* or arc flash suit hood Arc-rated jacket, parka, rainwear, or hard hat liner (AN) **Protective Equipment** Hard hat Safety glasses or safety goggles (SR) Hearing protection (ear canal inserts) Heavy duty leather gloves *(See Note 1.)* Leather work shoes (AN)
2	**Arc-Rated Clothing, Minimum Arc Rating of 8 cal/cm²** *(See Note 3.)* Arc-rated long-sleeve shirt and pants or arc-rated coverall Arc-rated flash suit hood or arc-rated face shield *(See Note 2)* and arc-rated balaclava Arc-rated jacket, parka, rainwear, or hard hat liner (AN) **Protective Equipment** Hard hat Safety glasses or safety goggles (SR) Hearing protection (ear canal inserts) Heavy duty leather gloves *(See Note 1.)* Leather work shoes
3	**Arc-Rated Clothing Selected so That the System Arc Rating Meets the Required Minimum Arc Rating of 25 cal/cm²** *(See Note 3.)* Arc-rated long-sleeve shirt (AR) Arc-rated pants (AR) Arc-rated coverall (AR) Arc-rated arc flash suit jacket (AR) Arc-rated arc flash suit pants (AR) Arc-rated arc flash suit hood Arc-rated gloves *(See Note 1.)* Arc-rated jacket, parka, rainwear, or hard hat liner (AN)

COMMENTARY TABLE 4.2 *Continued*

Hazard/Risk Category	Protective Clothing and PPE
	Protective Equipment
	Hard hat
	Safety glasses or safety goggles (SR)
	Hearing protection (ear canal inserts)
	Leather work shoes
4	**Arc-Rated Clothing Selected so That the System Arc Rating Meets the Required Minimum Arc Rating of 40 cal/cm²** *(See Note 3.)*
	Arc-rated long-sleeve shirt (AR)
	Arc-rated pants (AR)
	Arc-rated coverall (AR)
	Arc-rated arc flash suit jacket (AR)
	Arc-rated arc flash suit pants (AR)
	Arc-rated arc flash suit hood
	Arc-rated gloves *(See Note 1.)*
	Arc-rated jacket, parka, rainwear, or hard hat liner (AN)
	Protective Equipment
	Hard hat
	Safety glasses or safety goggles (SR)
	Hearing protection (ear canal inserts)
	Leather work shoes

AN: as needed (optional). AR: as required. SR: selection required.

Notes:

(1) If rubber insulating gloves with leather protectors are required by Table 130.7(C)(15)(a) and Table 130.7(C)(15)(b), additional leather or arc-rated gloves are not required. The combination of rubber insulating gloves with leather protectors satisfies the arc flash protection requirement.

(2) Face shields are to have wrap-around guarding to protect not only the face but also the forehead, ears, and neck, or, alternatively, an arc-rated arc flash suit hood is required to be worn.

(3) *Arc rating* is defined in Article 100 and can be either the arc thermal performance value (ATPV) or energy of break open threshold (EBT). ATPV is defined in ASTM F 1959, *Standard Test Method for Determining the Arc Thermal Performance Value of Materials for Clothing*, as the incident energy on a material, or a multilayer system of materials, that results in a 50 percent probability that sufficient heat transfer through the tested specimen is predicted to cause the onset of a second-degree skin burn injury based on the Stoll curve, in cal/cm². EBT is defined in ASTM F 1959 as the incident energy on a material or material system that results in a 50 percent probability of breakopen. Arc rating is reported as either ATPV or EBT, whichever is the lower value.

Source: NFPA 70E, 2012, Table 130.7(C)(16).

In addition to electrical safety requirements and procedures, other precautions such as lockout/tagout procedures should be exercised to prevent injury from the accidental starting of mechanical equipment, such as motors and pumping equipment. For more information on system impairment and lockout/tagout procedures, see Chapter 15 of NFPA 25.

References Cited in Commentary

ASTM International, 100 Barr Harbor Drive, P.O. Box C700, West Conshohocken, PA 19428-2959.

ASTM F 1505, *Standard Specification for Insulated and Insulating Hand Tools,* 2007.

ASTM F 1959, *Standard Test Method for Determining the Arc Thermal Performance Value of Materials for Clothing,* 2006.

Fire Protection Research Foundation, 1 Batterymarch Park, Quincy, MA 02169-7471.

"Fire Pump Field Data Collection and Analysis," Gayle Pennel, Aon Risk Solutions, January 2012.

National Fire Protection Association, 1 Batterymarch Park, Quincy, MA 02169-7471.

NFPA 3, *Recommended Practice for Commissioning and Integrated Testing of Fire Protection and Life Safety Systems,* 2012 edition.

NFPA 13, *Standard for the Installation of Sprinkler Systems,* 2013 edition.

NFPA 70E®, Standard for Electrical Safety in the Workplace®, 2012 edition.

U.S. Government Printing Office, Washington, DC 20402.

Title 29, Code of Federal Regulations, Part 1910.146, "Permit-Required Confined Spaces."

Title 29, Code of Federal Regulations, Part 1910, Subpart Z, "Toxic and Hazardous Substances."

Title 29, Code of Federal Regulations, Part 1926, "Safety and Health Regulations for Construction."

Title 29, Code of Federal Regulations, Part 1910.1200.

ITM NFPA 25 — INSPECTION TESTING MAINTENANCE

5

SPRINKLER SYSTEMS

Chapter 5 provides requirements for the inspection, testing, and maintenance of fire sprinkler systems, the most common type of fire protection systems installed in the built environment. Sprinkler systems are exceptionally reliable, but like any mechanical system, they require attention to remain operational. Unlike other types of building systems, such as plumbing or HVAC, sprinkler systems are relatively static in that they are normally not flowing water. Periodic inspection, testing, and maintenance are necessary to ensure that the system and its components will function as intended in the event of a fire emergency. Since many sprinkler system failures can be attributed to closed valves, it is imperative that inspection frequencies be observed and that prescribed items be inspected to ensure reliability of the systems.

5.1 General

5.1.1 Minimum Requirements.

5.1.1.1 This chapter shall provide the minimum requirements for the routine inspection, testing, and maintenance of sprinkler systems.

5.1.1.2 Table 5.1.1.2 shall be used to determine the minimum required frequencies for inspection, testing, and maintenance.

TABLE 5.1.1.2 *Summary of Sprinkler System Inspection, Testing, and Maintenance*

Item	*Frequency*	*Reference*
Inspection		
Gauges (dry, preaction, and deluge systems)	Weekly/quarterly	5.2.4.2, 5.2.4.3, 5.2.4.4
Control valves		Table 13.1.1.2
Waterflow alarm devices	Quarterly	5.2.5
Valve supervisory signal devices	Quarterly	5.2.5
Supervisory signal devices (except valve supervisory switches)	Quarterly	5.2.5
Gauges (wet pipe systems)	Quarterly	5.2.4.1
Hydraulic nameplate	Quarterly	5.2.6
Buildings	Annually (prior to freezing weather)	4.1.1.1
Hanger/seismic bracing	Annually	5.2.3
Pipe and fittings	Annually	5.2.2

(continues)

TABLE 5.1.1.2 Continued

Item	Frequency	Reference
Sprinklers	Annually	5.2.1
Spare sprinklers	Annually	5.2.1.4
Information sign	Annually	5.2.8
Fire department connections		Table 13.1.1.2
Valves (all types)		Table 13.1.1.2
Obstruction, internal inspection of piping	5 years	14.2
Heat trace	Per manufacturer's requirements	5.2.7
Test		
Waterflow alarm devices		
Mechanical devices	Quarterly	5.3.3.1
Vane and pressure switch–type devices	Semiannually	5.3.3.2
Valve supervisory signal devices		Table 13.1.1.2
Supervisory signal devices (except valve supervisory switches)		Table 13.1.1.2
Main drain		Table 13.1.1.2
Antifreeze solution	Annually	5.3.4
Gauges	5 years	5.3.2
Sprinklers (extra-high or greater temperature solder type)	5 years	5.3.1.1.1.4
Sprinklers (fast-response)	At 20 years and every 10 years thereafter	5.3.1.1.1.3
Sprinklers	At 50 years and every 10 years thereafter	5.3.1.1.1
Sprinklers	At 75 years and every 5 years thereafter	5.3.1.1.1.5
Sprinklers (dry)	At 10 years and every 10 years thereafter	5.3.1.1.1.6
Sprinklers (in harsh environments)	5 years	5.3.1.1.2
Valves (all types)		Table 13.1.1.2
Valve status test		13.3.1.2.1
Maintenance		
Valves (all types)		Table 13.1.1.2
Low-point drains (dry pipe system)		13.4.4.3.2
Sprinklers and automatic spray nozzles protecting commercial cooking equipment and ventilation systems	Annually	5.4.1.9
Investigation		
Obstruction		14.3

Critical Point

Some of the inspection activities listed in Table 5.1.1.2 can be performed by property owners, their designated representatives, or property maintenance personnel, with a minimal amount of training.

In addition to the requirements of this chapter, many of the requirements of Chapter 13, Valves, Valve Components, and Trim; Chapter 14, Obstruction Investigation; and Chapter 15, Impairments, apply to a sprinkler system. Table 5.1.1.2 summarizes the specific minimum requirements of sprinkler system inspection, testing, and maintenance. The "Reference" column includes NFPA 25 paragraph numbers so the user can refer to details on the inspection, testing, or maintenance to be performed.

5.1.2 Valves and Connections. Valves and fire department connections shall be inspected, tested, and maintained in accordance with Chapter 13.

5.1.3 Obstruction Investigations. The procedures outlined in Chapter 14 shall be followed where there is a need to conduct an obstruction investigation.

Case In Point

Some of the inspection activities listed in Table 5.1.1.2 can be performed by property owners, their designated representatives, or property maintenance personnel, with a minimal amount of training. For example, the weekly/monthly inspection of gauges and control valves can be performed with some basic training, and the quarterly inspection for the hydraulic nameplate is intended simply to verify that it is attached to the system riser and is legible. The more technical inspection tasks and the testing and maintenance activities required by Table 5.1.1.2 should be performed by a qualified contractor, service provider, or facilities personnel, since these activities require specialized training. (See 3.3.34 for the definition of the term *qualified* and related commentary.)

In some cases, actions can be performed by the owner or the owner's representative(s) while the contractor or service provider is present and supervising the activities. Some authorities having jurisdiction (AHJs) or governmental agencies require special licensing of the service provider or individual inspector, so it is recommended to always check with the local fire authority before performing any of the required inspection, testing, and maintenance (ITM) functions prescribed.

Critical Point

Some AHJs or governmental agencies require special licensing of the service provider or individual inspector, so it is recommended to always check with the local fire authority before performing any of the required ITM functions prescribed.

5.1.4 Impairments. The procedures outlined in Chapter 15 shall be followed where an impairment to protection occurs.

The inspection and testing functions generally produce a pass/fail result. However, some failed items do not necessarily trigger the impairment procedures detailed in Chapter 15. The definitions of the terms *deficiency* (3.3.7), *critical deficiency* (3.3.7.1), and *noncritical deficiency* (3.3.7.2) provide guidance to differentiate between items requiring immediate attention and those that can be repaired within a reasonable timeframe.

One example of a noncritical deficiency would be a missing hydraulic calculations placard on a hydraulically calculated sprinkler system, which is required by 5.2.6 to be verified quarterly. Despite missing this required component, the mechanical function of the sprinkler system would not be impacted, and therefore, this would be considered a noncritical deficiency. Where there is a concern over whether a deficient item should be considered critical or noncritical, it would be prudent to discuss such items with the local AHJ, owner, and insurance provider.

•

5.1.5 Hose Connections. Hose connections shall be inspected, tested, and maintained in accordance with Chapters 6 and 13.

5.2* Inspection

A.5.2 The provisions of the standard are intended to apply to routine inspections. In the event of a fire, a post-fire inspection should be made of all sprinklers within the fire area. In situations where the fire was quickly controlled or extinguished by one or two sprinklers, it might be necessary only to replace the activated sprinklers. Care should be taken that the replacement sprinklers are of the same make and model or that they have compatible performance characteristics (*see 5.4.1.2*). Soot-covered sprinklers should be replaced because deposits can result in corrosion of operating parts. In the event of a substantial fire, special consideration should be given to replacing the first ring of sprinklers surrounding the operated sprinklers because of the potential for excessive thermal exposure, which could weaken the response mechanisms.

Chapter 5 includes inspection requirements of varying frequencies, such as weekly, quarterly, or annually, for various system components such as sprinklers, piping, gauges, and hangers. Keeping track of which of these inspections have been conducted and what deficiencies were noted is critical to maintaining a system in good working order. While NFPA 25 does not contain

Critical Point

The definitions of the terms *deficiency* (3.3.7), *critical deficiency* (3.3.7.1), and *noncritical deficiency* (3.3.7.2) provide guidance to differentiate between items requiring immediate attention and those that can be repaired within a reasonable timeframe.

System Tagging

These photos show a cracked adapter for a nonmetallic piping system. The crack is only noticeable with the use of a magnifying glass. It would not be expected that an inspector would see this damaged adapter from the floor level, so it must be assumed that this condition was identified due to a leak or dripping caused by the damage. This would be considered a critical deficiency.

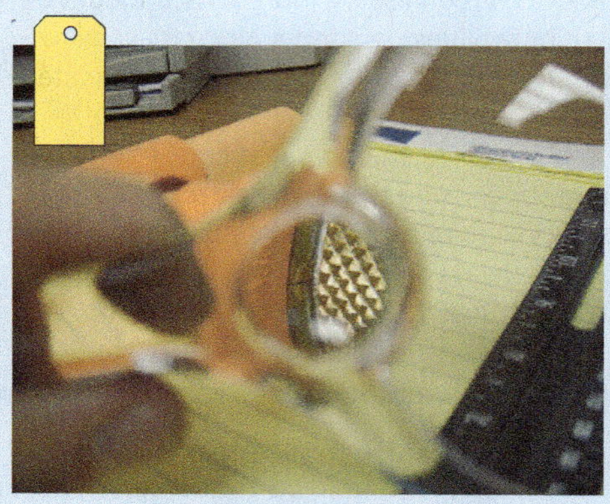

(Photos courtesy of Byron Blake and SimplexGrinnell)

Noncritical Deficiency | Critical Deficiency | Impairment

a specific form that must be filled out, having a form or series of forms that can be used to record the information obtained while conducting an inspection is important. Exhibit 5.1 is an example of a generic form that can be used to record a sprinkler inspection task required by Chapter 5. The form can be used for any inspection task identified in Table 5.1.1.2 to identify the frequency and the information associated with the inspection that is being conducted.

5.2.1 Sprinklers.

5.2.1.1* Sprinklers shall be inspected from the floor level annually.

A.5.2.1.1 The conditions described in this section can have a detrimental effect on the performance of sprinklers by affecting water distribution patterns, insulating thermal elements, delaying operation, or otherwise rendering the sprinkler inoperable or ineffectual.

Severely loaded or corroded sprinklers should be rejected as part of the visual inspection. Such sprinklers could be affected in their distribution or other performance characteristics not addressed by routine sample testing. Lightly loaded or corroded sprinklers could be permitted for continued use if samples are selected for testing based on worst-case conditions and the samples successfully pass the tests.

FAQ How do I inspect sprinklers that are located in a warehouse with high ceilings where it is difficult to determine the condition of the sprinklers?

Paragraph 5.2.1.1 requires a visual inspection of sprinklers to reveal obvious signs of damage, leakage, loading, or corrosion. Inspections are conducted from floor level, because it is usually impractical to get closer to the sprinklers for a more in-depth inspection, and the use of ladders is of limited benefit when compared to the cost. In buildings with higher ceilings, a flashlight or binoculars can assist in the inspection of sprinklers, piping, hangers, and other components. When other work is being done at the ceiling level using ladders or lifts, personnel could take advantage of the opportunity of being closer to the sprinklers to inspect the system.

AUTOMATIC SPRINKLER SYSTEMS: INSPECTION OF WET PIPE SPRINKLER SYSTEMS

Property Name: ______________________ **Inspector:** ______________
Property Address: ______________________ **Contract No.:** ______________
Phone Number: ______________________ **Date:** ______________

This Report Covers: ❑ **Monthly** ❑ **Quarterly** ❑ **Annual** ❑ **Five-Year**

Inspections
Monthly

❑ Yes ❑ No ❑ N/A Gauges—normal water pressure maintained

Control Valves

❑ Yes ❑ No ❑ N/A In the correct (open or closed) position
❑ Yes ❑ No ❑ N/A Sealed, locked, or supervised
❑ Yes ❑ No ❑ N/A Accessible
❑ Yes ❑ No ❑ N/A Free from damage or leaks
❑ Yes ❑ No ❑ N/A Proper signage

Alarm Valves

❑ Yes ❑ No ❑ N/A Free of damage
❑ Yes ❑ No ❑ N/A Accessible
❑ Yes ❑ No ❑ N/A Retard chamber/alarm drains not leaking

Quarterly

❑ Yes ❑ No ❑ N/A Alarm devices—free of damage
❑ Yes ❑ No ❑ N/A Hydraulic data nameplate—securely attached to riser/legible

Fire Department Connections

❑ Yes ❑ No ❑ N/A Visible and accessible
❑ Yes ❑ No ❑ N/A Coupling/swivels operate correctly
❑ Yes ❑ No ❑ N/A Plugs/caps are in place
❑ Yes ❑ No ❑ N/A Gaskets are not damaged
❑ Yes ❑ No ❑ N/A Identification signs are in place
❑ Yes ❑ No ❑ N/A Ball drip valve is functional

Pressure Reducing Valve

❑ Yes ❑ No ❑ N/A In the open position/not leaking
❑ Yes ❑ No ❑ N/A Maintaining downstream pressure
❑ Yes ❑ No ❑ N/A In good condition

NFPA Water-Based Fire Protection Systems Forms 1 of 2

EXHIBIT 5.1 *Sample Form for Automatic Sprinkler Systems: Inspection of Wet Pipe Sprinkler Systems.*

Annual

Sprinklers (visible)

❑ Yes ❑ No ❑ N/A No damage or leaks
❑ Yes ❑ No ❑ N/A Free of corrosion, foreign material, or paint
❑ Yes ❑ No ❑ N/A Installed in proper orientation
❑ Yes ❑ No ❑ N/A Fluid in glass bulbs
❑ Yes ❑ No ❑ N/A Spare sprinklers—proper number and type. Complete with wrench?
❑ Yes ❑ No ❑ N/A Hangers and seismic bracing—not damaged or loose

Pipes and Fittings (visible)

❑ Yes ❑ No ❑ N/A In good condition/no external corrosion
❑ Yes ❑ No ❑ N/A No leaks or mechanical damage
❑ Yes ❑ No ❑ N/A Correct allignment—no external loads
❑ Yes ❑ No ❑ N/A Hose racks—inspected per NFPA 1962
❑ Yes ❑ No ❑ N/A Building—wet piping not exposed to freezing temperatures

Five-Year

❑ Yes ❑ No ❑ N/A Alarm valve interior including strainers, filters, and restriction orifice
❑ Yes ❑ No ❑ N/A Check valve— internal moves freely, in good condition
❑ Yes ❑ No ❑ N/A Obstruction inspection—no foreign or obstructing material found

Comments

Signature: ______________________ Date: ______________

License/Certification No.: ______________________

NFPA Water-Based Fire Protection Systems Forms 2 of 2

EXHIBIT 5.1 Continued.

Inspectors who use a flashlight or binoculars must be careful about "false positive" results, which can occur with glass bulb sprinklers when the light reflects off the white printing on the bulb. This is also a point of contention with some building owners who mistakenly see the use of additional resources not mandated by the code, such as a flashlight or binoculars, as an effort to generate service work, as opposed to an effort to ensure a properly working sprinkler system. As with most issues, a little communication goes a long way in helping the owner understand why it is so important to look closely at the sprinklers.

FAQ How much corrosion or loading is acceptable for sprinklers?

As noted above, the inspection points required by Chapter 5 are pass/fail criteria and not subjective. It is not possible to determine how much loading or corrosion will impact a sprinkler's activation time with the naked eye. When there is a concern over whether sprinklers will perform as intended due to loading or corrosion, representative samples can be tested following the requirements of 5.3.1 to determine if replacement is warranted. It should also be noted that brass and bronze will discolor over time and such discoloration is not usually considered corrosion or loading. As discussed previously, an owner might not understand the need for testing to ensure proper sprinkler performance, but a bit of communication and education will go a long way.

5.2.1.1.1* Sprinklers shall not show signs of leakage; shall be free of corrosion, foreign materials, paint, and physical damage; and shall be installed in the correct orientation (e.g., upright, pendent, or sidewall).

ITM Deficiency, Impairment, or Hazard Evaluation?

ANSWER: Hazard Evaluation

While performing the annual visual inspection of sprinklers required by 5.2.1.1, an inspector observes several overhead doors without sprinkler protection below them. The inspector has specific knowledge that NFPA 13, *Standard for the Installation of Sprinkler Systems,* requires sprinklers under overhead doors when they create an obstruction more than 4 ft (1.2 m) wide. When open, these doors will extend more than 4 ft (1.2 m) from the walls, and therefore, they should have sprinklers under them.

(Courtesy of Byron Blake and SimplexGrinnell)

While the NFPA 25 inspector is required to inspect the system and identify any issues that would impair the system from functioning properly, the inspector is not required to verify the adequacy of the design of the system, as stated in 1.1.3.1. It is unusual for an NFPA document to state that the inspector is required to test the system but does not need to fully understand all aspects of the system, such as the system's design. However, some users of this document believe inspectors should know the design and installation requirements for every system covered by NFPA 25 as well. The NFPA 25 technical committee recognizes that no one person can possibly know all of the design and installation requirements in all of the different editions of all of the water-based fire protection system installation documents. However, if the inspector has specific knowledge of design and installation requirements and identifies what could be an incorrect installation or an area lacking sprinkler protection, a recommendation for a hazard evaluation should be made to the property owner or the owner's authorized representative.

The presence of paint or corrosion on a sprinkler can impact not only the activation characteristics of the devices but also the spray pattern development. Exhibit 5.2 shows a severely corroded sprinkler, while Exhibit 5.3 shows a sprinkler with a painted bulb. Exhibit 5.4 shows a sprinkler that has been covered with a spray-applied foam insulation product. All three of these sprinklers should be replaced immediately, without any attempt to remove the corrosion, paint, or foam.

__EXHIBIT 5.2__ Severely Corroded Sprinkler. (Courtesy of Wayne Automatic Fire Sprinkler, Inc.)

__EXHIBIT 5.3__ Painted Sprinkler. (Courtesy of Josh Elvove)

__EXHIBIT 5.4__ Sprinkler Encased in Foam Insulation. (Courtesy of Reliable Automatic Sprinkler Company, Inc.)

EXHIBIT 5.5 *Glued Concealer Plate. (Courtesy of Josh Elvove)*

A common deficiency involving concealed sprinklers is the application of caulking or glue to affix the concealer plate to the underside of the ceiling. NFPA 13 was modified in the 2013 edition to explicitly state that no glues, caulking compounds, or epoxies be used on concealer plates. It is not always easy to identify this condition from the floor level; however, telltale signs include discoloration around the concealer plate or any foreign material that appears to be adjoining the concealer plate and the ceiling, as shown in Exhibit 5.5.

A.5.2.1.1.1 Sprinkler orientation includes the position of the deflector in relation to the ceiling slope. The deflector is generally required to be parallel to the slope of the ceiling. The inspection should identify any corrections made where deficiencies are noted, for example, pipe with welded outlets and flexible grooved couplings that have "rolled" out of position.

System Tagging

A sprinkler that is leaking to this degree must be considered an impairment. In Table A.3.3.7, a distinction is made between "spraying or running" water and "dripping" water. Spraying water, which is what is shown in this photo, would be considered an impairment. A simple drip or a bead of water seeping out of the sprinkler might be considered a critical deficiency. The sprinkler could be damaged and might not activate as it is intended to. This condition would require immediate action because the functionality of the device is in question.

(Courtesy of Byron Blake and SimplexGrinnell)

Noncritical Deficiency

Impairment

Critical Deficiency

System Tagging

NFPA 25 requires, as part of the sprinkler inspection, a review for the proper orientation of sprinklers. In this case, the deflector of a horizontal sidewall sprinkler is not parallel to the ceiling. The inspector must use some judgment as to whether or not the degree of misalignment would result in ineffective system operation. A sprinkler that is misaligned will typically leave an excessive shadow area, which is why this is often considered a critical deficiency. However, where the misalignment is negligible, an inspector might consider this a noncritical deficiency.

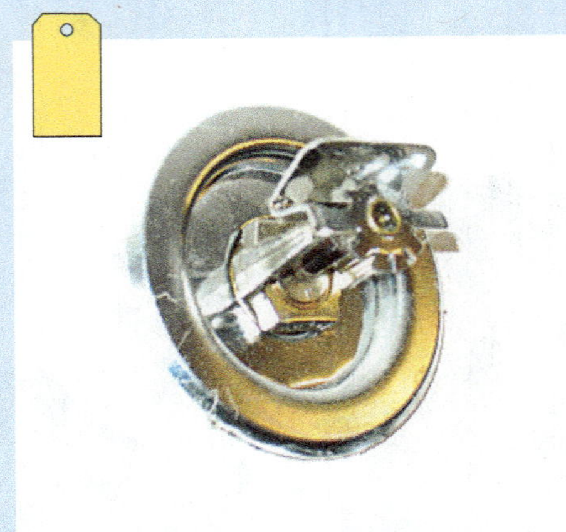

(Courtesy of Byron Blake and SimplexGrinnell)

Noncritical Deficiency

Impairment

Critical Deficiency

FAQ Is paint permitted on a sprinkler?

NFPA 25 does not permit paint on a sprinkler. This requirement is further clarified in 6.2.6.2 of NFPA 13, *Standard for the Installation of Sprinkler Systems,* which stipulates that sprinklers can be painted only by the manufacturer and that any other painted sprinklers must be replaced. In addition, NFPA 25, in 5.2.1.1.1, stipulates that a sprinkler must be free of paint other than that applied by the manufacturer. The only painting that is permitted is on the cover plate of concealed sprinklers and the frame of fusible link–type sprinklers as indicated in the listing for the sprinkler. It is important to inspect sprinklers from opposing vantage points. Often sprinklers are painted only on the side that faced the painter as the ceiling or structure was sprayed.

5.2.1.1.2 Any sprinkler that shows signs of any of the following shall be replaced:

(1) Leakage
(2)* Corrosion
(3) Physical damage
(4) Loss of fluid in the glass bulb heat-responsive element
(5)* Loading
(6) Painting unless painted by the sprinkler manufacturer

A.5.2.1.1.2(2) Corrosion found on the seat, or built up on the deflector that could affect the spray pattern, or a buildup on the operating elements that could affect the operation can have a detrimental effect on the performance of the sprinkler. Light surface corrosion on the boss, frame arms, and/or the deflector, and/or surface discoloration, not impacting the operation of the sprinkler should not warrant replacement. A degree of judgment should be exercised in the determination of the extent, location, and character of corrosion that would necessitate replacement.

A.5.2.1.1.2(5) In lieu of replacing sprinklers that are loaded with a coating of dust, it is permitted to clean sprinklers with compressed air or by a vacuum provided that the equipment does not touch the sprinkler.

FAQ Are there situations in which sprinklers can be cleaned of foreign material instead of being replaced?

Sprinklers that have a light coating of dust or dirt can be cleaned with a vacuum or a blast of compressed air, and they can remain in place, provided the sprinklers are not touched or agitated in the process. Exhibit 5.6 shows a sprinkler covered with some dust and cobwebs that could be cleaned with a vacuum or compressed air. However, if the vacuum or compressed air does not clear the debris, the sprinkler would need to be replaced. Under no circumstances should sprinklers be cleaned with bleach, ammonia, or any other household cleaner. Furthermore, attempting to remove paint with a razor blade or utility knife can further compromise the device and is not permitted.

Sprinklers that are heavily loaded with any contaminant such as dirt, dust, grease, or paint must be replaced. Painted sprinklers are never permitted to be cleaned and/or reinstalled, because the potential of damaging the assembly is too great. A "light" overspray or loading can be tolerated when a representative sample is tested to verify that the sprinklers will operate as intended (see A.5.2.1.1).

Sprinklers that are leaking or that have been damaged must be replaced without testing. Dissolved minerals and other residues in the water can solidify as the sprinkler leaks, hampering the operation of the sprinkler by changing internal clearances or acting like an adhesive, preventing parts from moving as intended. If corrosion is a significant problem, special corrosion-resistant sprinklers can be used.

Critical Point

Painted sprinklers are never permitted to be cleaned and/or reinstalled, because the potential of damaging the assembly is too great.

System Tagging

Loaded and painted heads can be tagged several ways in accordance with Table A.3.3.7. A single standard spray sprinkler in a nonresidential area will commonly be classified as a critical deficiency. However, this rule is not hard and fast, as a further consideration of the occupancy might be necessary.

A single painted sprinkler in a school or daycare facility, while not classified as residential, might necessitate an impairment tag based on the nature of the occupants. A condition where multiple sprinklers in the same compartment are painted or heavily loaded, such as the ones found in the bucket in this photo, would be considered an impairment.

(Photos courtesy of Byron Blake and SimplexGrinnell)

Noncritical Deficiency — Critical Deficiency — Impairment

EXHIBIT 5.6 Light Sprinkler Loading that Can Be Cleaned with Compressed Air. (Courtesy of Byron Blake and SimplexGrinnell)

Exhibit 5.7 through Exhibit 5.13 provide examples of sprinklers that require replacement. In Exhibit 5.7, the sprinkler is clearly corroded, which is an indicator that the other sprinklers in this space should also be carefully examined even if they do not appear initially to have any damage.

Exhibit 5.8 shows a sprinkler with a damaged deflector. Damaged deflectors should not be overlooked, although they can be difficult to notice from floor level. What might appear to be only slight damage to the deflector can have a drastic impact on the distribution of the spray pattern and the density of the water that is delivered.

EXHIBIT 5.7 Corroded Sprinkler. (Courtesy of Wiginton Fire Systems)

EXHIBIT 5.8 Damaged Sprinkler Deflector. (Courtesy of Wiginton Fire Systems)

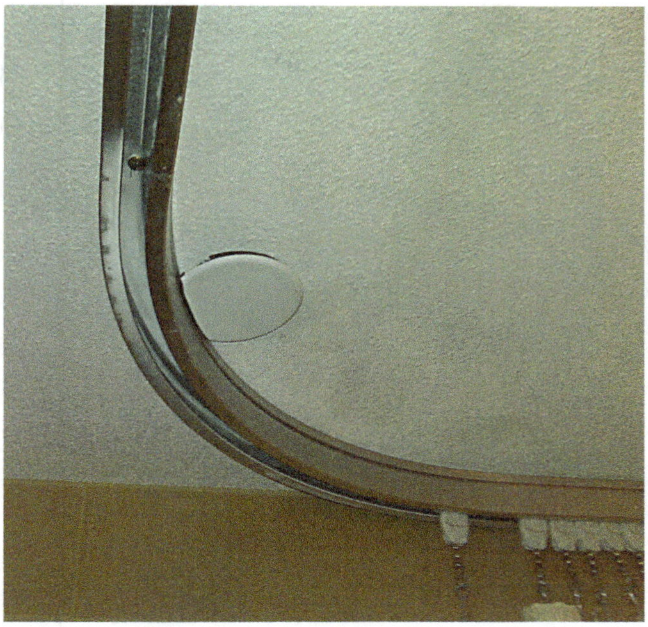

EXHIBIT 5.9 *Damaged Concealed Sprinkler.*

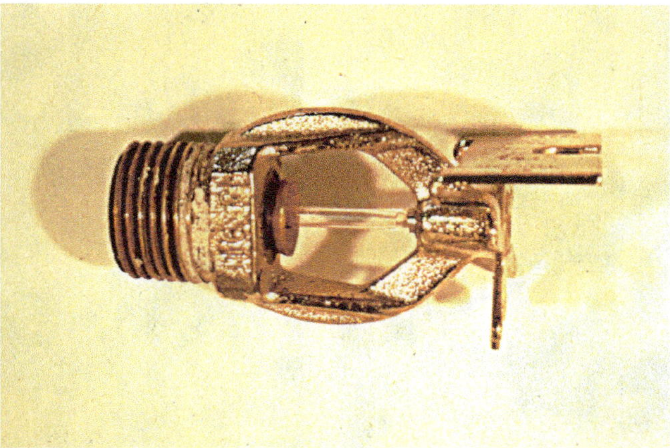

EXHIBIT 5.10 *Glass Bulb Sprinkler Without Fluid. (Courtesy of National Fire Sprinkler Association)*

Exhibit 5.9 shows a cover plate that has been cut, which is a clear violation of the listing of a concealed sprinkler. In this case, a privacy curtain track was installed retroactively around the sprinkler. Apparently, the track could not be moved and the sprinkler was in the way.

Glass bulb sprinklers operate when the liquid inside the bulb expands until the internal pressure is sufficient to break the glass. Missing liquid or a crack in the bulb will prevent the buildup of pressure inside the bulb and prevent the sprinkler from operating.

In some cases, rough handling of glass bulb sprinklers during shipment or installation can cause minute fractures in the bulb, allowing the fluid to leak out. Exhibit 5.10 shows a glass bulb sprinkler that contains no liquid inside the glass bulb. This sprinkler will not operate until the glass melts, which is well beyond the anticipated temperature rating and response time for this type of sprinkler. If a crack develops in the bulb and the liquid leaks out, the operating temperature of the sprinkler is no longer related to the liquid but rather to the melting temperature of the glass bulb. The melting point of glass is somewhere between 2600°F to 2900°F (1425°C to 1600°C), which is considerably higher than the temperature at which a sprinkler should operate.

Paint overspray is a common sprinkler problem. Exhibit 5.11 shows a sprinkler that has been hit by overspray from a careless or uninformed painter. Paint spray on the fusible link will inhibit the reaction of the link to heat and slow the operation of the sprinkler. Paint globs in the deflector could impact the spray pattern.

Exhibit 5.12 and Exhibit 5.13 show a loaded sprinkler. In this case, the sprinkler was not covered during a spray-on application process. As a result, the sprinkler deflector has been fouled, which will affect water distribution, and the glass bulb has been coated, which will delay operation.

Exhibit 5.14 shows a leaky sprinkler that was "repaired" and reinstalled in a sprinkler system. The machine bolt is not a fusible element, and this sprinkler will not operate in a fire emergency. Sprinklers should never be repaired; they must be replaced instead.

Exhibit 5.15 shows a sprinkler that is leaking in a freezer environment. The leak has caused ice to build up and form all around the sprinkler. The sprinkler either has a defect or has been damaged in some way to cause the leak, and it must be replaced.

EXHIBIT 5.11 Paint Overspray on a Sprinkler.

EXHIBIT 5.12 Loaded Sprinkler. (Courtesy of Wiginton Fire Systems)

EXHIBIT 5.13 Close-Up of Loaded Sprinkler. (Courtesy of Wiginton Fire Systems)

EXHIBIT 5.14 Sprinkler Leak Stopped and Sprinkler Returned to Service.

EXHIBIT 5.15 Frozen Sprinkler. (Courtesy of Shaun K. Wrightson, Koffel Associates, Inc.)

System Tagging

A single residential sprinkler that is painted or heavily loaded would be considered an impairment. Table A.3.3.7 makes a distinction between classifications in residential and nonresidential occupancies, highlighting the importance of not just looking at a single component but getting the complete picture of the facility and the impact of the observed condition in that particular facility.

- Noncritical Deficiency
- Critical Deficiency
- Impairment

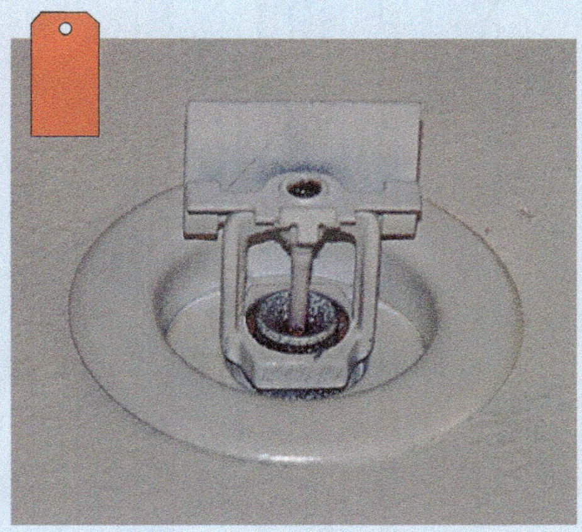

(Courtesy of Byron Blake and SimplexGrinnell)

The requirement in 6.2.1.1 of NFPA 13 states than when any sprinkler has been removed for any reason, it cannot be reinstalled. This rule applies to any sprinkler of any age that is removed for any purpose.

5.2.1.1.3* Any sprinkler that has been installed in the incorrect orientation shall be corrected by repositioning the branchline, drop, or sprig, or shall be replaced.

Exhibit 5.16 through Exhibit 5.18 show examples of sprinklers installed in the incorrect orientation.

EXHIBIT 5.16 *Upright Sprinkler Incorrectly Installed in the Pendent Position.*

EXHIBIT 5.17 *Pendent Sprinkler Incorrectly Installed in the Upright Position.*

EXHIBIT 5.18 *Sidewall Sprinkler Incorrectly Installed Upside Down.*

EXHIBIT 5.19 *Sidewall Sprinkler in the Middle of the Room. (Courtesy of M. Steven Welsh, Koffel Associates, Inc.)*

A.5.2.1.1.3 An example of a sprinkler needing replacement would be a pendent sprinkler installed in the upright position or vice versa.

Exhibit 5.19 shows a sidewall sprinkler that is now in open space in an office area. In this instance, a wall had been removed, and rather than replacing the sprinkler with a pendent, the sprinkler was left as previously installed in the wall, which is incorrect. This sprinkler should be replaced in accordance with NFPA 13, taking into consideration the spacing requirements and the adjacent sprinkler. This is an example of the type of issue that should be addressed when making modifications to the building compartmentation scheme, as noted in 4.1.5 and 4.1.6.

The requirement in 5.2.1.1.3 would not apply where a sprinkler has been intentionally installed in a position other than its listing, such as a pendent sprinkler in the upright position pointed up into the fan motor of a cooling tower.

System Tagging

This heavily corroded fitting and section of pipe would be considered a noncritical deficiency. This portion of the system is not actively leaking, which is what drives its classification. However, if there was a significant leak, the condition could be considered critical or even an impairment.

Noncritical Deficiency

Critical Deficiency

Impairment

(Courtesy of Byron Blake and SimplexGrinnell)

ITM Deficiency, Impairment, or Hazard Evaluation?

ANSWER: Hazard Evaluation

This photo shows a flexible fitting connecting from a branch line outlet to the suspended ceiling below. The listings for flexible fittings limit the number of allowable turns in the fitting as well as the radius of such turns. Excessive bending and looping of the flexible piping will cause the friction losses to skyrocket, which makes the assembly hydraulically inefficient and prevents the correct flow and pressure combination from being discharged during a fire event. That being said, this condition would not be noted on an NFPA 25 inspection form, because inspections above the ceiling are outside of the scope of the standard. This is the type of item that, if identified, could be brought to the attention of the owner on a separate list of hazard or hazard evaluation issues that might require an additional inspection beyond the scope of NFPA 25, as noted in the commentary in Chapter 4.

(Courtesy of Josh Elvove)

5.2.1.1.4* Sprinklers installed in concealed spaces such as above suspended ceilings shall not require inspection.

A.5.2.1.1.4 Examples include spaces above ceilings, whether the ceilings are lay-in tile or gypsum board, areas under theater stages, pipe chases, and other inaccessible areas, even if access panels or hatches are provided into the areas.

The requirement in 5.2.1.1.4 provides an exemption from the inspection requirements for sprinklers in concealed spaces on the basis of the two factors that follow:

1. It is extremely costly and impractical to inspect sprinklers in these spaces.
2. The sprinklers are in an inaccessible location, which means that they are less likely to be damaged, painted, or loaded than those in an open space.

However, when work is performed in concealed areas, it is prudent for the owner to have the sprinklers in these spaces inspected after the other work is completed.

5.2.1.1.5 Sprinklers installed in areas that are inaccessible for safety considerations due to process operations shall be inspected during each scheduled shutdown.

5.2.1.1.6 Escutcheons and coverplates for recessed, flush, and concealed sprinklers shall be replaced if found missing during the inspection.

Escutcheons and cover plates for recessed, flush, and concealed sprinklers are part of the listed sprinkler assembly and must be replaced when found missing. However, missing escutcheons would not normally be considered a critical deficiency. Exhibit 5.20 shows a sprinkler with an incomplete escutcheon assembly. While some of the hot gases from a fire might escape into the

EXHIBIT 5.20 *Sprinkler with Missing Escutcheon.*

plenum, the impact on activation time is negligible, which is why this is commonly referred to as a noncritical deficiency.

5.2.1.1.7 Escutcheons for pendent sprinklers that are not recessed, flush, or concealed shall not be required to be replaced if found missing during the inspection.

5.2.1.2* The minimum clearance to storage as described in 5.2.1.2.1 through 5.2.1.2.6 shall be maintained below all sprinkler deflectors.

Obstructions to sprinkler distribution patterns, such as those referred to in 5.2.1.2, can hamper the effectiveness of sprinklers. Obstructions that are less than 18 in. (457 mm) below the sprinkler have a greater impact on distribution patterns than do obstructions located further away. NFPA 13 provides guidance on these types of obstructions.

Obstructions below the sprinkler deflectors can include the installation of new equipment (installed since the last inspection) at the ceiling, the storage of stock on the top shelves in a retail store, or the installation of ceiling-mounted signs or decorations. The owner must make certain that storage, decorations, and other potential obstructions do not create a disruption to the sprinkler flow. NFPA 13 establishes the requirements for clearances to obstructions and differentiates between obstructions that impact sprinkler spray pattern development and those that obstruct the spray pattern from reaching the floor. According to NFPA 13, the top of storage is never permitted to be closer than 18 in. (457 mm) to the sprinkler deflector.

Obstructions to sprinklers range from the common to the unusual. Exhibits 5.21 and 5.22 show examples of the more common obstructions. Exhibit 5.23 shows an example of a more unusual obstruction.

Exhibit 5.21 shows a situation that occurred when signage in an airport was changed during a renovation of the terminal. The location of the sign was not evaluated against the provisions in NFPA 13 for the obstruction of spray pattern development. As a result, the sign is located too close to the sprinkler and would be considered an obstruction to the spray pattern. Exhibit 5.22 shows storage piled too close to the sprinklers, which is a common obstruction for sprinklers. Exhibit 5.23 shows a situation where a stainless steel bucket was fashioned to collect water from a leaking sprinkler.

A.5.2.1.2 NFPA 13, *Standard for the Installation of Sprinkler Systems,* in the storage definitions defines clearance as the distance from the top of storage to the ceiling sprinkler

EXHIBIT 5.21 *Typical Obstruction: Renovation.*

EXHIBIT 5.22 *Typical Obstruction: Storage.*

EXHIBIT 5.23 *Unusual Obstruction: Collection Device for a Leaking Sprinkler.*

deflectors. Other obstruction rules are impractical to enforce under this standard. However, if obstructions that might cause a concern are present, the owner is advised to have an engineering evaluation performed.

5.2.1.2.1* Unless greater distances are required by 5.2.1.2.2, 5.2.1.2.3, or 5.2.1.2.4, or lesser distances are permitted by 5.2.1.2.6, clearance between the deflector and the top of storage shall be 18 in. (457 mm) or greater.

A.5.2.1.2.1 The 18 in. (457 mm) clearance rule generally applies to standard pendent, upright and sidewall spray sprinklers, extended coverage upright and pendent sprinklers, and residential sprinklers.

5.2.1.2.2 Where standards other than NFPA 13, *Standard for the Installation of Sprinkler Systems*, specify greater clearance to storage minimums, they shall be followed.

5.2.1.2.3* Clearance between the deflector and the top of storage shall be 36 in. (914 mm) or greater for special sprinklers.

A.5.2.1.2.3 The special sprinklers that the minimum 36 in. (915 mm) clearance rule generally applies to includes large drop sprinklers, CMSA sprinklers, and early suppression fast-response (ESFR) sprinklers.

5.2.1.2.4 Clearance from the top of storage to sprinkler deflectors shall be 36 in. (914 mm) or greater where rubber tires are stored.

5.2.1.2.5 In-rack sprinklers shall not be required to meet the obstruction criteria and clearance from storage requirements.

5.2.1.2.6* Clearance between the deflector and the top of storage shall be permitted to be less than 18 in. (457 mm) where shown to be permitted by the installation standard.

A.5.2.1.2.6 The purpose of maintaining a minimum clearance is to ensure water discharge is not obstructed. There are certain installations where this can be achieved by other means. Examples include library stacks, record storage, and where sprinklers are installed in aisles in between storage shelving. Clearance is also not needed for shelving along perimeter walls since this does not cause an obstruction. NFPA 13 allows a clearance less than 18 in. (457 mm) where full-scale fire tests demonstrate an acceptable sprinkler discharge pattern. Also, where sufficient shielding of the sprinkler spray pattern has resulted in an increase in the hazard classification to Extra Hazard Group 2, a clearance less than 18 in. (457 mm) might be acceptable.

5.2.1.3* Storage closer to the sprinkler deflector than permitted by the clearance rules of the installation standard described in 5.2.1.2.1 through 5.2.1.2.4 shall be corrected.

A.5.2.1.3 Sprinkler spray patterns should not be obstructed by temporary or nonpermanent obstructions such as signs, banners, or decorations. While it is impractical for an inspector to know all of the various obstruction rules for all the different types of sprinklers, the inspector can observe when temporary or nonpermanent obstructions have been installed that could block or obstruct a

ITM Deficiency, Impairment, or Hazard Evaluation?

(Courtesy of M. Steven Welsh, Koffel Associates, Inc.)

ANSWER: Hazard Evaluation

Obstructions to sprinkler discharge are some of the most difficult observations to deal with in NFPA 25. While it appears that there is no wear and tear on the sprinkler, what is not certain is whether or not there is a violation of the NFPA 13 obstruction and spacing rules. Is the obstruction a permanent architectural or structural feature, or is it a mechanical or electrical component that was added after the sprinkler system was installed? Is the obstruction temporary and easily removed or moved away from the sprinkler?

While the proximity of the insulation to the sprinkler shown in this photo is an obvious problem, it is not specifically addressed in NFPA 25, and therefore, this situation should be reported as a recommendation for a hazard evaluation.

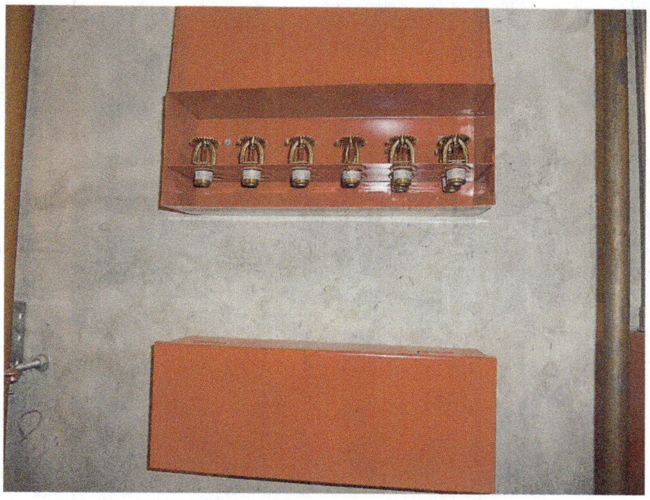

EXHIBIT 5.24 Spare Sprinkler Cabinet.

sprinkler's spray pattern. Temporary or nonpermanent obstructions that appear to be obstructions to sprinkler spray patterns should be removed or repositioned so they are not an obstruction.

The 2014 edition of NFPA 25 has taken a broader look at the storage requirements from NFPA 13 and how they should be reviewed as part of a sprinkler inspection. The requirements in 5.2.1.2 and all of its subsections, along with 5.2.1.3, provide a comprehensive look at the various storage requirements and how they should be treated.

5.2.1.4 The supply of spare sprinklers shall be inspected annually for the following:

(1) The correct number and type of sprinklers as required by 5.4.1.5
(2) A sprinkler wrench for each type of sprinkler as required by 5.4.1.5.5

It is necessary to maintain a supply of spare sprinklers, including all of the types and temperature ratings installed in the system (except for dry sprinklers, as covered in 5.4.1.5.3), to ensure systems can be returned to service immediately following a fire event or where a sprinkler has been damaged. It is important to verify that the spare sprinklers that are provided match the characteristics of the sprinklers installed and that the manufacturer's recommended wrenches that are appropriate for these sprinklers are also provided. For remote locations, it might be prudent to provide more spare sprinklers than those prescribed. A pipe or crescent wrench should never be used to install sprinklers, as these can stress the sprinkler frame and cause the sprinkler to leak. Exhibit 5.24 shows a spare sprinkler cabinet complete with two of each type of installed sprinkler and a sprinkler wrench in the lower cabinet.

(3) The list of spare sprinklers as required by 5.4.1.5.6

5.2.2* Pipe and Fittings. Sprinkler pipe and fittings shall be inspected annually from the floor level.

A.5.2.2 The conditions described in 5.2.2 can have a detrimental effect on the performance and life of pipe by affecting corrosion rates or pipe integrity or otherwise rendering the pipe ineffectual.

The inspection required by 5.2.2 includes checking for the conditions described in 5.2.2.1 and 5.2.2.2. These conditions are identified by a visual inspection from floor level, because it is usually impractical to get closer to the piping for a more detailed inspection. The use of ladders or lifts is of limited benefit when compared to the cost. As noted with the inspection of sprinklers, a flashlight or binoculars can assist in the inspection of piping in buildings with high ceilings.

System Tagging

This coupling is leaking a significant amount of water. In Table A.3.3.7, a distinction is made between "spraying or running water" and "slowly dripping water." Spraying or running water, which is what is shown in this photo, would be considered an impairment. A simple drip or a bead of water seeping out of the fitting would be considered a critical deficiency.

Noncritical Deficiency

Critical Deficiency

Impairment

(Courtesy of Byron Blake and SimplexGrinnell)

Where problems are suspected, such as those indicated by a wet spot on the floor, or when other work is being done at the ceiling level using ladders or lifts, the inspector could take advantage of the opportunity of being closer to the piping to inspect the system.

5.2.2.1* Pipe and fittings shall be in good condition and free of mechanical damage, leakage, and corrosion.

FAQ Is the inspector required to verify piping pitch during the annual inspection?

It can be difficult sometimes to determine the condition of piping corrosion, and mechanical damage might not be evident from floor level. However, Exhibit 5.25 provides an example of a more obvious problem with corrosion and the need for repair.

EXHIBIT 5.25 *Sprinkler Pipe with Obvious Corrosion and Leaking. (Courtesy of Wiginton Fire Systems)*

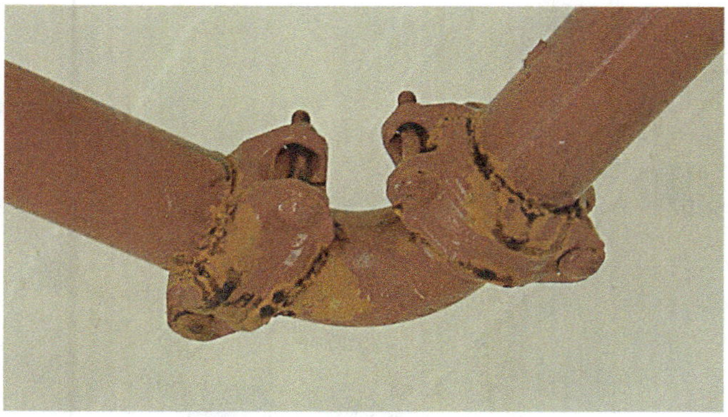

EXHIBIT 5.26 *Elbow with Leakage and Corrosion. (Courtesy of Josh Elvove)*

EXHIBIT 5.27 *Piping with Interior Corrosion. (Courtesy of Wiginton Fire Systems)*

It should be noted that installation issues such as piping pitch, pipe size, and type of pipe and fittings are not covered by the annual inspection required by 5.2.2. Where conditions indicate a potential problem, such as level or trapped sections of piping observed while making repairs to a system or section of a system, the owner should be advised that further investigation might be needed to avoid further problems related to this condition. Exhibit 5.26 shows an elbow that shows signs of leakage and corrosion. If not corrected, this condition could worsen and lead to a system impairment.

It is important to remember that the external condition of the pipe is not an indicator of the internal condition of the pipe. Many owners or their designated representatives will not conduct the internal assessment of sprinkler piping that is required by Chapter 14 because there were no pipe segments reported to show external corrosion during the annual inspection. Exhibit 5.27 shows piping that shows no signs of exterior corrosion, but the pipe interior is showing significant corrosion. The absence of exterior corrosion is not a valid reason for not conducting an internal assessment of the piping.

Critical Point

Where conditions indicate a potential problem, such as level or trapped sections of piping observed while making repairs to a system or section of a system, the owner should be advised that further investigation might be needed to avoid further problems related to this condition.

System Tagging

This pipe and its multiple fittings are showing signs of corrosion, most likely caused by a coupling that was leaking at some point. The condition as shown is not actively leaking, and therefore, it would be considered a noncritical deficiency.

Noncritical Deficiency

Critical Deficiency

Impairment

(Courtesy of Byron Blake and SimplexGrinnell)

System Tagging

In this case, the sprinkler remains intact despite the massive amount of corrosion on the welded outlet. Because the sprinkler does not appear to be of such a condition that activation would be delayed, this would be considered a noncritical deficiency. If the sprinkler itself is corroded and it appears that activation time or spray pattern development might be affected, this would become a critical deficiency.

- Noncritical Deficiency
- Critical Deficiency
- Impairment

(Courtesy of Byron Blake and SimplexGrinnell)

A.5.2.2.1 Surface corrosion not impacting the integrity of the piping strength or raising concern of potential leakage should not warrant the replacement of piping. A degree of judgment should be exercised in the determination of the extent of corrosion that would necessitate replacement.

Exhibit 5.28 shows a coupling that has a significant amount of rust and corrosion due to a constant leak. Even though they are showing a small amount of surface corrosion, it is not the intent to replace the two pieces of pipe that are connected by the fitting. The coupling is most likely the culprit and should be replaced, and the pipe simply scrubbed clean.

EXHIBIT 5.28 *Coupling Corrosion Caused by Leak. (Courtesy of M. Steven Welsh, Koffel Associates, Inc.)*

5.2.2.2 Sprinkler piping shall not be subjected to external loads by materials either resting on the pipe or hung from the pipe.

FAQ Is it permissible for small coaxial cable to be wrapped around sprinkler pipe?

Small coaxial cable is not permitted to be wrapped around sprinkler pipe. While the weight of the cable might be negligible, anything that is not a part of the sprinkler system should not be attached to piping. The design of sprinkler piping, fittings, and their supports takes into account the weight of the pipe and water and the normal forces anticipated during sprinkler operation. Although there are safety factors included in the design, the additional weight from other loads is not considered in the design of the system. In order to be certain that the sprinkler piping is not subjected to undo stress, the standard has adopted a zero-tolerance stance in regard to materials that are not a part of the system. Therefore, equipment, signs, decorations, or other materials should not be hung from, or otherwise supported by, the sprinkler piping. In addition to the potential for failure from the added load, attachments to sprinkler piping can also obstruct sprinklers or hinder their operation. The only exception is that listed linear wire detection systems, which are commonly used for activation of preaction or deluge systems, can be attached to sprinkler piping where listed for such use. Exhibit 5.29 provides an example of cable wrapped around sprinkler piping that is not permitted. Exhibit 5.30 shows clothing hanging from a sprinkler guard, which places additional loading on the sprinkler piping, the guard, and the fitting.

5.2.2.3* Pipe and fittings installed in concealed spaces such as above suspended ceilings shall not require inspection.

A.5.2.2.3 Examples include some floor/ceiling or roof/ceiling assemblies, areas under theater stages, pipe chases, and other inaccessible areas.

EXHIBIT 5.29 *Cabling Not Permitted to Be Wrapped Around Sprinkler Piping.*

EXHIBIT 5.30 *Overloaded Sprinkler Piping. (Courtesy of Josh Elvove)*

The requirement in 5.2.2.3 provides an exemption from the inspection requirements for pipe and fittings in concealed spaces on the basis of the two factors that follow:

1. It is costly and impractical to inspect piping in these spaces.
2. The inaccessibility of such piping means it is less likely to be damaged or otherwise affected than piping in an open space.

When other work is performed in concealed areas, it is prudent for the owner to have the pipe and fittings in these spaces inspected after the other work is completed.

5.2.2.4 Pipe and fittings installed in areas that are inaccessible for safety considerations due to process operations shall be inspected during each scheduled shutdown.

5.2.3* Hangers and Seismic Braces. Sprinkler pipe hangers and seismic braces shall be inspected annually from the floor level.

The visual inspection of hangers and seismic braces is conducted from floor level, as specified in 5.2.3, because it is usually impractical to get closer to the hangers and braces for a more detailed inspection. The use of ladders or lifts is of limited benefit when compared to the cost. A flashlight or binoculars can assist in the inspection of hangers and braces in buildings with high ceilings.

Where problems are suspected, or when other work is being done at the ceiling level using ladders or lifts, the inspector might take advantage of the opportunity of being closer to the hangers and braces to inspect the system. However, similar to inspecting piping, it is not always

System Tagging

A bent hanger rod will not typically have an impact on the functionality of the system, which is why it would typically be considered a noncritical deficiency. If several consecutive hangers are damaged or have failed, it might increase the likelihood for system failure and it would be a different classification.

Noncritical Deficiency

Critical Deficiency

Impairment

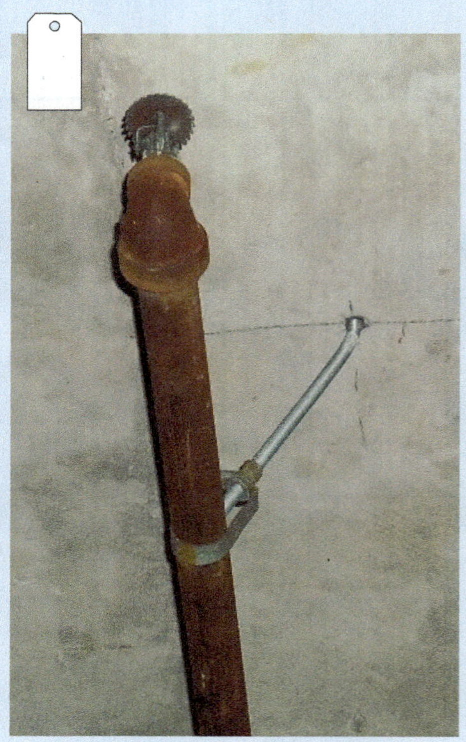

(Courtesy of Byron Blake and SimplexGrinnell)

EXHIBIT 5.31 *Hanger Rod Detached from the Ceiling Anchor.*

EXHIBIT 5.32 *Improper Use of a Closet Coat Rod Hanger to Support Pipe.*

EXHIBIT 5.33 *Improper Use of a Board to Support Pipe.*

necessary to get close to the hanger or bracing to identify problems. Exhibit 5.31 through Exhibit 5.33 show examples of clear violations of the requirements for hangers. Exhibit 5.31 shows a hanger that has come loose from its connection to the structure. Exhibit 5.32 shows piping being supported by a closet rod hanger that has been used to replace the missing pipe hanger, and Exhibit 5.33 shows a piece of wood that has been used to replace a hanger.

A.5.2.3 The conditions described in this section can have a detrimental effect on the performance of hangers and braces by allowing failures if the components become loose.

| **5.2.3.1** Hangers and seismic braces shall not be damaged, loose, or unattached.

The requirement in 5.2.3.1 is another example of the standard limiting the scope of the inspection to the functionality of the component. Issues such as the spacing, sizing, or type of hanger or seismic brace are not included in the inspection required by 5.2.3. This requirement does not impose the need to physically check that the seismic brace is tight, since the inspection is conducted from the floor per 5.2.3.

ITM NFPA 25

ITM Deficiency, Impairment, or Hazard Evaluation?

(Courtesy of Byron Blake and SimplexGrinnell)

ANSWER: Hazard Evaluation

A missing hanger for a sprinkler system is not something that an inspector conducting an annual inspection would be looking for. The spacing of hanging and bracing system components would be considered an item that would be part of a hazard evaluation.

If, during an inspection, an inspector sees a hanger that has been pulled from the ceiling as evidenced by the hanger ring and threaded rod dangling from the piping, this would be considered a noncritical deficiency. In extreme instances, where the deflection is to the point where it looks like couplings or fittings could fail or are showing signs of leaking, it could be upgraded to a critical deficiency.

5.2.3.2 Hangers and seismic braces that are damaged, loose, or unattached shall be replaced or refastened.

When hangers or seismic braces are loose, other hangers or braces must support the load. If enough supports need repair, overloading and failure of other hangers or braces can occur. Damaged hangers or braces can also fail. When such conditions are found, the damaged or loose supports must be replaced or refastened as appropriate.

5.2.3.3* Hangers and seismic braces installed in concealed spaces such as above suspended ceilings shall not require inspection.

Paragraph 5.2.3.3 provides an exemption from the inspection requirements for hangers and seismic braces in concealed spaces on the basis of the two factors that follow:

1. It is costly and impractical to inspect the hangers and braces in these spaces.
2. The inaccessibility of such hangers and braces means they are less likely to be damaged or otherwise affected than hangers and braces in open spaces.

However, when other work is performed in concealed areas, it is a good idea for the owner to have the hangers and braces in these spaces inspected after the other work is completed.

A.5.2.3.3 Examples of hangers and seismic braces installed in concealed areas include some floor/ceiling or roof/ceiling assemblies, areas under theater stages, pipe chases, and other inaccessible areas.

5.2.3.4 Hangers and seismic bracing installed in areas that are inaccessible for safety considerations due to process operations shall be inspected during each scheduled shutdown.

5.2.4 Gauges.

5.2.4.1* Gauges on wet pipe and deluge sprinkler systems shall be inspected quarterly to ensure that they are in good condition and that normal water supply pressure is being maintained.

The effectiveness of sprinklers depends on the availability and reliability of the water supply, and an inspection of the gauge on the sprinkler system riser, as required by 5.2.4.1, can identify potential problems with the water supply (see Exhibit 5.34). For example, if the gauge reads significantly less than it did in previous inspections, there might be a problem with the water supply, such as low tank levels in gravity systems, high water demand by new users, municipal or private pumps that are out of service, or closed sectional valves in gridded waterworks. If the gauge reads close to zero, it is likely that there is a shut valve or other problem that is preventing water from reaching the system.

Critical Point

The effectiveness of sprinklers depends on the availability and reliability of the water supply, and an inspection of the gauge on the sprinkler system riser, as required by 5.2.4.1, can identify potential problems with the water supply.

EXHIBIT 5.34 *Pressure Gauge.*

Any abnormal pressure readings should be investigated, because they might be the result of a damaged gauge, which must be repaired or replaced. The time it takes to restore normal pressure is also noteworthy, since a partially closed valve might permit normal pressure to be restored but over a longer period of time. Note that a normal pressure reading does not guarantee that no problems exist. A valve can be almost totally shut with the gauge still indicating a normal static pressure. A shut valve or other such impairment is usually discovered with the main drain test if the residual pressure falls dramatically below the expected reading. Gauges that show signs of mechanical damage, such as the ones shown in Exhibit 5.35 and Exhibit 5.36, are typically considered noncritical deficiencies, and should be replaced in a timely fashion.

EXHIBIT 5.35 *Damaged Gauge. (Courtesy of Wiginton Fire Systems)*

EXHIBIT 5.36 *Gauge with Dislodged Needle. (Courtesy of Byron Blake and SimplexGrinnell)*

A.5.2.4.1 Due to the high probability of a buildup of excess pressure, gridded wet pipe systems should be provided with a relief valve not less than ¼ in. (6.3 mm) in accordance with NFPA 13, *Standard for the Installation of Sprinkler Systems.*

It is normal, though, that the pressure above the alarm or system check valve is typically higher than that of the water supply as a result of trapped pressure surges.

Excessive pressures on the system side of a check valve might indicate a problem with pressure build-up as a result of solar heating (i.e., thermal expansion). Pressure relief valves are required on all wet systems installed, beginning with the 2010 edition of NFPA 13. Excessive pressures might indicate a malfunctioning relief valve and should be investigated. If excessive pressures are encountered on non-gridded systems, it might be prudent for the owner to install a pressure relief valve, even though one might not have been required when the system was originally installed.

5.2.4.2 Gauges on dry and preaction systems shall be inspected weekly to ensure that normal air or nitrogen and water pressures are being maintained.

5.2.4.3 Where air pressure supervision is connected to a constantly attended location, gauges shall be inspected monthly.

5.2.4.4* For dry pipe or preaction systems protecting freezers with two air pressure gauges on the air line(s) between the compressor and the dry pipe or preaction valve, the air pressure gauge near the compressor shall be compared weekly to the pressure gauge above the dry pipe or preaction valve.

A.5.2.4.4 See Figure A.5.2.4.4.

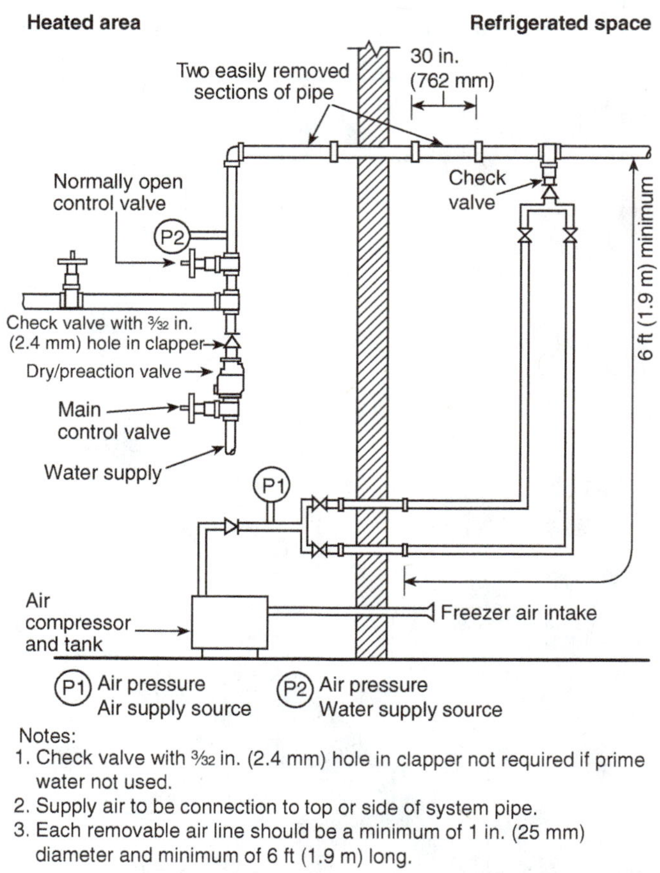

FIGURE A.5.2.4.4 *Refrigerator Area Sprinkler System Used to Minimize Chances of Developing Ice Plugs.*

The arrangement of the air supply for a dry pipe system, such as the one depicted in Figure A.5.2.4.4, is designed to prevent moisture in the make-up air for the dry system from condensing and freezing in the main sprinkler piping by allowing any moisture (ice) to accumulate in the air supply line. When the gauge near the compressor reads higher than the air pressure in the system, there is usually an ice blockage in the air line. Switching to the alternate line maintains the correct air pressure in the system, allowing the blocked line to be disconnected, the ice removed, and the line reinstalled without removing the dry pipe valve from service.

It should be noted that beginning with the 2013 edition of NFPA 13, the system air pressure can be provided with nitrogen or other approved gas. Such arrangements can include a pressure regulator or nitrogen generator. These alternative compressed gas systems should be inspected in the same manner as any traditional air supply, including any special recommendations of the manufacturer or supplier.

5.2.4.4.1 When the gauge near the compressor is reading higher than the gauge near the dry pipe valve, the air line in service shall be taken out of service, and the alternate air line opened to equalize the pressure.

5.2.4.4.2 The air line taken out of service shall be internally inspected, shall have all ice blockage removed, and shall be reassembled for use as a future alternate air line.

5.2.5 Waterflow Alarm and Supervisory Signal Initiating Device. Waterflow alarm and supervisory signal initiating devices shall be inspected quarterly to verify that they are free of physical damage.

5.2.6* Hydraulic Design Information Sign. The hydraulic design information sign shall be inspected quarterly to verify that it is provided, attached securely to the sprinkler riser, and is legible.

A.5.2.6 The hydraulic design information sign should be secured to the riser with durable wire, chain, or equivalent. *(See Figure A.5.2.6.)*

This system as shown on ____________ company
print no. ____________ dated ________
for ______________________________
at ____________ contract no. ________
is designed to discharge at a rate of ____________
gpm per ft^2 (L/min per m^2) of floor area over a maximum
area of ____________ ft^2 (m^2) when supplied
with water at a rate of ____________ gpm (L/min)
at ____________ psi (bar) at the base of the riser.
Hose stream allowance of ____________
gpm (L/min) is included in the above.

FIGURE A.5.2.6 *Sample Hydraulic Design Information Sign.*

ITM Deficiency, Impairment, or Hazard Evaluation?

(Courtesy of Byron Blake and SimplexGrinnell)

ANSWER: ITM Deficiency

At first glance, this hydraulic design information sign might appear to meet the requirements of 5.2.6 because it is provided and it is securely attached to the riser. It is the final requirement of this section that causes this to be a deficiency. Yes, the preprinted lettering on this sign is legible, but the hydraulic information that is required to be filled in on the sign is not legible, because it does not exist.

Although this is considered a noncritical deficiency because it does not hinder the ability of the fire protection system to function in a fire event, the information on this sign is important for hazard and water supply evaluations. The most recent set of working drawings, the hydraulic calculations, and/or other installation documents required by NFPA 13 will need to be reviewed to obtain the needed information. In the absence of the latest installation documents, a hazard evaluation might have to be performed and the required documentation recreated. Once the hydraulic design information has been determined, a new sign should be provided, or this sign filled in using a permanent marker.

Paragraph 5.2.6 requires that the hydraulic design information sign (also called a nameplate or placard) be inspected on a quarterly basis. NFPA 13 requires a hydraulic design information sign on hydraulically designed systems so that the design criteria and system demand can be readily determined. The hydraulic design information sign can provide useful information to the owner. If the design information sign is missing, the owner should contact a design professional to determine the demand for the system, which can be written on a new design information sign. The details are also documented on the approved plans and hydraulic calculations, but these plans can be misplaced and might not be available when the property changes owners. A hydraulic design information sign that is securely fastened to the riser can provide the details when these other data are missing (see Exhibit 5.37). If the sign becomes loose or is difficult to read, it must be repaired or replaced.

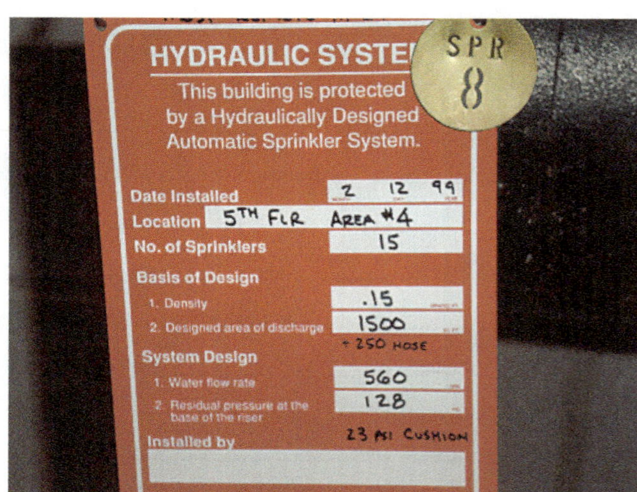

EXHIBIT 5.37 *Hydraulic Design Information Sign.*

A hydraulic design information sign will not be present on systems where the piping diameter was determined by the pipe schedule method. In order to avoid confusion on whether the sign is missing or simply not required, the 2014 edition of NFPA 25 requires pipe schedule systems to be provided with a sign indicating this basis of design.

5.2.6.1 A hydraulic design information sign that is missing or illegible shall be replaced.

5.2.6.2 A pipe schedule system shall have a hydraulic design information sign that reads "Pipe Schedule System."

5.2.7 Heat Tracing. Heat tracing shall be inspected and maintained in accordance with manufacturer's requirements.

5.2.8 Information Sign. The information sign required by 4.1.9 shall be inspected annually to verify that it is provided, securely attached, and legible.

The information sign required by 4.1.9 is in fact required to be added or replaced where missing. It is a rare retrofit requirement to the extent that at a minimum, information related to the four specific items noted in 4.1.9.2 must be provided on system risers so that the owner, owner's representative, or service provider is aware of the presence of these system attachments.

5.2.9* General Information Sign. The general information sign required by NFPA 13 shall be inspected annually to verify that it is provided, securely attached, and legible.

A.5.2.9 The information sign should be secured with wire, chain, or equivalent to each system control valve, antifreeze loop, and auxiliary system control valve indicating the information required by 4.1.9.

The general information sign required by 5.2.9 of NFPA 25 and Section 25.6 of NFPA 13 was introduced in the 2007 edition of NFPA 13 and, as such, should not be expected to be present on systems installed prior. Where the information is known, it is prudent to provide the sign for older systems so that future parties evaluating the system might have easy access to the data provided.

5.3 Testing

5.3.1* Sprinklers.

A.5.3.1 The sprinkler field service testing described in this section is considered routine testing. Nonroutine testing should be conducted to address unusual conditions not associated with the routine test cycles mandated within this standard. Due to the nature of nonroutine testing, specific tests cannot be identified in this standard. The type of tests to be conducted and the number and location of samples to be submitted should be appropriate to the problem discovered or being investigated and based on consultation with the manufacturer, listing agency, and the authority having jurisdiction.

Where documentation of the installation date is not available, the start date for the in-service interval should be based upon the sprinkler's manufacture date.

5.3.1.1* Where required by this section, sample sprinklers shall be submitted to a recognized testing laboratory acceptable to the authority having jurisdiction for field service testing.

Sprinklers are extremely reliable and can last as long as the building in which they are installed. However, unlike plumbing, electrical, and HVAC systems, sprinkler systems sit idle for years if no fire emergency occurs. As a result of this idleness, sprinkler system operation and the operation of the sprinkler itself cannot be verified without proper testing.

Paragraph 5.3.1.1 specifies that routine testing is needed to identify potential problems that would otherwise go unnoticed. Sprinklers that have visible signs of damage, that are leaking, or that are loaded, corroded, or painted are required to be replaced without testing. The test frequency varies based on the sprinkler type or its use.

As sprinklers age, the frequency with which they must be tested increases. For example, standard-response sprinklers are required to be tested after 50 years of service. They must then be retested every 10 years, and after 75 years the testing frequency increases to every 5 years.

FAQ What is the basis for determining whether sprinklers should be tested or simply replaced?

As sprinklers age, they may fail more frequently. As stated in 5.3.1.3, where one sprinkler fails the test, all sprinklers represented by that sample must be replaced. With the improvements being made in sprinkler technology, it might be desirable to replace the aging sprinklers rather than continue testing. In addition, if the sample area is small, it might be most cost efficient to replace the sprinklers. A cost-benefit analysis is necessary to determine if replacement is the best strategy.

The sprinklers being tested undergo a procedure known as a plunge test. The sprinkler is inserted (or plunged) into a device known as a plunge test apparatus where it is exposed to an airflow that has a controlled velocity and temperature (see Exhibit 5.38, top). The temperature in the device is considerably higher than the operating temperature of the sprinkler. The sprinkler is pressurized with 5 psi (0.4 bar) of air pressure. The amount of time taken for the fusible element or glass bulb to activate is measured. If the sprinkler fails to operate in the specified amount of time, the sprinkler fails the test and all sprinklers represented by the test sprinkler (see Exhibit 5.38, bottom) must be replaced.

EXHIBIT 5.38 *Plunge Test Apparatus (top) and Test Sprinkler (bottom).*

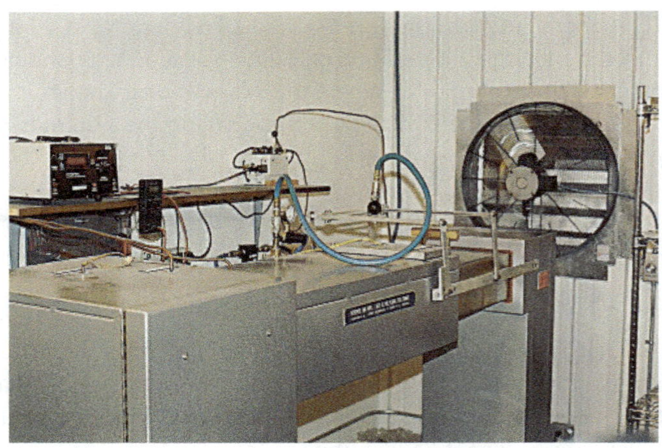

A.5.3.1.1 Sprinklers should be first given a visual inspection for signs of mechanical damage, cleaning, painting, leakage in service, or severe loading or corrosion, all of which are considered causes for immediate replacement. Devices that have passed the visual inspection should then be laboratory tested for sensitivity and functionality. The waterway should clear when sensitivity/functionality tested at 5 psi (0.4 bar) or the minimum listed operating pressure for dry sprinklers.

Thermal sensitivity should be not less than that permitted in post-corrosion testing of new sprinklers of the same type.

Sprinklers that have been in service for a number of years should not be expected to have all of the performance qualities of a new sprinkler. However, if there is any question about their continued satisfactory performance, the sprinklers should be replaced.

See Figure A.5.3.1.1.

FIGURE A.5.3.1.1 *Sprinkler Operating Element Identification.*

5.3.1.1.1 Where sprinklers have been in service for 50 years, they shall be replaced or representative samples from one or more sample areas shall be tested.

FAQ Where are the requirements found for testing 1 percent of piping and sprinklers when they reach 50 years old?

Paragraph 5.3.1.1.1 addresses the removal and testing of standard spray sprinklers after a service period of 50 years. Piping is not involved in this particular test. The correct quantity of sprinklers based on the total from 5.3.1.2 is removed and sent to an approved testing lab for evaluation of the response time index (RTI). The subsections of 5.3.1 should be reviewed carefully, because the test frequency varies based on the sprinkler type and use. Paragraph 5.3.1.2 addresses the number of sprinklers that must be removed for testing.

5.3.1.1.1.1 Test procedures shall be repeated at 10-year intervals.

5.3.1.1.1.2 Sprinklers manufactured prior to 1920 shall be replaced.

5.3.1.1.1.3* Sprinklers manufactured using fast-response elements that have been in service for 20 years shall be replaced or representative samples shall be tested and then retested at 10-year intervals.

Quick-response sprinklers, as defined in 3.3.40.13, are a relatively recent advancement in sprinkler technology, appearing in the industry in the early 1980s. The 20-year (rather than the typical 50-year) testing frequency requirement in 5.3.1.1.1.3 is included in the standard because the long-term performance of these relatively new sprinklers is not yet known. The reduced frequency is intended to identify potential problems early in the life cycle of quick-response sprinklers, rather than waiting for the 50-year test required of sprinklers whose performance is well known. As more quick-response sprinklers are tested over the next few years, the Committee on Inspection, Testing, and Maintenance of Water-Based Systems intends to monitor their performance and adjust the test frequency requirement if no significant problems are identified.

To determine the age of sprinklers in a system, the inspector often examines the sprinklers in the spare sprinkler cabinet. However, if a building has undergone renovations, or if the spare sprinklers have been replaced over the years, they might not be the same age as those in the system. In addition, the age of the sprinklers might not be the same as the sprinklers' years in service. In some cases, sprinklers could be a couple of years old before they are actually installed in a system. Therefore, the inspector should not rely exclusively on examining the spare sprinklers to determine the age of the sprinklers in the system, but rather, they should review the building completion documents or other data to determine when the system went into service.

FAQ Are glass bulb sprinklers an exception to the requirement in 5.3.1.1.1.3?

Although glass bulb sprinklers have a different operating mechanism and other types of failure for these sprinklers are possible, such as corrosion around the seat, contamination, and loading, there is no exception for testing glass bulb–type sprinklers at any other testing frequency.

A.5.3.1.1.1.3 Sprinklers defined as fast response have a thermal element with an RTI of 50 (meters-seconds)$^{1/2}$ or less. A quick-response sprinkler, residential sprinkler, and early suppression fast-response (ESFR) sprinklers are examples of fast-response sprinklers.

5.3.1.1.1.4* Representative samples of solder-type sprinklers with a temperature classification of extra high [325°F (163°C)] or greater that are exposed to semicontinuous to continuous maximum allowable ambient temperature conditions shall be tested at 5-year intervals.

A.5.3.1.1.1.4 Due to solder migration caused by the high temperatures to which these devices are exposed, it is important to test them every 5 years. Because of this phenomenon, the operating temperature can vary over a wide range.

5.3.1.1.1.5 Where sprinklers have been in service for 75 years, they shall be replaced or representative samples from one or more sample areas shall be submitted to a recognized testing laboratory acceptable to the authority having jurisdiction for field service testing and repeated at 5-year intervals.

5.3.1.1.1.6* Dry sprinklers that have been in service for 10 years shall be replaced or representative samples shall be tested and then retested at 10-year intervals.

FAQ Do dry sprinklers have to be tested, or can they be replaced?

Replacing all the dry-type sprinklers in a sample area, rather than testing them, is always an option. Experience with dry-type sprinklers has indicated a much higher failure rate than with standard sprinklers and a greater susceptibility to corrosion both internally, when moisture condenses inside the device, and externally. Corrosion at the water seal and at the weep hole at the bottom of the sprinkler has been reported in sprinklers older than 10 years. These sprinklers are usually installed in harsher environments, further compounding this problem.

FAQ Does 5.3.1.1.1.6 apply to listed dry sprinklers, or does it apply to all sprinklers installed in a dry pipe system?

Paragraph 5.3.1.1.1.6 refers to the listed dry-type sprinkler. The requirement does not apply to standard spray sprinklers installed on a dry pipe system.

Case In Point

The failure rate of dry-type sprinklers in service for 10 years is approximately 50 percent. Because of this high failure rate, it is important that these sprinklers be identified and tested as required. Dry-type sprinklers are custom-made in exact lengths, and therefore NFPA 13 does not require spare dry-type sprinklers in the spare sprinkler cabinet unless all the lengths are the same. When the age of a building or system appears to be approaching 10 years, or if the age is unknown, the inspector should perform a close inspection of a sampling of the sprinklers themselves to determine the date on the sprinklers. In cases where only a few dry sprinklers are installed or where corrosion is noted, it may be more cost-effective to replace the sprinklers rather than test them.

The 10-year threshold requirement was added to NFPA 25 in the 2002 edition, because not all of the conditions that cause failure are well understood, and the frequency of failure is higher for sprinklers that have been in service for more than 10 years. Design changes have been made to dry sprinklers, and it is expected that there will be an improvement in their performance.

A.5.3.1.1.1.6 See 3.3.40.4.

5.3.1.1.2* Where sprinklers are subjected to harsh environments, including corrosive atmospheres and corrosive water supplies, on a 5-year basis, either sprinklers shall be replaced or representative sprinkler samples shall be tested.

A.5.3.1.1.2 Examples of these environments are paper mills, packing houses, tanneries, alkali plants, organic fertilizer plants, foundries, forge shops, fumigation areas, pickle and vinegar works, stables, storage battery rooms, electroplating rooms, galvanizing rooms, steam rooms of all descriptions including moist vapor dry kilns, salt storage rooms, locomotive sheds or houses, driveways, areas exposed to outside weather, around bleaching equipment in flour mills, all portions of cold storage areas, and portions of any area where corrosive vapors prevail. Harsh water environments include water supplies that are chemically reactive.

FAQ Are walk-in refrigerators and walk-in freezers considered "harsh environments" and, therefore, subject to this standard?

Paragraph A.5.3.1.1.2 provides examples of areas that are considered to be harsh environments, including "cold storage areas." While A.5.3.1.1.2 does not further describe cold storage areas, a walk-in refrigerator or walk-in freezer could fall into this category.

5.3.1.1.3 Where historical data indicate, longer intervals between testing shall be permitted.

5.3.1.2* A representative sample of sprinklers for testing per 5.3.1.1.1 shall consist of a minimum of not less than four sprinklers or 1 percent of the number of sprinklers per individual sprinkler sample, whichever is greater.

The requirement in 5.3.1.2 for a minimum sample of four sprinklers or 1 percent of the total of sprinklers installed is intended to balance the cost of testing with the likelihood of identifying a possible problem. The sample should be somewhat random and should be representative of the sprinklers installed in the system. For example, sprinklers should be selected from different floors or areas of the building and not selected simply because they are more accessible than other sprinklers. In addition, the selection should take into consideration the age and types of sprinklers as well as the environmental conditions to which they are subjected.

The inspector and/or the owner can determine which groups of sprinklers the sprinkler sample represents. Keep in mind that if a single sprinkler from the sample fails the plunge test, all the sprinklers that the sample represents must be replaced. The sample can represent an entire system or one floor of a multi-story building.

FAQ Can the sprinklers in the spare sprinkler cabinet be used to comply with 5.3.1.2?

Note that only sprinklers that have been exposed to service conditions must be tested. The sprinklers in the spare sprinkler cabinet, for example, have not been exposed to service conditions and might not reveal any deficiencies.

A.5.3.1.2 Within an environment, similar sidewall, upright, and pendent sprinklers produced by the same manufacturer could be considered part of the same sample, but additional sprinklers would be included within the sample if produced by a different manufacturer.

5.3.1.3 Where one sprinkler within a representative sample fails to meet the test requirement, all sprinklers within the area represented by that sample shall be replaced.

The number of sprinklers selected for testing as specified in 5.3.1.2 is a minimum and will not guarantee that all problems will be discovered. As a result, when a single sprinkler fails, there is a high probability that other sprinklers will also fail. Therefore, 5.3.1.3 requires that all of the sprinklers represented by the sample be replaced.

5.3.1.3.1 Manufacturers shall be permitted to make modifications to their own sprinklers in the field with listed devices that restore the original performance as intended by the listing, where acceptable to the authority having jurisdiction.

5.3.2* Gauges.

A.5.3.2 The normal life expectancy of a gauge is between 10 and 15 years. A gauge can be permitted to have a reading with an error of ±3 percent of the maximum (full-scale) gauge reading. For example, a gauge having 200 psi (13.8 bar) maximum radius installed on a system with 60 psi (4.1 bar) normal pressure can be permitted if the gauge reads from 54 psi to 66 psi (3.7 bar to 4.5 bar).

The lower gauge pictured in Exhibit 5.39 has a maximum radius of 300 psi (20.7 bar). If the normal pressure on the system into which it is installed is 125 psi (8.6 bar) as shown, the actual pressure in the system can vary from 116 psi to 134 psi (8 bar to 9.2 bar). This range is acceptable.

5.3.2.1 Gauges shall be replaced every 5 years or tested every 5 years by comparison with a calibrated gauge.

5.3.2.2 Gauges not accurate to within 3 percent of the full scale shall be recalibrated or replaced.

FAQ Does NFPA 25 require that sprinkler system pressure gauges be tested for accuracy or calibration? If so, how often do they have to be tested?

Gauges are required by 5.3.2.1 to be tested every 5 years by comparison with a calibrated gauge. Gauges not accurate to within 3 percent of the full scale are required to be recalibrated or replaced.

Many of the inspections and tests prescribed in NFPA 25 rely on the accuracy of pressure gauges. It is not the intent of 5.3.2 to require that each and every gauge be individually calibrated. It is acceptable to compare the reading of one gauge with that of a calibrated gauge. Other gauges on systems that are installed in similar positions and elevations (e.g., on an

EXHIBIT 5.39 System Pressure Gauge.

adjacent riser) that show similar readings are acceptable. The 5-year period is based on the installed date as opposed to the manufacture date of a gauge, so it is acceptable to mark the installed date with a permanent marker on a label or on the body of a gauge when it is originally installed or calibrated.

5.3.2.3 Where multiple system risers are supplied by a common water supply source with gauges located at the same elevation, and the gauges for all systems read within 3 percent of the other(s), only one gauge shall be required to be tested to determine if replacement is required.

5.3.3 Waterflow Alarm Devices.

See 13.2.6 for the recommended operational procedure.

5.3.3.1 Mechanical waterflow alarm devices including, but not limited to, water motor gongs, shall be tested quarterly.

5.3.3.2* Vane-type and pressure switch–type waterflow alarm devices shall be tested semiannually.

A.5.3.3.2 Data concerning reliability of electrical waterflow switches indicate no appreciable change in failure rates for those tested quarterly and those tested semiannually. Mechanical motor gongs, however, have additional mechanical and environmental failure modes and need to be tested more often.

Critical Point

Each type of waterflow alarm has its own failure modes, and some are more susceptible than others to environmental conditions.

The waterflow alarm is a key part of the sprinkler system and must be tested. Paragraphs 5.3.3.1 and 5.3.3.2 take into account that each type of waterflow alarm has its own failure modes and that some are more susceptible than others to environmental conditions. For example, water motor gongs (see Exhibit 5.40, top) might be affected more by environmental conditions, such

EXHIBIT 5.40 *Water Motor Alarm Gong (top) and Water Motor Gong with Wasp Nests (bottom).*

as bird or wasp nests or freezing (see Exhibit 5.40, bottom), than would be an electric alarm initiating device such as a pressure switch or vane-type flow switch. For more information on alarm devices, see the commentary following 13.2.6.

5.3.3.3 Testing waterflow alarm devices on wet pipe systems shall be accomplished by opening the inspector's test connection.

5.3.3.3.1 Where freezing weather conditions or other circumstances prohibit use of the inspector's test connection, the bypass connection shall be permitted to be used.

Critical Point

Care should always be taken prior to discharging water to ensure proper drainage to an acceptable location where no property will be damaged.

The inspector's test connection should be arranged to facilitate testing. The valve should be readily accessible and the discharge should be directed so that the flow of water does not cause damage or unsafe conditions. An inspector's test valve that cannot be accessed, or conditions where the flow discharges to a location that does not allow testing, are situations that should be corrected. Care should always be taken prior to discharging water to ensure proper drainage to an acceptable location where no property will be damaged.

5.3.3.4 Fire pumps shall not be taken out of service during testing unless constantly attended by qualified personnel or all impairment procedures contained in Chapter 15 are followed.

It is vital that water supplies remain in service during testing for the following reasons:

1. Turning off fire pumps creates an impairment.
2. Sprinkler systems designed to be supplied by a fire pump might not perform properly without the flow and pressure the pump provides.
3. Testing the waterflow device with the fire pump in service simulates the conditions in which it is expected to operate.

Testing with the fire pump turned on exercises the pump and ensures that adequate water will be available should a fire occur during the test. It is required that an attendant be present in the pump room during operation. When it is necessary to shut down water supplies during testing, extra care and coordination are required to ensure that the pump or other water supply can be promptly returned to operation in case of emergency. It will also be necessary to apply all of the requirements in Chapter 15 of NFPA 25 whenever a pump is taken out of service. See NFPA 20, *Standard for the Installation of Stationary Pumps for Fire Protection*, for more information on fire pump operation.

Critical Point

Testing with the fire pump turned on exercises the pump and ensures that adequate water will be available should a fire occur during the test.

5.3.3.5* Testing waterflow alarm devices on dry pipe, preaction, or deluge systems shall be accomplished by using the bypass connection.

A.5.3.3.5 Opening the inspector's test connection can cause the system to trip accidentally.

5.3.4* Antifreeze Systems. Annually, before the onset of freezing weather, the antifreeze solution shall be tested using the following procedure:

(1) Using installation records, maintenance records, information from the owner, chemical tests, or other reliable sources of information, the type of antifreeze in the system shall be determined.
  (a) If the type of antifreeze is found to be a type that is no longer permitted, the system shall be drained completely and replaced with an acceptable solution.
  (b) If the type of antifreeze cannot be reliably determined, the system shall be drained completely and replaced with an acceptable solution.
(2) If the antifreeze is not replaced in accordance with step 1, test samples shall be taken at the top of each system and at the bottom of each system.
  (a) If the most remote portion of the system is not near the top or the bottom of the system, an additional sample shall be taken at the most remote portion.
  (b) If the connection to the water supply piping is not near the top or the bottom of the system, an additional sample shall be taken at the connection to the water supply.
(3) The specific gravity of each solution shall be checked using a hydrometer with a suitable scale or a refractometer having a scale calibrated for the antifreeze solution.
(4) If any of the samples exhibits a concentration in excess of what is permitted by NFPA 25, the system shall be emptied and refilled with a new acceptable solution. If a concentration greater than what is currently permitted by NFPA 25 was necessary to keep the fluid from freezing, alternate methods of preventing the pipe from freezing shall be employed.
(5) If any of the samples exhibits a concentration lower than what is necessary to keep the fluid from freezing, the system shall be emptied and refilled with a new acceptable solution.

Testing Procedure Alert

ITM NFPA 25

In recent years, the sprinkler industry found out how dangerous antifreeze can be in a sprinkler system when concentrations of the solution are too high. It is imperative that the exact type of solution and the concentration of the solution is known before even considering leaving an antifreeze solution in a system. All antifreeze solutions in existing systems must comply with 5.3.4.2.1. See the Case In Point on p. 157 for additional information. For more information about testing antifreeze systems, please refer to the detailed testing procedure at the end of this chapter.

A.5.3.4 Sampling from the top and bottom of the system helps to determine if the solution has settled. Antifreeze solutions are heavier than water. If the antifreeze compound is separating from the water due to poor mixing, it will exhibit a higher concentration in the lower portion of the system than in the upper portion of the system. If the concentration is acceptable near the top, but too low near the water connection, it might mean that the system is becoming diluted near the water supply. If the concentration is either too high or too low in both the samples, it might mean that the wrong concentration was added to the system.

Two or three times during the freezing season, test samples can be drawn from test valve B as shown in Figure 7.6.2.1(1) of NFPA 13, especially if the water portion of the system has been drained for maintenance or repairs. A small hydrometer can be used so that a small sample is sufficient. Where water appears at valve B, or where the sample indicates that the solution has become weakened, the entire system should be emptied and refilled with acceptable solution as previously described.

See Figure A.5.3.4 for expected minimum air temperatures in 48 of the United States and parts of Canada where the lowest one-day mean temperature can be used as one method of

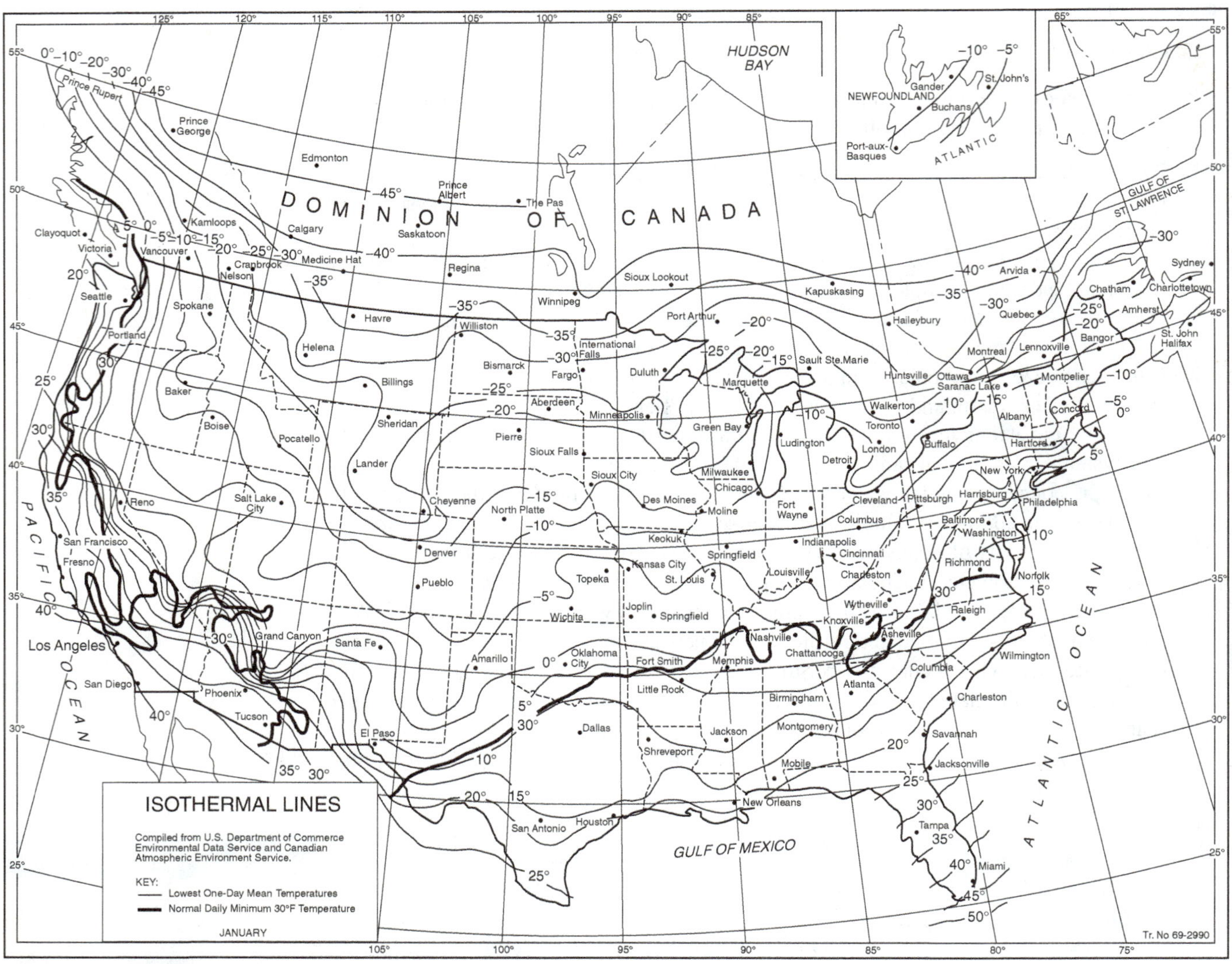

Source: Compiled from United States Weather Bureau records.
For SI units, °C = 5/9 (°F −32); 1 mi = 1.609 km.

FIGURE A.5.3.4 *Isothermal Lines — Lowest One-Day Mean Temperature (°F). [24:Figure A.10.5.1]*

Case In Point

After a series of fires where the presence of antifreeze in the system negatively contributed to the fire, the NFPA sprinkler committees responsible for NFPA 13 and NFPA 25 prepared a series of Tentative Interim Amendments (TIAs) to the standards to reflect the most up-to-date information on antifreeze.

This included a series of TIAs that were approved by the NFPA Standards Council following the printing of the 2011 edition of NFPA 25. These TIAs were written to the past edition of the standard (2011 edition), but the content provided in them was also picked up by the NFPA 25 technical committee in the 2014 revision cycle.

The revisions to the testing section for antifreeze systems require that a sample of the antifreeze solution be tested every year before the onset of freezing weather. The first photo shows an antifreeze solution test outlet and a test sample being collected and analyzed. Another photo shows an antifreeze loop with a test outlet and an expansion tank. With the use of a refractometer, such as the one shown in the final photo, the percentage of water and antifreeze in a given solution can be determined. The importance of this test cannot be understated, because it provides the owner and inspector with the concentration of the solution in the system. Certain concentrations of antifreeze have been proven in full-scale testing to significantly increase the heat release rate of a fire.

Antifreeze Testing. Obtaining an Antifreeze Sample by Draining a Small Quantity of Antifreeze from the Antifreeze System (top). Applying the Antifreeze Test Sample to a Refractometer (middle). Reading the Refractometric Index of the Antifreeze Sample by Holding the Refractometer Up to a Natural Light Source (bottom).

Antifreeze Loop. (Courtesy of Josh Elvove)

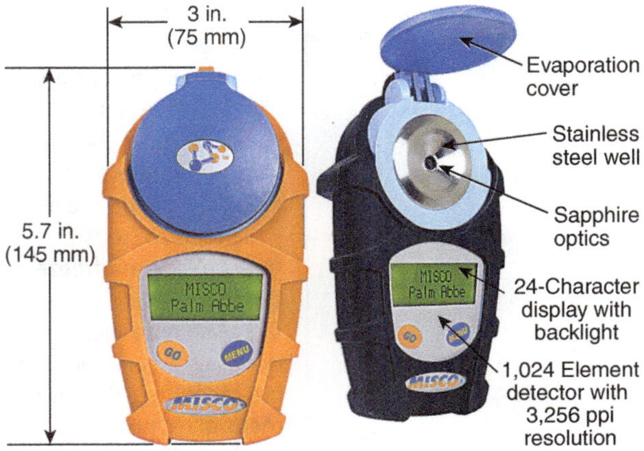

Digital Refractometer. (Courtesy of MISCO Refractometer, Cleveland, OH)

determining the minimum reasonable air temperature. In situations where the piping containing the antifreeze solution is protected in some way from exposure to the outside air, higher minimum temperatures can be anticipated.

Where systems are drained in order to be refilled, it is not typically necessary to drain drops. Most systems with drops have insufficient volume to cause a problem, even if slightly higher concentration solutions collect in the drops. For drops in excess of 36 in. (915 mm), consideration should be given to draining drops if there is evidence that unacceptably high concentrations of antifreeze have collected in these long drops.

When emptying and refilling antifreeze solutions, every attempt should be made to recycle the old solution with the antifreeze manufacturer rather than discard it.

•

5.3.4.1 The use of antifreeze solutions shall be in conformity with state and local health regulations.

5.3.4.1.1* Listed CPVC sprinkler pipe and fittings shall be protected from freezing with glycerine only.

A.5.3.4.1.1 Where inspecting antifreeze systems employing listed CPVC piping, the solution should be verified to be glycerine based.

5.3.4.1.1.1 The use of diethylene, ethylene, or propylene glycols shall be specifically prohibited.

Case In Point

As mentioned in the commentary associated with 5.3.4 and A.5.3.4, ascertaining the correct antifreeze concentration through a test sample is critical, as 5.3.4.2.1 provides several thresholds based on the determined concentration.

First, this requirement identifies that solutions with concentrations of antifreeze in excess of 40 percent propylene glycol and 50 percent glycerine should never be permitted in a sprinkler system. This concept is supported by full-scale testing using both residential and standard spray sprinklers where it was found that concentrations in excess of these values were prone to large scale ignition under certain conditions, such as pressure, ceiling height, or K-factor.

Secondly, antifreeze solutions at or below 30 percent propylene glycol and 38 percent glycerine are permitted to remain in the system, as testing summarized in an FM Technical Report, "K-25 Suppression Mode Sprinkler Protection for Areas Subject to Freezing," shows that these solutions are diluted to the point where no increase in heat release rate was noted when the solution was discharged onto a fire.

These two thresholds leave a gray area between 30 percent and 40 percent propylene glycol and 38 percent and 50 percent glycerine where in some cases these solutions might be acceptable and in others they might not be. The determination of whether or not a particular solution within these ranges is acceptable would be made by the AHJ, based on the data provided in a deterministic risk. The term *deterministic* is important because it requires the individual preparing the assessment to assume there is a fire. If a risk assessment was conducted in which the author stated that, "There won't be a fire, so I don't have to worry about the concentration of the antifreeze solution," then it would not be considered a deterministic approach and, therefore, would not be valid.

Also note that the technical committee has approved language that would phase out traditional antifreeze solutions by 2022. This allows combustible solutions to be used in the interim, but recognizes that using a combustible liquid in a life safety system is not ideal. This time allotment gives property owners time to budget for system modifications that consider alternate methods of freeze protection and also allows the manufacturers time to develop listed noncombustible solutions.

5.3.4.2 Except as permitted by 5.3.4.2.1 and 5.3.4.2.2, all antifreeze systems shall utilize listed antifreeze solutions.

The 2013 editions of NFPA 13 and NFPA 13R, *Standard for the Installation of Sprinkler Systems in Low-Rise Residential Occupancies,* allow the use of only listed antifreeze solutions. This is due to the unpredictability of traditional antifreeze solutions, as identified in the full-scale testing presented in "Antifreeze Solutions Supplied through Spray Sprinklers: Final Report," from the Fire Protection Research Foundation (FPRF). NFPA 25 also adopted this philosophy as a baseline requirement, but it contains several exceptions for existing systems, as identified in 5.3.4.2.1.

5.3.4.2.1* For systems installed prior to September 30, 2012, listed antifreeze solutions shall not be required until September 30, 2022, where all of the following conditions are met:

(1)* The concentration of the antifreeze solution shall be limited to 50 percent glycerine by volume or 40 percent propylene glycol by volume.
(2) Newly introduced solutions shall be factory premixed antifreeze solutions (chemically pure or United States Pharmacopeia 96.5 percent).
(3)* Antifreeze systems with concentrations in excess of 30 percent propylene glycol and 38 percent glycerine shall be permitted based upon an approved deterministic risk assessment prepared by a qualified person approved by the authority having jurisdiction.

A.5.3.4.2.1 All antifreeze systems installed after September 30, 2012, are assumed to meet the minimum requirements of NFPA 13, 2013 edition. For systems installed after September 30, 2012, that do not meet the requirements of the 2013 edition of NFPA 13, consideration should be given to applying 5.3.4.2.1.

A.5.3.4.2.1(1) The use of factory premixed solutions is required because solutions that are not mixed properly have a possibility of separating from the water, allowing the pure concentrate (which is heavier than water) to drop out of solution and collect in drops or low points of the system. Such concentrations are combustible and could present problems during fires. The properties of glycerine are shown in Table A.5.3.4.2.1(1).

TABLE A.5.3.4.2.1(1) *Properties of Glycerine and Propylene Glycol*

Material	*Solution (% by Volume)*	*Specific Gravity at 77°F (25°C)*	*Freezing Point °F*	*°C*
Glycerine (C.P. or U.S.P. grade)	0	1.000	32	0
	5	1.014	31	−0.5
	10	1.029	28	−2.2
	15	1.043	25	−3.9
	20	1.059	20	−6.7
	25	1.071	16	−8.9
	30	1.087	10	−12
	35	1.100	4	−15.5
	40	1.114	−2	−19
	45	1.130	−11	−24
	50	1.141	−19	−28
Propylene glycol	0	1.000	32	0
	5	1.004	26	−3
	10	1.008	25	−4
	15	1.012	22	−6

(continues)

TABLE A.5.3.4.2.1(1) Continued

Material	Solution (% by Volume)	Specific Gravity at 77°F (25°C)	Freezing Point °F	°C
	20	1.016	19	−7
	25	1.020	15	−10
	30	1.024	11	−12
	35	1.028	2	−17
	40	1.032	−6	−21

The freeze points provided in Table A.5.3.4.2.1(1) might not be appropriate for all manufactured solutions at the concentrations noted. The installer should confirm the amount of freeze protection that the selected solution provides and not simply assume that it will line up with the values in these tables.

A.5.3.4.2.1(3) Antifreeze solutions with a maximum concentration of 38 percent glycerine or 30 percent propylene glycol do not require a deterministic hazard analysis. The risk assessment should be prepared by individual(s) who can demonstrate an ability to prepare a risk assessment by education and experience and who can demonstrate an understanding of the issues associated with antifreeze sprinkler systems, including the available related fire tests. For additional information regarding the risk assessment process, documentation to be submitted, and the AHJ's role, refer to NFPA 551, *Guide for the Evaluation of Fire Risk Assessments,* and the SFPE *Engineering Guide: Fire Risk Assessment.*

Propylene glycol and glycerine antifreeze solutions discharged from sprinklers have the potential to ignite under certain conditions. Research testing has indicated that several variables may influence the potential for large-scale ignition of the antifreeze solution discharged from a sprinkler. These variables include, but are not limited to, the concentration of antifreeze solution, sprinkler discharge characteristics, inlet pressure at the sprinkler, ceiling height, and size of fire at the time of sprinkler discharge. All relevant data and information should be carefully reviewed and considered in the deterministic risk assessment. As appropriate, the risk assessment should consider factors such as the following:

(1) Occupancy use group per NFPA 13
(2) Ceiling height
(3) Antifreeze solution concentration and type
(4) Maximum system pressure (normal static pressures)
(5) Sprinkler type, including K-factor
(6) Potential and actual fuel load (Christmas trees)
(7) Type of structure (construction types)
(8) Size of structure
(9) Ability of the sprinkler system to control the fire
(10) Occupied spaces versus unoccupied spaces such as trash enclosures and dust collectors as follows:
    (a) Adjacent occupancies (spaces adjacent to the area protected by antifreeze systems)
    (b) Separation between areas protected with an antifreeze system and other areas
    (c) Ventilation of areas protected with an antifreeze system to prevent damage to adjacent areas
    (d) Duration of antifreeze discharge

Tests summarized in Table A.5.3.4.2.1(3) show that large-scale ignition of the sprinkler spray did not occur in tests with 50 percent glycerine and 40 percent propylene glycol antifreeze solutions discharging onto a fire having a nominal heat release rate (HRR) of 1.4 MW.

A deterministic risk assessment that demonstrates that the heat release rate for reasonably credible fire scenarios will be less than 1.4 MW at the time of sprinkler activation should be acceptable. The risk assessment should also address issues associated with management of change, such as change in occupancy and temporary fuel loads. A natural Christmas tree can result in an HRR well above 1.4 MW at the time of sprinkler activation. In addition to the variables identified previously, the deterministic risk assessment should include occupancy, quantity of solution, impact on life safety, and potential increase in heat release rate.

The following is a list of research reports that have been issued by the Fire Protection Research Foundation (FPRF) related to the use of antifreeze in sprinkler systems that should be considered in the development of the deterministic risk assessment:

(1) *Antifreeze Systems in Home Fire Sprinkler Systems — Literature Review and Research Plan,* Fire Protection Research Foundation, June 2010.
(2) *Antifreeze Systems in Home Fire Sprinkler Systems — Phase II Final Report,* Fire Protection Research Foundation, December 2010.
(3) *Antifreeze Solutions Supplied through Spray Sprinklers — Interim Report,* Fire Protection Research Foundation, February 2012.

Table A.5.3.4.2.1(3) provides an overview of the testing conducted by the FPRF.

TABLE A.5.3.4.2.1(3) *FPRF Testing Summary*

Topic	*Information*
Scope of sprinklers tested	The following sprinklers were used during the residential sprinkler research program described in the report dated December 2010: (1) Residential pendent style having nominal K-factors of 3.1, 4.9, and 7.4 gpm/psi$^{1/2}$ (2) Residential concealed pendent style having a nominal K-factor of 4.9 gpm/psi$^{1/2}$ (3) Residential sidewall style having nominal K-factors of 4.2 and 5.5 gpm/psi$^{1/2}$ The following sprinklers were used during the spray sprinkler research program described in the report dated February 2012: (1) Residential pendent style having a nominal K-factor of 3.1 gpm/psi$^{1/2}$ (2) Standard spray pendent style having nominal K-factors of 2.8, 4.2, 5.6, and 8.0 gpm/psi$^{1/2}$ (3) Standard spray concealed pendent style having a nominal K-factor of 5.6 gpm/psi$^{1/2}$ (4) Standard spray upright style having a nominal K-factor of 5.6 gpm/psi$^{1/2}$ (5) Standard spray extended coverage pendent style having a nominal K-factor of 5.6 gpm/psi$^{1/2}$
Antifreeze solution concentration	<50% glycerine and <40% propylene glycol antifreeze solutions — solutions were not tested. 50% glycerine and 40% propylene glycol antifreeze solutions — large-scale ignition of the sprinkler spray did not occur in tests with sprinkler discharge onto a fire having a nominal heat release rate (HRR) of 1.4 MW. Large-scale ignition of the sprinkler spray occurred in multiple tests with sprinkler discharge onto a fire having a nominal HRR of 3.0 MW.

(continues)

TABLE A.5.3.4.2.1(3) Continued

Topic	*Information*
	55% glycerine and 45% propylene glycol antifreeze solutions — large-scale ignition of the sprinkler spray occurred in tests with sprinkler discharge onto a fire having a nominal HRR of 1.4 MW. >55% glycerine and >45% propylene glycol antifreeze solutions — large-scale ignition of the sprinkler spray occurred in tests with sprinkler discharge onto a fire having an HRR of less than 500 kW. 70% glycerine and 60% propylene glycol antifreeze solutions — maximum antifreeze solution concentrations tested.
Sprinkler inlet pressure	Large-scale ignition of the sprinkler discharge spray was not observed when the sprinkler inlet pressure was 50 psi or less for tests using 50% glycerine or 40% propylene glycol.
Ceiling height	When discharging 50% glycerine and 40% propylene glycol antifreeze solutions onto fires having an HRR of 1.4 MW, no large-scale ignition of the sprinkler spray was observed with ceiling heights up to 20 ft. When discharging 50% glycerine and 40% propylene glycol antifreeze solutions onto fires having an HRR of 3.0 MW, large-scale ignition of the sprinkler spray was observed at a ceiling height of 20 ft.
Fire control	The test results described in the test reports of December 2010 and February 2012 indicated that discharging glycerine and propylene glycol antifreeze solutions onto a fire can temporarily increase the fire size until water is discharged. As a part of the residential sprinkler research described in report dated December 2010, tests were conducted to evaluate the effectiveness of residential sprinklers to control fires involving furniture and simulated furniture. The results of these tests indicated that 50% glycerine and 40% propylene glycol antifreeze solutions demonstrated the ability to control the furniture-type fires in a manner similar to water. For standard spray–type sprinklers, no tests were conducted to investigate the ability of these sprinklers to control the types and sizes of fires that these sprinklers are intended to protect.

The deterministic risk assessment should address, at a minimum, the items listed in A.5.3.4.2.1(3). The results of the testing conducted by the FPRF should be reviewed in preparing and reviewing the deterministic risk assessments. These reports provide a plethora of information on how antifreeze solutions can contribute to a fire under certain combinations of variables.

5.3.4.2.2 Premixed antifreeze solutions of propylene glycol exceeding 30 percent concentration by volume shall be permitted for use with ESFR sprinklers where the ESFR sprinklers are listed for such use in a specific application.

5.3.4.3 The antifreeze solution shall be tested at its most remote portion and where it interfaces with the wet pipe system.

5.3.4.4 When antifreeze systems have a capacity larger than 150 gal (568 L), tests at one additional point for every 100 gal (379 L) shall be made.

5.3.4.4.1 If the results indicate an incorrect freeze point at any point in the system, the system shall be drained and refilled with new premixed antifreeze.

5.3.4.4.2 For premixed solutions, the manufacturer's instructions shall be permitted to be used with regard to the number of test points and refill procedure.

5.4 Maintenance

5.4.1 Sprinklers.

5.4.1.1 Where a sprinkler has been removed for any reason, it shall not be reinstalled.

FAQ Can a building owner reinstall an existing sprinkler after removing the sprinkler and/or the drop nipple from the branch piping for maintenance or repairs?

Paragraph 5.4.1.1 prohibits reinstalling any sprinklers removed for any reason. Some installing contractors interpret this to mean that if the sprinkler drop assembly is removed without removing the sprinkler from the drop, it can be re-used. This section and the related section of the 2013 edition of NFPA 13 are intended to require that new sprinklers be installed because of concerns for damage to the device when handled during the disassembly. The intent is not to allow sprinklers to be removed and then reinstalled under any circumstances.

5.4.1.2* Replacement sprinklers shall have the proper characteristics for the application intended, which include the following:

(1) Style
(2) Orifice size and K-factor
(3) Temperature rating
(4) Coating, if any
(5) Deflector type (e.g., upright, pendent, sidewall)
(6) Design requirements

There are many types of sprinklers, a number of which have a similar appearance yet have different performance characteristics. Paragraph 5.4.1.2 identifies sprinkler characteristics that must be matched to the application.

Care must be taken to ensure that replacement sprinklers have the same characteristics as those they are replacing so that the design performance of the system is maintained. This is especially important for special sprinklers, as noted in 5.4.1.3. When the same make and model of a special sprinkler is no longer manufactured, persons with appropriate expertise and knowledge should be consulted when selecting a replacement sprinkler.

FAQ Must a sprinkler be replaced with a sprinkler having the exact temperature rating?

The replacement sprinkler is not required to have the exact temperature rating of the sprinkler being replaced, but it should be of the temperature range appropriate for the environment installed, based on Table 6.2.5.1 of NFPA 13.

Critical Point

Care must be taken to ensure that replacement sprinklers have the same characteristics as those they are replacing so that the design performance of the system is maintained.

A.5.4.1.2 To help in the replacement of like sprinklers, unique sprinkler identification numbers (SINs) are provided on all sprinklers manufactured after January 1, 2001. The SIN accounts for differences in orifice size, deflector characteristics, pressure rating, and thermal sensitivity.

5.4.1.2.1* Spray sprinklers shall be permitted to replace old-style sprinklers.

A.5.4.1.2.1 Old-style sprinklers are permitted to replace existing old-style sprinklers. Old-style sprinklers should not be used to replace standard sprinklers without a complete engineering review of the system. The old-style sprinkler is the type manufactured before 1953. It discharges approximately 40 percent of the water upward to the ceiling, and it can be installed in either the upright or pendent position.

5.4.1.2.2 Replacement sprinklers for piers and wharves shall comply with NFPA 307, *Standard for the Construction and Fire Protection of Marine Terminals, Piers, and Wharves.*

5.4.1.3 Only new, listed sprinklers shall be used to replace existing sprinklers.

5.4.1.4* Special and quick-response sprinklers as defined by NFPA 13, *Standard for the Installation of Sprinkler Systems*, shall be replaced with sprinklers of the same orifice, size, temperature range and thermal response characteristics, and K-factor.

A.5.4.1.4 It is imperative that any replacement sprinkler have the same characteristics as the sprinkler being replaced. If the same temperature range, response characteristics, spacing requirements, flow rates, and K-factors cannot be obtained, a sprinkler with similar characteristics should be used, and the system should be evaluated to verify the sprinkler is appropriate for the intended use. With regard to response characteristics, matching identical response time index (RTI) and conductivity factors are not necessary unless special design considerations are given for those specific values.

Prior to the 2008 edition of NFPA 25, the standard required that sprinklers be replaced by a sprinkler of the exact make and model, in addition to having the same performance characteristics. This is impractical, so the standard now simply requires that sprinklers be replaced with types having equal performance characteristics, such as orifice size, K-factor, temperature rating, and thermal response.

5.4.1.5* A supply of at least six spare sprinklers shall be maintained on the premises so that any sprinklers that have operated or been damaged in any way can be promptly replaced.

A.5.4.1.5 A minimum of two sprinklers of each type and temperature rating installed should be provided.

ITM Deficiency, Impairment, or Hazard Evaluation?

(Courtesy of Byron Blake and SimplexGrinnell)

ANSWER: ITM Deficiency

The spare sprinkler cabinet shown in this photo is required to be inspected annually to meet the requirements of 5.2.1.4. In this case, the cabinet is missing almost all of the spare parts required by 5.4.1.5, 5.4.1.5.5, and 5.4.1.5.6. A spare stock of sprinklers must include a minimum of six sprinklers for any protected facility, and at least two of each type and temperature rating is to be provided. A proper sprinkler wrench is also required, as well as a list of the sprinklers installed in the property.

The spare sprinkler requirements in the 2014 edition of NFPA 25 were correlated with the spare sprinkler requirements in the 2013 editions of NFPA 13 and NFPA 13R. Now all three documents have the exact same requirements.

This cabinet shown is missing the list of sprinklers installed in the property, so the most recent set of working drawings required by NFPA 13 will need to be reviewed to determine how many of each type of sprinkler is installed in the building or facility. In the absence of adequate working drawings, a complete visual examination of all of the sprinklers installed in the property must be performed, including those in concealed spaces and above ceilings.

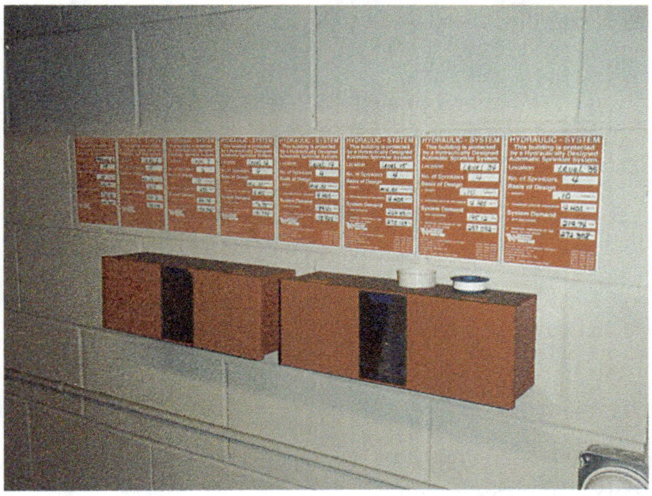

EXHIBIT 5.41 Sprinkler Cabinet with Hydraulic Information Signs. (Courtesy of Wayne Automatic Fire Sprinkler, Inc.)

Following an incident where a sprinkler has operated, either due to a fire or mechanical damage, it is important to minimize system impairments. Thus, 5.4.1.5 specifies that a supply of spare sprinklers must be available so that sprinklers can be replaced quickly following a small fire or accidental discharge. Note that the supply must be "on the premises" but need not be in each building of a complex or specifically at the sprinkler system riser. For example, a supply of spare sprinklers for a complex can be stored in a central location, provided that the supply is accessible and does not substantially delay the replacement of sprinklers following an incident. (A sprinkler cabinet with spare sprinklers is shown in Exhibit 5.24.)

5.4.1.5.1 The sprinklers shall correspond to the types and temperature ratings of the sprinklers in the property.

5.4.1.5.2 The sprinklers shall be kept in a cabinet located where the temperature in which they are subjected will at no time exceed 100°F (38°C).

Spare sprinklers must be maintained at normal temperatures to prevent their operation and also to prevent the phenomenon of "cold flow," which can result in premature operation. The 100°F (38°C) maximum temperature specified in 5.4.1.5.2 is a safe storage temperature for sprinklers of all temperature ratings. Exhibit 5.41 shows a series of well-maintained spare sprinkler cabinets with hydraulic information signs posted above.

5.4.1.5.3 Where dry sprinklers of different lengths are installed, spare dry sprinklers shall not be required, provided that a means of returning the system to service is furnished.

Dry sprinklers are manufactured to specific lengths. Due to the pitching of pipe or various ceiling heights, the dry sprinklers on a system are often of varying lengths, so it is impractical to have spare sprinklers for each specific length. However, if all the dry sprinklers on a system are the same length, spare dry sprinklers should be kept in the cabinet. For example, freezer boxes with dry sprinklers supplied by a wet system can all be the same length, and it is practical and reasonable to provide the spare dry sprinklers.

5.4.1.5.4 The stock of spare sprinklers shall include all types and ratings installed and shall be as follows:

(1) For protected facilities having under 300 sprinklers — no fewer than 6 sprinklers
(2) For protected facilities having 300 to 1000 sprinklers — no fewer than 12 sprinklers
(3) For protected facilities having over 1000 sprinklers — no fewer than 24 sprinklers

5.4.1.5.5* One sprinkler wrench as specified by the sprinkler manufacturer shall be provided in the cabinet for each type of sprinkler installed to be used for the removal and installation of sprinklers in the system.

A.5.4.1.5.5 One sprinkler wrench design can be appropriate for many types of sprinklers, and multiple wrenches of the same design should not be required.

Special wrenches as prescribed by the manufacturer must be kept in the sprinkler cabinet so that sprinklers can be properly replaced following an incident or in the event that inspection or testing indicates damage. There should be an appropriate wrench provided for each sprinkler type installed in the system.

5.4.1.5.6 A list of the sprinklers installed in the property shall be posted in the sprinkler cabinet.

5.4.1.5.6.1* The list shall include the following:

(1) Sprinkler identification number (SIN) if equipped; or the manufacturer, model, orifice, deflector type, thermal sensitivity, and pressure rating
(2) General description
(3) Quantity of each type to be contained in the cabinet
(4) Issue or revision date of the list

Starting with the 2007 edition, NFPA 13 has required a list of the sprinklers installed on the premises to be posted in the sprinkler cabinet. The list is required to include the Sprinkler Identification Number (SIN), a general description, the quantity of each sprinkler to be maintained in the cabinet, and the date of the list. It would be prudent for an owner to provide this list for projects completed under earlier editions of NFPA 13. The information can be found on the Contractor's Material Test Certificate or the as-built drawings that are part of the original records. If these records are not available, a survey of the installed sprinklers could be conducted by a qualified individual.

A.5.4.1.5.6.1 The minimum information in the list contained in the spare sprinkler cabinet should be marked with the following:

(1) General description of the sprinkler, including upright, pendent, residential, ESFR, and so forth
(2) Quantity of sprinklers to be maintained in the spare sprinkler cabinet

An example of the list is shown in Figure A.5.4.1.5.6.1.

Sprinklers Contained in this Cabinet			
Sprinkler Identification, SIN	General Description	Temperature Rating, °F	Sprinkler Quantity Maintained
TY9128	Extended Coverage, K-25, upright	155	6
VK425	Concealed pendent residential	145	6
Issued: 10/3/05	Revised:		

FIGURE A.5.4.1.5.6.1 *Sample List.*

5.4.1.6* Sprinklers shall not be altered in any respect or have any type of ornamentation, paint, or coatings applied after shipment from the place of manufacture.

A.5.4.1.6 Corrosion-resistant or specially coated sprinklers should be installed in locations where chemicals, moisture, or other corrosive vapors exist.

5.4.1.7 Sprinklers and automatic spray nozzles used for protecting commercial-type cooking equipment and ventilating systems shall be replaced annually.

5.4.1.7.1 Where automatic bulb-type sprinklers or spray nozzles are used and annual examination shows no buildup of grease or other material on the sprinklers or spray nozzles, such sprinklers and spray nozzles shall not be required to be replaced.

5.4.1.8 Protective Coverings.

5.4.1.8.1* Sprinklers protecting spray areas and mixing rooms in resin application areas installed with protective coverings shall continue to be protected against overspray residue so that they will operate in the event of fire.

As noted in the commentary following A.5.2.1.1.2(5), paint or overspray can prevent the proper operation of sprinklers. Paragraph 5.4.1.8 requires sprinklers to be protected from such overspray.

A.5.4.1.8.1 Typical sandwich bags purchased in a grocery store are generally plastic, not cellophane. Plastic bags have a tendency to shrink and adhere to the sprinkler prior to sprinkler activation, creating the potential for disruption of sprinkler spray patterns. Bags placed over sprinklers need to be true cellophane or paper.

5.4.1.8.2 Sprinklers installed as described in 5.4.1.8.1 shall be protected using cellophane bags having a thickness of 0.003 in. (0.076 mm) or less or thin paper bags.

Testing has shown that lightweight cellophane or paper bags will not adversely affect the operation of the sprinkler. However, sprinklers protected by lightweight cellophane or paper bags might require more frequent inspection than the annual inspection outlined in 5.2.1.1, to prevent excessive buildup on the bags. Depending on the use of the spray coating area, the inspection and subsequent replacement of the bags might need to be done daily. The use of plastic bags is not permitted due to concerns for the potential of a plastic bag to shrink prior to sprinkler activation and disrupt the discharge pattern. Exhibit 5.42 illustrates a sprinkler protected from overspray by a cellophane bag.

EXHIBIT 5.42 *Sprinkler Protected from Overspray with a Cellophane Bag.*

5.4.1.8.3 Coverings shall be replaced periodically so that heavy deposits of residue do not accumulate.

5.4.2* Dry Pipe Systems. Dry pipe systems shall be kept dry at all times.

A ruptured dry system as the result of water accumulating or left in the system is one of the most frequent and expensive consequences of an owner failing to enact a regular maintenance program. NFPA 25 addresses this with guidance on the use and frequency of operating dry auxiliary drains (see A.13.4.4.3.2).

A.5.4.2 Conversion of dry pipe systems to wet pipe systems on a seasonal basis causes corrosion and accumulation of foreign matter in the pipe system and loss of alarm service.

Moisture is one of the main causes of corrosion in sprinkler piping. Alternating from a dry system to a wet system allows moisture to enter the system and can accelerate corrosion rates. In addition, the waterflow alarm on most dry systems will need to be removed from service if the system is left wet. To minimize corrosion and prevent freezing of sprinkler piping, it is required that these systems always remain dry, and it is recommended that they be reset and drained following operation to minimize moisture inside the system. Exhibit 5.43 shows rust and scale removed from a dry pipe system.

EXHIBIT 5.43 Rust and Scale Removed from Dry Pipe System.

5.4.2.1 During nonfreezing weather, a dry pipe system shall be permitted to be left wet if the only other option is to remove the system from service while waiting for parts or during repair activities.

5.4.2.2 Refrigerated spaces or other areas within the building interior where temperatures are maintained at or below 40°F (4.0°C) shall not be permitted to be left wet.

5.4.2.3 Air driers shall be maintained in accordance with the manufacturer's instructions.

5.4.2.4 Compressors used in conjunction with dry pipe sprinkler systems shall be maintained in accordance with the manufacturer's instructions.

5.4.3* Marine Systems. Sprinkler systems that are normally maintained using fresh water as a source shall be drained and refilled, then drained and refilled again with fresh water following the introduction of raw water into the system.

A.5.4.3 Certain sprinkler systems, such as those installed aboard ships, are maintained under pressure by a small freshwater supply but are supplied by a raw water source following system activation. In these systems, the effects of raw water are minimized by draining and refilling with freshwater. For systems on ships, flushing within 45 days or the vessel's next port of call, whichever is longer, is considered acceptable.

5.5 Component Action Requirements

Component replacement tables offer guidance when system components are adjusted, repaired, rebuilt, or replaced. It is not necessary in each case to require a complete acceptance test for each component when maintenance is performed.

5.5.1 Whenever a component in a sprinkler system is adjusted, repaired, reconditioned, or replaced, the actions required in Table 5.5.1 shall be preformed.

TABLE 5.5.1 *Summary of Component Replacement Action Requirements*

Component	*Adjust*	*Repair/ Recondition*	*Replace*	*Required Action*
Water Delivery Components				
Pipe and fittings affecting less than 20 sprinklers	X	X	X	Inspect for leaks at system working pressure
Pipe and fittings affecting more than 20 sprinklers	X	X	X	Hydrostatic test in conformance with NFPA 13, *Standard for the Installation of Sprinkler Systems*
Sprinklers, less than 20	X		X	Inspect for leaks at system working pressure
Sprinklers, more than 20	X		X	Hydrostatic test in conformance with NFPA 13
Fire department connections	X	X	X	See Chapter 13
Antifreeze solution	X		X	Inspect freezing point of solution Inspect for leaks at system working pressure
Alarm and Supervisory Components				
Vane-type waterflow	X	X	X	Operational test using inspector's test connection
Pressure switch–type waterflow	X	X	X	Operational test using the inspector's test connection or alarm bypass test valve
Water motor gong	X	X	X	Operational test using inspector's test connection

(continues)

TABLE 5.5.1 Continued

Component	*Adjust*	*Repair/ Recondition*	*Replace*	*Required Action*
High and low air pressure switch	X	X	X	Operational test of high and low settings
Valve supervisory signal initiating device	X	X	X	Test for conformance with NFPA 13 and/or *NFPA 72, National Fire Alarm and Signaling Code*
Detection system (for deluge or preaction system)	X	X	X	Operational test for conformance with NFPA 13 and/or *NFPA 72*
Status-Indicating Components				
Gauges			X	Verify at 0 bar (0 psi) and system working pressure
Testing and Maintenance Components				
Air compressor	X	X	X	Operational test for conformance with NFPA 13
Automatic air maintenance device	X	X	X	Operational test for conformance with NFPA 13
Main drain	X	X	X	Main drain test
Auxiliary drains	X	X	X	Inspect for leaks at system working pressure; main drain test
Inspector's test connection	X	X	X	Inspect for leaks at system working pressure; main drain test
Structural Components				
Hanger/seismic bracing	X	X	X	Inspect for conformance with NFPA 13
Pipe stands	X	X	X	Inspect for conformance with NFPA 13
Informational Components				
Identification signs	X	X	X	Inspect for conformance with NFPA 13 and this standard
Hydraulic design information sign	X	X	X	Inspect for conformance with NFPA 13 and this standard
General information sign	X	X	X	Inspect for conformance with this standard

•

5.5.2 Where the original installation standard is different from the cited standard, the use of the appropriate installing standard shall be permitted.

5.5.3 These actions shall not require a design review, which is outside the scope of this standard.

References Cited in Commentary

National Fire Protection Association, 1 Batterymarch Park, Quincy, MA 02169-7471.

NFPA 13, *Standard for the Installation of Sprinkler Systems*, 2013 edition.

NFPA 13R, *Standard for the Installation of Sprinkler Systems in Low-Rise Residential Occupancies*, 2013 edition.

NFPA 20, *Standard for the Installation of Stationary Pumps for Fire Protection*, 2013 edition.

Fire Protection Research Foundation, 1 Batterymarch Park, Quincy, MA 02169-7471.

"Antifreeze Solutions Supplied through Spray Sprinklers: Final Report," February 2012.

FM Global, 1301 Atwood Avenue, P.O. Box 7500, Johnston, RI 02919.

"K-25 Suppression Mode Sprinkler Protection for Areas Subject to Freezing," FM Technical Report J.L. 0003004619, 2010.

Antifreeze Solution Test Procedure

Purpose

The purpose of completing a test of the antifreeze solution is to determine that the concentration of the solution is within the appropriate range for the application across the limits of the installed system, including minimum freeze protection for the location and maximum concentration for the specific antifreeze solution being used. In order to complete this testing, the type of antifreeze solution within the system must be known.

Tools/Equipment

A suitable means to measure or determine the specific gravity of the antifreeze solution and the associated concentration by volume and freeze point for the solution is required. This might include the use of either a hydrometer or refractometer having a scale calibrated for the specific antifreeze solution being tested. These two types are devices are discussed below.

Hydrometer

Hydrometers are designed to measure the specific gravity of the solution by assessing the buoyancy of certain items within the solution. This can generally include a measurement of the number of balls or disks that float or the tilt of a swing arm immersed in a sample of solution drawn into a sampling chamber, as is common with "automotive" style hydrometers.

Refractometer

Refractometers measure the refraction or "bending" of light as it passes through the solution sample, which can then be converted to a refractive index that is used to accurately determine a percent by volume concentration and freeze point for the solution. The refractive index is also temperature dependent and must be corrected as well for the temperature of the solution. With the small sample size, the temperature is essentially that of the test equipment, since the sample almost immediately takes on the temperature of the much larger piece of test equipment. Fortunately, many of the available refractometers are provided with automatic temperature compensation, making this task automatic rather than having to make corrections manually.

Procedure Steps

1. Prior to any testing, notify the fire department and/or the alarm monitoring company, as well as the facility representatives, that testing is going to be conducted.
2. Determine the required temperature protection needed for the location. One method by which this might be accomplished for exterior system locations is through the use of the one-day mean low temperatures from Figure A.5.3.4 for the locality under consideration. A determination of the type of antifreeze solution within the system must first be made. This might be directly obtained from a variety of sources, including but not limited to the signage provided at the system riser, original installation records, maintenance records, or information provided by the owner. However, in many cases, information might not be available or the reliability of the information that is available cannot be confirmed. In these instances, a chemical analysis is necessary to determine the specific type of antifreeze solution within the system and to establish a baseline for future system testing. This analysis is best done by extracting and submitting a sample(s) for chemical analysis to a qualified testing laboratory.
3. After a determination of the type of antifreeze solution is made, the following is to be verified:
   a. For Early Suppression Fast-Response (ESFR) antifreeze systems, the ESFR sprinkler must be listed for use with the specific antifreeze solution used.
   b. For systems other than ESFR systems installed after September 29, 2012, the antifreeze system must use a listed antifreeze solution for the specific usage conditions at hand. Note that at the time of this writing, no such listed antifreeze solution is currently available.
   c. For chlorinated polyvinyl chloride (CPVC) piping systems installed prior to September 30, 2012, the antifreeze solution must only be glycerine.
   d. For all other piping systems installed prior to September 30, 2012, the antifreeze solution must be glycerine or propylene glycol. The use of diethylene glycol and ethylene glycol is strictly prohibited.

A determination of the use of an antifreeze solution, other than those allowable solution types specified herein, requires replacement of the antifreeze solution or other alternative means of freeze protection. In such case, no additional concentration testing of the solution need be conducted prior to corrective action being taken.

4. Test samples for evaluation of the antifreeze concentration must be collected from the following locations:
   a. Top (uppermost elevation) of the antifreeze system piping.
   b. Bottom (lowermost elevation) of the antifreeze system piping.
   c. If the physically most remote portion of the system is not located at either 4a or 4b above, then an additional sample is to be collected at this location.
   d. If the connection to the water supply piping is not located at either 4a or 4b above, then an additional sample is to be collected at this location.
   e. For systems that exceed a capacity of 150 gal (568 L), one additional test point per 100 gal (379 L) is to be sampled for each 100 gal (379 L) or portion thereof above 150 gal (568 L).

The collection of samples across the spatial range on the system piping installation is to ensure that the solution has not separated due to settlement or become diluted and that it is reasonably uniform in concentration throughout. The collection process at the indicated locations might require the removal of a fitting or sprinkler at certain locations to obtain the sample of the solution, or it might be obtained from locations such as the inspector's test connection, drain connections, or test ports at the antifreeze loop. Where an excessive quantity of antifreeze solution is removed, additional premixed antifreeze solution of the appropriate concentration might need to be added to the system through the fill cup prior to returning the system control valve to the open position. After any control valve closure, a valve status test must be completed downstream of the closed valve in accordance with 13.3.1.2.1.

5. Once collected, each of the samples can then be either field tested or, if being sent in for chemical analysis as part of Item 2 above, the concentration can be tested at the laboratory. Field testing can be completed by determining the physical properties of the solution using an appropriate hydrometer or refractometer having a scale calibrated for the specific antifreeze solution being tested.
    a. Hydrometer Method: Once the measurements for specific gravity are taken and corrected as required, the results should be recorded and a determination made as to the percent by volume concentration and freeze point available from the solution, with these results also being recorded. At a reference temperature of 77°F (25°C), the approximate percent by volume and freeze point can be determined using Table A.5.3.4.2.1(1). If the reference temperature for the hydrometer is different than 77°F (25°C), then an alternative source of the information must be used.
    b. Refractometer Method: Once the measurements for specific gravity and concentration by volume and/or freeze point are taken and corrected if necessary for temperature, the results should be recorded. If not directly provided by the device, a determination must be made as to the percent by volume concentration and freeze point available from the solution, with these results also being recorded. At a reference temperature of 77°F (25°C), the approximate percent by volume and freeze point can be determined using Table A.5.3.4.2.1(1). If the reference temperature for the hydrometer is different than 77°F (25°C), then an alternative source of the information must be used.
6. Upon completion of all testing, notify the fire department and/or the alarm monitoring company, as well as the facility representatives, that testing is complete, and reset fire alarm system as necessary.

Evaluation of Results

Where a determination is made that an acceptable level of freeze protection cannot be obtained without exceeding the maximum concentrations of 40 percent by volume for propylene glycol [-6°F (-21°C)] or 50 percent by volume for glycerine [-19°F (-28°C)] then these antifreeze solutions are not a viable option and an alternative means of freeze protection must be provided. These temperature limits do not apply to listed ESFR systems using antifreeze, which must comply with the limits of their listing.

Where a determination is made that an inappropriate antifreeze solution type has been used within the system or where the content cannot be determined, the solution must be replaced or an alternative means of freeze protection must be provided. This could include the use of an antifreeze solution with an ESFR system for which it is not listed, the use of a non-listed antifreeze system that is installed after September 29, 2012, the use of other than a glycerine solution within a CPVC piping system installed prior to September 30, 2012, or the use of other than a propylene glycol or glycerine solution in other piping systems installed prior to September 30, 2012.

Assess the freezing point for the solution collected from all of the sampled locations to determine if they are sufficient for protection against the anticipated low temperature for the location. Where the freeze protection for any of the samples is inadequate, the system must be drained and refilled with a proper concentration to protect the system, using a factory premixed antifreeze solution appropriate for the system type. An exception to this would be where the required concentration for freeze protection exceeds 40 percent by volume for propylene glycol or 50 percent by volume for glycerine; then the use of antifreeze is not a viable solution and an alternative means of freeze protection must be provided. Where the required concentration for freeze protection is greater than 30 percent by volume for propylene glycol or 38 percent by volume for glycerine but less than the values above, the facility must have a deterministic risk analysis conducted by a qualified individual to demonstrate an acceptable risk or an alternative means of freeze protection must be provided.

Assess the maximum concentration of the solution by volume collected from all sampled locations. Where any of the samples has a concentration greater than 40 percent by volume for propylene glycol or 50 percent by volume for glycerine, the use of antifreeze solution must be drained and refilled with the proper concentration required for freeze protection. The exception would be where the required concentration for freeze protection exceeds these values; then the use of antifreeze is not a viable solution requiring the use of an alternative means of freeze protection. Additionally, where the required concentration for freeze protection is greater than 30 percent by volume for propylene glycol or 38 percent by volume for glycerine but less than the values above, the facility must have a deterministic risk analysis conducted by a qualified individual to demonstrate an acceptable risk or an alternative means of freeze protection must be provided.

The methods and practices outlined above represent only one way of complying with the test procedure required by 5.3.4 in NFPA 25.

ITM NFPA 25 — INSPECTION TESTING MAINTENANCE

6

STANDPIPE AND HOSE SYSTEMS

Chapter 6 of NFPA 25 covers the inspection, testing, and maintenance of standpipe systems. The operating condition of standpipe systems is important because standpipe systems are critical equipment in buildings where fire-fighting operations must be conducted internally due to the size or height of the building. A standpipe failure can have catastrophic consequences to both building occupants and first responders who rely on the system for their safety.

Standpipe systems may or may not be equipped with hose and nozzles. When hose and nozzles are part of the system, they must be tested and maintained as required by NFPA 1962, *Standard for the Care, Use, Inspection, Service Testing, and Replacement of Fire Hose, Couplings, Nozzles, and Fire Hose Appliances,* along with Chapter 6 and Chapter 13 of this standard.

Exhibit 6.1 shows a properly racked hose and nozzle. This installation demonstrates a properly maintained standpipe system hose rack assembly.

EXHIBIT 6.1 *Hose Rack Assembly in Proper Condition. (Courtesy of Josh Elvove)*

6.1 General

Critical Point

For work beyond simple and basic inspections, a qualified inspection service or contractor should perform all testing and maintenance activities listed in Table 6.1.1.2.

Some of the fire protection systems inspections listed in Table 6.1.1.2 can be performed by someone with minimal training and in many cases can be conducted by the property owner or owner's representative. These inspections could include items such as the operating position of the control valve or obvious signs of pipe damage or leaking.

However, for work beyond simple and basic inspections, a qualified inspection service or contractor should perform all testing and maintenance activities listed in Table 6.1.1.2.

6.1.1 Minimum Requirements.

6.1.1.1 This chapter shall provide the minimum requirements for the routine inspection, testing, and maintenance of standpipe and hose systems.

6.1.1.2 Table 6.1.1.2 shall be used to determine the minimum required frequencies for inspection, testing, and maintenance.

6.1.2 Table 6.1.2 shall be used for the inspection, testing, and maintenance of all classes of standpipe and hose systems.

TABLE 6.1.1.2 *Summary of Standpipe and Hose Systems Inspection, Testing, and Maintenance*

Item	*Frequency*	*Reference*
Inspection		
Control valves		Table 13.1.1.2
Pressure-regulating devices		Table 13.1.1.2
Piping	Annually	6.2.1
Hose connections		Table 13.1.1.2]
Cabinet	Annually	NFPA 1962
Gauges	Weekly/quarterly	6.2.2
Hose	Annually	NFPA 1962
Hose storage device	Annually	NFPA 1962
Hose nozzle	Annually and after each use	NFPA 1962
Hydraulic design information sign	Annually	6.2.3
Hose valves		Table 13.1.1.2
Hose connection		Table 13.1.1.2
Test		
Waterflow alarm devices		Table 13.1.1.2
Valve supervisory devices		Table 13.1.1.2
Supervisory signal devices (except valve supervisory switches)		Table 13.1.1.2
Hose storage device	Annually	NFPA 1962
Hose	5 years/3 years	NFPA 1962
Pressure control valve		Table 13.1.1.2
Pressure-reducing valve		Table 13.1.1.2
Hydrostatic test	5 years	6.3.2
Flow test	5 years	6.3.1
Main drain test		Table 13.1.1.2
Hose valves		Table 13.1.1.2
Hose connections		Table 13.1.1.2
Valve status test		13.3.1.2.1
Maintenance		
Hose connections	Annually	Table 6.1.2
Valves (all types)	Annually/as needed	Table 13.1.1.2
Hose valves		Table 13.1.1.2

TABLE 6.1.2 Standpipe and Hose Systems

Component/Checkpoint	*Corrective Action*
Hose Connections	
Cap missing	Replace
Fire hose connection damaged	Repair
Valve handles missing	Replace
Cap gaskets missing or deteriorated	Replace
Valve leaking	Close or repair
Visible obstructions	Remove
Restricting device missing	Replace
Manual, semiautomatic, or dry standpipe — valve does not operate smoothly	Lubricate or repair
Piping	
Damaged piping	Repair
Control valves damaged	Repair or replace
Missing or damaged pipe support device	Repair or replace
Damaged supervisory signal initiating device	Repair or replace
Hose	
Inspect	Remove and inspect the hose, including gaskets, and rerack or rereel at intervals in accordance with NFPA 1962, *Standard for the Care, Use, Inspection, Service Testing, and Replacement of Fire Hose, Couplings, Nozzles, and Fire Hose Appliances*
Mildew, cuts, abrasions, and deterioration evident	Replace with listed lined, jacketed hose
Coupling damaged	Replace or repair
Gaskets missing or deteriorated	Replace
Incompatible threads on coupling	Replace or provide thread adapter
Hose not connected to hose rack nipple or valve	Connect
Hose test outdated	Retest or replace in accordance with NFPA 1962
Hose Nozzle	
Hose nozzle missing	Replace with listed nozzle
Gasket missing or deteriorated	Replace
Obstructions	Remove
Nozzle does not operate smoothly	Repair or replace
Hose Storage Device	
Difficult to operate	Repair or replace
Damaged	Repair or replace
Obstruction	Remove
Hose improperly racked or rolled	Remove
Nozzle clip in place and nozzle correctly contained?	Replace if necessary
If enclosed in cabinet, will hose rack swing out at least 90 degrees?	Repair or remove any obstructions
Cabinet	
Inspect overall condition for corroded or damaged parts	Repair or replace parts; replace entire cabinet if necessary
Difficult to open	Repair
Cabinet door will not open fully	Repair or move obstructions
Door glazing cracked or broken	Replace
If cabinet is break-glass type, is lock functioning properly?	Repair or replace
Glass break device missing or not attached	Replace or attach
Not properly identified as containing fire equipment	Provide identification
Visible obstructions	Remove
All valves, hose, nozzles, fire extinguisher, etc., easily accessible	Remove any material not related

NFPA 1962 provides the requirements for fire hose and must be used where standpipe systems contain hose. For example, small hose must be stored unracked, unreeled, or unrolled and inspected annually. In addition, the hose must be removed and service tested 5 years from the date it was manufactured and every 3 years thereafter. The inspection provider should have a copy of NFPA 1962 to be certain that the requirements are met.

6.1.3 Checkpoints and corrective actions outlined in Table 6.1.2 shall be followed to determine that components are free of corrosion, foreign material, physical damage, tampering, or other conditions that adversely affect system operation.

6.1.4 Valves and fire department connections shall be inspected, tested, and maintained in accordance with Chapter 13.

6.1.5 The procedures outlined in Chapter 14 shall be followed where there is a need to conduct an obstruction investigation.

6.1.6 Where the inspection, testing, and maintenance of standpipe and hose systems results or involves a system that is out of service, the impairment procedures outlined in Chapter 15 shall be followed.

6.1.7 Where approved by the authority having jurisdiction, existing hose shall be permitted to be removed and shall not be recorded as a deficiency.

Guidance on the installation and use of small hose in buildings has evolved over recent years. Fire department personnel are normally trained not to use the fire hose that is supplied in a building or as part of a standpipe, due to a lack of verified maintenance or testing of the appliances provided. As a result, most fire departments bring their own hose into a building and connect it to the standpipe hose valves. In addition, the universal guidance is for untrained building occupants not to use hose and nozzles in fire-fighting activities. As a result, it is left to the fire department to decide on the disposition of existing hose, and many fire departments are requiring or approving the removal of small hose from standpipe systems.

•

ITM NFPA 25

ITM Deficiency, Impairment, or Hazard Evaluation?

(Courtesy of Byron Blake and SimplexGrinnell)

ANSWER: ITM Deficiency

There are several problems with the 2½ in. (65 mm) hose valve shown in this picture, and two of them are NFPA 25 deficiencies. The missing handle on the hose valve is a deficiency as described in Table 6.1.2 and must be replaced. The pile of boxes in front of the hose valve is creating an obstruction and needs to be removed per Table 6.1.2. All versions of NFPA 25 require that standpipe hoses be inspected and tested in accordance with NFPA 1962. If the hose is missing from a hose rack as is shown here, the inspector should report the missing hose as a deficiency. However, in the 2011 edition of NFPA 25 a new subsection, 6.1.7, was added that allows hoses to be permanently removed where approved by the authority having jurisdiction (AHJ), and the absence of the hose would not be reported as a deficiency. This subsection, which is also in the current edition, recognizes that many local fire departments will not use an existing hose when responding to a fire emergency, so NFPA 25 does not require the building owner to maintain it. The owner does need to verify to the inspector that permission was given by the AHJ to remove the hose.

System Tagging

Exterior corrosion can be seen on the hose outlet threading and welded outlet. Most facility managers would simply clean the threading and joint with a wire brush and apply a fresh coat of paint. For inspectors who have no maintenance responsibilities, they might call this a noncritical deficiency due to the light corrosion and recommend a remediation plan.

Noncritical Deficiency

Critical Deficiency

Impairment

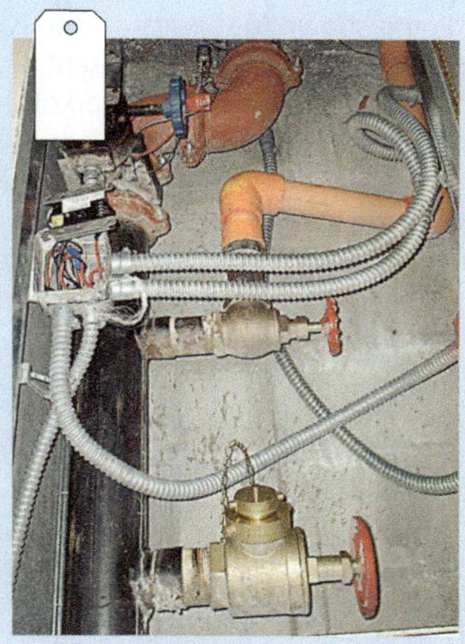

(Courtesy of Byron Blake and SimplexGrinnell)

6.2 Inspection

6.2.1 Components. Components of standpipe and hose systems shall be visually inspected annually or as specified in Table 6.1.1.2.

FAQ When were the inspection frequencies of components changed from quarterly to annually?

Inspection frequencies of piping and hose connections were changed from quarterly to annually in the 2008 edition, because the committee determined that this equipment does not require such rigorous inspections. The inspection frequencies for pressure-regulating devices and alarms, addressed in Chapter 13, were not changed because those devices require different inspection frequencies. Pressure-regulating devices are constantly working to assure that the inactive system components are not subject to conditions outside of the parameters of their design. For this reason, more frequent inspections are required for those system components as the impact of those inspections carry over to other system components.

6.2.2 Gauges.

6.2.2.1 Gauges on automatic wet and semiautomatic dry standpipe systems shall be inspected quarterly to ensure that they are in good condition and that normal water supply pressure is being maintained.

6.2.2.2 Gauges on automatic dry standpipe systems shall be inspected weekly to ensure that normal air or nitrogen and water pressure are being maintained.

6.2.2.3 Where air pressure supervision is connected to a constantly attended location, gauges shall be inspected monthly.

6.2.3* Hydraulic Design Information Sign. The hydraulic design information sign for standpipe systems shall be inspected annually to verify that it is provided, attached securely, and legible.

A hydraulic information sign is required by NFPA 14, *Standard for the Installation of Standpipe and Hose Systems,* regardless of whether the standpipe system piping is sized by hydraulic

Case In Point

The piping shown in the first photo is severely corroded and could fail if pressurized. Such a failure under pressure could present a safety hazard to the end user or to nearby building occupants. A simple visual observation of this system should reveal to the inspector that it has not been inspected for some time.

Even minor corrosion, which is caused by the interaction between dissimilar metals as shown in the second photo, should be noted when conducting an inspection, since this equipment is vital to fire department operations.

Severely Corroded Manual Dry Sandpipe.

Fire Hose Valve with Dissimilar Metals. (Courtesy of M. Steven Welsh, Koffel Associates, Inc.)

ITM Deficiency, Impairment, or Hazard Evaluation?

(Courtesy of Byron Blake and SimplexGrinnell)

ANSWER: ITM Deficiency

Table 6.1.1.2 does not include any reference for inspecting waterflow alarm devices. However, 6.2.1 requires all components of a standpipe system to be inspected annually. During the annual inspection, this waterflow switch was found without a cover; the wiring was disconnected, the terminal block was removed, and there was significant corrosion in the switch. The purpose of an inspection is to verify that the items being inspected appear in operating condition and are free of physical damage. In this case, the switch is missing the cover as well as other internal components and is not in operating condition. This condition should be considered a critical deficiency.

calculations or pipe schedule. This is not the case for sprinkler systems that are designed in accordance with NFPA 13, *Standard for the Installation of Sprinkler Systems*. The sign is required by NFPA 14 to be placed at the water control valve for automatic and semiautomatic systems or at an approved location for manual systems. NFPA 14 also requires that the following signs related to the system hydraulics be posted:

- A sign at the fire department connection indicating the pressure required at the inlets to deliver the system demand [**14:** 6.4.5.2.2]
- Where a fire pump is provided, a sign in the vicinity of the pump indicating the minimum pressure and flow required at the pump discharge flange to meet the system demand [**14:** 6.7]

The signs should be legible and securely attached.

A.6.2.3 The design information sign should be secured with durable wire, chain, or equivalent to the water supply control valve for automatic or semiautomatic standpipe systems and at an approved location for manual systems. See Figure A.6.2.3 for sample hydraulic information sign.

Location of the two hydraulically most remote hose connections: ____________

Design flow rate for the connections identified above: ____________

Design residual inlet and outlet pressures for the connections identified above: ____________

Design static pressure and design system demand (i.e., flow and residual pressure) at the system control valve, or at the pump discharge flange where a pump is installed, and at each fire department connection: ____________

FIGURE A.6.2.3 *Sample Hydraulic Sign. [**14:** Figure A.6.8]*

6.2.3.1 A hydraulic design information sign that is missing or illegible shall be replaced.

6.2.3.2 A standpipe system that was not sized by hydraulic design shall have a hydraulic design information sign that reads "Pipe Schedule System."

6.3 Testing

Where water damage is a possibility, an air test shall be conducted on the system at 25 psi (1.7 bar) prior to introducing water to the system.

Extreme care is needed when testing with pressurized air due to the greater amount of energy stored in compressed air that can be released if a portion of the system fails catastrophically. Pressures in excess of the 25 psi (1.7 bar) recommended should therefore be avoided for safety concerns. This section relates to standpipe systems that are normally dry, such as manual dry or semiautomatic dry standpipe systems. Operationally, manual systems rely on the fire pump provided by a fire engine or truck, with the fire fighter on the hose nozzle communicating with the engineer at the engine controls for the proper level of pressure regulation and flow during fire-fighting operations.

Since an automatic dry system is already pressurized with air, additional periodic air testing is not prescribed. Similarly, wet systems of all types do not require an air test, because leaks will be readily apparent in those systems.

Critical Point

Extreme care is needed when testing with pressurized air due to the greater amount of energy stored in compressed air that can be released if a portion of the system fails catastrophically.

6.3.1 Flow Tests.

Testing Procedure Alert

ITM NFPA 25

The purpose of conducting a flow test for the standpipe system is to ensure the adequacy of the water supply as well as the capability of the system to meet the original design flow and pressure criteria for the performance of the system. The design criteria to which the test data must be compared will vary depending on which edition of NFPA 14 the system was designed to; therefore, a preliminary assessment of the system is necessary. For more information on conducting the standpipe flow test, see the detailed testing procedure at the end of this chapter.

6.3.1.1* A flow test shall be conducted every 5 years on all Class I and Class III standpipe systems to verify that the required flow and pressure are available at the hydraulically most remote hose valve outlet(s) while flowing the standpipe system demand.

A.6.3.1.1 The hydraulically most remote hose connections in a building are generally at a roof manifold, if provided, or at the top of a stair leading to the roof. In a multizone system, the testing means is generally at a test header at grade or at a suction tank on higher floors.

NFPA 25 limits the 5-year flow test to automatic systems. This flow test is conducted to verify that the water supply for the system is still available and that all of the devices that could restrict flow in the standpipe are operating properly. These devices can include improperly installed or repaired check valves, gate valves, and pressure-regulating fire hose valves, or physical failures in the pipe network that might render the supply unavailable.

The test is intended to verify the design flow rate and pressure under functional flow conditions. Design pressure varies depending on the age of the system. For example, systems installed prior to 1993 should be tested to verify 65 psi (4.5 bar) at the topmost outlet. Systems installed after 1993 should be tested to verify 100 psi (6.9 bar) at the topmost outlet. It might be necessary to confirm the age of the system and what edition of NFPA 14 was in effect at the time of the original installation. Regardless of the pressure requirement, the test flow is based on a flow of 500 gpm (1892 L/min) for the most demanding riser and 250 gpm (946 L/min) for each additional riser up to a maximum of 1250 gpm (4731 L/min) for buildings that are partially or nonsprinklered, and 1000 gpm (3785 L/min) for buildings that are fully sprinklered. Note that NFPA 14 requires an additional outlet at the top of the most demanding standpipe for testing purposes. This requirement is intended to facilitate testing of the standpipe system at the design flow of 500 gpm (1892 L/min) [250 gpm × 2 = 500 gpm (946 L/min × 2 = 1892 L/min)]. The total required flow rate is required to be provided simultaneously for this test.

Case In Point

Usually, the most remote hose connections, as referred to in 6.3.1.1, are located on the roof or upper floors. Testing hose connections in such locations requires coordination to conduct the test safely, prevent water damage, and ensure proper water disposal. In some cases, special drain risers might be needed to remove the water generated by the test. If the test cannot be conducted at the most remote outlet, other arrangements can be made where acceptable to the AHJ. In cases where the drain riser provided is undersized to accept the full flow rates required, hoses can be run between two standpipes and arranged so that opposing standpipes can be used as the drain riser for the first. Where fire pumps are provided as a water supply for standpipes, the annual fire pump test required by Chapter 8 can often be accomplished at the same time as the standpipe's 5-year flow test.

It should be noted that special equipment might be required to measure both the flow rate (gpm or L/min) and the residual pressure (psi or bar) concurrently, especially where adjustments are needed for field adjustable pressure-regulating fire hose valves.

6.3.1.1.1 Where a flow test of the hydraulically most remote outlet(s) is not practical, the authority having jurisdiction shall be consulted for the appropriate location for the test.

6.3.1.2* The standpipe system demand shall include 500 gpm (1892 L/min) for the first standpipe and 250 gpm (946 L/min) for each additional standpipe until the total system demand is simultaneously flowing.

A.6.3.1.2 When the standpipe system was accepted, NFPA 14 required that each additional standpipe be flowed to simulate the hydraulic calculations. Typically, the lowest hose valve was used to create this simultaneous flow so hoses wouldn't have to be run all the way down each standpipe.

6.3.1.2.1* The 250 gpm (946 L/min) required from each additional standpipe shall be allowed to be flowed from the most convenient hose valve on that standpipe.

A.6.3.1.2.1 Since the pressures at each standpipe aren't required to be balanced by NFPA 14 or this standard, any hose valve on the standpipe can be flowed to achieve the additional 250 gpm (950 L/min) needed. It might be more convenient to use a hose valve on an upper level rather than the lowest one on the standpipe.

6.3.1.2.2* Where the 250 gpm (946 L/min) cannot be flowed from each additional standpipe, the authority having jurisdiction shall determine where the additional flow can be taken.

A.6.3.1.2.2 In some instances it isn't reasonable to attach a hose to a standpipe to provide this additional flow point. The authority having jurisdiction can allow the additional flow be made at other outlets on the standpipe system, such as from another standpipe, or from the fire pump test header. Although the results of having the flow points somewhere else in the standpipe system won't match the hydraulic calculations, the test will still prove that the most remote standpipe can provide the necessary flow and pressure required for fire department use while simultaneously flowing the full system demand.

6.3.1.3 The standpipe system demand shall be based on the design criteria in effect at the time of the installation.

FAQ What are some of the possible design pressures based on the date of installation?

The requirement in 6.3.1.3 requires the system to be flow tested and pressure tested according to the design criteria used for the standpipe system design. Depending on the edition of NFPA 14 the system was designed to, the hydraulic demand and permissible system pressures can vary greatly. Testing the system to the wrong demand and pressure could provide inaccurate results or damage the system. In the 1950s, standpipe systems were designed to flow water at a pressure of 50 psi (3.5 bar). The 65 psi (4.5 bar) pressure requirement was introduced into NFPA 14 in the early 1970s and was later revised to the current pressure of 100 psi (6.9 bar). In order to test to the proper pressure, the testing agent should remember these pressure differences to determine the date of installation of the standpipe system and, if possible, which edition of NFPA 14 was used in the design. Obviously, a standpipe system that was designed to operate at 65 psi (4.5 bar) most likely will not flow water at 100 psi (6.9 bar).

6.3.1.3.1 Where the standpipe system demand cannot be determined, the authority having jurisdiction shall determine the standpipe system demand.

6.3.1.3.2 The actual test method(s) and performance criteria shall be discussed in advance with the authority having jurisdiction.

6.3.1.4 Standpipes, sprinkler connections to standpipes, or hose stations equipped with pressure-reducing valves or pressure-regulating valves shall have these valves inspected, tested, and maintained in accordance with the requirements of Chapter 13.

6.3.1.5 A main drain test shall be performed on all standpipe systems with automatic water supplies in accordance with the requirements of Chapter 13.

6.3.1.5.1 The test shall be performed at the low point drain for each standpipe or the main drain test connection where the supply main enters the building (when provided).

6.3.1.5.2 Pressure gauges shall be provided for the test and shall be maintained in accordance with 5.3.2.

6.3.2 Hydrostatic Tests.

6.3.2.1* Hydrostatic tests of not less than 200 psi (13.8 bar) pressure for 2 hours, or at 50 psi (3.4 bar) in excess of the maximum pressure, where maximum pressure is in excess of 150 psi (10.3 bar), shall be conducted every 5 years on manual standpipe systems and semiautomatic dry standpipe systems, including piping in the fire department connection.

The 2011 edition of the standard eliminated the requirement for a hydrostatic test on automatic dry systems. Automatic dry systems are supervised under constant air pressure and the dry valve is subject to annual trip testing. However, semiautomatic dry systems use a deluge valve and are not supervised; thus, the need for the hydrostatic test remains.

FAQ Why is a hydrostatic test required for a manual dry standpipe system?

The requirement in 6.3.2.1 requires a hydrostatic test every 5 years for manual standpipe systems and semiautomatic dry standpipe systems to ensure the structural integrity of the system and detect potential failures before they are discovered during an actual emergency. Problems associated with piping integrity in wet systems are usually detected by leaks. Similarly, leaks in automatic dry systems are detected by the loss of air pressure and are subject to the 3-year test for air leakage as required by the standard (see 13.4.4.2.9). Pneumatic testing is required to detect leakage within manual dry and semiautomatic standpipe systems, which are more susceptible to corrosion due to the combination of air and moisture in the system. Undetected leaks can lead to failures when the dry standpipe is pressurized.

A.6.3.2.1 The intent of 6.3.2.1 is to ascertain whether the system retains its integrity under fire conditions. Minimum leakage existing only under test pressure is not cause for repair.

6.3.2.1.1 Manual wet standpipes that are part of a combined sprinkler/standpipe system shall not be required to be tested in accordance with 6.3.2.1.

It is not the intent of NFPA 25 to require a hydrostatic test on manual wet standpipes that are part of a combined system, since leaks in this type of system are usually detected immediately.

•

6.3.2.2 The hydrostatic test pressure shall be measured at the low elevation point of the individual system or zone being tested.

6.3.2.2.1 The inside standpipe piping shall show no leakage.

6.3.3 Waterflow Alarm and Supervisory Alarm Devices.

See 13.2.6 for the recommended operational procedure.

6.3.3.1 Where provided, waterflow alarm and supervisory alarm devices shall be tested in accordance with 13.2.6 and 13.3.3.5.

6.3.3.2 Where freezing conditions necessitate a delay in testing, tests shall be performed as soon as weather allows.

6.3.4* Gauges.

A.6.3.4 The normal life expectancy of a gauge is between 10 and 15 years. A gauge can be permitted to have a reading with an error of ±3 percent of the maximum (full-scale) gauge reading. For example, a gauge having 200 psi (13.8 bar) maximum radius installed on a system with 60 psi (4.1 bar) normal pressure can be permitted if the gauge reads from 54 psi to 66 psi (3.7 bar to 4.5 bar).

6.3.4.1 Gauges shall be replaced every 5 years or tested every 5 years by comparison with a calibrated gauge.

6.3.4.2 Gauges not accurate to within 3 percent of the full scale shall be recalibrated or replaced.

System Tagging

A gauge that has lost its seal or experienced mechanical damage, as shown in these photos, does not impact system performance and should be considered a noncritical deficiency.

(Photos courtesy of Byron Blake and Simplex Grinnell)

Noncritical Deficiency | Critical Deficiency | Impairment

6.4 Maintenance

6.4.1 Maintenance and repairs shall be in accordance with 6.1.3 and Table 6.1.2.

6.4.2 Equipment that does not pass the inspection or testing requirements shall be repaired and tested again or replaced.

6.5 Component Action Requirements

Component replacement tables provide guidance when system components are adjusted, repaired, rebuilt, or replaced. It is not necessary in each case to require a complete acceptance test for each component when maintenance is performed. It is important to keep in mind that corrective action must be completed in accordance with the applicable design standard. Quick

EXHIBIT 6.2 *Fire Hose Valve Connection with Duct Tape. (Courtesy of M. Steven Welsh, Koffel Associates, Inc.)*

fixes to undesirable field conditions, such as the duct tape provided in lieu of a cap as shown in Exhibit 6.2, are not appropriate, because they might only cause further deficiencies or potential impairments down the line.

6.5.1 Whenever components in standpipe and hose systems are adjusted, repaired, reconditioned, or replaced, the actions required in Table 6.5.1 shall be performed.

TABLE 6.5.1 *Summary of Component Replacement Action Requirements*

Component	*Adjust*	*Repair*	*Replace*	*Required Action*
Water Delivery Components				
Control valves	X	X	X	See Chapter 13
Hose valve pressure-regulating devices	X	X	X	See Chapter 13
System pressure-regulating devices	X	X	X	See Chapter 13
Piping	X	X	X	Hydrostatic test in conformance with NFPA 14, *Standard for the Installation of Standpipe and Hose Systems*
Fire hose			X	No action required
Fire hose		X		Perform hydrostatic test in accordance with NFPA 1962
Hose valve	X	X	X	See Chapter 13
Fire department connections	X	X	X	See Chapter 13
Backflow prevention device	X	X	X	See Chapter 13
Alarm and Supervisory Components				
Vane-type waterflow	X	X	X	Operational test using inspector's test connection
Pressure switch–type waterflow	X	X	X	Operational test using inspector's test connection
Water motor gong	X	X	X	Operational test using inspector's test connection
Valve supervisory device	X	X	X	Operational test for receipt of alarms and verification of conformance with NFPA 14 and/or NFPA 72, *National Fire Alarm and Signaling Code*

TABLE 6.5.1 Continued

Component	*Adjust*	*Repair*	*Replace*	*Required Action*
Status-Indicating Components				
Gauges			X	Verify at 0 psi (0 bar) and system working pressure
System Housing and Protection Components				
Cabinet	X	X	X	Verify compliance with NFPA 14
Hose storage rack	X	X	X	Verify compliance with NFPA 14
Testing and Maintenance Components				
Drain riser	X	X	X	Inspect for leaks while flowing from connection above the repair
Auxiliary drains	X	X	X	Inspect for leaks at system working pressure
Main drain	X	X	X	Inspect for leaks and residual pressure during main drain test
Structural Components				
Hanger/seismic bracing	X	X	X	Verify conformance with NFPA 14
Pipe stands	X	X	X	Verify conformance with NFPA 14
Informational Components				
Identification signs	X	X	X	Verify conformance with NFPA 14
Hydraulic placards	X	X	X	Verify conformance with NFPA 14

6.5.2 Where the original installation standard is different from the cited standard, the use of the appropriate installing standard shall be permitted.

•

6.5.3 These actions shall not require a design review, which is outside the scope of this standard.

References Cited in Commentary

National Fire Protection Association, 1 Batterymarch Park, Quincy, MA 02169-7471.

NFPA 13, *Standard for the Installation of Sprinkler Systems*, 2013 edition.

NFPA 14, *Standard for the Installation of Standpipe and Hose Systems*, 2013 edition.

NFPA 1962, *Standard for the Care, Use, Inspection, Service Testing, and Replacement of Fire Hose, Couplings, Nozzles, and Fire Hose Appliances*, 2013 edition.

Class I and Class III (Single Zone) Standpipe Flow

Purpose

The purpose of conducting a flow test for the standpipe system is to ensure the adequacy of the water supply as well as the capability of the system to meet the original design flow and pressure criteria for the system performance. The specific required design criteria for standpipe systems have changed through the various editions of NFPA 14, as well as for the specific type of installed standpipe. Therefore, a determination of the original design criteria as well as the type of standpipe must be established as a baseline against which the test results can be assessed. The types of systems include automatic wet, automatic dry, semiautomatic dry, manual wet, and manual dry. Definitions for each of these standpipe system types are provided by 3.3.15 of NFPA 14 and can be used in making a determination of the type of installed standpipe.

The specified minimum design pressure at the hydraulically most remote outlet for Class I and Class III standpipe systems was typically 65 psi (4.5 bar) for systems installed prior to 1993 and 100 psi (6.9 bar) for systems installed thereafter; however, the specific edition of NFPA 14 under which the system was installed and the required pressures mandated by the original should be confirmed. This information regarding the system type and specific design criteria might be directly obtained from a variety of sources including, but not limited to, the signage provided at the system riser, original installation records, maintenance records, information provided by the owner, and information provided by the AHJ.

It should be noted that this edition of NFPA 25 does not directly require the testing of a Class II standpipe system; however, the testing of such would be suggested using the following procedure with a minimum flow rate of 100 gpm (379 L/min) at the hydraulically most remote hose connection with a minimum residual pressure of 65 psi (4.5 bar).

Tools/Equipment

1. Sufficient hose lengths with the appropriate thread connection to allow for the discharge of the standpipe design flow to an appropriate location from the standpipe connections being tested is required. Typically 2½ in. (65 mm) hose with a coupling having a thread pattern matching that of the hose outlet are used for this purpose. The number and length(s) of hose will depend on the specific location and the distance from the tested outlet to a suitable location, as well as the number of outlets being tested.
2. A sufficient number of test nozzles to permit flow from the number of standpipe outlets to be tested simultaneously is required. Test nozzles can include UL play pipes or other approved test nozzles. The flow characteristics of the test nozzles must be known.
3. A pitot tube together with a test-pressure gauge[1], suitable for the pressures to be expected is required. [Usually, a 60 psi (4 bar) gauge is satisfactory.]
   - A pitot-tube-and-gauge assembly is indispensable for conducting flow tests from nozzles. The small opening at the end of the tube — not more than 1/16 in. (1.6 mm) in diameter — is inserted in the center of the stream in a direct line with the flow, at a distance in front of the opening equal to one-half the opening diameter. Velocity pressure is registered on the gauge attached to the tube.
   - Some available test nozzles include an integral pitot tube or pitotless flow measuring feature with an attached test gauge[1] allowing for direct measurement of flows without the use of a separate pitot tube.
4. A means to measure the operating pressure at the hydraulically most remote hose connections under flow conditions is required. Typically, a pressure gauge installed at the top of the standpipe or at the flowing outlet is used.
5. For automatic dry pipe standpipe systems, a timing device is needed to measure the duration between the opening of the most remote hose valve connection and water delivery.
6. Various wrenches and tools needed to facilitate the necessary connections are required.

Procedure Steps

Automatic Wet / Automatic Dry / Semiautomatic Dry Standpipe

1. Prior to any testing, notify the fire department and/or the alarm monitoring company, as well as the facility representatives, that testing is going to be conducted. An approval of the test method and performance criteria must be established with the AHJ prior to conduct of the test.
2. Ensure that the system is in the normal operating position.
3. For semiautomatic dry standpipe systems, locate or establish a point of connection for the introduction of pressurized air to the standpipe system on the dry side of the deluge valve, and pressurize the system piping to 25 psi (1.7 bar) to verify the integrity of the system piping, including piping leaks and open hose valves. No specific acceptance criteria are prescribed by NFPA 25 as to allowable leakage rate. Judgment should be applied to assess conditions that might be indicative of gross leakage raising concern for system integrity.
4. Set up a flow test connection from the hydraulically most remote standpipe hose valve connection(s). This usually includes the connection and routing of fire hose from the hose valve connection to a proper discharge location, including the connection to a suitable discharge device to allow for measurement of the flow rate. For other than horizontal standpipes, this will require a flow rate of 500 gpm (1893 L/min) from the most remote standpipe. This can be accomplished through the use of a roof manifold or connection to the upper most connection(s) within the stairway

---

1. Test-quality gauges should be used in accordance with ASME B40.1, *Gauges — Pressure Indicating Dial Type — Elastic Element, Grade 1A*, having an accuracy of ±1%. The use of quality test gauges produces results that are considered reasonably accurate within the scope of the test procedure. Care should be taken to protect the gauges from rough handling. Gauges should be calibrated at least annually by means of a dead-weight or calibrated tester throughout the range of operation before a test series is begun. Calibration sheets should be kept for each gauge and correction factors affixed to the back of each gauge.

where no roof manifold is provided. For a horizontal standpipe, a flow rate of 250 gpm (946 L/min) is required from the hydraulically most remote hose connection. A means must be established to measure the pressure at the hydraulically most remote hose connection during the conduct of the flow test. Typical installations include the installation of a pressure gauge at the top of the standpipe riser. However, where no pressure gauge is installed, the use of an adaptor with a pressure gauge or other alternative arrangement at or near the flowing outlet might be used.

5. Set up a flow connection for each of the additional standpipe risers (or hose connections for a horizontal standpipe) needed to achieve the required test flow. This usually includes the connection and routing of fire hose from the hose valve connection to a proper discharge location, including the connection to a suitable discharge device to allow for measurement of the flow rate. This will require the flow rate of an additional 250 gpm (946 L/min) from each additional standpipe riser (or additional hose connection for horizontal standpipes) up to a maximum flow rate of 1000 gpm (3785 L/min) for fully sprinklered buildings, 1250 gpm (4731 L/min) for nonsprinklered or partially sprinklered buildings, and 750 gpm (2840 L/min) for horizontal standpipe systems. The connection locations for these flows can be made at the most convenient point for flow purposes on the additional standpipe (or hose connection for horizontal standpipe) being used. Where the flow from these additional locations cannot be accomplished, other locations can be used as permitted by the AHJ. Such alternative flow locations might include other standpipes or the fire pump test header.

6. Check the area surrounding the discharge point for the connected discharge devices to ensure that there are no obvious conditions that would prevent water from being discharged safely or cause direct damage in the immediate vicinity, taking appropriate caution to ensure personnel and property are protected from the resultant high-pressure hose stream discharge. The use of a flow diffuser can aid in breaking the directed stream and managing resultant energy of the discharge. Where the discharge of water is to an area subject to potential freezing conditions, the facility representative should be advised of the potential for icing conditions.

7. For semiautomatic dry standpipes, initiate the means of manual operation of the deluge valve.

8. Initiate the flow test by slowly opening the hose valve(s) on the hydraulically most remote standpipe connection, allowing for the establishment of the automatic operation of any fire pumps or tripping of the dry pipe valve for automatic dry standpipe systems, followed by simultaneously opening additional hose valve(s) or flow locations as established above. For automatic dry standpipe systems, record the time that the first hose valve is fully opened and the elapsed time for water delivery to this most remote hose connection. The flow rate from each is to be measured and adjusted as required. The flow rate can be adjusted by metering the flow rate at the hose valve to obtain the required flow, with multiple flow points generally requiring multiple passes of adjustment to obtain the required flow rate at each outlet. Record the flow rate from each of the flowing connections.

9. Allow the flow to stabilize, take pressure readings at the hydraulically most remote hose connection, and record the results.

10. Slowly close each of the hose valves to avoid water hammer.

11. Shut down any fire pump(s) that has operated.

12. Remove all attached test equipment.

13. For dry systems, completely drain the system piping.

14. Restore the system to normal operating condition.

15. Upon completion of all testing, notify the fire department and/or the alarm monitoring company, as well as the facility representatives, that testing is complete, and reset the fire alarm system as necessary.

Manual Wet / Manual Dry Standpipe

1. Prior to any testing, notify the fire department and/or the alarm monitoring company, as well as the facility representatives, that testing is going to be conducted. An approval of the test method and performance criteria must be established with the AHJ prior to conducting the test.

2. Ensure that the system is in the normal operating position.

3. For manual dry standpipe systems, locate or establish a point of connection for the introduction of pressurized air to the standpipe system, and pressurize the system piping to 25 psi (1.7 bar) to verify the integrity of the system piping, including piping leaks or open hose valves. No specific acceptance criteria are prescribed by NFPA 25 as to allowable leakage rate. Judgment should be applied to assess conditions that might be indicative of gross leakage raising concern for system integrity.

4. Establish a temporary means of supplying pressurized water through the fire department connection. This might be accomplished by means of the fire department pumper truck or a portable pump having a comparable capacity to that of the fire department's pumper truck. This might be supplied with water from a fire hydrant on the municipal system or other alternate water supply. The typical fire department operating procedure is to charge the fire department connection to a pressure of 150 psi (10.3 bar) and maintain that during the flow testing; however, original design criteria might have established alternative pressure requirements.

5. Set up a flow test connection from the hydraulically most remote standpipe hose valve connection(s). This usually includes the connection and routing of fire hose from the hose valve connection to a proper discharge location, including the connection to a suitable discharge device to allow for measurement of the flow rate. For other than horizontal standpipes, this will require a flow rate of 500 gpm (1893 L/min) from the most remote standpipe. This can be accomplished through the use of a roof manifold or connection to the uppermost connection(s) within the

Class I and Class III (Single Zone) Standpipe Flow

stairway where no roof manifold is provided. For a horizontal standpipe, a flow rate of 250 gpm (946 L/min) is required from the hydraulically most remote hose connection. A means must be established to measure the pressure at the hydraulically most remote hose connection during the flow test. Typical installations include the installation of a pressure gauge at the top of the standpipe riser; however, where no pressure gauge is installed, the use of an adaptor with a pressure gauge or other alternative arrangement at or near the flowing outlet might be used.

6. Set up a flow connection for each of the additional standpipe risers (or hose connections for a horizontal standpipe) needed to achieve the required test flow. This usually includes the connection and routing of fire hose from the hose valve connection to a proper discharge location, including the connection to a suitable discharge device to allow for measurement of the flow rate. This will require the flow rate of an additional 250 gpm (946 L/min) from each additional standpipe riser (or additional hose connection for horizontal standpipes) up to a maximum flow rate of 1000 gpm (3785 L/min) for fully sprinklered buildings, 1250 gpm (4731 L/min) for nonsprinklered or partially sprinklered buildings, and 750 gpm (2840 L/min) for horizontal standpipe systems. The connection locations for these flows can be made at the most convenient point for flow purposes on the additional standpipe (or hose connection for horizontal standpipe) being used. Where the flow from these additional locations cannot be accomplished, other locations can be used as permitted by the AHJ. Such alternative flow locations might include other standpipes or the fire pump test header.

7. Check the area surrounding the discharge point for the connected discharge devices to ensure that there are no obvious conditions that would prevent water from being discharged safely or cause direct damage in the immediate vicinity, and take appropriate caution to ensure personnel and property are protected from the resultant high-pressure hose stream discharge. Where the discharge of water is to an area subject to potential freezing conditions, the facility representative should be advised of the potential for icing conditions.

8. Establish the fire department operating pressure at the fire department pumper or temporary pump connection to the fire department connection.

9. Initiate the flow test by slowly opening the hose valve(s) on the hydraulically most remote standpipe connection, followed by simultaneously opening additional hose valve(s) or flow locations as established above. The flow rate from each is to be measured and adjusted as required to obtain the required flow rate from each. The flow rate can be adjusted by metering the flow rate at the hose valve to obtain the required flow, with multiple flow points generally requiring multiple passes of adjustment to obtain the required flow rate at each outlet. Additionally, the operating pressure at the fire department pumper or temporary pump connection must be adjusted to maintain the established pressure for the connection. Record the flow rate from each of the flowing connections and the operating pressure at the fire department pumper or temporary pump.

10. Allow the flow to stabilize, take pressure readings at the hydraulically most remote hose connection, and record the results.

11. Slowly close each of the hose valves to avoid water hammer.

12. Shut down the fire department pumper or temporary pump.

13. Remove all attached test equipment.

14. For dry systems, completely drain the system piping.

15. Restore the system to normal operating condition.

16. Upon completion of all testing, notify the fire department and/or the alarm monitoring company, as well as the facility representatives, that testing is complete, and reset the fire alarm system as necessary.

Evaluation of Results

For semiautomatic dry and manual dry standpipe systems that are required to be air pressure tested at 25 psi (1.7 bar), an evaluation of the degree of leakage must be assessed. While no specific acceptance criteria are prescribed by NFPA 25 as to allowable leakage rate, an experienced level of judgment should be applied to assess if gross leakage raising concern for system integrity is evident. Where such conditions exist, the system should be investigated and corrective action taken prior to conducting the flow test of the system.

For automatic dry standpipe systems, the water delivery time to the hydraulically most remote hose valve should be compared with previous tests to reveal substantial delays for the water delivery time of dry pipe systems, since significant differences from one test to another are an indication of possible operational problems with the system. A drastic increase in the water delivery time from one full flow trip test of a dry pipe system to the next is of major concern, because such an increase is most likely due to corrosion of the system piping. Such conditions should be investigated and appropriate corrective action should be taken as necessary. NFPA 25 does not require a specific water delivery time for these systems.

An evaluation of the test results from the flow test must be compared to that established for the original design of the system. This would typically require that the available pressure at the hydraulically most remote hose connection while flowing the full design demand be either 100 psi (6.9 bar) or 65 psi (4.5 bar), depending on the specific edition of NFPA 14 under which the system was designed and installed. System performance falling below these criteria should be investigated and appropriate corrective action should be taken as necessary.

The procedure(s) outlined above is not mandated by NFPA 25 and represents only one approach to conducting this test.

ITM NFPA 25 — INSPECTION TESTING MAINTENANCE

7 PRIVATE FIRE SERVICE MAINS

Chapter 7 of NFPA 25 addresses the inspection, testing, and maintenance of private underground mains and their appurtenances. The term *private* indicates that the piping and equipment is located on private property and is neither owned by nor maintained by the local water district or purveyor. The flow testing discussed in this chapter is intended to apply to hydrants located on private property, although the procedures are the same for testing hydrants on public property. Often private firms or individuals are not permitted to operate equipment owned by the water district, so it is best to contact the owner of the system or equipment before conducting any test or operating valves or hydrants. In regions impacted by the Clean Water Act, it might be necessary to capture or treat the water discharged during hydrant flow tests. This can include de-chlorination or other precautionary measures to avoid washing debris into local storm drain piping. NFPA 291, *Recommended Practice for Fire Flow Testing and Marking of Hydrants,* provides guidance on public and private fire hydrant markings and color codes, which can be helpful in determining if a hydrant is public or private. NFPA 291 also provides guidance on how to conduct hydrant flow tests, in terms of which hydrants to use to read static and residual pressures, and which hydrants to flow for the best test results.

Critical Point

NFPA 291 provides guidance on public and private fire hydrant markings and color codes, which can be helpful in determining if a hydrant is public or private.

Exhibit 7.1 illustrates the boundary between public and private equipment and systems. Generally, when piping enters private property, the property owner is responsible for inspection, testing, and maintenance of piping and related equipment. Piping and equipment on the public side of the property line generally are the responsibility of the water purveyor.

7.1 General

The purpose of Section 7.1 mirrors the basic purpose of the standard; that is, "to provide requirements that ensure a reasonable degree of protection for life and property from fire through minimum inspection, testing, and maintenance methods. . . ." It should be recognized that when inspecting equipment such as private fire service mains, a visual examination of the equipment is usually not possible. Much of the inspection, therefore, centers on the inspection, testing, and maintenance of the attached equipment, such as hydrants, hose houses, and monitor nozzles. Testing of this equipment is necessary to assess the condition of the underground piping. Camera equipment can be helpful in discovering internal corrosion or obstructive materials. However, observing changes in flow and pressure during testing is the best way to determine if further investigation is needed.

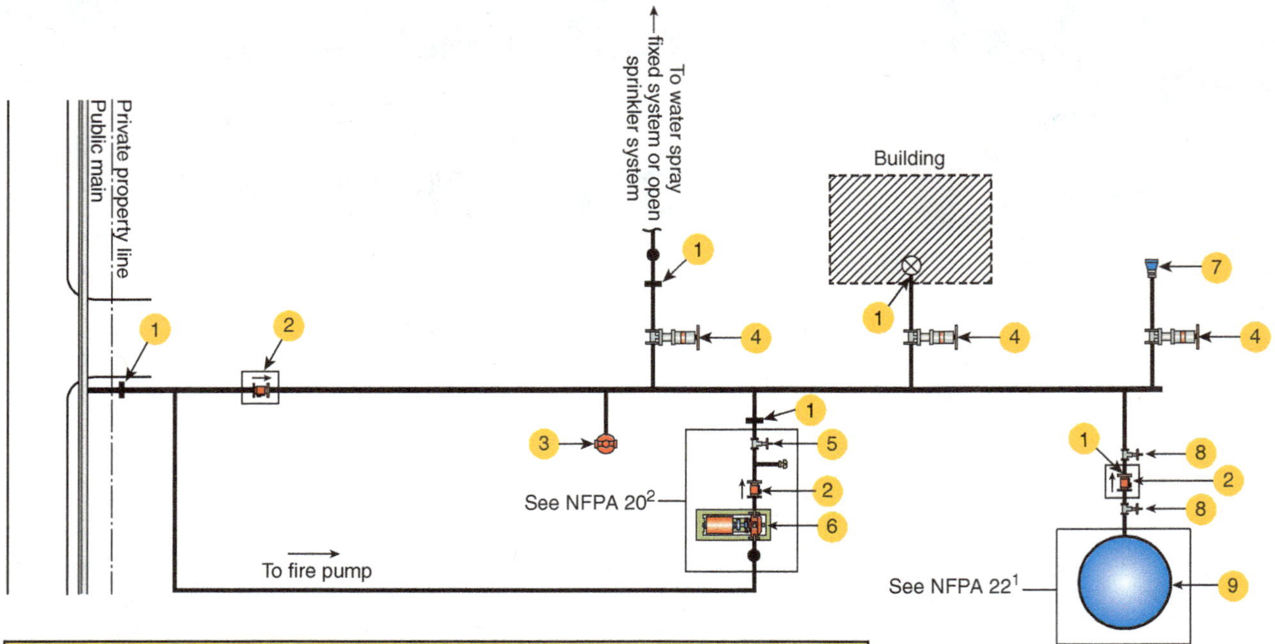

Legend

1 End of private fire service main
2 Check valve
3 Hydrant
4 Post indicating valve
5 Pump discharge valve
6 Fire pump
7 Monitor nozzle
8 Control valve
9 Water tank

Notes

The piping (aboveground or buried) shown is specific as to the end of the private fire service main, and this schematic is only for illustrative purposes beyond the end of the fire service main. Details of valves and their location requirements are covered in the specific standard involved.

1. See NFPA 22, *Standard for Water Tanks for Private Fire Protection.*
2. See NFPA 20, *Standard for the Installation of Stationary Pumps for Fire Protection.*

EXHIBIT 7.1 *Public/Private Equipment Boundary. (Courtesy of Stephan Laforest)*

FAQ Is the intent of NFPA 25 to require that the inspection, testing, and maintenance requirements of fire hydrants apply to both private and public hydrants?

Section 7.1 specifically states that this chapter applies to private fire service mains and appurtenances. The equipment that is covered in this chapter is designed and installed in accordance with NFPA 24, *Standard for the Installation of Private Fire Service Mains and Their Appurtenances.* Neither standard is intended to apply to public systems. However, on-site flow testing of hydrants and main drain flow testing of fire sprinkler or standpipe systems can be an indicator of a need to investigate the public water main system valves and piping as well. Where there is degradation in the test results from year to year or where the difference between the test results compared to the original results is significant (i.e., 10 percent or more), further investigation is normally needed.

7.1.1 Minimum Requirements.

7.1.1.1 This chapter shall provide the minimum requirements for the routine inspection, testing, and maintenance of private fire service mains and their appurtenances.

7.1.1.2 Table 7.1.1.2 shall be used to determine the minimum required frequencies for inspection, testing, and maintenance.

TABLE 7.1.1.2 *Summary of Private Fire Service Main Inspection, Testing, and Maintenance*

Item	*Frequency*	*Reference*
Inspection		
Hose houses	Quarterly	7.2.2.7
Hydrants (dry barrel and wall)	Annually and after each operation	7.2.2.4
Monitor nozzles	Semiannually	7.2.2.6
Hydrants (wet barrel)	Annually and after each operation	7.2.2.5
Mainline strainers	Annually and after each significant flow	7.2.2.3
Piping (exposed)	Annually	7.2.2.1
Piping (underground)	See 7.2.2.2	7.2.2.2
Test		
Monitor nozzles	Flow, annually (range and operation)	7.3.3
Hydrants	Flow, annually	7.3.2
Piping (exposed and underground) (flow test)	5 years	7.3.1
Valve status test		13.3.1.2.1
Maintenance		
Mainline strainers	Annually and after each operation	7.2.2.3
Hose houses	Annually	7.2.2.7
Hydrants	Annually	7.4.2
Monitor nozzles	Annually	7.4.3

7.1.2 Valves and Connections. Valves and fire department connections shall be inspected, tested, and maintained in accordance with Chapter 13.

7.1.3 Obstruction Investigations. The procedures outlined in Chapter 14 shall be followed where there is a need to conduct an obstruction investigation.

7.1.4 Fire Hose. Fire hose shall be maintained in accordance with NFPA 1962, *Standard for the Care, Use, Inspection, Service Testing, and Replacement of Fire Hose, Couplings, Nozzles, and Fire Hose Appliances.*

7.1.5 Impairments. The procedures outlined in Chapter 15 shall be followed wherever such an impairment to protection occurs.

Impairments such as those referred to in 7.1.5 involve removing a system from service. Impairments can be of an emergency nature, such as a broken fire service main, or they can be preplanned, as in cases of physical modification to the system or routine maintenance of the equipment. In either case, an impairment program must be in place to minimize the length of time the system is impaired and to verify that systems and valves are properly returned to service upon completion of the work.

An impairment program is critical for any work associated with a private fire service main, because impairment of these systems often involves shutting water supply control valves to one or more water-based fire protection systems or fire hydrants. Exhibit 7.2 shows one means of identifying hydrants that have been impaired.

As mentioned in the commentary on Chapter 1, shut valves are the most common cause of sprinkler system failure (see Commentary Table 1.1).

Critical Point

An impairment program must be in place to minimize the length of time the system is impaired and to verify that systems and valves are properly returned to service upon completion of the work.

EXHIBIT 7.2 *Hydrant with an "Out of Service" Collar Indicating an Impaired Condition.*

7.2 Inspection and Corrective Action

7.2.1 General. Private fire service mains and their appurtenances shall be inspected at the intervals specified in Table 7.1.1.2.

Generally, inspectors should check each inspection point as listed in Table 7.1.1.2. The intent is to perform a quick visual examination to verify that the system and components appear to be in good working order. As referenced in Chapter 1, it is important to verify that water supply control valves are in the full open position. The findings of the inspections conducted in accordance with Section 7.2 must be recorded and maintained by the owner. Exhibit 7.3 is a sample form that can be used to not only record the inspection-related tasks of Section 7.2 but also the maintenance and testing tasks required by the other sections of Chapter 7.

7.2.2* Procedures. All procedures shall be carried out in accordance with the manufacturer's instructions, where applicable.

A.7.2.2 The requirements in 7.2.2 outline inspection intervals, conditions to be inspected, and corrective actions necessary for private fire service mains and associated equipment.

7.2.2.1 Exposed Piping.

ITM NFPA 25 — ITM Deficiency, Impairment, or Hazard Evaluation?

(Courtesy of Byron Blake and SimplexGrinnell)

ANSWER: ITM Deficiency

The hydrant shown here has been struck or backed over by a vehicle. Pictures from different angles show no obvious damage to the hydrant itself, although based on the position of the hydrant, the following can be determined:

1. The barrel is cracked or broken below ground level.
2. The shaft is bent and will not operate.

Table 7.2.2.4 and Table 7.2.2.5 both indicate that if the hydrant barrel is cracked, it needs to be replaced. It is interesting to note that NFPA 25 does not specifically address situations like this one, where the component is obviously damaged, but the damage is not visible.

PRIVATE FIRE SERVICE MAINS: INSPECTION, TESTING, AND MAINTENANCE OF PRIVATE FIRE SERVICE MAINS

Name of Property: ______________________ Inspector: ______________

Address: ______________________ Contract No.: ______________

Phone Number: ______________________ Date: ______________

This Report Covers: ❑ Quarterly ❑ Semi-Annual ❑ Annual ❑ Five-Year

Inspections

Quarterly — **Hose Houses**

❑ Yes ❑ No ❑ N/A Accessible
❑ Yes ❑ No ❑ N/A Free from damage or leaks
❑ Yes ❑ No ❑ N/A No missing equipment

Semi-annual — **Monitor Nozzles**

❑ Yes ❑ No ❑ N/A Not leaking
❑ Yes ❑ No ❑ N/A Free of damage
❑ Yes ❑ No ❑ N/A Free of corrosion

Annual — **Hydrants (Dry Barrel & Wall Type)**

❑ Yes ❑ No ❑ N/A Accessible
❑ Yes ❑ No ❑ N/A Barrel is free of water/ice
❑ Yes ❑ No ❑ N/A Barrel drains properly
❑ Yes ❑ No ❑ N/A Not leaking
❑ Yes ❑ No ❑ N/A Barrel is free of cracks
❑ Yes ❑ No ❑ N/A Outlets are not excessively tight
❑ Yes ❑ No ❑ N/A Nozzle threads are not worn
❑ Yes ❑ No ❑ N/A Operating nut is not worn
❑ Yes ❑ No ❑ N/A Operating wrench is available

Hydrants (Wet Barrel)

❑ Yes ❑ No ❑ N/A Accessible
❑ Yes ❑ No ❑ N/A Not leaking
❑ Yes ❑ No ❑ N/A Barrel is free of cracks
❑ Yes ❑ No ❑ N/A Outlets not excessively tight
❑ Yes ❑ No ❑ N/A Nozzle threads are not worn
❑ Yes ❑ No ❑ N/A Operating nut is not worn
❑ Yes ❑ No ❑ N/A Operating wrench is available

NFPA NFPA Water-Based Fire Protection Systems Forms 1 of 2

EXHIBIT 7.3 Sample Form for Inspection, Testing, and Maintenance of Private Fire Service Mains.

Mainline Stringers

❑ Yes ❑ No ❑ N/A Not plugged or fouled
❑ Yes ❑ No ❑ N/A Free of corrosion

Exposed Piping

❑ Yes ❑ No ❑ N/A Not leaking
❑ Yes ❑ No ❑ N/A Free of damage/corrosion
❑ Yes ❑ No ❑ N/A Hangers intact/not damaged

Test

Annual

Monitor Nozzles

❑ Yes ❑ No ❑ N/A Flow test until all foreign material has cleared (not less than one minute)
❑ Yes ❑ No ❑ N/A Operated through full range of motion

Hydrants

❑ Yes ❑ No ❑ N/A Flow test until all foreign material has cleared (not less than one minute)
❑ Yes ❑ No ❑ N/A Barrel drains within 60 minutes

5-Year

Piping

❑ Yes ❑ No ❑ N/A Flow test piping at rate anticipated during a fire
❑ Yes ❑ No ❑ N/A Flow test results comparable to previous test

Maintenance

Annual

Mainline Strainers

❑ Yes ❑ No ❑ N/A Clean strainer basket by back flushing or basket removal

Hose House

❑ Yes ❑ No ❑ N/A Verify usable condition of house and all components

Hydrants

❑ Yes ❑ No ❑ N/A Lubricate stems, caps, plugs, and threads
❑ Yes ❑ No ❑ N/A Accessible (no snow, ice, or other material)
❑ Yes ❑ No ❑ N/A Protected from damage

Monitor Nozzles

❑ Yes ❑ No ❑ N/A Lubricate per manufacturer's recommendations

Comments

Signature: ______________________________ Date: ______________

License/Certification No.: ______________________________

NFPA® NFPA Water-Based Fire Protection Systems Forms 2 of 2

EXHIBIT 7.3 *Continued.*

7.2.2.1.1 Exposed piping shall be inspected annually.

7.2.2.1.2 Piping shall be inspected, and the necessary corrective action shall be taken as specified in Table 7.2.2.1.2.

TABLE 7.2.2.1.2 *Exposed Piping*

Condition	*Corrective Action*
Leaks	Repair
Physical damage	Repair or replace
Corrosion	Clean or replace and coat with corrosion protection
Restraint methods	Repair or replace

7.2.2.1.3 Piping installed in areas that are inaccessible for safety considerations due to process operations shall be inspected during each scheduled shutdown.

7.2.2.2* Underground Piping.

A.7.2.2.2 Generally, underground piping cannot be inspected on a routine basis. However, flow testing can reveal the condition of underground piping and should be conducted in accordance with Section 7.3.

7.2.2.3* Mainline Strainers. Mainline strainers shall be inspected and cleaned after each system flow exceeding that of a nominal 2 in. (50 mm) orifice and shall be removed and inspected annually for failing, damaged, and corroded parts, with the necessary corrective action taken as specified in Table 7.2.2.3.

TABLE 7.2.2.3 *Mainline Strainers*

Condition	*Corrective Action*
Plugging or fouling	Clean
Corrosion	Replace or repair

A.7.2.2.3 Any flow in excess of the flow through the main drain connection should be considered significant.

A strainer is installed on a system where foreign material is likely to be present in the water supply and might obstruct an orifice. For example, the presence of rocks, pebbles, leaves, and sediment in a raw water source, such as a pond, would necessitate a strainer.

Strainers must be installed where sprinkler or nozzle orifices are smaller than ⅜ in. (10 mm) in diameter. The concern is increased where higher volumes of water are used for fire protection purposes, as opposed to volumes typically encountered in a plumbing system. In fire pump installations, the velocity of the water could be sufficient to dislodge and carry material to an orifice. In this case, the strainer serves to "filter" the water. During inspection, the strainer flushing valve should be opened and allowed to flow until the water is clear. In cases of severe sediment, the strainer basket should be removed and cleaned. Exhibit 7.4 shows a typical fire protection system strainer.

EXHIBIT 7.4 *Fire Protection Mainline Strainer.*

7.2.2.4 Dry Barrel and Wall Hydrants. Dry barrel and wall hydrants shall be inspected annually and after each operation, with the necessary corrective action taken as specified in Table 7.2.2.4.

TABLE 7.2.2.4 *Dry Barrel and Wall Hydrants*

Condition	***Corrective Action***
Inaccessible	Make accessible
Barrel contains water or ice (presence of water or ice could indicate a faulty drain, a leaky hydrant valve, or high groundwater table)	Repair and drain; for high groundwater it could be necessary to plug the drain and pump out the barrel after each use
Improper drainage from barrel	Repair drain
Leaks in outlets or at top of hydrant	Repair or replace gaskets, packing, or parts as necessary
Cracks in hydrant barrel	Repair or replace
Tightness of outlet caps	Lubricate if necessary; tighten if necessary
Worn outlet threads	Repair or replace
Worn hydrant operating nut	Repair or replace
Availability of operating wrench	Make sure wrench is available

The dry barrel hydrants (see Figure A.3.3.12.1) are also known as frost-proof hydrants. These hydrants are used where there is a chance that temperatures will drop below freezing. The valve controlling the water is located below the frost line between the foot piece and barrel of the hydrant. The barrel of the hydrant is dry, and water is admitted upon opening the operating nut. A drain valve at the base of the barrel is open when the main valve is closed, allowing residual water in the barrel to drain out.

System Tagging

The hydrant shown in this photo has had the cap removed or stolen, and it was not replaced. This hydrant was exposed to harsh conditions for several seasons, and threading has deteriorated to the point where a hose line could not be attached. In this instance, the hydrant can no longer accept a hose, and therefore, it would not be considered functional during a fire. This condition should be considered an impairment of that hydrant.

Furthermore, this hydrant is not labeled with an out of service tag; therefore, when responding personnel arrive on scene, they would have a false indication of a potential manual source of fire suppression. This can be problematic as it can require responding personnel to modify operational tactics on the fly, costing them several minutes in response time.

(Courtesy of Byron Blake and SimplexGrinnell)

Noncritical Deficiency　Critical Deficiency　Impairment

ITM Deficiency, Impairment, or Hazard Evaluation?

ANSWER: ITM Deficiency

The hydrant shown here apparently is used frequently, because the operating nut is worn so badly that it is completely round. Hydrants are required in 7.2.2.4 and 7.2.2.5 to be inspected annually and after each operation. Table 7.2.2.4 and Table 7.2.2.5 describe the conditions the inspector is looking for and the corrective action that needs to be taken to fix each specific problem.

In this case, a worn hydrant operating nut is a condition described in each table, and the required corrective action is to repair or replace the operating nut. Although it will be difficult to open this hydrant in a fire event, if an adjustable hydrant wrench or pipe wrench is available it can be done, and the condition should be recorded as a critical deficiency.

(Courtesy of Byron Blake and SimplexGrinnell)

7.2.2.5 Wet Barrel Hydrants. Wet barrel hydrants shall be inspected annually and after each operation, with the necessary corrective action taken as specified in Table 7.2.2.5.

TABLE 7.2.2.5 *Wet Barrel Hydrants*

Condition	*Corrective Action*
Inaccessible	Make accessible
Leaks in outlets or at top of hydrant	Repair or replace gaskets, packing, or parts as necessary
Cracks in hydrant barrel	Repair or replace
Tightness of outlet caps	Lubricate if necessary; tighten if necessary
Worn outlet threads	Repair or replace
Worn hydrant operating nut	Repair or replace
Availability of operating wrench	Make sure wrench is available

ITM Deficiency, Impairment, or Hazard Evaluation?

ANSWER: ITM Deficiency

The hydrant shown here was opened before the annual flow test required by 7.3.2 in order to perform the inspections required by 7.2.2.4 and 7.2.2.5, which require hydrants to be inspected annually and after each operation. Table 7.2.2.4 and Table 7.2.2.5 describe the conditions the inspector is looking for and the corrective action that needs to be taken to fix each specific problem.

In this case, a hydrant that leaks from the outlets or top is a condition described in each table. The required corrective action is to repair or replace gaskets, packing, or parts as necessary. Although it leaks, this hydrant can still be used in a fire event and the condition should be recorded as a critical deficiency.

(Courtesy of Byron Blake and SimplexGrinnell)

EXHIBIT 7.5 Wet Barrel Hydrant.

Paragraph 7.2.2.5 addresses the inspection of wet barrel hydrants, also known as "California" hydrants, such as the one shown in Exhibit 7.5. These hydrants are sometimes used where the temperature remains above freezing. A wet barrel hydrant usually has a compression valve at each outlet, but it might have another valve in the bonnet that controls the water flow to all outlets. As with any flow testing, the inspector should verify where water will go before any valves or hydrants are opened. Extra care should be used in newly landscaped areas that could be washed out by a normal flow test, because such actions could result in owners or occupants losing the desire to comply with ITM requirements provided in this standard.

7.2.2.6 Monitor Nozzles. Monitor nozzles shall be inspected semiannually, with the necessary corrective action taken as specified in Table 7.2.2.6.

TABLE 7.2.2.6 *Monitor Nozzles*

Condition	*Corrective Action*
Leakage	Repair
Physical damage	Repair or replace
Corrosion	Clean or replace, and lubricate or protect as necessary

Monitor nozzles are provided where large amounts of combustible materials, such as log piles, lumber piles, flammable or combustible liquids, or railway cars are stored in yards and it is necessary to provide a means of quickly delivering large volumes of water to control fires. See Exhibit 7.6 for an example of a monitor nozzle.

EXHIBIT 7.6 Monitor Nozzle.

7.2.2.7 Hose Houses. Hose houses shall be inspected quarterly, with the necessary corrective action taken as specified in Table 7.2.2.7.

TABLE 7.2.2.7 *Hose Houses*

Condition	*Corrective Action*
Inaccessible	Make accessible
Physical damage	Repair or replace
Missing equipment	Replace equipment

A hose house is typically located in an industrial environment to provide manual fire-fighting equipment outside to protect buildings, exposures, or process equipment that might be exposed to a fire hazard. Such equipment generally includes hose, nozzles, wrenches, gated wyes, and hose couplings.

7.3 Testing

7.3.1* Underground and Exposed Piping Flow Tests. Underground and exposed piping serving hydrants shall be flow tested at minimum 5-year intervals.

Testing Procedure Alert

ITM NFPA 25

It is not uncommon for underground piping and other appurtenances associated with private service mains to become damaged due to the environment where the equipment is located. This can lead to a degradation in the water supply that is delivered to the systems downstream. The underground flow test is intended to identify reductions in flow, among other things. For a procedure for conducting this test, please refer to the detailed testing procedure at the end of this chapter.

A.7.3.1 Full flow tests of underground piping can be accomplished by methods including, but not limited to, flow through yard hydrants, fire department connections once the check valve has been removed, main drain connections, and hose connections. The flow test should be conducted in accordance with NFPA 291.

A hydrant flow test, which helps determine the internal condition of piping, should be compared to the most recent test conducted, as well as to the original test. If deterioration is noted, additional testing, such as a hydraulic gradient analysis, should be conducted. All data gathered during either test should be recorded on report forms.

Critical Point

A hydrant flow test, which helps determine the internal condition of piping, should be compared to the most recent test conducted, as well as to the original test.

7.3.1.1 Any flow test results that indicate deterioration of available waterflow and pressure shall be investigated to the complete satisfaction of the authority having jurisdiction to ensure that the required flow and pressure are available for fire protection.

7.3.1.2 Where underground piping supplies individual fire sprinkler, standpipe, water spray, or foam-water sprinkler systems and there are no means to conduct full flow tests, tests generating the maximum available flows shall be permitted.

Case In Point

The flows expected during a fire are not just those used by automatic suppression systems such as sprinklers. Many jurisdictions use fire flows from the adopted fire code, which could include much higher flows that take into consideration manual fire-fighting tactics. It might be necessary to consult the fire department to ascertain what amount of flow was accounted for in the design of the system. This is particularly important if the private fire service main(s) supplies only manual fire-fighting equipment such as hydrants. If the flow test required by 7.3.1.1 indicates a potential problem, additional information can be obtained by conducting an engineering analysis or hydraulic gradient analysis. For further information on how to conduct a hydraulic gradient analysis, see Section 15 in the 20th edition of the *Fire Protection Handbook*.

7.3.2 Hydrants. Hydrants shall be tested annually to ensure proper functioning.

Testing Procedure Alert

ITM NFPA 25

The purpose of completing a hydrant test is to exercise the hydrant valves through the full range of operation, clear the hydrant of any built-up debris, and to ensure proper operation of the barrel drainage for dry barrel and wall hydrants. Hydrant operation is often critical for manual fire-fighting operations, and this test provides a level of assuredness that the hydrants will be available for fire department use. For a process for conducting the test required by 7.3.2, please refer to the detailed testing procedure at the end of this chapter.

7.3.2.1 Each hydrant shall be opened fully and water flowed until all foreign material has cleared.

7.3.2.2 Flow shall be maintained for not less than 1 minute.

7.3.2.3 After operation, dry barrel and wall hydrants shall be observed for proper drainage from the barrel.

The dry barrel hydrant is often used in areas where there are concerns about freezing temperatures. Proper drainage is important, because if the water does not drain out of the barrel promptly, the potential for freezing increases. (See the commentary following 7.2.2.4 for more information.) One common cause of drainage failure or slow drainage is plugged weep holes, which are easily repaired by excavating around the hydrant and cleaning out the drainage holes.

7.3.2.4 Full drainage shall take no longer than 60 minutes.

7.3.2.5 Where soil conditions or other factors are such that the hydrant barrel does not drain within 60 minutes, or where the groundwater level is above that of the hydrant drain, the hydrant drain shall be plugged and the water in the barrel shall be pumped out.

7.3.2.6 Dry barrel hydrants that are located in areas subject to freezing weather and that have plugged drains shall be identified clearly as needing pumping after operation.

7.3.3 Monitor Nozzles.

7.3.3.1 Monitor nozzles that are mounted on hydrants shall be tested as specified in 7.3.2.

7.3.3.2 All monitor nozzles shall be oscillated and moved throughout their full range annually to ensure proper operability.

A monitor nozzle can be a fixed-position device or a directional device capable of being manually or automatically rotated either from side to side or up and down. The requirement in 7.3.3.2 addresses the necessity of ensuring the device maintains full movement in all directions. (See the commentary following Table 7.2.2.6 for more explanation of monitor nozzles and their use.)

7.4 Maintenance

7.4.1 General. All equipment shall be maintained in proper working condition, consistent with the manufacturer's recommendations.

7.4.2 Hydrants.

7.4.2.1 Hydrants shall be lubricated annually to ensure that all stems, caps, plugs, and threads are in proper operating condition.

7.4.2.2* Hydrants shall be kept free of snow, ice, or other materials and protected against mechanical damage so that free access is ensured.

A.7.4.2.2 The intent of 7.4.2.2 is to maintain adequate space for use of hydrants during a fire emergency. The amount of space needed depends on the configuration as well as the type and size of accessory equipment, such as hose, wrenches, and other devices that could be used.

FAQ What measures can be taken to indicate the locations of hydrants in case of snow burial?

The requirement in 7.4.2.2 that hydrants be kept free of snow is sometimes difficult to meet. For example, a hydrant near the street runs the risk of being buried by a snowplow. Where the potential exists for snow burial, extra measures should be taken to ensure that hydrants can be easily located, such as with the use of flags, indicating poles, banners, or other measures, as shown in Exhibit 7.7. Note that in addition to snow, vegetation can also obscure a hydrant location. In that case, the vegetation should be cleared away completely.

EXHIBIT 7.7 *Hydrant with Indicating Pole.*

7.4.3 Monitor Nozzles. Monitor nozzles shall be lubricated annually to ensure proper operating condition.

7.5 Component Action Requirements

Component replacement tables offer the user guidance of the standard when system components are adjusted, repaired, rebuilt, or replaced. It is not necessary in each case to require a complete acceptance test for each component when maintenance is performed.

Critical Point

Component replacement tables offer the user guidance of the standard when system components are adjusted, repaired, rebuilt, or replaced.

7.5.1 Whenever a component in a private fire service system is adjusted, repaired, reconditioned, or replaced, the action required in Table 7.5.1 shall be performed.

7.5.2 Where the original installation standard is different from the cited standard, the use of the appropriate installing standard shall be permitted.

•

7.5.3* Where a main drain is not provided, other equivalent means of flow testing shall be permitted.

A.7.5.3 Private fire service mains might not include a main drain connection; therefore, other equivalent means of flow such as an installed fire hydrant can be used.

7.5.4 The actions of 7.5.1 shall not require a design review, which is outside the scope of this standard.

TABLE 7.5.1 Summary of Component Replacement Action Requirements

Component	Adjust	Repair/ Recondition	Replace	Test Criteria
Water Delivery Components				
Pipe and fittings (exposed)	X	X	X	Hydrostatic test in conformance with NFPA 24, *Standard for the Installation of Private Fire Service Mains and Their Appurtenances*
Pipe and fittings (underground)				Flush in conformance with NFPA 24 or NFPA 20, as appropriate
Hydrants	X	X	X	Hydrostatic test in conformance with NFPA 24 Waterflow in conformance with NFPA 24 Inspect for proper drainage
Monitor nozzles	X	X	X	Hydrostatic test in conformance with NFPA 24 Flush in conformance with NFPA 24
Mainline strainers	X	X	X	Flow test downstream of strainer
Fire department connection	X	X	X	See Chapter 13
Alarm and Supervisory Components				
Valve supervisory device	X	X	X	Operational test for conformance with NFPA 24 and/or *NFPA 72, National Fire Alarm and Signaling Code*
System-Indicating Components				
Gauges			X	Verify at 0 psi (0 bar) and system working pressure
System Housing and Protection Components				
Hose houses	X	X	X	Verify integrity of hose house and hose house components
Hose		X		Repair and test hose in accordance with NFPA 1962
Hose			X	No action required
Structural Components				
Thrust blocks	X	X	X	Test at system working pressure
Tie rods	X	X	X	Test at system working pressure
Retainer glands	X	X	X	Test at system working pressure
Informational Components				
Identification signs	X	X	X	Verify conformance with NFPA 24

References Cited in Commentary

American Society of Mechanical Engineers, Three Park Avenue, New York, NY 10016-5990.

ASME B40.1, *Gauges — Pressure Indicating Dial Type — Elastic Element*, Grade 1A, 2005 edition.

National Fire Protection Association, 1 Batterymarch Park, Quincy, MA 02169-7471.

Cote, A. E. ed., *Fire Protection Handbook®*, 20th edition, 2008.

Cote, A. E. ed., *Fire Protection Handbook®*, 19th edition, 2003.

NFPA 20, *Standard for the Installation of Stationary Pumps for Fire Protection*, 2013 edition.

NFPA 22, *Standard for Water Tanks for Private Fire Protection*, 2013 edition.

NFPA 24, *Standard for the Installation of Private Fire Service Mains and Their Appurtenances*, 2013 edition.

NFPA 291, *Recommended Practice for Fire Flow Testing and Marking of Hydrants*, 2013 edition.

U.S. Government Printing Office, Washington, DC 20402.

P.L. 92-500, Federal Water Pollution Control Amendments of 1972, Clean Water Act, October 18, 1972.

Testing Procedure for 7.3.1

Underground Flow Test

Purpose

The purpose of completing a test of the underground piping is to assess the adequacy of the connected water supply and to detect significant reductions that might adversely impact the system operation. This assessment is accomplished by comparing the results of the current test to previous baseline test data as well as to the system demands supplied by the underground piping.

Tools/Equipment

1. A ruler with 1⁄16 in. (1.6 mm) divisions is required. [A minimum 6 in. (50 mm) rule is recommended.]
2. A pitot tube (see Testing Exhibit 7.1), together with a test-pressure gauge[1], suitable for the pressures to be expected is required. [Usually, a 60 psi (4 bar) gauge is satisfactory.]

   A pitot-tube-and-gauge assembly, such as the one shown in Testing Exhibit 7.1, is indispensable for conducting flow tests from hydrants and nozzles. The small opening at the end of the tube — not more than 1⁄16 in. (1.6 mm) in diameter — is inserted in the center of the stream in a direct line with the flow, at a distance in front of the opening equal to one-half the opening diameter. Velocity pressure is registered on the gauge attached to the tube.

TESTING EXHIBIT 7.1 *Typical Pitot Tube Assembly.*

Some available test nozzles include an integral pitot tube or pitotless flow-measuring feature with an attached test gauge[1] allowing for direct measurement of flows without the use of a separate pitot tube. Where used, a sufficient number of test nozzles are needed to achieve the desired flow rate.

3. A 2½ in. (65 mm) hydrant gauge cap with fittings (see Testing Exhibit 7.2) [assuming 2½ in. (65 mm) outlets], together with a test-pressure gauge[1] suitable for the pressures to be expected, with 1 psi (0.07 bar) graduations is required. [Usually, a 200 psi (13.8 bar) gauge is satisfactory.]

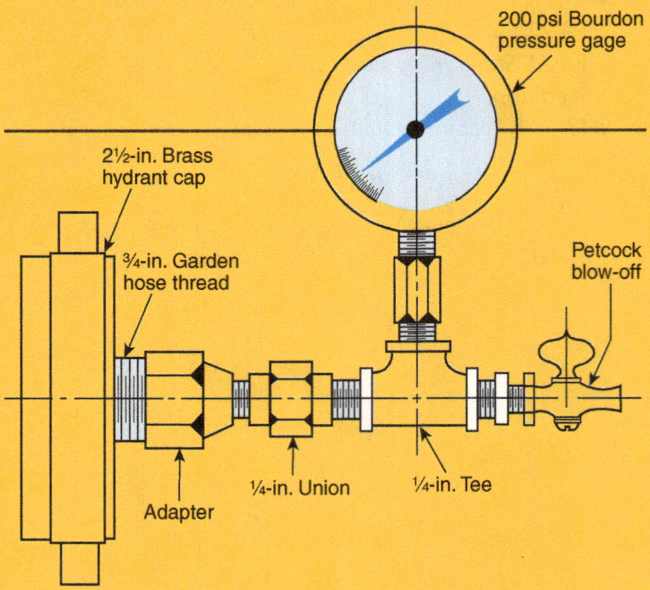

TESTING EXHIBIT 7.2 *Typical Hydrant Cap and Gauge. (Source: NFPA Fire Protection Handbook® 2003, Figure 10.6.2)*

In Testing Exhibit 7.2, note the petcock blow-off. Its operation permits air from the hydrant barrel to be vented when the hydrant is opened and air to re-enter when the hydrant is closed. Failure to open the petcock during hydrant closing might subject the pressure gauge to a partial vacuum, which could introduce errors in future gauge readings.

4. A hydrant wrench compatible with the operating nut of the hydrants being tested is required. Adjustable wrenches, pipe wrenches, and parrot wrenches should not be used, because these could damage the operating nut.
5. A form for recording test data, including a sketch of the test location showing hydrants and any other salient features, is required. All test data should be recorded in a neat, systematic fashion, along with any system operating conditions that might affect the test results.

Procedure Steps

1. Prior to any testing, notify the water department, the fire department and/or the alarm monitoring company (if needed), and the facility representatives that testing is going to be conducted.
2. Establish a test plan with the AHJs and obtain required approvals.

1. Test-quality gauges should be used in accordance with ASME B40.1, *Gauges — Pressure Indicating Dial Type — Elastic Element, Grade 1A,* having an accuracy of ±1%. The use of quality test gauges produces results that are considered reasonably accurate within the scope of the test procedure. Care should be taken to protect the gauges from rough handling.

Gauges should be calibrated at least annually by means of a dead-weight or calibrated tester throughout the range of operation before a test series is begun. Calibration sheets should be kept for each gauge and correction factors affixed to the back of each gauge.

3. Identify the test hydrants (non-flowing and flowing). In a single-direction feed water supply, the test hydrant should be the one nearest the source. In a gridded-type system where the water flow comes from multiple directions, the location of the test and flow hydrants is not as critical. In any case, the test should be conducted in the vicinity of the required point of connection.

   Based on the underground piping configuration, Testing Exhibit 7.3 provides the recommended arrangement of flow (*F*) and residual (*R*) hydrants. The hydrants used to measure flow and pressure vary, based on the directional flow of the water in relationship to the hydrants being tested, along with the piping configuration.

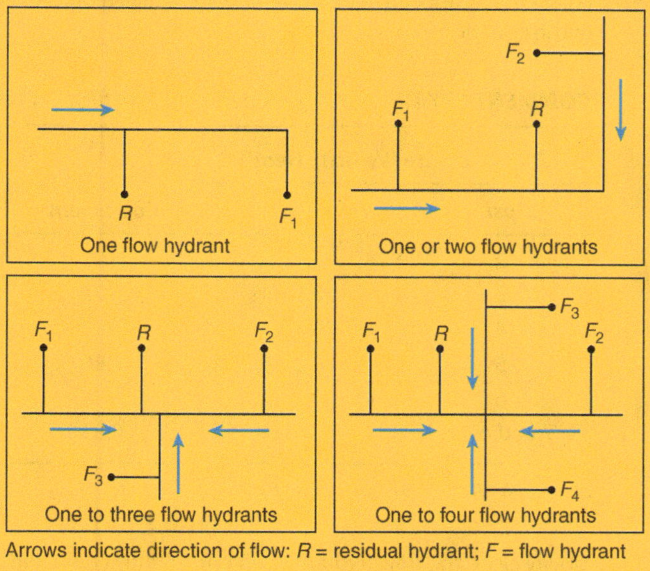

TESTING EXHIBIT 7.3 *Suggested Flow Test Arrangements. (Source: NFPA 291, 2013, Figure 4.3.4)*

4. Check the area surrounding the flow (*F*) hydrant(s) to ensure that there are no obvious conditions that would prevent water from being discharged safely or cause direct damage in the immediate vicinity. Take appropriate caution to ensure personnel and property are protected from the resultant high-pressure hose stream discharge. The use of a flow diffuser can aid in breaking the directed stream, managing resultant energy of the discharge, and directing the flow in a specific direction. Where the discharge of water is to an area subject to potential freezing conditions, the facility representative should be advised of the potential for icing conditions.

5. Remove the residual hydrant (*R*) cap and flush. Flushing the hydrant reduces the possibility of damage to the test equipment or the tester. Once a clean, steady stream of water is observed, the hydrant valve should be slowly closed. Note that the flow of fire hydrants supplied by a fire pump will result in the operation of the fire pump.

6. Install the hydrant gauge cap and pressure gauge on the residual hydrant (*R*) with the petcock open (see Testing Exhibit 7.2). Ensure that the non-gauged caps on the residual hydrant (*R*) are secured and tight.

7. Open the residual hydrant (*R*), obtain a flow from the petcock, then close the petcock.

8. Record the static pressure (i.e., no other hydrants flowing) from the pressure gauge on the residual hydrant (*R*).

9. Remove the cap(s) from the flow hydrant(s) (*F*) and measure the diameter of the opening. Then determine the hydrant coefficient (see Testing Exhibit 7.4).

TESTING EXHIBIT 7.4 *Verifying Hydrant Outlet Coefficient.*

   The hydrant coefficient can be determined by inserting one hand into the hydrant opening and comparing the shape of the outlet to the shapes shown in Testing Exhibit 7.5.

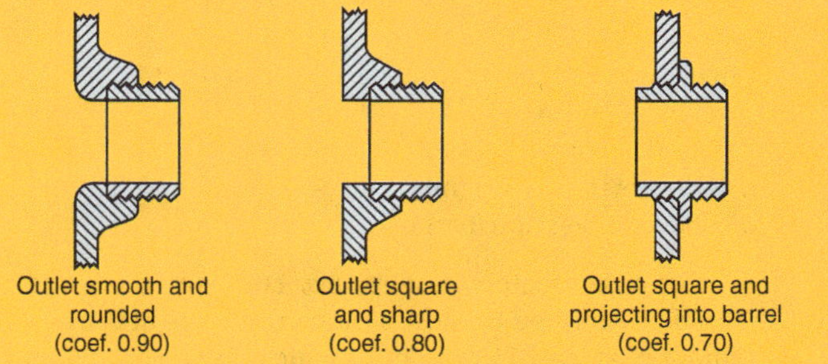

TESTING EXHIBIT 7.5 *Three General Types of Hydrant Outlets and Their Coefficients of Discharge. (Source: NFPA 291, 2013, Figure 4.7.1)*

   The use of a test nozzle and/or diffuser can also be used to provide an accurate coefficient for use in determining the

Underground Flow Test

flow rate. Where used, these devices should be installed on the flow outlets for the hydrant. Ensure that the non-flowing caps on the flow hydrant (*F*) are secured and tight.

10. Fully open the flow hydrant(s) (*F*) and wait for a steady stream of water and stabilized pressure reading at the residual hydrant (*R*).

11. Measure flowing pressure at each flowing outlet by inserting a pitot tube and gauge assembly (see Testing Exhibit 7.1) into the center of the stream at a distance of one-half the opening diameter out from the face of the hydrant outlet, while holding the center line of the pitot orifice at a right angle to the face of the hydrant outlet. The air chamber of the pitot tube should be kept elevated.

    Where a test nozzle is attached that includes an integral pitot tube or pitotless flow-measuring feature, the flowing pressures can be read directly from the pressure gauge on the attached device.

12. Record the pitot gauge reading(s) while the pressure reading at the residual hydrant (*R*) is recorded simultaneously.

    The pressure recorded by the pitot tube assembly is the velocity pressure, which is used to calculate flow. Converting the pitot pressure to flow can be accomplished by using the flow listed in the flow tables in NFPA 291. Alternatively, the flow can be calculated by inserting the outlet coefficient (*c*), outlet diameter (*d*), and velocity pressure (*p*) into the following formula:

$$Q = 29.84cd^2\sqrt{p}$$

where:
Q = flow in gpm
c = coefficient of discharge
d = diameter of outlet (in.)
p = flowing pressure (psi)

In metric units, the formula is as follows:

$$Q = 0.0666cd^2\sqrt{p}$$

where:
Q = flow in L/min
c = coefficient of discharge
d = diameter of outlet (mm)
p = flowing pressure (bar)

To obtain satisfactory test results, sufficient discharge should be achieved to cause a drop in pressure at the test hydrant of at least 25 percent of static pressure or to flow the total system demand. Should additional flow rates be desired, additional outlets or hydrants can be opened and pressures measured accordingly.

Pitot readings of less than 10 psi (0.7 bar) or over 30 psi (2.1 bar) at any open hydrant should be avoided. To keep within these pressure limits, the rate of flow can be controlled by throttling the hydrant, opening a second outlet, or both. However, the flow hydrants should be opened sufficiently so that hydrant drains are closed. Water continuously discharged through the drains tends to erode the soil from the base of the hydrant.

The use of the larger pumper connection on a hydrant for testing should be avoided unless flow and pressure are strong enough to produce a full stream. When pumper outlets are used, a proper coefficient of discharge must be determined, based on the extent to which the orifice is completely filled with water. This is accomplished by applying an additional correction factor to the calculated flow rate using the proper pumper outlet coefficient from Table 4.8.2 from NFPA 291. See Commentary Table 7.1. This coefficient is applied in addition to the coefficient applied in the equation above.

COMMENTARY TABLE 7.1 *Pumper Outlet Coefficients*

Pitot Pressure (Velocity Head)		
psi	***bar***	***Coefficient***
2	0.14	0.97
3	0.21	0.92
4	0.28	0.89
5	0.35	0.86
6	0.41	0.84
7 and over	0.48 and over	0.83

Source: NFPA 291, 2013, Table 4.8.2.

13. Slowly close the flow hydrant(s) (*F*) one at a time to avoid water hammer until all hydrant flow is ceased.

14. Record the return static pressure on the residual hydrant (*R*).

15. Allow any fire pumps that might have automatically operated to run for 10 minutes, for an electric fire pump, or 30 minutes, for a diesel fire pump. Follow this with a shutdown of the fire pump and restoration of automatic operating condition.

16. Remove any installed test nozzles and/or diffusers.

17. Check the flow hydrant(s) (*F*) barrel for proper drainage.

18. Replace the hydrant cap(s) on the flow hydrant(s) (*F*).

19. Close the residual hydrant (*R*) and open the petcock on the hydrant gauge cap. The petcock should be opened prior to complete closure to avoid the development of a vacuum.

20. Remove the hydrant gauge cap from the residual hydrant (*R*) and check the barrel for proper drainage.

21. Replace the hydrant cap on the residual hydrant (*R*).

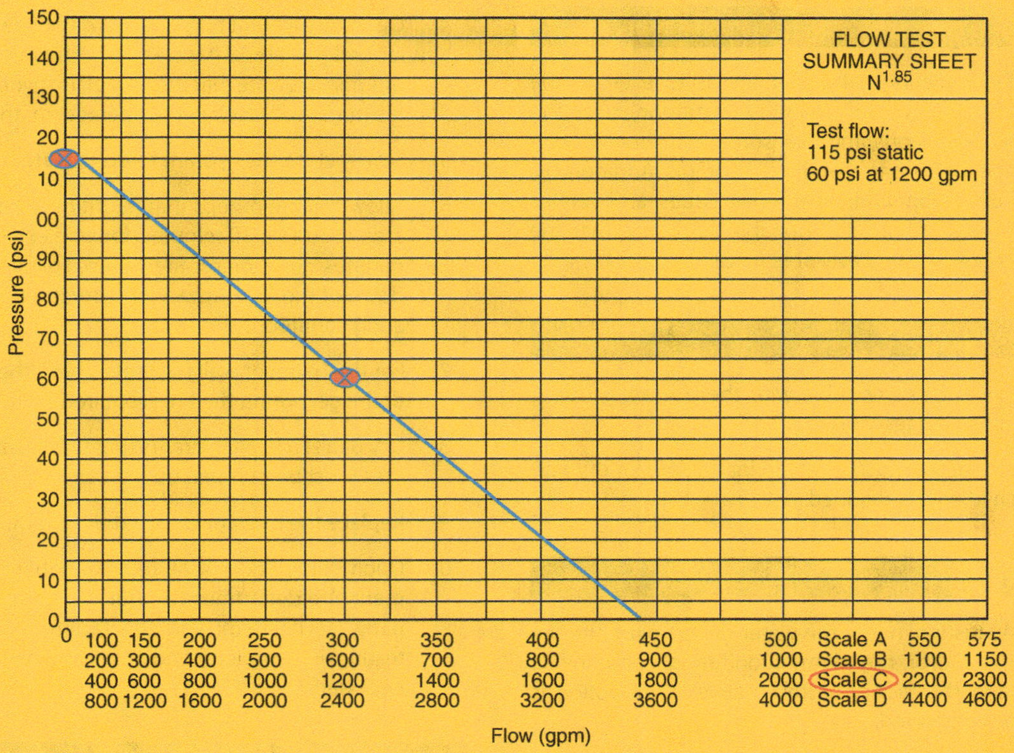

TESTING EXHIBIT 7.6 *Water Supply Test Plotted on $N^{1.85}$ Graph Paper.*

22. Upon completion of all testing, notify the water department, fire department and/or the alarm monitoring company (as needed), and the facility representatives that testing is complete, and reset fire alarm system as necessary.

Evaluation of Results

The test data should be plotted on $N^{1.85}$ graph paper. See Testing Exhibit 7.6. The data is plotted by drawing a line between the two points represented by the recorded static pressure along the vertical axis at 0 flow on the horizontal axis and the recorded residual pressure along the vertical axis at the recorded total flow rate of all flowing outlets.

A comparative evaluation should be made against the original test data, if available, or previous test data. A significant degradation in performance should warrant further investigation, such as a hydraulic gradient analysis, to determine the source of the reduced system capacity. For further information regarding the conduct of a hydraulic gradient analysis, see the *Fire Protection Handbook*, 20th edition, Section 15. Additionally, the supplied system demands, such as fire sprinkler, standpipe, or hydrants that are connected to the underground piping and are being tested, should be plotted on $N^{1.85}$ graph paper. Where those demands fall above the plotted water supply curve, an unacceptable condition is represented requiring further investigation, and corrective action should be taken as necessary.

The procedure(s) outlined above is not mandated by NFPA 25 and represents only one approach to conducting this test.

Testing Procedure for 7.3.2

Hydrant Test

Purpose

The purpose of completing a hydrant test is to exercise the hydrant valves through the full range of operation, clear the hydrant of any built-up debris, and to ensure proper operation of the barrel drainage for dry barrel and wall hydrants. Additionally, the test verifies operation, addresses repair issues, and verifies reliability of the hydrant with a reasonable assurance that the water supply is connected and available for use.

Tools/Equipment

1. A hydrant wrench compatible with the operating nut of the hydrants being tested is required.
2. A timing device to measure the drainage time for dry barrel or wall hydrants is required.

Procedure Steps

1. Prior to any testing, notify the water department, the fire department and/or the alarm monitoring company (if needed), and the facility representatives that testing is going to be conducted.
2. Check the area surrounding the hydrant to ensure that there are no obvious conditions that would prevent water from being discharged safely or cause direct damage in the immediate vicinity. Take appropriate caution to ensure personnel and property are protected from the resultant high-pressure hose stream discharge. The use of a flow diffuser can aid in breaking the directed stream and managing resultant energy of the discharge. Where the discharge of water is to an area subject to potential freezing conditions, the facility representative should be advised of the potential for icing conditions.
3. Remove a hydrant cap from the hydrant. Ensure the non-flowing caps are secured and tight.
4. Open the hydrant fully and allow the flow to continue until all foreign material has cleared, with a minimum flow period of 1 minute. Note that the flow of fire hydrants supplied by a fire pump will result in the operation of the fire pump.
5. Slowly close the hydrant to avoid water hammer.
6. Allow any fire pumps that might have automatically operated to run for 10 minutes, for an electric fire pump, or 30 minutes, for a diesel fire pump. Follow this with a shutdown of the fire pump and restoration of automatic operating condition.
7. For dry barrel or wall hydrants, check the barrel for proper drainage, noting the time required for full drainage.
8. Dry barrel or wall hydrants that fail to drain within 60 minutes must be pumped out.
9. Replace the hydrant cap on the hydrant.
10. Upon completion of all testing, notify the water department, fire department and/or the alarm monitoring company (as needed), and the facility representatives that testing is complete, and reset the fire alarm system as necessary.

Evaluation of Results

The hydrant should cycle through the full range of operation without excessive operating force or disengagement and should fully stop the hydrant flow upon closure.

Dry barrel and wall hydrants must fully drain within a period of not more than 60 minutes. Hydrants that do not properly drain must be pumped out to ensure protection of the hydrant against freeze-up.

The hydrant should produce a substantial water flow when fully open. Obvious degradation in the flow capacity of the hydrant should be investigated, and corrective action should be taken as necessary.

The procedure(s) outlined above is not mandated by NFPA 25 and represents only one approach to conducting this test.

8

FIRE PUMPS

Chapter 8 addresses the inspection, testing, and maintenance of fire pumps and related equipment. In most cases where fire pumps are installed, the fire protection systems they connect to would be ineffective in fire control if the pumps and connected drivers were not fully functioning. Therefore, proper maintenance is critical to help ensure constant system readiness at all times.

Fire pumps and their associated equipment also constitute a major investment, ranging in cost from $20,000 to more than $1,000,000 when suction tanks are required.

Fire pumps are typically required for effective fire protection system operations. When pumps that were put in place to protect a facility must be removed from service for repairs, the facility can become unprotected for the duration of the repair. In some of these situations, the protection of property and life is considered impaired, resulting in a shutdown of operations mandated by the authority having jurisdiction (AHJ) until repairs are made. This shutdown could result in a major business interruption loss in addition to costly pump repairs. Thus, failure to properly maintain fire pumps and related equipment can quickly result in significant direct and indirect costs.

Critical Point

Failure to properly maintain fire pumps and related equipment can quickly result in significant direct and indirect costs.

8.1* General

A.8.1 A fire pump assembly provides waterflow and pressure for private fire protection. The assembly includes the water supply suction and discharge piping and valving; pump; electric, diesel, or steam turbine driver and control; and the auxiliary equipment appurtenant thereto.

8.1.1 Minimum Requirements.

8.1.1.1 This chapter shall provide the minimum requirements for the routine inspection, testing, and maintenance of fire pump assemblies.

FAQ What is the testing frequency for a fire pump that only services a standpipe system in a municipal or private office building?

All stationary fire pumps must be tested weekly or monthly, depending on the driver type and a complete annual test per the requirements of Table 8.1.1.2, regardless of the type of system they are connected to or building occupancy they are installed in.

Exhibit 8.1 shows a typical horizontal split-case fire pump installation, and Exhibit 8.2 shows a typical vertical turbine fire pump installation.

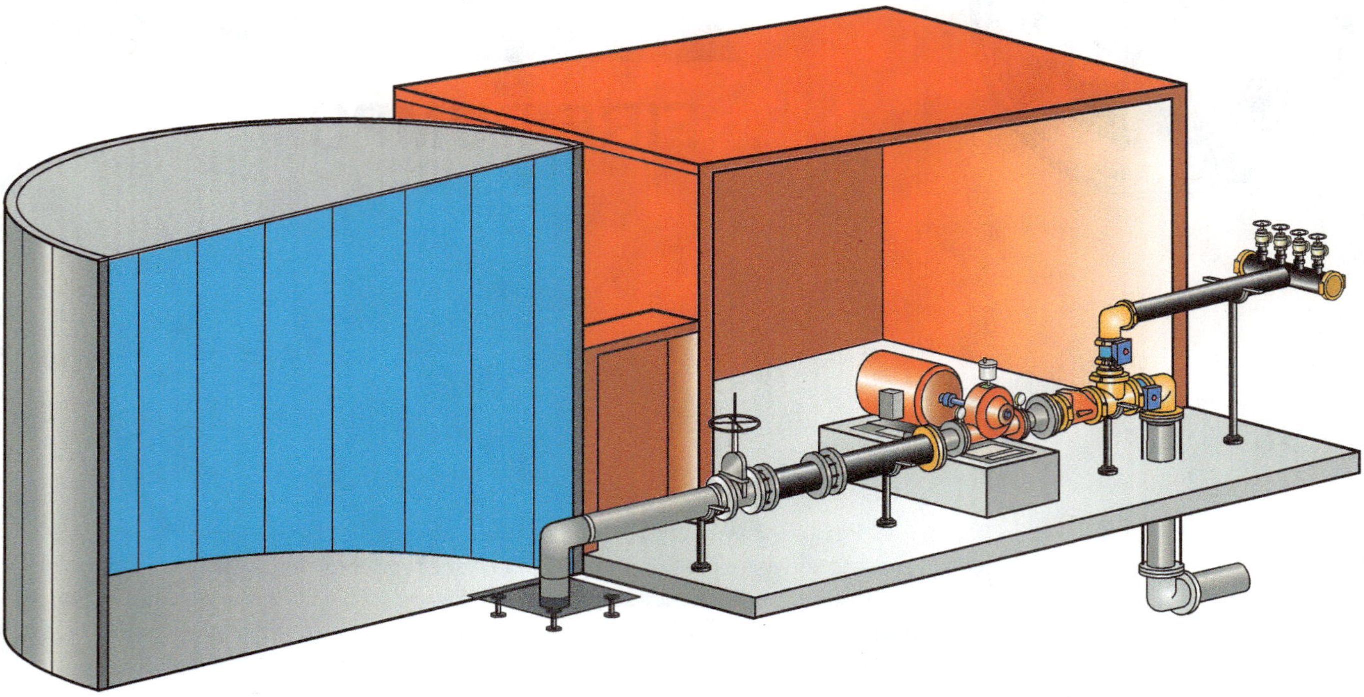

EXHIBIT 8.1 Horizontal Split-Case Fire Pump Assembly. (Courtesy of Stephan Laforest)

EXHIBIT 8.2 Vertical Turbine Fire Pump Assembly. (Source: NFPA 20, 2013, Figure A.7.2.2.1)

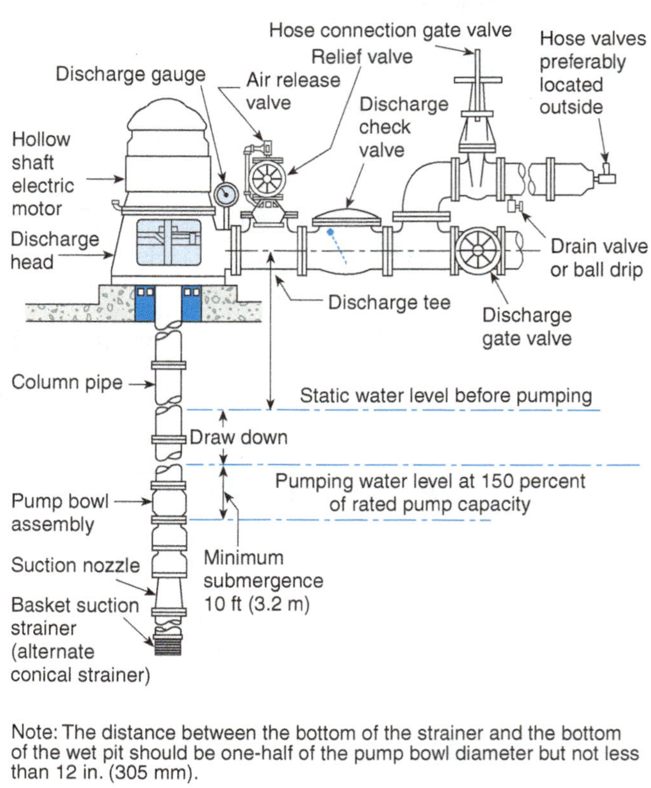

Note: The distance between the bottom of the strainer and the bottom of the wet pit should be one-half of the pump bowl diameter but not less than 12 in. (305 mm).

8.1.1.2 Table 8.1.1.2 shall be used to determine the minimum required frequencies for inspection, testing, and maintenance.

TABLE 8.1.1.2 Summary of Fire Pump Inspection, Testing, and Maintenance

Item	*Frequency*	*Reference*
Inspection		
Pump house, heating ventilating louvers	Weekly	8.2.2(1)
Fire pump system	Weekly	8.2.2
Test		
Pump operation		
No-flow condition		8.3.1
Diesel engine–driven fire pump	Weekly	
Electric motor–driven fire pump	See 8.3.1.2	
Flow condition	Annually	8.3.3
Fire pump alarm signals	Annually	8.3.3.5
Maintenance		
Hydraulic	Annually	8.5
Mechanical transmission	Annually	8.5
Electrical system	Varies	8.5
Controller, various components	Varies	8.5
Motor	Annually	8.5
Diesel engine system, various components	Varies	8.5

8.1.2 Alternative Inspection, Testing, and Maintenance Procedures. In the absence of manufacturer's recommendations for preventive maintenance, Table 8.1.2 shall be used for alternative requirements.

The manufacturer's maintenance recommendations per equipment manuals, user guides, and/or other product documentation should be followed for preventive maintenance. The information contained in this material is important for the long-term care of the various components and should be provided to owners with all new installations by the installing contractor. For new installations, it is important that all manufacturers' manuals and informational material for the pump and the associated equipment remain in the possession of the owner for reference in maintaining the equipment. This material should be kept in a location accessible for regular reference but not where it could be subject to environmental damage or decay, as might be the case in some pump rooms.

Manufacturers will typically replace missing or damaged documents or provide electronic files. In most cases, requirements are manufacturer and model specific, so it is best to confirm that acquired documentation is specific to the manufacturer, model, and serial number of the specific equipment being maintained.

In the rare case that information is not available through the manufacturer or manufacturer's representative, Table 8.1.2 provides guidelines for general minimum preventative maintenance tasks including frequencies.

Critical Point

The manufacturer's maintenance recommendations per equipment manuals, user guides, and/or other product documentation should be followed for preventive maintenance.

8.1.3 Valves and Connections. Valves and fire department connections shall be inspected, tested, and maintained in accordance with Chapter 13.

8.1.4 Obstruction Investigations. The procedures outlined in Chapter 14 shall be followed where there is a need to conduct an obstruction investigation.

TABLE 8.1.2 Alternative Fire Pump Inspection, Testing, and Maintenance Procedures

Complete as Applicable	*Visual Inspection*	*Inspect*	*Change*	*Clean*	*Test*	*Frequency*
Pump System						
Pump bearings		X				Annually
Lubricate pump bearings			X			As needed
Inspect pump shaft end play		X				Annually
Inspect accuracy of pressure gauges and sensors		X	X			Annually (replace or recalibrate when 5% out of calibration)
Inspect pump coupling alignment		X				Annually
Wet pit suction screens		X		X		After each pump operation
Mechanical Transmission						
Lubricate coupling			X			Annually
Lubricate right-angle gear drive			X			Annually
Electrical System						
Exercise isolating switch and circuit breaker					X	Monthly
Trip circuit breaker (if mechanism provided)					X	Annually
Operate manual starting means (electrical)					X	Semiannually
Inspect and operate emergency manual starting means (without power)	X				X	Annually
Tighten electrical connections as necessary		X				Annually
Lubricate mechanical moving parts (excluding starters and relays)		X				Annually
Calibrate pressure switch settings		X				Annually
Grease motor bearings			X			Annually or as needed
Voltmeter and ammeter for accuracy (5%)		X				Annually
Any corrosion on printed circuit boards (PCBs)	X					Annually
Any cracked cable/wire insulation	X					Annually
Any leaks in plumbing parts	X					Annually
Any signs of water on electrical parts	X					Annually
Diesel Engine System						
Fuel						
Tank level	X	X				Weekly
Tank float switch	X				X	Weekly
Solenoid valve operation	X				X	Weekly
Strainer, filter, or dirt leg, or combination thereof				X		Quarterly
Water and foreign material in tank				X		Annually
Water in system		X		X		Weekly
Flexible hoses and connectors	X					Weekly
Tank vents and overflow piping unobstructed		X			X	Annually
Piping	X					Annually
Lubrication system						
Oil level	X	X				Weekly
Oil change			X			50 hours or annually
Oil filter(s)			X			50 hours or annually
Lube oil heater		X				Weekly
Crankcase breather	X		X	X		Quarterly

TABLE 8.1.2 *Continued*

Complete as Applicable	*Visual Inspection*	*Inspect*	*Change*	*Clean*	*Test*	*Frequency*
Cooling system						
Level	X	X				Weekly
Antifreeze protection level					X	Semiannually
Antifreeze		X				Annually
Adequate cooling water to heat exchanger		X				Weekly
Rod out heat exchanger				X		Annually
Water pump(s)	X					Weekly
Condition of flexible hoses and connections	X	X				Weekly
Jacket water heater		X				Weekly
Inspect duct work, clean louvers (combustion air)	X	X	X			Annually
Water strainer				X		Quarterly
Exhaust system						
Leakage	X	X				Weekly
Drain condensate trap		X				Weekly
Insulation and fire hazards	X					Quarterly
Excessive back pressure					X	Annually
Exhaust system hangers and supports	X					Annually
Flexible exhaust section	X					Semiannually
Battery system						
Electrolyte level		X				Weekly
Terminals clean and tight	X	X				Quarterly
Case exterior clean and dry	X	X				Monthly
Specific gravity or state of charge					X	Monthly
Charger and charge rate	X					Monthly
Equalize charge		X				Monthly
Clean terminals				X		Annually
Cranking voltage exceeds 9 volts on a 12 volt system or 18 volts on a 24 volt system		X				Weekly
Electrical system						
General inspection	X					Weekly
Tighten control and power wiring connections		X				Annually
Wire chafing where subject to movement	X	X				Quarterly
Operation of safeties and alarms		X			X	Semiannually
Boxes, panels, and cabinets				X		Semiannually
Circuit breakers or fuses	X	X				Monthly
Circuit breakers or fuses			X			Biennially
Voltmeter and ammeter for accuracy (5%)		X				Annually
Any corrosion on printed circuit boards (PCBs)	X					Annually
Any cracked cable/wire insulation	X					Annually
Any leaks in plumbing parts	X					Annually
Any signs of water on electrical parts	X					Annually

Obstructions and damage related to microbiologically influenced corrosion (MIC), which is discussed in Supplement 1, are not typically associated with fire pump installations. However, a number of cases have occurred in which biological activity that was associated with MIC has resulted in obstructive growth and pipe wall damage in piping within several feet of fire pump discharge flange. Contrary to what might be expected, the increased velocity and flow rate at fire pump discharge does not eliminate the risk of system damage from MIC in this area. If questionable flange leaks or reduced fire pump flow performance is detected and not determined to be an impeller or driver issue, the discharge piping on the fire pump should be visually inspected per Chapter 14.

8.1.5* Auxiliary Equipment. The pump assembly auxiliary equipment shall include the following:

(1) Pump accessories as follows:
  (a) Pump shaft coupling

A pump shaft coupling, as shown in Exhibit 8.3, is used to connect a driver and a pump shaft together to transmit driver torque and rotation to the pump. These couplings allow for minor angular and parallel driver-to-pump misalignment from base plate flexibility and temperature changes.

  (b) Automatic air release valve

Automatic float-operated air release valves are required to be located at the high point on centrifugal fire pump casings to remove air that can become trapped inside the pump casing. Trapped air can cause pump cavitation, which can quickly result in permanent impeller damage if not released. On some old installations, the air release valve might be manually operated. Manually operated air release valves are allowed by the current standard but only for manually operated pumps. Exhibit 8.4 shows an air release valve mounted on top of the fire pump casing.

  (c) Pressure gauges

Pressure gauges, such as the two shown in Exhibit 8.5, are located on the suction and discharge pipes on a horizontal split case pump. These gauges are required to be located near the pump casing by NFPA 20, *Standard for the Installation of Stationary Pumps for Fire Protection*. There are also specific gauge sizes and gauge pressure ratings that are required based on specific fire system suction pressure and pump working pressures. A pressure gauge is also required as a

EXHIBIT 8.3 *Pump Shaft Coupling Under a Protective Cage.*

EXHIBIT 8.4 *Air Release Valve.*

EXHIBIT 8.5 *Pump Casing with Air Release Valve and Suction and Discharge Pressure Gauges.*

component of the engine side of a diesel driver heat exchanger to verify cooling water flow and that the water pressure is within the specifications of the driver heat exchanger. See 8.2.2(4)(j) for a more detailed discussion.

Heat exchanger ratings normally range from 30 psi to 60 psi (2.1 bar to 4.1 bar). Because the cooling water is taken off of the pump discharge, it is easy to overpressurize the cooling water heat exchanger, which can result in catastrophic failure.

(d) Circulation relief valve (not used in conjunction with diesel engine drive with heat exchanger)

The circulation relief valve, often referred to as the automatic relief valve, is also sometimes confused with the pressure relief valve, which is much larger and has a different function. The circulating relief valve, as the name implies, is designed to open slightly below churn pressure and provide impeller, packing, and bearing cooling by relieving water that is circulating around the pump casing water. This valve opens to relieve some of the water heated in the casing by friction, which then allows a small flow of cooler water to pass from the suction pipe through the casing.

Critical Point

If circulation relief valves are incorrectly set too low or fail to operate, pumps can operate without casing relief, but within a short time they will overheat from frictional heat.

If circulation relief valves are incorrectly set too low or fail to operate, pumps can operate without casing relief but within a short time they will overheat from frictional heat. In cases where this has occurred, the steam created raises temperatures to levels that operate pump room sprinklers. At this point, the damage to fire pumps often requires a complete and costly impeller replacement. Adequate flow through the circulating relief valves should be the top priority each time a pump is churned, because these valves are usually spring-operated and they can fail with the slightest bit of foreign material in the water supply or hard water buildup in the valve body. NFPA 20 requires these valves to "discharge to drain," but the intent is that they are piped in such a way to an "open drain" where flow can be verified with each pump churn.

Circulating relief valves are typical only installed on electric-driven pumps, since most diesel-driven pumps have heat exchanger cooling supplied by a water connection to the pump discharge pipe that serves the same purpose as the circulation relief valve. Per NFPA 20, these valves must be sized at 0.75 in. (19 mm) for pump capacities of 2500 gpm (9462 L/min) or less and 1 in. (25 mm) for all larger pump size. Exhibit 8.6 shows a typical circulation relief valve, while Exhibit 8.7 shows a diesel-driven pump with a cooling water line connection.

EXHIBIT 8.6 *Circulation Relief Valve Installed on Electric-Driven Pump.*

EXHIBIT 8.7 *Cooling Water Line Connection Installed on Diesel-Driven Pump, Negating the Need for a Circulating Relief Valve.*

Case In Point

The purpose of the circulating relief valve is to create enough flow to maintain water temperature in the pump casing. Accordingly, the optimum circulating relief valve setting is based on the maximum temperature that a pump casing should experience, to avoid damage to the packing glands, impeller, and bearings. Because the water in the pump casing should never exceed 140°F (60°C), this valve must be open enough to prevent the pump from reaching this temperature at churn. However, the valve setting is really not determined by pressure but by flow, which is dependent on the brake horsepower (BHP) of the pump running at churn and not on the capacity of the pump. This valve setting can be roughly calculated with assumptions such as the churn BHP of a pump equals approximately 50 percent maximum pump BHP, and that heat is approximately 80 percent dissipated energy.

Assumptions would also have to be made on the maximum water temperature in order to calculate heat absorption. Going through this exercise, a safe rough estimate would be a minimum flow of about 16 gpm (60 L/min). In fact, neither the calculation nor the valve settings are usually critical since the design standard, NFPA 20, requires a larger relief valve for larger pumps. NFPA 20 requires a ¾ in. (19 mm) casing relief valve for pumps with a capacity up to 2500 gpm (9462 L/min), and a 1 in. (25 mm) valve for larger pumps. A ¾ in. (19 mm) valve set to fully open discharges about 30 gpm (114 L/min), which is easily enough to cool a 100 hp to 200 hp pump. For a 2500 gpm (9462 L/min) pump, this is approximately 1 percent of the rated flow if wide open. Setting a casing relief valve on smaller pumps, such as a 500 gpm (1892 L/min) pump, might take some consideration. A relief valve flowing 30 gpm (114 L/min) could actually relieve a significant portion of the fire pump's capacity. In this case, it would be recommended to set the valve to discharge much less.

(2) Pump test device(s)

As indicated in 8.1.5(2), each pump must have the means to be flow tested annually and after any overhauls or repairs. The most common method to meet this requirement is to install a fire pump test header. Per NFPA 20, test headers, such as the ones shown in Exhibit 8.8, are sized such that there will be at least one hose valve for each 250 gpm (946 L/min) of pump flow capacity.

The second most common method is to install a flowmeter (see Exhibit 8.9). Flowmeters are typically installed on a bypass pipe or test header with discharge back to a water storage tank or reservoir. The benefit of this test method is that it can be completed without the added time or resources that are needed to connect test hoses and play pipes. Tools, such as a pitot tube to measure flow, also are not required. With this method, wasting water and risking damage from play pipe discharge to the ground are also avoided.

However, when using such devices in a "closed loop," every third year the annual test must be conducted using hose streams or the flowmeter bypass to discharge into a drain or suction reservoir.

Critical Point

As indicated in 8.1.5(2), each pump must have the means to be flow tested annually and after any overhauls or repairs.

EXHIBIT 8.8 *Examples of Fire Pump Test Headers. (Left: Courtesy of John Munno and Arthur J. Gallagher Risk Management Services, Inc.)*

EXHIBIT 8.9 *Fire Pump Flowmeter.*

Case In Point

The need for a main pump relief valve and subsequent required sizing are dictated by NFPA 20. In the 2013 edition, the requirements dictating when a pressure relief valve is required have been changed from previous editions. Although they previously were required for all engine-driven fire pumps due to concern over speed governors failing and creating an overspeed (system overpressurization) condition, there have not been any documented cases of this occurring. Thus, they now are only required as follows:

1. For standard diesel engine–driven pumps, where 1.21 × net-rated shutoff (churn) plus elevation-adjusted maximum pump suction supply static pressure exceeds any system component listings' maximum allowable pressure [typically 175 psi (12.1 bar)]
2. For pressure-limiting diesel engine–driven pumps, where elevation-adjusted maximum total discharge churn pressure at rated speed exceeds any system component listings' maximum allowable pressure [typically 175 psi (12.1 bar)]
3. For electric motor–driven pumps with variable speed pressure limiting controllers, where elevation-adjusted maximum total discharge churn pressure at rated speed exceeds any system component listings' maximum allowable pressure [typically 175 psi (12.1 bar)]

Flowmeters must be listed, and they often need factory calibration. Stream-straightening devices on test headers, such as play pipes, should also be listed. If a test header or flowmeter cannot be used, a site fire hydrant or standpipe can be used to connect hoses and nozzles to conduct the testing.

(3) Pump relief valve and piping (where maximum pump discharge pressure exceeds the rating of the system components or the driver is of variable speed)

A relief valve, which is sometimes called a main relief valve or pressure relief valve, is designed to open at a predetermined pressure [usually 175 psi (1.2 bar)] to protect the piping and other components in the fire protection system from overpressurization. See Exhibit 13.31 for an illustration of a pressure relief valve. This should not be confused with a pressure reducing valve shown in Exhibit 13.32. Confusion can occur if the abbreviation of PRV is used, which typically is used for reducing valves and not relief valves.

(4) Alarm sensors and indicators

The alarm sensors and indicators listed in 8.1.5(4) are the required supervisory signals addressing pump readiness or operation that require immediate attention per NFPA 20 installation requirements. This requirement encompasses not just the fire pump driver and controller but all alarms associated with devices to keep the fire pump fully operational, including those for valve tamper switches, controller alarms for the "off" position, failure to start, tanks fuel levels (for diesel drivers), and loss of phase for electric pumps. Per NFPA 20 requirements, these alarms must be of an audible level to be clearly heard in the pump room while the pump is running and in every controller position except "off." In addition, if pump rooms are not constantly attended, the alarms listed in NFPA 20 also must transmit to a constantly attended location and annunciate either with an audible or visible indication. These requirements should guide testing per NFPA 25.

It should be noted that while most remote annunciation signals can be grouped into a common "trouble" signal, the signals for "pump running" and "pump off" must indicate separately from all other signals. A common error in new pump installations and pumps that have been damaged or rewired is the crossing of wires and the poor panel description programming of these signals. One such example illustrating this is when a "pump running" signal appears on a security center screen as a simple "valve trouble alarm," and thus, does not get the immediate action required. For this reason, alarms should be tested from the driver to the remote annunciation location when possible.

The quality and training of attendants at the annunciator location should also be regularly reviewed to verify that they meet the definition of the term *qualified*, in relation to understanding the significance and actions required when various fire pump system signals are received.

(5) Right-angle gear sets (for engine-driven vertical shaft turbine pumps)

Right-angle gear drives (see Exhibit 8.10) are built with a cooling oil reservoir. That reservoir must be checked regularly per the manufacturer's requirements to verify oil levels are acceptable to provide proper cooling and prevent overheating. Oil temperatures should typically be maintained below 200°F (93.3°C) and preferably below 135°F (57°C) to avoid premature breakdown.

(6) Pressure maintenance (jockey) pump and accessories

The pressure maintenance pump, or jockey pump, as shown in Exhibit 8.11, is a low-flow, high-pressure pump, installed in parallel with a primary pump on automatic fire pump systems (see Exhibit 8.12). Pressure maintenance pumps are installed to prevent the unplanned operation of primary pumps when system pressures fluctuate or drop slightly, such as from small fitting leaks in large underground water main systems. These pumps typically are rated to slightly exceed system pressure created by the primary pump but with the capability of only providing 40 gpm to 75 gpm (150 L/min to 284 L/min). Thus, the discharge of a pressure maintenance pump is not intended to supply a fire sprinkler system and typically can only supply the requirements of a single sprinkler.

EXHIBIT 8.10 *Right-Angle Gear Drive Attached to a Diesel Engine Driver.*

EXHIBIT 8.11 *Pressure Maintenance (Jockey) Pump.*

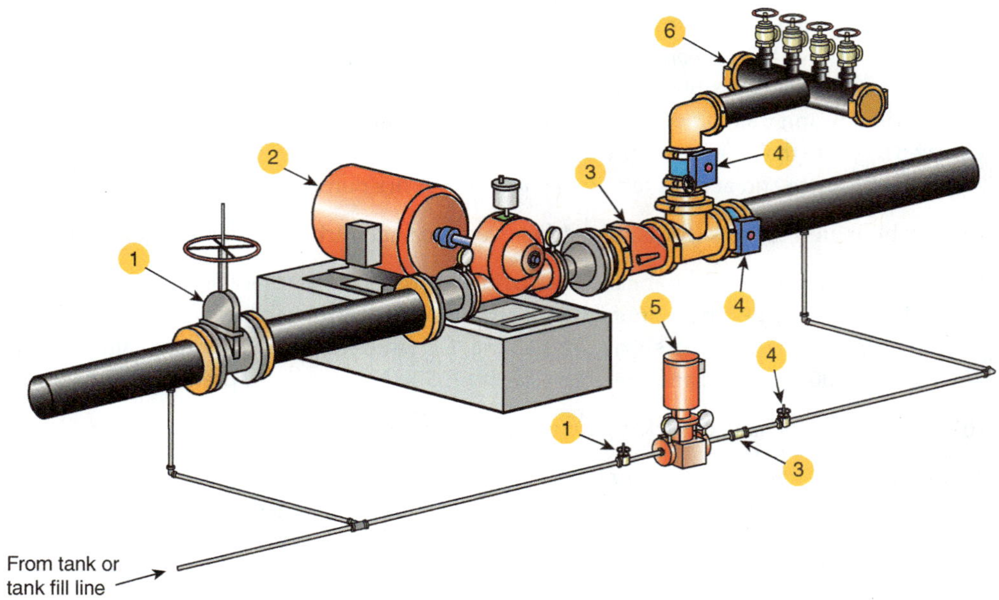

Legend			
1	OS&Y gate valve	4	OS&Y gate valve or indicating butterfly valve
2	Fire pump	5	Jockey pump
3	Check valve	6	Hose header

EXHIBIT 8.12 *Required Jockey and Fire Pump Installation Arrangement. (Courtesy of Stephan Laforest)*

Pressure maintenance pumps are actuated by dedicated controllers and system pressure-sensing lines, so the system pressure decreases when one or more sprinklers activate. Under these circumstances, as the system pressure drops, the pressure switch (see Exhibit 8.13) in the pressure maintenance pump controller senses the pressure drop and activates the pump. Since the pressure maintenance pump is not designed to keep pace with the flowing sprinkler(s), the system pressure continues to drop.

The pressure switch shown in Exhibit 8.13 is located in the fire pump controller and has high- and low-pressure settings to start and stop the associated pump. The pressure switch is internally mounted in each controller and has specific settings based on the water supply. These can also be fully electronic and part of a controller master display only.

The following is an example of recommended pressure switch settings from A.14.2.5(4)(f) of NFPA 20:

Main pump start is set at maximum churn − 5 psi (0.4 bar).

Main pump stop is set at main pump start setting + 15 psi (1 bar).

Jockey pump start is set at main pump start setting − 10 psi (0.7 bar).

Jockey pump stop is set equal to main pump stop setting.

In fact, there is usually at least a 5 psi (0.4 bar) pressure differential between the pressure maintenance pump and the fire pump stop settings. Pressure switch settings less than 5 psi (0.4 bar) can cause unintentional starting of the fire pump.

Some jurisdictions and insurance companies often require fire pumps only be stopped manually and that the automatic stop relay in the controller be removed or bypassed. This is

EXHIBIT 8.13 Mercoid-Type Pressure Switch.

due to concerns that a fire pump could repeatedly cycle on and off during a fire and eventually not restart from an off sequence. NFPA 20 allows both an automatic and manual fire pump stop setting, except in cases where a pump constitutes a sole site fire protection water source. In these cases, NFPA 25 requires a manual stop only.

The jockey pump should be sized to be able to maintain the system pressure above the starting pressure of the main fire pump, while accounting for small leaks in the system. Such units are not required to be listed for fire service but should be listed as general electrical pump devices. Flow tests of pressure maintenance pumps and controllers are not required in NFPA 25, but they should be started and run through a normal on-off cycle during the annual fire pump test to verify set points. Conditions where a primary pump appears to be running excessively should be investigated. This problem could indicate an incorrect jockey pump setting or service wear that might require replacement.

In addition to dedicated controllers, each pressure maintenance pump and primary pump is required to have independent, dedicated pressure-sensing lines (arranged as shown in Exhibit 8.14). Each sensing line is required to have two check valves or ground-face unions installed, to act as pressure snubbers and limit fire pump start-stop cycles due to momentary pressure surges, as shown in Exhibit 8.15.

Critical Point

Some jurisdictions and insurance companies often require fire pumps only be stopped manually and that the automatic stop relay in the controller be removed or bypassed.

A.8.1.5 Types of centrifugal fire pumps include single and multistage units of horizontal or vertical shaft design. Listed fire pumps have rated capacities of 25 gpm to 5000 gpm (95 L/min to 18,925 L/min), with a net pressure range from approximately 40 psi to 400 psi (2.75 bar to 27.6 bar).

Multistage pumps have a single shaft and casing but more than one impeller inside the casing. For example, a four-stage pump has four impellers on the same shaft and in the casing that take suction from the same source. Impellers can be installed in both parallel and series arrangements. A parallel multistage arrangement, as shown in Exhibit 8.16, is used to boost flow beyond that of a single-stage (single-impeller) pump. A series multistage arrangement, as shown in Exhibit 8.17, is used to boost pressure beyond that of a single-stage pump. This arrangement is typically seen in high-rise applications. Testing of these types of pumps is typically completed in the same manner as for single-stage pumps. One exception to this is multistage, multi-outlet pumps, where a single casing can have multiple outputs in the casing, each providing an independent discharge. These are being used in Europe where a single fire pump can supply multiple levels of a high-rise structure without pressure zoning. Exhibit 8.18 shows a single-shaft, two-impeller horizontal split case pump.

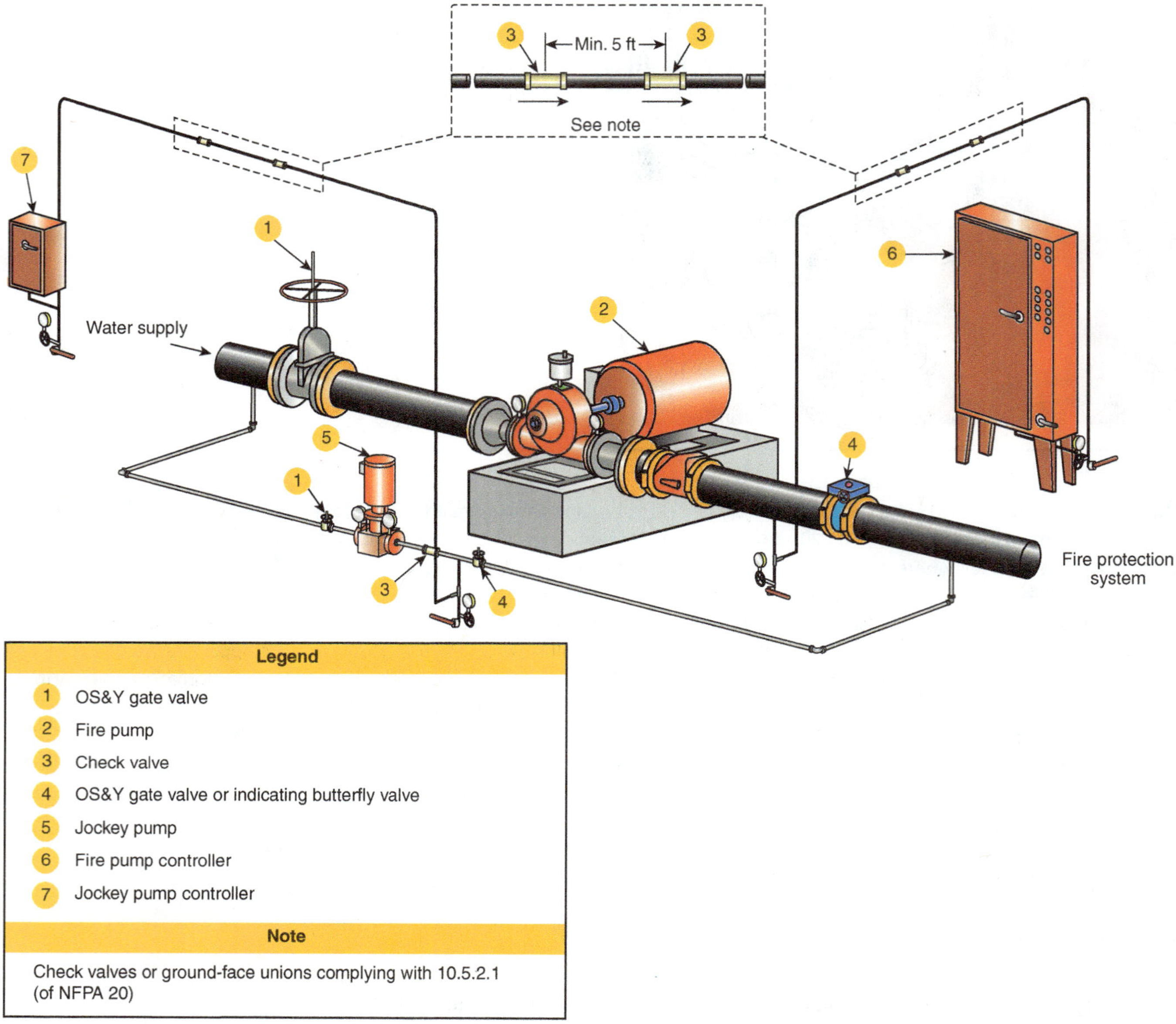

EXHIBIT 8.14 *Required Pressure Maintenance Pump and Fire Pump Arrangement with Dedicated Pressure-Sensing Lines. (Courtesy of Stephan Laforest)*

(1) *Horizontal Split Case.* This pump has a double suction impeller with an inboard and outboard bearing and is used with a positive suction supply. A variation of this design can be mounted with the shaft in a vertical plane. *[See Figure A.8.1.5(a).]*
(2) *End Suction and Vertical In-Line.* This pump can have either a horizontal or vertical shaft with a single suction impeller and a single bearing at the drive end. *[See Figure A.8.1.5(b).]*
(3) *Vertical Shaft, Turbine Type.* This pump has multiple impellers and is suspended from the pump head by a column pipe that also serves as a support for the shaft and bearings. This pump is necessary where a suction lift is needed, such as from an underground reservoir, well, river, or lake. *[See Figure A.8.1.5(c).]*

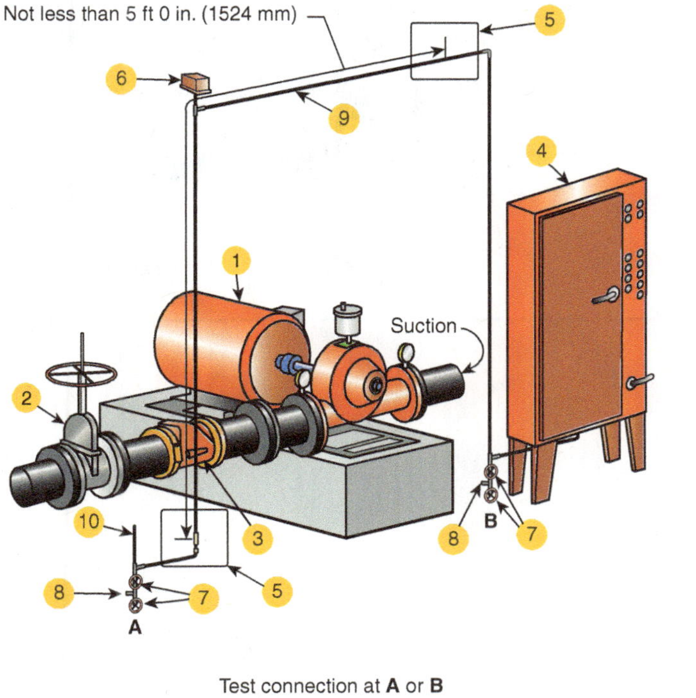

Legend

1. Fire pump
2. Indicating control valve
3. Check valve
4. Control panel (with pressure switch)
5. Bronze check valves with 3/32 in. (2 mm) orifice in clapper
6. Supplemental air chamber or pulsation damper—might be needed if water pulsation causes erratic operation of the pressure switch or the recorder
7. ½ in. (15 mm) globe valves
8. ¼ in. (mm) plug
9. Not less than ½ in. (15 mm) brass pipe with brass fittings or equivalent
10. Connect to a tapped boss or other suitable outlet between the indicating control valve and check valve

Notes

(1) Solenoid drain valve used for engine-driven fire pumps can be at A, B, or inside controller enclosure.
(2) If water is clean, ground-face unions with noncorrosive diaphragms drilled for 3/32 in. (2 mm) orifices can be used in place of the check valves.

EXHIBIT 8.15 *Typical Fire Pump Piping Sensing Line Arrangement Showing Snubber Placement. (Courtesy of Stephan Laforest)*

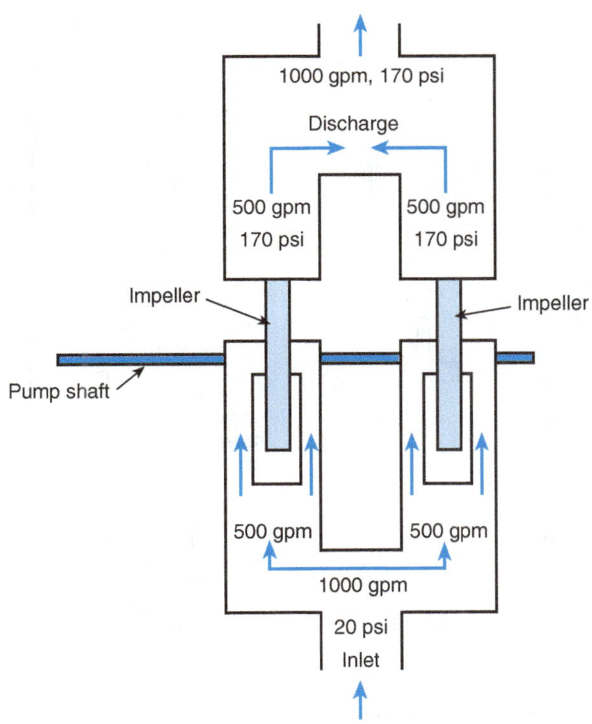

For SI units, 1 psi = 0.0689 bar; 1 gpm = 3.785 L/min

EXHIBIT 8.16 *Two-Stage Pump in Parallel. (Source: Pumps for Fire Protection Systems, 2002, Figure 3.9)*

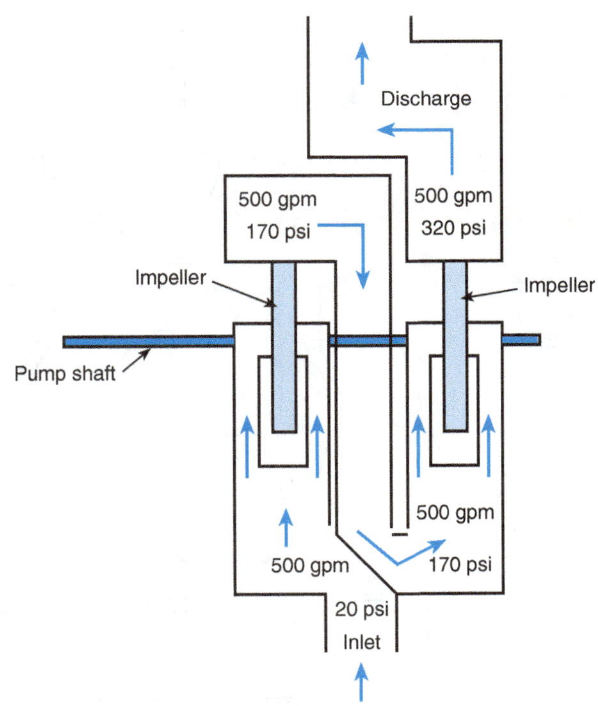

For SI units, 1 psi = 0.0689 bar; 1 gpm = 3.785 L/min

EXHIBIT 8.17 *Two-Stage Pump in Series. (Source: Pumps for Fire Protection Systems, 2002, Figure 3.10)*

EXHIBIT 8.18 Two Pump Impellers on a Single Shaft with Shrouds, Vanes, and Eye. (Source: Pumps for Fire Protection Systems, 2002, Figure 3.3)

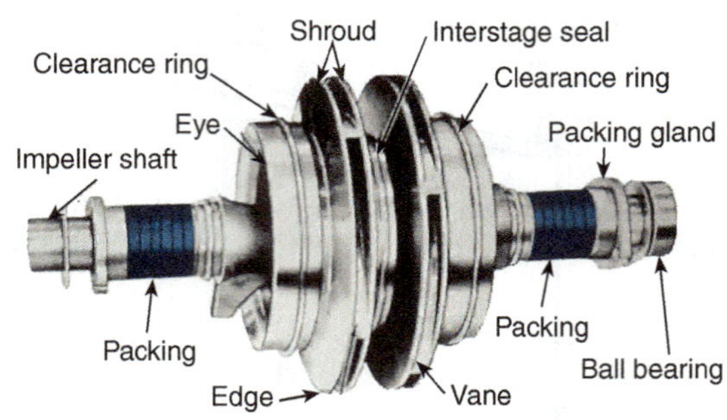

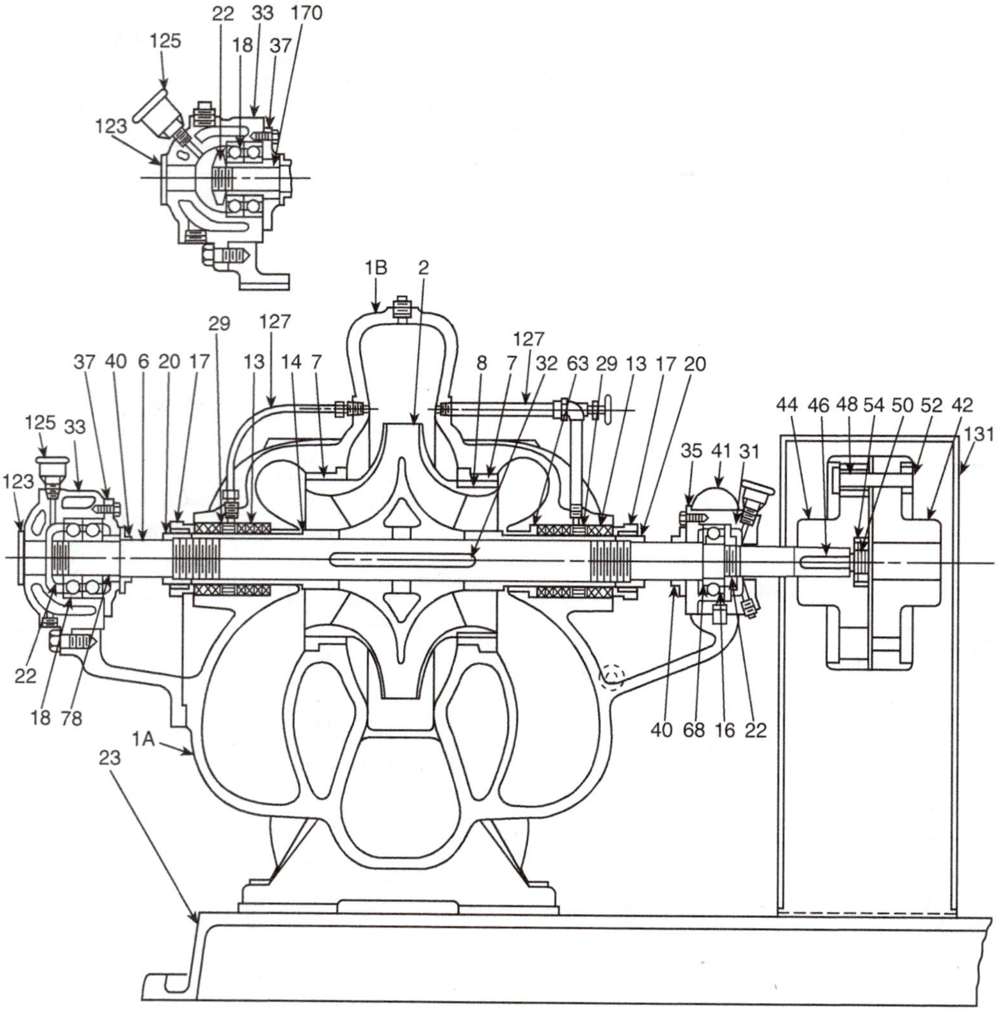

FIGURE A.8.1.5(a) *Impeller Between Bearings, Separately Coupled, Single-Stage Axial (Horizontal) Split Case. (Courtesy of the Hydraulic Institute, Parsippany, NJ, www.Pumps.org.)*

NFPA 20 allows positive displacement pumps for water mist and foam system applications. For details on the inspection, testing, and maintenance of these pumps, the pump manufacturer should be contacted.

8.1.6 Water Supply to Pump Suction.

8.1.6.1 The suction supply for the fire pump shall provide the required flow at a gauge pressure of 0 psi (0 bar) or higher at the pump suction flange to meet the system demand.

The pressure observed at the suction gauge while the pump is operating should be positive, even when the fire pump is operating at its overload condition. The suction gauge pressure is permitted by NFPA 20 to drop to −3 psi (−0.2 bar) when the supply is a suction tank with its base at or above the same elevation as the pump, provided the lowest water tank level and all maximum fire system demands and durations have been properly supplied. However, it should also be noted that when taking water from a public water system, most municipal water companies limit the residual pressure on the system to 20 psi (1.4 bar) due to concerns of collapsing water mains and drastically increasing water turbidity. But this is generally the minimum required at the city connection. Many times, gauges at the location of a pump test could be 10 psi to 20 psi (0.7 bar to 1.4 bar) above that at the city connection due to pressure losses across backflow apparatus and numerous valves and other reasons. Thus, in these cases, an additional pressure gauge should be attached to a hydrant or the city side of the site backflow apparatus and examined so the accurate public water supply pressure can be monitored for compliance with any minimum amounts stipulated by the municipality.

8.1.6.2 Those installations for which NFPA 20, *Standard for the Installation of Stationary Pumps for Fire Protection*, permitted negative suction gauge pressures at the time of pump installation, where the system demand still can be met by the pump and water supply, shall be considered to be in compliance with 8.1.6.

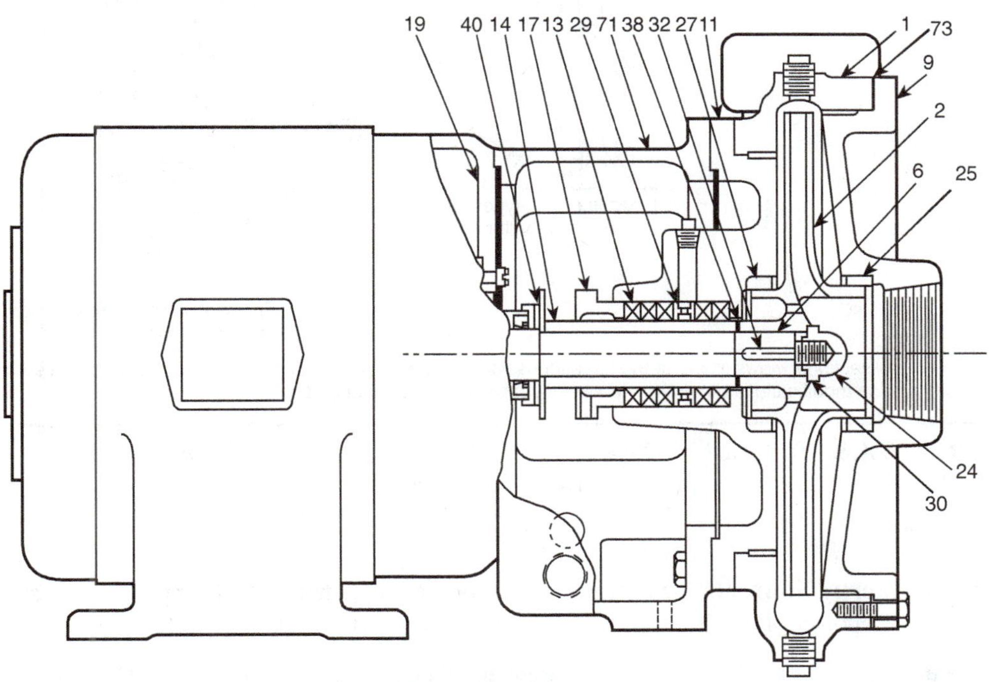

The numbers used in this figure do not necessarily represent standard part numbers used by any manufacturer.

FIGURE A.8.1.5(b) *Overhung Impeller, Close-Coupled, Single-Stage, End Suction. (Courtesy of the Hydraulic Institute, Parsippany, NJ, www.Pumps.org.)*

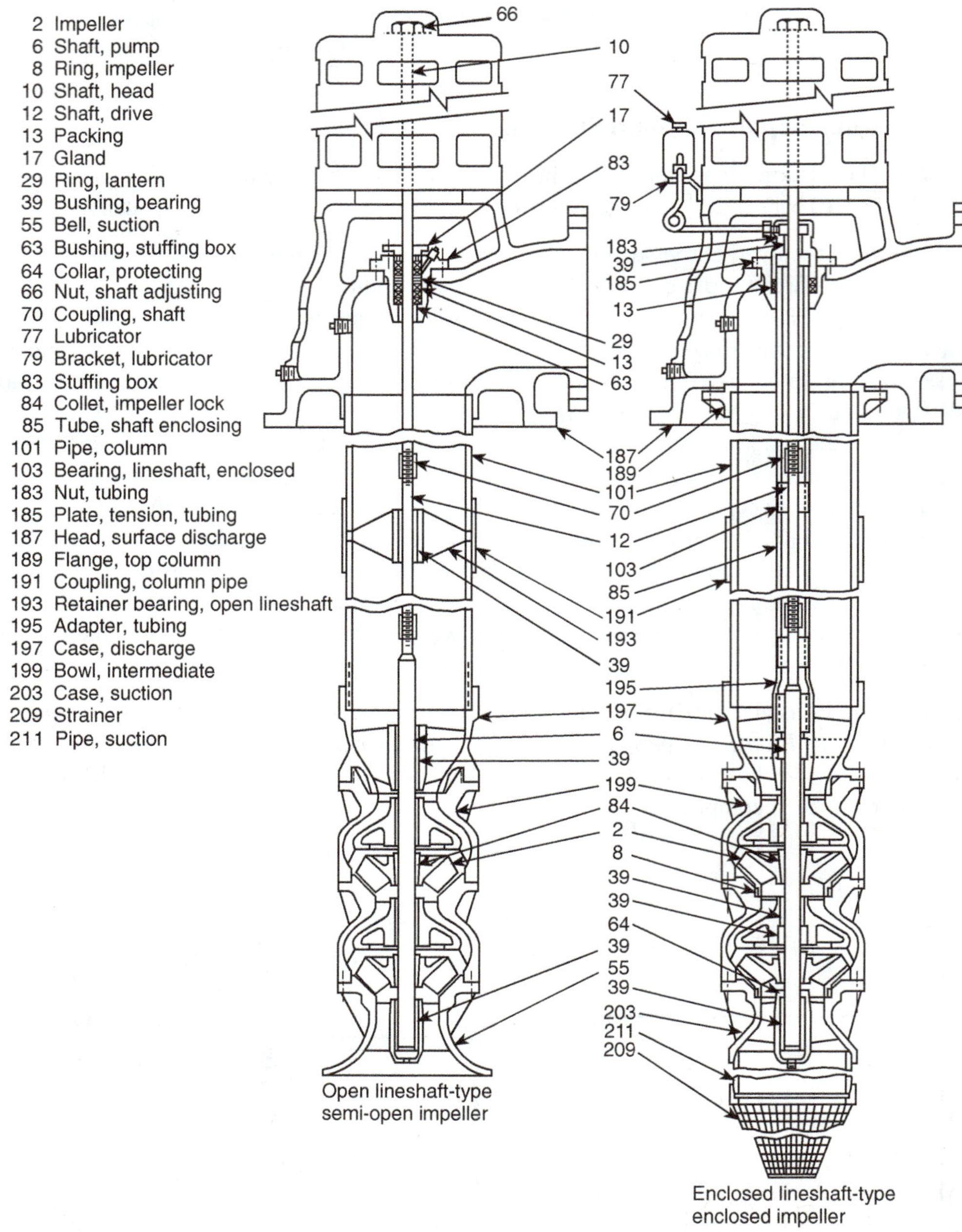

The cross-sectional views illustrate the largest possible number of parts in their proper relationship and some construction modifications but do not necessarily represent recommended design.

FIGURE A.8.1.5(c) *Turbine-Type, Vertical, Multistage, Deep Well. (Courtesy of the Hydraulic Institute, Parsippany, NJ, www.Pumps.org.)*

8.1.7 Energy Source. The energy sources for the pump driver shall supply the necessary brake horsepower of the driver so that the pump meets system demand.

The energy sources referred to in 8.1.7 are electric power, steam, and diesel fuel. Other fuels currently are not allowed for newly installed fire pump installations.

8.1.8 Driver. The pump driver shall not overload beyond its rating (including any service factor allowance) when delivering the necessary brake horsepower.

8.1.9* Controller. Automatic and manual controllers for applying the energy source to the driver shall be capable of providing this operation for the type of pump used.

The two types of controllers approved for electric fire pump service are limited-service and full-service controllers. For installations with across-the-line starting of squirrel-cage type motors 30 hp or less and 600 V or less, limited-service controllers can be used. These controllers are typically less expensive and less complex than standard controllers and are used in cases where pumps are connected to large, high-capacity water mains with very low pressures. The other type of controller approved for electric fire pump service is a full-service controller, which is seen in most installations.

A.8.1.9 Controllers include air-, hydraulic-, or electric-operated units. These units can take power from the energy source for their operation, or the power can be obtained elsewhere. Controllers used with electric power sources can apply the source to the driver in one (across-the-line) or two (reduced voltage or current) steps. Controllers can be used with automatic and manual transfer switches to select the available electric power source where more than one is provided.

The most common reason for installing multiple electric power supplies, as referred to in A.8.1.9, is in cases where one or more of the supplies do not offer the reliability required by NFPA 20. When testing fire pumps, each power source must be tested using the transfer switch to determine its adequacy for supplying the fire pump under full load. Testing each pump on each power supply under full load is critical in determining true reliability and redundancy.

Critical Point

Testing each pump on each power supply under full load is critical in determining true reliability and redundancy.

8.1.10 Impairments. The procedures outlined in Chapter 15 shall be followed where an impairment to protection occurs.

8.2 Inspection

8.2.1 The purpose of inspection shall be to verify that the pump assembly appears to be in operating condition and is free from physical damage.

Exhibit 8.19 shows a pump casing that has been severely damaged. In this instance, the contractor replaced a valve in the suction piping and used a mechanical assist on the piping connected to the fire pump to reconnect the piping to the valve.

EXHIBIT 8.19 *Crack in Pump Casing. (Courtesy of Damon Pietraz, Underwood Fire Equipment)*

ITM Deficiency, Impairment, or Hazard Evaluation?

(Courtesy of Byron Blake and SimplexGrinnell)

ANSWER: ITM Deficiency

This photo shows a serious problem with the fire pump test header. The normal visual inspection probably did not reveal this problem, since the control valve to the test header is in a typical closed position and the hose valves are cracked but not broken apart. However, when the control valve to the test header was opened to begin the annual fire pump flow test, the damaged valves began spraying water. Subsection 8.2.1 requires that the pump assembly be free from physical damage, and as described in 8.1.5(2), the pump test device is part of the pump assembly auxiliary equipment.

While this break is serious, it should not be considered an impairment because the pump assembly itself will function in a fire event. The repairs can be made without affecting the pump when the test header control valve is shut.

8.2.2* The pertinent visual observations specified in the following checklists shall be performed weekly:

While NFPA 25 does not mandate that a specific inspection form must be used, it is important to document that each item required to be inspected has been addressed at the proper frequency. Exhibit 8.20 is a sample form that could be used to document the items identified in 8.2.2.

(1) Pump house conditions as follows:
  (a) Heat is adequate, not less than 40°F (4°C) for pump room with diesel engine–driven pumps with engine heaters.
  (b) Heat is adequate, not less than 70°F (21°C) for pump room with diesel engine–driven pumps without engine heaters.

Water is present in the fire pump piping and sprinkler system that protects the fire pump room, so it is necessary to prevent temperatures that could freeze the water in these systems by providing heat in the pump house.

In the case of diesel engines, it is also necessary to provide a sufficiently cooled environment in summer months for the engine to properly operate. Diesel engines are derated 1 percent for each 10°F (5.6°C) above 77°F (25°C) ambient temperature. Thus, at significantly elevated temperatures, diesel drivers might not be capable of delivering the necessary horsepower to drive a pump at its required speed. High temperatures can also be an issue in electric pump rooms where controllers can be damaged. Many controllers have maximum operating temperatures of 104°F (40°C). Some offer a 122°F (50°C) option but, in either case, fire pump buildings with metal deck roofs in areas subject to high temperatures could quickly see elevated temperatures without air conditioning.

FIRE PUMPS WEEKLY INSPECTION

This form covers a 3-month period.

Year: ______________________ **System:** ______________________

Location: ______________________

Y = Satisfactory **N** = Unsatisfactory (explain below) **N/A** = Not applicable

Date: ______________ Inspector: ______________________

	Y	N	N/A	Comments
Heat in pump room is 40°F (4.4°C) or higher	❑	❑	❑	
Intake air louvers in pump room appear operational	❑	❑	❑	
Pump suction, discharge, and bypass valves are open	❑	❑	❑	
No piping or hoses leak	❑	❑	❑	
Fire pump leaking one drop of water per second at seals	❑	❑	❑	
Suction line pressure is normal	❑	❑	❑	
System line pressure is normal	❑	❑	❑	
Suction reservoir is full	❑	❑	❑	
Controller pilot light (power on) is illuminated	❑	❑	❑	
Transfer switch normal power light is illuminated	❑	❑	❑	
Isolating switch for standby power is closed	❑	❑	❑	
Reverse-phase alarm light is not illuminated	❑	❑	❑	
Normal-phase rotation light is illuminated	❑	❑	❑	
Oil level in vertical motor sight glass is normal	❑	❑	❑	
Diesel fuel tank is at least ⅔ full	❑	❑	❑	
Controller selector switch is in "auto" position	❑	❑	❑	
Voltage readings for batteries (2) are normal	❑	❑	❑	
Charging current readings are normal for batteries	❑	❑	❑	
Pilot lights for batteries are "on" or battery failure pilot lights are "off"	❑	❑	❑	
All alarm pilot lights are "off"	❑	❑	❑	
Record engine running time from meter	❑	❑	❑	
Oil level is normal in right-angle gear-drive pumps	❑	❑	❑	
Crankcase oil level is normal	❑	❑	❑	
Cooling water level is normal	❑	❑	❑	
Electrolyte level in batteries is normal	❑	❑	❑	
Battery terminals are free of corrosion	❑	❑	❑	
Water-jacket heater is operational	❑	❑	❑	
For steam-driven pumps, steam pressure is normal	❑	❑	❑	
Examine exhaust system for leaks	❑	❑	❑	
Check lube oil heater for operation (diesel pumps)	❑	❑	❑	
Drain condensate trap of cooling system	❑	❑	❑	
Check for water in diesel fuel tank	❑	❑	❑	

NFPA® NFPA Water-Based Fire Protection Systems Forms

EXHIBIT 8.20 *Sample Form for Fire Pumps Weekly Inspection.*

EXHIBIT 8.21 Ventilation Louvers.

(c) Ventilating louvers are free to operate.

Paragraph 8.2.2(1)(c) requires a weekly pump house ventilating louver inspection to verify their operational condition. These louvers are needed to maintain fresh air for proper diesel system combustion and room air temperature control. In cold climates, the inlet air might need to be heated to prevent small piping and gauges from freezing. See Exhibit 8.21.

(2) Pump system conditions as follows:

(a) Pump suction and discharge and bypass valves are fully open.

NFPA 13, *Standard for the Installation of Sprinkler Systems,* requires that all valves controlling water supplies be supervised. Although Chapter 13 of NFPA 25 indicates that electronically monitored control valves and/or locked valve inspections can be decreased from weekly to monthly inspection frequencies, this section supersedes that requirement for control valves associated with fire pumps. All valves directly associated with the fire pump (i.e., in the pump room) must be part of the weekly visual inspection.

(b) Piping is free of leaks.

Piping leaks in suction and discharge piping, most frequently found at flanges or couplings as shown in Exhibit 8.22, can be considered impairments or can lead to impaired systems, depending on the severity of the leak.

Frequently, inexperienced personnel might view the water in the drip pocket as a leak. If the packing glands are tightened to the point where no water is allowed to drip, then the packing gland will dry out and fail. Exhibit 8.23 illustrates a packing gland that has been properly adjusted, because water can be seen in the drip pocket.

(c) Suction line pressure gauge reading is within acceptable range.

As required by 8.2.2(2)(c), an inspection of pump system conditions includes a check to ensure that the suction line pressure gauge reading is within an acceptable range. Where the water supply is taken from a tank, the gauge should be positive at a pressure approximately equal to the water height in the tank multiplied by 0.433 psi/ft. As an example, a pump attached to a tank that has 20 ft (6.1 m) of water above the point of the pipe connection should have

EXHIBIT 8.22 *Leaking Coupling in Fire Pump Piping. (Courtesy of Indonesian Fire and Rescue Foundation)*

EXHIBIT 8.23 *Fire Pump Packing Gland and Drip Pocket.*

approximately 8.5 psi (0.6 bar) on the suction gauge [0.433 psi/ft × 20 ft = 8.7 psi (0.6 bar)]. Where the supply is from a public source, the gauge should equal the public water supply pressure, adjusted for elevation changes.

(d) System line pressure gauge reading is within acceptable range.

Fire pumps boost pressure only when they are running. Therefore, when a pump is not running, the pressure on the system line pressure (pump discharge) gauge should be the same as the suction pressure gauge.

(e) Suction reservoir has the required water level.

An altitude float located on the exterior of the tank is usually used to check the pressure in the suction reservoir. These devices should always be moved slightly at inspection to ensure that the sliding gauge has not become stuck in the full position. In some installations, a pressure gauge that reads in feet of water or psi is installed in the pump room. See Chapter 9 for inspection, testing, and maintenance requirements for suction tanks.

In the case of suction from a well or underground reservoir, a different method must be used to check the water level. It involves using an air line that extends below the water surface and a pressure gauge that, when air is blown into the tubing, reads how far below grade the water is. See A.7.3.5.3 and Figure A.7.3.5.3 of NFPA 20, for a detailed description of this method.

(f) Wet pit suction screens are unobstructed and in place.

The size and shape of the wet pit are governed by the Hydraulic Institute standards. The following must be checked during the inspection required by 8.2.2(2)(f):

- Trash rack
- Two sets of screens
- Pump suction strainer at the bottom of the submerged pump assembly

A small amount of debris can negatively affect pump operation by damaging the pump impeller or by causing obstruction of the pump and the piping around the pump.

(g) Waterflow test valves are in the closed position.

Paragraph 4.16.2 of NFPA 20 requires test header control valves be supervised in the closed position. That requirement is meant to protect against pump operation caused by small header cap leaks or unauthorized use of the header. The test header must be protected from freezing, so water must be fully drained following each use after the control valve is fully closed. Frequently, hose valves are found broken from freezing if they have not been properly drained following their use.

(3) Electrical system conditions as follows:
  (a) Controller pilot light (power on) is illuminated.
  (b) Transfer switch normal pilot light is illuminated.

Transfer switches are installed only with electric motors that have an alternative electric power supply installed. Exhibit 8.24 through Exhibit 8.26 offer three views of a fire pump controller with an attached power transfer switch.

  (c) Isolating switch is closed — standby (emergency) source.

Isolation switches on each source of power to an electric motor drive must be closed when the pump is in the operating or standby condition.

EXHIBIT 8.24 *Fire Pump Controller with Transfer Switch.*

EXHIBIT 8.25 *Close-Up of Controller Indicators.*

EXHIBIT 8.26 *Close-Up of Transfer Switch Indicators.*

(d) Reverse phase alarm pilot light is off, or normal phase rotation pilot light is on.

Once the electric motor/controller installation is completed and the phases have been checked, there is little chance that reverse phase will be a problem, unless work is conducted on the electric supply system.

(e) Oil level in vertical motor sight glass is within acceptable range.
(f) Power to pressure maintenance (jockey) pump is provided.

In addition to the items listed in 8.2.2(3)(a) through 8.2.2(3)(f), the general condition of the electrical components should be observed and recorded. Potential problems that can be discovered include rodent nesting, plugged motor vents, broken parts, unlocked controller doors, and improperly labeled electrical panel boards. If the pump room is equipped with electric heat, the thermostat should be checked to determine that it is operating in cold weather and that the temperature is set at normal room temperature. Any electric controls for ventilation should also be verified that they are fully operational.

(4) Diesel engine system conditions as follows:
(a) Fuel tank is at least two-thirds full.

The fuel tank is designed to hold an 8-hour supply of fuel for the diesel engine. The quantity of fuel needed is based on a formula of 1 gal per horsepower (5.07 L per kW), plus a 10 percent volume adjustment for expansion and sump). If an engine does not consume this quantity of fuel — and many do not — the fuel storage can be adjusted for the actual demand. Fuel in the tank should be consumed within 1 year. Fuel kept longer than 1 year has an increased risk of becoming contaminated with biological growth and can clog the engine fuel filters, preventing the engine from running.

The bottom 5 percent of a fuel tank is reserved for collecting water or other contaminants. Therefore, this fuel should be removed on an annual basis, as is required by NFPA 25. Also, the bottom of the fuel tank should be observed to make sure that it is above the level of the fuel injectors so that if the fuel pump fails, the engine can still operate. Exhibit 8.27 and Exhibit 8.28 show a diesel fuel tank and fuel gauge.

EXHIBIT 8.27 *Fuel Tank for a Diesel Fire Pump (with Driver Supply Line Installed to Allow a 5 Percent Sump as Required).*

EXHIBIT 8.28 *Fuel Level Indicator for a Diesel Fire Pump.*

(b) Controller selector switch is in auto position.
(c) Batteries' (2) voltage readings are within acceptable range.
(d) Batteries' (2) charging current readings are within acceptable range.
(e) Batteries' (2) pilot lights are on or battery failure (2) pilot lights are off.
(f) All alarm pilot lights are off.
(g) Engine running time meter is reading.
(h) Oil level in right angle gear drive is within acceptable range.
(i) Crankcase oil level is within acceptable range.
(j) Cooling water level is within acceptable range.

A pressure gauge is also required as a component of the engine side of a diesel driver heat exchanger, to verify cooling water flow and that the water pressure is within the specifications of the driver heat exchanger. Engine heat exchanger ratings typically range from 30 psi to 60 psi (2.1 bar to 4.1 bar). Thus cooling water, which comes off of the pump discharge main piping typically at over 125 psi (8.6 bar), could quickly cause permanent damage to the exchanger if not properly regulated or if the regulator fails.

Older fire pumps typically only have one regulator installed on the main line. Thus, during emergency operation where cooling water flows only through the bypass line, pressure regulation must be done manually with the bypass line quarter turn valves and by monitoring the pressure gauge. Exhibit 8.29 through Exhibit 8.31 illustrate common engine driver cooling arrangements.

(k) Electrolyte level in batteries is within acceptable range.

The electrolyte level in a battery is normal when the battery is full of water up to the ring under the cell cap. A lower water level exposes the battery to an accumulation of hydrogen in the cell, which can lead to an explosion when the pump is started and there is a spark in the battery. Proper personal protective equipment (PPE) should be worn when filling batteries with electrolyte. (See Commentary Table 4.2 for PPE selection.)

Diesel fire pumps are required by NFPA 20 to have two battery banks. During the cranking cycle, the fire pump controller will alternate cranking from one bank to the other, with a 15-second rest period between attempts until the engine starts or until it has signaled through

EXHIBIT 8.29 *Close-Up of Fire Pump Cooling Line.*

EXHIBIT 8.30 *Typical Engine Driver Cooling Water Pressure Arrangement with Single Regulator.*

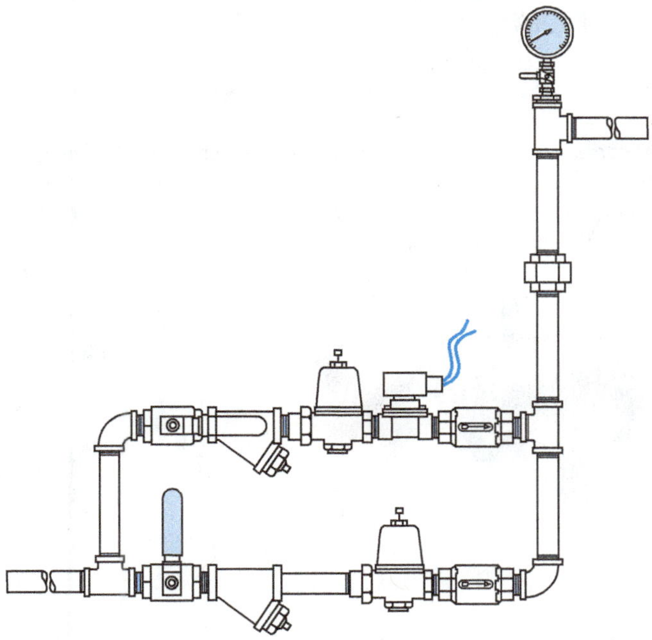

EXHIBIT 8.31 *Spring-Loaded Check Valve Arrangement. [Source: NFPA 20, 2013, Figure A.11.2.8.5.3.8(A)]*

EXHIBIT 8.32 *Diesel Fire Pump Batteries.*

a 3-minute cycle with six attempts on each battery. Both batteries must be maintained fully charged at all times. Exhibit 8.32 shows the two battery banks for a diesel fire pump.

(l) Battery terminals are free from corrosion.

(m) Water-jacket heater is operating.

(5)* Steam system conditions: Steam pressure gauge reading is within acceptable range.

A.8.2.2 See Table A.8.2.2 and Figure A.8.2.2.

A.8.2.2(5) Visual indicators other than pilot lights can be used for the same purpose.

TABLE A.8.2.2 *Observations — Before Pumping*

Item	*Before Pump Is Operated*
Horizontal pumps	1. Inspect drip pockets under packing glands for proper drainage. Standing water in drip pockets is the most common cause of bearing failure. 2. Inspect packing adjustment — approximately one drop per second is necessary to keep packing lubricated. 3. Observe suction and discharge gauges. Readings higher than suction pressure indicate leakage back from system pressure through either the fire pump or jockey pump.

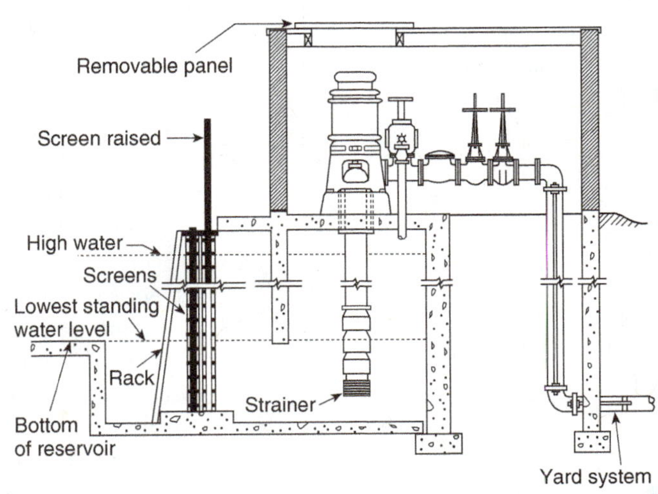

FIGURE A.8.2.2 *Wet Pit Suction Screen Installation.*

System Tagging

The batteries shown in this first photo are showing early signs of corrosion. The batteries are not at the point of failure and not corroded to the point where the system will not work; however, they are in need of some maintenance. In these instances, inspectors might consider this observed condition to be a noncritical deficiency.

The terminal on the battery shown in this next photo is severely corroded, which would be considered a critical deficiency. Many facility managers who conduct inspections would simply clean the terminal when they observed this condition; however, it is important to keep in mind that not all individuals conducting NFPA 25 inspections have the authority to engage in maintenance or to correct any observed conditions. The final photo shows a new battery with clean terminals.

Noncritical Deficiency

Critical Deficiency

Impairment

(Photos courtesy of Byron Blake and SimplexGrinnell)

ITM Deficiency, Impairment, or Hazard Evaluation?

(Courtesy of Indonesian Fire and Rescue Foundation)

ANSWER: Hazard Evaluation

This photo shows several installation errors on the suction side of this centrifugal fire pump. NFPA 20 has always prohibited the installation of elbows in the horizontal plane this close to the suction flange, and the suction reducer must be an eccentric type to avoid air pockets. Some might also question the installation of a strainer in the suction piping. The current installation as shown could affect the performance of the pump by introducing trapped air into the casing, likely increasing the maintenance required on the pump due to the uneven wear on bearings, sleeves, and other internal parts. However, with all of these apparent problems, these are not NFPA 25 deficiencies or impairments.

While the NFPA 25 inspector is required to inspect the system and identify any issues that would impair the system from functioning properly, the inspector is not required to verify the adequacy of the design of the system, as stated in 1.1.3.1. It is unusual for an NFPA document to state that the inspector is required to test the system but does not need to fully understand all aspects of the system, such as the system's design. However, some users of this document believe inspectors should know the design and installation requirements for every system covered by NFPA 25 as well. The NFPA 25 technical committee recognizes that no one can possibly know all of the design and installation requirements for all of the different editions of all of the water-based fire protection system installation documents. However, if the inspector has specific knowledge of design and installation requirements and identifies what could be an incorrect installation, a recommendation for a hazard evaluation should be made to the property owner or the owner's authorized representative.

8.3* Testing

A.8.3 The purpose of testing the pump assembly is to ensure automatic or manual operation upon demand and continuous delivery of the required system output. An additional purpose is to detect deficiencies of the pump assembly not evident by inspection.

8.3.1 Frequency.

The weekly testing requirement in 8.3.1 has generated significant discussion in the fire protection industry over the years. Weekly testing of fire pumps can be costly due to the time required. Many owners do not have qualified personnel and must contract the tests to a qualified contractor or service provider, which can cost $5,000 to $10,000 or more annually. Electricity and diesel fuel are expensive and further add to the costs. In addition, it is difficult to apply a one-size-fits-all approach to testing such a critical piece of equipment given all the possible configurations of the equipment and the various controllers. For this reason, the standard permits a performance-based testing schedule beyond the weekly frequency recommended that is subject to approval by the authority having jurisdiction (AHJ). Performance-based testing requires a review of the historical data for the particular system to determine what should be the appropriate performance-based test frequency.

The decision by the technical committee to reduce the frequency of the no-flow test for electric motor–driven fire pumps from weekly to monthly in the 2011 edition of NFPA 25 came after significant review and discussion of data presented by owners and maintainers of large numbers of fire pumps. The data submitted appeared to indicate that a change to monthly testing from weekly testing of electric motor–driven fire pumps had no significant change in the failure rates with these types of pumps and drivers. However, there was no substantive data to justify reducing the testing frequency for diesel engine–driven pumps.

Critical Point

The decision by the technical committee to reduce the frequency of the no-flow test for electric motor–driven fire pumps from weekly to monthly in the 2011 edition of NFPA 25 came after significant review and discussion of data presented by owners and maintainers of large numbers of fire pumps.

8.3.1.1* A non-flow test shall be conducted for diesel engine–driven fire pumps without recirculating water back to the pump suction on a test frequency in accordance with 8.3.1.1.1 or 8.3.1.1.2.

A.8.3.1.1 Fire pump systems should be designed so that the pressure relief valve has a minimum flow (to verify pressure relief valve is properly set and operating) at churn, and only allows a larger flow under abnormal conditions (i.e., engine overspeed or failure of a variable speed pressure limiting control). The fire pump imparts more energy into the water when recirculating the water through the pump than when the pump is operating at churn (no flow). NFPA 20 requires a circulation relief valve downstream of the pressure relief valve whenever the pressure relief valve is piped back to the pump suction. Improperly installed and/or operating circulation relief valves can result in unacceptably high water temperature, especially when recirculating the water to the pump suction. High water temperatures can affect the operation of a diesel engine drive. Modern engines, due to EPA requirements, are more sensitive to cooling water temperatures.

8.3.1.1.1 Except as permitted in 8.3.1.1.2, a weekly test frequency shall be required.

Exhibit 8.33 shows a nonmandatory sample form that can be used to document the weekly operating test required by 8.3.1.1.1.

8.3.1.1.2* The test frequency shall be permitted to be established by an approved risk analysis.

The data collected as part of the FPRF project concluded that diesel pumps were more than 99 percent effective when tested weekly, compared to only 96 percent effective when tested monthly. The data showed that diesel pumps were not quite as effective as electric-driven pumps when tested on a monthly basis, which prompted the technical committee to keep the baseline requirement for operating tests of diesel fire pumps at a weekly frequency. The data

Case In Point

In the 2014 edition of NFPA 25, this monthly testing frequency has been moved back to weekly testing. This change was based on a review of results from a 2012 report, "Fire Pump Field Data Collection and Analysis," prepared by the Fire Protection Research Foundation (FPRF). This project aimed to identify, collect, and analyze available data to validate, or revalidate, optimum testing frequency for fire pumps.

Several companies that assisted in the project provided a total of 32 data sets with 3396 non-flow tests on 79 fire pumps using pass/fail criteria forms only. A total of 112 failures were reported in these tests. One company provided a total of 6 data sets with 749 non-flow tests on 17 fire pumps using pass/fail/repair forms. A total of 20 pumps needing repairs and 29 failures were reported in these tests. Comments regarding maintenance issues that do not directly affect the fire pump starting and operating were recorded on 122 tests. Additionally, one company provided electronic records in a different format on 983 tests on 41 fire pumps. A total of 103 tests identified repairs were needed, and 11 failures were reported.

The sample size for non-flow weekly testing was deemed adequate to provide a meaningful fire pump failure rate range with 95 percent confidence. For electric motor pumps, there were 19 failures in 1547 tests, for a failure rate of 1.23 percent and a 95 percent confidence in a failure range between 0.67 percent and 1.79 percent. For diesel engine pumps, there were 98 failures in 4855 tests, for a failure rate of 2.02 percent and a 95 percent confidence in a failure range between 1.61 percent and 2.42 percent. When electric motor and diesel engine pump data are combined, there are 117 failures in 6402 tests, for a failure rate of 1.83 percent, a standard deviation of 0.17 percent, and a 95 percent confidence in a failure range between 1.49 percent and 2.16 percent. The sample size for no-flow monthly testing provided a 95 percent confidence failure range of between 0 percent and 7.9 percent for monthly testing, which does not allow meaningful analysis of the monthly test data.

The data solidly justified re-establishing a weekly test requirement for electric fire pumps in NFPA 25.

reviewed did show that some pumps were still highly effective when tested monthly, and as such, the 2014 edition of NFPA 25 allows for an approved risk analysis to be used as a mechanism for modifying the frequency to allow for a monthly test or some other frequency supported by the analysis.

A.8.3.1.1.2 The risk analysis should be prepared and reviewed by qualified people. Increased test frequencies might be desirable when high impact losses could result from an uncontrolled fire. Examples where increased fire pump test frequencies can be considered could include high piled storage facilities and buildings where the predominant occupancy is protected by an extra hazard density sprinkler system.

Test frequency has been a heavily discussed and researched topic for several years, and is still continuing to be researched. A set of data was submitted in 2008 by a group of owners and maintainers of large numbers of fire pumps. This data was presented to the committee as indicating a decreased test frequency on electric fire pumps did not "significantly" impact "reliability"; however, "reliability" as used in the discussion of the data presentation was actually the failure rate, and did not take into account the effect of test frequency on the fire pump reliability (i.e., the time between failure and discovery of the failure affects reliability). Subsequently, the NFPA Research Council commissioned research, and the resultant "Fire Pump Field Data Collection and Analysis Report" in 2011 (available for download at www.nfpa.org\Foundation) reported that electric fire pumps tested weekly had a failure rate of approximately

FIRE PUMPS WEEKLY OPERATING TESTS

System: ______________________

Location: ______________________

Y = Satisfactory **N** = Unsatisfactory (explain below) **N/A** = Not applicable

Date: ______________ Inspector: ______________________

	Y	N	N/A	Comments
Operate fire pump for 10 minutes (30 minutes for diesel pump)	❑	❑	❑	
Check packing gland tightness (slight leak at no flow)	❑	❑	❑	
Record suction pressure from gauge in psi (bar)	❑	❑	❑	
Record discharge pressure from gauge in psi (bar)	❑	❑	❑	
Adjust gland nuts if necessary	❑	❑	❑	
Check for unusual noise or vibration	❑	❑	❑	
Check packing boxes, bearings, or pump casing for overheating	❑	❑	❑	
Record pump starting pressure	❑	❑	❑	
Observe time for motor to accelerate to full speed (diesel and steam pumps)	❑	❑	❑	
For reduced-voltage or reduced-current starting, record time controller is on first step	❑	❑	❑	
Record time pump runs after starting for pumps having automatic stop feature	❑	❑	❑	
Record time for diesel engine to crank	❑	❑	❑	
Record time for diesel engine to reach running speed	❑	❑	❑	
Check oil pressure gauge, speed indicator, water and oil temperatures while engine is running	❑	❑	❑	
Check heat exchanger for cooling water flow	❑	❑	❑	
Record steam pressure for steam-operated pumps	❑	❑	❑	
Check water tank float switch	❑	❑	❑	
Check solenoids for proper operation	❑	❑	❑	
Operate speed governor (internal combustion engine only)	❑	❑	❑	
Check steam trap (steam turbine only)	❑	❑	❑	
Check steam relief valve (steam turbine only)	❑	❑	❑	
Check controller alarms	❑	❑	❑	
Record any notes that the inspector believes to be significant in the corresponding action's comments field.				

NFPA Water-Based Fire Protection Systems Forms

EXHIBIT 8.33 *Sample Form for Fire Pumps Weekly Operating Tests.*

0.64 per year. Assuming a failure rate independent of the test frequency, and assuming that on the average the impairment occurs at the midpoint of the test interval, this failure rate provides approximately 99.4 percent reliability with weekly testing and approximately 97.3 percent reliability with monthly testing. Diesel engine fire pumps tested weekly had a failure rate of approximately 1.02 per year. Assuming a failure rate independent of the test frequency and assuming that on the average the impairment occurs at the midpoint of the test interval, this failure rate provides approximately 99.1 percent reliability with weekly testing and approximately 96.0 percent reliability with monthly testing.

Based on this data, the lower reliability has not been determined to be acceptable for all facilities. Decisions to decrease test frequency must be based on more than cost savings. A reliability/risk analysis to decrease test frequency should take into account the risk associated with life safety, property values, hazards, and business interruption at the protected property. Fire pump redundancy can impact overall fire system reliability and be used in a reliability/risk analysis.

8.3.1.2* A non-flow test shall be conducted for electric motor–driven fire pumps without recirculating water back to the pump suction on a test frequency in accordance with 8.3.1.2.1, 8.3.1.2.2, 8.3.1.2.3, or 8.3.1.2.4.

A.8.3.1.2 For pressure relief valve operation, see 8.3.2.4.

8.3.1.2.1 Except as permitted in 8.3.1.2.2 and 8.3.1.2.3, a weekly test frequency shall be required for the following electric fire pumps:

(1) Fire pumps that serve fire protection systems in high rise buildings that are beyond the pumping capacity of the fire department
(2) Fire pumps with limited service controllers
(3) Vertical turbine fire pumps
(4) Fire pumps taking suction from ground level tanks or a water source that does not provide sufficient pressure to be of material value without the pump

8.3.1.2.2 A monthly test frequency shall be permitted for electric fire pumps not identified in 8.3.1.2.1.

The majority of electric-driven fire pumps will still be required to have a monthly operating test, which is consistent with the modification made in the 2011 edition. The difference between the 2011 and 2014 editions is that not all electric-driven pumps can take advantage of the reduced operating test frequency. Section 8.3.1.2.1 lists several types of installations where the operating test frequency for an electric-driven pump must revert back to a weekly frequency. These installations are considered to be the installations that either have a higher risk associated with them or a higher failure rate.

8.3.1.2.3* A monthly test frequency shall be permitted for electric fire pump systems having a redundant fire pump.

A.8.3.1.2.3 For systems where multiple fire pumps are required to meet the system demand, a one-for-one redundancy is not necessary (i.e., one backup pump for two or more primary pumps meets the intent of this section).

8.3.1.2.4* The test frequency shall be permitted to be established by an approved risk analysis.

Not unlike the diesel operating pump test described above, the frequency for an electric-driven pump operating test can also be modified based on a risk analysis.

A.8.3.1.2.4 The risk analysis should be prepared and reviewed by qualified people. Increased test frequencies can be desirable when high impact losses could result from an uncontrolled fire. Examples where increased fire pump test frequencies can be considered could include high piled storage facilities and buildings where the predominant occupancy is protected by an extra hazard density sprinkler system.

One of the most controversial topics in the ITM industry over the past few years is determining the appropriate operating test frequency for fire pumps. Irrespective of what the appropriate frequency is, it is important that the building owner, the facility manager, or the inspector understands how to conduct this no-flow test. The testing procedures vary depending on whether the pump is electric driven or diesel driven. Procedures for each of these pump types are provided at the end of this chapter.

Testing Procedure Alert

ITM NFPA 25

8.3.2 No-Flow Condition.

8.3.2.1 A test of fire pump assemblies shall be conducted without flowing water.

FAQ Is it an acceptable procedure to crack open a test header valve to provide enough water flow to keep the pump casing cool, even though 8.3.2.1 specifically states that the test "shall be conducted without flowing water"?

The reason that 8.3.2.1 references the associated test "without flowing water" is to indicate that it is not the intent of the standard to require a flow test on a weekly basis. Paragraph 8.3.2.1 is intended to verify that the pump will start and will not overheat. Paragraph 8.3.2.5 allows the circulation relief valve to open to flow water as a cooling measure. Allowing additional water flow to prevent overheating is not a requirement of the standard. Flow from the circulation relief valve should be sufficient to prevent overheating of the pump, and the flowing of additional water is a waste and should be avoided.

8.3.2.2 The test shall be conducted by starting the pump automatically.

FAQ What is meant by an automatic start?

Paragraph 8.3.2.2 requires that pumps be started automatically, rather than by the use of the "start" button on the front panel of the fire pump controller. The pump must be started by drawing water from the sensing line to simulate a pressure drop in the system. As the pressure drops, the pressure switch (see Exhibit 8.13) will sense this drop in pressure and should start the pump automatically. Using the start button on the fire pump controller does not constitute an automatic start.

Exhibit 8.34 shows a pressure-sensing line connection that can be used to draw water to simulate a pressure drop in the pressure-sensing line. This pressure drop should cause the jockey pump, and ultimately the fire pump, to start automatically.

8.3.2.3 The electric pump shall run a minimum of 10 minutes.

When a pump is started, a great deal of heat is generated in the pump windings from the energy needed to bring the pump up to speed. Paragraph 8.3.2.3 requires that the electric motor be run for 10 minutes, which is seen as the minimum time for the motor windings to cool back down after starting. Repeatedly starting and running a motor-driven fire pump for less than 10 minutes each time it is started could significantly shorten the motor's life span.

Another reason for the 10-minute requirement is that it allows time to check the pump packing and bearings to determine if they are overheating or leaking excessively. It is important to inspect the pump bearings, as failed bearings can lead to larger issues. Exhibit 8.35 shows an impeller that was destroyed due to the outboard bearing failing, which caused the impeller to come into contact with the pump casing.

Exhibit 8.36 shows a stone in the impeller of a fire pump. The stone was revealed by weekly testing.

EXHIBIT 8.34 *Pressure-Sensing Line.*

EXHIBIT 8.35 *Impeller Failure Due to Failed Bearing. (Courtesy of Damon Pietraz, Underwood Fire Equipment)*

EXHIBIT 8.36 *Some Faults Found in Testing: Stone in Impeller (left) and Size of Stone (right). (Courtesy of John Jensen)*

8.3.2.4 The diesel pump shall run a minimum of 30 minutes.

FAQ Why does NFPA 25 require the diesel pump to be run at churn for 30 minutes each week?

Paragraph 8.3.2.4 requires that a diesel fire pump operate for 30 minutes each week. This requirement is intended to allow the pump and driver to reach operating temperature, which is the point most experts agree that overheating problems can be detected. The 30-minute operating time is also intended to consume enough fuel to prevent the fuel from stagnating. Running the driver long enough to get up to rated temperature also helps prevent wet stacking. Wet stacking is a condition where unburned fuel residue collects in diesel engine cylinders and exhaust. Over a prolonged period, this buildup can seriously degrade engine performance, increase fuel consumption, and ultimately cause engine failure. Exhibit 8.37 shows a diesel engine that has failed due to excess exhaust pressure.

EXHIBIT 8.37 Engine Failure Due to Excessive Exhaust Pressure. (Courtesy of Damon Pietraz, Underwood Fire Equipment)

8.3.2.5 A valve installed to open as a safety feature shall be permitted to discharge water.

The following are three types of valves that might be found in a pump room that are designed to open as a safety feature that discharges water:

- Circulating (automatic) relief valve, for cooling electric-driven pump casings
- Pressure relief valve (sometimes called the "main" relief valve), for preventing system overpressurization
- Heat exchanger cooling line automatic valve, on diesel drivers

During a weekly test, all three valves, if installed, must be verified to be operational, but the valve primarily referred to in 8.3.2.5 is the pressure (or "main") relief valve (see Exhibit 13.31). This valve is installed to limit overpressurization on the system downstream of the pump. It is important that this valve be set so that it discharges water only at or above the rating of the piping downstream of the pump. This valve, if present, should cause water to discharge while the pump is operating at churn, meaning that water will be discharging for 30 minutes each week. However, this can be controlled by adjusting valve trim settings to allow only a small flow. A common complaint about 30-minute weekly runs is the amount of water wasted when the valve is open, which if adjusted incorrectly could total more than 500 gpm (1893 L/min) for the 30-minute weekly run. There are also provisions, which did not exist in previous editions, to allow this valve to be piped back to a suction water source for water conservation.

8.3.2.6 An automatic timer that meets 8.3.2.6.1 through 8.3.2.6.3 shall be permitted to be substituted for the starting procedure.

The requirements in 8.3.2.6 address the acceptable methods used for pump tests. These do not intend to supersede the requirement in 8.3.2.7 that someone must be present and watching the pump system each time it operates.

8.3.2.6.1 A solenoid valve drain on the pressure control line shall be the initiating means for a pressure-actuated controller.

8.3.2.6.2 In a pressure-actuated controller, performance of this program timer shall be recorded as a pressure drop indication on the pressure recorder.

8.3.2.6.3 In a non-pressure-actuated controller, the test shall be permitted to be initiated by means other than a solenoid valve.

8.3.2.7 Qualified personnel shall be in attendance whenever the pump is in operation.

8.3.2.7.1* The use of the automatic timer allowed in 8.3.2.6 shall not eliminate the requirement of 8.3.2.7 to have qualified personnel present during test.

Critical Point

Qualified personnel are required by 8.3.2.7 to be in attendance whenever a fire pump is operating.

The definition of the term *qualified* is a point of debate in the fire protection community (and in most committees tasked with defining the word). There is disagreement because the word could have several meanings depending on where it is used. For example, a professional qualified to complete a weekly fire pump test might not be qualified to complete an annual flow test, and analyzing the data from an annual flow test might require someone else's qualification and expertise. Qualified personnel are required by 8.3.2.7 to be in attendance whenever a fire pump is operating. In this instance, the intent is to have a person present that has the qualifications to know if something visually appears wrong with the fire pump system (such as excessive cooling water temperature, overpressurization, or low oil temperature) and both the ability and authority to quickly address the issue.

Often, the presence of qualified personnel is thought to affect the decision of using an automatic timer, not only for starting a fire pump but for automatically shutting it down as well.

Some jurisdictions and insurance companies require that fire pumps only be stopped manually and that the automatic stop relay in the controller be removed or bypassed. This is due to concerns that a fire pump could repeatedly cycle on and off in an actual fire and eventually not restart from an "off" sequence. NFPA 20 allows both an automatic and manual fire pump stop setting, except in cases where a pump constitutes a sole site fire protection water source. In these cases, this standard requires a manual stop only.

A.8.3.2.7.1 An automatic timer allows a person who has been instructed on what to watch for and record during this test to monitor the test and request assistance should any issues arise.

8.3.2.8 The pertinent visual observations or adjustments specified in the following checklists shall be conducted while the pump is idle:

(1) Record the system suction and discharge pressure gauge readings

The system suction and discharge pressure gauge readings are taken both when the pump is running and when the pump is shut off. For diesel engine–driven pumps, it is recommended that the cooling water pressure be recorded to ensure that the heat exchanger is not being overpressurized due to a failed or improperly adjusted cooling water line pressure regulator.

(2) For pumps that use electronic pressure sensors to control the fire pump operation, record the current pressure and the highest and the lowest pressure shown on the fire pump controller event log
(3) If the highest or lowest pressure is outside of the expected range, record all information from the event log that helps identify the abnormality

8.3.2.9* The pertinent visual observations or adjustments specified in the following checklists shall be conducted while the pump is running:

(1) Pump system procedure as follows:
    (a) Record the pump starting pressure from the pressure switch or pressure transducer
    (b) Record the system suction and discharge pressure gauge readings
    (c) Inspect the pump packing glands for slight discharge
    (d) Adjust gland nuts if necessary
    (e) Inspect for unusual noise or vibration
    (f) Inspect packing boxes, bearings, or pump casing for overheating

(g) Record pressure switch or pressure transducer reading and compare to the pump discharge gauge
(h) For pumps that use electronic pressure sensors to control the fire pump operation, record the current pressure and the highest and the lowest pressure shown on the fire pump controller event log
(i) For electric motor and radiator cooled diesel pumps, check the circulation relief valve for operation to discharge water

(2) Electrical system procedure as follows:
(a) Observe the time for motor to accelerate to full speed
(b) Record the time controller is on first step (for reduced voltage or reduced current starting)
(c) Record the time pump runs after starting (for automatic stop controllers)

(3) Diesel engine system procedure as follows:
(a) Observe the time for engine to crank

The engine controller on diesel-driven pump systems will alternate battery supplies with each attempt-to-start sequence. There should be a pilot light on the controller that indicates which bank of batteries is being used for the present start cycle. If the engine is difficult to start, the cause should be investigated, and maintenance or repair may be necessary to correct the problem. Excessive cranking time causes the batteries to discharge, and delays in the engine starting sequence can affect the delivery of needed water or liquids to the fire protection system.

(b) Observe the time for engine to reach running speed

The engine should come up to speed within 10 seconds by NFPA 20 requirements. A tachometer is required instrumentation on a listed diesel fire pump driver, but handheld units, such as the one shown in Exhibit 8.38, are often used as a means for measuring engine running speed and are typically more accurate.

(c) Observe the engine oil pressure gauge, speed indicator, water, and oil temperature indicators periodically while engine is running
(d) Record any abnormalities

Abnormalities that should be recorded include excessively low or high driver speed, low oil pressure, high temperature, high cooling water pressure for diesel engines using discharge water for cooling, leaking hoses, or strange vibrations. The weekly checklist should have a location to record these items and, more importantly, an active management system to note and address discrepancies immediately.

(e) Inspect the heat exchanger for cooling waterflow

Radiator-cooled fire pump drivers are rarely used, and they are typically used only in locations where water is highly restricted. Such pumps do not have a connection off the pump for a heat exchanger, due to the radiator. (See Exhibit 8.39.)

Heat exchangers typically discharge into an open drain in the pump room or to an outside location, in which case there should also be a sight glass or pressure gauge inside the pump room so the water flow can be observed directly. There should not be any antifreeze or discoloration in the heat exchanger water discharge when visually observed.

(4) Steam system procedure as follows:
(a) Record the steam pressure gauge reading
(b) Observe the time for turbine to reach running speed

A.8.3.2.9 See Table A.8.3.2.9.

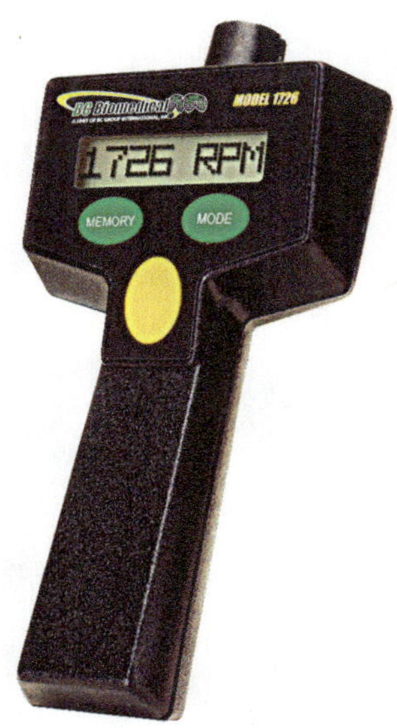

EXHIBIT 8.38 *Handheld Tachometer Used to Measure Motor Speed in rpm. (Courtesy of BC Group International)*

EXHIBIT 8.39 Fire Pump Driver with Radiator.

TABLE A.8.3.2.9 Observations — While Pumping

Item	While Pump Is Operating
Horizontal pumps	1. Read suction and discharge gauges — difference between these readings indicates churn pressure, which should match churn pressure as shown on fire pump nameplate. 2. Observe packing glands for proper leakage for cooling of packing. 3. Observe discharge from casing relief valve — adequate flow keeps pump case from overheating.
Vertical pumps	1. Read discharge gauge — add distance to water level in feet (or meters) and divide by 2.31 to compute psi (30.47 to compute bar). This total must match churn pressure as shown on fire pump nameplate. 2. Observe packing glands for proper leakage for cooling of packing. 3. Observe discharge from casing relief valve — adequate flow keeps pump case from overheating.
Diesel engines	1. Observe discharge of cooling water from heat exchanger — if not adequate, inspect strainer in cooling system for obstructions. If still not adequate, adjust pressure-reducing valve for correct flow. 2. Inspect engine instrument panel for correct speed, oil pressure, water temperature, and ammeter charging rate. 3. Inspect battery terminal connections for corrosion and clean if necessary. 4. After pump has stopped running, inspect intake screens, if provided; replace diesel system pressure recorder chart; and rewind if necessary.

While the pump is running, observations that include the items listed in Table A.8.3.2.9 should be made and a record of the test prepared. Exhibit 8.40 shows operating personnel recording test data during the weekly test. The weekly checklist should have a location to record these items and, more importantly, an active management system to note and address discrepancies in a timely manner.

EXHIBIT 8.40 *Weekly Test.*

8.3.3 Annual Flow Testing.

8.3.3.1* An annual test of each pump assembly shall be conducted by qualified personnel under no-flow (churn), rated flow, and 150 percent of the pump rated capacity flow of the fire pump by controlling the quantity of water discharged through approved test devices.

It is specifically required that the annual flow test be conducted by qualified personnel. The definition of the term *qualified* is a point of debate in the fire protection community (and in most committees tasked with defining the word). There is disagreement because the term could have several meanings depending on where it is used. For example, a professional qualified to complete a weekly fire pump churn test might not be qualified to complete an annual flow test, and analyzing the data from an annual flow test might require someone else's qualification and expertise. See Supplement 2 for annual pump test requirements and the required analysis.

In most instances, annual testing requires the owner to contract with a service provider or contractor to conduct the test. In some jurisdictions, a licensed contractor is required to conduct the test. The annual flow test generally follows the same procedure as was used for the original acceptance test.

During the flow test, a minimum of the following three points must be measured and plotted on graph paper:

1. *Pump churn.* The no-flow condition, sometimes called shutoff or churn, is the point where water is not flowing through test apparatus of play pipes. The normal water flowing at churn is restricted to diesel engine cooling water, casing relief, and/or the main relief valve.

Testing Procedure Alert

ITM NFPA 25

Unlike the no-flow test, there is no controversy associated with the frequency at which the fire pump flow test must be conducted. The annual flow test is an important test because it serves to confirm that the pump's performance has not degraded significantly when compared to the certified shop test curve. For more information on conducting the annual flow test, refer to Supplement 2, which provides an in-depth look at the data analysis associated with determining the pump's performance relative to when it is installed. Also, see the detailed testing procedures at the end of this chapter for guidance on conducting these tests for various arrangements.

2. *Rated capacity.* The rated capacity is the volume of water at the pressure indicated on the pump nameplate.
3. *Overload.* The overload test point is where the pump discharge is flowed to 150 percent of rated capacity flow. At this point, pressure is measured and should meet the rated pressure stamped on the pump nameplate or manufacturer's pump curve.

Exhibit 8.41 illustrates the annual test at churn. In this example, calibrated test gauges have been installed. Exhibit 8.42 shows measuring flow at rated capacity. Exhibit 8.43 illustrates the annual test at overload. A difference can be observed in the quantity of water of the annual test at overload versus rated capacity (from Exhibit 8.42).

A.8.3.3.1 Minimum flow for a pump is the churn pressure.

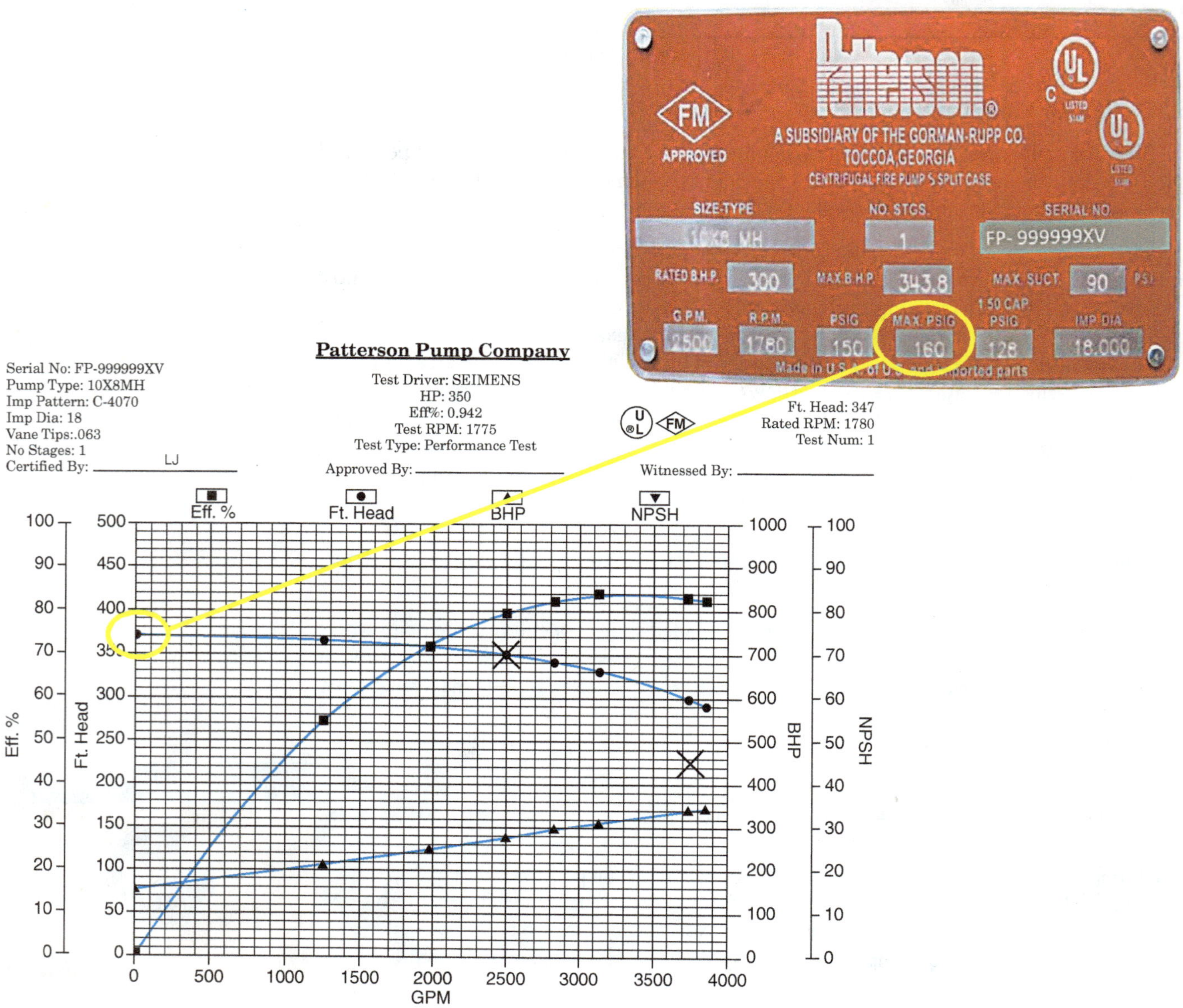

EXHIBIT 8.41 *Annual Test at Churn.*

8.3.3.1.1 If available suction supplies do not allow flowing of 150 percent of the rated pump capacity, the fire pump shall be tested to the maximum allowable discharge.

Two reasons a suction supply does not allow for the flow equal to 150 percent of the rated pump capacity are as follows:

1. Drainage in the area around the pump test is not of capacity to allow a continued test.
2. Low fire pump suction pressure drops to a point where discharge flow cannot safely be made to reach 150 percent of rated flow.

In case 1, the proper drainage as required by the design standard should be installed, and the test should be completed when the required drainage is in place. Another approach to addressing insufficient drainage is to allow the area to drain between test points, such as testing at the 100 percent point, waiting for water to drain in the area, and then testing at the 150 percent point.

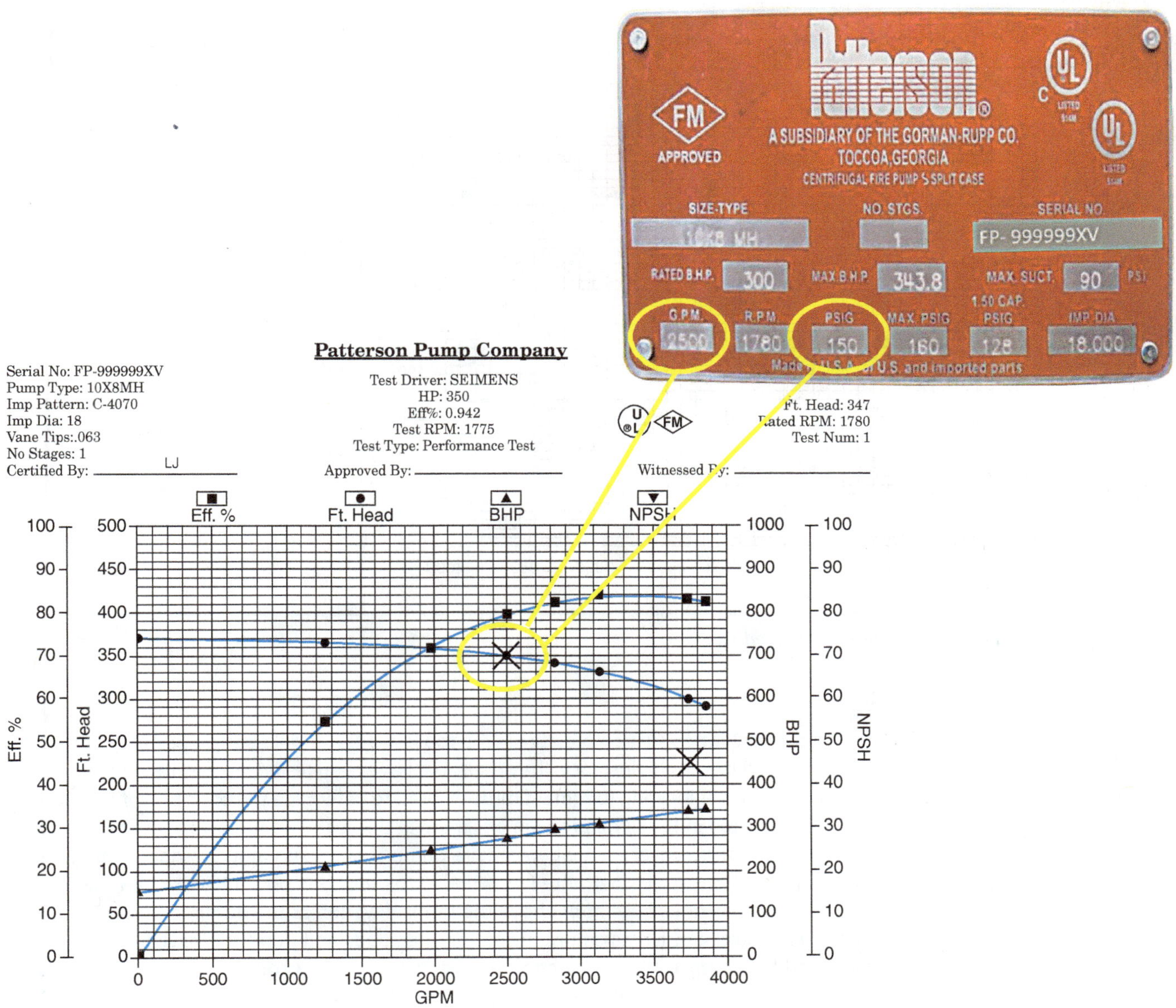

EXHIBIT 8.42 Discharge from Annual Fire Pump Test at Rated Capacity.

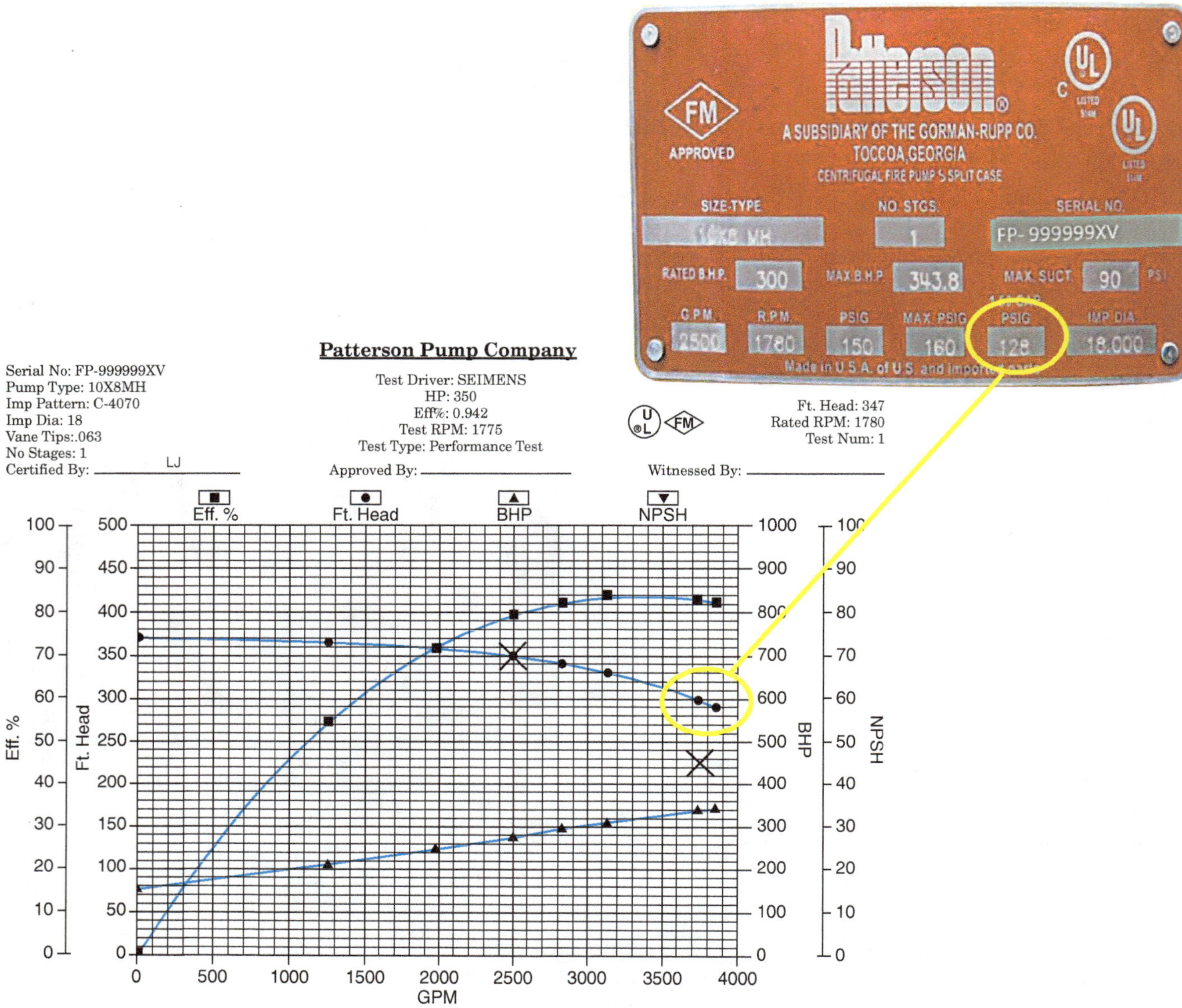

EXHIBIT 8.43 *Discharge from Annual Test at Overload.*

For case 2, when suction pressures drop too low, a test should be stopped and plotted with the data available. In this instance, the points that have been tested (at least three) must be plotted, and some engineering judgment must be used to determine the condition of the pump.

As indicated in 8.1.6, the pressure observed at the suction gauge while the pump is operating should be positive, even when the fire pump is operating at its overload condition. The suction gauge pressure is permitted by NFPA 20 to drop to -3 psi (-0.2 bar) when the supply is a suction tank with its base at or above the same elevation as the pump, provided the lowest water tank level and all maximum fire system demands and durations have been acceptably supplied. However, it should also be noted that when taking water from a public water system, most municipal water companies' purveyors limit the residual pressure on the system to 20 psi (1.4 bar) due to concerns of collapsing water mains and drastically increasing water turbidity.

But this is generally the minimum required at the city connection. Many times, gauges at the location of a pump test could be 10 psi to 20 psi (0.7 bar to 1.4 bar) above that at the city connection due to such issues as pressure losses across backflow apparatus and numerous valves. Thus, in these cases, an additional pressure gauge should be attached to a hydrant or the city side of the site backflow apparatus and examined so the accurate public water supply pressure can be monitored for compliance with any minimum amounts stipulated by the municipality.

FAQ If a pump test must be stopped early per 8.3.3.1.1, is it still considered an acceptable test, and can an acceptable annual NFPA 25 test be reported to the local jurisdiction?

To meet the requirements of an acceptable pump test, the criteria in 8.3.6.3 (with the 5 percent "pass/fail" requirement for speed-adjusted flow performance) must be completed. If a test is stopped early, as allowed per 8.3.3.1.1, then all that can be determined is how the pump performed along the test curve up to the point the that test was stopped. If the test was stopped close to the 150 percent flow point, engineering judgment with some extrapolation could likely be used to indicate if the test was the equivalent of a successful annual test. However, there is a second pump test data analysis requirement in 8.3.6.1. The test flow would have to be long enough to allow a complete analysis to confirm there is enough water flow and pressure to supply all connected fire protection systems. If the test could not be made with enough flow to verify that system demands are met, the test cannot be considered complete. A fire pump test report could then only have assumptions based on extrapolations of flow curves and engineering judgment. See Supplement 2 for more information.

8.3.3.1.2* The annual test shall be conducted as described in 8.3.3.1.2.1, 8.3.3.1.2.2, or 8.3.3.1.2.3.

A.8.3.3.1.2 The method described in 8.3.3.1.2.3 is not considered as complete as those in 8.3.3.1.2.1 and 8.3.3.1.2.2, because it does not test the adequacy of the water supply for compliance with the requirements of 8.1.6 at the suction flange.

When testing fire pumps, the inspector should closely examine beneath the coupling guard for filings, which can indicate a misaligned coupling. When the pump is running, excessive vibration can also be an indication of coupling misalignment.

Testing Procedure Alert

While fire pumps are operated on a weekly or monthly basis as part of the testing required by 8.3.1, those tests do not require the flow of water and are primarily intended to exercise the pump. The annual flow test required by 8.3.3 provides the current flow characteristics and allows for a comparison to the pump's original capabilities. For more information on conducting a flow test, see the testing procedure at the end of this chapter.

8.3.3.1.2.1 Use of Pump Discharge via Hose Streams.

(A) Pump suction and discharge pressures and the flow measurements of each hose stream shall determine the total pump output.

(B) Care shall be taken to prevent water damage by verifying there is adequate drainage for the high pressure water discharge from hoses.

The impact of the test discharge water on the property will vary, depending on the topography of the property. The contractor administering the test should have an understanding of where the discharge flow will drain to and the potential impact on the property, adjacent property, and roadways.

8.3.3.1.2.2 Use of Pump Discharge via Bypass Flowmeter to Drain or Suction Reservoir. Pump suction and discharge pressures and the flowmeter measurements shall determine the total pump output.

8.3.3.1.2.3 Use of Pump Discharge via Bypass Flowmeter to Pump Suction (Closed-Loop Metering).

(A) Pump suction and discharge pressures and the flowmeter measurements shall determine the total pump output.

(B) When testing includes recirculating water back to the fire pump suction, the temperature of the recirculating water shall be monitored to verify that it remains below temperatures that could result in equipment damage as defined by the pump and engine manufacturers.

8.3.3.1.3 Where the annual test is conducted periodically in accordance with 8.3.3.1.2.3, a test shall be conducted every 3 years in accordance with 8.3.3.1.2.1 or 8.3.3.1.2.2 in lieu of the method described in 8.3.3.1.2.3.

FAQ Why is a flow test required every 3 years, and is the roof manifold permitted to be used for this test?

Fire pump flow tests are required annually. However, if the pump is equipped with a closed-loop metering device, a flow test must be conducted every 3 years, which allows water to be discharged. The purpose of this test is to verify the adequacy of the water supply, which the closed-loop method does not do. Fire hydrants, standpipes, roof manifolds, or other devices that have outlets with the capacity to discharge water at the peak flow typically are acceptable for the pump test.

8.3.3.1.4 Where 8.3.3.1.2.2 or 8.3.3.1.2.3 is used, the flowmeter shall be adjusted immediately prior to conducting the test in accordance with the manufacturer's instructions.

8.3.3.1.4.1 If the test results are not consistent with the previous annual test, 8.3.3.1.2.1 shall be used.

Flowmeters are used only once a year and, as a result, must be adjusted and calibrated prior to conducting the annual flow test. If a test using a flowmeter shows a material deviation from the previous year's test, the pump is required to be retested using hose streams.

8.3.3.1.4.2 If testing in accordance with 8.3.3.1.2.1 is not possible, a flowmeter calibration shall be performed and the test shall be repeated.

8.3.3.2 The pertinent visual observations, measurements, and adjustments specified in the following checklists shall be conducted annually while the pump is running and flowing water under the specified output condition:

(1) At no-flow condition (churn) as follows:

(a) Inspect the circulation relief valve for operation to discharge water

The circulation relief valve opens when the pump is running to allow a small amount of water to discharge, usually into a drain box or outside the pump room. The intent is to allow water to enter the pump casing for cooling purposes. Circulation relief valves are usually pressure operated (i.e., spring-loaded) and can fail easily when a small amount of obstructing material or corrosion enters the valve. The circulation relief valve should be set to open at the pump-rated pressure.

(b) Inspect the pressure relief valve (if installed) for proper operation

The main or pressure relief valve is used to maintain the system pressure below the maximum system rated pressure, which is usually 175 psi (12.1 bar). If a diesel engine reaches a runaway or overspeed condition, this valve is used in case a speed governor fails to control the pressure. The overspeed governor is set at 120 percent of rated engine speed. Using the pump affinity laws, the formula to determine the overspeed governor setting is as follows:

$$H_1 = H_2\left(\frac{N_1}{N_2}\right)^2$$

where:

H_1 = Head at test speed (ft or m)

H_2 = Head at rated speed (ft or m)

N_1 = Test speed (rpm)

N_2 = Rated speed (rpm)

For example, a rated speed of 1750 rpm, a rated pressure of 125 psi (8.6 bar), and a test speed of 2100 rpm gives us the following:

$$H_1 = 125\left(\frac{2100}{1750}\right)^2 = 180 \text{ psi } (12.4 \text{ bar})$$

A pressure of 180 psi (12.4 bar) is above the rating of the pipe.

The overspeed governor should be tested annually to make sure it will operate in an emergency situation. For pipe and fittings rated for 175 psi (12.1 bar), the setting of the main relief valve should be no higher than this pressure.

(2) At each flow condition as follows:

(a) Record the electric motor voltage and current (all lines)

The electric motor voltage and current data should be recorded only by someone trained and qualified in electrical hazards and equipped with the necessary safety equipment as outlined in *NFPA 70E®, Standard for Electrical Safety in the Workplace®*. Voltage and amperage should be recorded for each phase of the electrical circuits at each flow condition, including shutoff.

(b) Record the pump speed in rpm

The pump speed should be recorded using a calibrated strobe tachometer or handheld rpm counter placed on the end of the pump shaft. The newer engine electronic tachometers are as accurate as any portable unit and are considered adequate for fire pump testing purposes. Electric motors run at their synchronous speed and, unlike diesel engines, their speed will not vary drastically with load.

(c) Record the simultaneous (approximately) readings of pump suction and discharge pressures and pump discharge flow

(3)* For electric motor–driven pumps, do not shut down the pump until it has run for 10 minutes

A.8.3.3.2(3) It is not the intent to discharge water for the full 1-hour test duration, provided all flow tests can be conducted in less time and efforts are taken to prevent the pump from overheating.

(4) For diesel motor–driven pumps, do not shut down the pump until it has run for 30 minutes

8.3.3.3* For installations having a pressure relief valve, the operation of the relief valve shall be closely observed during each flow condition to determine whether the pump discharge pressure exceeds the normal operating pressure of the system components.

The pressure setting of the pressure relief valve must be set at the working pressure of the system, which is usually 175 psi (12.1 bar) or less. When the pressure relief valve is set to open just below the rated pressure of the pipe network, it prevents excessive strain on the system by not allowing the system components to experience pressures in excess of the normal operating pressure.

A.8.3.3.3 A pressure relief valve that opens during a flow condition is discharging water that is not measured by the recording device(s). It can be necessary to temporarily close the pressure relief valve to achieve favorable pump test results. At the conclusion of the pump test, the

Critical Point

The pressure setting of the pressure relief valve must be set at the working pressure of the system, which is usually 175 psi (12.1 bar) or less.

pressure relief valve must be readjusted to relieve pressures in excess of the normal operating pressure of the system components.

If the pressure relief valve is open during the flowing conditions due to the fact that the pressure is too high for the components in the fire protection system, the discharge control valve should be closed prior to closing the pressure relief valve to make sure that the fire protection system is not overpressurized. After the test, the valve must be opened again.

8.3.3.3.1* The pressure relief valve shall also be observed during each flow condition to determine whether the pressure relief valve closes at the proper pressure.

A.8.3.3.3.1 A pressure relief valve that is open during a flow condition will affect test results.

8.3.3.3.2 The pressure relief valve shall be closed during flow conditions if necessary to achieve minimum rated characteristics for the pump and reset to normal position at the conclusion of the pump test.

8.3.3.3.2.1 When it is necessary to close the relief valve to achieve minimum rated characteristics for the pump, the pump discharge control valve shall be closed if the pump churn pressure exceeds the system rated pressure.

8.3.3.3.3 When pressure relief valves are piped back to the fire pump suction, the temperature of the recirculating water shall be monitored to verify that it remains below temperatures that could result in equipment damage as defined by the pump and engine manufacturers.

8.3.3.4 For installations having an automatic transfer switch, the following test shall be performed to ensure that the overcurrent protective devices (i.e., fuses or circuit breakers) do not open:

(1) Simulate a power failure condition while the pump is operating at peak load
(2) Verify that the transfer switch transfers power to the alternate power source
(3) Verify that the pump continues to perform at peak horsepower load on the alternate power source for 10 minutes for an alternate utility or 30 minutes if the alternate power source is a standby generator set
(4) Remove the power failure condition and verify that, after a time delay, the pump is reconnected to the normal power source

8.3.3.5* Alarm conditions shall be simulated by activating alarm circuits at alarm sensor locations, and all such local or remote alarm indicating devices (visual and audible) shall be observed for operation.

All of the conditions designated as "alarm conditions" in NFPA 20 are considered "tamper" or "trouble" conditions, as per the definitions found in *NFPA 72®, National Fire Alarm and Signaling Code,* and all require immediate action. For fire pump alarm conditions required by 8.3.3.5, immediate local attention is required, but fire department response is not required.

A.8.3.3.5 It is not the intent to verify that all the alarm conditions required by NFPA 20 (e.g., low oil pressure, high coolant temperature, failure of engine to start, engine overspeed) transmit individually to a remote location, as long as these alarms can be individually verified at the fire pump controller.

8.3.3.6* Safety. Section 4.9 shall be followed for safety requirements while working near electric motor–driven fire pumps.

Working near electric motor–driven pumps requires licensed or qualified personnel using the appropriate safety equipment, such as gloves, protective clothing, and a face shield.

Due to the voltages present in a typical fire pump controller, *NFPA 70E* considers it a motor control center (MCC). Therefore, protective equipment in the form of flash protection for the face, rated gloves, and noncombustible shirt and trousers must be worn when opening the controller. (See Exhibit 4.12 for the safety equipment needed to work on a fire pump controller, and see Commentary Table 4.2 for PPE selection.)

A.8.3.3.6 See also *NFPA 70E* for additional safety guidance.

8.3.3.7* Suction Screens. After the waterflow portions of the annual test or fire protection system activations, the suction screens shall be inspected and cleared of any debris or obstructions.

A.8.3.3.7 During periods of unusual water supply conditions such as floods, inspection should be on a daily basis.

The "unusual water supply conditions" mentioned in A.8.3.3.7 include floods and times of low water. In times of low water, the suction supply should be inspected to make sure there is adequate water to cover the pump bowls so that cavitation does not occur.

If the pump takes suction from a well, the water level needs to be checked regularly during times of drought or unusual water demand to make sure there is adequate water available for the pump. If the water level is too low, it might be necessary to extend the well deeper or add additional sections of pipe in the well. In extreme cases, additional bowl sections and a larger pump driver might need to be added.

8.3.3.8* Where engines utilize electronic fuel management control systems, the backup electronic control module (ECM) and the primary and redundant sensors for the ECM shall be tested annually.

A.8.3.3.8 ECM and Sensor Testing. To verify the operation of the alternate ECM with the stop, the ECM selector switch should be moved to the alternate ECM position. Repositioning of this should cause an alarm on the fire pump controller. Then the engine is started; it should operate normally with all functions. Next, the engine is shut down, switched back to the primary ECM, and restarted briefly to verify that correct switchback has been accomplished.

To verify the operation of the redundant sensor, with the engine running, the wires are disconnected from the primary sensor. There should be no change in the engine operation. The wires are then reconnected to the sensor, then disconnected from the redundant sensor. There should be no change in the engine operation. The wires should next be reconnected to the sensor. This process is repeated for all primary and redundant sensors on the engines. It should be noted whether disconnecting and reconnecting of wires to the sensors can be done while the engine is not running, then starting the engine after each disconnecting and reconnecting of the wires to verify engine operation.

8.3.4 Diesel Fuel Testing and Maintenance.

8.3.4.1 Diesel fuel shall be tested for degradation no less than annually.

Most diesel fuels available have a shelf life of about 1 year. Fuel oil stored for more than 1 year can become highly susceptible to oxidation and biological growth. The amount of time that a fuel spends at a fueling dispatch station cannot be controlled, so shelf life of the fuel is a significant problem.

In an effort to address this issue, additives are mixed with fuels to extend the shelf life of the diesel fuel. Trying to predict how long fuel can be stored is a complex problem with many variables, but commercial power plants have experienced satisfactory storage periods of up to 1 year when stability additives were used. Fuel additives are typically either biocides and/or stabilizer additives, and they are used commonly throughout the fuel industry to prevent degradation.

Stabilizers are used to inhibit oxidation of the fuel oil. When diesel fuel oxidizes, it precipitates insoluble particulates and gum, which can clog fuel filters and foul diesel engine injectors. Biocide additives are used to prohibit biological growth in the water/fuel interface.

Critical Point

Fuel oil stored for more than 1 year can become highly susceptible to oxidation and biological growth.

A consistent fuel draining program as discussed in A.8.5.1, backed by a biocide treatment program, can significantly lower the risk of having fuel oil contaminated with biological growth.

It is important to realize that biocides and stabilizers are not effective in correcting existing fuel contamination problems. Extreme caution should be taken not to treat a tank that is already in service with a known problem. Stabilizers will not reduce the particulates in fuel; it will only prevent more particulate from forming in the fuel. Furthermore, if a diesel fuel tank has large microbiological colonies, treating the fuel tank with a biocide while it is in service could result in engine failure if the colonies are killed and float to the engine suction line. If a tank is known to have severe to moderate biological growth problems, the tank should be taken out of service and cleaned prior to adding a biocide per 8.3.4.2.

8.3.4.1.1* Fuel degradation testing shall comply with ASTM D 975-11b, *Standard Specification for Diesel Fuel Oils,* or ASTM D 6751-11b, *Standard Specification for Biodiesel Fuel Blend Stock (B100) for Middle Distillate Fuels*, as approved by the engine manufacturer, using ASTM D 7462-11, *Standard Test Method for Oxidation Stability of Biodiesel (B100) and Blends of Biodiesel with Middle Distillate Petroleum Fuel (Accelerated Method).*

Critical Point

The best maintenance for diesel fuel oil is consumption of the fuel.

The best maintenance for diesel fuel oil is consumption of the fuel. It is extremely advantageous to establish a driver test schedule that will "turn over" the fuel tank's full contents at least once a year. With a 30-minute weekly run required, fuel in diesel driver tanks should be fully consumed annually. Commercial distillate fuels, including diesel fuel, are subject to various degrees of degradation. In the 2014 edition of NFPA 25, diesel fuel is now required to be analyzed annually to ensure that fuel supplying diesel fire pump drivers will not cause engine failure.

A.8.3.4.1.1 Commercial distillate fuel oils used in modern diesel engines are subject to various detrimental effects from storage. The origin of the crude oil, refinement processing techniques, time of year, and geographical consumption location all influence the determination of fuel blend formulas. Naturally occurring gums, waxes, soluble metallic soaps, water, dirt, blends, and temperature all contribute to the degradation of the fuel as it is handled and stored. These effects begin at the time of fuel refinement and continue until consumption. Proper maintenance of stored distillate fuel is critical for engine operation, efficiency, and longevity.

Storage tanks should be kept water-free. Water contributes to steel tank corrosion and the development of microbiological growth where fuel and water interface. This and the metals of the system provide elements that react with fuel to form certain gels or organic acids, resulting in clogging of filters and system corrosion. Scheduled fuel maintenance helps to reduce fuel degradation. Fuel maintenance filtration can remove contaminants and water and maintain fuel conditions to provide reliability and efficiency for standby fire pump engines. Fuel maintenance and testing should begin the day of installation and first fill.

8.3.4.2* If diesel fuel is found to be deficient in the testing required in 8.3.4.1.1, the fuel shall be reconditioned or replaced, the supply tank shall be cleaned internally, and the engine fuel filter(s) shall be changed.

A.8.3.4.2 Where environmental or fuel quality conditions result in degradation of the fuel while stored in the supply tank, from items such as water, micro-organisms and particulates, or destabilization, active fuel maintenance systems permanently installed on the fuel storage tanks have proven to be successful at maintaining fuel quality. An active fuel maintenance system will maintain the fuel quality in the tank, therefore preventing the fuel from going through possible cycles of degradation, risking engine reliability, and then requiring reconditioning.

8.3.4.2.1 After the restoration of the fuel and tank in 8.3.4.2, the fuel shall be retested every 6 months until experience indicates the fuel can be stored for a minimum of 1 year without degradation beyond that allowed in 8.3.4.1.1.

8.3.4.3 When provided, active fuel maintenance systems shall be listed for fire pump service.

8.3.4.3.1 Maintenance of active fuel maintenance systems shall be in accordance with the manufacturer's recommendations.

8.3.4.3.2 Maintenance of active fuel maintenance systems shall be performed at a minimum annual frequency for any portion of the system that the manufacturer does not provide a recommended maintenance frequency.

8.3.4.3.3 Fuel additives shall be used and maintained in accordance with the active fuel maintenance system manufacturer's recommendations.

8.3.5 Positive Displacement Pumps. [20:14.2.6.4.3]

8.3.5.1 Except as provided in 8.3.5.1 through 8.3.5.7, positive displacement pumps shall be tested in accordance with 8.3.1 through 8.3.3.

8.3.5.2 The pump flow for positive displacement pumps shall be tested and determined to meet the specified rated performance criteria where only one performance point is required to establish positive displacement pump acceptability. [**20:**14.2.6.4.3.1]

8.3.5.3 The pump flow test for positive displacement pumps shall be accomplished using a flowmeter or orifice plate installed in a test loop back to the supply tank, to the inlet side of a positive displacement water pump, or to drain. [**20:**14.2.6.4.3.2]

8.3.5.4 The flowmeter reading or discharge pressure shall be recorded and shall be in accordance with the pump manufacturer's flow performance data. [**20:**14.2.6.4.3.3]

8.3.5.5 If orifice plates are used, the orifice size and corresponding discharge pressure to be maintained on the upstream side of the orifice plate shall be made available to the authority having jurisdiction. [**20:**14.2.6.4.3.4]

8.3.5.6 Flow rates shall be as specified while operating at the system design pressure. Tests shall be performed in accordance with HI 3.6, *Rotary Pump Tests*. [**20:**14.2.6.4.3.5]

8.3.5.7 Positive displacement pumps intended to pump liquids other than water shall be permitted to be tested with water; however, the pump performance will be affected, and manufacturer's calculations shall be provided showing the difference in viscosity between water and the system liquid. [**20:**14.2.6.4.3.6]

8.3.6 Other Tests.

8.3.6.1 Engine generator sets supplying emergency or standby power to fire pump assemblies shall be tested routinely in accordance with NFPA 110, *Standard for Emergency and Standby Power Systems*.

8.3.6.2 Automatic transfer switches shall be tested routinely and exercised in accordance with NFPA 110.

8.3.6.3 Tests of appropriate environmental pump room space conditions (e.g., heating, ventilation, illumination) shall be made to ensure proper manual or automatic operation of the associated equipment.

8.3.6.4* Parallel and angular alignment of the pump and driver shall be inspected during the annual test, and any misalignment shall be corrected.

When inspecting and testing fire pumps, the inspector should closely examine beneath the coupling guard for filings, which can indicate a misaligned coupling. When the pump is running, excessive vibration can also be an indication of coupling misalignment. Exhibit 8.44 shows a pump's shaft that has failed at the internal bearing. The cause of the failure was determined to be due to misalignment.

Couplings are required to be listed by NFPA 20 due to several past catastrophic failures of several types of elastomeric and polymeric couplings. This listing requirement was added to the

EXHIBIT 8.44 *Broken Shaft Due to Misalignment. (Courtesy of Damon Pietraz, Underwood Fire Equipment)*

1996 edition of NFPA 20, but approved models did not become available by manufacturers until several years later. Thus, pumps installed in the early 1990s to meet NFPA 20 requirements might not have been installed with approved couplings, although they were required by the standard. Speed, torque, and specific load ratings are part of the individual listings for these products.

Pump shaft couplings are typically protected by a cage or cover that meets the requirements of ANSI B15.1, *Safety Standard for Mechanical Power Transmission Apparatus,* as required by 4.12.8 of NFPA 20 and as shown in Exhibit 8.3.

Pump alignment is required to be checked annually per 8.3.5.4. Coupling alignment requirements are dictated by both the pump and coupling manufacturer specifications. The allowable amount of misalignment can vary by type of pump, driver, coupling, model, and size. Irrespective of which type of coupling is used, proper maintenance is imperative, because a failed coupling can lead to an impaired pump. Exhibit 8.45 shows a shaft that has failed due to misalignment where a metallic coupling was used. Exhibit 8.46 shows a failed elastopolymer coupling that sheared in half due to misalignment.

EXHIBIT 8.45 *Misalignment Caused Broken Shaft (with Metallic Coupling). (Courtesy of Damon Pietraz, Underwood Fire Equipment)*

EXHIBIT 8.46 *Coupling Sheared in Half Due to Misalignment. (Courtesy of Damon Pietraz, Underwood Fire Equipment)*

Most couplings now come almost fully enclosed, and thus, guards must be removed for inspection. The important things to look for to make sure a pump is aligned and to minimize chances of failure are as follows:

- *Flatness of the foundation and base plate.* If the base is not leveled and properly grouted, the pump will lose alignment over time.
- *Pipe strain.* If the pump is aligned but the pump flanges are being pulled in one direction or other, the pump will lose alignment. The best way to check this is to remove the bolts from the pump flanges to see if the piping needs to be pulled over to get the bolts back in.
- *Misinstalled set screws.* Incorrectly installed set screws are not limited to pumps already in service. There have been instances in which pumps have been shipped directly from the manufacturer with the set screws incorrectly installed.

Critical Point

When inspecting and testing fire pumps, the inspector should closely examine beneath the coupling guard for filings, which can indicate a misaligned coupling.

A.8.3.6.4 If pumps and drivers were shipped from the factory with both machines mounted on a common baseplate, they were accurately aligned before shipment. All baseplates are flexible to some extent and, therefore, must not be relied on to maintain the factory alignment. Realignment is necessary after the complete unit has been leveled on the foundation and again after the grout has set and foundation bolts have been tightened. The alignment should be inspected after the unit is piped and reinspected periodically. To facilitate accurate field alignment, most manufacturers either do not dowel the pumps or drivers on the baseplates before shipment or, at most, dowel the pump only.

After the pump and driver unit has been placed on the foundation, the coupling halves should be disconnected. The coupling should not be reconnected until the alignment operations have been completed.

The purpose of the flexible coupling is to compensate for temperature changes and to permit end movement of the shafts without interference with each other while transmitting power from the driver to the pump.

There are two forms of misalignment between the pump shaft and the driver shaft:

(1) *Angular misalignment.* Shafts with axes concentric but not parallel
(2) *Parallel misalignment.* Shafts with axes parallel but not concentric

The faces of the coupling halves should be spaced within the manufacturer's recommendations and far enough apart so that they cannot strike each other when the driver rotor is moved hard over toward the pump. Due allowance should be made for wear of the thrust bearings. The necessary tools for an approximate inspection of the alignment of a flexible coupling are a straight edge and a taper gauge or a set of feeler gauges.

A check for angular alignment is made by inserting the taper gauge or feelers at four points between the coupling faces and comparing the distance between the faces at four points spaced at 90 degree intervals around the coupling *[see Figure A.8.3.6.4(a)]*. The unit will be in angular alignment when the measurements show that the coupling faces are the same distance apart at all points.

A check for parallel alignment is made by placing a straight edge across both coupling rims at the top, bottom, and at both sides *[see Figure A.8.3.6.4(b)]*. The unit will be in parallel alignment when the straight edge rests evenly on the coupling rim at all positions. Allowance might be necessary for temperature changes and for coupling halves that are not of the same outside diameter. Care must be taken to have the straight edge parallel to the axes of the shafts.

Angular and parallel misalignment are corrected by means of shims under the motor mounting feet. After each change, it is necessary to recheck the alignment of the coupling halves. Adjustment in one direction might disturb adjustments already made in another direction. It should not be necessary to adjust the shims under the pump.

The permissible amount of misalignment will vary with the type of pump and driver; and coupling manufacturer, model, and size. [**20:** A.6.5]

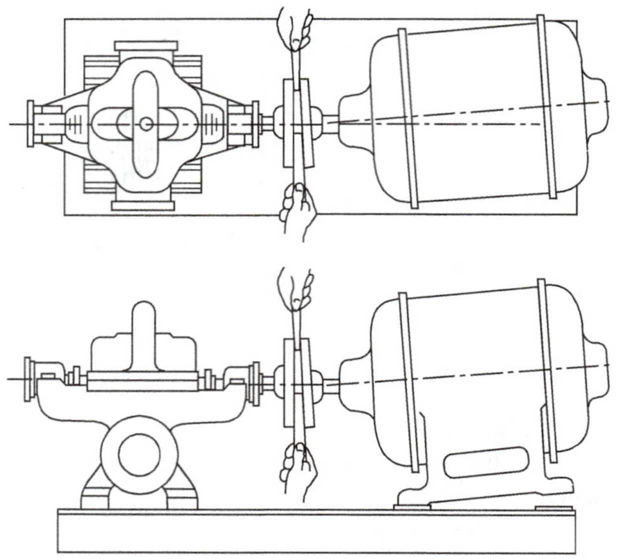

FIGURE A.8.3.6.4(a) Checking Angular Alignment. (Courtesy of the Hydraulic Institute, Parsippany, NJ, www.Pumps.org.)

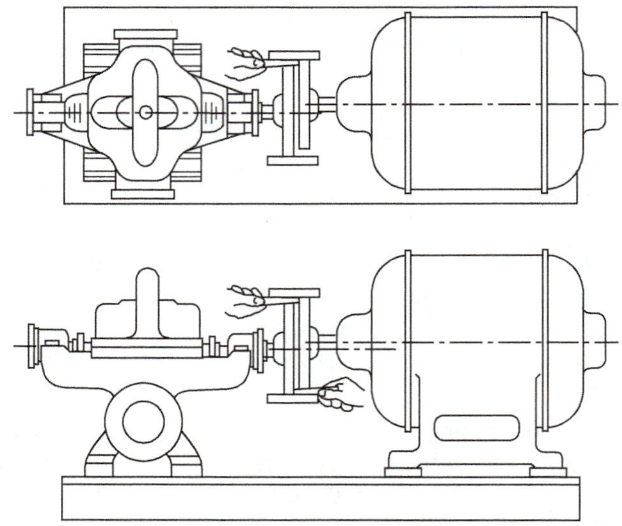

FIGURE A.8.3.6.4(b) Checking Parallel Alignment. (Courtesy of the Hydraulic Institute, Parsippany, NJ, www.Pumps.org.)

8.3.7 Test Results and Evaluation.

8.3.7.1* Interpretation.

A.8.3.7.1 Where the information is available, the test plot should be compared with the original acceptance test plot. It should be recognized that the acceptance test plot could exceed the minimum acceptable pump requirements as indicated by the rated characteristics for the pump. While a reduction in output is a matter of concern, this condition should be evaluated in light of meeting the rated characteristics for the pump. *[See Figure A.8.3.7.3(1)(a).]*

The test equipment should be of high quality and accuracy. All equipment should have been calibrated within the last 12 months by an approved calibration facility. Where possible, the calibration facility should provide documentation indicating the instrument reading against the calibrated reading. Instruments that pass the calibration test should be labeled by the calibration facility with the name of the facility and the date of the test.

Pressure gauges should have an accuracy not greater than 1 percent of full scale. To prevent damage to a pressure gauge utilizing a Bourdon tube mechanism, it should not be used where the expected test pressure is greater than 75 percent of the test gauge scale. Some digital gauges can be subjected to twice the full-scale pressure without damage. The manufacturer's recommendations should be consulted for the proper use of the gauge. To be able to easily read an analog gauge, the diameter of the face of the analog gauge should be greater than 3 in. (76 mm). Pressure snubbers should be used for all gauges to minimize needle fluctuation. All gauges used in the test should be such that a gauge with the lowest full-scale pressure is used. For example, a 300 psi (20.7 bar) gauge should not be used to measure a 20 psi (1.4 bar) pitot pressure.

Equipment other than pressure gauges, such as volt/ammeters, tachometers, and flowmeters, should be calibrated to the manufacturer's specifications. The readings from equipment with this level of accuracy and calibration can be used without adjustment for accuracy.

8.3.7.1.1 The interpretation of the test results shall be the basis for determining performance of the pump assembly.

8.3.7.1.2 Qualified individuals shall interpret the test results.

8.3.7.2 Engine Speed.

8.3.7.2.1 Theoretical factors for correction to the rated speed shall be applied where determining the compliance of the pump per the test.

NFPA 25 requires the use of affinity laws for correction to the rated pump speed. This correction is needed as the pump ratings were established under ideal factory conditions and most pump performance will vary from the factory ratings under field conditions. Uncorrected pump performance must be sufficient to supply the required system demand as required in 8.3.7.7, to assure that the active systems are properly supported with appropriate pressure and flow.

8.3.7.2.2 Increasing the engine speed beyond the rated speed of the pump at rated condition shall not be permitted as a method for meeting the rated pump performance.

8.3.7.3 The fire pump assembly shall be considered acceptable if either of the following conditions is shown during the test:

(1)* The test is no less than 95 percent of the pressure at rated flow and rated speed of the initial unadjusted field acceptance test curve, provided that the original acceptance test curve matches the original certified pump curve by using theoretical factors.

A.8.3.7.3(1) See Figure A.8.3.7.3(1)(a) and Figure A.8.3.7.3(1)(b).

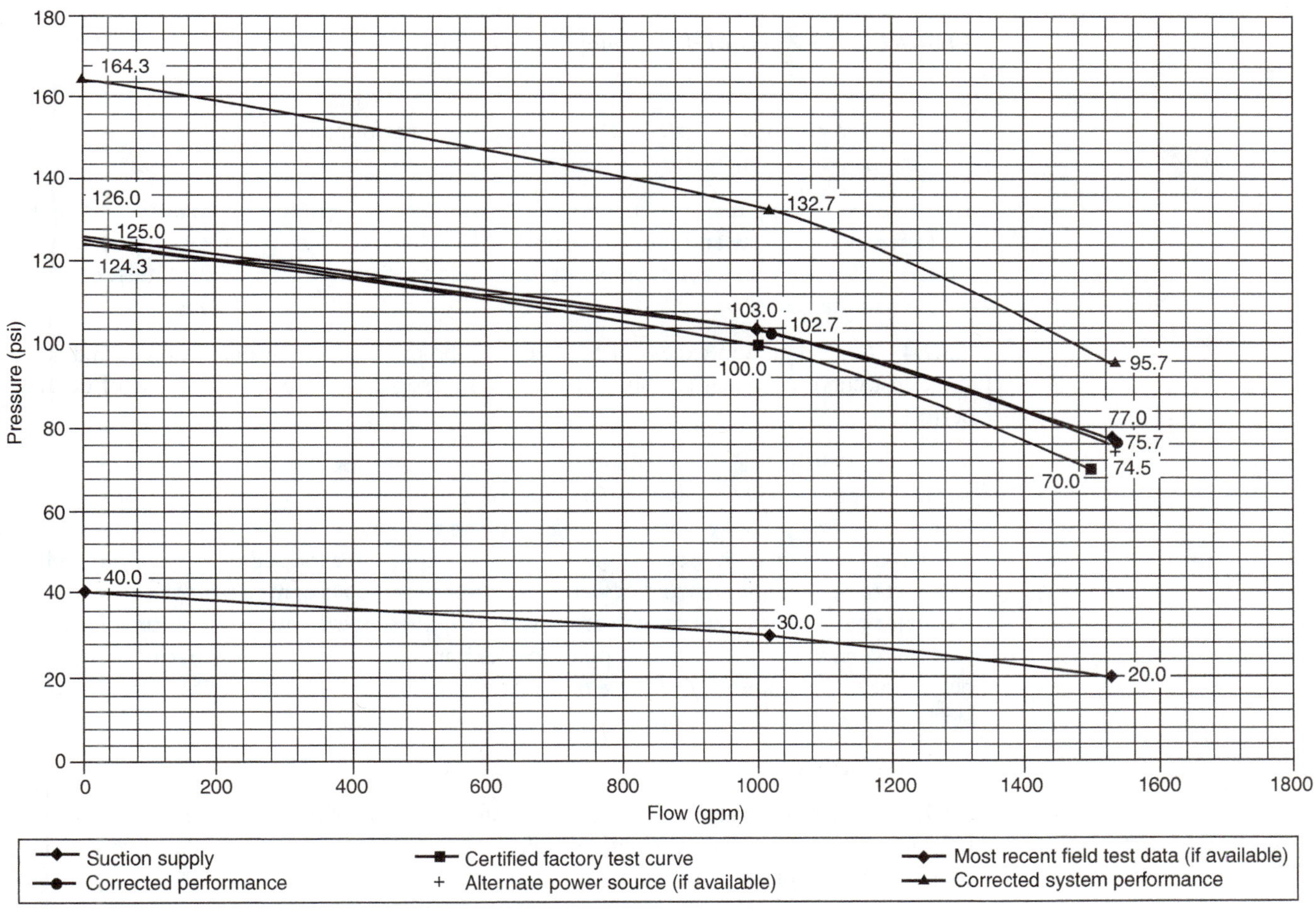

***FIGURE A.8.3.7.3(1)(a)** Fire Pump Performance Curve — Corrected Data.*

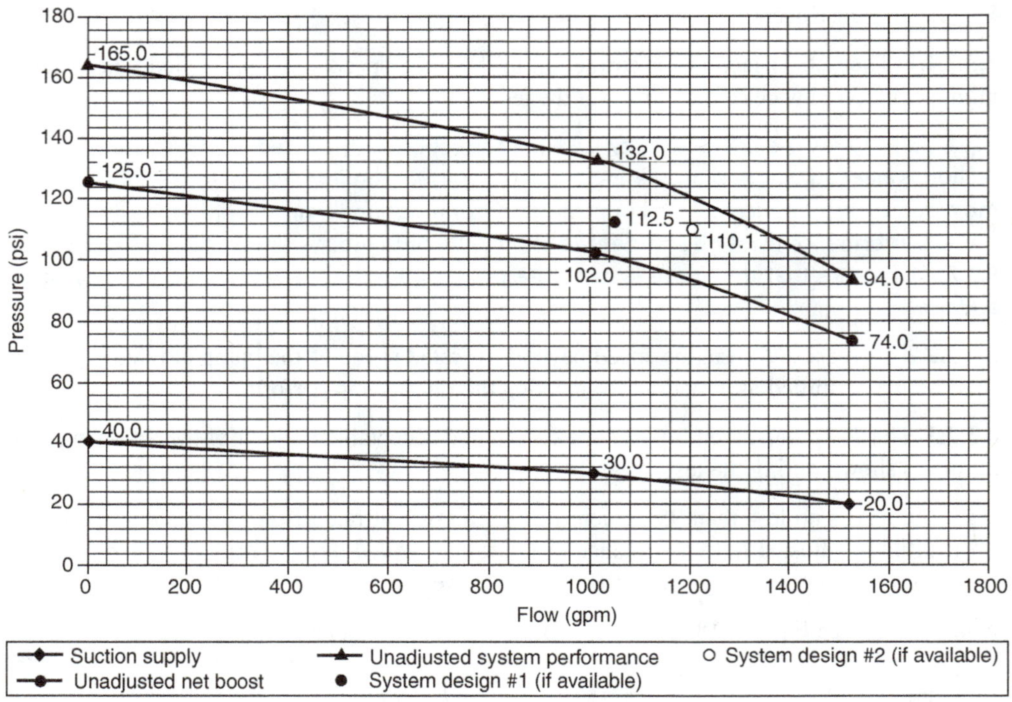

FIGURE A.8.3.7.3(1)(b) *Fire Pump Performance Curve — Unadjusted Data.*

(2) The fire pump is no less than 95 percent of the performance characteristics as indicated on the pump nameplate.

The annual fire pump test is intended to evaluate the net fire pump performance. This net performance includes flow and pressure. Figure A.8.3.7.3(1) illustrates that the pump is evaluated based on discharge minus suction pressure. Variations of more than 5 percent must be investigated as to the cause.

8.3.7.4* Degradation in excess of 5 percent of the pressure of the initial unadjusted acceptance test curve or nameplate shall require an investigation to reveal the cause of degraded performance.

If the pump pressure falls more than 5 percent below that shown on the nameplate or the initial pump acceptance test, an investigation must be undertaken. History shows that the most common cause of degraded pump performance is the presence of a shut or partially shut control valve in the water supply. Booster pumps taking water from a looped or gridded network can be affected by sectional valves that were left closed after repairs or modifications were made by the water purveyor. In high-growth areas, water supplies may deteriorate over time from more user demand, and this can impact pump performance at rated or peak capacity. If water supplies appear to be normal, the cause may rest within the pump itself, and the pump should be evaluated by qualified personnel.

A.8.3.7.4 See Annex C.

8.3.7.5 Current and voltage readings whose product does not exceed the product of the rated voltage and rated full-load current multiplied by the permitted motor service factor shall be considered acceptable.

8.3.7.6 Voltage readings at the motor within 5 percent below or 10 percent above the rated (i.e., nameplate) voltage shall be considered acceptable.

8.3.7.7 The pump performance shall be evaluated using the unadjusted flow rates and pressures to ensure the pump can supply the system demand as supplied by the owner.

•

8.4 Reports

8.4.1 Any abnormality observed during inspection or testing shall be reported promptly to the property owner or designated representative.

8.4.2 Test results and any documented performance issues shall be recorded and retained for comparison purposes in accordance with Section 4.3.

Although no formal recording process is mandated by the standard, there are several key pieces of information that should be recorded so that the pump efficiency and performance can be plotted if needed. This information includes the percent of rated pump discharge, suction and discharge pressure, flow, and pump operating speed. Supplement 5 contains sample forms that provide a means for recording this information.

Critical Point

Although no formal recording process is mandated by the standard, there are several key pieces of information that should be recorded so that the pump efficiency and performance can be plotted if needed.

8.4.3 All time delay intervals associated with the pump's starting, stopping, and energy source transfer shall be recorded.

8.5 Maintenance

8.5.1* A preventive maintenance program shall be established on all components of the pump assembly in accordance with the manufacturer's recommendations or Table 8.1.2.

A.8.5.1 It is important to provide proper bearing lubrication and to keep bearings clean. Some bearings are the sealed type and need no relubrication. Couplings with rubber drive parts do not need lubrication; other types generally do. The following practices are recommended:

(1) Lubricant fittings should be cleaned before relubricating with grease.
(2) The proper amount of lubricant should be used. Too much lubricant results in churning, causing excessive power loss and overheating.
(3) The correct lubricant should be used.

Engine Maintenance. Engines should be kept clean, dry, and well lubricated. The proper oil level in the crankcase should be maintained.

Battery Maintenance. Only distilled water should be used in battery cells. Plates should be kept submerged at all times. An automatic battery charger is not a substitute for proper maintenance of the battery and charger. Periodic inspection ensures that the charger is operating correctly, the water level in the battery is adequate, and the battery is holding its proper charge.

An adequate water level in battery cells is established when the water level reaches the ring at the bottom of the cell fill connection. The use of distilled water in battery cells is also critical. Exhibit 8.47 shows a battery that was not properly maintained because the battery cells were refilled with tap water. Tap water contains impurities, which bridged the gap between the lead cells, causing the battery to fail. If there is any space in the top of the cell, hydrogen can accumulate in that space. If that occurs, an explosion could result when the pump starts.

It is important to make sure that the proper amount and type of lubricants are used to maintain bearings. Exhibit 8.48 shows a broken drive shaft that was caused by seized outboard bearings, while Exhibit 8.49 shows an over-greased outboard bearing.

Fuel Supply Maintenance. The fuel storage tank should be kept at least two-thirds full. Fuel should be maintained free of water and foreign material by draining water and foreign material from the tank sump annually. This necessitates draining approximately 5 gal (19 L).

EXHIBIT 8.47 *Battery Failure Due to Improper Maintenance. (Courtesy of Damon Pietraz, Underwood Fire Equipment)*

EXHIBIT 8.48 *Broken Pump Shaft Due to Seized Bearings. (Courtesy of Damon Pietraz, Underwood Fire Equipment)*

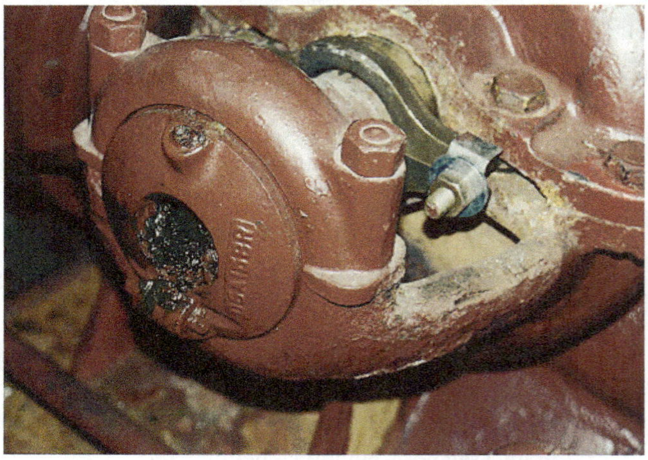

EXHIBIT 8.49 *Over-Greased Outboard Bearing. (Courtesy of Damon Pietraz, Underwood Fire Equipment)*

Fuel that is kept for longer than 1 year could contain biological growth that can clog the engine fuel filters. Therefore, fuel should be rotated or filtered annually and new biological inhibitors added. For more information on testing of diesel fuel, see the commentary following 8.3.4.1.1.

Temperature Maintenance. The temperature of the pump room, pump house, or area where engines are installed should never be less than the minimum recommended by the engine manufacturer. The manufacturer's temperature recommendations for water and oil heaters should be followed.

8.5.2 Records shall be maintained on all work performed on the pump, driver, controller, and auxiliary equipment.

8.5.3 The preventive maintenance program shall be initiated immediately after the pump assembly has passed acceptance tests.

As with every installation standard, once commissioning activity has been completed, the scope of the installation standard is no longer applicable and the scope (and frequencies) of the maintenance standard begins.

Critical Point

As with every installation standard, once commissioning activity has been completed, the scope of the installation standard is no longer applicable and the scope (and frequencies) of the maintenance standard begins.

8.6 Component Replacement Testing Requirements

Component replacement tables provide guidance to the user of the standard when system components are adjusted, repaired, rebuilt, or replaced. It is not necessary in each case to require a complete acceptance test for each component when maintenance is performed.

8.6.1 Whenever a component in a fire pump is adjusted, repaired, rebuilt, or replaced, the tests required to restore the system to service shall be performed in accordance with Table 8.6.1.

TABLE 8.6.1 *Summary of Component Replacement Testing Requirements*

Component	*Adjust*	*Repair*	*Rebuild*	*Replace*	*Test Criteria*
Fire Pump System					
Entire pump assembly				X	Perform acceptance test in accordance with NFPA 20, *Standard for the Installation of Stationary Pumps for Fire Protection*
Impeller/rotating assembly		X		X	Perform acceptance test in accordance with NFPA 20
Casing		X		X	Perform acceptance test in accordance with NFPA 20 with alignment inspection
Bearings				X	Perform annual test in accordance with 8.3.3
Sleeves				X	Perform annual test in accordance with 8.3.3
Wear rings				X	Perform annual test in accordance with 8.3.3
Main shaft		X		X	Perform annual test in accordance with 8.3.3
Packing	X			X	Perform test in accordance with 8.3.2
Mechanical Transmission					
Gear right angle drives		X	X	X	Perform acceptance test in accordance with NFPA 20
Drive coupling	X	X	X	X	Perform test in accordance with 8.3.3 with alignment inspection
Electrical System/Controller					
Entire controller				X	Perform acceptance test in accordance with NFPA 20
Electronic component or module that can prevent the controller from starting or running			X	X	Perform acceptance test in accordance with NFPA 20
Electronic component or module that will not prevent the controller from starting or running			X	X	Perform weekly test in accordance with NFPA 25
Plumbing part				X	Perform weekly test in accordance with NFPA 25
Isolating switch				X	Perform test in accordance with 8.3.2 and exercise six times

(continues)

TABLE 8.6.1 Continued

Component	Adjust	Repair	Rebuild	Replace	Test Criteria
Circuit breaker	X				Perform six momentary starts in accordance with NFPA 20
Circuit breaker				X	Perform a 1-hour full-load current test in accordance with 8.3.3, including six starts at peak load
Electrical connections	X				Perform test in accordance with 8.3.2
Main contactor		X		X	Perform test in accordance with 8.3.3 with six starts
Power monitor				X	Perform six operations of the circuit breaker/isolation switch disconnect (cycle the power on/off)
Start relay				X	Perform test in accordance with 8.3.2 with six starts
Pressure switch	X			X	Perform test in accordance with 8.3.2 and exercise six times automatically
Pressure transducer	X			X	Perform six automatic no-load starts
Manual start or stop switch				X	Perform six operations under load
Transfer switch — load carrying parts		X	X	X	Perform a 1-hour full-load current test, six starts at peak horsepower load, and transfer from normal power to emergency power and back one time
Transfer switch — no-load parts		X	X	X	Perform six no-load operations of transfer of power
Electric Motor Driver					
Electric motor		X	X	X	Perform acceptance test in accordance with 8.3.3, including alignment tests
Motor bearings				X	Perform annual test in accordance with 8.3.3
Incoming power conductors				X	Perform a 1-hour full-load current test including six starts at peak load
Diesel Engine Driver					
Entire engine			X	X	Perform acceptance test in accordance with NFPA 20
Fuel transfer pump	X		X	X	Perform test in accordance with 8.3.2
Fuel injector pump or ECM	X			X	Perform test in accordance with 8.3.3
Fuel system filter		X		X	Perform test in accordance with 8.3.2
Combustion air intake system		X		X	Perform test in accordance with 8.3.2
Fuel tank		X		X	Perform test in accordance with 8.3.2
Cooling system		X	X	X	Perform test in accordance with 8.3.3
Batteries		X		X	Perform start/stop sequence in accordance with NFPA 25
Battery charger		X		X	Perform test in accordance with 8.3.2
Electric system		X		X	Perform test in accordance with 8.3.2
Lubrication filter/oil service		X		X	Perform test in accordance with 8.3.2
Steam Turbines					
Steam turbine		X		X	Perform acceptance test in accordance with NFPA 20
Steam regulator or source upgrade		X		X	Perform acceptance test in accordance with NFPA 20

TABLE 8.6.1 Continued

Component	*Adjust*	*Repair*	*Rebuild*	*Replace*	*Test Criteria*
Positive Displacement Pumps					
Entire pump				X	Perform acceptance test in accordance with NFPA 20
Rotors				X	Perform annual test in accordance with 8.3.3
Plungers				X	Perform annual test in accordance with 8.3.3
Shaft				X	Perform annual test in accordance with 8.3.3
Driver		X	X	X	Perform acceptance test in accordance with NFPA 20
Bearings				X	Perform annual test in accordance with 8.3.3
Seals				X	Perform test in accordance with 8.3.2
Pump House and Miscellaneous Components					
Baseplate		X			Perform test in accordance with 8.3.2 with alignment inspection
Baseplate				X	Perform test in accordance with 8.3.3 with alignment inspection
Foundation		X	X	X	Perform test in accordance with 8.3.2 with alignment inspection
Suction/discharge pipe		X		X	Perform visual inspection in accordance with 8.2.2
Suction/discharge fittings		X		X	Perform visual inspection in accordance with 8.2.2
Suction/discharge valves		X	X	X	Perform operational test in accordance with 13.3.3.1

Table 8.6.1 is intended to be used only in the absence of the pump manufacturer's recommendations for maintenance. It is best to obtain the manufacturer's operations and maintenance manuals for a fire pump, because some manufacturers might have special requirements particular to their equipment. Failure to follow such requirements could result in damage to the pump or its components and could void any warranties.

8.6.2 NFPA 20 shall be consulted for the minimum requirements for design, installation, and acceptance testing.

8.6.3 Replacement parts shall be provided that will maintain the listing for the fire pump component assembly whenever possible.

8.6.3.1 If the part is no longer available from the original equipment manufacturer, then an approved like part shall be permitted to be used.

References Cited in Commentary

ANSI Publications. American National Standards Institute, Inc., 25 West 43rd Street, 4th Floor, New York, NY 10036.

ANSI B15.1, *Safety Standard for Mechanical Power Transmission Apparatus*, 2000.

National Fire Protection Association, 1 Batterymarch Park, Quincy, MA 02169-7471.

NFPA 13, *Standard for the Installation of Sprinkler Systems*, 2013 edition.
NFPA 20, *Standard for the Installation of Stationary Pumps for Fire Protection*, 2013 edition.
NFPA 70E®, Standard for Electrical Safety in the Workplace®, 2012 edition.
NFPA 72®, National Fire Alarm and Signaling Code, 2013 edition.

Isman, K. E. and M.T. Puchovsky, *Pumps for Fire Protection Systems*, 2002 edition.

Fire Protection Research Foundation, 1 Batterymarch Park, Quincy, MA 02169-7471.

"Fire Pump Field Data Collection and Analysis," April 2012.

No-Flow Condition Fire Pump (Non-Variable Speed) Test

Purpose

The purpose of completing a no-flow condition test of a fire pump is to verify that the fire pump will start under automatic conditions and run without overheating or development of unacceptable operating conditions. In meeting this objective, a variety of normal operating conditions are to be observed during the test.

Tools/Equipment

1. A handheld tachometer is required to measure the operating speed of the fire pump.
2. The various wrenches and tools necessary to facilitate any required adjustments during the test are required.
3. A timing device is needed to measure the fire pump run time duration.
4. A form for recording test data is required.

Procedure Steps

Electric-Driven Fire Pump

1. Prior to any testing, notify the fire department and/or the alarm monitoring company (as needed), as well as the facility representatives, that testing is going to be conducted.
2. The inspection activities required by Section 8.2 should be completed prior to any testing of the fire pump to ensure its readiness for testing.
3. Review the fire pump assembly nameplates noting the voltage rating, rated speed, and churn pressure for the unit.
4. Prior to operation of the fire pump, record the suction and discharge pressures. For fire pump controllers that use electronic pressure sensors, record the current pressure reading as well as the highest and lowest pressures indicated on the fire pump controller event log. Where these values are outside of the expected range, a record of the entire event log must be made and further investigation of the condition conducted. Appropriate corrective action should be taken prior to conducting the no-flow condition test.
5. Check the area surrounding any relief valve or cooling water discharge outlets to ensure that there are no obvious conditions that would prevent water from being discharged safely or cause direct damage in the immediate vicinity. Where the discharge of water is to an area subject to potential freezing conditions, the facility representative should be advised of the potential for icing conditions.
6. Consider closing the fire pump discharge control valve prior to conducting the test, to limit the exposure of the connected systems to the resultant pressure surge during the start-up of the fire pump.
7. For pressure-actuated fire pump controllers, simulate an automatic start for the fire pump by creating a pressure drop in the sensing line to the fire pump controller. This is typically accomplished by slowly opening the drain valve on the sensing line located near the fire pump controller until the fire pump starts automatically. The use of the "start" button on the fire pump controller is not acceptable for the purpose of simulating an automatic start. Note the start time of the fire pump to measure the run time duration and record the starting pressure.

   For pressure-actuated fire pump controllers that use an automatic timer, an automatic opening of a solenoid valve in the sensing line to the fire pump controller might be used to simulate the automatic start of the fire pump. These systems must include a record of the pressure drop on the pressure recorder for the controller. For non-pressure-actuated fire pump controllers, the automatic start can be simulated by other means.

   In all cases, the attendance of qualified personnel is required whenever the fire pump is in operation.
8. Immediately check the operating speed of the motor, noting the time required to reach rated speed. The motor should reach rated speed within 20 seconds. The measurement can be taken with a handheld tachometer. Note that the use of a strobe-type handheld tachometer requires advance preparation prior to the test for proper measurement, including the application of a reflective tape on the shaft and/or removal of protective covers that shield the rotating shaft.
9. For reduced-voltage start or reduced-current start fire pump controllers, record the time the controller remains in the first step voltage. This transition is characterized by a distinct change in the sound of the motor rotation speed as the controller voltage/current goes to full output. The maximum time to transition to full voltage/current should not exceed 10 seconds.
10. Check the fire pump packing gland for a slight discharge (slow drip) of water, adjusting the packing gland nuts as needed to achieve approximately 1 drip per second. For safety purposes, the adjustment of the packing gland should be made while the pump is not running. Care should be exercised to ensure the glands are not tightened to the point of breaking.
11. Monitor the fire pump operation for any unusual noise, vibration, or other signs of malfunction.
12. Verify that the operation of the circulation (casing) relief valve has a steady stream of water to ensure proper cooling of the pump case.
13. If the fire pump is equipped with a main pressure relief valve, verify the operation of the pressure relief valve such that outlet pressures do not exceed the pressure rating of the piping downstream of the fire pump. Usually, this rating is 175 psi (12.1 bar); however, some systems are designed for higher pressures.
14. Record the suction and discharge pressures. Note that for vertical turbine pumps, only the discharge pressure is recorded.

15. Record the pressure at the fire pump controller pressure switch or pressure transducer, and compare it with the discharge pressure recorded above.

16. Check the packing gland box, shaft bearings, and pump casing for overheating approximately every 5 minutes during the test. The packing gland box and shaft bearings might be warm to the touch, and the pump casing should remain cool.

17. Allow the fire pump to continue operating uninterrupted for 10 minutes, rechecking for overheating periodically during the test, and then shut down manually at the end of the 10-minute duration. Note that some fire pump controllers might include automatic run timers that automatically shut down the pump after a specified run time. For these controllers, check the run time for the fire pump to ensure the required 10 minutes has elapsed.

18. If closed, reopen the fire pump discharge control valve and conduct a valve status test downstream of the closed valve in accordance with 13.3.1.2.1.

19. Restore the fire pump to the automatic operating position.

20. Upon completion of all testing, notify the fire department and/or the alarm monitoring company (as needed), as well as the facility representatives that testing is complete. Reset fire alarm system as necessary.

Diesel-Driven Fire Pump

1. Prior to any testing, notify the fire department and/or the alarm monitoring company (as needed), as well as the facility representatives, that testing is going to be conducted.

2. The inspection activities required by Section 8.2 should be completed prior to any testing of the fire pump to ensure its readiness for testing.

3. Review the fire pump assembly nameplates, noting the rated speed and churn pressure for the unit.

4. Prior to operation of the fire pump, record the suction and discharge pressures. For fire pump controllers that use electronic pressure sensors, record the current pressure reading as well as the highest and lowest pressures indicated on the fire pump controller event log. Where these values are outside of the expected range, a record of the entire event log must be made and further investigation of the condition conducted. Appropriate corrective action should be taken prior to conducting the test.

5. Check the area surrounding any relief valve or cooling water discharge outlets to ensure that there are no obvious conditions that would prevent water from being discharged safely or cause direct damage in the immediate vicinity. Where the discharge of water is to an area subject to potential freezing conditions, the facility representative should be advised of the potential for icing conditions.

6. Consider closing the fire pump discharge control valve prior to conducting the test, to limit the exposure of the connected systems to the resultant pressure surge during the start-up of the fire pump.

7. For pressure-actuated fire pump controllers, simulate an automatic start for the fire pump by creating a pressure drop in the sensing line to the fire pump controller. This is typically accomplished by slowly opening the drain valve on the sensing line located near the fire pump controller until the fire pump starts automatically. The use of the "start" button on the fire pump controller is not acceptable for the purpose of simulating an automatic start. Note the start time of the fire pump to measure the run time duration and record the starting pressure.

   For pressure-actuated fire pump controllers that use an automatic timer, an automatic opening of a solenoid valve in the sensing line to the fire pump controller might be used to simulate the automatic start of the fire pump. These systems must include a record of the pressure drop on the pressure recorder for the controller.

   For non-pressure-actuated fire pump controllers, the automatic start can be simulated by other means.

   In all cases, the attendance of qualified personnel is required whenever the fire pump is in operation.

8. Observe the amount of time required for the diesel engine to crank. Delays in starting of the engine should be investigated further, and corrective action should be taken as necessary. Typically, the controller will attempt three 15-second crank cycles before registering a failure-to-start trouble condition.

9. Immediately check the operating speed of the diesel engine, noting the time required to reach rated speed. The engine should reach rated speed within 20 seconds. The measurement can be taken from the engine instrument panel and can be verified by a handheld tachometer if necessary. Note that the use of a strobe-type handheld tachometer requires advance preparation prior to the test for proper measurement, including the application of a reflective tape on the shaft and/or removal of protective covers that shield the rotating shaft.

10. Observe the engine instrument panel to ensure that the engine oil pressure, operating speed, water and oil temperature, and charging rate are within the acceptable range, rechecking approximately every 5 minutes during the test.

11. Check the fire pump packing gland for a slight discharge (slow drip) of water, adjusting the packing gland nuts as needed to achieve approximately 1 drip per second. For safety purposes, the adjustment of the packing gland should be made while the pump is not running. Care should be exercised to ensure the glands are not tightened to the point of breaking.

12. Monitor the fire pump operation for any unusual noise, vibration, or other signs of malfunction.

13. For radiator-cooled diesel fire pumps, verify that the operation of the circulation (casing) relief valve has a steady stream of water to ensure proper cooling of the pump case.

14. For heat exchanger–cooled diesel fire pumps, verify that the heat exchanger has a proper flow of cooling water.

No-Flow Condition Fire Pump (Non-Variable Speed) Test *(Continued)*

15. If the fire pump is equipped with a main pressure relief valve, verify the operation of the pressure relief valve such that outlet pressures do not exceed the pressure rating of the piping downstream of the fire pump. Usually, this rating is 175 psi (12.1 bar); however, some systems are designed for higher pressures.
16. Record the suction and discharge pressures. Note that for vertical turbine pumps, only the discharge pressure is recorded.
17. Record the pressure at the fire pump controller pressure switch or pressure transducer, and compare it with the discharge pressure recorded above.
18. Check the packing gland box, shaft bearings, and pump casing for overheating approximately every 5 minutes during the test. The packing gland box and shaft bearings might be warm to the touch, and the pump casing should remain cool.
19. Allow the fire pump to continue operating for 30 minutes, rechecking for overheating and abnormal readings on the engine instrument panel periodically during the test, and then shut down manually. Any observed abnormalities are to be recorded. Note that some fire pump controllers might include automatic run timers that automatically shut down the pump after a specified run time. For these controllers, check the run time for the fire pump to ensure the required 30 minutes has elapsed.
20. If closed, reopen the fire pump discharge control valve and conduct a valve status test downstream of the closed valve in accordance with 13.3.1.2.1.
21. Inspect and clean any installed intake screens.
22. Where the fire pump controller is so equipped, replace any pressure recorder charts and rewind as necessary.
23. Restore the fire pump to the automatic operating position.
24. Upon completion of all testing, notify the fire department and/or the alarm monitoring company (as needed), as well as the facility representatives, that testing is complete. Reset the fire alarm system as necessary.

Evaluation of Results

An evaluation of the recorded test pressure should be conducted by comparing the test results with previous test results and by assessing the churn pressure achieved during the conduct of the test with the nameplate data on the fire pump unit. The churn pressure for a centrifugal pump is determined by subtracting the recorded suction pressure from the discharge pressure. For a vertical turbine pump, the churn pressure is determined by adding the elevation pressure difference between the water level in the wet pit or well supplying the pump and the discharge gauge to the recorded discharge pressure on the pump. The elevation pressure difference is determined by multiplying the height differential in feet between the water level in the wet pit or well and the discharge gauge by 0.434 to determine a pressure difference in psi (height differentials in meters are converted to bar using a 0.0981 multiplier). A drop in churn pressure of more than 5 percent below the nameplate data should be investigated, and corrective action should be taken as necessary. It should be noted that the operation of a relief valve during the test will not result in true churn/no-flow conditions and should be considered in evaluating the results of the pressure readings for the fire pump.

Note that, at any time during the test, should abnormal conditions be observed that might result in personal injury or damage to the equipment or property, the fire pump should be immediately shut down and the test terminated until the cause can be investigated and corrective action taken as necessary.

Additionally, all abnormal conditions observed during the test, including those specified in the testing procedures above, should be referred for further investigation and corrective action taken as necessary. Minor field adjustments to correct for packing gland leakage rate, casing relief or cooling water discharge, or other situations might be conducted as part of the testing procedure.

The procedure(s) outlined above is not mandated by NFPA 25 and represents only one approach to conducting this test.

Annual Fire Pump (Non-Variable Speed) Flow Test

Purpose

The purpose of completing an annual flow test is to verify the adequacy of the fire pump performance through its range of operating conditions. In meeting this objective, a variety of normal operating conditions and performance measures are to be observed during the test.

Testing Via Hose Streams

Tools/Equipment

1. A handheld tachometer[1] is required to measure the operating speed of the fire pump.
2. A pitot tube together with a test-pressure gauge[2], suitable for the pressures to be expected, is required. [Usually, a 60 psi (4 bar) gauge is satisfactory.]

   A pitot-tube-and-gauge assembly is indispensable for conducting flow tests from nozzles. The small opening at the end of the tube — not more than 1/16 in. (1.6 mm) in diameter — is inserted in the center of the stream, in a direct line with the flow, at a distance in front of the opening equal to one-half the opening diameter. Velocity pressure is registered on the gauge attached to the tube.

   Some available test nozzles include an integral pitot tube or pitotless flow-measuring feature with an attached test gauge[2] allowing for direct measurement of flows without the use of a separate pitot tube.
3. Test-pressure gauges[2] suitable for measurement of the expected suction and discharge pressures on the fire pump are required. Note that the pressure gauge used on the suction side of the fire pump should be a compound pressure and vacuum gauge having pressure range of at least twice the static pressure available from the suction supply. The pressure gauge used on the discharge side of the fire pump should have a range of at least twice the rated working pressure of the fire pump.
4. A sufficient number of test nozzles are required, including hoses for connections where needed, to permit the measurement of a flow rate up to 150 percent of the rated capacity of the fire pump. [Typically 50 ft (15.2 m) lengths of lined 2½ in. (65 mm) hose are used for this purpose.] Test nozzles can include UL play pipes or other approved test nozzles. The flow characteristics (i.e., discharge coefficients) of the test nozzles must be known.
5. The various wrenches and tools necessary to facilitate required adjustments during the test and for the installation of the required test equipment are required.
6. A multimeter[1] and/or clamp-on ammeter[1] is required for taking voltage and amperage readings.
7. A timing device is required to measure the fire pump run time duration.
8. The required personal protective equipment for the job is needed. See Commentary Table 4.2.
9. A form for recording test data is required.

Procedure Steps

Electric-Driven Fire Pump

1. Prior to any testing, notify the fire department and/or the alarm monitoring company (as needed), as well as the facility representatives, that testing is going to be conducted.
2. The inspection activities required by Section 8.2 should be completed prior to any testing of the fire pump to ensure its readiness for testing.
3. Complete a check of the parallel and angular alignment of the fire pump and driver in accordance with 8.3.5.4.
4. Review the fire pump assembly nameplates, noting the following:
   a. Voltage rating
   b. Rated speed
   c. Rated capacity
   d. Churn pressure
   e. Rated pressure
   f. Overload pressure
5. Where the pump churn pressure will exceed the rated pressure for the system components, the fire pump discharge control valve must be closed prior to the operation of the fire pump to limit the exposure of the connected systems to the resultant elevated pressures at churn. For all other systems, consider closing the fire pump discharge control valve prior to conducting the test, to limit the exposure of the connected systems to the resultant pressure surge during the start-up of the fire pump.
6. Remove the installed pressure gauges on the suction and discharge side of the fire pump, and install the appropriate calibrated pressure gauges at each location. Note that for a vertical turbine fire pump there will only be a discharge pressure gauge.
7. Install sufficient flow measuring equipment, fed from the test header, to allow for the overload flow rate (150 percent

---

1. All test equipment should be calibrated by an approved testing laboratory within 12 months prior to the test.

2. Test-quality gauges should be used in accordance with ASME B40.1, *Gauges — Pressure Indicating Dial Type — Elastic Element, Grade 1A* having an accuracy of ±1%. The use of quality test gauges produces results that are considered reasonably accurate within the scope of the testing procedure. Care should be taken to protect the gauges from rough handling.

Gauges should be calibrated at least annually by means of a dead-weight or calibrated tester throughout the range of operation before a test series is begun. Calibration sheets should be kept for each gauge and correction factors affixed to the back of each gauge.

Annual Fire Pump (Non-Variable Speed) Flow Test *(Continued)*

flow) required to test the fire pump. This might include the use of hose connected to the test header and extending to flow measurement devices such as UL play pipes or other test nozzles that are adequately secured or protected against movement. In some cases, the connection of the flow measurement devices directly to the test header might be possible. The use of a flow diffuser can aid in preventing movement, as well as breaking the directed stream and managing resultant energy of the discharge.

8. Check the area surrounding any relief valve or cooling water discharge outlets as well as the discharge points for the fire pump flow to ensure that there are no obvious conditions that would prevent water from being discharged safely or cause direct damage in the immediate vicinity. Take appropriate caution to ensure personnel and property are protected from the resultant high-pressure hose stream discharge. The use of a flow diffuser can aid in breaking the directed stream and managing resultant energy of the discharge. Where the discharge of water is to an area subject to potential freezing conditions, the facility representative should be advised of the potential for icing conditions.

9. For testing using hose-connected flow measuring devices or UL play pipes directly connected to the test header, ensure that the individual hose connection valves on the test header are in the closed position and slowly open the main test header control valve pressurizing the test header. For other flow-measuring devices directly connected to the test header, ensure that the main test header control valve is closed. Then fully open a single hose connection valve readying the arrangement for the first flow point with the flow being controlled by the main test header control valve. Any time the use of an additional hose connection valve is required with this latter arrangement, the additional hose connection valves are to be fully opened and the flow again controlled by the adjustment of the main test header control valve rather than the individual hose connection valves.

10. For pressure-actuated fire pump controllers where the fire pump does not automatically start as a result of completing Step 9 above, simulate an automatic start for the fire pump by creating a pressure drop in the sensing line to the fire pump controller. This is typically accomplished by slowly opening the drain valve on the sensing line located near the fire pump controller until the fire pump starts automatically. The use of the "start" button on the fire pump controller is not acceptable for the purpose of simulating an automatic start. Note the start time of the fire pump to measure the run time duration, and record the starting pressure.

    For pressure-actuated fire pump controllers that use an automatic timer, an automatic opening of a solenoid valve in the sensing line to the fire pump controller might be used to simulate the automatic start of the fire pump. These systems must include a record of the pressure drop on the pressure recorder for the controller.

    For non-pressure-actuated fire pump controllers, the automatic start can be simulated by other means.

11. Check the fire pump packing gland for a slight discharge (slow drip) of water, adjusting the packing gland nuts as needed to achieve approximately 1 drip per second. For safety purposes, the adjustment of the packing gland should be made while the pump is not running. Care should be exercised to ensure the glands are not tightened to the point of breaking.

12. Monitor the fire pump operation for any unusual noise, vibration, or other signs of malfunction.

13. Verify that the operation of the circulation (casing) relief valve has a steady stream of water to ensure proper cooling of the pump case.

14. Check the packing gland box, shaft bearings, and pump casing for overheating approximately every 5 minutes during the test. The packing gland box and shaft bearings might be warm to the touch, and the pump casing should remain cool throughout the test.

15. If the fire pump is equipped with a main pressure relief valve, verify the operation of the pressure relief valve such that outlet pressures do not exceed the pressure rating of the piping downstream of the fire pump. Usually, this rating is 175 psi (12.1 bar); however, some systems are designed for higher pressures.

16. Record the electric motor voltage and current on all phases (lines). This can be done with a multimeter and/or clamp-on ammeter; however, these readings should only be taken by individuals trained and qualified in electrical hazards and equipped with the needed safety equipment as outlined in *NFPA 70E®, Standard for Electrical Safety in the Workplace.* Also see 8.3.3.6 and Section 4.8, along with the associated commentary, for additional guidance. Alternatively, some newer fire pump controllers include a display of the voltage and amperage readings on the front of the controller.

17. Record the suction and discharge pressures. Note that for vertical turbine pumps, only the discharge pressure is recorded.

18. Record the operating speed (rpm) of the motor. The measurement can be taken with a handheld tachometer. Note that the use of a strobe-type handheld tachometer requires advance preparation prior to the test for proper measurement, including the application of a reflective tape on the shaft and/or removal of protective covers that shield the rotating shaft.

19. For systems equipped with a main pressure relief valve as described in Step 15 above, adjust the pressure relief valve to temporarily stop the relief valve operation and repeat Step 16 through Step 18 above, then re-adjust the main relief valve to the normal operating position, and continue with Step 20 below.

20. Data recorded in Step 16 through Step 19 should be noted as the churn test point.

21. Initiate the flow from the test header by either slowly opening one of the hose connection control valves on the test header or by slowly opening the main test header control valve, depending on the arrangement selected in Step 9 above. As the valve is opened, take flow measurements to determine the flow rate, adjusting the valve position until a measured flow rate equal to the rated capacity from the nameplate on the fire pump is achieved. Where the flow rate with the valve fully open is inadequate to meet this rated capacity, additional hose connection valves must be opened and adjustments made until the total measured flow rate equals the rated capacity. The flow rate should be stabilized as much as possible prior to taking the final flow measurements. When using multiple outlets, the final measurement on all outlets cannot accurately be made until all adjustments have been completed. See the discussion regarding flow measurement calculations near the end of the this testing procedure.

22. Repeat Step 16 through Step 19 above, then continue with Step 23 below. The data recorded should be noted as the rated (or 100 percent flow) test point.

23. Continue to open hose valves and/or the main test header control valve, depending on the arrangement selected in Step 9 above. Take flow measurements to determine the flow rate, adjusting the position of the valve(s) until a total measured flow rate equal to the 150 percent of the rated capacity from the nameplate on the fire pump is achieved, or until the maximum available flow rate available from the water supply is met, whichever is lower. The flow rate should be stabilized as much as possible prior to taking the final flow measurements. When using multiple outlets, the final measurement on all outlets cannot accurately be made until all adjustments have been completed. Care must be taken to ensure that the suction pressure does not fall below an acceptable level as the flow rate is increased to 150 percent. NFPA 20, *Standard for the Installation of Stationary Pumps for Fire Protection,* allows a pressure at the suction flange of 0 psi (0 bar) for fire pumps connected to municipal supplies and -3 psi (-0.2 bar) for fire pumps taking suction from a grade level tank. Local water authorities might require a higher maintained pressure [commonly 20 psi (1.4 bar)] at the point of connection to the municipal supply.

24. Repeat Step 16 through Step 19 above, then continue with Step 25 below. The data recorded should be noted as the overload (or 150 percent flow) test point.

25. For installations including an automatic transfer switch, simulate a power failure condition and verify the following:

    a. That the transfer switch transitions to an alternate power source with continued operation of the fire pump

    b. That the operation of the pump continues for 10 minutes when the alternate power supply is from a separate electrical connection, or 30 minutes when the alternate power supply is from a standby generator set

    Restore the normal power source and verify that the transfer switch transitions back to the normal power supply with continued operation of the fire pump.

26. Shut down the discharge flow by slowly closing the main test header control valve to avoid water hammer.

27. If the run time for the fire pump has not reached 10 minutes, then allow the fire pump to continue operating until a total run time of 10 minutes has elapsed, and then shut down manually upon completion of the test.

28. Verify the closure of the circulation relief valve and pressure relief valve (if installed) upon completion of the test.

29. Plot test the data against the manufacturer's specifications *before* removing test equipment, to verify that a retest is not required due to bad data recording or other reasons.

30. Remove all attached test equipment from the test header, restore the closed position of all hose connection valves, and reinstall hose caps. In locations subject to freezing, check that the test header properly drains or pumps out as necessary.

31. Remove the calibrated pressure gauges, and reinstall the original pressure gauges on the suction and discharge side of the fire pump.

32. Inspect and clean any installed intake screens.

33. Initiate a fire pump phase reversal supervisory signal by jumping across the monitored points in the fire pump controller (typically labeled on a wiring diagram posted on the inside of the fire pump controller) to simulate an activation. Verify that all of the fire pump supervisory signals, such as fire pump running, loss of power, or phase reversal, have been indicated at the fire pump controller as well as being transmitted to any connected fire alarm control panels.

34. If closed, reopen the fire pump discharge control valve and conduct a valve status test downstream of the closed valve in accordance with 13.3.1.2.1.

35. Restore the fire pump to the automatic operating position.

36. Upon completion of all testing, notify the fire department and/or the alarm monitoring company (as needed), as well as the facility representatives, that testing is complete. Reset the fire alarm system as necessary. A verification of the receipt of initiated signals with the fire department and/or the alarm monitoring company, where so connected, would be appropriate as part of this effort.

Diesel-Driven Fire Pump

1. Prior to any testing, notify the fire department and/or the alarm monitoring company (as needed), as well as the facility representatives, that testing is going to be conducted.

2. The inspection activities required by Section 8.2 should be completed prior to conducting any testing of the fire pump to ensure its readiness for testing.

Annual Fire Pump (Non-Variable Speed) Flow Test *(Continued)*

3. Complete a check of the parallel and angular alignment of the fire pump and driver in accordance with 8.3.5.4.
4. Review the fire pump assembly nameplates, noting the following:
   a. Rated speed
   b. Rated capacity
   c. Churn pressure
   d. Rated pressure
   e. Overload pressure
5. Where the pump churn pressure will exceed the rated pressure for the system components, the fire pump discharge control valve must be closed prior to the operation of the fire pump to limit the exposure of the connected systems to the resultant elevated pressures at churn. For all other systems, consider closing the fire pump discharge control valve prior to the test to limit the exposure of the connected systems to the resultant pressure surge during the start-up of the fire pump.
6. Remove the installed pressure gauges on the suction and discharge side of the fire pump, and install the appropriate calibrated pressure gauges at each location. Note that for a vertical turbine fire pump, there will only be a discharge pressure gauge.
7. Install sufficient flow measuring equipment, fed from the test header, to allow for the overload flow rate (150 percent flow) required to test the fire pump. This might include the use of hose connected to the test header and extending to flow measurement devices such as UL play pipes or other test nozzles that are adequately secured or protected against movement. In some cases, the connection of the flow measurement devices directly to the test header might be possible. The use of a flow diffuser can aid in preventing movement, as well as breaking the directed stream and managing resultant energy of the discharge.
8. Check the area surrounding any relief valve or cooling water discharge outlets as well as the discharge points for the fire pump flow to ensure that there are no obvious conditions that would prevent water from being discharged safely or cause direct damage in the immediate vicinity. Take appropriate caution to ensure personnel and property are protected from the resultant high-pressure hose stream discharge. The use of a flow diffuser can aid in breaking the directed stream and managing resultant energy of the discharge. Where the discharge of water is to an area subject to potential freezing conditions, the facility representative should be advised of the potential for icing conditions.
9. For testing using hose-connected flow measuring devices or UL play pipes directly connected to the test header, ensure that the individual hose connection valves on the test header are in the closed position and slowly open the main test header control valve pressurizing the test header. For other flow-measuring devices directly connected to the test header, ensure that the main test header control valve is closed, Then fully open a single hose connection valve, readying the arrangement for the first flow point, with the flow being controlled by the main test header control valve. Any time the use of an additional hose connection valve is required with this latter arrangement, the additional hose connection valves are to be fully opened and the flow again controlled by the adjustment of the main test header control valve rather than the individual hose connection valves.
10. For pressure-actuated fire pump controllers where the fire pump does not automatically start as a result of completing Step 9 above, simulate an automatic start for the fire pump by creating a pressure drop in the sensing line to the fire pump controller. This is typically accomplished by slowly opening the drain valve on the sensing line located near the fire pump controller until the fire pump starts automatically. The use of the "start" button on the fire pump controller is not acceptable for the purpose of simulating an automatic start. Note the start time of the fire pump to measure the run time duration, and record the starting pressure.

    For pressure-actuated fire pump controllers that use an automatic timer, an automatic opening of a solenoid valve in the sensing line to the fire pump controller might be used to simulate the automatic start of the fire pump. These systems must include a record of the pressure drop on the pressure recorder for the controller.

    For non-pressure-actuated fire pump controllers, the automatic start can be simulated by other means.
11. Observe the engine instrument panel to ensure that the engine oil pressure, operating speed, water and oil temperature, and charging rate are within the acceptable range, rechecking approximately every 5 minutes.
12. Check the fire pump packing gland for a slight discharge (slow drip) of water, adjusting the packing gland nuts as needed to achieve approximately 1 drip per second. For safety purposes, the adjustment of the packing gland should be made while the pump is not running. Care should be exercised to ensure the glands are not tightened to the point of breaking.
13. Monitor the fire pump operation for any unusual noise, vibration, or other signs of malfunction.
14. For radiator-cooled diesel fire pumps, verify that the operation of the circulation (casing) relief valve has a steady stream of water to ensure proper cooling of the pump case.
15. For heat exchanger–cooled diesel fire pumps, verify that the heat exchanger has a proper flow of cooling water.
16. Check the packing gland box, shaft bearings, and pump casing for overheating approximately every 5 minutes during the test. The packing gland box and shaft bearings might be warm to the touch, and the pump casing should remain cool throughout the test.
17. If the fire pump is equipped with a main pressure relief valve, verify the operation of the pressure relief valve such

that outlet pressures do not exceed the pressure rating of the piping downstream of the fire pump. Usually, this rating is 175 psi (12.1 bar); however, some systems are designed for higher pressures.

18. Record the suction and discharge pressures. Note that for vertical turbine pumps, only the discharge pressure is recorded.

19. Record the operating speed (rpm) of the motor. The measurement can be taken with a handheld tachometer. Note that the use of a strobe-type handheld tachometer requires advance preparation prior to the test for proper measurement, including the application of a reflective tape on the shaft and/or removal of protective covers that shield the rotating shaft.

20. For systems equipped with a main pressure relief valve as described in Step 17 above, adjust the pressure relief valve to temporarily stop the relief valve operation and repeat Step 18 and Step 19 above, then re-adjust the main relief valve to the normal operating position, and continue with Step 21 below.

21. The data recorded in Step 18 through Step 20 should be noted as the churn test point.

22. Initiate the flow from the test header by either slowly opening one of the hose connection control valves on the test header or by slowly opening the main test header control valve, depending on the arrangement selected in Step 9 above. As the valve is opened, take flow measurements to determine the flow rate, adjusting the valve position until a measured flow rate equal to the rated capacity from the nameplate on the fire pump is achieved. Where the flow rate with the valve fully open is inadequate to meet this rated capacity, additional hose connection valves must be opened and adjustments made until the total measured flow rate equals the rated capacity. The flow rate should be stabilized as much as possible prior to taking the final flow measurements. When using multiple outlets, the final measurement on all outlets cannot accurately be made until all adjustments have been completed. See the discussion of flow measurement calculations near the end of this testing procedure.

23. Repeat Step 18 through Step 20 above, then continue with Step 24 below. The data recorded should be noted as the rated (or 100 percent flow) test point.

24. Continue to open hose valves and/or the main test header control valve, depending on the arrangement selected in Step 9 above. Take flow measurements to determine the flow rate, adjusting the position of the valve(s) until a total measured flow rate equal to the 150 percent of the rated capacity from the nameplate on the fire pump is achieved or until the maximum available flow rate available from the water supply is met, whichever is lower. The flow rate should be stabilized as much as possible prior to taking the final flow measurements. When using multiple outlets, the final measurement on all outlets cannot accurately be made until all adjustments have been completed. Care must be taken to ensure that the suction pressure does not fall below an acceptable level as the flow rate is increased to 150 percent. NFPA 20 allows a pressure at the suction flange of 0 psi (0 bar) for fire pumps connected to municipal supplies and -3 psi (-0.2 bar) for fire pumps taking suction from a grade level tank. Local water authorities might require a higher maintained pressure [commonly 20 psi (1.4 bar)] at the point of connection to the municipal supply.

25. Repeat Step 18 and Step 19 above, then continue with Step 26 below. Data recorded should be noted as the overload (or 150 percent flow) test point.

26. Shut down the discharge flow by slowly closing the main test header control valve to avoid water hammer.

27. If the run time for the fire pump has not reached 30 minutes, then allow the fire pump to continue operating until a total run time of 30 minutes has elapsed, and then shut down manually upon completion of the test.

28. For engines that use an electronic fuel management control system, complete the following additional tests:

    a. Move the electronic control module (ECM) selector switch from the primary to the alternate position.

    b. Verify an alarm signal at the fire pump controller.

    c. Manually start the fire pump and verify normal operation.

    d. Manually shut down the fire pump.

    e. Reposition the ECM switch back to the primary position.

    f. Manually start the fire pump and verify the normal operation.

    g. With the fire pump still operating, disconnect the wire connection for each of the primary sensors for the ECM, verify the continued operation of the fire pump, and then reconnect the wires. Repeat the process on that same sensor removing the wires from the redundant sensor. This process should be completed for all of the primary and redundant sensors on the engine.

    h. Upon completion of the testing of all primary and redundant sensors, shut down the fire pump manually.

29. Verify the closure of the circulation relief and pressure relief valves (if installed) upon completion of the test.

30. Plot the test data against the manufacturer's specifications *before* removing test equipment, to verify a retest is not required due to bad data recording or other reasons.

31. Remove all attached test equipment from the test header, restore the closed position of all hose connection valves, and reinstall hose caps. In locations subject to freezing, check that the test header properly drains or pumps out as necessary.

Annual Fire Pump (Non-Variable Speed) Flow Test *(Continued)*

32. Remove the calibrated pressure gauges, and reinstall the original pressure gauges on the suction and discharge side of the fire pump.
33. Inspect and clean any installed intake screens.
34. Initiate the various fire pump supervisory signals, such as low oil pressure, high coolant temperature, failure to start, or engine overspeed, to simulate an activation at the individual sensors. Verify that all of the fire pump supervisory signals, such as pump running, low oil pressure, high coolant temperature, failure to start, or engine overspeed, have been indicated at the fire pump controller as well as being transmitted to any connected fire alarm panels.
35. If closed, reopen the fire pump discharge control valve and conduct a valve status test downstream of the closed valve in accordance with 13.3.1.2.1.
36. Restore the fire pump to the automatic operating position.
37. Upon completion of all testing, notify the fire department and/or the alarm monitoring company (as needed), as well as the facility representatives, that testing is complete. Reset the fire alarm system as necessary. A verification of the receipt of initiated signals with the fire department and/or the alarm monitoring company, where so connected, would be appropriate as part of this effort.

Testing Via Flow Meter

Tools/Equipment

1. A handheld tachometer[1] is required to measure the operating speed of the fire pump.
2. Test-pressure gauges[2] suitable for measurement of the expected suction and discharge pressures on the fire pump are required. The pressure gauge used on the suction side of the fire pump should be a compound pressure and vacuum gauge having pressure range of at least twice the static pressure available from the suction supply. The pressure gauge used on the discharge side of the fire pump should have a range of at least twice the rated working pressure of the fire pump.
3. The various wrenches and tools necessary to allow for required adjustments during the test and for the installation of required test equipment are required.
4. A multimeter[1] and/or clamp-on ammeter[1] is required for taking voltage and amperage readings.
5. A timing device is needed to measure the fire pump run time duration.
6. The required personal protective equipment for the job is needed. See Commentary Table 4.2.
7. A form for recording test data is required.

Procedure Steps

Electric-Driven Fire Pump

1. Prior to any testing, notify the fire department and/or the alarm monitoring company (as needed), as well as the facility representatives, that testing is going to be conducted.
2. The inspection activities required by Section 8.2 should be completed prior to any testing of the fire pump to ensure its readiness for testing.
3. Complete a check of the parallel and angular alignment of the fire pump and driver in accordance with 8.3.5.4.
4. Review the fire pump assembly name plates noting the following:
   a. Voltage rating
   b. Rated speed
   c. Rated capacity
   d. Churn pressure
   e. Rated pressure
   f. Overload pressure
5. Where the pump churn pressure will exceed the rated pressure for the system components, the fire pump discharge control valve must be closed prior to the operation of the fire pump to limit the exposure of the connected systems to the resultant elevated pressures at churn. For all other systems, consider closing the fire pump discharge control valve prior to conducting the test, to limit the exposure of the connected systems to the resultant pressure surge during the start-up of the fire pump.
6. Remove the installed pressure gauges on the suction and discharge side of the fire pump, and install the appropriate calibrated pressure gauges at each location. Note that for a vertical turbine fire pump there will only be a discharge pressure gauge.
7. Open the flow meter inlet control valve and ensure the flow meter outlet valve is closed.
8. Check the area surrounding any relief valve or cooling water discharge outlets as well as the discharge points for the fire pump flow to ensure that there are no obvious conditions that would prevent water from being discharged safely or cause direct damage in the immediate vicinity. Take appropriate caution to ensure personnel and property are protected from the resultant high-pressure discharge. Where the discharge of water is to an area subject to potential freezing conditions, the facility representative should be advised of the potential for icing conditions.
9. For pressure-actuated fire pump controllers, simulate an automatic start for the fire pump by creating a pressure

drop in the sensing line to the fire pump controller. This is typically accomplished by slowly opening the drain valve on the sensing line located near the fire pump controller until the fire pump starts automatically. The use of the "start" button on the fire pump controller is not acceptable for the purpose of simulating an automatic start. Note the start time of the fire pump to measure the run time duration and record the starting pressure.

For pressure-actuated fire pump controllers that use an automatic timer, an automatic opening of a solenoid valve in the sensing line to the fire pump controller might be used to simulate the automatic start of the fire pump. These systems must include a record of the pressure drop on the pressure recorder for the controller.

For non-pressure-actuated fire pump controllers, the automatic start can be simulated by other means.

10. Check the fire pump packing gland for a slight discharge (slow drip) of water, adjusting the packing gland nuts as needed to achieve approximately 1 drip per second. For safety purposes, the adjustment of the packing gland should be made while the pump is not running. Care should be exercised to ensure the glands are not tightened to the point of breaking.

11. Monitor the fire pump operation for any unusual noise, vibration, or other signs of malfunction.

12. Verify that the operation of the circulation (casing) relief valve has a steady stream of water to ensure proper cooling of the pump case.

13. Check the packing gland box, shaft bearings, and pump casing for overheating approximately every 5 minutes during the test. The packing gland box and shaft bearings might be warm to the touch, and the pump casing should remain cool throughout the test.

14. If the fire pump is equipped with a main pressure relief valve, verify the operation of the pressure relief valve such that outlet pressures do not exceed the pressure rating of the piping downstream of the fire pump. Usually, this rating is 175 psi (12.1 bar); however, some systems are designed for higher pressures.

15. Record the electric motor voltage and current on all phases (lines). This can be done with a multimeter and/or clamp-on ammeter; however, these readings should only be taken by individuals trained in and qualified in electrical hazards and equipped with the needed safety equipment as outlined in *NFPA 70E®, Standard for Electrical Safety in the Workplace*. Also see 8.3.3.6 and Section 4.8, along with the associated commentary, for additional guidance. Alternatively, some newer fire pump controllers include a display of the voltage and amperage readings on the front of the controller.

16. Record the suction and discharge pressures. Note that for vertical turbine pumps, only the discharge pressure is recorded.

17. Record the operating speed (rpm) of the motor. The measurement can be taken with a handheld tachometer. Note that the use of a strobe-type handheld tachometer requires advance preparation prior to the test for proper measurement, including the application of a reflective tape on the shaft and/or removal of protective covers that shield the rotating shaft.

18. For systems equipped with a main pressure relief valve as described in Step 14 above, adjust the pressure relief valve to temporarily stop the relief valve operation and repeat Step 15 through Step 17 above, then re-adjust the main relief valve to the normal operating position, and continue with Step 19 below.

19. The data recorded in Step 15 through Step 18 should be noted as the churn test point.

20. Initiate the flow through the flow meter by slowly opening the flow meter throttling valve on the outlet side of the flow meter. Throttle the valve position to achieve a reading on the flow meter equal to the rated capacity from the nameplate on the fire pump. If the flow is circulated back to the fire pump suction, then the temperature of the re-circulated water must be monitored under flow conditions throughout the remainder of the test, to ensure that temperatures remain below that which would result in equipment damage, as defined by the pump and engine manufacturers.

21. Repeat Step 15 through Step 18 above, then continue with Step 23 below. The data recorded should be noted as the rated (or 100 percent flow) test point.

22. Continue to open the throttling valve to achieve a reading on the flow meter equal to the 150 percent of the rated capacity from the nameplate on the fire pump or until the maximum available flow rate available from the water supply is met, whichever is lower. Care must be taken to ensure that the suction pressure does not fall below an acceptable level as the flow rate is increased to 150 percent. NFPA 20 allows a pressure at the suction flange of 0 psi (0 bar) for fire pumps connected to municipal supplies and -3 psi (-0.2 bar) for fire pumps taking suction from a grade level tank. Local water authorities might require a higher maintained pressure [commonly 20 psi (1.4 bar)] at the point of connection to the municipal supply.

23. Repeat Step 15 through Step 18 above, then continue with Step 24 below. The data recorded should be noted as the overload (or 150 percent flow) test point.

24. For installations including an automatic transfer switch, simulate a power failure condition and verify the following:

    a. That the transfer switch transitions to an alternate power source with continued operation of the fire pump

    b. That the operation of the pump continues for 10 minutes when the alternate power supply is from a separate electrical connection, or 30 minutes when the alternate power supply is from a standby generator set

Annual Fire Pump (Non-Variable Speed) Flow Test *(Continued)*

Restore the normal power source and verify that the transfer switch transitions back to the normal power supply with continued operation of the fire pump.

25. Shut down the discharge flow by slowly closing the throttling valve to avoid water hammer.
26. If the run time for the fire pump has not reached 10 minutes, then allow the fire pump to continue operating until a total run time of 10 minutes has elapsed, and then shut down manually upon completion of the test.
27. Verify the closure of the circulation relief valve and pressure relief valve (if installed) upon completion of the test.
28. Plot the test data against the manufacturer's specifications *before* removing test equipment, to verify a retest is not required due to bad data recording or other reasons.
29. Return the flow meter control valve to the closed position.
30. Remove the calibrated pressure gauges, and reinstall the original pressure gauges on the suction and discharge side of the fire pump.
31. Inspect and clean any installed intake screens.
32. Initiate a fire pump phase reversal supervisory signal by jumping across the monitored points in the fire pump controller (typically labeled on a wiring diagram posted on the inside of the fire pump controller) to simulate an activation. Verify that all of the fire pump supervisory signals, such as fire pump running, loss of power, or phase reversal, have been indicated at the fire pump controller as well as being transmitted to any connected fire alarm control panels.
33. If closed, reopen the fire pump discharge control valve and conduct a valve status test downstream of the closed valve in accordance with 13.3.1.2.1.
34. Restore the fire pump to the automatic operating position.
35. Upon completion of all testing, notify the fire department and/or the alarm monitoring company (as needed), as well as the facility representatives, that testing is complete. Reset the fire alarm system as necessary. A verification of the receipt of initiated signals with the fire department and/or the alarm monitoring company, where so connected, would be appropriate as part of this effort.

Diesel-Driven Fire Pump

1. Prior to any testing, notify the fire department and/or the alarm monitoring company (as needed), as well as the facility representatives, that testing is going to be conducted.
2. The inspection activities required by Section 8.2 should be completed prior to conducting any testing of the fire pump to ensure its readiness for testing.
3. Complete a check of the parallel and angular alignment of the fire pump and driver in accordance with 8.3.5.4.
4. Review the fire pump assembly nameplates, noting the following:
    a. Rated speed
    b. Rated capacity
    c. Churn pressure
    d. Rated pressure
    e. Overload pressure
5. Where the pump churn pressure will exceed the rated pressure for the system components, the fire pump discharge control valve must be closed prior to the operation of the fire pump to limit the exposure of the connected systems to the resultant elevated pressures at churn. For all other systems, consider closing the fire pump discharge control valve prior to the test to limit the exposure of the connected systems to the resultant pressure surge during the start-up of the fire pump.
6. Remove the installed pressure gauges on the suction and discharge side of the fire pump, and install the appropriate calibrated pressure gauges at each location. Note that for a vertical turbine fire pump, there will only be a discharge pressure gauge.
7. Open the flow meter inlet control valve and ensure the flow meter outlet valve is closed.
8. Check the area surrounding any relief valve or cooling water discharge outlets as well as the discharge points for the fire pump flow to ensure that there are no obvious conditions that would prevent water from being discharged safely or cause direct damage in the immediate vicinity. Take appropriate caution to ensure personnel and property are protected from the resultant high-pressure discharge. Where the discharge of water is to an area subject to potential freezing conditions, the facility representative should be advised of the potential for icing conditions.
9. For pressure-actuated fire pump controllers, simulate an automatic start for the fire pump by creating a pressure drop in the sensing line to the fire pump controller. This is typically accomplished by slowly opening the drain valve on the sensing line located near the fire pump controller until the fire pump starts automatically. The use of the "start" button on the fire pump controller is not acceptable for the purpose of simulating an automatic start. Note the start time of the fire pump to measure the run time duration, and record the starting pressure.

    For pressure-actuated fire pump controllers that use an automatic timer, an automatic opening of a solenoid valve in the sensing line to the fire pump controller might be used to simulate the automatic start of the fire pump. These systems must include a record of the pressure drop on the pressure recorder for the controller.

    For non-pressure-actuated fire pump controllers, the automatic start can be simulated by other means.

10. Observe the engine instrument panel to ensure that the engine oil pressure, operating speed, water and oil temperature, and charging rate are within the acceptable range, rechecking approximately every 5 minutes.

11. Check the fire pump packing gland for a slight discharge (slow drip) of water, adjusting the packing gland nuts as needed to achieve approximately 1 drip per second. For safety purposes, the adjustment of the packing gland should be made while the pump is not running. Care should be exercised to ensure the glands are not tightened to the point of breaking.

12. Monitor the fire pump operation for any unusual noise, vibration, or other signs of malfunction.

13. For radiator-cooled diesel fire pumps, verify that the operation of the circulation (casing) relief valve has a steady stream of water to ensure proper cooling of the pump case.

14. For heat exchanger–cooled diesel fire pumps, verify that the heat exchanger has a proper flow of cooling water.

15. Check the packing gland box, shaft bearings, and pump casing for overheating approximately every 5 minutes during the test. The packing gland box and shaft bearings might be warm to the touch, and the pump casing should remain cool throughout the test.

16. If the fire pump is equipped with a main pressure relief valve, verify the operation of the pressure relief valve such that outlet pressures do not exceed the pressure rating of the piping downstream of the fire pump. Usually, this rating is 175 psi (12.1 bar); however, some systems are designed for higher pressures.

17. Record the suction and discharge pressures. Note that for vertical turbine pumps, only the discharge pressure is recorded.

18. Record the operating speed (rpm) of the motor. The measurement can be taken with a handheld tachometer. Note that the use of a strobe-type handheld tachometer requires advance preparation prior to the test for proper measurement, including the application of a reflective tape on the shaft and/or removal of protective covers that shield the rotating shaft.

19. For systems equipped with a main pressure relief valve as described in Step 16 above, adjust the pressure relief valve to temporarily stop the relief valve operation and repeat Step 17 and Step 18 above, then re-adjust the main relief valve to the normal operating position, and continue with Step 20 below.

20. The data recorded in Step 17 through Step 19 should be noted as the churn test point.

21. Initiate the flow through the flow meter by slowly opening the flow meter throttling valve on the outlet side of the flow meter. Throttle the valve position to achieve a reading on the flow meter equal to the rated capacity from the nameplate on the fire pump. If the flow is circulated back to the fire pump suction, then the temperature of the re-circulated water must be monitored under flow conditions throughout the remainder of the test, to ensure that temperatures remain below that which would result in equipment damage, as defined by the pump and engine manufacturers.

22. Repeat Step 17 through Step 19 above, then continue with Step 23 below. The data recorded should be noted as the rated (or 100 percent flow) test point.

23. Continue to open the throttling valve to achieve a reading on the flow meter equal to the 150 percent of the rated capacity from the nameplate on the fire pump, or until the maximum available flow rate available from the water supply is met, whichever is lower. Care must be taken to ensure that the suction pressure does not fall below an acceptable level as the flow rate is increased to 150 percent. NFPA 20 allows a pressure at the suction flange of 0 psi (0 bar) for fire pumps connected to municipal supplies and -3 psi (-0.2 bar) for fire pumps taking suction from a grade level tank. Local water authorities might require a higher maintained pressure [commonly 20 psi (1.4 bar)] at the point of connection to the municipal supply.

24. Repeat Step 17 through Step 19 above, and then continue with Step 25 below. The data recorded should be noted as the overload (or 150 percent flow) test point.

25. Shut down the discharge flow by slowly closing the throttling valve to avoid water hammer.

26. If the run time for the fire pump has not reached 30 minutes, then allow the fire pump to continue operating until a total run time of 30 minutes has elapsed, and then shut down manually upon completion of the test.

27. For engines that use an electronic fuel management control system complete the following additional tests:

    a. Move the electronic control module (ECM) selector switch from the primary to the alternate position.

    b. Verify an alarm signal at the fire pump controller.

    c. Manually start the fire pump and verify normal operation.

    d. Manually shut down the fire pump.

    e. Reposition the ECM switch back to the primary position.

    f. Manually start the fire pump and verify the normal operation.

    g. With the fire pump still operating, disconnect the wire connection for each of the primary sensors for the ECM, verify the continued operation of the fire pump, and then reconnect the wires. On that same sensor, repeat the process removing the wires from the redundant sensor. This process should be completed for all of the primary and redundant sensors on the engine.

    h. Upon completion of the testing of all primary and redundant sensors, shut down the fire pump manually.

Annual Fire Pump (Non-Variable Speed) Flow Test *(Continued)*

28. Verify the closure of the circulation relief and pressure relief valves (if installed) upon completion of the test.
29. Plot the test data against the manufacturer's specifications *before* removing test equipment, to verify a retest is not required due to bad data recording or other reasons.
30. Return the flow meter control valve to the closed position.
31. Remove the calibrated pressure gauges, and reinstall the original pressure gauges on the suction and discharge side of the fire pump.
32. Inspect and clean any installed intake screens.
33. Initiate the various fire pump supervisory signals, such as low oil pressure, high coolant temperature, failure to start, or engine overspeed, to simulate an activation at the individual sensors. Verify that all of the fire pump supervisory signals, such as pump running, low oil pressure, high coolant temperature, failure to start, or engine overspeed, have been indicated at the fire pump controller as well as being transmitted to any connected fire alarm panels.
34. If closed, reopen the fire pump discharge control valve and conduct a valve status test downstream of the closed valve in accordance with 13.3.1.2.1.
35. Restore the fire pump to the automatic operating position.
36. Upon completion of all testing, notify the fire department and/or the alarm monitoring company (as needed), as well as the facility representatives, that testing is complete. Reset the fire alarm system as necessary. A verification of the receipt of initiated signals with the fire department and/or the alarm monitoring company, where so connected, would be appropriate as part of this effort.

Flow Measurement Calculation

The pressure recorded by the pitot tube assembly is velocity pressure, which is used to calculate flow. Converting the pitot pressure to flow can be accomplished by using manufacturer supplied tables for the specific flow-measuring device, or it can be calculated by inserting the outlet coefficient (c), outlet diameter (d), and velocity pressure (p) into the following formula:

$$Q = 29.84cd^2\sqrt{p}$$

where:
Q = flow in gpm
c = coefficient of discharge
d = diameter of outlet (in.)
p = flowing pressure (psi)

In metric units, the formula is as follows:

$$Q = 0.0666cd^2\sqrt{p}$$

where:
Q = flow in L/min
c = coefficient of discharge
d = diameter of outlet (mm)
p = flowing pressure (bar)

Pitot readings of less than 10 psi (0.7 bar) or more than 30 psi (2.1 bar) at the flow nozzle should be avoided. To keep within these pressure limits, the rate of flow can be controlled by throttling the control valve, opening additional outlets, or both.

Evaluation of Results

For a complete analysis on evaluating the results of the annual flow test, refer to Supplement 2 of this handbook.

Note that, at any time during the test, should abnormal conditions be observed that might result in personal injury or damage to the equipment/property, the fire pump should be immediately shut down and the test terminated until the cause can be investigated and corrective action taken as necessary.

References

American Society of Mechanical Engineers, Three Park Avenue, New York, NY 10016-5990.

ASME B40.1, *Gauges — Pressure Indicating Dial Type — Elastic Element,* Grade 1A, 2005.

National Fire Protection Association, 1 Batterymarch Park, Quincy, MA 02169-7471.

NFPA 20, *Standard for the Installation of Stationary Pumps for Fire Protection, 2013 edition.*

NFPA 70E®, Standard for Electrical Safety in the Workplace, 2012 edition.

The procedure(s) outlined above is not mandated by NFPA 25 and represents only one approach to conducting this test.

Testing Procedures for 8.3.3.1.2

Annual Flow Test

Purpose

The following is a description of the testing procedures necessary to conduct the annual test required in 8.3.3.1.2. This test allows the owner/inspector to determine the fire pump's capability based on its current operating condition. The owner/inspector can then compare the current data to the certified shop test curve to quantify the pump degradation since its installation.

Tools/Equipment

The test equipment required should be in good working condition and of such quality as to not be questioned should a flow test fail. All gauges used in the test (i.e., those for the suction and discharge sides of the pump as well as the pitot tube) should have recent calibration within the past 6 to 12 months. Gauges should be calibrated at least annually by means of a dead-weight or calibrated tester throughout the range of operation. Calibration sheets should be kept for each gauge and correction factors affixed to the back of each gauge before a test series is begun. The flow characteristics of the test nozzles should be known as well. While listed play pipes have standard flow performance curves, some proprietary flow devices, such as hose monster test nozzles, have equipment specific flow charts. If flow meters are used, they require pre-test calibration. Typical tools equipment used in annual pump flow testing are as follows:

1. A handheld tachometer[1] is required to measure the operating speed of the fire pump and verify that on the engine.
2. A pitot tube together with a test-pressure gauge[2],suitable for the pressures to be expected, is required. [Usually, a 60 psi (4 bar) gauge is satisfactory.] Test-quality gauges should be used in accordance with ASME B40.1, *Gauges — Pressure Indicating Dial Type — Elastic Element, Grade 1A,* having an accuracy of ±1 percent. The use of quality test gauges produces results that are considered reasonably accurate within the scope of the testing procedure. Care should be taken to protect the gauges from rough handling. While pitot tubes for fire protection flow testing are not listed, the gauges used should be listed and should be acceptable for water or water/air systems and not only air systems. A pitot-tube-and-gauge assembly, such as the one shown in Exhibit 7.1, is indispensable in conducting flow tests from hydrants and nozzles. The small opening at the end of the tube — not over 1⁄16 in. (1.6 mm) in diameter — is inserted in the center of the stream, in a direct line with the flow, at a distance in front of the opening equal to one-half the opening diameter. Velocity pressure is registered on the gauge attached to the tube. Some available test nozzles, including an integral pitot tube, are pitotless flow measuring features with an attached test gauge allowing for direct measurement of flows without the use of a separate pitot tube.
3. Test-pressure gauges suitable for measurement of expected the suction and discharge pressures on the fire pump are required. Note that the gauge used on the suction side of the fire pump should be a compound pressure and vacuum gauge having a pressure range of at least twice the static pressure available from the suction supply. The gauge used on the discharge side of the fire pump should have a range of at least twice the rated working pressure of the fire pump.
4. A sufficient number of test nozzles are required, including hoses for connection to test headers where needed [typically 50 ft (15.2 m) lengths of lined 2½ in. (65 mm) hose might be used for this purpose], to permit the measurement of a flow rate up to 150 percent of the rated capacity of the fire pump. Test nozzles can include UL play pipes or other approved test nozzles. The flow characteristics of the test nozzles must be known.
5. The various wrenches and other specific tools necessary to allow for required adjustments during the test and for the installation of required test equipment are required.
6. For electric pump tests, a multimeter and/or clamp-on ammeter is required for taking voltage and amperage readings.
7. A timing device is required to measure the duration of the fire pump run (on acceptance tests).
8. A form for recording test data is required.
9. While not required for the actual pump test data collections, approved hearing protection and personal protective equipment (PPE) are typically required. If PPE is not handy, then the start of pump testing can often be delayed while the PPE is being located.

Positive Displacement Pump Flow Tests

The pump flow for positive displacement pumps is completed to meet the specified rated performance criteria. Typically, one performance point is required to establish positive displacement pump performance. Many positive displacement pumps also have special applications, such as a component in foam or water mist systems. For these types of pumps, manufacturers' data should be referenced. The pump flow test for foam positive displacement pumps typically is completed using a flowmeter or orifice plate installed in a test loop back to the foam concentrate tank or the inlet side of a water pump. The flowmeter reading or discharge pressure should be recorded and should be in accordance with the pump manufacturer's flow performance data. If orifice plates are used, the orifice size and corresponding discharge pressure to be maintained on the

---

1. All test equipment should be calibrated by an approved testing laboratory within 12 months prior to the test.

2. Test-quality gauges should be used in accordance with ASME B40.1, *Gauges — Pressure Indicating Dial Type — Elastic Element, Grade 1A* having an accuracy of ±1%. The use of quality test gauges produces results that are considered reasonably accurate within the scope of the testing procedure. Care should be taken to protect the gauges from rough handling.

Gauges should be calibrated at least annually by means of a dead-weight or calibrated tester throughout the range of operation before a test series is begun. Calibration sheets should be kept for each gauge and correction factors affixed to the back of each gauge.

Annual Flow Test *(Continued)*

upstream side of the orifice plate should be made available to the authority having jurisdiction (AHJ). Flow rates should be as specified while operating at the system design pressure.

In addition to manufacturers' recommendations, systems with positive displacement pumps should reference ANSI/Hydraulic Institute 3.6, *Rotary Pump Tests,* and Chapter 12, Chapter 13, and Annex C of NFPA 750, *Standard on Water Mist Fire Protection Systems.* ANSI 3.6 applies to industrial rotary positive displacement pumps, and it includes detailed procedures on the setup and methods for conducting hydrostatic test and performance tests of such pumps.

Steam-Driven Turbine Fire Pumps

While Chapter 13 of NFPA 20 is dedicated to these types of pumps, they are rarely installed today due to maintenance and reliability concerns both on the driver and steam source. A steam turbine driver for a fire pump should always be kept "warmed up" to permit instant operation at full rated speed. The automatic starting of the turbine should not be dependent on any manual valve operation or period of low-speed operation, and this should be verified as part of the test. Steam turbines are provided with governors to maintain a predetermined speed, with some adjustment for higher or lower speeds. Desired speeds below this range can be obtained by adjusting the main throttle valve.

Vertical Turbine and Horizontal Split Case Pump Flow Tests

As required by 8.3.3.1, pump flow and pressure, as well as drive shaft speed (rpm), is measured at a minimum of three points: churn (no flow), rated (100 percent flow), and peak/overload (150 percent flow).

Procedure Steps

Pre-Flow Activities/Planning

1. If the site has a fire alarm system, notify the alarm monitoring company and place the fire alarm system on test.
2. Notify all facility representatives that testing is going to be conducted. In some cases, the amount of water flowing on the ground might require extra maintenance personnel. Or, suction pressure requirements could affect site production equipment (i.e., site cooling towers could trip off line due to low supply pressure, and so forth.)
3. Notify the local fire department and/or any other AHJ that might want to attend, such as property insurance representatives.
4. Complete the required activities in Section 8.2 to ensure pump system readiness for full flow testing. This maintenance should include checking battery electrolyte levels and specific gravity, inspecting cable conditions, and checking for any corrosion on pipes and other system equipment.
5. Review the fire pump assembly nameplates, noting the following:
    a. Voltage rating (electric pumps)
    b. Rated speed
    c. Rated capacity (flow)
    d. Churn pressure
    e. Rated pressure
    f. Overload pressure
6. Ensure all valves are in the proper position. Normally, pump discharge valves that supply site fire systems or fire system loops should be kept fully open throughout both weekly testing and the annual pump test. These are closed sometimes due to concern that the "high pump pressure" or "pressure fluctuations" might damage the fire system where pipe integrity is a concern. However, if fire mains are subject to damage due to pump pressures, it is best to discover this during an annual test rather than an actual fire. However, where the pump churn pressure will significantly exceed the rated pressure for the system, the fire pump discharge control valve might be closed prior to the operation of the fire pump to limit the exposure of the connected systems. Consideration might be given to closure of the fire pump discharge control valve prior to conducting the test, to limit the exposure of the connected systems to the resultant pressure surge during the start-up of the fire pump. Another method to avoid potential water hammer affects is to partially open a test header outlet at start-up, and then close it for the churn pressure reading.
7. Replace suction and discharge pressure gauges with the appropriate calibrated test gauges (for a vertical turbine fire pump there will only be a discharge gauge).
8. Install flow test equipment on the test header as required, to allow for the maximum flow rate. This might include the use of hoses connected to the test header and extending them to flow measurement devices, such as listed play pipes or other test nozzles.
9. Confirm that all test equipment, such as play pipes, hoses, play pipe test stands, and so forth, are adequately secured or protected against movement. Manually securing play pipes (i.e., by hand holding) is not acceptable and could cause serious personal damage. Swivel elbows, if installed on test headers, must be locked in place with set screws to avoid spinning during testing, which could cause serious personal damage. In some cases, the connection of flow test equipment such as play pipes can be made directly to the test header. The use of a flow diffuser can aid in preventing movement as well as breaking the directed stream.
10. For testing using hose-connected flow-measuring devices or UL play pipes directly connected to the test header, ensure that the individual hose connection valves on the test header are in the closed position, and slowly open the main test header control valve pressurizing the test header. For other flow measuring devices directly connected to the test header, ensure that the main test header control valve is closed. Then fully open a single hose connection valve, readying the arrangement for the first flow point,

with the flow being controlled by the main test header control valve. Any time the use of an additional hose connection valve is required with this latter arrangement the additional hose connection valves are to be fully opened and the flow again controlled by the position of the main test header control valve rather than the individual hose connection valves.

11. Check the area surrounding the test per 8.3.3.1.2.1(B) for potential uncontrolled water flow that could cause damage — plugged drain pipes, local construction with open pits, berms subject to hydraulic erosion, and so forth. Check any relief valve or cooling water discharge outlets as well as the discharge points for the fire pump flow to ensure that there are no obvious conditions that would prevent water from being discharged safely or cause direct damage in the immediate vicinity. Where the discharge of water is to an area subject to potential freezing conditions, the facility representative should be advised of the potential for icing conditions.
12. Check the fire pump packing glands for a slight discharge (slow drip) of water, adjusting the packing gland nuts as needed to achieve approximately 1 drip per second. If the packing is completely dry before start-up, it could overheat and fail once the pump is turned for the pump test, The overheating of bearings could also quickly occur. For safety purposes, the adjustment of the packing gland should be made while the pump is not running. Care should be exercised to ensure the glands are not tightened to the point of breaking.
13. For vertical turbine pumps, make sure the pump is completely primed.
14. In cases where a handheld tachometer will be used, verify where the point readings will be taken from and that reflective tape is installed as required. Some handheld tachometers require advance preparation prior to the test for proper measurement, such as the application of a reflective tape on the shaft. Care should be taken not to apply reflective tape on a shaft that already has it applied, because false "double readings" might occur. Since a reading will be required at each point during the actual test, care should be taken of where the location tape is applied, considering the proximity to casing spray, hot engine exhaust, accessibility to the sides of a rotating shaft, and so forth.

Flow Test Activities/Planning

1. For pressure-actuated fire pump controllers, simulate an automatic start for the fire pump by creating a pressure drop in the sensing line to the fire pump controller. This is typically accomplished by slowly opening the drain valve on the sensing line located near the fire pump controller until the fire pump starts automatically. The use of the "start" button on the fire pump controller is not acceptable for the purpose of simulating an automatic start. Note the start time of the fire pump, and record the starting pressure.

   For pressure-actuated fire pump controllers that use an automatic timer, an automatic opening of a solenoid valve in the sensing line to the fire pump controller might be used to simulate the automatic start of the fire pump. These systems must include a record of the pressure drop on the pressure recorder for the controller.

   For non-pressure-actuated fire pump controllers, the automatic start can be simulated by other means.
2. Check the fire pump packing gland for a slight discharge (slow drip) of water, adjusting the packing gland nuts as needed to achieve approximately 1 drip per second. For safety purposes, the adjustment of the packing gland should be made while the pump is not running. Care should be exercised to ensure the glands are not tightened to the point of breaking.
3. Monitor the fire pump operation for any unusual noise, vibration, or other signs of malfunction.
4. For electric-driven pumps, verify that the operation of the circulation (casing) relief valve has a steady stream of water to ensure proper cooling of the pump case.
5. Check the packing gland box, shaft bearings, and pump casing for overheating approximately every 5 minutes during the test. The packing gland box and shaft bearings might be warm to the touch, but the pump casing should remain cool throughout the test.
6. If the fire pump is equipped with a main pressure relief valve, verify the operation of the pressure relief valve such that outlet pressures do not exceed the pressure rating of the piping downstream of the fire pump. Usually, this rating is 175 psi (12.1 bar); however, some systems might be designed for higher pressures.
7. For electric-driven pumps, motor voltage and current on all phases (lines) should be recorded. This can be done with a multimeter and clamp-on ammeter; however, these readings should only be taken by individuals trained and qualified in electrical hazards and equipped with the needed safety equipment as outlined in *NFPA 70E®, Standard for Electrical Safety in the Workplace.* Also see 8.3.3.6 and Section 4.8, along with the associated commentary, for additional guidance.
8. Record the suction and discharge pressures. Note that only the discharge pressure is recorded for vertical turbine pumps.
9. Record the operating speed (rpm) of the motor.
10. The data recorded in Step 8 and Step 9 should be noted as the churn test point.
11. Initiate the flow of the fire pump by slowly opening either hose control valves on the test header or by slowly opening the main test header control valve, depending on the arrangement selected. As the valve is opened, take flow measurements to determine the flow rate, adjusting each valve position until a combined flow rate equal to the rated capacity from the nameplate on the fire pump is achieved.

Annual Flow Test *(Continued)*

Where the flow rate with the valve fully open is inadequate to meet this rated capacity, additional hose valves must be opened and adjustments made until the total measured flow rate equals the rated capacity. For systems equipped with a main pressure relief valve, this should now be verified as closed and not flowing water. If it is flowing water, adjust the pressure relief valve to temporarily close.

12. Repeat Step 8 and Step 9.
13. The data recorded in Step 11 and Step 12 should be noted as the rated (or 100 percent flow) test point.
14. Continue to open hose valves and/or the main test header control valve, as in Step 11, and take flow measurements to determine the flow rate, adjusting the valve(s) position until a flow rate equal to the 150 percent of the rated capacity from the nameplate on the fire pump is achieved, or until the maximum available flow rate is achieved, whichever is lower. Care must be taken to ensure that the suction pressure (at the point of the city connection) does not fall below an acceptable level. NFPA 20, *Standard for the Installation of Stationary Pumps for Fire Protection,* allows a suction pressure of 0 psi (0 bar) for fire pumps connected to municipal supplies and -3 psi (-0.2 bar) for fire pumps taking suction from a grade level tank. Some local water authorities might require a higher maintained pressure [typically 20 psi (1.4 bar) where this is a regulation], as the flow rate is increased to 150 percent.
15. Repeat Step 8 and Step 9.
16. The data recorded in Step 14 and Step 15 should be noted as the overload (or 150 percent flow) test point.
17. Shut down the flow by slowly closing the main test header control valve to avoid water hammer.
18. If the run time for the fire pump has not reached 10 minutes, then allow the fire pump to continue operating until a total run time of 10 minutes is reached, rechecking for overheating periodically during the test, and then shut down manually upon completion of the test.
19. Plot the test data against the manufacturer's specifications *before* removing test equipment, to verify a retest is not required due to bad data recording and other reasons.
20. Remove all attached test equipment from the test header, restore the closed position of all hose connection valves, and reinstall hose caps. In locations subject to freezing, check that the test header properly drains or pumps out as necessary.
21. Remove the calibrated gauges, and reinstall the original gauges on the suction and discharge side of the fire pump. Any gauges found out of calibration should be replaced.
22. Inspect and clean any installed intake screens.
23. For electric-driven pumps, verify a fire pump phase reversal supervisory signal by jumping across the monitored points in the fire pump controller (typically labeled on a wiring diagram posted on the inside the fire pump controller) to simulate an activation.
24. Verify that all of the fire pump supervisory signals (e.g., fire pump running, loss of power, phase reversal) required by NFPA 20 and NFPA 25 (see 8.3.3.5) have been indicated at the fire pump controller as well as being transmitted to any connected fire alarm panels.
25. Verify the main pressure relief valve is properly set and adjusted.
26. Verify fire pump start-stop settings (see below).
27. If closed, reopen the fire pump discharge control valve, and conduct a valve status test downstream of the closed valve in accordance with 13.3.1.2.1.
28. Restore the fire pump to the automatic operating position.
29. Complete a coupling alignment check.
30. Upon completion of all testing, notify the fire department and/or the alarm monitoring company (as needed), as well as the facility representatives, that testing is complete. Reset the fire alarm system as necessary.
31. Complete the pump test data analysis to determine compliance with 8.3.3.1 and 8.3.6.1. See Supplement 2.

At each flow test point, the quantity of water discharging from the fire pump test assembly should be stabilized as best as possible before taking pitot readings. When using multiple outlets for pitot readings, final readings on all outlets cannot accurately be made until all adjustments are completed. Due to hydraulic principles, each time one test header outlet is adjusted, the flow on others will be affected. That is, opening an additional test header will cause the flow (and related pitot readings) on others flowing to drop to a different number than the reading taken before the additional valve was opened.

Fire Pump Operational Settings

The fire pump system, when started by pressure drop, should be arranged as follows:

1. The jockey pump stop point should equal the pump churn pressure plus the minimum static supply pressure.
2. The jockey pump start point should be at least 10 psi (0.7 bar) less than the jockey pump stop point.
3. The fire pump start point should be 5 psi (0. 4 bar) less than the jockey pump start point. Use 10 psi (0.7 bar) increments for each additional pump.
4. Where minimum run timers are provided, the pump will continue to operate after attaining these pressures. The final pressures should not exceed the pressure rating of the system.
5. Where the operating differential of pressure switches does not permit these settings, the settings should be as close as equipment will permit. The settings should be established by pressures observed on test gauges.

Evaluation of Results

Supplement 2 of this handbook provides an in-depth review of interpreting the results of the annual flow test and determining if the pump has seen more than a 5 percent degradation when compared to the certified shop test curve. For more information, see Supplement 2 of this handbook.

References

American Society of Mechanical Engineers, Three Park Avenue, New York, NY 10016-5990.

ASME B40.1, *Gauges — Pressure Indicating Dial Type — Elastic Element, Grade 1A,* 2005.

National Fire Protection Association, 1 Batterymarch Park, Quincy, MA 02169-7471.

NFPA 20, *Standard for the Installation of Stationary Pumps for Fire Protection,* 2013 edition.

NFPA 70E®, Standard for Electrical Safety in the Workplace®, 2012 edition.

NFPA 750, *Standard on Water Mist Fire Protection Systems,* 2010 edition.

Hydraulic Institute, 6 Campus Drive First Floor North, Parsippany, NJ 07054-4406.

ANSI/Hydraulic Institute 3.6, *Rotary Pump Tests,* 2010 edition.

The procedure(s) outlined above is not mandated by NFPA 25 and represents only one approach to conducting this test.

ITM NFPA 25 — INSPECTION TESTING MAINTENANCE

9

WATER STORAGE TANKS

Chapter 9 of NFPA 25 covers the inspection, testing, and maintenance activities necessary to ensure that water storage tanks will provide the proper quantity and, in some cases, pressure, in the event of a fire. Water storage tanks can be made of various materials such as wood, steel, concrete, fiberglass-reinforced plastic (FRP), and rubberized fabric. Each type of water storage tank, such as suction, gravity, or bladder tanks, can have vastly different system components.

Water storage tanks are frequently used where an adequate quantity or volume of water is not readily available. Depending on the needs of the system that the tanking is supporting and the geographical and topographical location of building, the type of tank that is used will vary. Accordingly, the ITM tasks that must be executed will also vary, because the components will vary based on the type of tank involved.

Some tanks, such as pressure tanks, serve dual purposes by providing the necessary quantity of water and the needed pressure to propel the water through a fire protection system. Section 4.31 of NFPA 20, *Standard for the Installation of Stationary Pumps for Fire Protection,* gives requirements for "break tanks," which basically have the following three purposes:

1. Serve as a backflow prevention device between the city water supply and the fire pump suction
2. Eliminate pressure fluctuations in the city water supply and provide a steady suction pressure to the fire pump
3. Augment the city water supply when the volume of water available from the city is inadequate for the fire protection system demand

FAQ What is the purpose of a break tank?

A break tank, as its name suggests, provides a break or separation between a fire pump and a city water supply. Traditionally, a break tank is a small tank intended to isolate the fire pump from the water supply. However, a break tank can also be quite large.

Gravity tanks, as referred to in NFPA 22, *Standard for Water Tanks for Private Fire Protection,* rely on gravity to provide the 0.433 psi/ft (0.09 bar/m) of pressure necessary. Suction tanks rely on fire pumps to pressurize fire protection systems.

Regardless of the type or arrangement, water storage tanks are a critical part of the water-based fire protection system, and they must be properly maintained if they are to function when needed.

The requirements in Chapter 9, as well as those in NFPA 22, apply only to private tanks dedicated to fire protection and installed on private property. Storage tanks installed on public property serving public domestic water systems are not within the scope of the standard.

9.1* General

A.9.1 One source of information on the inspection and maintenance of steel gravity and suction tanks is the AWWA *Manual of Water Supply Practices — M42 Steel Water-Storage Tanks*, Part III and Annex C.

Critical Point

The appropriate size of a dedicated water storage tank is determined by multiplying the water demand of the fire protection system it serves by the discharge duration specified in the installation standard.

In many cases, water storage tanks are the primary — if not the only — source of water for firefighting purposes. Requirements for the design, construction, and operation of water storage tanks can be found in NFPA 22.

The appropriate size of a dedicated water storage tank is determined by multiplying the water demand of the fire protection system it serves by the discharge duration specified in the installation standard. The total capacity is measured from the discharge pipe to the overflow pipe. Exhibit 9.1 shows a steel ground-level tank that constitutes the sole source of water for a small manufacturing plant.

Not all tanks are required to be designed around the total system demand. Break tanks can also be used to provide suction to a fire pump; however, their capacity is often less than the fire protection system demand, which is calculated as flow rate multiplied by the required flow duration. Break tanks are normally used to avoid cavitation of the main upon initial startup of the fire pump, and they are usually sized for a 20-minute water supply duration or less. Where a system design depends on an automatic fill valve to replace the water in the tank to achieve the required duration of the system demand (flow rate), inspection and testing of the refill mechanism is critically important.

Exhibit 9.1 *Steel Ground-Level Tank.*

Exhibit 9.2 *Elevated Water Storage Tank.*

Exhibit 9.3 *Wooden Gravity Tanks. (Courtesy of Hall-Woolford Tank Co., Inc.)*

Exhibit 9.2 shows an elevated storage tank with a riser more than 3 ft (0.9 m) in diameter. In some areas of the United States, the size of the riser will dictate the need for heating.

The use of wooden gravity (elevated) tanks for fire protection has decreased significantly over the past 100 years, although they are still found in many older cities (see Exhibit 9.3). These wooden tanks were installed high in the building (usually on or above the roof), feeding a sprinkler system by gravity. Exhibit 9.4 shows a wooden gravity tank under construction. Exhibit 9.5 shows a fiberglass-reinforced plastic (FRP) tank. Exhibit 9.6 illustrates a ground-level suction tank and the discharge pipe connected to the bottom of the tank in a valve pit.

The use of steel pressure tanks, as shown in Exhibit 9.7, has declined largely due to space considerations. These tanks require a large volume of space to provide sufficient water capacity for a sprinkler system. Smaller-volume pressure tanks are now being provided, in part due to the introduction of residential sprinkler systems and the corresponding lower duration values they typically require when installed in accordance with NFPA 13D, *Standard for the Installation of Sprinkler Systems in One- and Two-Family Dwellings and Manufactured Homes*.

An embankment-supported fabric suction tank, as illustrated in Exhibit 9.8, is designed much like a waterbed. Water is enclosed in fabric that is given its shape by earthen berms located around the tank. The fabric reduces the likelihood of marine life attaching to the walls of the tank.

Exhibit 9.4 *Wooden Gravity Tank Under Construction. (Courtesy of Hall-Woolford Tank Co., Inc.)*

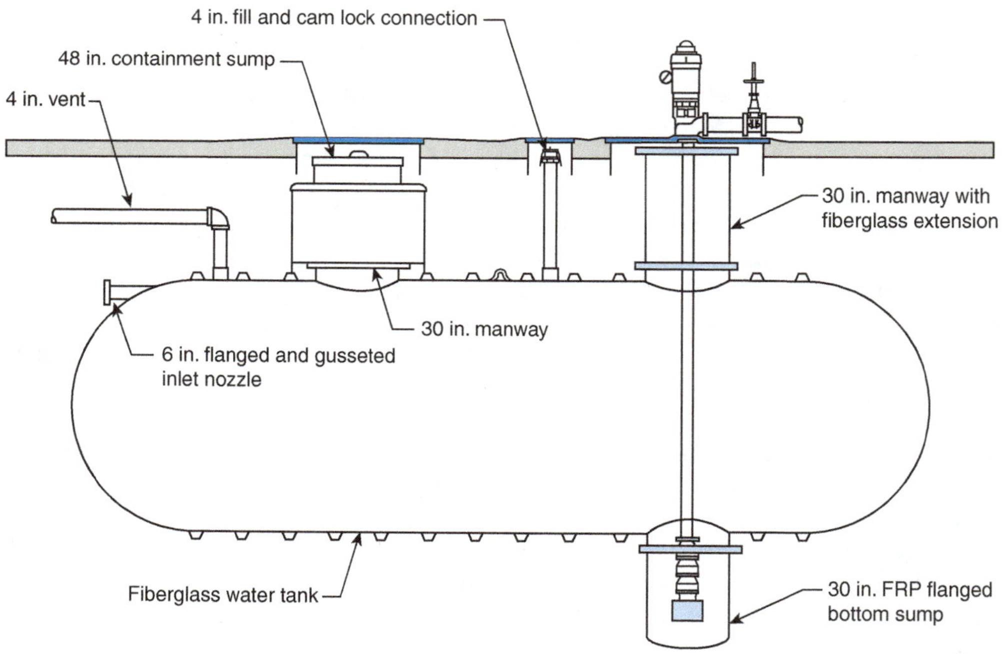

For SI units, 1 in. = 25.4 mm.

Exhibit 9.5 *Fiberglass-Reinforced Plastic (FRP) Tank.*

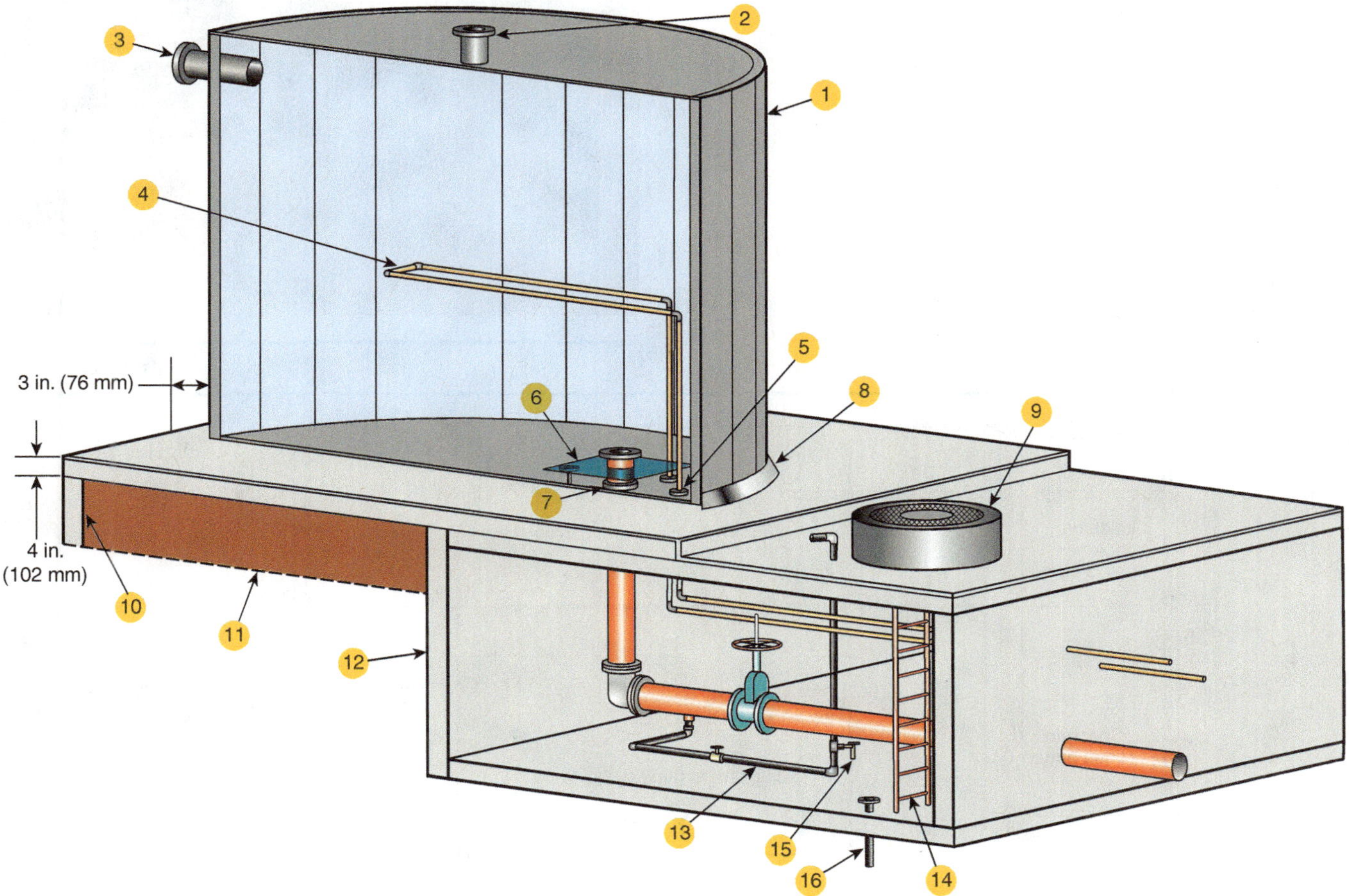

Legend

1 Pump suction tank
2 Screened vent
3 Stub overflow pipe
4 Steam coil for heating
5 Extra-heavy couplings welded to bottom of tank
6 Vortex plate
7 Watertight lead slip joint
8 Flashing around tank
9 Manhole with cover
10 Concrete ring wall
11 Sand or concrete pad (depending on soil condition)
12 Valve pit
13 Drain pipe
14 Ladder
15 Drain cock
16 Valve pit drain suction tank

Exhibit 9.6 *Ground-Level Suction Tank. (Courtesy of Stephan Laforest)*

Exhibit 9.7 *Steel Pressure Tank.*

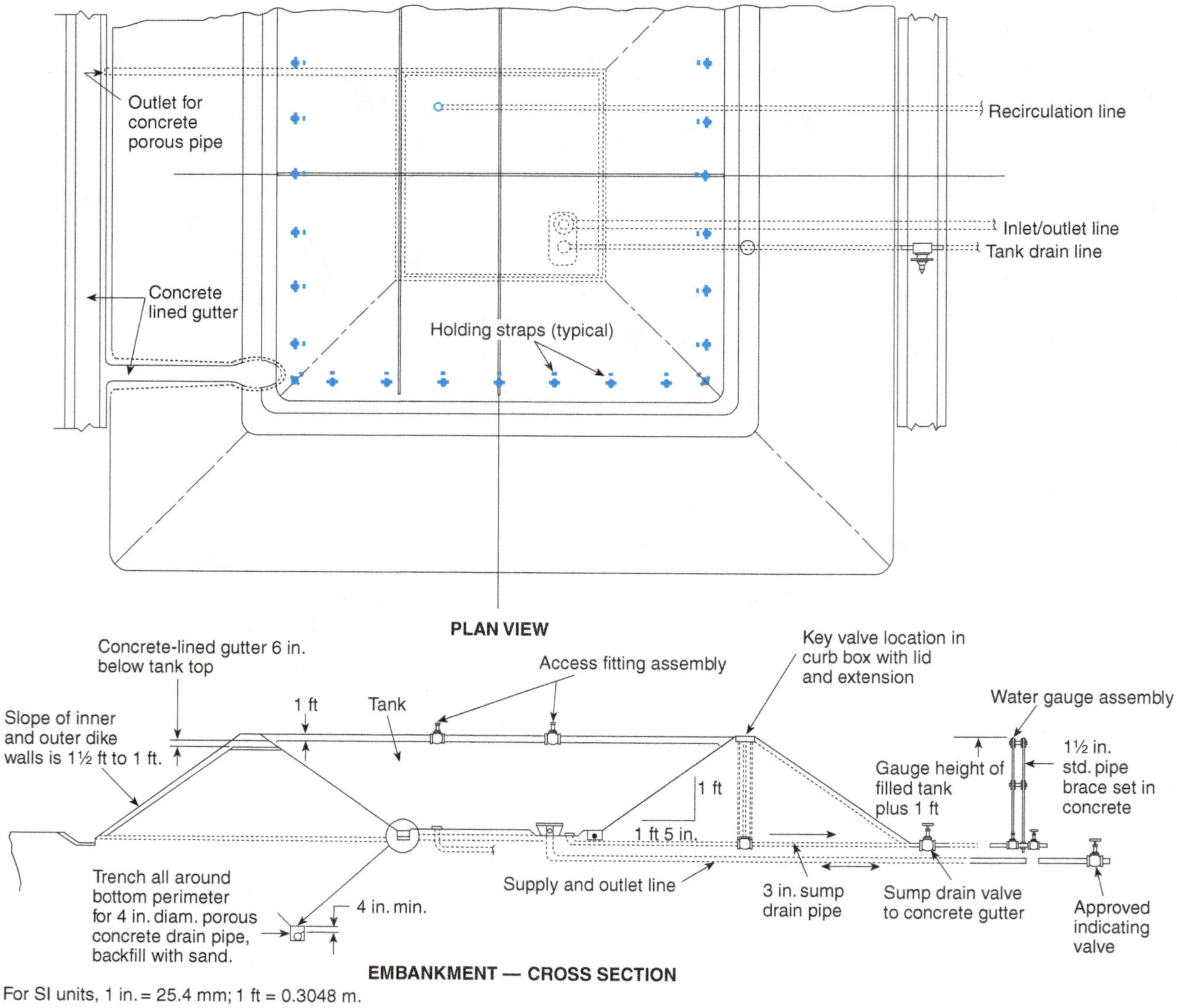

Exhibit 9.8 *Embankment-Supported Fabric Suction Tank. [Source: NFPA 22, 2013, Figure B.1(e)]*

9.1.1 Minimum Requirements.

9.1.1.1 This chapter shall provide the minimum requirements for the routine inspection, testing, and maintenance of water storage tanks dedicated to fire protection use.

The private fire service tanks referenced in this chapter are those tanks that are dedicated only to fire protection use. It is necessary to specify this because, in the past, this chapter has been incorrectly applied to municipal water storage tanks and private tanks that supply both domestic and fire protection water. The inspection, testing, and maintenance of storage tanks for domestic, potable water is not within the scope of this standard.

9.1.1.2 Table 9.1.1.2 shall be used to determine the minimum required frequencies for inspection, testing, and maintenance.

TABLE 9.1.1.2 *Summary of Water Storage Tank Inspection, Testing, and Maintenance*

Item	*Frequency*	*Reference*
Inspection		
Water temperature — low temperature alarms connected to constantly attended location	Monthly	9.2.4.2
Water temperature — low temperature alarms not connected to constantly attended location	Weekly	9.2.4.3
Heating system — tanks with supervised low temperature alarms connected to constantly attended location	Weekly*	9.2.3.1
Heating system — tanks without supervised low temperature alarms connected to constantly attended location	Daily*	9.2.3.2
Control valves		Table 13.1.1.2
Water level — tanks equipped with supervised water level alarms connected to constantly attended location	Quarterly	9.2.1.1
Water level — tanks without supervised water level alarms connected to constantly attended location	Monthly	9.2.1.2
Air pressure — tanks that have their air pressure source supervised	Quarterly	9.2.2.1
Air pressure — tanks without their air pressure source supervised	Monthly	9.2.2.2
Tank — exterior	Quarterly	9.2.5.1
Support structure	Quarterly	9.2.5.1
Catwalks and ladders	Quarterly	9.2.5.1
Surrounding area	Quarterly	9.2.5.2
Hoops and grillage	Annually	9.2.5.4
Painted/coated surfaces	Annually	9.2.5.5
Expansion joints	Annually	9.2.5.3
Interior — tanks without corrosion protection	3 years	9.2.6.1.1
Interior — all other tanks	5 years	9.2.6.1.2
Temperature alarms — connected to constantly attended location	Monthly*	9.2.4.2
Temperature alarms — not connected to constantly attended location	Weekly*	9.2.4.3
Check valves		Table 13.1.1.2
Test		
Tank heating system	Prior to heating season	9.3.2
Low water temperature alarms	Monthly*	9.3.3
High temperature limit switches	Monthly*	9.3.4
Water level alarms	Semiannually	9.3.5
Level indicators	5 years	9.3.1
Pressure gauges	5 years	9.3.6
Valve status test		13.3.1.2.2.1

(continues)

TABLE 9.1.1.2 Continued

Item	Frequency	Reference
Maintenance		
Water level	—	9.4.2
Control valves	—	Table 13.1.1.2
Embankment-supported coated fabric (ESCF)	—	9.4.6
Check valves	—	13.4.2.2

*Cold weather/heating season only.

9.1.2 Valves and Connections. Valves and fire department connections shall be inspected, tested, and maintained in accordance with Chapter 13.

9.1.3 Obstruction Investigations. The procedures outlined in Chapter 14 shall be followed where there is a need to conduct an obstruction investigation.

9.1.4 Impairments. The procedures outlined in Chapter 15 shall be followed where an impairment to protection occurs.

The inspection, testing, and maintenance of water storage tanks can involve or result in a system that is out of service. In cases where a tank is the sole source of supply to a fire protection system, it is recommended that an alternative water supply be arranged while maintenance is performed on the tank.

An alternative source of water can take the form of temporary hose connections to a nearby hydrant, the temporary placement of a tanker truck, portable storage tanks, or hose connections and a portable fire pump to a body of water, either natural or manmade. Exhibit 9.9 shows portable tanks that were used as an alternative water supply during the renovation of supply piping from the prime water source of a fire protection system in Yellowstone National Park.

•

9.2 Inspection

There are several different types of inspections that must be conducted to be assured that a water storage tank is in good working order. Exhibit 9.10 is a sample form that can be used to record all of the different inspection tasks outlined in Section 9.2 and Table 9.1.1.2.

Exhibit 9.9 *Portable Storage Tanks as an Alternative Source of Water. (Courtesy of National Park Service)*

INSPECTION OF WATER STORAGE TANKS

Name of Property: ____________________ **Inspector:** ____________

Address: ____________________ **Contract No.:** ____________

Phone Number: ____________________ **Date:** ____________

Inspection

Daily — **(Cold weather/heating season only)**

❑ Yes ❑ No ❑ N/A Water temperature (when not supervised)
❑ Yes ❑ No ❑ N/A Heating system (when not supervised)

Weekly — **(Cold weather/heating season only)**

❑ Yes ❑ No ❑ N/A Water temperature (supervised systems)
❑ Yes ❑ No ❑ N/A Heating system (supervised systems)

Monthly — **Control Valves**

❑ Yes ❑ No ❑ N/A In the correct (open or closed) position
❑ Yes ❑ No ❑ N/A Sealed, locked, or supervised
❑ Yes ❑ No ❑ N/A Accessible
❑ Yes ❑ No ❑ N/A Free from damage or leaks
❑ Yes ❑ No ❑ N/A Proper signage
❑ Yes ❑ No ❑ N/A Water level (unsupervised) full
❑ Yes ❑ No ❑ N/A Air pressure (unsupervised) proper level

Weekly

❑ Yes ❑ No ❑ N/A Water level (supervised) full
❑ Yes ❑ No ❑ N/A Air pressure (supervised) proper level
❑ Yes ❑ No ❑ N/A Tank (exterior) free of damage or signs of weakening (including support structure, catwalks, and ladders)

Tank Surrounding Area

❑ Yes ❑ No ❑ N/A Free of combustibles
❑ Yes ❑ No ❑ N/A Free of material that could accelerate corrosion or rot
❑ Yes ❑ No ❑ N/A Free of ice buildup
❑ Yes ❑ No ❑ N/A Embankments (for coated fabric tank) free of erosion

NFPA® NFPA Water-Based Fire Protection Systems Forms 1 of 2

Exhibit 9.10 Sample Form for Inspection of Water Storage Tanks.

Annual

❑ Yes ❑ No ❑ N/A Hoops and grillage are in good condition
❑ Yes ❑ No ❑ N/A Painted/coated surfaces are in good condition
❑ Yes ❑ No ❑ N/A Expansion joints not cracked or leaking

Five Year/Three Year

Interior Inspection

❑ Yes ❑ No ❑ N/A Silt has been removed for underwater evaluation
❑ Yes ❑ No ❑ N/A Interior surfaces free of pitting, corrosion, spalling, or other forms of deterioration
❑ Yes ❑ No ❑ N/A Interior is free of waste material, aquatic growth, and debris
❑ Yes ❑ No ❑ N/A Interior coating is intact
❑ Yes ❑ No ❑ N/A Tank floor is free of dents
❑ Yes ❑ No ❑ N/A Heating system and components are in good condition
❑ Yes ❑ No ❑ N/A Anti-vortex plate is in good condition and is not obstructed
❑ Yes ❑ No ❑ N/A Check valve (internal) moves freely in good condition

Comments

Signature: ______________________________ Date: ______________

License/Certification No.: ______________________________

NFPA NFPA Water-Based Fire Protection Systems Forms 2 of 2

Exhibit 9.10 *Continued.*

NFPA 25 does not require that this specific form be used. This is provided as just one example of an acceptable inspection form.

9.2.1 Water Level.

9.2.1.1* The water level in tanks equipped with supervised water level alarms that are connected to a constantly attended location shall be inspected quarterly.

A.9.2.1.1 More frequent inspections should be made where extreme conditions, such as freezing temperatures or arid climate, can increase the probability of adversely affecting the stored water.

Supervisory water level alarms installed on tanks provide notification that the tank water level is above or below an acceptable level. The water level of the tank is the main concern as opposed to the condition of the water. For convenience, inspection of the condition of the water can take place concurrently with the water level inspection.

The water level in a tank is determined by measuring the amount of water in the tank between the overflow and discharge outlet. This quantity is determined by the calculated flow of the fire protection system for a specified duration, such as 30, 60, or 90 minutes. The level should never be lower than 3 in. to 4 in. (76 mm to 102 mm) below the designated fire service level.

Exhibit 9.11 and Exhibit 9.12 show two different means to determine the quantity of water within a tank. Exhibit 9.11 shows the indicator running the full height of a gravity tank that is under construction. Exhibit 9.12 shows a sight glass on a steel pressure tank.

Critical Point

The water level in a tank is determined by measuring the amount of water in the tank between the overflow and discharge outlet.

Exhibit 9.11 *Wood Tank with a Liquid Level Gauge.*

Exhibit 9.12 *Sight Glass on a Steel Pressure Tank. (Courtesy of National Fire Sprinkler Association)*

9.2.1.2 The water level in tanks not equipped with supervised water level alarms connected to a constantly attended location shall be inspected monthly.

9.2.2 Air Pressure.

9.2.2.1 The air pressure in pressure tanks that have their air pressure source supervised in accordance with *NFPA 72, National Fire Alarm and Signaling Code*, shall be inspected quarterly.

According to NFPA 22, a pressure tank must be kept two-thirds full of water and a minimum air pressure of at least 75 psi (5.2 bar) must be maintained. As the last of the water leaves the pressure tank, the residual pressure shown on the gauge cannot be less than 0. The air pressure must be sufficient to provide not less than 15 psi (1.03 bar) at the highest sprinkler under the main roof of the building.

Critical Point

The need for a tank heating system is based on the type of tank construction and the lowest one-day mean temperature as determined by isothermal lines.

9.2.2.2 The air pressure in pressure tanks with a nonsupervised air pressure source shall be inspected monthly.

9.2.3 Heating System.

9.2.3.1 Tank heating systems installed on tanks equipped with supervised low water temperature alarms that are connected to a constantly attended location shall be inspected weekly.

ITM Deficiency, Impairment, or Hazard Evaluation?

(Courtesy of Byron Blake and SimplexGrinnell)

ANSWER: ITM Deficiency and Possible Impairment

This photo shows the interior of a water storage tank. It was taken while an inspector was standing on the ladder on the outside of the tank and looking through the access hatch down the ladder that is inside the tank. The water level in this ground storage tank is below what is required for the fire protection system. Notice that the water level is well below the tank water level switch and tank temperature switch.

The water level in tanks with supervised water level alarms is required by 9.2.1.1 to be inspected quarterly. This requirement involves inherent safety risks such as climbing the ladder on the exterior of the tank and accessing a confined space, but it can be performed safely by using the proper safety equipment and following the required procedures.

In addition to the quarterly inspection, the water level alarm is required in 9.3.5 to be tested semiannually, and the low water temperature alarm is required in 9.3.3 to be tested monthly during cold weather. In the case of this photo, the water level switch should have signaled the low water condition. One method used to test this switch is to manually move the float up and down to simulate water levels in the tank. To test the thermostat in the low water temperature alarm switch, a manufacturer simply states that the device must be exposed to temperatures of 40°F and 140°F (4°C and 60°C). One way to test low temperature is to insert the probe into a bag of ice.

The low water condition found in this tank should be recorded as a critical deficiency if the tank is still partially full, which would allow the fire protection system to function in a fire event, but with a reduced duration. If the tank is empty, the water tank unit is impaired. If this tank is the sole water supply source, the systems downstream would not function at all during a fire event, and the tank should be considered impaired. The impairment procedures in Chapter 15 should be implemented immediately.

Case In Point

NFPA 22 recognizes the following heating systems:

- Steam water heaters
- Gas-fired water heaters
- Oil-fired water heaters
- Coal-burning water heaters
- Electric water heaters
- Vertical steam radiators
- Hot water
- Steam coils inside tanks
- Direct discharge of steam
- Solar heating

The requirements for each of these systems are discussed in Section 16.3 of NFPA 22. Regardless of which method is used to heat the tank, the intent is to ensure reliability by keeping the tank free of ice. An ice plug in a riser pipe, for example, can obstruct flow and could break the pipe. In addition, ice in or on the tank structure can cause collapse of the tank.

Generally, only tanks that are subject to freezing will have heating systems. The need for a tank heating system is based on the type of tank construction and the lowest one-day mean temperature as determined by isothermal lines. The accompanying map shows the isothermal lines of the United States. For example, a suction tank in Atlanta, Georgia, does not need to be heated. An elevated tank with a riser of 3 ft (0.9 m) or less in diameter in the same city must be heated in the riser portion only. See NFPA 22 to determine whether a heating system is needed for a particular tank.

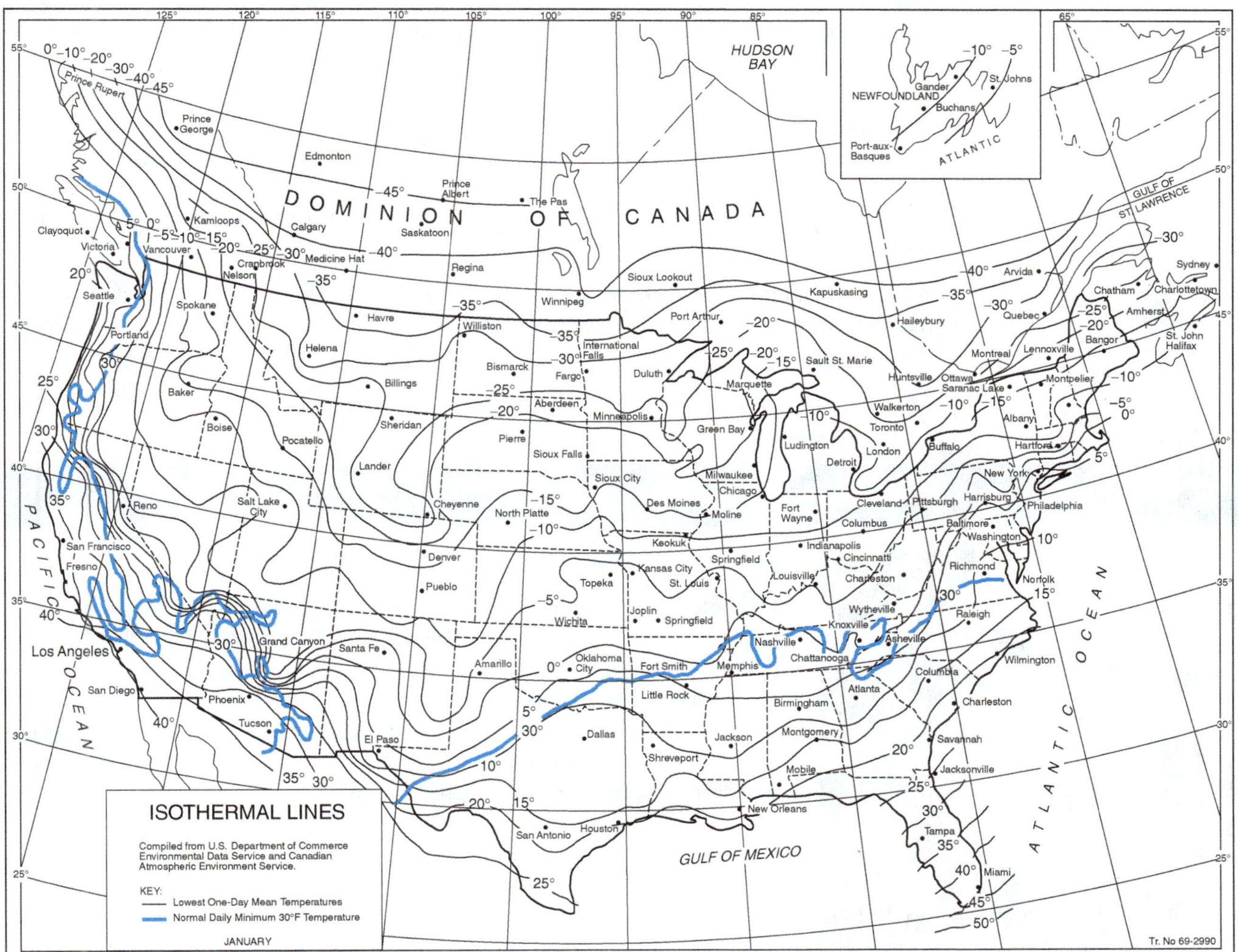

Source: Compiled from United States Weather Bureau records.
For SI units, °C = 5/9 (°F −32); 1 mi = 1.609 km.

Isothermal Lines — Lowest One-Day Mean Temperature. (Source: NFPA 22, 2013 edition, Figure 16.1.4)

9.2.3.2 Tank heating systems without a supervised low temperature alarm connected to a constantly attended location shall be inspected daily during the heating season.

9.2.4 Water Temperature.

9.2.4.1 The temperature of water in tanks shall not be less than 40°F (4.0°C).

9.2.4.2 The temperature of water in tanks with low temperature alarms connected to a constantly attended location shall be inspected and recorded monthly during the heating season when the mean temperature is less than 40°F (4.0°C).

9.2.4.3 The temperature of water in tanks without low temperature alarms connected to a constantly attended location shall be inspected and recorded weekly during the heating season when the mean temperature is less than 40°F (4.0°C).

9.2.5 Exterior Inspection.

9.2.5.1* The exterior of the tank, supporting structure, vents, foundation, and catwalks or ladders, where provided, shall be inspected quarterly for signs of obvious damage or weakening.

The inspection required by 9.2.5.1 is a visual inspection, as indicated by the words "obvious damage." An exterior visual inspection should reveal loose or missing bolts, excessive corrosion, and cracking and peeling paint, among other items.

The roof vent provides ventilation above the maximum water level and also keeps birds and other animals out of the tank. The vent may be equipped with a perforated plate having ⅜ in. (9.5 mm) perforations or a corrosion-resistant screen. The perforations must be kept clean to maintain proper ventilation.

For FRP tanks, the exterior inspection should verify that the tank is protected from freezing, mechanical, and ultraviolet (UV) damage.

A.9.2.5.1 Lightning protection systems, where provided, should be inspected, tested, and maintained in accordance with NFPA 780, *Standard for the Installation of Lightning Protection Systems*.

ITM Deficiency, Impairment, or Hazard Evaluation?

(Courtesy of Byron Blake and SimplexGrinnell)

ANSWER: ITM Deficiency

This photo was taken during an inspection of a ground storage tank as required by Section 9.2, and it reveals a couple of problems that should be identified as deficiencies. First, the ladder is required to be inspected for signs of obvious damage or weakening as described in 9.2.5.1. This ladder is not only damaged but part of it is missing, and this should be noted as a deficiency. Second, the area surrounding the tank does not comply with the conditions described in 9.2.5.2, because there is brush that could present a fire exposure hazard.

9.2.5.2 The area surrounding the tank and supporting structure, where provided, shall be inspected quarterly to ensure that the following conditions are met:

(1) The area is free of combustible storage, trash, debris, brush, or material that could present a fire exposure hazard.
(2) The area is free of the accumulation of material on or near parts that could result in accelerated corrosion or rot.
(3) The tank and support are free of ice buildup.
(4) The exterior sides and top of embankments supporting coated fabric tanks are free of erosion.

Critical Point

An inspection of the area surrounding the tank and supporting structure is similar to an exposure evaluation provided during an insurance inspection.

An inspection of the area surrounding the tank and supporting structure is similar to an exposure evaluation provided during an insurance inspection. It must be determined whether any surrounding material presents an exposure hazard to the tank or support structure. This exposure hazard can be from any of the following issues:

- Fires [see 9.2.5.2(1)]
- Corrosion [see 9.2.5.2(2)]
- Increased loads due to ice [see 9.2.5.2(3)]
- Erosion of the berm or embankment, as in the case of an embankment-supported coated fabric tank — [see 9.2.5.2(4)]

9.2.5.3 Expansion joints, where provided, shall be inspected annually for leaks and cracks.

9.2.5.4 The hoops and grillage of wooden tanks shall be inspected annually.

9.2.5.5 Exterior painted, coated, or insulated surfaces of the tank and supporting structure, where provided, shall be inspected annually for signs of degradation.

9.2.6 Interior Inspection.

Exhibit 9.13 shows the interior of a tank revealing silt and a noncompliant vortex plate. This material was found inside a 1 million gal (3,785,411 L) suction tank at a power plant. This tank supplies two 2000 gpm (7570 L/min) fire pumps. The photo demonstrates the importance of performing an interior inspection of water tanks supplying fire protection systems. See Exhibit 9.15 for a proper anti-vortex plate.

When an interior inspection of a tank is performed, it is critical to comply with proper safety precautions, including confined space entry requirements.

Critical Point

When an interior inspection of a tank is performed, it is critical to comply with proper safety precautions, including confined space entry requirements.

Exhibit 9.13 *Interior of Tank Requiring an Inspection.*

FAQ Is it necessary to drain a tank to perform an interior inspection?

It is not necessary to drain the tank for the inspection. One means of accomplishing a tank inspection without draining the tank is by using a certified commercial diver. An alternative approach would be to use remote video equipment if acceptable to the authority having jurisdiction (AHJ).

9.2.6.1 Frequency.

9.2.6.1.1* The interior of steel tanks without corrosion protection shall be inspected every 3 years.

A.9.2.6.1.1 To aid in the inspection and evaluation of test results, it is a good idea for the property owner or designated representative to stencil the last known date of an interior paint job on the exterior of the tank in a conspicuous place. A typical place is near one of the manways at eye level.

9.2.6.1.2* The interior of all other types of tanks shall be inspected every 5 years.

FAQ What is the inspection requirement for internal inspections of tanks used for fire protection?

A 3-year internal inspection is required for tanks that do not have corrosion protection (i.e., cathodic protection or equivalent). A 5-year internal inspection is required for all other tanks.

A.9.2.6.1.2 If written verification of interior corrosion protection for a tank per NFPA 22, *Standard for Water Tanks for Private Fire Protection*, cannot be provided by the building owner, the interior of the tank should be inspected every 3 years.

9.2.6.2 Where interior inspection is made by means of underwater evaluation, silt shall first be removed from the tank floor.

Critical Point

Underwater inspections are desirable because draining, inspecting, and refilling a tank can be time consuming and costly and can create a lengthy system impairment.

Underwater interior inspections were introduced in the 2002 edition of NFPA 25. Underwater inspections are desirable because draining, inspecting, and refilling a tank can be time consuming and costly and can create a lengthy system impairment. Over time, silt will accumulate on the bottom of a tank. This accumulation not only poses a threat to a diver, but it can also hide flaws in the tank.

It is important to note that NFPA 22 requires a fill connection that is designed to fill the tank within 8 hours. Therefore, draining, inspecting, and refilling can take longer than a normal work shift. Section 15.5 of NFPA 25 requires supplemental protection and special precautions for any impairment in excess of 10 hours in a 24-hour period. As a result, interior inspections can involve considerable planning and expense.

Also, it is important to verify that the diver is a certified commercial diver who is qualified for confined space entry. It is not advisable for certified recreational divers to attempt to perform internal tank inspections. Prior to entering the tank, it is critical to lock out the fire pump systems and observe all the requirements of Chapter 15 during this inspection. Failure to observe these basic safety practices can result in serious injury and even loss of life for the diver.

Exhibit 9.14 illustrates the presence of a potential obstruction — a fish — inside a steel tank. The photo was taken during an internal inspection.

9.2.6.3 The tank interior shall be inspected for signs of pitting, corrosion, spalling, rot, other forms of deterioration, waste materials and debris, aquatic growth, and local or general failure of interior coating.

9.2.6.4 Steel tanks exhibiting signs of interior pitting, corrosion, or failure of coating shall be tested in accordance with 9.2.7.

Exhibit 9.14 Potential Obstruction Inside Steel Tank. (Courtesy of Conrady Consultant Services)

9.2.6.5* Tanks on ring-type foundations with sand in the middle shall be inspected for evidence of voids beneath the floor.

A.9.2.6.5 This inspection can be performed by looking for dents on the tank floor. Additionally, walking on the tank floor and looking for buckling of the floor will identify problem areas.

Ground-level suction tanks are typically set on crushed stone, sand, or concrete foundations. A concrete ring wall at least 2½ ft (760 mm) high and 10 in. (250 mm) thick surrounds the tank foundation. This ring typically projects 6 in. (152 mm) above grade, but it should be inspected carefully because the sand can shift beneath the tank and create a void. Leaks from the tank bottom can also cause subsurface erosion. Without adequate support at the void, failure of the tank bottom can occur.

9.2.6.6 The heating system and components including piping shall be inspected.

9.2.6.7 The anti-vortex plate shall be inspected for deterioration or blockage.

An anti-vortex plate, such as the one illustrated in Exhibit 9.15, is provided on a tank to reduce the likelihood of introducing air pockets into the suction line. The introduction of an air pocket into a pump suction line could result in cavitation of the pump, which in turn could damage the pump casing and cause a reduction in pump performance.

9.2.7 Tests During Interior Inspection. Where a drained interior inspection of a steel tank is required by 9.2.6.4, the following tests shall be conducted:

(1) Evaluation of tank coatings shall be made in accordance with the adhesion test of ASTM D 3359, *Standard Test Methods for Measuring Adhesion by Tape Test*, generally referred to as the "cross-hatch test."
(2) Dry film thickness measurements shall be taken at random locations to determine the overall coating thickness.
(3) Nondestructive ultrasonic readings shall be taken to evaluate the wall thickness where there is evidence of pitting or corrosion.
(4) Interior surfaces shall be spot wet-sponge tested to detect pinholes, cracks, or other compromises in the coating. Special attention shall be given to sharp edges such as ladder rungs, nuts, and bolts.

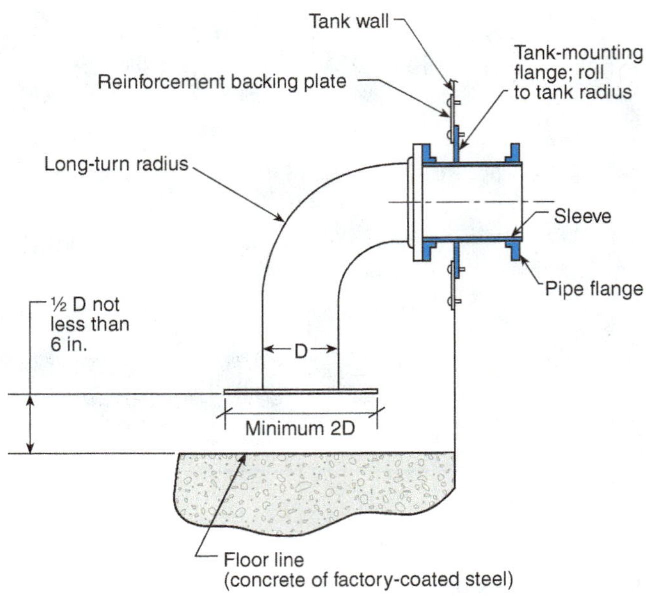

For SI units, 1 in. = 25.4 mm.

Exhibit 9.15 *Anti-Vortex Plate. [Source: NFPA 22, 2013, Figure B.1(p)]*

(5) Tank bottoms shall be tested for metal loss and/or rust on the underside by use of ultrasonic testing where there is evidence of pitting or corrosion. Removal, visual inspection, and replacement of random floor coupons shall be an acceptable alternative to ultrasonic testing.

(6) Tanks with flat bottoms shall be vacuum-box tested at bottom seams in accordance with test procedures found in NFPA 22, *Standard for Water Tanks for Private Fire Protection*.

Critical Point

All the tests required by 9.2.7 are intended to identify any failure points or weakened areas of the tank.

All the tests required by 9.2.7 are intended to identify any failure points or weakened areas of the tank. The list identifies a specific and thorough series of tests to be conducted for a drained tank. Many of the deficiencies noted during this inspection are not going to be detected by a simple visual inspection. Therefore, it is critical to complete this more thorough internal investigation so that any areas that could cause catastrophic tank failure can be identified and corrected.

9.3 Testing

9.3.1* Level indicators shall be tested every 5 years for accuracy and freedom of movement.

Testing Procedure Alert

ITM NFPA 25

While it is important to look at the manufacturer's instructions for testing procedures associated with water level indicators, there are many steps in the process of conducting the test mandated by 9.3.1 that apply to many tank arrangements. It is important to confirm that the indicator can move freely and has not become inadvertently "stuck" in the wrong position. For a detailed testing procedure to execute the test required by 9.3.1, see the Testing Procedures for 9.3.1 at the end of this chapter.

A.9.3.1 The testing procedure for listed mercury gauges is as follows.

To determine that the mercury gauge is accurate, the gauge should be tested every 5 years as follows [steps (1) through (7) coincide with Figure A.9.3.1]:

(1) Overflow the tank.
(2) Close valve F. Open test cock D. The mercury will drop quickly into the mercury pot. If it does not drop, there is an obstruction that needs to be removed from the pipe or pot between the test cock and the gauge glass.
(3) If the mercury does lower at once, close cock D and open valve F. If the mercury responds immediately and comes to rest promptly opposite the "FULL" mark on the gauge board, the instrument is functioning properly.
(4) If the mercury column does not respond promptly and indicate the correct reading during the test, there probably are air pockets or obstructions in the water connecting pipe. Open cock D. Water should flow out forcibly. Allow water to flow through cock D until all air is expelled and rusty water from the tank riser appears. Close cock D. The gauge now

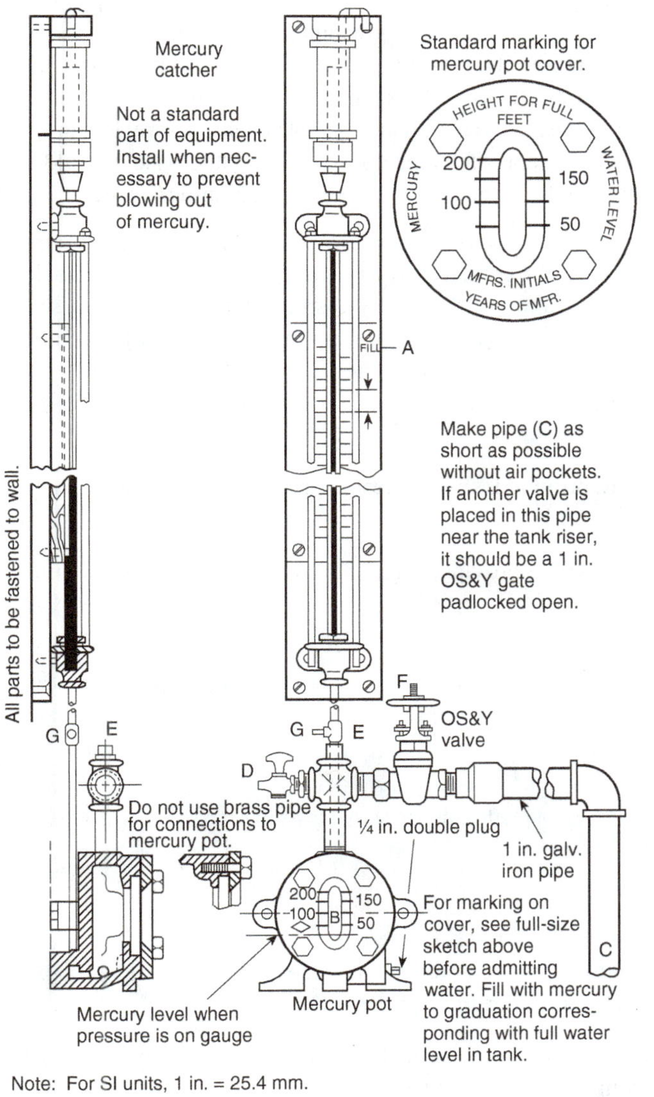

***FIGURE A.9.3.1** Mercury Gauge.*

likely will read correctly. If air separates from the water in the 1 in. (25 mm) pipe due to being enclosed in a buried tile conduit with steam pipes, the air can be removed automatically by installing a ¾ in. (20 mm) air trap at the high point of the piping. The air trap usually can be installed most easily in a tee connected by a short piece of pipe at E, with a plug in the top of the tee so that mercury can be added in the future, if necessary, without removing the trap. If there are inaccessible pockets in the piping, as where located below grade or under concrete floors, the air can be removed only through petcock D.

(5) If, in step (4), the water does not flow forcibly through cock D, there is an obstruction that needs to be removed from the outlet of the test cock or from the water pipe between the test cock and the tank riser.

(6) If there is water on top of the mercury column in the gauge glass, it will provide inaccurate readings and should be removed. First, lower the mercury into the pot as in step (2). Close cock D and remove plug G. Open valve F very slowly, causing the mercury to rise slowly and the water above it to drain through plug G. Close valve F quickly when mercury appears at plug G, but have a receptacle ready to catch any mercury that drains out. Replace plug G. Replace any escaped mercury in the pot.

(7) After testing, leave valve F open, except under the following conditions: If it is necessary to prevent forcing mercury and water into the mercury catcher, the controlling valve F can be permitted to be closed when filling the tank but should be left open after the tank is filled. In cases where the gauge is subjected to continual fluctuation of pressure, it could be necessary to keep the gauge shut off except when it needs to be read. Otherwise, it could be necessary to remove water frequently from the top of the mercury column as in step (5).

A level indicator is a device used to measure the water level in a tank. Although mercury gauges are often used for this purpose, they are not permitted on new installations due to the health risks associated with ingestion of mercury vapors and mercury absorption through human skin. Therefore, they will be found only on older tanks. A mercury gauge should not be adjusted until a Material Safety Data Sheet (MSDS) (see Exhibit 4.7) is obtained from the property owner and all precautions have been followed. Electronic tank water level indicators should be tested in accordance with the manufacturer's instructions.

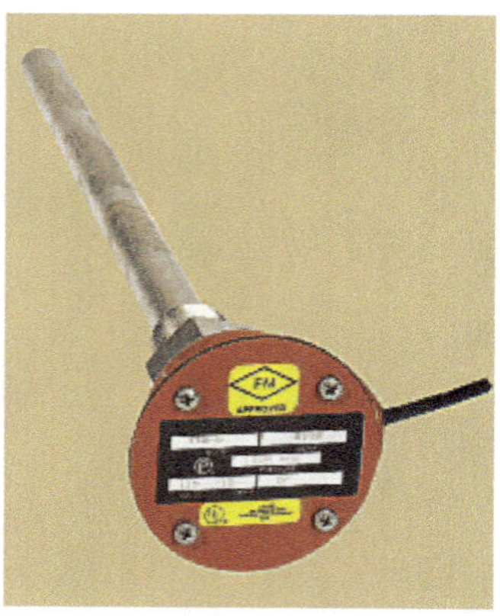

Exhibit 9.16 Electronic Water Temperature Sensor. (Courtesy of Potter Electric Signal Company, LLC)

9.3.2 The tank heating system, where provided, shall be tested prior to the heating season to make certain it is in the proper working order.

9.3.3 Low water temperature signals, where provided, shall be tested monthly (cold weather only).

Testing Procedure Alert

ITM NFPA 25

Freezing water in a storage tank can be extremely problematic for several reasons. First, it will limit or, in some cases, eliminate the water that can be delivered to the system (e.g., sprinkler or standpipe system) that is trying to draw it. Second, it can lead to significant damage of the tank and its appurtenances, or it can cause a fire pump to run "dry" because no water is being drawn from the tank. For a process for conducting low temperature signal tests in accordance with 9.3.3, see the Testing Procedures for 9.3.3 at the end of this chapter.

Critical Point

As the water level decreases, the likelihood of freezing increases. Therefore, testing must be done more frequently during cold weather.

As the water level decreases, the likelihood of freezing increases. Therefore, testing must be done more frequently during cold weather. The water temperature sensor shown in Exhibit 9.16 is a common way to monitor water temperature during cold weather periods.

9.3.4* High water temperature limit switches on tank heating systems, where provided, shall be tested monthly whenever the heating system is in service.

A.9.3.4 The manufacturer's instructions should be consulted for guidance on testing. In some situations, it might not be possible to test the actual initiating device. In such cases, only the circuitry should be tested.

The purpose of completing a high temperature limit switch test is to ensure that an initiated high temperature condition will cause the shutdown of the attached heating system. There are several ways to conduct this test, one of which is provided in the Testing Procedures for 9.3.4 at the end of this chapter.

Testing Procedure Alert

ITM NFPA 25

9.3.5* High and low water level signals shall be tested semiannually.

Low water level alarms must be tested monthly during cold weather periods as opposed to the semiannual test addressed in 9.3.5 (see 9.3.3 and related commentary). Exhibit 9.17 illustrates a water level indicator.

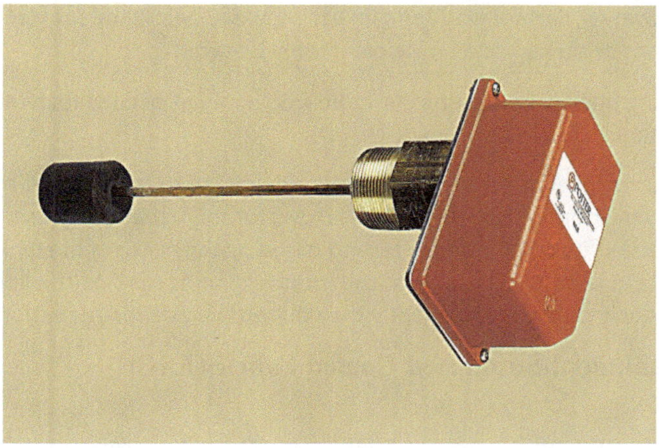

Exhibit 9.17 *Electronic Water Level Indicator. (Courtesy of Potter Electric Signal Company, LLC)*

Not unlike the level indicator test required by 9.3.1, this test focuses on making sure there is sufficient water in the tank. Rather than looking at the indicator, however, this test focuses on the functionality of the system components that send a signal to indicate that the tank is experiencing an issue with the water level. See the Testing Procedures for 9.3.5 at the end of this chapter for guidance in conducting this test.

Testing Procedure Alert

ITM NFPA 25

A.9.3.5 See A.9.3.4.

9.3.6* **Pressure Gauges.**

A.9.3.6 See A.5.3.2.

9.3.6.1 Pressure gauges shall be tested every 5 years with a calibrated gauge in accordance with the manufacturer's instructions.

9.3.6.2 Gauges not accurate to within 3 percent of the scale of the gauge being tested shall be recalibrated or replaced.

9.4 Maintenance

9.4.1 Voids discovered beneath the floors of tanks shall be filled by pumping in grout or accessing the sand and replenishing.

Voids can result from slow leaks, which, if left untreated, could cause further subsoil erosion and result in failure of the tank bottom. This issue is addressed in greater detail in A.9.2.6.5 and its associated commentary.

9.4.2 The tank shall be maintained full or at the designed water level.

The water level should never be lower than 3 in. to 4 in. (76 mm to 102 mm) below the designated fire service level.

9.4.3 The hatch covers in the roofs and the door at the top of the frostproof casing shall always be kept securely fastened with substantial catches as a protection against freezing and windstorm damage.

9.4.4 No waste materials, such as boards, paint cans, trim, or loose material, shall be left in the tank or on the surface of the tank.

Any material left in or on the tank could block the waterway. Steel tanks are easy to inspect for the presence of such material. However, certain wooden tanks have space below the roof (see Exhibit 9.18) in which old paint cans or similar items were commonly placed.

9.4.5 Silt shall be removed during interior inspections or more frequently as needed to avoid accumulation to the level of the tank outlet.

Critical Point

The actual frequency required for removal of silt or other debris accumulation, as required by 9.4.5, is permitted to be identified over time.

The actual frequency required for removal of silt or other debris accumulation, as required by 9.4.5, is permitted to be identified over time. The requirement allows for this variation by using the term "or more frequently as needed" to customize the testing based on the needs of the tank. For example, if a raw water source is used to fill the tank, the introduction of silt into the line presents a problem that must be dealt with more frequently than during the interior inspection.

9.4.6 Maintenance of Embankment-Supported Coated Fabric (ESCF) Suction Tanks.

Embankment-supported coated fabric tanks present a unique design and installation challenge. The tank manufacturer's maintenance requirements must be followed, in addition to those prescribed by the standard. Standard inspection requirements for these tanks, such as not allowing waste materials, not allowing large accumulations of ice to collect on top of the tank, and inspection of the berms for erosion, are outlined in the standard. However, specific maintenance requirements, such as painting or refurbishing the top surface of the tank, and interior inspection frequencies and methods, must come from the manufacturer.

9.4.6.1 The maintenance of ESCF tanks shall be completed in accordance with this section and the tank manufacturer's instructions.

9.4.6.2 The exposed surfaces of ESCF tanks shall be cleaned and painted every 2 years or in accordance with the manufacturer's instructions.

9.5 Automatic Tank Fill Valves

Automatic tank fill valves are usually spring-loaded, diaphragm-type valves that sense pressure in the tank and open and close to maintain a constant water level in the tank. The valve manufacturer's instructions should be followed for testing, maintenance, and adjustment. Exhibit 9.19 shows an automatic tank fill valve.

9.5.1 Inspection.

| **9.5.1.1** Automatic tank fill valves shall be inspected in accordance with Table 9.5.1.1.

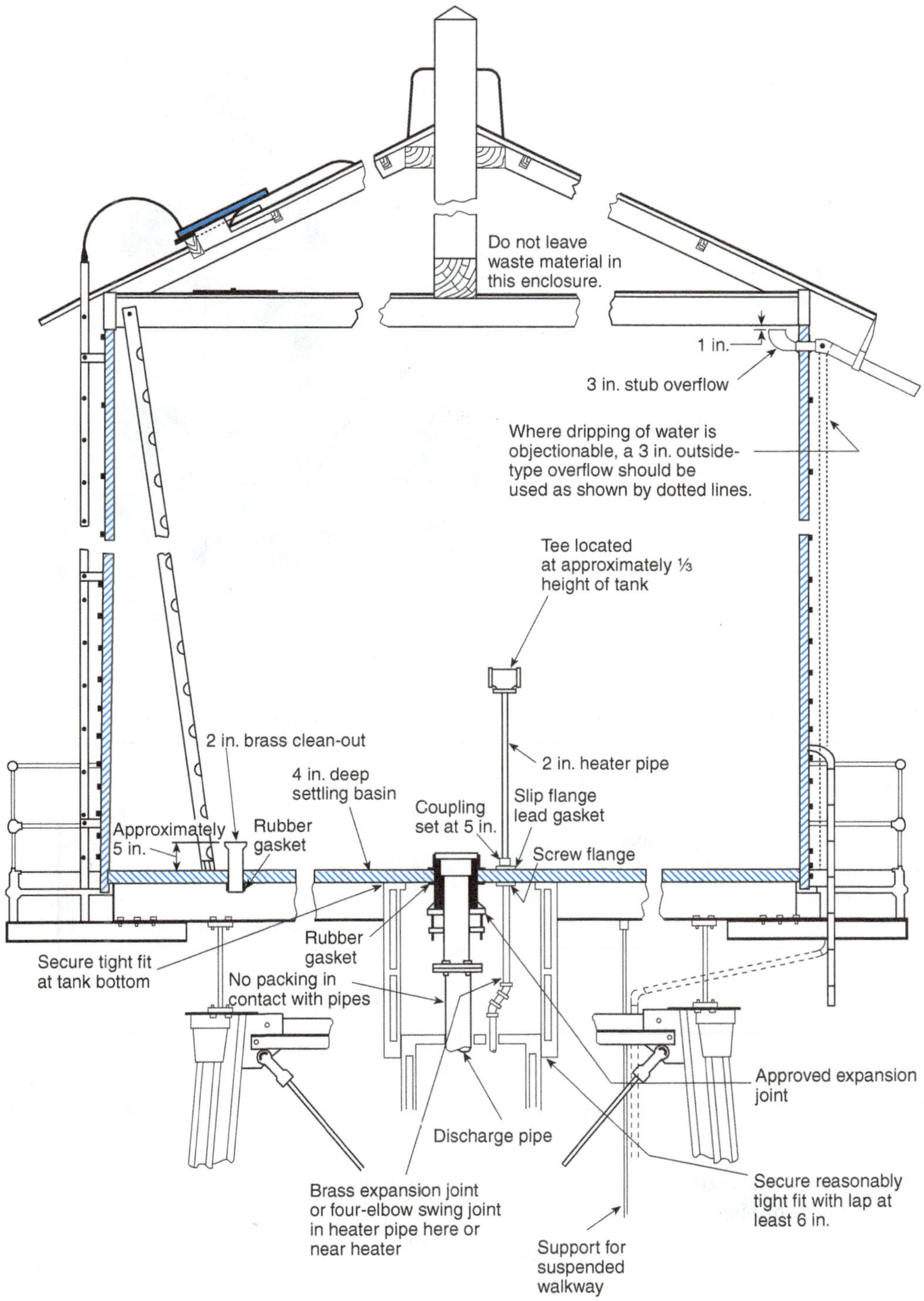

Exhibit 9.18 *Wooden Tank with Space Below the Roof. [Source: NFPA 22, 2013, Figure B.1(i)]*

Exhibit 9.19 *Automatic Tank Fill Valve.*

TABLE 9.5.1.1 *Summary of Automatic Tank Fill Valve Inspection and Testing*

Item	*Frequency*	*Reference*
Inspection		
Strainers, filters, orifices (inspect/clean)	5 years	13.4.1.2
Enclosure (during cold weather)	Daily/weekly	13.4.3.1.1
Exterior	Monthly	13.4.3.1.6
Interior	Annually/5 years	13.4.3.1.7
Test		
Automatic tank fill valve	Annually	

9.5.1.1.1 OS&Y isolation valves that are a part of the automatic fill valves shall be inspected in accordance with Chapter 13.

9.5.1.2 Valves secured with locks or electrically supervised in accordance with applicable NFPA standards shall be inspected monthly.

9.5.1.3 The enclosure shall be inspected to verify that it is heated and secured.

9.5.2 Maintenance.

9.5.2.1 Maintenance of all automatic tank fill valves shall be conducted by a qualified person following the manufacturer's instructions in accordance with the procedure and policies of the authority having jurisdiction.

9.5.2.2 Rubber parts shall be replaced in accordance with the frequency required by the authority having jurisdiction and the manufacturer's instructions.

9.5.2.3 Strainers shall be cleaned quarterly.

9.5.3 Testing. All automatic tank fill valves shall be tested yearly in accordance with the following:

(1) The valve shall be actuated automatically by lowering the water level in the tank.
(2) The refill rate shall be measured and recorded.

Testing Procedure Alert

ITM NFPA 25

While 9.3.1 and 9.3.5 focus on testing various components of the tank system to confirm there is sufficient water, 9.5.3 is aimed at making sure that where a low level signal is received, the refill valve activates appropriately and provides the correct refill rate for the tank. For more information on this test, including a detailed procedure for conducting it, please see the Testing Procedures for 9.5.3 at the end of this chapter.

9.6 Component Action Requirements

Component replacement tables offer guidance to the user of the standard when system components are adjusted, repaired, rebuilt, or replaced. It is not necessary in each case to require a complete acceptance test for each component when maintenance is performed.

Critical Point

Component replacement tables offer guidance to the user of the standard when system components are adjusted, repaired, rebuilt, or replaced.

9.6.1 Whenever a component in a water storage tank is adjusted, repaired, reconditioned, or replaced, the action required in Table 9.6.1 shall be performed.

TABLE 9.6.1 Summary of Component Replacement Action Requirements

Component	*Adjust*	*Repair/ Recondition*	*Replace*	*Test Criteria*
Tank Components				
Tank interior		X	X	Remove debris Verify integrity in conformance with NFPA 22, *Standard for Water Tanks for Private Fire Protection*
Tank exterior		X	X	Verify integrity in conformance with NFPA 22
Support structure		X	X	Verify integrity in conformance with NFPA 22
Heating system	X	X	X	Verify heating system is in conformance with NFPA 22
Catwalks and ladders	X	X	X	Verify integrity in conformance with NFPA 22
Hoops and grillage	X	X	X	Verify integrity in conformance with NFPA 22
Expansion joints	X	X	X	Verify integrity in conformance with NFPA 22
Overflow piping	X	X	X	Verify integrity in conformance with NFPA 22
Insulation		X	X	Verify integrity in conformance with NFPA 22

(continues)

TABLE 9.6.1 Continued

Component	Adjust	Repair/ Recondition	Replace	Test Criteria
Alarm and Supervisory Components				
High and low water level	X	X	X	Operational test for conformance with NFPA 22 and/or *NFPA 72, National Fire Alarm and Signaling Code*, and the design water levels
Water temperature	X	X	X	Operational test for conformance with NFPA 22 and/or *NFPA 72*
Enclosure temperature	X	X	X	Operational test for conformance with NFPA 22 and/or *NFPA 72*
Valve supervision	X	X	X	Operational test for conformance with NFPA 22 and/or *NFPA 72*
Fill and Discharge Components				
Automatic fill valves				See Chapter 13
Valves	X	X	X	See Chapter 13
Status Indicators				
Level indicators	X	X	X	Verify conformance with NFPA 22
Pressure gauges			X	Verify at 0 psi (0 bar) and at system working pressure

9.6.2 Where the original installation standard is different from the cited standard, the use of the appropriate installing standard shall be permitted.

•

9.6.3 These actions shall not require a design review, which is outside the scope of this standard.

References Cited in Commentary

National Fire Protection Association, 1 Batterymarch Park, Quincy, MA 02169-7471.

NFPA 13D, *Standard for the Installation of Sprinkler Systems in One- and Two-Family Dwellings and Manufactured Homes*, 2013 edition.

NFPA 22, *Standard for Water Tanks for Private Fire Protection*, 2013 edition.

NFPA 20, *Standard for the Installation of Stationary Pumps for Fire Protection*, 2013 edition.

Testing Procedures for 9.3.1

Water Tank Level Indicator Test (Float Type)

Purpose

The purpose of completing a test of the float-type water tank level indicator is to ensure that the indicator is accurate and moves freely throughout the full range of operation. Some tanks might also employ other types of level indicators such as sight glass tubes, altitude pressure gauges, or mercury gauges. The testing of these devices should be in accordance with the manufacturer's instructions, with specific criteria being provided in A.9.3.1 for the testing of mercury gauges. The float-type level indicator typically includes an internal float, which rises and drops with the water level inside the tank, connected through a cable and pulley system to an external target on the shell of the tank. The movement of the internal float with the water level causes a corresponding movement of the target on the exterior of the tank, which can be visually observed against an adjacent scale marking. Generally these include a drop in the target position as the water level rises; however, some devices might include a compound pulley system that results in an upward movement of the external target to correspond directly with the water level in the tank. This testing can be coordinated with the drainage of a tank that is part of an internal tank inspection conducted as required by 9.2.6.1.

Tools/Equipment

1. The various wrenches and tools necessary to facilitate the removal of such equipment as drain flanges are required.

Procedure Steps

1. Prior to any testing, notify the fire department and/or the alarm monitoring company (if needed), as well as the facility representatives, that testing is going to be conducted. Make the necessary arrangements for a preplanned impairment in accordance with Section 15.5.
2. Close the control valves on any automatic fill line connection and the control valve on any outlet connections from the tank.
3. Visually check the position of the water level target to determine the designated position of the water within the tank, and confirm that the actual water level within the tank matches this designation.
4. Locate a drain connection for the tank and check the area surrounding the discharge point to ensure that there are no obvious conditions that would prevent water from being discharged safely or cause direct damage in the immediate vicinity. Where the discharge of water is to an area subject to potential freezing conditions, the facility representative should be advised of the potential for icing conditions.
5. Initiate the drainage of the tank by opening the located drain connection and observe the water level target for movement as the water level begins to drop. Make periodic checks of the target position against the provided scale to ensure the accuracy of the scale designation as the tank is drained.
6. Once the tank is completely drained, closely check the float, cable, pulleys, guides, and external target for signs of corrosion or deterioration, noting any conditions that might inhibit free movement of the system. Note that internal entry into the tank requires special precautions for confined space entry.
7. Close any tank drain connections that were opened in Step 5 above.
8. Open any control valves that were closed in the automatic fill line in Step 2 above, or open the manual refill line to initiate a refill of the tank and observe the water level target for movement as the water level begins to rise. Make periodic checks of the target position against the provided scale to ensure the accuracy of the scale as the tank is refilled.
9. Open any control valves that were closed in the tank outlet connections in Step 2 above.
10. Upon completion of all testing, notify the fire department and/or the alarm monitoring company (as needed), as well as the facility representatives, that testing is complete, and reset the fire alarm system as necessary.

Alternative Approaches

The drainage of a water storage tank is not mandated by 9.2.6.1, so alternative methods to the described process above could be used. This could include the use of a certified commercial diver or remote video equipment to complete an internal inspection of the water tank where acceptable to the authority having jurisdiction (AHJ). Rather than checking the movement of the float with a change in actual water level, a certified commercial diver might manually cycle the system through a simulated movement by pulling the float to the bottom of the tank and returning it to the surface or by using some other alternative means to submerge the float.

These methods provide the benefit of not requiring drainage of the water tank and associated impairment of the system; however, the freedom of movement of the system cannot be fully assessed against an actual response of the system from a change in water level in the tank. This can be checked to some extent by conducting a partial drainage of the tank to ensure initiation of movement through the starting range for the system, but this will not check the full movement of the target.

Evaluation of Results

The water level indicator should provide an accurate indication of the actual water level within the tank throughout the full range of operation. The components of the water level indicator should not show signs of excessive corrosion, deterioration, or other conditions that inhibit the free movement of the system. Any noted inaccuracies or conditions that inhibit free movement should be investigated and corrective action should be taken as necessary.

The procedure(s) outlined above is not mandated by NFPA 25 and represents only one approach to conducting this test.

Water Tank Low Temperature Signal Test

Purpose

The purpose of completing a low temperature signal test is to ensure that an initiated low temperature supervisory switch will transmit a signal to the appropriate monitoring location. The actual testing of the low temperature supervisory switch would require the exposure of the switch to a temperature at or below its designated operating temperature. For a water storage tank, this would require the removal of the switch from the tank, which is not normally possible without drainage of the tank. As a result, the testing of the circuitry between the installed switch and the monitoring location is to be conducted. Some manufacturers might provide additional testing instructions for the internal circuitry of the switch.

Consideration might be given to removal and testing of the actual operation of the switch whenever the tank is drained for inspection or maintenance to ensure proper operation.

Tools/Equipment

1. The various tools necessary to facilitate the removal of cover plates and disconnection of wire connections are required.
2. A multimeter is required for additional manufacturers' recommended testing, as needed.

Procedure Steps

1. Prior to any testing, notify the fire department and/or the alarm monitoring company (as needed), as well as the facility representatives, that testing is going to be conducted.
2. Check the fire alarm control panel, remote annunciator, or other supervising panel to determine that the low temperature signal is not currently active on the system.
3. Remove the cover plate on the low temperature supervisory switch, exposing the wiring connection of the unit.
4. Disconnect one leg of the wired connection of the low temperature supervisory switch to simulate an open condition.
5. Verify the receipt of a low temperature supervisory signal at the fire alarm control panel, remote annunciator, or other supervising panel.
6. Complete additional testing of the switch as recommended by the manufacturer.
7. Reconnect the wiring connection for the low temperature supervisory switch and reinstall the cover plate.
8. Upon completion of all testing, notify the fire department and/or the alarm monitoring company (as needed), as well as the facility representatives, that testing is complete, and reset the fire alarm system as necessary. A verification of the receipt of initiated signals with the fire department and/or the alarm monitoring company, where so connected, would be appropriate as part of this effort.

Evaluation of Results

The transmission of the low temperature supervisory signal is verified by receipt at the fire alarm control panel, remote annunciator, or other supervising panel upon simulation of an activation of the low temperature supervisory switch. Failure to transmit a proper signal requires further investigation and corrective action as necessary.

The procedure(s) outlined above is not mandated by NFPA 25 and represents only one approach to conducting this test.

Water Tank High Temperature Limit Switch Test

Purpose

The purpose of completing a high temperature limit switch test is to ensure that an initiated high temperature condition will cause the shutdown of the attached heating system. The actual testing of the high temperature limit switch would require the exposure of the switch to a temperature at or above its designated operating temperature. For a water storage tank this would require the removal of the switch from the tank, which is not normally possible without draining the tank. As a result, the testing of the circuitry to cause a shutdown of the heating system is to be conducted. Some manufacturers might provide additional testing instructions for the internal circuitry of the switch.

Consideration might be given to removal and testing of the actual operation of the switch whenever the tank is drained for inspection or maintenance to ensure proper operation.

Tools/Equipment

1. The various tools necessary to facilitate the removal of cover plates and disconnection of wire connections are required.
2. A multimeter is required for additional manufacturers' recommended testing as needed.

Procedure Steps

1. Prior to any testing, notify the fire department and/or the alarm monitoring company (as needed), as well as the facility representatives, that testing is going to be conducted.
2. Check the heating system to ensure that it is in operation prior to initiating the test.
3. Remove the cover plate on the high temperature limit switch, exposing the wiring connection of the unit.
4. Disconnect one leg of the wired connection of the high temperature limit switch to simulate an open condition.
5. Verify the shutdown of the power or fuel source to the heating system.
6. Complete additional testing of the switch as recommended by the manufacturer.
7. Reconnect the wiring connection for the high temperature limit switch and reinstall the cover plate.
8. Upon completion of all testing, notify the fire department and/or the alarm monitoring company (as needed), as well as the facility representatives, that testing is complete, and reset the fire alarm system as necessary. An additional reset of the heating system might be required.

Evaluation of Results

The shutdown of the power or fuel source to the heating system upon simulation of a high temperature limit switch activation verifies the operation of the shutdown circuitry of the heating system from the connection to the high temperature limit switch to the heating system. Failure of the power or fuel source to shutdown requires further investigation and corrective action as necessary.

The procedure(s) outlined above is not mandated by NFPA 25 and represents only one approach to conducting this test.

Non-Pressure Water Tank High/Low Water Level Signal Test

Purpose

The purpose of completing a high/low water level signal test is to ensure that high and low water conditions (as monitored) are initiated and transmitted to the appropriate monitoring location. Some manufacturers might provide additional testing instructions for the internal circuitry of the switch.

Tools/Equipment

1. The various wrenches and tools necessary to facilitate the drainage of water from the tank and overfill are required.
2. A tape measure or other measuring means is required to measure the change in water level required to initiate the high/low water level signals.

Procedure Steps

1. Prior to any testing, notify the fire department and/or the alarm monitoring company (as needed), as well as the facility representatives, that testing is going to be conducted.
2. Check the fire alarm panel, remote annunciator, or other supervising panel to determine that a high/low water level signal is not currently active on the system.
3. Note the normal operating water level in the tank.
4. For tanks that include a high water level switch, complete Step 5 through Step 9. Otherwise, skip to Step 10.
5. Open the manual fill valve for the tank and monitor the fire alarm panel, remote annunciator, or other supervising panel to verify the initiation and transmission of the high water level signal. Ensure that signal is initiated prior to a rise in water level of more than 12 in. (300 mm) from the normal operating water level or overflow from the tank, whichever comes first.
6. Shut down the manual fill valve.
7. Locate a drain connection for the tank, and check the area surrounding the discharge point to ensure that there are no obvious conditions that would prevent water from being discharged safely or cause direct damage in the immediate vicinity. Where the discharge of water is to an area subject to potential freezing conditions, the facility representative should be advised of the potential for icing conditions.
8. Drain water from the tank down to the normal operating level.
9. Verify that the high water level signal is restored at the fire alarm panel, remote annunciator, or other supervising panel, resetting as necessary.
10. For tanks that include a low water level switch, complete Step 11 through Step 16. Otherwise, skip to Step 17.
11. Shut down the control valve for any automatic fill lines to the tank.
12. Locate a drain connection for the tank, and check the area surrounding the discharge point to ensure that there are no obvious conditions that would prevent water from being discharged safely or cause direct damage in the immediate vicinity. Where the discharge of water is to an area subject to potential freezing conditions, the facility representative should be advised of the potential for icing conditions.
13. Initiate drainage of water from the tank, monitor the fire alarm panel, remote annunciator, or other supervising panel, and verify the initiation and transmission of the low water level signal. Ensure that the signal is initiated prior to a drop in water level of more than 12 in. (300 mm) from the normal operating water level.
14. Shut down the drainage of the tank.
15. Open the manual fill valve, or open any control valves that were shut as a part of Step 11 above, and refill to the normal operating water level.
16. Verify that the low water level signal is restored at the fire alarm panel, remote annunciator, or other supervising panel, and reset them as necessary.
17. Upon completion of all testing, notify the fire department and/or the alarm monitoring company (as needed), as well as the facility representatives, that testing is complete, and reset the fire alarm system as necessary. A verification of the receipt of initiated signals with the fire department and/or the alarm monitoring company, where so connected, would be appropriate as part of this effort.

Evaluation of Results

The proper operation of the high/low water level signals is verified by the initiation and transmission of appropriate signals at the fire alarm panel, remote annunciator, or other supervising panel under the simulated high and low water conditions, as well as restoration upon return to the normal operating water level. The initiation of the high and low level water signals are to operate at not more than 12 in. (300 mm) above or below the normal operating water level. Failure of the high/low switches to initiate within the specified water levels and transmit a signal or to restore requires further investigation and corrective action as necessary.

The procedure(s) outlined above is not mandated by NFPA 25 and represents only one approach to conducting this test.

Testing Procedures for 9.5.3

Water Tank Automatic Fill Valve Test

Purpose

The purpose of completing a test on the automatic fill valve is to ensure operation of the fill valve through the connected control system and to verify the refill rate of the installation. Some manufacturers might provide additional testing instructions for the fill valve.

Tools/Equipment

1. The various wrenches and tools necessary to facilitate the drainage of water from the tank and overfill are required.
2. A timing device is required to measure the time for refill of the tank.
3. A tape measure or other measuring means is required to determine the tank dimensions and measure the water fill height.

Procedure Steps

1. Prior to any testing, notify the fire department and/or the alarm monitoring company (as needed), as well as the facility representatives, that testing is going to be conducted.
2. Locate a drain connection for the tank, and check the area surrounding the discharge point to ensure that there are no obvious conditions that would prevent water from being discharged safely or cause direct damage in the immediate vicinity. Where the discharge of water is to an area subject to potential freezing conditions, the facility representative should be advised of the potential for icing conditions.
3. Initiate drainage of water from the tank, and monitor the automatic fill valve for actuation.
4. Shut down the drainage of the tank.
5. Close the control valve to the automatic fill line.
6. Continue drainage of the tank allowing a sufficient quantity of water to drain to provide for a measurable refill time for calculation of a refill rate. Depending on the size of the tank, a drop of 1 ft (0.3 m) to 5 ft (1.5 m) should be sufficient. Measure and record the drop in water level from the normal operating level.
7. Open the control valve to the automatic fill line.
8. Upon establishment of a full flow rate from the automatic fill valve, record a start time and establish a water level in the tank. Continue to monitor the refill process by recording the elapsed time to reach a specific increase in water level. Depending on the size of the tank, a rise of 1 ft (0.3 m) to 5 ft (1.5 m) should be sufficient.
9. Allow the automatic fill valve operation to return the tank to the normal operating level.
10. Upon completion of all testing, notify the fire department and/or the alarm monitoring company (as needed), as well as the facility representatives, that testing is complete, and reset the fire alarm system as necessary.

Evaluation of Results

The proper operation of the automatic fill valve is verified by the actuation of the valve upon a drop in water level. Failure of the automatic fill valve to operate and refill the tank to normal operating level and then shutoff requires further investigation and corrective action as necessary.

The determination of the fill rate is calculated by a timed average fill rate by using the measured time to refill a known height. An example of this calculation for a round vertical water tank is as follows. Calculate the fill rate from the automatic fill valve using the following formula, which applies to a round vertical water tank:

$$Q = \frac{352.5H \cdot d^2}{t}$$

Where:

Q = fill rate (gpm)

H = tank height filled during the measured time duration t (ft)

d = tank diameter (ft)

t = measured time duration (seconds)

Where the automatic fill valve flow rate is intended to provide make-up water for tanks designed to meet a capacity less than the full required duration, a direct evaluation of the adequacy of this fill rate would require knowledge of the original design criteria for the system. However, a comparative assessment can also be made against previous test results where available. A significant degradation in performance should warrant further investigation and corrective action as necessary. Fill rates below those known performance measures require further investigation and corrective action as necessary.

The procedure(s) outlined above is not mandated by NFPA 25 and represents only one approach to conducting this test.

ITM NFPA 25 — INSPECTION TESTING MAINTENANCE

10

WATER SPRAY FIXED SYSTEMS

Chapter 10 of NFPA 25 covers inspection, testing, and maintenance of water spray fixed systems. Water spray fixed systems are highly specialized systems that are typically installed under the scope of NFPA 15, *Standard for Water Spray Fixed Systems for Fire Protection*. These systems are used to protect equipment such as cable trays, belt conveyors, pumps, compressors, vessels, transformers, structures, and miscellaneous equipment. Water spray systems are also used on flammable and combustible liquid pool fires and are frequently used for exposure protection.

Water spray fixed systems use different types of detection systems, such as electronic (including smoke, heat, infrared, and linear cable detection), pneumatic rate-of-rise, and wet and dry pilot detection systems for actuation. Depending on the detection system used, inspection, testing, and maintenance (ITM) of water spray fixed systems may require specialized training beyond that required for conventional sprinkler systems.

> **Critical Point**
>
> Due to the nature of the hazards being protected, extreme care must be given to the maintenance of the system, because a seemingly small deviation from the original design can have catastrophic consequences.

The inspection and testing of a water spray fixed system as prescribed in Chapter 10 requires knowledge of NFPA 15. To determine if test results are acceptable, the original installation drawings and hydraulic calculations should be available for comparison. If these documents are not available, or if the inspector does not have sufficient knowledge of NFPA 15, the interpretation of test results might require the assistance of a professional engineer or an engineering technician. Due to the nature of the hazards being protected, extreme care must be given to the maintenance of the system, because a seemingly small deviation from the original design can have catastrophic consequences. Exhibit 10.1 shows a water spray fixed system being discharged as part of a test or routine inspection.

EXHIBIT 10.1 *Discharging System.*

10.1* General

A.10.1 The effectiveness and reliability of water spray fixed systems depends on maintenance of the integrity of hydraulic characteristics, water control valves, deluge valves and their fire detection/actuation systems, pipe hangers, and prevention of obstructions to nozzle discharge patterns.

Water spray fixed systems are most commonly used to protect processing equipment and structures, flammable liquid and gas vessels, piping, and equipment such as transformers, oil switches, and motors. They also have been shown to be effective on many combustible solids.

Many of the components and subsystems found in a water spray system require the same inspection, test, and maintenance procedures where they are used in automatic sprinkler systems and other fixed water-based fire protection systems. Other chapters of this standard should be consulted for particulars on required inspection and maintenance.

Water spray fixed systems are designed and installed in accordance with the requirements of NFPA 15. These systems normally employ an open piping network with open spray nozzles that discharge water in a predetermined pattern on a hazard area, structure, or vessel (see Exhibit 10.2 through Exhibit 10.4). Exhibit 10.2 illustrates a typical water spray system protecting

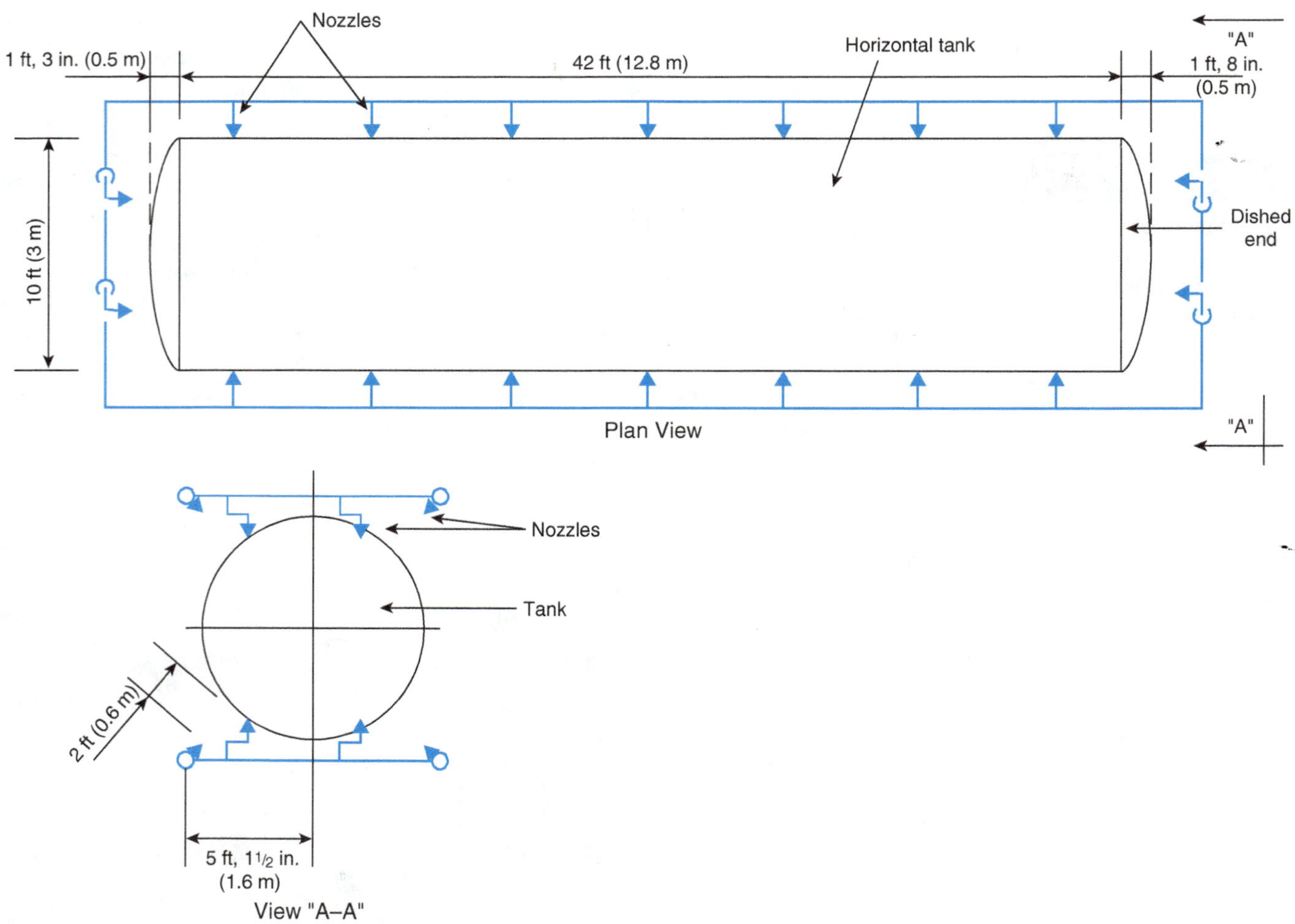

EXHIBIT 10.2 *Horizontal Chemical Tank Nozzle Arrangement. (Source: Adapted from Fire Protection Systems for Special Hazards, 2004, Figure 6.3)*

a horizontal chemical storage tank, with two levels of nozzles used to completely engulf the tank in water spray. This type of system design is typically used for fire extinguishment as well as exposure protection.

Exhibit 10.3 illustrates water spray protection for a pipe rack assembly of the type that is usually located in a chemical processing plant or refinery. The water spray nozzles are intended to spray directly on process piping and, in some cases, are used to protect the structural steel of the pipe rack itself. In this illustration, water spray nozzles are also employed to protect a process pump.

Exhibit 10.4 illustrates the use of water spray nozzles to protect a conveyor for coal or another type of material. In this application, nozzles are located along the conveyor belt to spray water either across the width of the belt or along the length of the belt. There is a detector at the roof of this enclosed conveyor that is used to activate the deluge valve in the water spray system. Often, the detection system is tied into the operational controls for a facility in addition to the suppression system, and if triggered, this can shut down conveyors or other pieces of industrial equipment.

The piping network is exposed to the atmosphere through the open spray nozzles, so these systems are subject to interior corrosion and blockages resulting from debris, such as insects nesting. Scale from the inside of the pipes can clog the water spray nozzles, affecting both the discharge pattern and density being applied to the surface of the hazard.

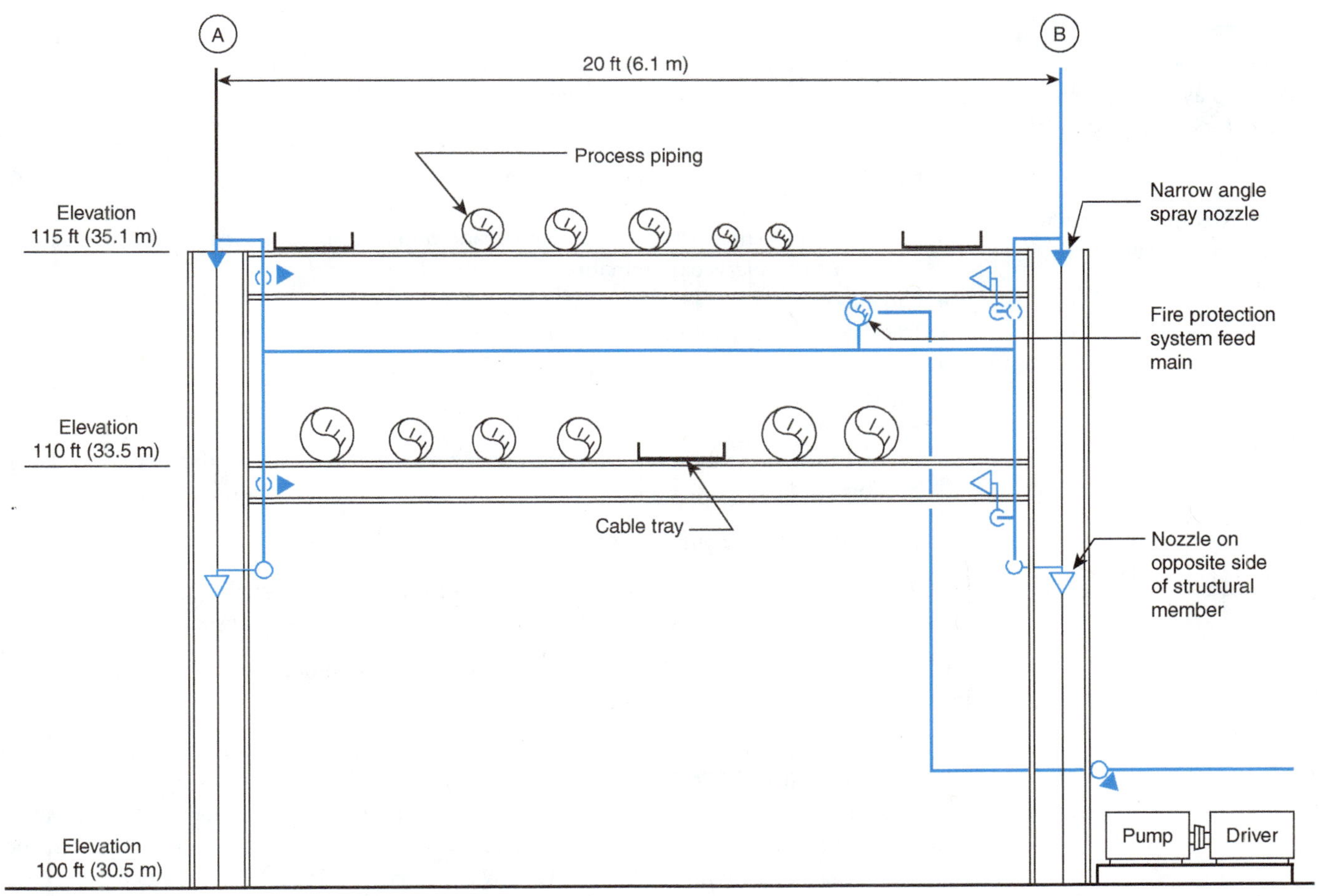

EXHIBIT 10.3 *Pipe Rack Nozzle Arrangement. (Source: Adapted from Fire Protection Systems for Special Hazards, 2004, Figure 6.16)*

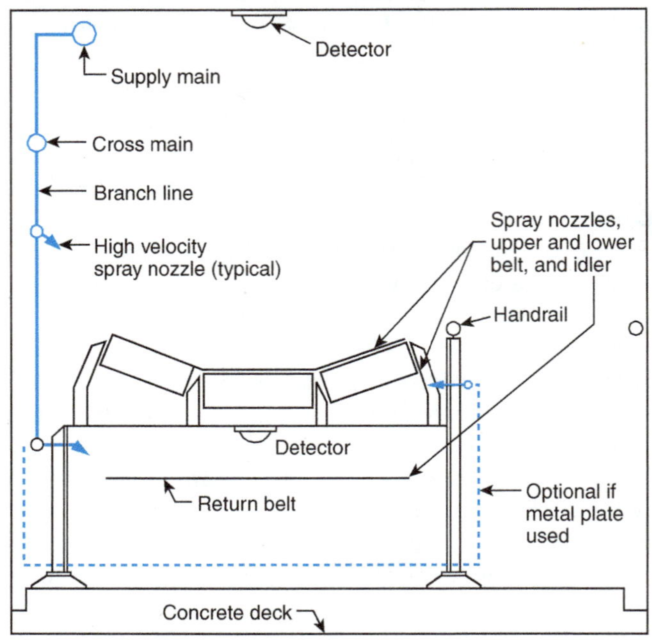

EXHIBIT 10.4 Typical Conveyor Belt Protection. [Source: NFPA 15, 2012, Figure A.7.2.3.3.1(a)]

Critical Point

Water spray fixed systems protect special hazards that require higher densities and direct impingement of water over the entire surface area of the equipment or structure.

Water spray fixed systems protect special hazards that require higher densities and direct impingement of water over the entire surface area of the equipment or structure. Examples of these special hazards include the following:

- Storage vessels containing hazardous chemicals or flammable and combustible liquids
- Piping and pumps involved in the processing or transfer of hazardous chemicals or flammable and combustible liquids
- Structural steel supporting vessels, piping, and other equipment used in the processing of hazardous chemicals or flammable and combustible liquids
- Cable trays and cable runs
- Transformers (see Exhibit 10.5 and Exhibit 10.6)
- Belt conveyors
- Turbine bearings
- Boiler fronts
- Other similar equipment involving the use of hazardous chemicals or flammable and combustible liquids

10.1.1 Minimum Requirements.

10.1.1.1 This chapter shall provide the minimum requirements for the routine inspection, testing, and maintenance of water spray protection from fixed nozzle systems only.

10.1.1.2 Table 10.1.1.2 shall be used to determine the minimum required frequencies for inspection, testing, and maintenance.

Many of the inspection, testing, and maintenance requirements in Table 10.1.1.2 reference other chapters within NFPA 25 or other NFPA documents. Those components in a water spray fixed system that are common to other water-based fire protection systems, such as waterflow alarms, fire pumps, water storage tanks, and valves, are addressed in other chapters. The detection systems are addressed in *NFPA 72®, National Fire Alarm and Signaling Code.* All other components specific to water spray fixed systems, such as the nozzles, pipe, fittings, hangers, supports, strainers, and manual releases, are addressed in the specific Chapter 10 paragraphs indicated in Table 10.1.1.2.

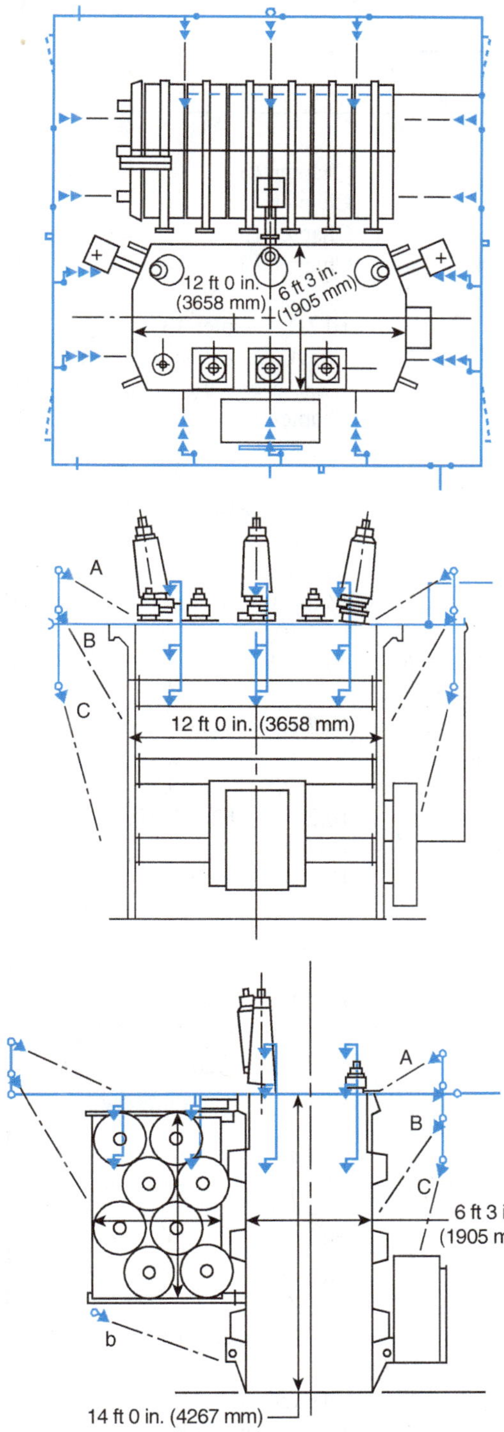

EXHIBIT 10.5 *Typical Transformer Water Spray System. (Source: NFPA 15, 2012, Figure A.7.4.4.1)*

EXHIBIT 10.6 *Water Spray System for Oil-Filled Electric Power Transformers.*

TABLE 10.1.1.2 Summary of Water Spray Fixed System Inspection, Testing, and Maintenance

Item	*Frequency*	*Reference*
Inspection		
Backflow preventer		Chapter 13
Check valves		Chapter 13
Control valves	Weekly (sealed)	Chapter 13
Control valves	Monthly (locked, supervised)	Chapter 13
Deluge valve		10.2.2, Chapter 13
Detection systems		*NFPA 72, National Fire Alarm and Signaling Code*
Detector check valves		Chapter 13
Drainage	Quarterly	10.2.8
Electric motor		10.2.9, Chapter 8
Engine drive		10.2.9, Chapter 8
Fire pump		10.2.9, Chapter 8
Fittings	Quarterly	10.2.4, 10.2.4.1
Fittings (rubber-gasketed)	Quarterly	10.2.4.1, A.10.2.4.1
Gravity tanks		10.2.10, Chapter 9
Hangers	Annually and after each system activation	10.2.4.2
Heat (deluge valve house)	Daily/weekly	10.2.1.5, Chapter 13
Nozzles	Annually and after each system activation	10.2.1.1, 10.2.1.2, 10.2.1.6, 10.2.5.1, 10.2.5.2
Pipe	Annually and after each system activation	10.2.1.1, 10.2.1.2, 10.2.4, 10.2.4.1
Pressure tank		10.2.10, Chapter 9
Steam driver		10.2.9, Chapter 8
Strainers	Manufacturer's instruction	10.2.7
Suction tanks		10.2.10, Chapter 9
Supports	Quarterly	10.2.1.1, 10.2.1.2, 10.2.4.2
Water supply piping		10.2.6.1, 10.2.6.2
UHSWSS — detectors	Monthly	10.4.2
UHSWSS — controllers	Each shift	10.4.3
UHSWSS — valves	Each shift	10.4.4
Operational Test		
Backflow preventer		Chapter 13
Check valves		Chapter 13
Control valves	Annually	13.3.3.1
Deluge valve		10.2.2, Chapter 13
Detection systems		*NFPA 72*
Detector check valve		Chapter 13
Electric motor		10.2.9, Chapter 8
Engine drive		10.2.9, Chapter 8
Fire pump		10.2.9, Chapter 8
Flushing	Annually	10.2.1.3, Section 10.3 (flushing of connection to riser, part of annual test)
Gravity tanks		10.2.10, Chapter 9
Main drain test	Annually	13.3.3.4
Manual release	Annually	10.2.1.3, 10.3.6
Nozzles	Annually	10.2.1.3, 10.2.1.6, Section 10.3

TABLE 10.1.1.2 Continued

Item	*Frequency*	*Reference*
Pressure tank		Section 10.2, Chapter 9
Steam driver		10.2.9, Chapter 8
Strainers	Annually	10.2.1.3, 10.2.1.7, 10.2.7
Suction tanks		10.2.10, Chapter 9
Waterflow alarm	Quarterly	Chapter 5
Water spray system test	Annually	Section 10.3, Chapter 13
Water supply flow test		7.3.1
UHSWSS	Annually	Section 10.4
Valve status test		13.3.1.2.1
Maintenance		
Backflow preventer		Chapter 13
Check valves		Chapter 13
Control valves	Annually	10.2.1.4, Chapter 13
Deluge valve		10.2.2, Chapter 13
Detection systems		*NFPA 72*
Detector check valve		Chapter 13
Electric motor		10.2.9, Chapter 8
Engine drive		10.2.9, Chapter 8
Fire pump		10.2.9, Chapter 8
Gravity tanks		10.2.10, Chapter 9
Pressure tank		10.2.6, Chapter 9
Steam driver		10.2.9, Chapter 8
Strainers	Annually	10.2.1.4, 10.2.1.6, 10.2.7
Strainers (baskets/screen)	5 years	10.2.1.4, 10.2.1.7, A.10.2.7
Suction tanks		10.2.10, Chapter 9
Water spray system	Annually	10.2.1.4, Chapter 13

10.1.2 Water Spray Protection. This chapter shall not cover water spray protection from portable nozzles, sprinkler systems, monitor nozzles, or other means of application.

As indicated in 10.1.1, Chapter 10 does not specifically describe the inspection, testing, and maintenance requirements applicable to systems with portable nozzles, monitor nozzles, or any means of application other than fixed nozzles. Nonetheless, many of the requirements in this chapter can be applied in the absence of another inspection, testing, and maintenance standard.

For example, monitor nozzles used in an aircraft hangar should be inspected, tested, and maintained in accordance with NFPA 409, *Standard on Aircraft Hangars.* However, monitor nozzles that are fed from a fire service underground main are covered in Chapter 7 of this standard.

Critical Point

Many of the requirements in this chapter can be applied in the absence of another inspection, testing, and maintenance standard.

10.1.3* Design and Installation. NFPA 15, *Standard for Water Spray Fixed Systems for Fire Protection,* shall be consulted to determine the requirements for design and installation, including acceptance testing.

FAQ Is the inspector required to have in-depth knowledge of NFPA 15?

The inspector of a water-based fire protection system is not required to have complete knowledge of the design requirements of the installation standard NFPA 15. However, that standard should be consulted as required by 10.1.3, and a certain level of knowledge is necessary to properly inspect and test a water spray fixed system. It is important to understand the intended design of the system so that the location or quantity of water does not accidentally get modified during what can be considered routine maintenance or component replacement.

For example, an inspector might not be able to determine if the nozzles are aligned correctly, because it depends on whether direct impingement over the entire surface of the structure, vessel, or equipment is required, or if the system is intended to protect the structure, vessel, or equipment from an exposure fire. An inspector familiar with NFPA 15 would be aware that, when protecting vessels to control burning, the nozzles are to be positioned to impinge directly on the vessel and on all areas around the vessel where a spill is likely to spread or accumulate; when protecting structural steel, water spray from the nozzles is required to impinge only on one side of the structural member, and rundown is anticipated.

A.10.1.3 Insulation acting in lieu of water spray protection is expected to protect a vessel or structure for the duration of the exposure. The insulation is to prevent the temperature from exceeding 850°F (454°C) for structural members and 650°F (393°C) for vessels. If the insulation is missing, the structure or vessel is not considered to be protected, regardless of water spray protection or insulation on other surfaces. To re-establish the proper protection, the insulation should be replaced or the water spray protection should be extended, using the appropriate density.

FAQ Why is it important to check insulation?

NFPA 15 allows encasing structural steel in fire-resistant insulating material approved by the authority having jurisdiction (AHJ), in lieu of protecting the steel with a water spray fixed system. Missing or damaged insulation has the same impact on the protection of the structure as does a totally clogged sprinkler nozzle. In cases of damaged or missing insulation, immediate corrective action should be taken. Therefore, it is important that the inspection of the water spray fixed system include a check of the integrity of any such insulation.

Critical Point

Missing or damaged insulation has the same impact on the protection of the structure as does a totally clogged sprinkler nozzle.

10.1.4 Obstruction Investigations. The procedures outlined in Chapter 14 shall be followed where there is a need to conduct an obstruction investigation.

10.1.5 Valves and Connections. Valves and fire department connections shall be inspected, tested, and maintained in accordance with Chapter 13.

10.1.6* Impairments. The procedures outlined in Chapter 15 shall be followed where an impairment to protection occurs.

A.10.1.6 The inspection, testing, and maintenance of water spray fixed systems can involve or result in a system that is out of service. Also see Chapter 15.

10.1.6.1 When a water spray fixed system or any portion thereof is out of service for any reason, notice shall be given to facility management, the local fire department, the on-site fire brigade, and other authorities having jurisdiction, as applicable.

Water spray fixed systems are typically employed in high-hazard or high-challenge areas. These are areas where the potential exists for fires of higher intensity than are ordinarily encountered in lower hazard occupancies such as office buildings or residential occupancies. These areas are also subject to more hazardous activities, such as welding, and the use of flammable or combustible liquids.

In addition to the system impairment requirements prescribed by Chapter 15, controls over hazardous activities and operations must be considered during any system impairment. The interconnection between the industrial processes being protected and the fire protection system should be considered during the design of the system and the preparation of the facility maintenance plan. When a system is shut down for maintenance or repair, the facility maintenance plan should account for the consequences of the planned system impairment and address the impaired condition.

Critical Point

When a system is shut down for maintenance or repair, the facility maintenance plan should account for the consequences of the planned system impairment and address the impaired condition.

10.1.6.2 A sign shall be posted at each fire department connection or system control valve indicating which portion of the system is out of service.

10.2 Inspection and Maintenance Procedures

10.2.1 Components. The components described in this section shall be inspected and maintained at the frequency specified in Table 10.1.1.2 and in accordance with this standard and the manufacturer's instructions.

As in other chapters of NFPA 25 that address systems, a table (Table 10.1.1.2) summarizes this chapter's inspection, testing, and maintenance requirements. The referenced sections provide specific information for each item and the associated activity.

10.2.1.1 Items in areas that are inaccessible for safety considerations due to factors such as continuous process operations and energized electrical equipment shall be inspected during each scheduled shutdown but not more than every 18 months.

The facility manager is obligated to notify the inspector when a shutdown occurs. In some cases, repairs or unscheduled maintenance will necessitate a shutdown, during which inspection, testing, and/or maintenance activities can be conducted. Where processes are rarely shut down, the facility manager must use reasonable judgment and take advantage of shutdowns when they occur. If inspection, testing, and maintenance have not been conducted for almost 18 months because no shutdown has occurred, a shutdown must be scheduled to allow these activities to take place.

10.2.1.2 Inspections shall not be required for items in areas with no provision for access and that are not subject to the conditions noted in 10.2.4.1, 10.2.4.2, and 10.2.5.1.

10.2.1.3 Items in areas that are inaccessible for safety considerations shall be tested at longer intervals in accordance with 13.4.3.2.2.4.

ITM Deficiency, Impairment, or Hazard Evaluation?

ANSWER: ITM Deficiency

(Courtesy of Jackie "JW" Ward and SimplexGrinnell)

The condition of excessive external corrosion shown in this photo was found on the piping installed in a cooling tower. Galvanized pipes and fittings were installed, and there is some minor corrosion occurring at the pipe threads and on the fittings. However, significant corrosion is occurring at the hanger locations and a sample is shown in the picture. The hanger was removed to show the extent of the corrosion.

Table 10.1.1.2 indicates that the piping of a water spray fixed system is to be inspected annually and after each activation of the system. The requirements in 10.2.1.1, 10.2.1.2, 10.2.4, 10.2.4.1 are referenced in the table, and 10.2.4.1 includes a list of the negative conditions the inspector is looking for. Item 2 in this list describes external conditions such as missing or damaged paint or coatings, rust, and corrosion. All of these conditions are found on the piping at all hanger locations during the inspection and should be recorded as a noncritical deficiency. However, time is of the essence to correct this deficiency, because the severity of corrosion could cause the pipe to start leaking and possibly rupture.

The cause of this accelerated corrosion at the hanger locations should be investigated as well. It would be appropriate for the inspector to recommend a hazard evaluation to determine the cause and any corrective action needed to correct it.

FAQ Why does NFPA 25 allow longer intervals between discharge tests for inaccessible systems?

Water spray fixed systems are often used to protect equipment in industrial occupancies and are intended to discharge a considerable amount of water, so performing a functional test requires coordination with the property owner or operator. Discharge tests must be planned carefully to avoid disruption of building or process operations, with further consideration given to potential damage to the building, equipment, or products. The standard allows a longer interval between discharge tests to facilitate this coordination for systems that are inaccessible.

NFPA 25 permits a second control valve to be installed above the deluge valve for the purpose of testing the valve and the detection system. Such tests are intended to provide only a partial test, and they do not meet the requirements to provide a full flow trip test where all nozzles can be verified for proper aim and position and internal piping obstructions can be identified.

10.2.1.4 Other maintenance intervals shall be permitted, depending on the results of the visual inspection and operating tests.

10.2.1.5 Deluge valve enclosures shall be inspected in accordance with the provisions of Chapter 13.

10.2.1.6 Nozzle strainers shall be removed, inspected, and cleaned during the flushing procedure for the mainline strainer.

Nozzle strainers are installed only with certain types of nozzles. When a nozzle strainer is present, it will usually need cleaning as required by 10.2.1.6, since it is intended to capture obstructing material and prevent the plugging of the nozzle. Exhibit 10.7 shows a typical fire protection strainer with a basket-type screen of corrosion-resistant metal.

10.2.1.7 Mainline strainers shall be removed and inspected every 5 years for damaged and corroded parts.

10.2.2 Deluge Valves. Deluge valves shall be inspected, tested, and maintained in accordance with Chapter 13.

10.2.3 Automatic Detection Equipment.

10.2.3.1 Automatic detection equipment shall be inspected, tested, and maintained in accordance with *NFPA 72, National Fire Alarm and Signaling Code.*

EXHIBIT 10.7 *Strainer with Basket-Type Screen.*

Detection systems, which include smoke, heat, infrared, and linear cable systems, are used as an actuating means for water spray fixed systems. Detection systems are not covered in detail in NFPA 25. They are usually installed in accordance with *NFPA 72* and should be maintained by a qualified electrician in accordance with that standard, as required by Chapter 14.

An electronic detection system functions as an actuation mechanism by sending a signal via a local fire alarm or deluge system panel to a solenoid valve mounted on the deluge system trim piping. Signaling the solenoid to open and release water pressure from the alarm line and deluge valve diaphragm allows the deluge valve to open. The solenoid can be configured to either energize to open or de-energize to open (providing fail-safe operation). Exhibit 10.8 shows a detection system being tested by applying a heat source to the solenoid valve.

EXHIBIT 10.8 *Testing a Heat Detector on a Water Spray System.*

A wet pilot system is composed of a system of ½ in. (15 mm) pipe with sprinklers attached, similar to a wet pipe sprinkler system. The system is pressurized to normal static pressure. Sprinklers in this case are used as detectors. When a sprinkler activates, pressure in the system is released to the atmosphere. This reduced pressure allows the deluge valve diaphragm to open the deluge valve.

Wet pilot systems are usually provided with an inspector's test connection in the form of a ½ in. (15 mm) ball valve. Opening the inspector's test connection releases pressure and activates the system. Wet pilot systems are used where freezing is not a concern.

Dry pilot systems operate in a similar manner, except that the system is filled with air, as in a dry pipe sprinkler system. When a sprinkler activates in a dry pilot system, air pressure is released and a device such as a dry pilot actuator opens and bleeds water from the diaphragm on the deluge valve, allowing the deluge valve to open.

A dry pilot system is normally equipped with an inspector's test connection in the form of a ½ in. (15 mm) ball valve. Dry pilot systems are typically provided where freezing is a concern.

EXHIBIT 10.9 *Dry Pilot Actuation Trim Arrangement.*

Inspection, testing, and maintenance of pilot sprinkler systems should be conducted according to the requirements for sprinkler systems in Chapter 5 of NFPA 25. Regardless of whether the pilot system is wet or dry, spare sprinklers should be provided, as discussed in 5.2.1.4.

The air pressure for a dry pilot system can be supplied from either plant air or from a nitrogen supply. Exhibit 10.9 illustrates the proper connections for air supplies.

10.2.3.2 Automatic fire detection equipment not covered by *NFPA 72* shall be inspected, tested, and maintained to ensure that the detectors are in place, securely fastened, and protected from corrosion, weather, and mechanical damage and that the communication wiring, control panels, or tubing system is functional.

10.2.4* System Components. System piping, fittings, hangers, and supports shall be inspected and maintained to ensure continuity of water delivery to the spray nozzles at full waterflow and design pressure.

Due to the forces exerted on the pipe and pipe supports from the water spray coming from a system of open nozzles, water spray fixed systems are prone to water hammer and pipe vibration. It is therefore extremely important to inspect the integrity of the piping, the hanger or supports, and the nozzles.

A.10.2.4 The operation of the water spray system is dependent on the integrity of the piping, which should be kept in good condition and free of mechanical damage. The pipe should not be used for support of ladders, stock, or other material. Where piping is subject to a corrosive atmosphere, a protective corrosion-resistant coating should be provided and maintained. Where the age or service conditions warrant, an internal examination of the piping should be made. Where it is necessary to flush all or part of the piping system, this work should be done by sprinkler contractors or other qualified workers.

In many cases, pipe in water spray fixed systems is installed near equipment, rather than in a more protected spot near the roof or ceiling, as in the case of sprinkler systems. Therefore, an inspection of piping is essential to ensure that pipe, fittings, and pipe supports are not damaged.

Usually, where piping is installed in corrosive atmospheres, the pipe, fittings, and pipe supports are galvanized. The galvanization protects against corrosion, which could lead to the eventual failure of the component. Inspections should be made to ensure that the galvanizing has not been removed through abrasion or any other means. As with sprinkler systems, water spray fixed systems must be supported in accordance with NFPA 13, *Standard for the*

Installation of Sprinkler Systems. NFPA 13 requires that pipe supports other than those illustrated in the standard be designed to support five times the weight of the water-filled pipe plus a 250 lb (113.4 kg) safety factor. Otherwise, additional weight from ladders, stock, or other material could place a load beyond that designed for the pipe support.

10.2.4.1* Piping and Fittings. System piping and fittings shall be inspected for the following:

(1) Mechanical damage (e.g., broken piping or cracked fittings)

Piping in a water spray fixed system is installed primarily in areas where production processes, machinery, and other equipment can come in contact with it. Therefore, it is more susceptible to mechanical damage and misalignment than piping in most other water-based fire protection systems.

Supporting detection system pipe or conduit is a common practice in the installation of water spray fixed systems. This practice is acceptable because the detection system is a component of the overall fire protection system. A release line hanger is typically used for hanging detection piping from the water piping. Angle iron or angle struts can also be used for this application.

No other items are permitted to be attached to or hung from the water spray fixed system piping.

(2) External conditions (e.g., missing or damaged paint or coatings, rust, and corrosion)

Many water spray fixed systems are installed outside or in atmospheres where corrosion is a common problem. Although it is required that piping be protected, rust, damaged paint, or other signs of corrosion frequently are found upon inspection.

(3) Misalignment or trapped sections

In many installations, water spray fixed system piping that surrounds equipment or vessels is installed in sections. This configuration allows the piping to be disassembled if the equipment or vessels need to be removed for repair or replacement. When such disassembly takes place, piping sections can be reinstalled incorrectly. The inspection should assess whether this has happened.

(4) Low-point drains (automatic or manual)

Water spray fixed systems are often installed outdoors, and if they are exposed to freezing temperatures, these systems can freeze. Low-point drains and/or trapped sections of pipe should be inspected, particularly prior to the onset of cold weather. The guidance provided in A.13.4.4.3.2 on removing water from a dry system is also good practice to follow for a water spray fixed system or any system that is expected to be maintained in a dry condition.

(5) Location of rubber-gasketed fittings

Mechanical damage to fittings is more prevalent in a water spray system than in a wet or dry pipe system, because the pipe and fittings can be exposed to the hazard that is protected by the water spray system. For example, if pipe and fittings are exposed to heat prior to system actuation, the sudden contraction of cool water entering a hot pipe could fracture the pipe or fitting. Inspection of a water spray system should include a check for this type of failure.

A.10.2.4.1 Rubber-gasketed fittings in the fire areas are inspected to determine whether they are protected by the water spray or other approved means. Unless properly protected, fire could cause loss of the rubber gasket following excessive leakage in a fire situation.

NFPA 15 allows the use of rubber-gasketed fittings in fire areas if those fittings are also protected by a water spray fixed system automatically controlled through the use of a detection system.

Spray from the open nozzles on the system is expected to keep the gaskets from melting or disfiguring during a fire. These fittings may be protected by an adjacent system, provided that system is activated by a detection system in the same fire area as the fittings.

Rubber gaskets are normally found in the following two types of fittings that can be installed in a water spray fixed system: flanged fittings and grooved fittings.

When using flanged fittings and/or joints, a rubber gasket, usually red or black, is sandwiched between every two flanges to prevent leakage. These gaskets can be either the ring type that will be visible inside the ring of bolts holding the flanges together, or the full-face type that extends beyond the bolts to the outer edge of the flanges. There is a metallic alternative to rubber gaskets that can be used with flanges. These metallic gaskets, which are normally silver or gray, do not need protection when installed in a fire area.

Grooved fittings use rubber gaskets that are installed around the outside of the pipe in a coupling. These gaskets are not normally visible during an inspection. To date, there is no alternative to rubber gaskets in grooved fittings. Therefore, a water spray fixed system, or other method in accordance with NFPA 15, must protect all grooved fittings or couplings in a fire area.

10.2.4.2* Hangers and Supports. Hangers and supports shall be inspected for the following and repaired or replaced as necessary:

(1) Condition (e.g., missing or damaged paint or coating, rust, and corrosion)
(2) Secure attachment to structural supports and piping
(3) Damaged or missing hangers

A.10.2.4.2 Hangers and supports are designed to support and restrain the piping from severe movement when the water supply operates and to provide adequate pipe slope for drainage of water from the piping after the water spray system is shut down. Hangers should be kept in good repair. Broken or loose hangers can put undue strain on piping and fittings, cause pipe breaks, and interfere with proper drainage of the pipe. Broken or loose hangers should be replaced or refastened.

10.2.5* Water Spray Nozzles.

A.10.2.5 Systems need inspection to ensure water spray nozzles effectively discharge water unobstructed onto surfaces to be protected from radiant heat (exposure protection) or onto flaming surfaces to extinguish or control combustion. Factors affecting the proper placement of water spray nozzles include the following:

(1) Changes or additions to the protected area that obstruct existing nozzles or require additional coverage for compliance
(2) Removal of equipment from the protected area that results in nozzle placement at excessive distances from the hazard
(3) Mechanical damage or previous flow tests that have caused nozzles to be misdirected
(4) A change in the hazard being protected that requires more or different nozzles to provide adequate coverage for compliance

Spray nozzles can be permitted to be placed in any position necessary to obtain proper coverage of the protected area. Positioning of nozzles with respect to surfaces to be protected, or to fires to be controlled or extinguished, should be guided by the particular nozzle design and the character of water spray produced. In positioning nozzles, care should be taken that the water spray does not miss the targeted surface and reduce the efficiency or calculated discharge rate.

Critical Point

Automatic water spray nozzles are intended to open automatically by operation of a heat-responsive element, which keeps the discharge orifice closed by the exertion of force on a cap (i.e., button or disc).

The following two basic types of nozzles are used on water spray fixed systems: automatic water spray nozzles and non-automatic water spray nozzles. The requirements of 10.2.5 apply to both types of nozzles.

Automatic water spray nozzles are intended to open automatically by operation of a heat-responsive element, which keeps the discharge orifice closed by the exertion of force on a cap (i.e., button or disc). When discharging water under pressure, the nozzle distributes the water in a specific, directional pattern similar to a sprinkler. In fact, automatic water spray nozzles are sometimes mistaken for sprinklers. Exhibit 10.10 shows typical automatic spray nozzles.

A non-automatic, or open, water spray nozzle is an open water discharge device which, when discharging water under pressure, will distribute the water in a specific, directional pattern. Some non-automatic nozzles resemble an operated sprinkler, while others appear very different. Exhibit 10.11 illustrates a low velocity non-automatic spray nozzle. Exhibit 10.12 shows a high velocity non-automatic spray nozzle in both its inert and activated states.

Critical Point

A non-automatic, or open, water spray nozzle is an open water discharge device which, when discharging water under pressure, will distribute the water in a specific, directional pattern.

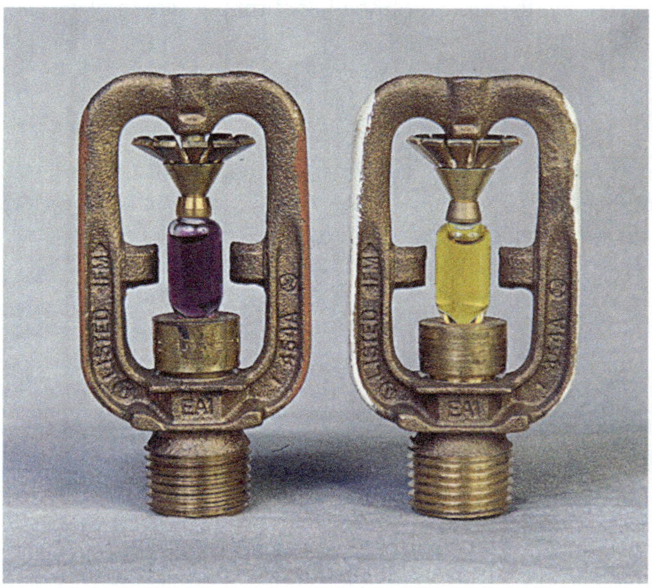

EXHIBIT 10.10 *Automatic Spray Nozzles.*

EXHIBIT 10.11 *Low Velocity Non-Automatic Water Spray Nozzles.*

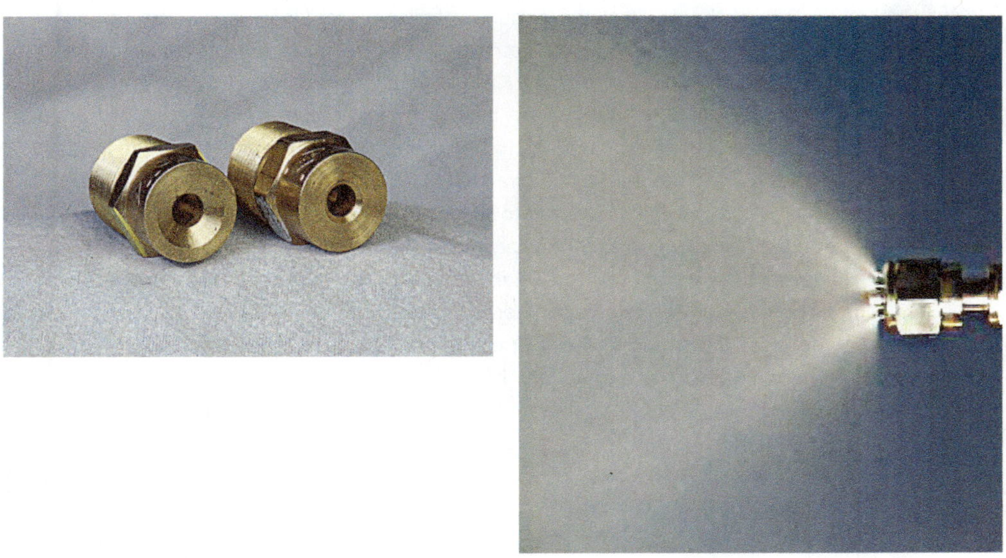

EXHIBIT 10.12 *High Velocity Non-Automatic Spray Nozzles (left) and Activated High-Velocity Non-Automatic Spray Nozzle (right).*

10.2.5.1 Water spray nozzles shall be inspected and maintained to ensure that they are in place, continue to be aimed or pointed in the direction intended, and are free from external loading and corrosion.

The requirements of 10.2.5.1 are important because water spray nozzles are subject to corrosion from the effects of weather or exposure to corrosive chemicals. When water spray nozzles are installed in harsh environments, corrosion protection in the form of wax or lead coatings applied by the nozzle manufacturer are permitted to be used.

A problem commonly seen with the use of grooved fittings is the misalignment of piping sections due to the thrust from the nozzles, which causes the pipe to rotate. If this condition is discovered during an inspection, nozzles should be realigned and additional restraints added.

10.2.5.2 Where caps or plugs are required, the inspection shall confirm they are in place and free to operate as intended.

FAQ Does NFPA 25 require systems to use plugs or caps?

Prior to an inspection, the inspector must determine if plugs or caps are required for the system, since not all systems require them. Bees and hornets have been known to build nests inside open nozzles and sprinklers, creating obstructions. Dust and dirt can also find their way into nozzles and piping. If nests or other obstructing material are found during an inspection on a system that does not require them, consideration should be given to the use of listed plastic

ITM Deficiency, Impairment, or Hazard Evaluation?

(Courtesy of Dave Smith and SimplexGrinnell)

ANSWER: ITM Deficiency or Impairment

Chapter 10 of NFPA 25 covers a group of systems installed to protect specialized hazards such as processing equipment and structures, flammable liquid and gas vessels, piping, and equipment such as transformers, oil switches, and motors. In this photo, the system protecting a transformer is being annually tested as required by Section 10.3.

One of the goals of the annual water spray test is to make sure the water spray nozzles "... continue to be aimed or pointed in the direction intended . . . ," as required in 10.2.5.1, and that ". . . water discharge patterns from all of the open spray nozzles shall be observed . . . to ensure that nozzles are correctly positioned, and to ensure that obstructions do not prevent discharge patterns from wetting surfaces to be protected," as required in 10.3.3.3.1.

In 10.1.3, the user is referred to NFPA 15 to determine the requirements for the design and installation as they apply to the requirements of NFPA 25. This does not mean that the design has to be verified or the installation has to be inspected to comply with NFPA 15. However, there are design objectives in NFPA 15 that must be understood and applied when inspecting and testing per NFPA 25.

For transformer protection, NFPA 15 requires ". . . complete water spray impingement on all exposed exterior surfaces." In this picture, the spray from the nozzles is not reaching the surfaces between the cooling fins on the side of the transformer facing the photographer. If there were small dry areas requiring minor adjustments to the alignment of the nozzles, then describing this as a deficiency would probably be appropriate. However, in this case an entire side of the transformer is not getting direct impingement, which would elevate this situation to the level of impairment.

plugs or nozzle caps as covered in 10.2.5.2. Only plastic plugs or caps listed for use with the nozzle can be used. Exhibit 10.13 shows an example of a nozzle with such a cap. The use of any plug or cap should be reviewed against the manufacturer's data to ensure that their inclusion does not prevent the system from operating as designed.

Critical Point

The use of any plug or cap should be reviewed against the manufacturer's data to ensure that their inclusion does not prevent the system from operating as designed.

10.2.5.3 Misaligned water spray nozzles shall be adjusted (aimed) by visual means, and the discharge patterns shall be inspected at the next scheduled flow test.

When adjusting misaligned water spray nozzles as required by 10.2.5.3, if it is not obvious in what direction the nozzle should be aimed, that information should be obtained from the as-built drawings for the system.

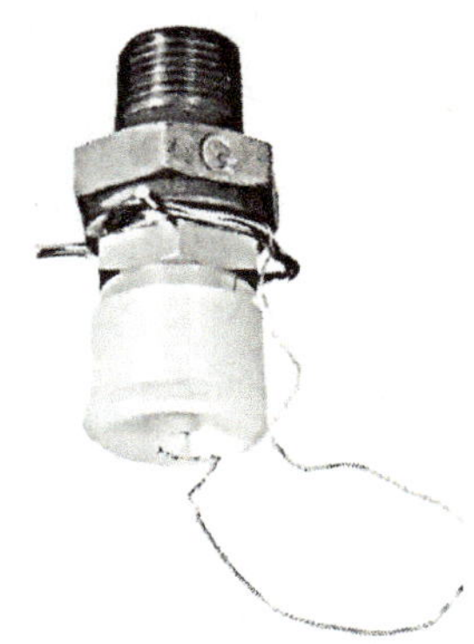

EXHIBIT 10.13 *Nozzle with Blow-Off Cap. (Source: Fire Protection Handbook, 2003, Figure 10.15.11)*

10.2.6 Water Supply.

10.2.6.1 The dependability of the water supply shall be ensured by regular inspection and maintenance, whether furnished by a municipal source, on-site storage tanks, a fire pump, or private underground piping systems.

10.2.6.2* Water supply piping shall be maintained free of internal obstructions.

A.10.2.6.2 Water supply piping should be free of internal obstructions that can be caused by debris (e.g., rocks, mud, tubercles) or by closed or partially closed control valves. See Chapter 5 for inspection and maintenance requirements.

It is important to ensure that water supplies remain free of obstructing materials because nozzles can be easily obstructed. Therefore, it is critical to comply with the inspection and maintenance requirements of Chapter 7.

10.2.7* Strainers.

A.10.2.7 Mainline strainers should be removed and inspected for damaged and corroded parts every 5 years.

10.2.7.1 Mainline strainers (basket or screen) shall be flushed until clear after each operation or flow test.

Listed strainers are equipped with flushing connections that allow back-flushing of the strainer basket (see Exhibit 10.7). If flushing as required by 10.2.7.1 reveals excessive obstructing material, the basket can be removed easily for visual inspection and cleaning.

10.2.7.2 Individual water spray nozzle strainers shall be removed, cleaned, and inspected after each operation or flow test.

10.2.7.3 All strainers shall be inspected and cleaned in accordance with the manufacturer's instructions.

10.2.7.4 Damaged or corroded parts shall be replaced or repaired.

10.2.8 Drainage. The area beneath and surrounding a water spray fixed system shall be inspected visually on a quarterly basis to ensure that drainage facilities, such as trap sumps and drainage trenches, are not blocked and retention embankments or dikes are in good repair.

10.2.9 Fire Pumps. Chapter 8 shall be followed for inspection and maintenance requirements.

10.2.10 Water Tanks (Gravity, Pressure, or Suction Tanks, or Reservoirs). Chapter 9 shall be followed for inspection and maintenance requirements.

10.3 Operational Tests

Unlike many of the operational tests required by NFPA 25, the operational tests of a water spray fixed system require the inspector to have information about the design of the system.

Critical Point

Unlike many of the operational tests required by NFPA 25, the operational tests of a water spray fixed system require the inspector to have information about the design of the system.

For example, the inspector might need to review the as-built working plans to determine the intended nozzle discharge patterns and to identify the surfaces that are intended to be wetted. As another example, the inspector will need the original system design pressures to properly evaluate the pressure readings required in 10.3.3.4. If this information is not available, an evaluation of the system design criteria might be necessary to adequately test the system.

While NFPA 25 does not mandate that a specific form be used to record the information obtained while conducting the tests outlined in Section 10.3, Exhibit 10.14 is one example of a form that can be used to document these tests.

10.3.1 Performance.

10.3.1.1 Frequency of system tests shall be in accordance with Table 10.1.1.2.

10.3.1.2 Water spray fixed systems shall be serviced in accordance with this standard and with the manufacturer's instructions.

•

10.3.2* Test Preparation. Precautions shall be taken to prevent damage to property during the test.

A.10.3.2 The property owner or designated representative should take care to prevent damage to equipment or the structure during the test. Damage could be caused by the system discharge or by runoff from the test site. It should be verified that there is adequate and unobstructed drainage. Equipment should be removed or covered as necessary to prevent damage. Means such as curbing or sandbagging should be used to prevent entry of the water.

A test of a water spray fixed system can result in the discharge of water into the environment. Proper procedures for the disposal of test water must be followed based on local, state, and national environmental policies.

10.3.3 Operational Test Performance. Operational tests shall be conducted to ensure that the water spray fixed systems respond as designed, both automatically and manually.

For procedures necessary to satisfy the operational test requirement in 10.3.3, see 13.4.3.2, which details the testing requirements for deluge valves.

10.3.3.1* Response Time.

A.10.3.3.1 Test methods are as follows:

(1) Some detection circuits can be permitted to be deliberately desensitized in order to override unusual ambient conditions. In such cases, the response required in 10.3.3.1 can be permitted to be exceeded.
(2) Testing of integrating tubing systems can be permitted to be related to this test by means of a standard pressure impulse test specified by the listing laboratory.
(3) One method of testing heat detection uses a radiant heat surface at a temperature of 300°F (149°C) and a capacity of 350 watts at a distance of 1 in. (25 mm) but not more than 2 in. (50 mm) from the nearest part of the detector. This method of testing with an electric test set should not be used in hazardous locations. Other test methods can be permitted to be employed, but the results should be obtained under these conditions.

10.3.3.1.1 Under test conditions, the heat detection systems, where exposed to a heat test source, shall operate within 40 seconds.

10.3.3.1.2 Under test conditions, the flammable gas detection system, where exposed to a standard test gas concentration, shall operate within the time frame specified in the system design.

10.3.3.1.3 These response times shall be recorded.

TESTING OF WATER SPRAY SYSTEMS

Name of Property: ______________________ **Inspector:** ______________

Address: ______________________ **Contract No.:** ______________

Phone Number: ______________________ **Date:** ______________

This Report Covers: ❑ **Quarterly** ❑ **Semi-annual** ❑ **Annual**

Test

Quarterly

❑ Yes ❑ No ❑ N/A Alarm devices—water motor gong

❑ Yes ❑ No ❑ N/A Main drain test—if sole supply is through a backflow preventer

Static psi ______________ Residual psi ______________

❑ Yes ❑ No ❑ N/A Do results differ by more than 10 percent from previous test?

❑ Yes ❑ No ❑ N/A Deluge valve priming water—test level

❑ Yes ❑ No ❑ N/A Low air alarm—test per manufacturer's instructions

Semi-annual

❑ Yes ❑ No ❑ N/A Supervisory switch functions

❑ Yes ❑ No ❑ N/A Alarm devices—inspector's test or bypass opened/observed waterflow

Annual

❑ Yes ❑ No ❑ N/A Main drain test

Static psi ______________ Residual psi ______________

❑ Yes ❑ No ❑ N/A Do results differ by more than 10 percent from previous test?

❑ Yes ❑ No ❑ N/A All control valves operated through full range of motion and returned to normal position

Full Flow Trip Test

❑ Yes ❑ No ❑ N/A Unobstructed discharge from all nozzles

❑ Yes ❑ No ❑ N/A Pressure reading at most remote nozzle: __________ psi

❑ Yes ❑ No ❑ N/A Nozzle spray patterns and direction verified

❑ Yes ❑ No ❑ N/A Air maintenance device functions correctly

System Response Time

❑ Yes ❑ No ❑ N/A Heat detection responded in __________ sec

❑ Yes ❑ No ❑ N/A Flammable gas detection responded in __________ sec

❑ Yes ❑ No ❑ N/A Mainline strainer flushed after trip test

❑ Yes ❑ No ❑ N/A Backflow preventer/backflow test

❑ Yes ❑ No ❑ N/A Detection system tested in accordance with *NFPA 72*

Comments

Signature: ______________________ Date: ______________

License/Certification No.: ______________________

NFPA NFPA Water-Based Fire Protection Systems Forms

EXHIBIT 10.14 *Sample Form for Testing of Water Spray Systems.*

Critical Point

The observation of water discharge patterns during the discharge test required by 10.3.3.3.1 is intended to verify not only that there are no obstructions but also that the water spray pattern attains complete impingement on the surface to be protected.

10.3.3.2 Discharge Time. The time lapse between operation of detection systems and water delivery time to the protected area shall be recorded.

10.3.3.3* Discharge Patterns.

A.10.3.3.3 Spray nozzles can be of different sizes and types. Some are more subject to internal obstructions than others.

10.3.3.3.1* The water discharge patterns from all of the open spray nozzles shall be observed to ensure that patterns are not impeded by plugged nozzles, to ensure that nozzles are correctly positioned, and to ensure that obstructions do not prevent discharge patterns from wetting surfaces to be protected.

The observation of water discharge patterns during the discharge test required by 10.3.3.3.1 is intended to verify not only that there are no obstructions but also that the water spray pattern attains complete impingement on the surface to be protected.

A.10.3.3.3.1 See 13.4.3.2.2.3.

10.3.3.3.1.1 Where the nature of the protected property is such that water cannot be discharged, the nozzles shall be inspected for proper orientation and the system tested with air to ensure that the nozzles are not obstructed.

10.3.3.3.2 Where obstructions occur, the piping and nozzles shall be cleaned and the system retested.

10.3.3.4 Pressure Readings.

10.3.3.4.1 Pressure readings shall be recorded at the hydraulically most remote nozzle to ensure the waterflow has not been impeded by partially closed valves or by plugged strainers or piping.

The proper functioning of a water spray nozzle is based on the nozzle pressure. To achieve the intended water spray pattern, pressures of either 15 psi (1 bar) or 30 psi (2.1 bar) are necessary for most nozzles. Evaluation of the most hydraulically demanding nozzle ensures that all other nozzles in the system will experience at least equivalent — if not higher — pressures.

10.3.3.4.2 A second pressure reading shall be recorded at the deluge valve to ensure the water supply is adequate.

10.3.3.4.3 Readings shall be compared to the hydraulic design pressures to ensure the original system design requirements are met and the water supply is adequate to meet the design requirements.

The residual pressure measured at the deluge valve should be compared to the calculated pressure as indicated in the hydraulic calculations. Any pressure in excess of that calculated pressure is cause for concern and should be sufficient to initiate a complete evaluation of the piping system.

10.3.3.4.3.1 Where the hydraulically most remote nozzle is inaccessible, nozzles shall be permitted to be checked visually without taking a pressure reading on the most remote nozzle.

10.3.3.4.3.2 Where the reading taken at the riser indicates that the water supply has deteriorated, a gauge shall be placed on the hydraulically most remote nozzle and the results compared with the required design pressure.

10.3.4 Multiple Systems. The maximum number of systems expected to operate in case of fire shall be tested simultaneously to inspect the adequacy of the water supply.

10.3.5 Manual Operation. Manual actuation devices shall be operated annually.

10.3.6 Return to Service. After the full flow test, the water spray system shall be maintained and returned to service in accordance with the manufacturer's instructions.

•

10.3.6.1 Low Point Drains.

10.3.6.1.1 To prevent freezing and corrosion, all low point drains in aboveground piping shall be opened, the pipe drained, and the valves closed and plugs replaced.

10.3.6.1.2 Where weep holes are provided in lieu of low-point drains, they shall be inspected to ensure they are clear and unobstructed.

The weep holes mentioned in 10.3.6.1.2 are usually holes of $^{3}/_{16}$ in. (5 mm) in diameter drilled into fittings to create an automatic drain for trapped sections of pipe. The presence of these drains might not be obvious, but once found they must be carefully inspected.

10.4 Ultra-High-Speed Water Spray System (UHSWSS) Operational Tests

An ultra-high-speed water spray system is a type of automatic water spray system in which water spray is rapidly applied to protect specific hazards in areas where deflagrations are anticipated. Ultra-high-speed water spray systems are intended to operate in less than 100 milliseconds. Specialized training and experience are necessary to work on these systems.

10.4.1 A full operational test, including measurements of response time, shall be conducted at intervals not exceeding 1 year.

10.4.1.1 Systems out of service shall be tested before being placed back in service.

10.4.2 All detectors shall be tested and inspected monthly for physical damage and accumulation of deposits on the lenses of optical detectors.

10.4.3 Controllers shall be inspected for faults at the start of each working shift.

10.4.4 Valves.

10.4.4.1 Valves on the water supply line shall be inspected at the start of each working shift to verify they are open.

10.4.4.2 Valves secured in the open position with a locking device or monitored by a signaling device that sounds a trouble signal at the deluge system control panel or other central location shall not require inspection.

10.4.5 Response Time.

10.4.5.1 The response time shall be verified during the operational test.

10.4.5.2 The response time shall be in accordance with the requirements of the system but not more than 100 milliseconds.

10.5 Component Action Requirements

FAQ Is an acceptance test required whenever maintenance is performed on a system component?

Component replacement tables provide guidance when system components are adjusted, repaired, rebuilt, or replaced. It is not necessary in each case to require a complete acceptance test for each component when maintenance is performed.

10.5.1 Whenever a component in a water spray fixed system is adjusted, repaired, reconditioned, or replaced, the action required in Table 10.5.1 shall be performed.

TABLE 10.5.1 *Summary of Component Replacement Action Requirements*

Component	*Adjust*	*Repair/ Recondition*	*Replace*	*Required Action*
Water Delivery Components				
Pipe and fittings	X	X	X	Operational flow test
Nozzles	X	X	X	Operational flow test
Manual release	X	X	X	(1) Operational test (2) Check for leaks at system working pressure (3) Test all alarms
Fire department connections				See Chapter 13
Alarm and Supervisory Components				
Pressure switch–type waterflow	X	X	X	Operational test using inspector's test connection
Water motor gong	X	X	X	Operational test using inspector's test connection
Valve supervisory device	X	X	X	Test for conformance with NFPA 15, *Standard for Water Spray Fixed Systems for Fire Protection*, and/or *NFPA 72, National Fire Alarm and Signaling Code*
Detection system	X	X	X	Operational test for conformance with NFPA 15 and/or *NFPA 72*
Status-Indicating Components				
Gauges			X	Verify at 0 psi (0 bar) and system working pressure
Testing and Maintenance Components				
Main drain	X	X	X	Full flow main drain test
Auxiliary drains	X	X	X	(1) Inspect for leaks at system working pressure (2) Main drain test
Structural Components				
Hanger/seismic bracing	X	X	X	Inspect for conformance with NFPA 15 and/or NFPA 13, *Standard for the Installation of Sprinkler Systems*
Pipe stands	X	X	X	Inspect for conformance with NFPA 15 and/or NFPA 13
Informational Components				
Identification signs	X	X	X	Inspect for conformance with NFPA 15

10.5.2 Where the original installation standard is different from the cited standard, the use of the appropriate installing standard shall be permitted.

•

10.5.3 The actions of 10.5.1 shall not require a design review, which is outside the scope of this standard.

References Cited in Commentary

National Fire Protection Association, 1 Batterymarch Park, Quincy, MA 02169-7471.

Cote, A. E., ed., *Fire Protection Handbook®*, 19th edition, 2003.

Hague, D. R., *Fire Protection Systems for Special Hazards*, 2004 edition.

NFPA 13, *Standard for the Installation of Sprinkler Systems*, 2013 edition.

NFPA 15, *Standard for Water Spray Fixed Systems for Fire Protection*, 2012 edition.

NFPA 72®, *National Fire Alarm and Signaling Code*, 2013 edition.

NFPA 409, *Standard on Aircraft Hangars*, 2011 edition.

11

FOAM-WATER SPRINKLER SYSTEMS

Chapter 11 of NFPA 25 covers the inspection, testing, and maintenance (ITM) of foam-water sprinkler and foam-water spray systems. Foam-water systems are used in extra hazard areas containing flammable or combustible liquids. As is the case with water spray fixed systems, the proper inspection, testing, and maintenance of foam-water systems requires specialized training and experience. In addition to the requirements in this chapter, the manufacturer's instructions for inspection, testing, and maintenance should be followed.

Foam-water sprinkler systems are installed to protect structures, equipment, and facilities where a potential hazard for a two-dimensional flammable liquid fire exists. These systems can provide prevention, extinguishment, control, and/or exposure protection, depending on the application. Typically, foam-water sprinkler systems are installed in aircraft hangars, petrochemical plants, tank farms, fuel-loading facilities, and power plants.

Protein, fluoroprotein, and aqueous film-forming concentrates or film-forming fluoroprotein foam concentrates are suitable for use with foam-water sprinklers. The latter type of foam concentrate also has proven suitable for use with standard sprinklers of the type referred to in NFPA 13, *Standard for the Installation of Sprinkler Systems*, where the system is provided with the necessary foam concentrate proportioning equipment. But foam products are tested to specific sprinklers, and care should be exercised to ensure that the chosen concentrate and discharge device installed are listed for use together.

Case In Point

The uses of foam for fire protection have increased greatly since its first use in the 1900s. Early foam development and its use as a fire extinguishing agent started in the petroleum industry, as people in the industry began to recognize the special hazard of combustible and flammable liquids. In the 1930s, the U.S. Navy realized the need for improvement on the cumbersome, early foam systems and initiated further research. The U.S. military and numerous petrochemical companies continued this research for several years. The concept of using foam in ceiling-mounted systems was researched, developed, and tested through the 1950s. In 1961, Factory Mutual Research Corporation (now FM Global) began conducting tests of high-expansion foam systems for possible protection in various industries. That same year, the National Board of Fire Underwriters and the U.S. Naval Research Laboratory showed the effectiveness of foam-water sprinklers in test demonstrations, and NFPA published NFPA 16, *Standard for the Installation of Foam-Water Sprinkler and Foam-Water Spray Systems,* which was the association's first standard on foam-water sprinklers. In 1969, testing conducted by FM Global and Grinnell (now SimplexGrinnell) showed the positive effects of foam-water deluge systems on jet fuel fires, resulting in the formation of the committee for NFPA 409, *Standard on Aircraft Hangars,* which recommended these systems over regular water-based sprinkler systems (Richardson 2003).

11.1 General

11.1.1 Minimum Requirements.

11.1.1.1 This chapter shall provide the minimum requirements for the routine inspection, testing, and maintenance of foam-water systems.

11.1.1.2 Table 11.1.1.2 shall be used to determine the minimum required frequencies for inspection, testing, and maintenance.

TABLE 11.1.1.2 *Summary of Foam-Water Sprinkler System Inspection, Testing, and Maintenance*

System/Component	*Frequency*	*Reference*
Inspection		
Discharge device location (sprinkler)	Annually	11.2.5
Discharge device location (spray nozzle)	Monthly	11.2.5
Discharge device position (sprinkler)	Annually	11.2.5
Discharge device position (spray nozzle)	Monthly	11.2.5
Foam concentrate strainer(s)	Quarterly	11.2.7.2
Drainage in system area	Quarterly	11.2.8
Proportioning system(s) — all	Monthly	11.2.9
Pipe corrosion	Annually	11.2.3
Pipe damage	Annually	11.2.3
Fittings corrosion	Annually	11.2.3
Fittings damage	Annually	11.2.3
Hangers/supports	Annually	11.2.4
Waterflow devices	Quarterly	11.2.1
Water supply tank(s)		Chapter 9
Fire pump(s)		Chapter 8
Water supply piping		11.2.6.1
Control valve(s)	Weekly/monthly	—
Deluge/preaction valve(s)		11.2.1, Chapter 13
Detection system	See *NFPA 72, National Fire Alarm and Signaling Code*	11.2.2
Test		
Discharge device location	Annually	11.3.2.6
Discharge device position	Annually	11.3.2.6
Discharge device obstruction	Annually	11.3.2.6
Foam concentrate strainer(s)	Annually	11.2.7.2
Proportioning system(s) — all	Annually	11.2.9
Complete foam-water system(s)	Annually	11.3.3
Foam-water solution	Annually	11.3.5
Manual actuation device(s)	Annually	11.3.4
Backflow preventer(s)	Annually	Chapter 13
Fire pump(s)	See Chapter 8	—
Waterflow devices	Quarterly/semiannually	11.3.1.3
Water supply piping	Annually	Chapter 10
Control valve(s)	See Chapter 13	—
Strainer(s) — mainline	See Chapter 10	11.2.7.1
Deluge/preaction valve(s)	See Chapter 13	11.2.1
Detection system	See *NFPA 72*	11.2.2
Backflow preventer(s)	See Chapter 13	—
Water supply tank(s)	See Chapter 9	—

TABLE 11.1.1.2 Continued

System/Component	*Frequency*	*Reference*
Water supply flow test	5 years	7.3.1
Valve status test		13.3.1.2.1
Maintenance		
Foam concentrate pump operation	Monthly	11.4.6.1, 11.4.7.1
Foam concentrate strainer(s)	Quarterly	Section 11.4
Foam concentrate samples	Annually	11.2.10
Proportioning system(s) standard pressure type		
Ball drip (automatic type) drain valves	5 years	11.4.3.1
Foam concentrate tank — drain and flush	10 years	11.4.3.2
Corrosion and hydrostatic test	10 years	11.4.3.3
Bladder tank type		
Sight glass	10 years	11.4.4.1
Foam concentrate tank — hydrostatic test	10 years	11.4.4.2
Line type		
Foam concentrate tank — corrosion and pickup pipes	10 years	11.4.5.1
Foam concentrate tank — drain and flush	10 years	11.4.5.2
Standard balanced pressure type		
Foam concentrate pump(s)	5 years *(see Note)*	11.4.6.2
Balancing valve diaphragm	5 years	11.4.6.3
Foam concentrate tank	10 years	11.4.6.4
In-line balanced pressure type		
Foam concentrate pump(s)	5 years *(see Note)*	11.4.7.2
Balancing valve diaphragm	5 years	11.4.7.3
Foam concentrate tank	10 years	11.4.7.4
Pressure vacuum vents	5 years	11.4.8
Water supply tank(s)	See Chapter 9	—
Fire pump(s)	See Chapter 8	—
Water supply	Annually	11.2.6.1
Backflow preventer(s)	See Chapter 13	—
Detector check valve(s)	See Chapter 13	—
Check valve(s)	See Chapter 13	—
Control valve(s)	See Chapter 13	—
Deluge/preaction valves	See Chapter 13	11.2.1
Strainer(s) — mainline	See Chapter 10	—
Detection system	See *NFPA 72*	11.2.2

Note: Also refer to manufacturer's instructions and frequency. Maintenance intervals other than preventive maintenance are not provided, as they depend on the results of the visual inspections and operational tests. For foam-water systems in aircraft hangars, refer to the inspection, test, and maintenance requirements of NFPA 409, *Standard on Aircraft Hangars*, Table 11.1.1.

11.1.2 Other System Components. Fire pumps, water storage tanks, and valves common to other types of water-based fire protection systems shall be inspected, tested, and maintained in accordance with Chapters 8, 9, and 13, respectively, and as specified in Table 11.1.1.2.

11.1.3 Foam-Water Systems.

FAQ What is the difference between the various systems covered in NFPA 13, NFPA 11, NFPA 16, and the systems covered by Chapter 11 of NFPA 25?

NFPA 11, *Standard for Low-, Medium-, and High-Expansion Foam,* addresses low-, medium-, and high-expansion foam systems, while NFPA 16 addresses foam-water systems. NFPA 11

Critical Point

Sprinkler operating pressure affects foam quality, discharge patterns, and fire extinguishment (or control) capabilities.

specifically indicates that foam-water sprinkler and spray systems, which are covered in NFPA 16, are outside of the scope of NFPA 25.

Chapter 11 of NFPA 25 addresses the foam-water system types and applications covered in NFPA 16. The NFPA 16 foam-water sprinkler systems addressed in this chapter are similar to the standard wet, dry, preaction, and deluge systems. The exception is that NFPA 16 systems discharge a foam-water extinguishing agent and, thus, are also equipped with added trim, including a foam concentrate connection, a proportioning device, and a foam concentrate storage tank.

Foam-water systems also use either aspirating foam-water nozzles or non-aspirating sprinklers in systems similar to a standard ceiling sprinkler system. Nozzles and sprinklers must be listed and tested for use with the specific foam used in the system. Specific foam concentrates typically are listed or approved with specific sprinklers. Part of the approval and listing is a minimum sprinkler operating pressure. Sprinkler operating pressure affects foam quality, discharge patterns, and fire extinguishment (or control) capabilities. Discharge pressures less than the specified minimum pressure should be corrected immediately; therefore, it is necessary to test under full flow conditions. In many cases, standard sprinklers that hold a double listing are used, meaning that they are used in both standard and foam-water systems.

Foam-water systems, such as NFPA 13 sprinkler systems, will run until a shutoff valve is closed. If the foam concentrate supply is exhausted, these systems continue to operate with water discharge only. And in some installations, they are designed to sequence the discharge between water and foam.

Systems addressed in NFPA 15, *Standard for Water Spray Fixed Systems for Fire Protection,* do not use foam concentrate. These systems use water only, but they were tested and are designed to provide protection for specific special hazards such as electrical hazards (including transformers, cable trays, and motors), ordinary combustibles, and certain combustible solids. NFPA 15 systems use listed spray nozzles, not sprinklers (except as activation pilots).

NFPA 750, *Standard on Water Mist Fire Protection Systems,* which is covered in Chapter 12 of NFPA 25, is the standard for water mist systems, which are entirely different than NFPA 15 water spray systems. Water mist systems also are tested and designed for specific special hazards and use sprinklers or nozzles with system-specific listings. Both water spray and water mist systems are also approved for use in the protection of some of the same hazards, such as motor protection. However, water mist systems are specifically excluded from the 2013 edition of NFPA 13.

11.1.3.1 This section shall apply to foam-water systems as specified in NFPA 16, *Standard for the Installation of Foam-Water Sprinkler and Foam-Water Spray Systems.*

11.1.3.2 This section shall not include systems detailed in NFPA 11, *Standard for Low-, Medium-, and High-Expansion Foam.*

NFPA 11 covers a different type of foam system from those addressed in NFPA 25. Where applicable, testing should include the requirements of Chapter 11 of NFPA 11, and maintenance should include the requirements of Chapter 12 of NFPA 11.

11.1.4 Foam-Water System.

11.1.4.1 If during routine inspection and testing the foam-water system is determined to have been altered or replaced (e.g., equipment replaced, relocated, or foam concentrate replaced), it shall be determined whether the design intent has been altered and whether the system operates properly.

Alterations or changes must be evaluated for compliance with NFPA 16. This evaluation might require referring to the original design and installation documents to determine if the system has been altered. The evaluation should be conducted by a qualified contractor, consultant, or engineer.

Case In Point

TYPES OF FOAM CONCENTRATES

Foam-water systems, as referenced in NFPA 13, use aspirating foam-water nozzles or non-aspirating nozzles with low expansion–type foam only. For foam-only systems, see NFPA 11.

Foam-water spray systems are equipped with a foam concentrate connection and proportioning device connected to a foam concentrate storage tank. The types of foam concentrates used in these systems include protein, fluoroprotein, aqueous film-forming foam or film-forming fluoroprotein, or alcohol-resistant foam concentrates. These types of concentrates are described below and have been compiled from information in NFPA 11.

Protein-Foam Concentrate. Products from a natural hydrolyzed protein, plus added stabilizers and inhibitors to protect against freezing, prevent corrosion of equipment and containers, resist bacterial decomposition, control viscosity, and otherwise ensure readiness for use under emergency conditions. These are typically diluted with water to form 3 percent to 6 percent solutions, depending on the type, and are compatible with certain dry chemicals.

Aqueous Film-Forming Foam (AFFF) Concentrate. Products with fluorinated surfactants plus stabilizers. These are typically diluted with water to a 1 percent, 3 percent, or 6 percent solution. The foam that forms acts as a barrier both to exclude air or oxygen and to develop an aqueous film on the fuel surface that is capable of suppressing the evolution of fuel vapors. These are suitable for use with dry chemicals.

Fluoroprotein-Foam (FP) Concentrate. Products similar to protein-foam concentrate but with a synthetic fluorinated surfactant additive. In addition to an air-excluding foam blanket, it also can deposit a vaporization-preventing film on the surface of a liquid fuel. These are typically diluted with water to form 3 percent to 6 percent solutions, depending on the type, and they are compatible with certain dry chemicals.

Film-Forming Fluoroprotein (FFFP) Foam Concentrate. Products with protein-based and fluorinated surfactants that produce a fluid aqueous film for suppressing hydrocarbon fuel vapors. This foam has stabilizers and inhibitors added to protect against freezing, corrosion, and bacterial decomposition for added resistance to prevent fuel pickup. These are typically diluted with water to a 3 percent or 6 percent solution and are dry chemical–compatible.

Alcohol-Resistant (AR) Foam Concentrate. Products used for fighting fires on water-soluble materials and other fuels destructive to regular AFFF or FFFP foams. It is also used for fires involving hydrocarbons. The three general types of alcohol-resistant foam concentrates are as follows:

1. Water-soluble natural polymers, such as protein or fluoroprotein concentrate, containing alcohol-insoluble materials that precipitate as an insoluble barrier in the bubble structure.
2. Water-soluble natural polymers, such as protein or fluoroprotein concentrates, containing a gelling agent that protects the foam from water-soluble fuels. This foam can also have film-forming and fluoroprotein characteristics on hydrocarbon fuels.
3. Synthetic concentrates containing gelling agents that surround the foam bubbles and form a protective raft on the surface of water-soluble fuels. These foams can also have film-forming characteristics on hydrocarbon fuels.

Alcohol-resistant aqueous film-forming foams (AR-AFFF) are foams resistant to the action of alcohols. These concentrates are typically used in up to 10 percent solutions, depending on the nature of the hazard to be protected and the type of concentrate.

NFPA 11 permits different brands of foam concentrate to be mixed if they are of the same type and the manufacturer provides data to the authority having jurisdiction (AHJ) demonstrating compatibility. Foam concentrates of different concentrations cannot be substituted.

11.1.4.1.1 Mechanical waterflow devices, including but not limited to water motor gongs, shall be tested quarterly.

11.1.4.1.2 Valve-type and pressure switch–type waterflow devices shall be tested semiannually.

11.1.4.1.3 Waterflow devices shall be inspected quarterly to verify that they are free of physical damage.

See 13.2.6 for the recommended operational procedure.

11.1.4.2 The inspection shall verify that all components, including foam concentrate discharge devices and proportioning equipment, are installed in accordance with their listing.

11.1.5 Obstruction Investigations. The procedures outlined in Chapter 14 shall be followed where there is a need to conduct an obstruction investigation.

11.1.6 Impairments. The procedures outlined in Chapter 15 shall be followed where an impairment to protection occurs.

11.2 Inspection

Systems shall be inspected in accordance with the frequency specified in Table 11.1.1.2.

11.2.1 Deluge Valves. Deluge valves shall be inspected in accordance with the provisions of Chapter 13.

11.2.2 Automatic Detection Equipment. Automatic detection equipment shall be inspected, tested, and maintained in accordance with *NFPA 72, National Fire Alarm and Signaling Code*, to ensure that the detectors are in place, securely fastened, and protected from corrosion, weather, and mechanical damage and that the communication wiring, control panels, or pneumatic tubing system is functional.

For more information on the automatic detection equipment addressed in 11.2.2, see the commentary following 10.2.3.

11.2.3 System Piping and Fittings. System piping and fittings shall be inspected for the following:

(1) Mechanical damage (e.g., broken piping or cracked fittings)
(2) External conditions (e.g., missing or damaged paint or coatings, rust, and corrosion)
(3) Misalignment or trapped sections
(4) Low-point drains (automatic or manual)
(5) Location and condition of rubber-gasketed fittings

11.2.4 Hangers and Supports. Hangers and supports shall be inspected for the following and repaired or replaced as necessary:

(1) Condition (e.g., missing or damaged paint or coating, rust, and corrosion)
(2) Secure attachment to structural supports and piping
(3) Damaged or missing hangers

11.2.5* Foam-Water Discharge Devices.

Discharge devices for foam-water spray systems can be either air-aspirating nozzles or non-air-aspirating sprinklers. Discharge devices for foam-water sprinkler systems, such as wet, dry, or preaction systems, are automatic in operation and must be non-air-aspirating and listed for use with the specific foam utilized. Exhibit 11.1 and Exhibit 11.2 show an air-aspirating discharge nozzle, while Exhibit 11.3 and Exhibit 11.4 show various deflector types for foam-water systems and foam-water spray nozzles.

Critical Point

Discharge devices for foam-water sprinkler systems, such as wet, dry, or preaction systems, are automatic in operation and must be non-air-aspirating and listed for use with the specific foam utilized.

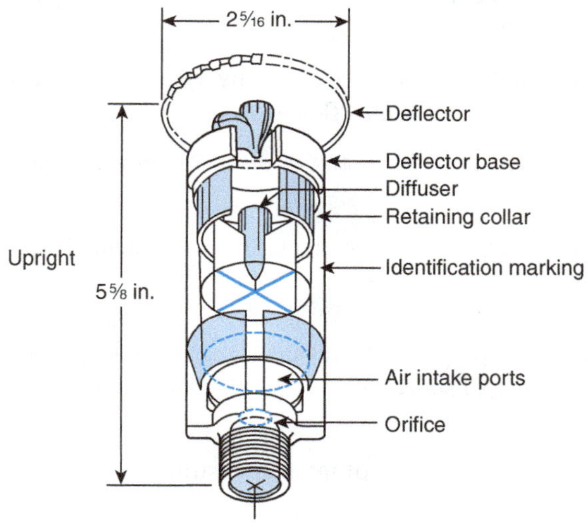

For SI units, 1 in. = 25.4 mm.

EXHIBIT 11.1 *Diagram of an Air-Aspirating Discharge Nozzle. (Courtesy of Kidde Fire Systems, Ashland, MA)*

EXHIBIT 11.2 *Air-Aspirating Discharge Nozzle.*

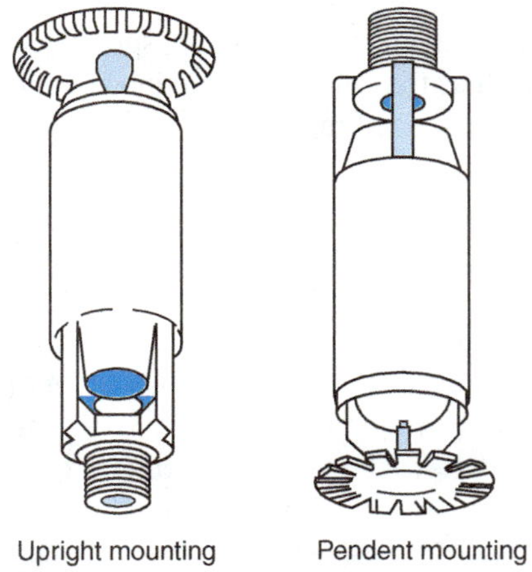

EXHIBIT 11.3 *Variations in Deflector Shape. (Source: NFPA 16, 2011, Figure A.3.3.6.1)*

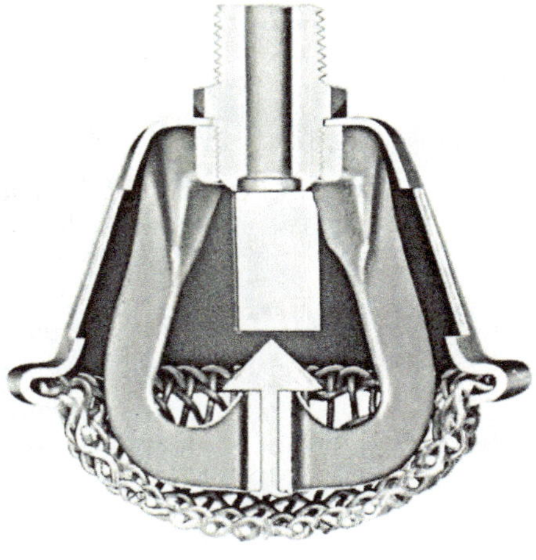

EXHIBIT 11.4 *Foam-Water Spray Nozzle. (Source: NFPA 16, 2011, Figure A.3.3.6.4)*

A.11.2.5 Directional-type foam-water discharge devices are quite often located in heavy traffic areas and are more apt to be dislocated compared to ordinary sprinkler locations. Of particular concern are low-level discharge devices in loading racks in and around low-level tankage and monitor-mounted devices that have been pushed out of the way for convenience. Inspection frequency might have to be increased accordingly.

11.2.5.1 Foam-water discharge devices shall be inspected visually and maintained to ensure that they are in place, continue to be aimed or pointed in the direction intended in the system design, and are free from external loading and corrosion.

11.2.5.2 Where caps or plugs are required, the inspection shall confirm they are in place and free to operate as intended.

11.2.5.3 Misaligned discharge devices shall be adjusted (aimed) by visual means, and the discharge patterns shall be inspected at the next scheduled flow test.

11.2.5.4* Inspection shall verify that unlisted combinations of discharge devices and foam concentrate have not been substituted.

A.11.2.5.4 Discharge devices are listed or approved for particular foam concentrates.

11.2.6 Water Supply.

11.2.6.1 The dependability of the water supply shall be ensured by regular inspection and maintenance, whether furnished by a municipal source, on-site storage tanks, a fire pump, or private underground piping systems.

11.2.6.2* Water supply piping shall be maintained free of internal obstructions.

A.11.2.6.2 Water supply piping should be free of internal obstructions that can be caused by debris (e.g., rocks, mud, tubercles) or by closed or partially closed control valves. See Chapter 5 for inspection and maintenance requirements.

11.2.7 Strainers.

11.2.7.1 Mainline and individual discharge device strainers (basket or screen) shall be inspected in accordance with the provisions of Chapter 10.

11.2.7.2 Foam concentrate strainers shall be inspected visually to ensure the blow-down valve is closed and plugged.

11.2.7.3 Baskets or screens shall be removed and inspected after each operation or flow test.

11.2.8 Drainage. The area beneath and surrounding a foam-water spray system shall be inspected to ensure that drainage facilities, such as trap sumps and drainage trenches, are not blocked and retention embankments or dikes are in good repair.

11.2.9* Proportioning Systems.

The following description of various types of proportioning systems is adapted from A.8.4 of NFPA 16.

In *vacuum induction*, a venturi device, called an inductor, decreases the pressure in the foam concentrate proportioning inlet tube to the system flow passageway to below the atmospheric pressure in which the foam is pushed by atmospheric pressure through a calibrated orifice. This system is limited to deluge systems, and the friction loss of the discharge system to the discharge devices must be considered for proper performance. Proportioning percentage can vary based on the discharge and supply characteristics as well as the foam concentrate. These systems require full system discharge flow testing to determine proper installation and sizing.

A *foam concentrate pump* discharges foam directly into the discharge system through a metering orifice specifically sized for the system performance that is being installed. This type of system is limited to deluge-type systems and must be flow tested at the desired maximum flow rate of the system. Each installation must be full system flow tested due to specific design parameters for a particular installation.

Balanced pressure proportioning with a foam pump or bladder tank uses a modified venturi proportioning device in the system water supply pipe. The waterflow to the system through the modified venturi causes a metered pressure drop in the foam concentrate inlet chamber. As the flow increases, the metered pressure loss increases, causing increased flow of foam concentrate through a calibrated orifice into the system water supply. With a foam pump system, a balancing valve is used to measure the supply pressure at the inlet of the modified venturi proportioner and balance the foam concentrate pressure down to equal that pressure. When using a bladder tank, the foam concentrate is stored inside a bladder within a pressure vessel where the supply water in the system supply is directed to the outside of the bladder, causing the foam concentrate to push out from the top or bottom of the tank and travel to the inlet of the modified venturi device. As the water pressure increases or decreases due to flow conditions, so does the foam concentrate pressure. In both the balanced pump system and the bladder tank system, when the inlet pressures of both foam concentrate and water supply match, the inlet calibrated orifice of the modified venturi device provides the proper mixture of foam to water as specified by the manufacturer. Also, these systems allow for system isolation test valves to be applied and divert test foam solution flow to a containment area that can be easily disposed of.

Positive pressure proportioning with a foam pump or pressure-controlled bladder tank system uses a system that supplies foam concentrate to a modified venturi at a higher pressure than the supply water pressure. The modified venturi includes a calibrated orifice that is sized to match the system equipment and the foam concentrate being used.

Exhibit 11.5 illustrates a hydraulically operated concentrate valve. As the deluge valve on the system activates, water pressure is applied to the valve, allowing it to open. This valve should also be included in the inspection, testing, and maintenance program for foam-water systems.

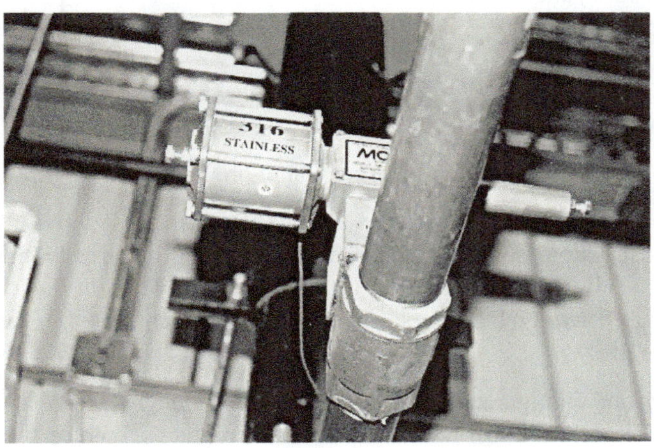

EXHIBIT 11.5 *Hydraulically Operated Concentrate Valve.*

A.11.2.9 Proportioning systems might or might not include foam concentrate pumps. If pumps are part of the proportioning system, the driver, pump, and gear reducer should be inspected in accordance with the manufacturer's recommendations, and the inspection can include items such as lubrication, fuel, filters, oil levels, and clutches.

11.2.9.1 The components of the various proportioning systems described in 11.2.9 shall be inspected in accordance with the frequency specified in Table 11.1.1.2.

11.2.9.2 Valves specified to be inspected shall be permitted to be open or closed, depending on specific functions within each foam-water system.

11.2.9.3 The position (open or closed) of valves shall be verified in accordance with specified operating conditions.

11.2.9.4* Inspection of the concentrate tank shall include verification that the quantity of foam concentrate satisfies the requirements of the original design.

A.11.2.9.4 In some cases, an adequate supply of foam liquid is available without a full tank. This is particularly true of foam liquid stored in nonmetallic tanks. If liquid is stored in metallic tanks, the proper liquid level should be one-half the distance into the expansion dome.

Critical Point

As specified in NFPA 16, the minimum quantity of foam concentrate required is determined by the following hydraulic calculation:

% concentration × amount of flow × duration = foam concentrate quantity

As specified in NFPA 16, the minimum quantity of foam concentrate required is determined by the following hydraulic calculation:

% concentration × amount of flow × duration = foam concentrate quantity

This minimum quantity should be maintained in the storage tank at all times.

11.2.9.5 Additional inspection requirements shall be performed as detailed for the proportioning systems specified in 11.2.9.

11.2.9.5.1 Standard Pressure Proportioner.

11.2.9.5.1.1* The pressure shall be removed before the inspection to prevent injury.

A.11.2.9.5.1.1 The standard pressure proportioner is a pressure vessel. Although under normal standby conditions this type of proportioning system should not be pressurized, some installations allow for inadvertent pressurization. Pressure should be removed before inspection.

Pressure can be removed by slowly opening the drain or vent valve.

11.2.9.5.1.2 The inspection shall verify the following:

(1) Ball drip valves (automatic drains) are free and opened.
(2) External corrosion on foam concentrate storage tanks is not present.

11.2.9.5.2 Bladder Tank Proportioner.

11.2.9.5.2.1* The pressure shall be removed before the inspection to prevent injury.

A.11.2.9.5.2.1 The bladder tank proportioner is a pressure vessel. Where inspecting for a full liquid tank, the manufacturer's instructions should be followed. If inspected incorrectly, the tank sight gauges could indicate a full tank when the tank actually is empty of foam liquid. Some foam liquids, due to their viscosity, might not indicate true levels of foam liquid in the tank where inspected via the sight glass.

CAUTION: Depending on system configuration, this type of proportioner system might be pressurized or nonpressurized under normal conditions. Pressure should be removed before inspection.

11.2.9.5.2.2 The inspection shall include the following:

(1) Water control valves to foam concentrate tank
(2) An inspection for external corrosion on foam concentrate storage tanks
(3) An inspection for the presence of foam in the water surrounding the bladder (annual)

11.2.9.5.3 Line Proportioner. The inspection shall include the following:

(1)* Strainers

A.11.2.9.5.3(1) See 11.2.7.1.

(2)* Verification that pressure vacuum vent is operating freely

A.11.2.9.5.3(2) See Figure A.3.3.32.

(3) An inspection for external corrosion on foam concentrate storage tanks

11.2.9.5.4 Standard Balanced Pressure Proportioner. The inspection shall include the following:

(1)* Strainers

ITM Deficiency, Impairment, or Hazard Evaluation?

ANSWER: Impairment

(Courtesy of David W. Smith and SimplexGrinnell)

This photo shows the actual rubber bladder on a foam-water sprinkler system pulled from the bladder tank after it was determined that foam concentrate had leaked from it. The bladder is separated or torn at the bottom of the neck, allowing the concentrate to migrate from the bladder into the water surrounding it and the water to enter the bladder, diluting the concentrate. In addition, the amount of the diluted concentrate in the bladder was about half of what it should have been. This condition was discovered while performing the annual inspection and test required by Chapter 11. A subsequent pressure test on the bladder and vacuum test on the tank shell indicated that the integrity of the bladder itself had failed.

The dilution and amount of the foam concentrate in the tank created an impaired condition. This portion of the system will no longer function as intended in a fire event. Instead of delivering the proper duration of pure foam concentrate to the proportioner, only a short duration of diluted concentrate will be proportioned into this foam-water system. The hazard being protected requires a certain concentration of a foam-water solution in accordance with NFPA 16 for proper fire extinguishment or control. Because of the failure of the bladder, this system will not provide that concentration level.

A.11.2.9.5.4(1) See 11.2.7.1.

(2)* Verification that pressure vacuum vent is operating freely

A.11.2.9.5.4(2) See Figure A.3.3.32.

(3) Verification that gauges are in good operating condition
(4) Verification that sensing line valves are open
(5) Verification that power is available to foam liquid pump

11.2.9.5.5 In-Line Balanced Pressure Proportioner. The inspection shall include the following:

(1)* Strainers

A.11.2.9.5.5(1) See 11.2.7.1.

(2)* Verification that pressure vacuum vent is operating freely

A.11.2.9.5.5(2) See Figure A.3.3.32.

(3) Verification that gauges are in good working condition
(4) Verification that sensing line valves at pump unit and individual proportioner stations are open
(5) Verification that power is available to foam liquid pump

11.2.9.5.6 Orifice Plate Proportioner. The inspection shall include the following:

(1)* Strainers

A.11.2.9.5.6(1) See 11.2.7.1.

(2)* Verification that pressure vacuum vent is operating freely

A.11.2.9.5.6(2) See Figure A.3.3.32.

Exhibit 11.6 illustrates an arrangement for orifice plate proportioning. This method uses a second deluge valve supplied by a foam concentrate pump. The diameter of the orifice is determined by hydraulic calculations and controls the flow of foam concentrate.

EXHIBIT 11.6 Orifice Plate Proportioning Arrangement.

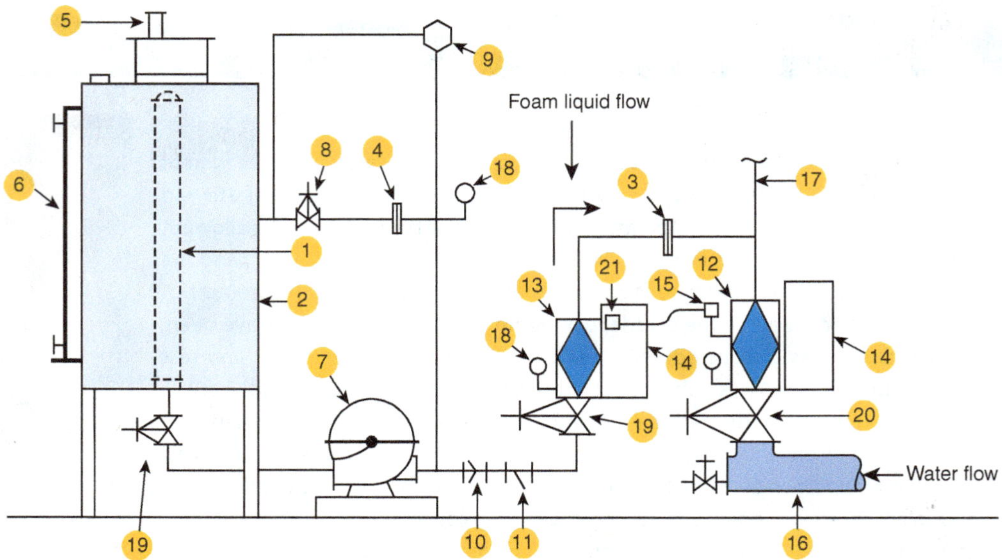

Legend

1	Strainer	12	Deluge valve
2	Foam liquid storage tank	13	Companion deluge valve
3	Flange with orifice plate (for calibration of foam liquid injection)	14	Deluge system control panel
4	Flange with orifice plate (for pump test calibration)	15	Deluge valve solenoid
5	Vent-vacuum breather	16	Water supply strainer
6	Sight glass	17	To system foam water solution
7	Foam liquid pump	18	Pressure indicating gauge
8	Re-circulating and test line gate valve (N.C.)	19	Foam liquid control valve (N.O.)
9	Relief valve	20	Water supply control valve (N.O.)
10	Check valve	21	Companion deluge valve solenoid
11	Foam liquid supply strainer		

N.C. = Normally closed.
N.O. = Normally open.

(3) Verification that gauges are in good working condition

(4) Verification that power is available to foam liquid pump

11.2.10 Foam Concentrate Samples. Samples shall be submitted in accordance with the manufacturer's recommended sampling procedures.

11.3* Operational Tests

Frequency of system tests shall be in accordance with Table 11.1.1.2.

A.11.3 Operational tests generally should be comprised of the following:

(1) A detection/actuation test with no flow to verify that all components such as automated valves, foam and water pumps, and alarms operate properly

(2) A water-only flow test to inspect piping continuity, discharge patterns, pressures, and line flushing
(3) A foam flow test to verify solution concentration
(4) Resetting of system to its normal standby condition, including draining of lines and filling of foam liquid tank

System tests must be conducted in accordance with Table 11.1.1.2. This testing includes annual testing of the foam concentrate, which helps ensure the reliability of the foam system.

Properly stored foam concentrates have an excellent record of effectiveness and longevity. Protein-type concentrates have an expected shelf life of approximately 7 to 10 years. Synthetic-type concentrates have an expected shelf life of 20 to 25 years. Factors affecting shelf life and effectiveness include wide temperature variations, high or low temperature extremes, evaporation, dilution, and contamination by foreign materials.

Manufacturers of foam concentrates provide laboratory testing services to analyze foam concentrate samples. The samples are tested for such chemical characteristics as pH level and specific gravity. They are also tested for contamination from water, oil, noncompatible foam concentrates, and possible microbiological organisms. In addition, changes in the concentrate due to storage temperature fluctuations and sediment are analyzed. Samples for testing should be taken according to the manufacturer's instructions.

A test connection is normally provided as a means of periodically checking the performance of the proportioners used in foam sprinkler systems. Typical test connections are illustrated in Figure A.11.3.2. If the proportioning controller is located after the main sprinkler valve, an additional supervised OS&Y valve is provided to isolate the sprinkler overhead during the proportioner test. This additional precaution is intended to eliminate problems caused by air cushions in wet pipe sprinkler systems or the servicing delays that result from charging and draining preaction or deluge sprinkler systems. The manufacturer's test procedures should be followed closely.

11.3.1* Test Preparation. Precautions shall be taken to prevent damage to property during the test.

FAQ What precautions must be taken to prevent property damage during a discharge test?

Some of the precautions required by 11.3.1 to prevent property damage during a discharge test might include the following:

- Checking related facility systems for proper operation, to ensure that the water supply to the deluge system has been inspected and tested.
- Making sure sump tanks are empty and operating properly.
- Inspecting any drainage provisions required to contain or collect the foam-water discharging from the system.
- Making sure the owner or the owner's authorized representative has protected all exposed equipment or property that could be damaged by the system when discharging.
- Making sure proper post-test cleanup has been arranged and planned before testing.
- Making sure environmental compliance has been confirmed and planned before testing. Requirements for cleanup and disposal are often enforced differently in different jurisdictions, regardless of material safety data sheets (MSDS) indications.

A.11.3.1 The property owner or designated representative should take care to prevent damage to equipment or the structure during the test. Damage could be caused by the system discharge or by runoff from the test site. It should be verified that there is adequate and unobstructed drainage. Equipment should be removed or covered as necessary to prevent damage. Means such as curbing or sandbagging should be used to prevent entry of the foam-water solution.

Critical Point

During maintenance tests, only a small amount of foam concentrate should be discharged to verify the correct concentration of foam in the foam-water solution.

During maintenance tests, only a small amount of foam concentrate should be discharged to verify the correct concentration of foam in the foam-water solution. Designated foam-water test ports can be designed into the piping system so that the discharge of foam-water solution can be directed to a controlled location, such as a container that would be transported to an approved disposal site by a licensed hazardous waste contractor. The remainder of the test should be conducted using water only.

11.3.1.1 Mechanical waterflow devices, including but not limited to water motor gongs, shall be tested quarterly.

11.3.1.2 Vane-type and pressure switch–type waterflow devices shall be tested semiannually.

11.3.1.3 Waterflow Devices. Waterflow devices shall be inspected quarterly to verify that they are free of physical damage.

See 13.2.6 for the recommended operational procedure.

11.3.2* Operational Test Performance.

See 13.2.6 for the recommended operational procedure.

A.11.3.2 An alternative method for achieving flow can be permitted to be an installation as shown in Figure A.11.3.2. This type of testing does not verify system pipe conditions or discharge device performance but only the water supply, foam concentrate supply, and proportioning accuracy.

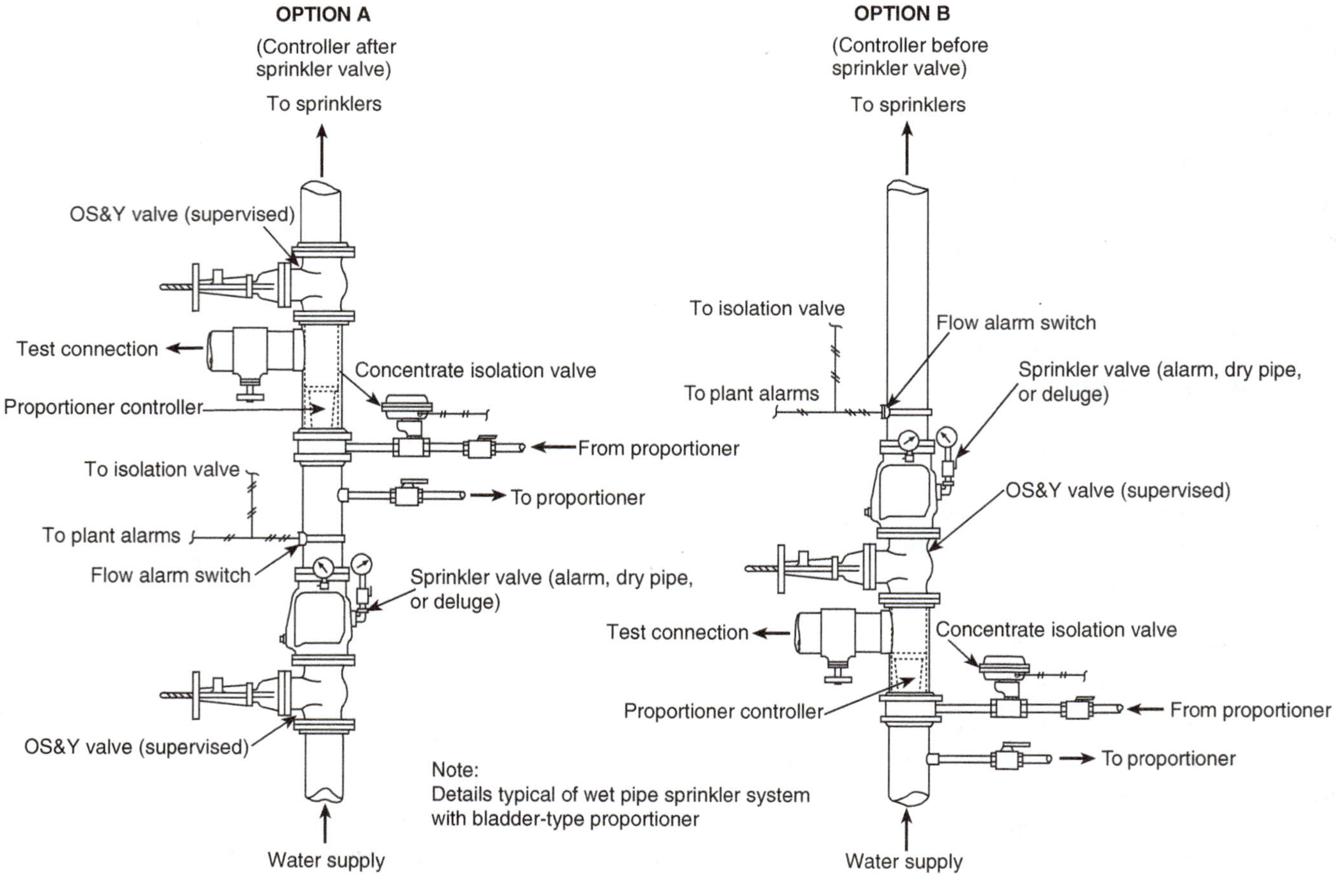

FIGURE A.11.3.2 Foam System/Test Header Combination.

Foam-water systems are typically used where the fire hazard is greater than that of a hazard that can be protected by a standard water-only sprinkler system. The fires that occur in these areas can grow rapidly, which is why confirming system functionality through testing is critical. For a procedure for conducting this operational test, please refer to the detailed testing procedure at the end of this chapter.

Testing Procedure Alert

ITM NFPA 25

11.3.2.1 Operational tests shall be conducted to ensure that the foam-water system(s) responds as designed, both automatically and manually.

Knowledge of the specialized equipment and test apparatus is necessary to perform an acceptable test. Therefore, the tests should be conducted by qualified persons who are knowledgeable about foam equipment and systems.

11.3.2.2 The test procedures shall simulate anticipated emergency events so the response of the foam-water system(s) can be evaluated.

11.3.2.3 Where discharge from the system discharge devices would create a hazardous condition or conflict with local requirements, an approved alternate method to achieve full flow conditions shall be permitted.

When an alternative method of achieving full flow conditions is permitted, full flow can be achieved from the discharge test connection at the system riser.

11.3.2.4 Response Time. Under test conditions, the automatic fire detection systems, when exposed to a test source, shall operate within the requirements of *NFPA 72, National Fire Alarm and Signaling Code*, for the type of detector provided, and the response time shall be recorded.

11.3.2.5 Discharge Time. The time lapse between operation of detection systems and water delivery time to the protected area shall be recorded for open discharge devices.

11.3.2.6 Discharge Patterns.

11.3.2.6.1 The discharge patterns from all of the open spray devices shall be observed to ensure that patterns are not impeded by plugged discharge devices and to ensure that discharge devices are correctly positioned and that obstructions do not prevent discharge patterns from covering surfaces to be protected.

11.3.2.6.2 Where obstructions occur, the piping and discharge devices shall be cleaned and the system retested.

11.3.2.6.3 Discharge devices shall be permitted to be of different orifice sizes and types.

11.3.2.7* Pressure Readings.

A.11.3.2.7 Specific foam concentrates typically are listed or approved with specific sprinklers. Part of the approval and listing is a minimum sprinkler operating pressure. Sprinkler operating pressure affects foam quality, discharge patterns, and fire extinguishment (control) capabilities. Discharge pressures less than this specified minimum pressure should be corrected immediately; therefore, it is necessary to test under full flow conditions.

11.3.2.7.1 Pressure readings shall be recorded at the highest, most remote discharge device.

11.3.2.7.2 A second pressure reading shall be recorded at the main control valve.

11.3.2.7.3 Readings shall be compared to the hydraulic design pressures to ensure the original system design requirements are met.

11.3.3 Multiple Systems. The maximum number of systems expected to operate in case of fire shall be tested simultaneously to inspect the adequacy of the water supply and concentrate pump.

ITM Deficiency, Impairment, or Hazard Evaluation?

ANSWER: Impairment

These three photos have to be viewed in a group to fully understand what they are showing. The foam concentrate line going from the foam concentrate tank to the proportioner was cut to determine the cause of a failed test. As can be seen, this piping was clogged with congealed foam concentrate.

During the annual system discharge test required by Table 11.1.1.2 and Section 11.3, a foam sample was taken as required in 11.3.5, and an attempt was made to verify the concentration. The use of a refractometer revealed that very little foam concentrate was mixing with the water, and the test was stopped. The investigation started at the tank and eventually led to this section of piping. At this point, the system was determined to be impaired, and the recommendation was made to the owner to begin the impairment procedures required in Chapter 15.

(Photos courtesy of David W. Smith and SimplexGrinnell)

11.3.4 Manual Actuation Devices. Manual actuation devices shall be tested annually.

11.3.5 Concentration Testing.

11.3.5.1 During the full flow foam test, a foam sample shall be taken.

11.3.5.2 The foam sample shall be inspected by refractometric or other methods to verify concentration of the solution.

11.3.5.3 Concentration shall be within 10 percent of the acceptance test results but in no case more than 10 percent below minimum design standards.

11.3.6 Return to Service. After the full flow test, the foam-water system shall be returned to service and the foam concentrate tank shall be replenished to design level.

Testing Procedure Alert

See the detailed testing procedure at the end of this chapter for more information.

11.4* Maintenance

A.11.4 The maintenance items specified in the body of this standard are in addition to the typical inspection and test procedures indicated. Foam-water systems are, as are all fire protection systems, designed to be basically maintenance free. There are, however, some areas that need special attention. Foam concentrate shelf life varies between liquids and is affected by factors such as heat, cold, dilution, contamination, and many others. As with all systems, common sense dictates those maintenance-sensitive areas that should be given attention. Routine testing and inspection generally dictate the need for additional maintenance items. Those maintenance items specified are key procedures that should be performed routinely.

11.4.1 Maintenance of foam-water systems shall be in accordance with the requirements of those chapters covering the specific component parts.

11.4.2 Foam Components. Maintenance of specific foam components shall be in accordance with 11.4.3 through 11.4.7.

11.4.3 Standard Pressure Proportioner.

11.4.3.1 The ball drip (automatic-type) drain valves shall be disassembled, cleaned, and reassembled.

11.4.3.2* The foam liquid storage tank shall be drained of foam liquid and flushed.

A.11.4.3.2 Foam concentrates tend to settle out over time. Depending on the specific characteristics of the foam concentrate, sedimentation accumulates in the bottom of the storage vessel. This sediment can affect proportioning and foam concentrate integrity. Some concentrates tend to settle out more rapidly than others. If the annual samples indicate excessive sediment, flushing the tank could be required more frequently.

11.4.3.3 Foam liquid shall be permitted to be salvaged and reused.

11.4.3.4 The foam liquid tank shall be inspected for internal and external corrosion and hydrostatically tested to the specified working pressure.

11.4.4 Bladder Tank Proportioner.

11.4.4.1 Sight glass, where provided, shall be removed and cleaned.

11.4.4.2* The foam concentrate tank shall be hydrostatically tested to the specified working pressure.

A.11.4.4.2 Where hydrostatically testing bladder tanks, the generation of a pressure differential across the diaphragm should not be allowed. The manufacturer should be consulted for specific procedures.

11.4.5 Line Proportioner.

11.4.5.1 The foam concentrate tank shall be inspected for internal corrosion.

11.4.5.2 Pickup pipes inside the tank shall be inspected for corrosion, separation, or plugging.

11.4.5.3 The foam concentrate tank shall be drained and flushed.

11.4.5.4 Foam concentrate shall be permitted to be salvaged and reused.

11.4.6 Standard Balanced Pressure Proportioner.

11.4.6.1 Pump Operation.

11.4.6.1.1 The foam concentrate pump shall be operated.

11.4.6.1.2 Foam concentrate shall be circulated back to the tank.

11.4.6.2 Servicing. Foam pumps, drive train, and drivers shall be serviced in accordance with the manufacturer's instructions and frequency but not at intervals of more than 5 years.

11.4.6.3 Flushing. The diaphragm balancing valve shall be flushed through the diaphragm section with water or foam concentrate until fluid appears clear or new.

11.4.6.4 Corrosion and Sediment.

11.4.6.4.1 The foam concentrate tank shall be inspected internally for corrosion and sediment.

11.4.6.4.2 Excessive sediment shall require draining and flushing of the tank.

11.4.7 In-Line Balanced Pressure Proportioner.

11.4.7.1 Pump Operation.

11.4.7.1.1 The foam concentrate pump shall be operated.

11.4.7.1.2 Foam concentrate shall be circulated back to the tank.

11.4.7.2 Servicing. Foam pumps, drive train, and drivers shall be serviced in accordance with the manufacturer's instructions and frequency but not at intervals of more than 5 years.

11.4.7.3 Flushing. The diaphragm balancing valve shall be flushed through the diaphragm section with water or foam concentrate until fluid appears clear or new.

11.4.7.4 Corrosion and Sediment.

11.4.7.4.1 The foam concentrate tank shall be inspected internally for corrosion and sediment.

11.4.7.4.2 Excessive sediment shall require draining and flushing of the tank.

11.4.8 Pressure Vacuum Vents. The procedures specified in 11.4.8.1 through 11.4.8.13 shall be performed on pressure vacuum vents every 5 years.

11.4.8.1 The vent shall be removed from the expansion dome.

11.4.8.2 The vent shall be inspected to ensure that the opening is not blocked and that dirt or other foreign objects do not enter the tank.

11.4.8.3 The vent bonnet shall be removed.

11.4.8.4 The vacuum valve and pressure valve shall be lifted out.

11.4.8.5 The vent body shall be flushed internally, and the vacuum valve and the pressure valve shall be washed thoroughly.

11.4.8.6 The vent shall be inspected to ensure that the screen is not clogged, and the use of any hard, pointed objects to clear the screen shall be avoided.

11.4.8.7 If the liquid has become excessively gummy or solidified, the vent body and parts shall be soaked in hot soapy water.

11.4.8.8 The vent body shall be turned upside down and drained thoroughly.

11.4.8.9 Parts shall be dried by placing them in a warm and dry area or by using an air hose.

11.4.8.10 Parts shall be sprayed with a light Teflon® coating, and the vent shall be reassembled.

11.4.8.11 The use of any type of oil for lubrication purposes shall not be permitted.

11.4.8.12 The vent bonnet shall be replaced, and the vent shall be turned upside down slowly a few times to ensure proper freedom of the movable parts.

11.4.8.13 The vent shall be attached to the liquid storage tank expansion dome.

11.5 Component Action Requirements

Component replacement tables offer guidance for situations when system components are adjusted, repaired, rebuilt, or replaced. It is not necessary in each case to conduct a complete acceptance test for each component when maintenance is performed.

11.5.1 Whenever a component in a foam-water sprinkler system is adjusted, repaired, reconditioned, or replaced, the action required in Table 11.5.1 shall be performed.

TABLE 11.5.1 *Summary of Component Replacement Action Requirements*

Component	*Adjust*	*Repair/ Recondition*	*Replace*	*Required Action*
Water Delivery Components				
Pipe and fittings on open head system	X	X	X	Operational flow test
Pipe and fittings on closed head system	X	X	X	Hydrostatic test in conformance with NFPA 16, *Standard for the Installation of Foam-Water Sprinkler and Foam-Water Spray Systems*
Discharge devices	X		X	(1) Inspect for leaks at system working pressure (2) Inspect for impairments at orifice
Fire department connections	X	X	X	See Chapter 13
Manual release	X	X	X	(1) Operational test (2) Inspect for leaks at system working pressure (3) Test all alarms
Foam Components				
Foam concentrate strainer(s)				See Chapter 13
Proportioning system(s)	X	X	X	Conduct flow test and inspect proportioning by refractometer test or equivalent
Water supply tank(s)				See Chapter 9
Foam concentrate	X		X	Submit 1 pint (473 mL) sample for laboratory analysis for conformance with manufacturer's specifications
Foam concentrate pump				See Chapter 8
Ball drip (automatic-type) drain valves				See Chapter 13
Foam concentrate tank	X	X	X	Inspect for condition; repair as appropriate
Bladder tank	X	X	X	Inspect water jacket for presence of foam concentrate
Alarm and Supervisory Components				
Vane-type waterflow	X	X	X	Operational test using inspector's test connection
Pressure switch–type waterflow	X	X	X	Operational test using inspector's test connection
Water motor gong			X	Operational test using inspector's test connection

(continues)

TABLE 11.5.1 Continued

Component	*Adjust*	*Repair/ Recondition*	*Replace*	*Required Action*
Valve supervisory device			X	Test for conformance with NFPA 16 and/or *NFPA 72, National Fire Alarm and Signaling Code*
Detection system	X	X	X	Operational test for conformance with NFPA 16 and/ or *NFPA 72*
Status-Indicating Components				
Gauges			X	Verify at 0 psi (0 bar) and system working pressure
Testing and Maintenance Components				
Main drain	X	X	X	Full flow main drain test
Auxiliary drains	X	X	X	Inspect for leaks at system working pressure
Inspector's test connection	X	X	X	Inspect for leaks at system working pressure
Structural Components				
Hanger/seismic bracing	X	X	X	Inspect for conformance with NFPA 16 and/or NFPA 13, *Standard for the Installation of Sprinkler Systems*
Pipe stands	X	X	X	Inspect for conformance with NFPA 16 and/or NFPA 13
Informational Components				
Valve signs	X	X	X	Inspect for conformance with NFPA 16 and/or NFPA 13
Hydraulic placards	X	X	X	Inspect for conformance with NFPA 16 and/or NFPA 13

11.5.2 Where the original installation standard is different from the cited standard, the use of the appropriate installing standard shall be permitted.

11.5.3 The actions of 11.5.1 and 11.5.2 shall not require a design review, which is outside the scope of this standard.

References Cited in Commentary

National Fire Protection Association, 1 Batterymarch Park, Quincy, MA 02169-7471.

NFPA 11, *Standard for Low-, Medium-, and High-Expansion Foam*, 2010 edition.

NFPA 13, *Standard for the Installation of Sprinkler Systems*, 2010 edition.

NFPA 15, *Standard for Water Spray Fixed Systems for Fire Protection*, 2012 edition.

NFPA 16, *Standard for the Installation of Foam-Water Sprinkler and Foam-Water Spray Systems*, 2011 edition.

NFPA 409, *Standard on Aircraft Hangars*, 2011 edition.

NFPA 750, *Standard on Water Mist Fire Protection Systems*, 2014 edition.

Richardson, L. K. (ed.), *History of Fire Protection Engineering*, 2003.

Testing Procedures for 11.3.2.2

Foam Water Operational Test

Purpose

The following procedure is recommended when conducting a full flow operational test of a foam-water deluge system. This procedure will vary slightly based on the equipment installed.

Tools/Equipment

Equipment including a pipette, graduated cylinder, and refractometer are used when conducting this test.

Procedure Steps

Step 1: Inspect the deluge valve.

A. Check the deluge valve, trim piping and valves, and associated riser piping and valves for proper arrangement.

B. Position all system and trim valves properly, including any low-point drains, in accordance with the system installation drawings or the valve schematic.

Step 2: Isolate the foam supply from the deluge system.

A. Shut the tank foam liquid supply valve.

B. Disconnect the electrical service to the foam concentrate pump(s) if applicable, or shut off the water pressure supply to the bladder tank.

Step 3: Check the deluge system for proper operation.

A. Run and test any fire pumps.

B. Open the test valve to trip the deluge valve.

C. Allow water to flow through the system, flushing all piping until the water runs clear.

D. Check for obstructed nozzles.

Step 4: Prepare the system for concentration test.

A. Test detectors.

B. Test manual actuation.

C. Test alarms and auxiliary equipment shutdowns.

Step 5: Return the foam tank and proportioning system to normal condition.

A. If a foam concentrate pump is used, check the oil level and re-energize the power supply to the pump.

B. Inspect the foam tank for proper setup and check the level of foam concentrate.

C. If a bladder tank is used, avoid damage to the bladder tank by carefully opening the water pressure supply to the tank in accordance with the manufacturer's procedures.

D. Set up the remaining proportioning equipment, and open the tank foam liquid supply valve.

Step 6: Make sample foam mixtures.

A. Both the foam concentrate and the water used for the mixtures must be taken from the site supply.

B. Mix in 100 mL graduated cylinders. If the same cylinder and pipette are used for both samples, clean them thoroughly before each use.

C. For 3 percent mixture, add 3 parts foam concentrate to 97 parts water. For 6 percent mixture, add 6 parts foam concentrate to 94 parts water.

D. Using a pipette, stir each sample until thoroughly mixed.

E. Using the pipette, place a small amount of the 3 percent mixture on the refractometer.

F. Thoroughly clean the refractometer and repeat for the 6 percent mixture.

G. Plot the refractometer reading on the graph. The graph should now include all information except the plot point from the actual system full flow discharge test sample and the end head residual pressure.

Step 7: Prepare the highest, most remote branch line for the end head residual pressure recording.

A. Remove the end head (discharge device) from the highest, most remote branch line.

B. Install ½ in. to ¼ in. (15 mm to 6 mm) adapter, or whatever size is needed.

C. Attach ¼ in. (15 mm) hose to the adapter. A 20 ft (6.1 m) hose is recommended unless a longer one is necessary.

D. Attach a pressure gauge with a 0–250 psi (0–17 bar) range and a valve to the end of the hose.

E. The valve will be used to take a foam sample and will then be shut to record the residual pressure.

Step 8: Prepare to perform the operational test of the system.

A. Using the system installation drawings and hydraulic calculations, determine the pressure required at the highest, most remote discharge outlet. Record this pressure on the test graph in the appropriate place.

B. Check fire pump(s), foam tank, foam concentrate piping, and proportioning system, to make sure all components are set up in normal mode.

C. Check all system valves to make sure they are in the normally open or normally closed position, as appropriate for the system discharge test.

D. Prepare the area for the full flow operational test.

E. Coordinate the efforts of the testing personnel before the test begins, to minimize the amount of time the system discharges.

F. Notify all testing and witnessing personnel that the test is about to begin.

Step 9: Trip the system and record the results.

A. Manually trip the system using the manual pull station at the deluge valve.

B. Allow the system to flow long enough for the discharge from the nozzles to stabilize before checking spray patterns and taking foam-water samples.

C. Visually inspect nozzles for proper discharge. Note any deficiencies.

D. Take three foam-water samples as follows: one from the valve attached to the highest, most remote discharge outlet (see Step 7E), and two from the foam blanket.

E. Shut off the valve attached to the highest, most remote discharge outlet.

F. Take residual pressure readings at the system deluge valve and at the gauge installed at the highest, most remote discharge outlet.

Step 10: Begin system shutdown procedures.

A. Shut off the foam supply. Allow only water to flow through the system until the system is flushed clear.

B. Shut the water supply valve to the deluge system.

C. Open system drains.

Step 11: Determine foam concentrate levels from the samples taken during the test.

A. Use the pipette to apply a drop of foam-water sample #1 to the refractometer, then take reading #1.

B. Clean the pipette and refractometer after each reading, and follow this procedure for foam-water samples #2 and #3.

C. Plot the single best result from the three foam-water samples on the graph.

D. Adjust the pressure recorded from the gauge connected to the end of the hose to that from the highest, most remote discharge outlet by determining the head pressure caused by the elevation difference between the gauge and the branch line fitting and subtracting it from the gauge reading.

For example, if the gauge at the end of the hose is 15 ft (4.57 m) below the branch line fitting, the pressure difference caused by elevation is calculated as follows:

$P_e = 15 \times 0.433$ psi $\quad (P_e = 4.57 \times 0.0978$ bar)

$P_e = 6.5$ psi $\quad (P_e = 0.45$ bar)

Subtracting the head pressure [6.5 psi (0.45 bar)] from the pressure recorded at the gauge at the end of the hose will provide the actual pressure at the highest, most remote discharge outlet. Record this pressure in the appropriate place on the graph.

Evaluation of Results

Analyze the test results against those from previous tests and the installation requirements. Where the results of this test do not compare favorably to previous tests, system modifications might be necessary.

The procedure(s) outlined above is not mandated by NFPA 25 and represents only one approach to conducting this test.

Testing Procedures for 11.3.5.2

Solution Concentration Testing

Purpose

The purpose of this test is to identify foam concentrate levels. The two methods shown below — the *refractometric* and the *conductivity* methods — are used to determine the required foam concentration levels.

Refractometric Method

The *refractometric method* uses a prism that has a much greater refractive index than the sample solution to be measured. Measurements are made at the interface of the prism and the sample solution. For a weak (low-concentration) solution, the difference between the refractive index of the prism and that of the solution is high. For a strong (high-concentration) solution, the difference between the refractive index of the prism and that of the sample is smaller. The following is a sample test procedure for foam concentration determination.

The foam solution concentration determination test is used to determine the percent concentration of a foam concentrate in the water being used to generate foam. It is typically used as a means of determining the accuracy of a system's proportioning equipment. If the level of foam concentrate injection varies widely from that of the design, it may abnormally influence the expansion and drainage foam quality values, which may, in turn, affect the foam's fire performance.

A handheld refractometer is used to measure the refractive index of the foam solution samples. This method is not particularly accurate for aqueous film-forming foam (AFFF) concentrate or alcohol-resistant AFFFs, since they typically exhibit very low refractive index readings. For this reason, the conductivity method might be preferred where these products are used.

A detailed procedure illustrating the use of the refractive index method, a0s found in Annex B of NFPA 16, follows.

Tools/Equipment

Prepare a base (calibration) curve using the following apparatus:

- Four 3.4 fl oz (100 mL) or larger plastic bottles with caps
- One measuring pipette [0.34 fl oz (10 mL)] or syringe [0.34 fl oz (10 mL)]
- One 3.4 fl oz (100 mL) or larger graduated cylinder
- Three plastic-coated magnetic stirring bars
- One handheld refractometer (American Optical Model 10400 or 10441, Atago NI, or equivalent)
- Standard graph paper
- Ruler or other straight edge

Procedures

1. Using water and foam concentrate from the system to be tested, make up three standard solutions using the 3.4 fl oz (100 mL) or larger graduated cylinders. These samples should include the nominal intended percentage of injection, the nominal percentage plus 1 percent, and the nominal percentage minus 1 percent.
2. Place the water in the graduated cylinders (leaving adequate space for the foam concentrate), and then carefully measure the foam concentrate samples into the water using the pipette or syringe. Use care not to pick up air in the foam concentrate samples.
3. Pour each measured foam solution from the graduated cylinder into a 3.4 fl oz (100 mL) or larger plastic bottle. Each bottle should be marked with the percent solution it contains. Add a plastic stirring bar to the bottle, cap it, and shake it thoroughly to mix the foam solution.
4. After thoroughly mixing the foam solution samples, take a refractive index reading of each sample. To do this, place a few drops of the solution on the refractometer prism, close the cover plate, and observe the scale reading at the dark field intersection. Since the refractometer is temperature compensated, it might take 10 to 20 seconds for the sample to be read properly. It is important to take all refractometer readings at ambient temperatures of 50°F (10°C) or above.
5. Using standard graph paper, plot the refractive index readings on one axis and the percent concentration readings on the other (see Testing Procedure Exhibit 11.3). This plotted curve will serve as the known baseline for the test series.
6. Set the solution samples aside in the event the measurements need to be checked.
7. *Sampling.* Collect foam solution samples from the proportioning system, being careful to ensure that the samples are taken at an adequate distance downstream of the proportioner being tested.

Evaluation of Results

Take refractive index readings of the sample and compare them to the plotted curve to determine the percentage of the samples. Testing Procedure Exhibit 11.1 shows a typical graph of refractive index versus foam concentrate.

Both the conductivity and refractometric methods of measuring foam concentrate involve comparing foam solution test samples to premeasured solutions that are plotted on a baseline graph of percent.

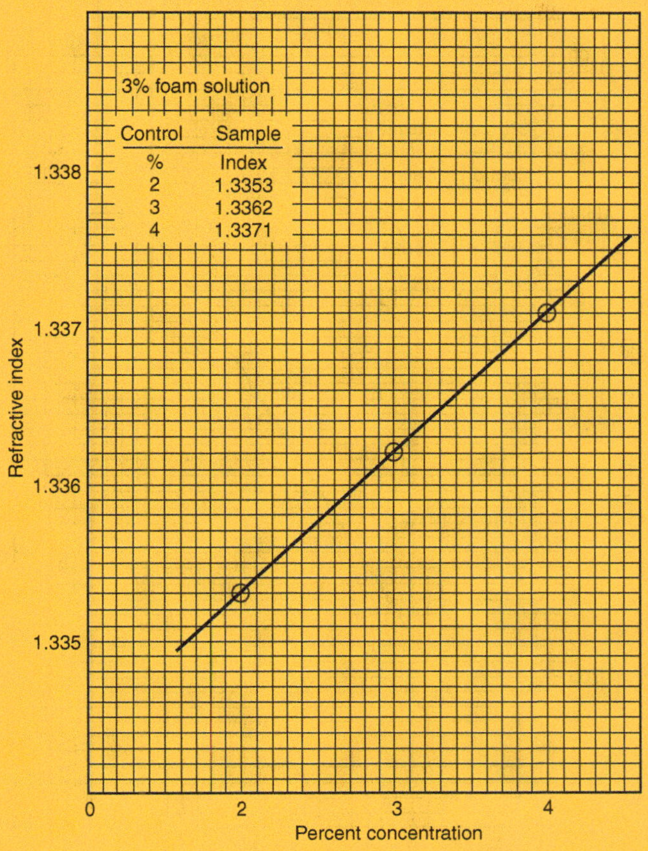

TESTING PROCEDURE EXHIBIT 11.1 *Typical Graph of Refractive Index Versus Foam Concentration. (Source: NFPA 11, 2010, Figure D.2.1.1.2)*

Conductivity Method

The conductivity method uses a conductivity meter (as shown in Testing Procedure Exhibit 11.2), which evaluates changes in electrical conductivity as foam concentrate is added to water.

The conductivity method is accurate for measuring foam concentrate percentage, provided there are substantial changes in conductivity, as foam concentrate is added to the water in relatively low percentages. Because it is highly conductive, the use of salt or brackish water might not be suitable for this method, due to small conductivity changes as foam concentrate is added. It will be necessary to make foam and water solutions in advance to determine if adequate changes in conductivity can be detected if the water source is salty or brackish.

A detailed procedure illustrating the use of the conductivity method, as found in Annex D of NFPA 11, follows.

Tools/Equipment

Prepare a base (calibration) curve using the following apparatus:

- Four 3.4 fl oz (100 mL) or larger plastic bottles with caps
- One measuring pipette [0.34 fl oz (10 mL)] or syringe [0.34 fl oz (10 mL)]
- One 3.4 fl oz (100 mL) or larger graduated cylinder
- Three plastic-coated magnetic stirring bars
- A portable, temperature-compensated conductivity meter (Omega Model CDH-70, VWR Scientific Model 23198-014, or equivalent)
- Standard graph paper
- Ruler or other straight edge

Testing Procedure Exhibit 11.2 shows the equipment needed for using the conductivity method.

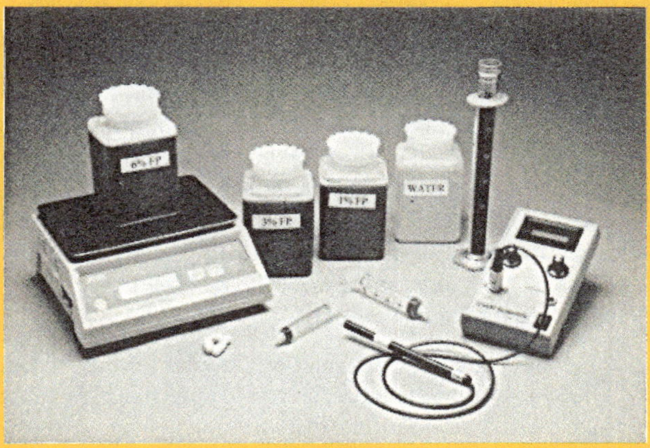

TESTING PROCEDURE EXHIBIT 11.2 *Equipment Needed for Conductivity Method of Proportioning Measurement. (Source: NFPA 11, 2010, Figure D.2.1.2)*

Procedure Steps

1. Using water and foam concentrate from the system to be tested, make up three standard solutions using the 3.4 fl oz (100 mL) or larger graduated cylinders. These samples should include the nominal intended percentage of injection, the nominal percentage plus 1 percent, and the nominal percentage minus 1 percent.
2. Place the water in the graduated cylinders (leaving adequate space for the foam concentrate), and then carefully measure the foam concentrate samples into the water using the pipette or syringe. Use care not to pick up air in the foam concentrate samples.
3. Pour each measured foam solution from the graduated cylinder into a 3.4 fl oz (100 mL) or larger plastic bottle. Each bottle should be marked with the percent solution it contains. Add a plastic stirring bar to the bottle, cap it, and shake it thoroughly to mix the foam solution.
4. After making the three foam solutions in this manner, measure the conductivity of each. Refer to the instructions that came with the conductivity meter to determine proper procedures for taking readings. It will be necessary to switch the meter to the correct conductivity range setting to obtain a proper reading.

Solution Concentration Testing *(Continued)*

5. Most synthetic-based foams used with fresh water result in foam solution conductivity readings of less than 2000 microsiemens. Protein-based foams in fresh water solutions generally produce conductivity readings in excess of 2000 microsiemens. Due to the temperature compensation feature of the conductivity meter, it may take a short time to obtain a consistent reading.
6. Once the solution samples have been measured and recorded, set the bottles aside for control sample reference. The conductivity readings then should be plotted on the graph paper (see Testing Procedure Exhibit 11.3). It is most convenient to place the foam solution percentage on the horizontal axis and the conductivity readings on the vertical axis.
7. Use a ruler or straightedge to draw a line that approximates connecting all three points. Although it might not be possible to hit all three points with a straight line, they should be very close. If not, repeat the conductivity measurements and, if necessary, make new control sample solutions until all three points plot in a nearly straight line. This plot will serve as the known base (calibration) curve to be used for the test series.
8. *Sampling.* Collect foam solution samples from the proportioning system, being careful to ensure that the samples are taken at an adequate distance downstream of the proportioner being tested. Using foam solution samples that are allowed to drain from expanded foam is not recommended, because doing so may produce misleading conductivity readings.

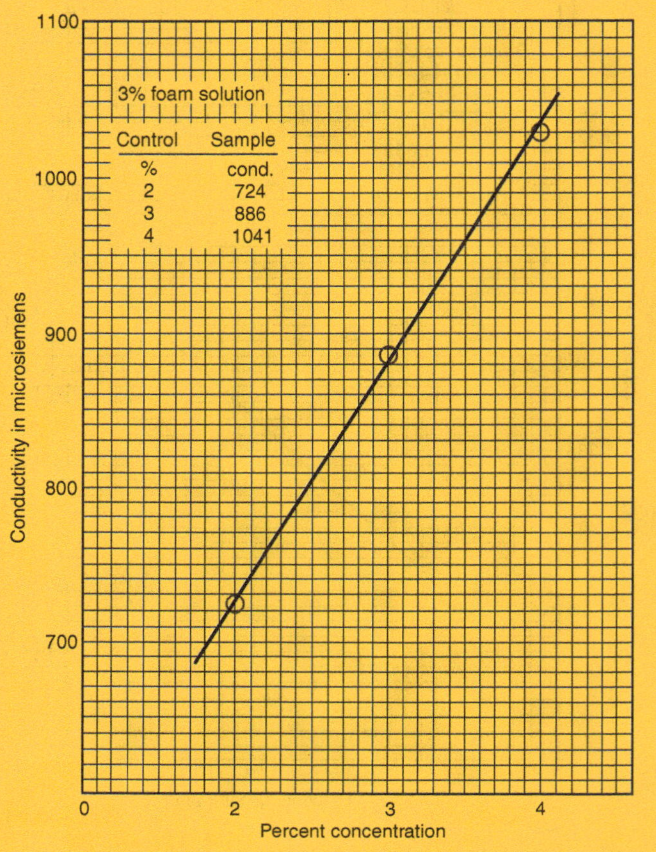

TESTING PROCEDURE EXHIBIT 11.3 *Typical Graph of Conductivity Versus Foam Concentration. (Source: NFPA 11, 2010, Figure D.2.1.2.2)*

Evaluation of Results

Once one or more samples have been collected, read the conductivity of each and find the corresponding percentage from the base curve prepared from the control sample solutions. Testing Procedure Exhibit 11.3 shows a typical graph of conductivity versus foam concentration.

Both the conductivity and refractometric methods of measuring foam concentrate involve comparing foam solution test samples to premeasured solutions that are plotted on a baseline graph of percent.

References

National Fire Protection Association, 1 Batterymarch Park, Quincy, MA 02169-7471.

NFPA 11, *Standard for Low-, Medium-, and High-Expansion Foam*, 2010 edition.

NFPA 16, *Standard for the Installation of Foam-Water Sprinkler and Foam-Water Spray Systems*, 2011 edition.

The procedure(s) outlined above is not mandated by NFPA 25 and represents only one approach to conducting this test.

ITM NFPA 25 — INSPECTION TESTING MAINTENANCE

12

WATER MIST SYSTEMS

Chapter 12 covers the inspection, testing, and maintenance (ITM) of water mist systems. These systems, which are simple when compared to today's complex water-based fire protection systems, were first used to protect ferries in the 1940s. Today, water mist systems are used widely in the marine sector as alternatives to sprinkler systems on passenger cruise ships. Water mist systems are also increasingly used in special hazards and equipment protection, such as for turbines, generators, industrial deep fat frying equipment, cleanroom production tools, and mobile mining equipment. They are also being installed for protection in historic buildings and computer rooms (i.e., for subfloor protection). Some are also approved for use to protect light hazard occupancies. As with any fire protection system, water mist systems require reliable, competent inspection, testing, and maintenance to ensure they will perform properly if a fire occurs.

FAQ Can water mist systems use the same components as a sprinkler system?

Water mist systems use different components and fire control concepts than the more commonly known water-based fire protection systems that comply with NFPA 13, *Standard for the Installation of Sprinkler Systems.* Some system configurations involve components borrowed from gaseous agent extinguishing systems, such as compressed gas cylinders, pressure regulators, and pneumatic control valves. Others use positive displacement pumps and high pressure piping or tubing not previously used in fire protection. Many inspectors familiar with ITM of fire pumps, sprinkler systems, and water spray or foam-deluge systems will be unfamiliar with some of the hardware used in a water mist system. Similarly, inspectors familiar with gaseous clean agent systems might not be familiar with pumps, controllers, filters, and water storage equipment.

The design and performance criteria of a water mist system also are different than those of a typical sprinkler system. NFPA 750, *Standard on Water Mist Fire Protection Systems,* is founded on the premise of performance-based design and mandates that the application density for each type of water mist system be individually determined. This determination is required to be completed by fire testing to a comprehensive, unbiased fire test protocol. And unlike NFPA 13, which dictates that all pumps, pipes, fittings, and components be of a certain type and pressure rating, NFPA 750 allows for a range of pressure arrangements, power sources, and component technologies. Some water mist systems exceed an operating pressure of more than 500 psi (34.5 bar). Thus NFPA 750 establishes only general design factors applicable to all water mist systems, such as hydraulic calculation procedures, acceptance testing, water quality, and duration of protection, as well as only generic requirements, such as those needed for corrosion control and mechanical strength.

Critical Point

Water mist systems use different equipment components and fire control concepts than the typical water-based fire protection systems that comply with NFPA 13.

Guidance is provided for testing agencies on characteristics of an adequate fire test protocol but relies on the manufacturers and the listing or approval agencies for details that might be unique to a particular technology or manufacturer. Special training by the manufacturer's representative is often necessary, especially where proprietary differences exist between water mist systems developed by different manufacturers.

The requirements for ITM of water mist systems in NFPA 25 are taken from Chapter 13 of NFPA 750. These requirements must be supplemented by the specific requirements contained in each manufacturer's design, installation, operation, and maintenance (DIOM) manual. A DIOM manual is required for all listed or approved water mist system equipment and should be readily available for reference at the premises where such systems are installed.

12.1 Inspection and Testing

12.1.1 Components and Systems.

Some of the components of water mist systems are similar to those of other water-based extinguishing systems, such as ancillary fire and smoke detection and activation systems. Detection and activation systems are addressed by *NFPA 72®, National Fire Alarm and Signaling Code.* All other components specific to water mist systems — such as special pump assemblies, gas cylinders, control valves, nozzles, pipe and stainless steel tubing, hangers, supports, strainers and filters, and manual releases — are addressed in specific paragraphs in this chapter.

FAQ Are water mist and sprinkler systems considered to be equivalent to each other and allowed to be installed to protect the same occupancies?

Water mist fire suppression systems are gaining use in various industries for special hazards and equipment protection. But whether they are equivalent to a typical fixed fire sprinkler system is currently under industry (and NFPA 750 committee) debate. The NFPA 13 technical committees reviewed this issue during the development of the 2013 edition and added language in NFPA 13, NFPA 13R, *Standard for the Installation of Sprinkler Systems in Low-Rise Residential Occupancies,* and NFPA 13D, *Standard for the Installation of Sprinkler Systems in One- and Two-Family Dwellings and Manufactured Homes,* that concluded that water mist systems are not covered by the sprinkler standards and the design practices associated with these systems must come from NFPA 750. The NFPA 750 committee considered adding language that would call water mist systems "equivalent" or an "alternative" to a sprinkler system; however, that language was removed at the committee's second meeting.

Currently, since there is no design procedure that would allow extrapolation of test results to different hazards and general hazard types, including those in NFPA 13, water mist systems are not being considered equivalent to typical sprinkler systems by the NFPA 13 or NFPA 750 technical committees. For land-based applications, they are primarily only used for property protection. Testing to determine the ability of water mist systems to protect life as well as property from fire will need to be qualified through some type of testing on various hazard types and hazard configurations in standard occupancies. For the purposes of inspections and testing, they are also different; these systems are configured more like special extinguishing systems and can have pressures in excess of 500 psi (34.5 bar).

Critical Point

A water mist system can be interlocked with other mechanical systems such as automatic door closers, emergency power shutdown, stopping or starting of ventilation fans, opening or closing of automatic fire dampers, stopping of production lines, or other ancillary actions integrated in a cause and effect protocol.

12.1.1.1 All components and systems shall be inspected and tested to verify that they function as intended.

A water mist system can be interlocked with other mechanical systems such as automatic door closers, emergency power shutdown, stopping or starting of ventilation fans, opening or closing of automatic fire dampers, stopping of production lines, or other ancillary actions integrated in a cause and effect protocol. In these cases, interlocked systems might need to be shut down temporarily for testing. Relevant plant operations personnel should be informed of any temporary shutdowns.

ITM Deficiency, Impairment, or Hazard Evaluation?

ANSWER: ITM Deficiency

Water tanks for water mist systems are to be inspected weekly or monthly for proper water level, and all valves and appurtenances are to be inspected semiannually as required by Table 12.1.2. The relief valve on this hydropneumatic tank was found to have a plug installed that will not allow the valve to operate to perform its intended function as required by 12.1.1.1. This valve is required by NFPA 750 as a safety device to relieve excess pressure.

The purpose of an inspection is to verify that the items being inspected appear in operating condition and are free of physical damage. In this case, the relief valve is not in operating condition, and this finding should be considered a critical deficiency. Although this condition will probably not have a material effect on the ability of the system to function as intended in a fire event, excess pressure in the tank could cause a failure of the tank, or the components connected to the tank, and presents a possible safety hazard to personnel.

(Courtesy of Zachary Magnone and Tyco Fire Protection Products)

Where such interlocks exist, the owner should ensure that the service contractor is familiar with the entire integrated system and associated interlocks before beginning system testing or maintenance. It should be verified that the temporary shutdown of the water mist system or associated interlocks will not inadvertently interrupt important site operations.

12.1.1.2 Water mist systems that are equipped with an additive system shall be tested with the specific additive system engaged or used during the acceptance testing.

Some water mist systems have used aqueous film-forming foam (AFFF) as an additive to increase suppression capabilities. Such additives have been used for bilge area and large mobile mining equipment systems. AFFF additives are particularly suited for semi-enclosed equipment and areas. The AFFF solution storage and injection system typically is supplied by pressurized nitrogen gas, and special valves must open in order to propel the foam solution into the water stream. The requirement in 12.1.1.2 recognizes that the additive feed line is an essential part of the system and must be included in any functional or operational tests. The actual additive, not a substitute, must be used. The testing must confirm that the correct quantity of additive is injected at the correct rate.

Details of the intended operation of all elements of the system, including the additive injection system, are described in the "written sequence of operations" specified in 11.4.5 of NFPA 750. This written description is required by NFPA 750 to be submitted with the system documentation.

12.1.2 Requirements. The components of typical water mist systems to be inspected and tested are provided in Table 12.1.2.

FAQ What is involved in inspecting a water mist system?

The term *inspection* is defined in Chapter 3. When "inspecting" a component, a qualified individual visually examines it to ascertain whether it appears to be in its normal operating state and is undamaged.

Testing involves actual operation of devices and is more involved than simply inspecting components. There are different levels of testing. Some components are tested without

Critical Point

Testing involves actual operation of devices and is more involved than simply inspecting components.

involving other components in the system. For example, a fire pump can be started to confirm that electric power or diesel fuel is available and that starting switches work. There are other tests, however, that cannot be performed without involving most elements of the system. Testing a pressure regulating valve or an automatic sectional control valve, for example, may require releasing compressed gas cylinders or flowing water into overhead piping.

The manufacturer of a water mist systems must produce a design, installation, operation, and maintenance (DIOM) manual as part of the listing or approval documentation for the system. The DIOM manual contains specific ITM instructions for specialized or proprietary components or system features, which must be followed in addition to the requirements of this chapter. A copy of the manufacturer's DIOM manual must be provided to the user upon completion of the installation per 11.4.3 of NFPA 750. The DIOM manual is required to provide enough information to allow the user or a third party to verify that the system has been designed and installed properly.

Critical Point

It might be necessary to coordinate the annual ITM activities of the water mist system with another activity that requires system shutdown, so that access can be gained to protected spaces for inspection of the piping and nozzles.

FAQ What qualifications are needed to service a water mist system?

Extensive expertise is required to service any water mist system, since it involves testing and maintenance of the system components, as well as draining the system, replacing gas cylinders, and putting the system back in operating order. The owner or occupant might employ a specialized contractor with the necessary training and experience to carry out the inspections. The ITM service contractor should provide certificates issued by the water mist system manufacturer confirming the training of the employees/inspectors who will inspect, test, or provide maintenance on a water mist system.

Table 12.1.2 is a detailed list of the types of components that might be included in water mist systems. Not all of the components listed are found in every water mist system. The table indicates the number of times per year each component should be inspected or tested to the degree indicated. The manufacturer's DIOM manual must be consulted to determine if more frequent inspection or testing is recommended.

In this commentary, each grouping of components in Table 12.1.2 is followed by additional explanatory information. Schematic diagrams in Exhibit 12.1 through Exhibit 12.3 illustrate different types of components and arrangements typically encountered in water mist systems.

It might be necessary to coordinate the annual ITM tasks for the water mist system with another activity that requires system shutdown, so that access can be gained to protected spaces for inspection of the piping and nozzles. For example, it is not normally possible to enter a turbine enclosure while it is running, but if it is shut down for routine maintenance, an inspection of the water mist system could be scheduled at the same time to minimize disruption.

The inspection service provider is directed to Chapter 12 of NFPA 750, which provides details on functionality testing of water mist systems. The requirement in 12.2.5 of NFPA 750 describes preliminary functional tests to be performed to confirm that detection devices, supervisory devices, automatic and manual releasing devices, and pneumatic and hydraulic components are functional. Auxiliary functions such as alarm functions, remote annunciators, air handling shutdown, and power shutdown are checked for operation in accordance with the system requirements. Depending on the nature of the repair that was performed to remediate the deficiency or impairment, some or all of the functional and operational tests of the system should be performed.

Critical Point

Repairs requiring specialized knowledge of the components should be undertaken only by a qualified contractor, service provider, or the manufacturer's representative.

Minor deficiencies can be repaired by the owner or owner's representative if qualified. Repairs requiring specialized knowledge of the components should be undertaken only by a qualified contractor, service provider, or the manufacturer's representative.

The following commentary is divided into sections with titles corresponding to the related requirements in Table 12.1.2.

Water Supply. A reliable water source is essential for any water-based fire protection system. Chapter 10 of NFPA 750 requires that the water supply be capable of supplying, for a specified

TABLE 12.1.2 Mist Systems

Item	*Task*	*Weekly*	*Monthly*	*Quarterly*	*Semi-annually*	*Annually*	*Other*
Water supply (general)	Inspect source pressure.			X			
	Inspect source quality (*first year).				X*	X	
	Test source pressure, flow, quantity, duration.					X	
Water storage tanks	Inspect water level (unsupervised).		X				
	Inspect water level (supervised).			X			
	Confirm sight glass valves are open.		X				
	Inspect tank gauges, pressure.			X			
	Inspect all valves, appurtenances.				X		
	Drain tank, inspect interior, and refill.					X	
	Inspect tank condition (corrosion).					X	
	Inspect water quality.					X	
	Inspect water temperature.						Extreme weather
Water storage cylinder (high pressure)	Inspect water level (load cells).				X		
	Inspect water level (unsupervised).			X			
	Inspect support frame/restraints.					X	
	Inspect vent plugs at refilling.					X	
	Inspect cylinder pressure on discharge.					X	
	Inspect filters on refill connection.					X	
Additive storage cylinders	Inspect general condition, corrosion.			X			
	Inspect quantity of additive agent.				X		
	Test quality of additive agent.					X	
	Test additive injection, full discharge test.					X	
Water recirculation tank	Inspect water level (unsupervised).		X				
	Inspect water level (supervised).			X			
	Inspect supports, attachments.					X	
	Test low water level alarm.					X	
	Inspect water quality, drain, flush, and refill.					X	
	Test operation of float-operated valve.					X	
	Test pressure at outlet during discharge.					X	
	Test backflow prevention device (if present).					X	
	Inspect and clean filters, strainers, cyclone separator.					X	
Compressed gas cylinders	Inspect support frame and cylinder restraints.			X			
	Inspect cylinder pressure (unsupervised).		X				
	Inspect cylinder pressure (supervised).			X			
	Confirm cylinder control valve is open.		X				

(continues)

TABLE 12.1.2 Continued

Item	*Task*	*Weekly*	*Monthly*	*Quarterly*	*Semi-annually*	*Annually*	*Other*
	Inspect cylinder capacity and pressure rating.					X	
	Inspect cylinder compliance specification.					X	
	Confirm compressed gas meets specifications (moisture, cylinder pressure).					X	
	Hydrostatic test cylinders.						5–12 years
Plant air, compressors, and receivers	Inspect air pressure (unsupervised).	X					
	Inspect air pressure (supervised).		X				
	Start compressor.	X					
	Check compressor/receiver capacity, changes.				X		
	Inspect compressed air moisture content.					X	
	Clean filters, moisture traps.				X		
	Test full capacity, duration, and any changes in other demands.					X	
Pumps and drivers	Inspection, testing, and maintenance in accordance with the requirements of NFPA 20, *Standard for the Installation of Stationary Pumps for Fire Protection*, and NFPA 25.	X	X	X	X	X	
Standby pump	Inspect and empty the moisture trap, oil injection (pneumatic).		X				
	Inspect compressed gas supply, inlet air pressure.		X				
	Inspect outlet water (standby) pressure.		X				
	Test start/stop pressure settings for standby pressure.			X			
Pneumatic valves	Inspect cylinder valves, master release valves.		X				
	Inspect all tubing associated with release valves.			X			
	Test solenoid release of master release valve.				X		
	Test manual release of master release valve.					X	
	Test operation of slave valves.					X	
	Reset all pneumatic cylinder release valves.					X	
	Test on-off cycling of valves intended to cycle.					X	

TABLE 12.1.2 Continued

Item	Task	Weekly	Monthly	Quarterly	Semi-annually	Annually	Other
System control valves	Inspection, testing, and maintenance in accordance with the requirements of NFPA 25.	X	X	X	X	X	
Control equipment	Inspection, testing, and maintenance in accordance with the requirements of *NFPA 72, National Fire Alarm and Signaling Code.*						
Water mist system piping and nozzles	Inspection, testing, and maintenance in accordance with NFPA 25. Inspect sample of nozzle screens and strainers *(see 10.5.1.4 of NFPA 750, Standard on Water Mist Fire Protection Systems).*	X	X	X	X	X	After discharge
Enclosure features, interlocks	Inspect enclosure integrity.				X		
Ventilation	Test interlocked systems (e.g., ventilation shutdown).					X	
	Test shutdown of fuel/lubrication systems.					X	

duration, the largest single hazard or group of hazards to be protected simultaneously. An annual waterflow test is required to confirm flow rate and pressure. Where the water supply is taken from a reliable large system, testing for "quantity" may not be required, but where it is taken from a potentially limited source such as a tank, the quantity of water available should be checked. The quantity available must be sufficient to meet the design flow rate for the specified duration of protection. The basic requirement in NFPA 750 is for a 30-minute duration of protection. That duration might be relaxed for pre-engineered systems.

If the minimum duration intended for the system is less than 30 minutes, this specification should be identified on the "system design information sign" required in 11.4.4 of NFPA 750 or in other system documentation. The intent is to ensure that all pertinent design information be displayed on the sign. If the design duration of protection is not indicated on the sign, the as-built design documents and acceptance test forms should be consulted.

The water supply must be taken from a source that is equivalent in quality to potable water, with respect to bacteria, particulates, and dissolved solids, or from a source of natural seawater. Marine vessels can use natural seawater after the potable water supply has been depleted. As vessels enter waters of different quality in different ports, however, bacteria-free and silt-free seawater is not always available. In some cases, the high chloride content of seawater would be damaging to stainless steel components, and switching to seawater would shorten the service life of such piping systems, so this step is not always recommended.

FAQ What types of water quality tests should be conducted during the first semiannual inspection?

Checking water quality requires sending at least one sample of water to a water quality testing laboratory. During the first semiannual inspection, at a minimum, tests should be conducted for turbidity, algae, *Legionella* bacteria, *E. coli* bacteria, and chloride content. For stainless steel piping systems, the chloride content of water should not exceed 50 ppm. The need for laboratory testing may be reduced in future inspections if the quality of the source water is found to be acceptable over at least two consecutive semiannual tests.

The quality of the source water should be checked semiannually during the first year. If no adverse conditions are noted, it is acceptable to test the water quality annually. If adverse conditions are found, remedial measures should be taken and the water quality should be tested monthly until it is assured that the quality has improved.

Water Storage Tanks. When maintaining water storage tanks, the owner or the designated representative should be concerned primarily with confirming the following issues about the tank:

1. It contains the correct quantity of water.
2. The water quality is acceptable.
3. The discharge valves are open.

It is also important to confirm that the ambient temperature at the tank location is within the acceptable range of 40°F to 130°F (4°C to 54°C) and that there are no signs of leakage or corrosion.

For water storage tanks that are not supervised by means of a low water level switch connected to an alarm panel, the water level in the tank must be verified visually on a weekly basis. A sight glass or other means of quickly verifying water level in a tank is required. If a tank is supervised by low water level switches connected to an alarm panel, a quarterly inspection will suffice.

Even if the quality of the source water used to fill a tank is shown to be adequate, the quality of stored water must be carefully monitored. The task list in Table 12.1.2 for storage tanks indicates that the tank's water quality should be checked annually. As indicated previously for the water source, at a minimum, a semiannual check should be made during the first year. It is not advisable to wait for 1 full year to evaluate whether the quality of the water in the storage tank is deteriorating. The two greatest concerns with respect to quality of stored water are bacteria and algae growth. Algae will plug screens and small nozzle orifices; bacteria such as *Legionella* can cause illness.

A visual check for algae growth and discoloration might be enough to identify the presence of fouling substances. However, where people could be exposed to the water mist taken from stored water, it is advisable to send water samples to a laboratory to test for *Legionella* and *E. coli* bacteria. *Legionella* bacteria have been identified in some water storage tanks for water mist systems, and ultraviolet (UV) antibacterial systems have been installed as a remedial measure. Efficient UV devices can be installed inside each storage tank, which eliminates the need for re-circulation piping. If present, ancillary devices associated with the tank must be included in the ITM task list.

Paragraph 10.5.1.5.3 of NFPA 750 requires that a filter or strainer be supplied downstream (on the system side) of any water storage reservoir or break tank with an air–water interface greater than 11 ft^2 (1 m^2). For a circular tank, this interface corresponds to a tank diameter of approximately 4 ft (1.2 m). (See Exhibit 12.1.) The objective of the strainer is to remove any algae that might grow in the tank before it enters the water mist system piping. The intent of NFPA 750 is to protect tanks with a "large" air–water interface. Listed water mist systems currently on the market that use a water tank ensure that water going into the tank is filtered. However, if algae grow in the water once the water is inside the tank, fibrous material will clog valves and nozzles in the water mist system if discharged. Notwithstanding the limit in NFPA 750, the potential exists for algae growth in stored water in tanks of diameters less than 4 ft (1.2 m).

An inspector should check for the presence of algae in stored water. A sample of water — approximately 1 qt (1 L) — should be removed via the tank clean-out access and examined under good lighting for the presence of algae. A water sample should be sent to a laboratory for bacteria l analysis and turbidity and chloride tests.

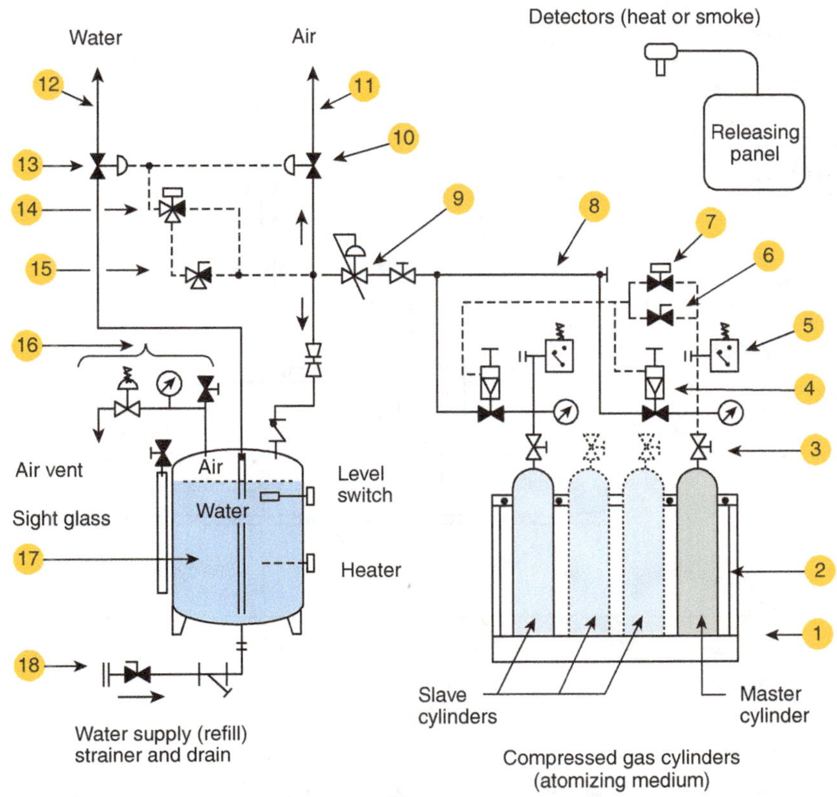

Legend			
1	Steel base and frame	10	Air-actuated globe valve
2	Compressed gas cylinders (atomizing medium): master and slave cylinders	11	Air line to twin-fluid nozzles (low pressure)
3	Cylinder control valve	12	Water line to twin-fluid nozzles (low pressure)
4	Pneumatic cylinder release valve	13	Air-actuated globe valve (cycle water line)
5	Pressure supervisory switch with burst disc	14	Low pressure solenoid valves (for operating air-actuated globe valves)
6	Manually operated master release valve	15	Manual release valve (opens globe valves)
7	Solenoid operated master release valve	16	Pressure gauge, pressure relief valve, and vent valve
8	High pressure tubing manifold	17	Low pressure rated water tank
9	Air pressure control valve (high to low pressure)	18	Drain and refill connection with strainer

EXHIBIT 12.1 *Gas-Driven System with Water Stored in Tank with Air-Water Interface Greater than 11 ft^2 (1 m^2).*

As indicated under preventive maintenance, once per year, the tank must be completely emptied, cleaned, inspected for corrosion, and refilled with filtered water.

Water Storage Cylinders. Water storage cylinders are modified compressed gas cylinders, with a special head assembly to allow water to enter and air to escape from the cylinder for refilling. Exhibit 12.2 is a schematic diagram of a water mist system using these water storage cylinders. The cylinders are typically lined with a corrosion-resistant coating. Refilling is done by connecting a fill line to an inlet and opening a vent port. The cylinder is full when water starts to flow out of the open vent port. Only an experienced technician should perform the filling operation. A permanent water supply connection with a filter and hose valve should be

EXHIBIT 12.2 *Pre-Engineered, Gas-Driven Single Fluid, High Pressure Water Mist System with Water Stored in High Pressure Cylinders.*

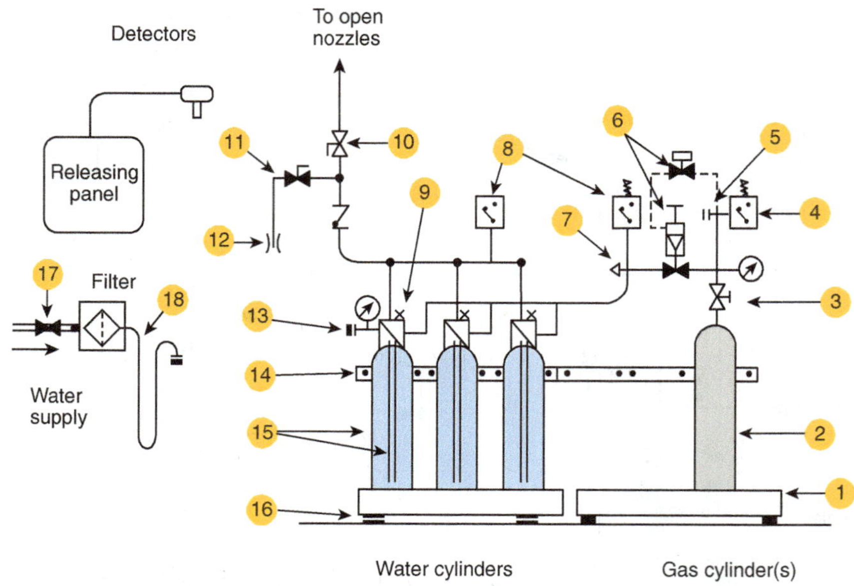

Legend

1. Steel base and frame
2. Compressed gas cylinders (driving medium)
3. Cylinder control valve
4. Pressure switch, supervise cylinder pressure
5. Burst disc
6. Solenoid operated master release valve
7. Micro-leakage valve
8. Pressure switches, alarm if system trips
9. Vent port, for filling water cylinders (fill until water discharges from open port)
10. Primary system or sectional control valve
11. Test connection and drain
12. Test orifice (alternative to full discharge)
13. Cylinder discharge header with filling port
14. Cylinder rack with restraints
15. Pressure-rated water = 150 cylinders with dip tube
16. Optional load cells to confirm water level in cylinders
17. Water supply valve, normally closed
18. Filter and hose with adaptor fitting for filling

mounted near the unit to facilitate refilling with filtered water. The intent of this provision is that concern about algae growth in water stored in cylinders is minimal because the air–water interface in a full cylinder is very small.

FAQ How can it be determined whether the water storage cylinders are full?

Some manufacturers provide a dipstick on each water storage cylinder to allow an inspector to easily verify that the tanks are full. However, with some water mist systems, once the cylinders are filled and the vent ports sealed, it is not possible to visually confirm that water storage cylinders are full of water. A written record of who last filled the cylinders must be maintained at the cylinder station. If the person doing a monthly or quarterly inspection is not the person who last filled the system, he or she must rely on the written record as the only evidence that the cylinders were at least full of water at some time in the recent past.

Some end users require placing the cylinder rack on load cells to continuously monitor the weight of the cylinders. This is the best way to supervise water level in such closed cylinders.

Additive Storage Cylinders. The most commonly used additive with water mist systems, AFFF, is used in some marine systems for bilge nozzles. ITM procedures for reservoirs of AFFF solution should comply with Chapter 11 of NFPA 25. Specialized water mist systems might incorporate additives to prevent freezing or to enhance extinguishing effectiveness. For the most part, these are small pre-engineered systems for which the additive is pre-mixed with water in the water storage container. ITM activities for such systems should follow the manufacturer's recommendations.

Water Recirculation Tanks. Water recirculation tanks are also referred to as "break tanks" and are commonly used on marine systems. The break tank is typically built into the frame of the pump assembly with direct connections to the suction inlets of the pumps. The water level is usually supervised, in that if the water level drops to a certain level, the "float-operated valve" will open and allow water from a permanent source to flow into the tank. It is preferable to have an additional low-level alarm for supervision of water level, to provide an alert to declining water level without having to wait for the refill switch to activate. Inspection and testing must include the water supply to the break tank, which might include a backflow prevention device and/or a seawater connection with cyclone separator. The water supply to the break tank must be tested for capacity: the minimum flow rate must equal or exceed the maximum output capacity of the pump assembly.

The quality of the water in the break tank should be checked frequently. If there is algae growth in the water and there is no strainer on the outlet from the tank to the suction inlet of the pumps, fibrous algae material can pass through the pumps and into the water mist system piping, where it will clog vital components. The possibility of algae growth can be reduced by frequent changeover of the water in the tank, which can occur with weekly startup of the pump assembly. Otherwise, fixed UV antibacterial units can be installed directly in the break tank.

Compressed Gas Cylinders. Exhibit 12.1 and Exhibit 12.2 illustrate two types of water mist systems that use compressed gas cylinders as the propellant or as the atomizing medium. Exhibit 12.1 shows a twin-fluid system, which includes a pressure-reducing valve to create a high pressure side and a low pressure side. The compressed air is reduced [from typically 2175 psi to 116 psi (150 bar to 8 bar)], which drives the water out of the water storage tank and acts as a propellant. This compressed air also goes to the air inlet of each of several twin-fluid nozzles, to act as the atomizing medium. The twin-fluid system is a low pressure water mist system; therefore, the water storage tank is not rated for high pressure. A similar arrangement is used for single-fluid low or intermediate pressure water mist systems.

Exhibit 12.2 represents a decaying pressure, high pressure water mist system. The compressed gas cylinder pressure is typically between 2175 psi and 2900 psi (150 bar and 200 bar), and the pressure is not regulated to a lower pressure. The compressed gas forces the water out of the high pressure water storage cylinders and into the water mist system piping. The pressure in the system piping typically starts at about 1450 psi (100 bar) and declines over approximately 10 minutes to about 290 psi (20 bar). Typical applications for a high pressure water mist system include turbine enclosures and machinery spaces.

Gas-driven water mist systems involve components not previously used with water-based fire protection systems. There are "master" and "slave" pneumatic releasing valves that are interconnected by compressed gas tubing. If the technician accidentally activates a slave valve instead of a master valve, it is possible that not all cylinders will discharge. The testing of compressed gas cylinders and their interconnections should be performed only by technicians with specialized training and experience with the particular equipment involved.

Plant Air, Compressors, and Receivers. Systems that use plant air tend to be small systems, such as a twin-fluid system for wet benches or fume hood protection. Generally, it is preferable to have a dedicated air supply for these systems, which can be better controlled than a compressed gas system serving other purposes. It is difficult for a fire protection ITM service provider to have access to test plant air compressors. The full capacity of the plant supply will have been verified during the original system acceptance tests. The plant supply should be inspected annually to confirm that assumptions about the plant capacity are still valid (e.g., that there are no new demands or operational constraints). A means should be devised for confirming the capacity of the air supply at the point of connection to the water mist system.

Pumps and Drivers. Large water mist systems used to protect entire passenger ships, or as equivalent sprinkler systems in buildings, may utilize a pumped water supply. The ITM procedures for standard fire pumps, as provided in Chapter 8 of NFPA 25, generally apply equally well to water mist systems involving pumps. Some additional requirements or cautions, however, should be observed for water mist systems that use positive displacement pumps.

When using positive displacement pumps to supply fire protection systems that have widely varying demands, the challenge is to provide a means of bypassing pump water flow that cannot be discharged from the number of open nozzles. The operation of flow-bypass valves (called "unloader" valves) is critically important to the operation of the system. Even if only a few thermally activated water mist nozzles operate, the pump assembly must deliver the proper flow to those nozzles while bypassing the unused portion of the flow. If many nozzles are opened by heat so that the entire output from the pump assembly is required, the unloader valves should not allow any water to flow to the bypass. The operation of the pump assembly for both small and maximum demands will have been verified at the acceptance test. It is important not to disturb any of the settings on the unloader valves in the course of the annual pump testing.

Exhibit 12.3 is a schematic of one type of high pressure pump assembly for an engineered water mist system. The assembly consists of two electric motors, each driving two small piston pumps. The total capacity of the pump assembly is the sum of the output of the four pumps. If there are not enough nozzles opened on the system to flow the full capacity of the assembly, the system pressure will rise until it causes one or more unloader valves to divert the pump flow to the bypass line. The pressure at which an unloader valve opens is pre-set at the time of the system design. Typical applications for this pump assembly include passenger ships, multi-zone industrial systems, and sprinkler-equivalent systems in buildings.

As shown with the tag numbered 9 in Exhibit 12.3, a test connection should be provided on the discharge manifold. The test connection should include a predesigned orifice device that will allow the full flow of the pump at the desired backpressure. Since the unloader valves respond to the system backpressure, it will not be possible to verify the proper operation of all contributing pumps without approximately matching the corresponding system pressure. A throttling valve can be installed on the discharge line to allow the backpressure in the system to be modified.

Some high pressure pump assemblies do not have unloader valves installed on each pump but, instead, have an equivalent large-volume pressure relief valve on the discharge manifold. (See the valve identified with tag number 8 in Exhibit 12.3.) Before conducting annual flow testing, the inspector must understand the intended operation of all pressure or flow control devices.

In Exhibit 12.3, the fire pump controller is shown as a pair of controllers. The FM Global listing for the pump assembly requires one controller per electric motor (per NFPA 20, *Standard for the Installation of Stationary Pumps for Fire Protection*). The controller is usually programmed to start or stop individual motors according to the demands on the system. Pressure and flow sensors are provided on the pump assembly, which are used in the system control logic program, to determine if another motor is needed or if one of several motors can be shut down. These programmable features are different from a standard fire pump installation, which usually involves only one motor and one pump.

The waterflow test results from the annual pump test need to be compared to the flow test results obtained during the acceptance testing of the system. This is relatively straightforward for low and intermediate pressure systems that use centrifugal-type pumps. As indicated in Chapter 8 of NFPA 25, the usual practice is to plot at least three points from the flow test on a pressure versus flow plot, subtract the water supply pressures from the test pressures, and draw

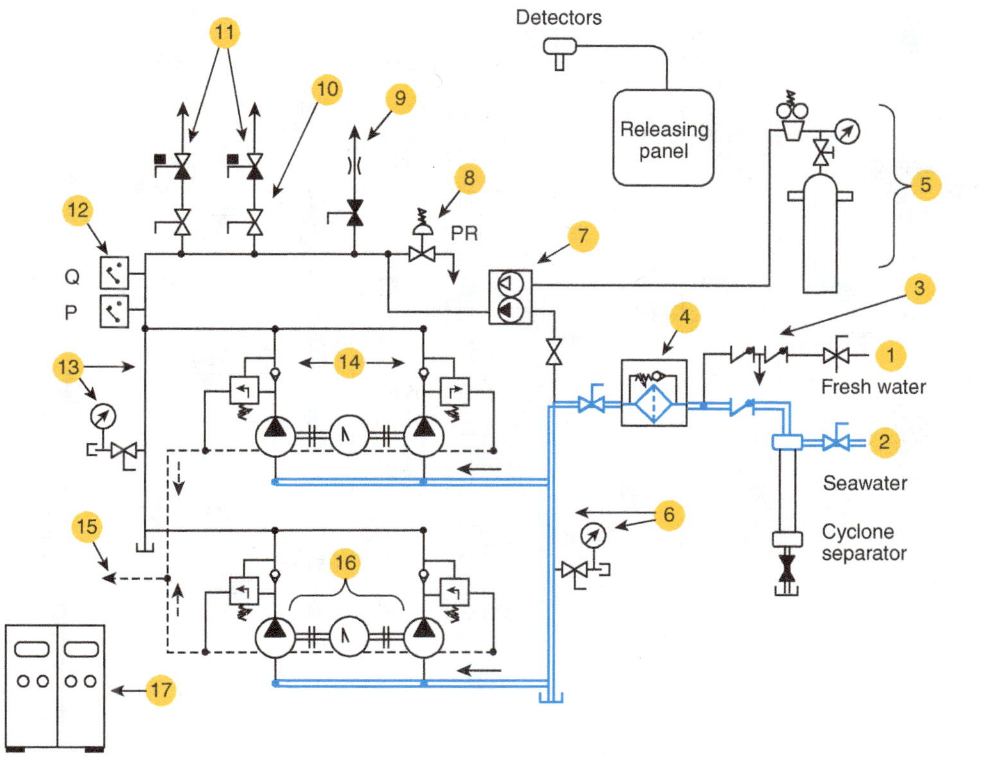

Legend

1 Fresh water supply	10 Isolation valve for sectional valves
2 Seawater supply with cyclone separator (marine systems only)	11 Solenoid actuated sectional control valves
3 Backflow device (optional)	12 Control sensors, pressure (P) and flow (Q), connected to programmable controller
4 Filters or screens with bypass	13 Discharge manifold with pressure gauge
5 Air-supply and regulator for standby pressure maintenance pump	14 Unloader valve (one per pump)
6 Suction manifold with NPSH gauge (+/–)	15 Unloader valve discharge bypass line (to drain)
7 Pneumatic standby pressure pump	16 Positive displacement pumps (2 per motor)
8 Pressure relief valve	17 Programmable pump controllers in bank (one per motor)
9 Test connection with flowmeter	

EXHIBIT 12.3 *Engineered Water Mist System with Positive Displacement Pump Assembly That Has Unloader Valves on Individual Pumps and Sectional Control Valves for Separate Zones.*

a smooth curve through the resulting points. The smooth curve should lie on or above the "factory" pump curve. If it does not, one concludes that some impairment has occurred.

The situation is more complex for a positive displacement pump or pump assembly. One reason is that the unloader valves or pressure relief valve on the manifold are usually set at less than the maximum possible pump pressure, and there may be at least a 5 percent variation in the settings of individual unloader valves. The pump curve for the output of an assembly of positive displacement pumps, with all pumps contributing, will invariably be less than the presumed unloader setting, and the flow will be less than the theoretical sum of the individual pumps. Exhibit 12.4 illustrates the theoretical pump curve for a positive displacement pump assembly with eight pumps and eight unloader valves set nominally at 1595 psi (110 bar).

EXHIBIT 12.4 *Composite Pump Curve for a Four-Motor, Eight-Pump Assembly, with All Eight Unloader Valves Set at 1595 psi (±72.5 psi) [110 bar (±5 bar)].*

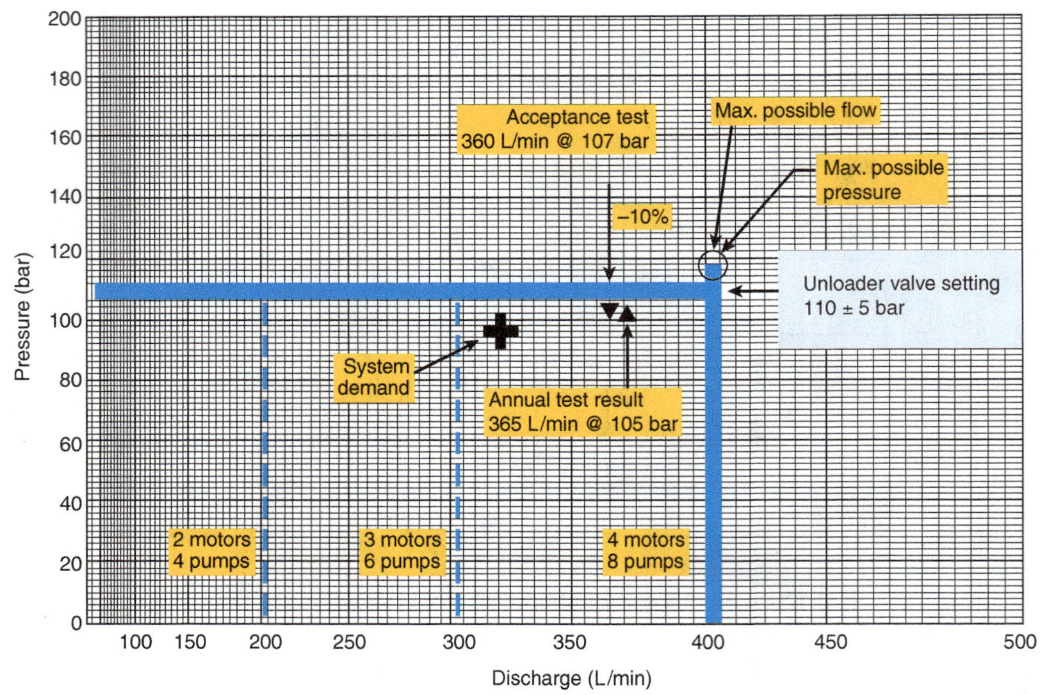

FAQ What factors determine whether the pump assembly is operating properly?

The key to determining whether the pump assembly is operating as it should is to ensure that the pressure achieved at maximum flow is within 5 percent of the pressure relief valve setting or the highest unloader valve setting and that the total flow is within 10 percent of the theoretical maximum flow rate, as shown in Exhibit 12.4.

Exhibit 12.4 illustrates one flow test comparison with the acceptance test result. During the acceptance test, a maximum flow/pressure result of 9 gpm at 1552 psi (360 L/min at 107 bar) was accepted. That result was 10 percent lower flow than the theoretical maximum of 10 gpm at 1552 psi (400 L/min at 107 bar), and 1552 psi (107 bar) was within the ±70 psi (±5 bar) precision of the unloader valves. During the annual flow test, a maximum flow rate of 9 gpm (365 L/min) was recorded at a pressure of 1524 psi (105 bar). This was an acceptable condition, since the flow was slightly greater than the acceptance test, and within 10 percent of the maximum; and the pressure was within the ±70 psi (±5 bar) precision of the unloader valve settings.

In the case of high pressure pumps operating at around 1450 psi (100 bar), the supply water pressure [typically less than 30 psi (2 bar)] does not show up on the scale. The pressure at the suction inlet to the pumps is nevertheless important. Positive displacement pumps are more sensitive to net positive suction head (NPSH) problems than centrifugal pumps. The need to filter all water going to the system adds hydraulic resistance to the suction inlet line. If filters are present on the suction line, friction losses in the suction supply might become too high. In some cases, a suction booster pump might be used to increase the suction line pressure to overcome friction loss through the filters. Starting of the positive displacement pumps must then be preceded by starting the suction booster pump.

It is important that filters be able to accommodate the maximum system demand flow rate for an extended period of time without clogging. Unlike the manual-fill systems suitable for pre-engineered systems (see Exhibit 12.2), which only need to accommodate the flow needed to refill a reservoir in a reasonable period of time, the filters on the suction supply to a pumped system must be sized for the maximum system flow rate. In Exhibit 12.3, a spring-loaded check valve is placed on a bypass around the filters (tag number 4). If the filters become plugged, the bypass line should automatically open to allow flow of unfiltered water to the pumps. This must be tested during the annual discharge test.

Gauges on the suction side of the pumps should be checked to ensure that they are capable of measuring the range of suction line pressures. If they are not, then a new gauge with the appropriate scale to read the NPSH will need to be installed. Other pumps use two banks of filters in parallel, with a pressure transducer connected to the pump controller. If the head loss across the first bank of filters becomes too high (as the filters become clogged), a solenoid valve opens and reroutes the flow through the redundant bank of filters.

The maintenance plan for a water mist pump system must be designed to evaluate the specific details of the suction supply to the pumps. It is recommended that temporary supplemental instrumentation be installed to measure performance factors during the annual discharge test.

Standby Pumps. ITM tasks for standby pumps on low or intermediate pressure water mist systems are covered in Chapter 8 of NFPA 25. High pressure water mist systems may use a pneumatic pump to maintain the system piping at an elevated pressure [e.g., 360 psi (25 bar)]. Inspection and testing of such pumps might not be covered in Chapter 8. Tag number 7 in Exhibit 12.3 shows the symbol for a pneumatic standby pump with dedicated cylinders of compressed gas. The pneumatic pump activates on a signal from the pump controller panel (tag number 17), which receives pressure and flow information from the switches identified by tag number 12. It is important to check the compressed gas supply and outlet pressure on a monthly basis. The owner's personnel can monitor the standby pump on a monthly basis, but the ITM service provider should perform a test of the unit during the annual servicing.

Pneumatic Valves. Exhibits 12.1 and 12.2 show compressed gas cylinders used for two types of gas-driven water mist systems. Special pneumatic valves are released by a solenoid valve to activate the systems. On some systems, there is a master valve on the first cylinder, which is electrically released on a signal from the releasing panel. A pneumatic line then connects the first releasing valve to additional valves on the remaining cylinders.

The integrity of pneumatic tubing connecting master and slave valves is seldom or never supervised. For that reason, pneumatic valves must be inspected carefully for possible problems with the control tubing. Generally, the system can only be determined to function properly at the time of an annual trip test of the system.

Some water mist systems are designed to "cycle" flow on and off for specified periods of time. For example, the system might be designed to discharge water for 60 seconds, stop for 40 seconds, and then resume discharging water mist for another 60 seconds. The cycle is repeated for the intended duration of protection, which is usually 10 minutes.

FAQ How should the total volume of water required for 10 minutes of protection be calculated for a cycling system?

The total volume of water required for 10 minutes of protection for a cycling system is based on the total number of minutes of actual discharge, not on the total elapsed time from the start of discharge.

Under ideal conditions, cycling systems can extinguish hydrocarbon fires in enclosures more quickly and with less total water usage than continuous discharge systems. They rely, however, on the confinement of heat and water vapor inside the enclosure to assist in extinguishing fire. During each "off" cycle at the start of discharge, a hydrocarbon fire might re-grow and increase the temperature in the enclosure and, hence, the amount of water vapor that can be held in the vapor state. By the third or fourth cycle, the enclosure is warm enough to sustain a high water vapor concentration, which reduces the oxygen concentration and results in extinguishing the fire. Therefore, all ventilation to an enclosure must be interlocked to stop upon activation of the water mist system. It is also critical to verify that the enclosure integrity is maintained for such systems.

The timing of the on and off cycles is controlled by a system control panel and specialized solenoid valves. The operation of the timing circuit should be tested during the annual inspection.

System Control Valves. As on any water-based fire protection system, all primary control valves must be supervised. Valves identified with tag number 10 in Exhibit 12.3 are lever-handled ball valves. They can be visually determined to be "open" and can be supervised by using a standard tamper switch arranged to indicate a movement of the lever handle.

Zone control valves operate in a manner similar to deluge valves on sprinkler and water spray systems. On high pressure systems, zone control valves are specially designed to be able to hold back high pressure water but open the control valve with a small solenoid valve typical of standard deluge systems, such as a 1 amp maximum draw. See valves identified with tag number 11 in Exhibit 12.3. Special steps are required to ensure that such valves are properly set to be released upon a signal from the releasing panel. During the annual inspection and testing, all solenoid valves and zone control valves must be tested to ensure that they operate properly.

It is particularly important to test the sectional control valves used in large water mist systems. These valves are electrically activated. They might be non-indicating, and it might not be clear how to open them manually. Even during a trip test, it is not possible to tell by visual inspection whether they are fully or partially open. On the high pressure systems, friction loss through sectional control valves can be very high. In order to limit friction losses within reasonable bounds, the total flow to a system might be divided among two or more sectional valves. This introduces the possibility that one of the valves in the group could fail to open, and the hydraulic performance of the system could be adversely affected. The only way to test sectional control valves is to open them under full pressure and flow water. In facilities where a full flow test can be conducted annually, the performance of sectional valves can be easily tested. If it is not possible to do a full system discharge, an alternative means of testing the functionality of the valves must be found. It might be necessary to install a test line on each individual zone in a water mist system, downstream of (on the system side of) each sectional control valve so that the valve can be tested.

Control Equipment. Water mist systems rely on a separate detection system to activate. The one exception is water mist systems that are sprinkler equivalents and use thermally activated (automatic) nozzles. Inspection and testing procedures, therefore, must include the detection system. The type of detection is often dictated by the listing for the system. For example, turbine enclosures and machinery spaces listed by FM Global are required to use thermal detectors. The listing also indicates that certain shutdown events must be synchronized with the release of the system. The ventilation system should be shut down, door holders released, dampers closed, and lubrication or fuel lines shut off. Obviously, the annual testing of a water mist system must confirm the proper operation of all events associated with release of the water mist system.

Where a cause and effect protocol has been developed to identify how the water mist system integrates with other process and safety operations, the protocol should be available to the water mist system inspector.

12.1.2.1 Water mist nozzles shall be inspected from the floor level annually.

Exhibit 12.5 through Exhibit 12.7 illustrate several different styles of nozzles that might be used on water mist systems.

12.1.2.1.1 Any water mist nozzle that shows signs of any of the following shall be replaced:

(1) Leakage
(2) Corrosion
(3) Physical damage
(4) Loss of fluid in the glass bulb heat responsive element

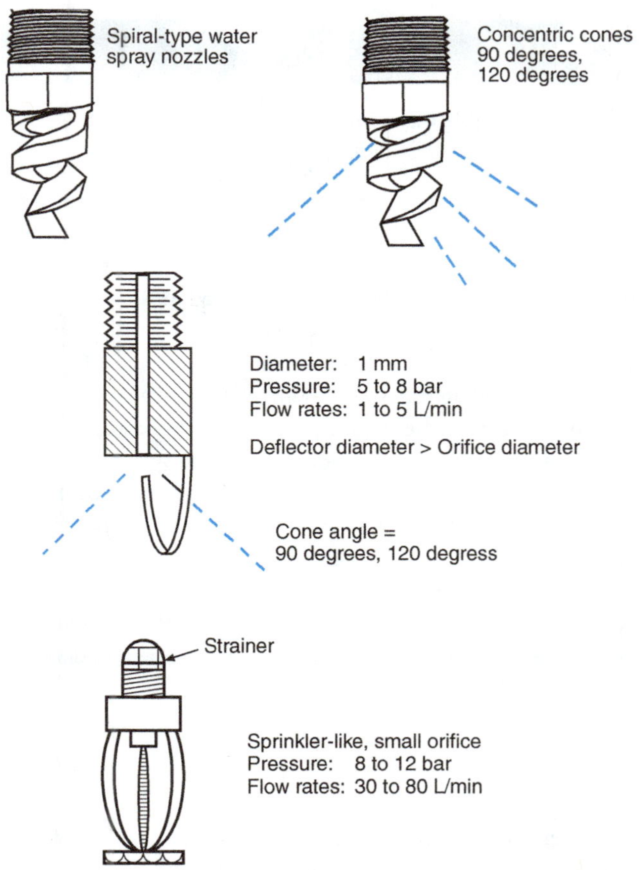

EXHIBIT 12.5 *Several Types of Impingement Nozzles for Fine Sprays. (Source: Fire Protection Handbook, 2008, Figure 16.8.12)*

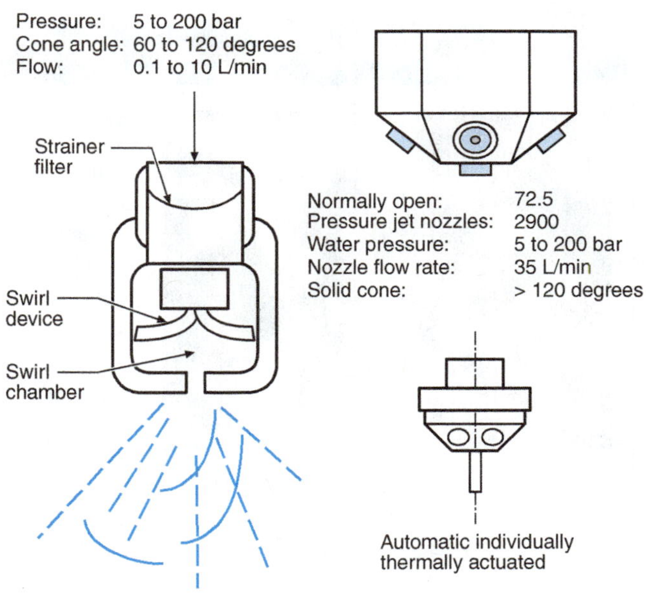

EXHIBIT 12.6 *Pressure Jet Nozzles. (Source: Fire Protection Handbook, 2008, Figure 16.8.13)*

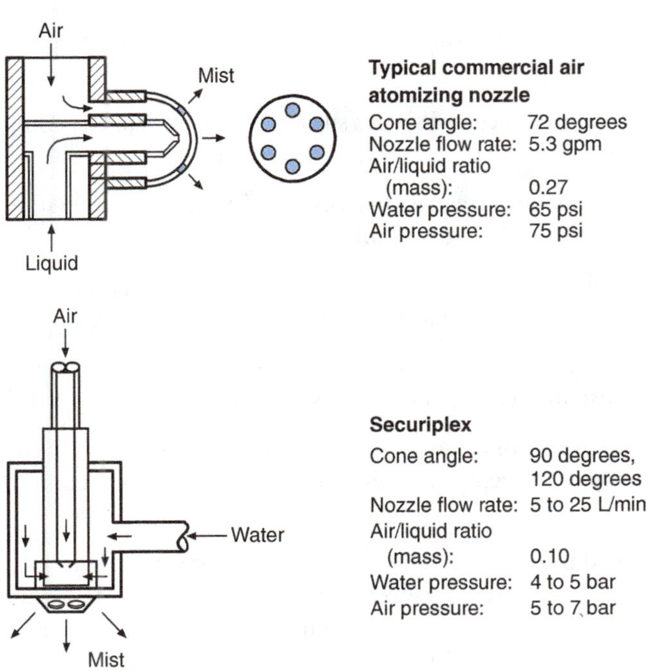

EXHIBIT 12.7 *Two Designs for Twin-Fluid Nozzles. (Source: Fire Protection Handbook, 2008, Figure 16.8.14)*

ITM Deficiency, Impairment, or Hazard Evaluation?

(Courtesy of Zachary Magnone and Tyco Fire Protection Products)

ANSWER: ITM Deficiency and Possible Impairment

Compressed gas cylinders for water mist systems are to be inspected monthly (if unsupervised) or quarterly (if supervised) for proper cylinder pressure as required by Table 12.1.2. The gauge shown in this photo indicates there is no pressure in the cylinder. Compressed gas cylinders can be used to pressurize a self contained water source, such as a tank or water cylinders, or can be used as an atomizing media to create the mist that is common in twin fluid systems.

Based on the number of compressed gas cylinders on the system, this finding could be considered either a critical deficiency or an impairment. If this is one of multiple cylinders and the others have the required pressure as indicated on the gauge, this could be considered a critical deficiency because the system will still function in a fire event. The lack of pressure from this one cylinder will adversely impact the system performance by lowering the duration time of the discharge, by lowering the discharge pressure, or both. However, if this is the only compressed gas cylinder for the system, then the system should be considered impaired and the impairment procedures in Chapter 15 should be implemented immediately.

(5)* Loading
(6) Painting, unless painted by the water mist nozzle manufacturer

A.12.1.2.1.1(5) In lieu of replacing water mist nozzles that are loaded with a coating of dust, it is permitted to clean the nozzles with compressed air or by a vacuum provided that the equipment does not touch the nozzle.

12.1.2.1.2 Any water mist nozzle that has been installed in the incorrect orientation shall be corrected by repositioning the branch line, drop, or sprig, or shall be replaced.

12.1.2.1.3 Water mist nozzles with glass bulbs shall be replaced if the bulbs are empty of fluid.

12.1.2.1.4 Water mist nozzles installed in concealed spaces such as above suspended ceilings shall not require inspection.

12.1.2.1.5 Water mist nozzles installed in areas that are inaccessible for safety considerations due to process operations shall be inspected during each scheduled shutdown.

12.1.2.1.6 Stock, furnishings, or equipment closer to the water mist nozzle than permitted by the clearance specified in the manufacturer's installation instructions shall be corrected.

12.1.2.2 The supply of spare automatic water mist nozzles shall be inspected annually for the correct number and type of water mist nozzles as required by NFPA 750.

12.1.2.3* Where required by this section, sample automatic water mist nozzles shall be submitted to a recognized testing laboratory acceptable to the authority having jurisdiction for functional testing.

A.12.1.2.3 Water mist nozzles should be first given a visual inspection for signs of mechanical damage, cleaning, painting, leakage in service, or severe loading or corrosion, all of which are causes for immediate replacement. Nozzles that have passed the visual inspection should be laboratory tested for sensitivity and functionality. Thermal sensitivity should be not less than that permitted in post-corrosion testing of new water mist nozzles of the same type. Water mist nozzles that have been in service for a number of years should not be expected to have all of the performance qualities of a new water mist nozzle. However, if there is any question about their continued satisfactory performance, the water mist nozzles should be replaced.

12.1.2.3.1 Automatic water mist nozzles that have been in service for 20 years shall be replaced, or representative samples shall be tested and then retested at 10-year intervals.

12.1.2.3.2* Where water mist nozzles are subjected to harsh environments, including corrosive atmospheres and corrosive water supplies, they shall be either replaced or representative samples tested on a 5-year basis.

A.12.1.2.3.2 These environments include outdoor weather conditions and portions of any area where corrosive vapors prevail. Harsh water environments include water supplies that are chemically reactive.

12.1.2.4* A representative sample of water mist nozzles for testing per 12.1.2.3 shall consist of a minimum of four water mist nozzles or 1 percent of the number of water mist nozzles per individual water mist nozzle sample, whichever is greater.

A.12.1.2.4 Within the area represented by the selected sample, water mist nozzles of the same design produced by the same manufacturer can be considered part of the same sample, but additional water mist nozzles would need to be selected if produced by a different manufacturer.

12.1.2.4.1 Where one water mist nozzle within a representative sample fails to meet the test requirement, all water mist nozzles within the area represented by that sample shall be replaced.

FAQ How long are water mist tanks allowed to be kept in service? How are they refilled?

As indicated in Table 12.1.2, high pressure compressed gas cylinders are permitted to be kept in service (if not discharged) for up to 12 years. Generally, cylinders that are discharged are replaced by new cylinders that are provided by a compressed gas supplier; therefore, it is seldom necessary to send cylinders out for hydrostatic testing. Some water mist systems, however, use customized compressed gas cylinders or fixed banks of cylinders that are intended to be refilled in place. A special fill connection should be provided on the manifold to permit refill in place. If the system is discharged after the first 5 years, the cylinders must be removed and sent for hydrostatic testing.

If refilling in place cannot be easily accomplished within 24 hours of a system release, replacement cylinders should be kept in stock. The spent cylinders should be removed and sent to a supplier to be refilled, and the stock replacement cylinders installed.

Exhibit 12.8 shows an assembly of high pressure nitrogen cylinders used as the driving medium for a water mist system. Each cylinder has its own low pressure alarm pressure switch. The manifold is designed to permit refill in place.

Different countries have different pressure ratings for cylinders. In Europe, Asia, and Australia, 2900 psi (200 bar) cylinders are standard. In North America, commercially available cylinders are typically 2250 psi (155 bar). It is important to check the actual cylinder volume and pressure whenever a new cylinder is installed. It is not uncommon to find a commercially supplied cylinder at, for example, 2175 psi (150 bar), about 3 percent less than the 2250 (155 bar) expected. A small difference (e.g., ~3 percent) probably will not significantly affect the overall performance of the system. The calculation to determine the number of cylinders needed for a given duration of discharge is based on the volume of the cylinders and the pressure. If the water mist system was designed on the basis of 50 liter cylinders at 2900 psi (200 bar) and similarly sized replacement cylinders were only 2250 (155 bar), the duration of protection will be reduced. Before any cylinders are replaced, the system documentation should be consulted to determine the minimum acceptable cylinder volume and pressure.

EXHIBIT 12.8 Compressed Gas Cylinders for Use as Propellant for a Water Mist System.

12.2 Maintenance

12.2.1 General.

12.2.1.1 Maintenance shall be performed to keep the system equipment operable or to make repairs.

FAQ How should the inspector notify the owner or the designated representative of any deficiency or impairment found in the system?

If the inspection and testing activity identifies a deficiency or impairment in the system, the ITM service provider must bring the problem to the attention of the owner or designated representative. The owner, or a representative of the owner who has some authority to act, should be verbally informed of the problem. Additionally, a signed inspection form should be submitted with the deficiency clearly marked. Immediate action should be taken to restore the system to its intended level of operation in as short a time as possible.

12.2.1.1.1 Mechanical waterflow devices, including but not limited to water motor gongs, shall be tested quarterly.

12.2.1.1.2 Vane-type and pressure switch–type waterflow devices shall be tested semiannually.

12.2.1.1.3 Waterflow devices shall be inspected quarterly to verify that they are free of physical damage.

12.2.1.1.4 Where required by the manufacturer, maintenance shall be performed on devices not described in this standard.

12.2.1.2 As-built system installation drawings, original acceptance test records, and device manufacturer's maintenance bulletins shall be retained to assist in the proper care of the system and its components.

ITM tasks associated with a water mist system should not be undertaken without the as-built drawings and original acceptance test records. The ITM service provider should also ensure that copies of the drawings and acceptance test records are kept with the new test record.

The ITM service provider is not required to confirm that the original design for a particular installation was appropriate for the hazard. However, information on the original design is provided on the system design information sign required by 11.4.4 of NFPA 750. The sign provides a description of the hazard protected that the maintenance service provider can check against the observed conditions.

At the completion of the system installation, the service provider is required to provide the owner with the as-built installation drawings, as noted in 11.4.2 of NFPA 750. The service provider should review these drawings with the owner and, if necessary, the designated representative, and note any changes in ventilation, compartment size, and obstructions, which could significantly alter the ability of the water mist system to achieve its intended performance.

Critical Point

At the completion of the system installation, the service provider is required to provide the owner with the as-built installation drawings, as noted in 11.4.2 of NFPA 750.

12.2.1.3 Replacement components shall be in accordance with the manufacturer's specifications and the original system design.

Certain components of water mist systems can have relatively long delivery times for replacement. If a component must be removed for repair, such work should be scheduled to minimize the amount of time the system will be out of service.

12.2.1.4 Spare components shall be accessible and shall be stored in a manner to prevent damage or contamination.

12.2.1.5* After each system operation, a representative sample of operated water mist nozzles in the activated zone shall be inspected.

A.12.2.1.5 The representative sample should include 10 percent of the water mist nozzles in the activated zone. If contamination of filters or strainers is found on inspection, it is recommended that all nozzles within the activated zone be inspected.

12.2.1.6 After each system operation due to fire, the system filters and strainers shall be cleaned or replaced.

12.2.1.7 Where applicable to the type of water mist system, scheduled maintenance shall be performed as outlined in Table 12.2.1.7.

TABLE 12.2.1.7 *Maintenance Frequencies*

Item	*Activity*	*Frequency*
Water tank	Drain and refill	Annually
System	Flushing	Annually
Strainers and filters	Clean or replace as required	After system operation

12.2.2 Water Mist Nozzles.

12.2.2.1* Replacement water mist nozzles shall have the proper characteristics for the application intended, including the nozzle model and temperature rating.

A.12.2.2.1 Each water mist nozzle has unique requirements for protection applications and end-use limitations.

12.2.2.2 Only new water mist nozzles shall be used to replace existing water mist nozzles.

12.2.2.3* A supply of spare water mist nozzles (never fewer than three) shall be maintained on the premises so that any water mist nozzles that have operated or been damaged in any way can be promptly replaced.

A.12.2.2.3 A minimum of two water mist nozzles of each type and temperature rating installed should be provided.

12.2.2.3.1 The water mist nozzles shall correspond to the manufacturer(s), models, and temperature ratings of the water mist nozzles in the property.

12.2.2.3.2 The automatic water mist nozzles shall be kept in a cabinet located where the temperature will at no time exceed 100°F (38°C).

12.2.2.3.3 The stock of spare thermally activated nozzles shall include all types and ratings installed and shall be as follows:

(1) For systems having fewer than 50 nozzles, not fewer than 3 nozzles
(2) For systems having 50 to 300 nozzles, not fewer than 6 nozzles
(3) For systems having 301 to 1000 nozzles, not fewer than 12 nozzles
(4) For systems having over 1000 nozzles, not fewer than 24 nozzles

12.2.2.4* Where required by the manufacturer, a special water mist nozzle wrench shall be provided and kept in the cabinet to be used for the removal and installation of nozzles.

A.12.2.2.4 Other types of wrenches could damage the water mist nozzles.

12.2.2.4.1 One water mist nozzle wrench shall be provided for each type of nozzle installed.

12.2.2.5 Protective Coverings.

12.2.2.5.1 Water mist nozzles protecting spray areas and mixing rooms in resin application areas installed with protective coverings shall continue to be protected against overspray residue so that they will operate in the event of fire.

12.2.2.5.2* Water mist nozzles installed as described in 12.2.2.5.1 shall be protected using cellophane bags having a thickness of 0.003 in. (0.076 mm) or less or thin paper bags.

A.12.2.2.5.2 Typical sandwich bags purchased in a grocery store are generally plastic, not cellophane. Plastic bags have a tendency to shrink and adhere to the nozzles prior to nozzle activation, creating the potential for disruption of nozzle spray pattern. Bags placed over nozzles need to be cellophane or paper.

12.2.2.5.3 Coverings shall be replaced periodically so that heavy deposits of residue do not accumulate.

12.2.2.6 Water mist nozzles shall not be altered in any respect or have any type of ornamentation, paint, or coatings applied after shipment from the manufacturer.

12.2.2.7 Automatic water mist nozzles used for protecting commercial-type cooking equipment and ventilating systems shall be replaced annually.

12.2.2.7.1 Where automatic water mist nozzles are used and annual examination shows no buildup of grease or other material on the nozzles, the nozzles shall not be required to be replaced.

12.3 Training

12.3.1 All persons who might be expected to inspect, test, maintain, or operate water mist systems shall be trained thoroughly in the functions they are expected to perform.

Most manufacturers of water mist systems equipment provide training on their equipment. Some manufacturers provide certificates for confirmation that specific employees have completed the specialized training.

In addition to familiarity with the various types of hardware encountered in water mist systems, it is important that the service provider be familiar with the features and limitations of the particular system. The service provider should obtain and study all of the system documentation provided before beginning testing or maintenance work on the system.

12.3.2 Refresher training shall be provided as recommended by the manufacturer or by the authority having jurisdiction.

New technology and procedures in water mist systems are being introduced with increasing speed, which necessitates that continuing education should be undertaken at a similar speed. The investment in ongoing training usually provides a significant return by helping to prevent large capital expenditures for replacing neglected equipment, with the ultimate return evident when a system operates as intended and prevents a catastrophic loss. In addition, personnel changes in an ITM service company can result in an employee who received specialized training leaving the company. It is intended that all employees who work on a water mist system receive the specialized training provided by the manufacturer.

Case In Point

Annex D of NFPA 750 provides a Water Mist System Questionnaire that is intended to collect information on the reliability of water mist systems. If a water mist system operates for any reason, either accidentally or due to fire, it is likely that the ITM service provider will be involved in restoring the system. It would be valuable if the service provider would consent to completing the questionnaire and send it to the NFPA staff liaison responsible for NFPA 750.

References Cited in Commentary

National Fire Protection Association, 1 Batterymarch Park, Quincy, MA 02169-7471.

Cote, A. E., ed., *Fire Protection Handbook®*, 20th edition, 2008.

NFPA 13, *Standard for the Installation of Sprinkler Systems*, 2013 edition.

NFPA 20, *Standard for the Installation of Stationary Pumps for Fire Protection*, 2013 edition.

NFPA 72®, National Fire Alarm and Signaling Code, 2013 edition.

NFPA 750, *Standard on Water Mist Fire Protection Systems*, 2010 edition.

13

VALVES, VALVE COMPONENTS, AND TRIM

Chapter 13 of NFPA 25 covers valves and valve components. Every type of water-based fire protection system has one or more valves that affect system operation. System and control valves form a critical part of any fire protection system; ever since the first fire protection systems were installed, both closed and partially closed valves have caused systems to not work as intended, resulting in major fire loss. A fully maintained and tested fire sprinkler or fire pump system will only work well as long as the valves controlling the system are fully open. Thus, a comprehensive inspection, testing, and maintenance (ITM) program is not complete without diligence to constantly ensure system valves and associated components operate correctly and are in the correct position.

This chapter was written to consolidate the ITM of all types of valves. Therefore, when inspecting, testing, and maintaining any water-based system, Chapter 13 should always be referenced along with the requirements for specific hardware or components covered in other chapters. Although system control valves are not necessarily complex, they do require specialized care by a qualified contractor to remain functional throughout their life cycle. And when a fire protection system control valve must be closed for any reason, the requirements in Chapter 15 must also be implemented.

This chapter also covers *trim*, which is a term not defined in any NFPA water-based fire protection system document including NFPA 13, *Standard for the Installation of Sprinkler Systems*, NFPA 20, *Standard for the Installation of Stationary Pumps for Fire Protection*, or NFPA 25. In fact, the term is primarily used in NFPA 25. The term is loosely used in NFPA 25 to reference equipment that is peripheral to the "primary" fire system components. For example, pressure gauges could be considered trim on alarm valves, sprinkler risers, and fire pumps. And alarm check valves have trim that might include pressure gauges, a main drain valve and piping, a water motor gong, and a water motor gong retard chamber.

Some trim has specific ITM requirements and some does not. As such, proper maintenance of trim is required because the trim is considered a component of the primary equipment that it supports. Some trim, such as a gauge, has requirements to be listed, and some trim does not, such as a main drain line. And some trim, such as a quarter-turn valve on a dry pipe valve, is not required to be specifically listed but might be listed as part of an assembly. Some components also must be listed but do not hold listings for fire service and vice-versa. Such is the case with jockey pumps that could be considered to be trim to a fire pump assembly.

13.1* General

A.13.1 *Alarm Valves.* Alarm valves are installed in water-based fire protection systems to sound a fire alarm when a flow of water from the system equals or exceeds the flow of a single discharge device. A retarding chamber, which minimizes false alarms due to surges and fluctuating water supply pressure, can be supplied with the alarm valve.

An alarm valve is a regular swing check valve with several trim options to allow for a water motor gong alarm connection. The alarm trim options provide pipe to a water motor alarm gong for the purpose of generating a waterflow alarm. The water motor gong retard chamber is used to snub water pressure surges, which could create false local alarms. In many systems, a water gong is replaced by an electrically operated annunciator, which may or may not be connected or operated by alarm check trip. Exhibit 13.1 shows a piping system that includes an alarm check valve with a hydraulically operated water motor gong as the alarm trim option (see also Exhibit 13.2).

Exhibit 13.3 shows a typical trip arrangement and a connection for an optional alarm pressure switch that could also be connected to annunciate an electric bell. This switch is shown in Exhibit 13.4.

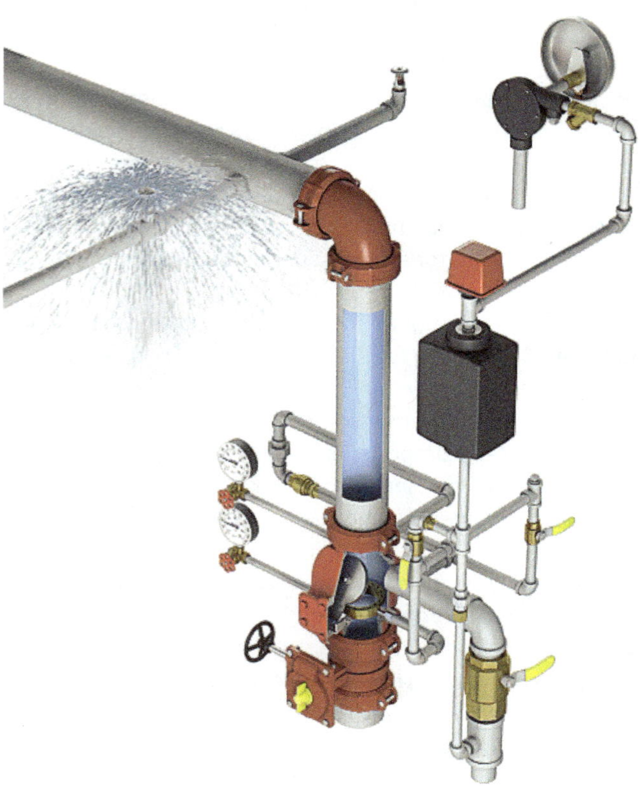

EXHIBIT 13.1 *Wet Pipe System Riser with Alarm Valve and Trim. (Courtesy of Jesse Cappon and Viking®)*

EXHIBIT 13.2 *Pressure Type Waterflow Alarm on Sprinkler System Riser Bypass Line.*

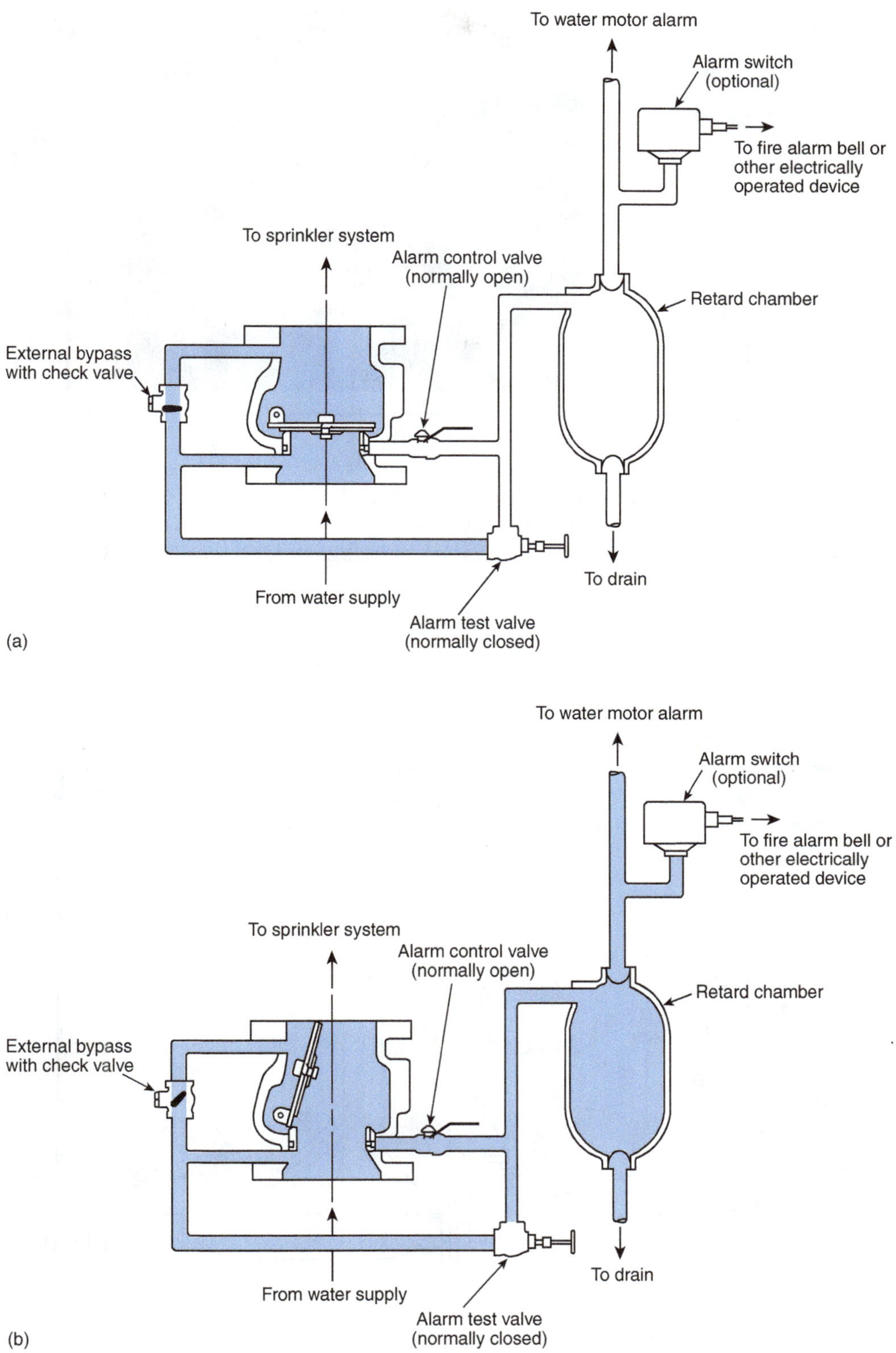

EXHIBIT 13.3 *Pressure Type Waterflow Alarm on Standard Alarm Check Valve: (a) Closed (Set) and (b) Open (Activated).*

EXHIBIT 13.4 Pressure Type Waterflow Alarm Switch. (Courtesy of Potter Electric Signal Company, LLC)

Backflow Prevention Devices. Backflow prevention devices are used to prevent water in a fire protection system from entering the public water supply due to a reverse flow of water, thermal expansion, hydraulic shock, back pressure, or back siphonage. [*See Figure A.13.1(a).*]

Backflow prevention devices, which are commonly called backflow assemblies when installed with valve trim, are not required for fire protection purposes by NFPA 13. They are typically required on fire systems connected to potable water sources by private and municipal water companies. These are installed as part of efforts to comply with the Safe Drinking Water Act, the 1974 federal law that requires that all suppliers (both public and private) guarantee their consumers receive clean drinking water and protection from water supply contamination.

Specific installation requirements are based on enforcement by local water authorities and can vary between municipalities. The type of backflow apparatus required to be installed is typically based on the type of fire systems installed on the premises, and the municipal enforcement is based on the perceived risk of contamination from the fire system to the municipal

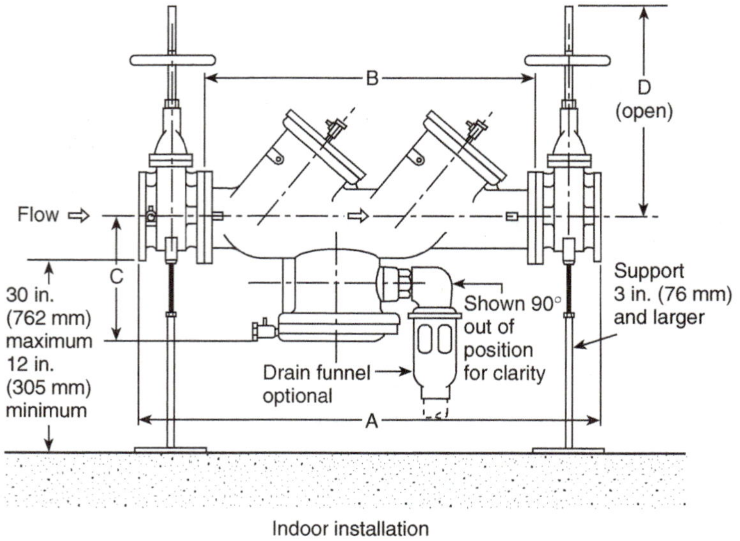

***FIGURE A.13.1(a)** Reduced-Pressure Backflow Preventers (left) and Double Check Valve Assemblies (right).*

water supply. For example, systems that have antifreeze or foam concentrates in the systems typically require a more sophisticated type of backflow apparatus. And this determination, where enforced, normally references the American Water Works Association Manual M14, *Recommended Practice for Backflow Prevention and Cross-Connection Control.*

There are two primary types of backflow prevention devices used in fire protection systems: a reduced-pressure (RP) type and a double check (DC) type. Both are required to be listed as complete assemblies that include shutoff valves and test cocks at each end of the apparatus. The double check type can sometimes include air gap trim, which provides an air gap barrier to virtually eliminate any chance of backflow.

A reduced-pressure backflow preventer consists of two independently acting, spring-loaded check valves with a hydraulically operating, mechanically independent, spring-loaded pressure differential relief valve between the check valves and below the first check valve. These provide a level of protection higher than a standard double check valve assembly, which consists only of two independently acting, spring-loaded check valves.

While NFPA 13 does not require the backflow preventer to be installed as part of a fire protection system, it does require that, if installed, a means be included downstream to flow test them regularly per NFPA 25. Poorly performing backflow check valve assemblies can result in an impaired fire protection sprinkler system. Backflow assemblies must be listed. Part of this listing for use in a fire protection system includes evaluations for durability, reliability, and flow characteristics. Each assembly has flow characteristics that show pressure drop related to flow. This must be evaluated with system demands, especially when installed on a pre-existing system.

Ball Valves. Ball valves are manually operated through their full range of open to closed positions with a one-quarter turn.

Ball valves cannot be used for controlling fire protection system water supplies or sprinkler system operation. NFPA 13 requires that listed indicating valves be used where controlling connections *to* water supplies and sprinkler systems. NFPA 13 also requires that all valves listed for fire service control not be designed to close in less than 5 seconds when operated at maximum speed. The purpose of this second requirement is to prevent water hammering and valve damage; this eliminates ball valves as they can quickly be closed with a quarter turn. Thus, ball valves are only used for trim on alarm valves or to control waterflow discharge *from* sprinkler systems on inspector's test connections, auxiliary drains, and main drain valves. Ball valves are not required to be listed, but some that are installed in inspector's tests connections are for use other than for fire system control. Exhibit 13.5 shows a ball valve commonly used on an inspector's test connection. Exhibit 13.6 shows a listed combination auxiliary drain and inspector's test valve.

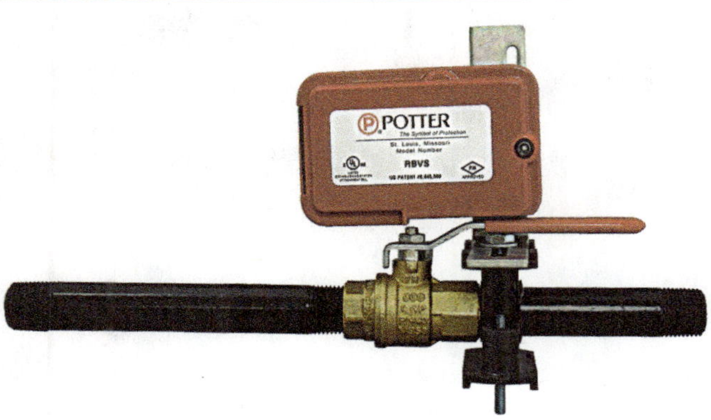

EXHIBIT 13.5 *Typical Ball Valve. (Courtesy of Potter Electric Signal Company, LLC)*

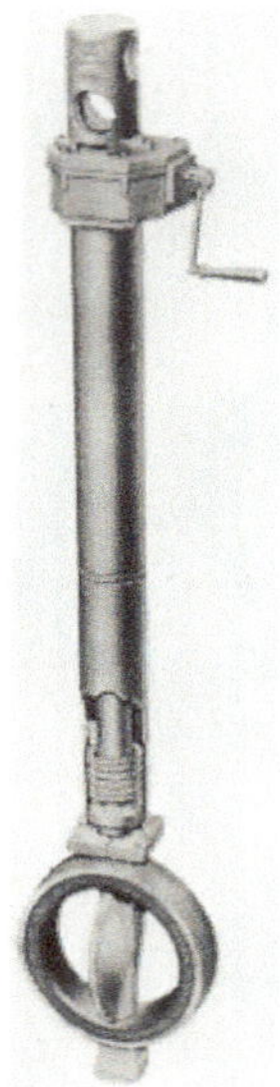

FIGURE A.13.1(b) *Butterfly Post Indicator Valve. (Courtesy of Henry Pratt Co.)*

EXHIBIT 13.6 *Listed Combination Auxiliary Drain and Inspector's Test Valve.*

Butterfly Valves. Butterfly valves are water supply control valves with gear operators to assist in opening and closing. Butterfly valves can be of the wafer or grooved-end type. [*See Figure A.13.1(b).*]

Butterfly valves, such as the one shown in Exhibit 13.7, are typically used as a lighter, less-expensive alternative to OS&Y valves. They are indicating and are usually equipped with a factory-installed tamper switch. There are specific restrictions to their use in NFPA 20 on suction supplies for fire pumps.

Check Valves. Check valves allow waterflow in one direction only. [*See Figure A.13.1(c).*]

FIGURE A.13.1(c) *Detector Check Valve.*

EXHIBIT 13.7 *Butterfly Valve.*

DCA. A double check assembly (DCA) consists of two independently operating spring-loaded check valves. The assembly includes two resilient-seated isolation valves and four test cocks required for testing.

DCDA. A double check detector assembly (DCDA) is hydraulically balanced to include a metered bypass assembly to detect system leakage. The main valve assembly and bypass assembly afford equal levels of backflow prevention and are each equipped with two resilient-seated isolation valves and four test cocks required for testing.

Deluge Valves. Deluge valves hold water at the valve until actuated by the operation of a detection system or manual release. [*See Figure A.13.1(d).*]

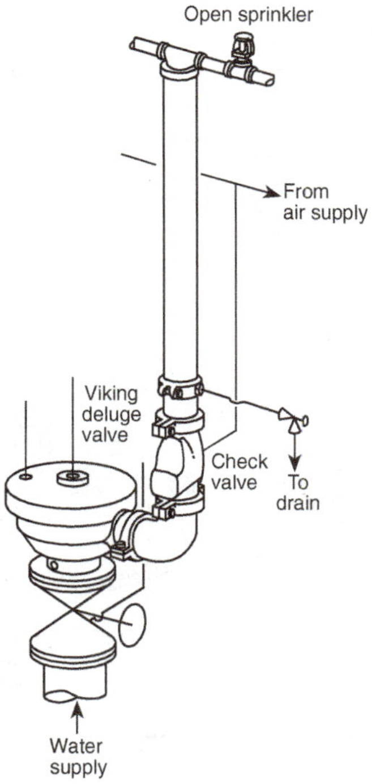

FIGURE A.13.1(d) *Deluge Valve.*

See Exhibit 13.8 for an illustration of a double check detector assembly (DCDA) valve. Backflow assemblies can be found mounted vertically, but they must be listed for installation in that orientation. Some types are also reconfigured for space requirements and other considerations, as shown in Exhibit 13.9 and Exhibit 13.10.

EXHIBIT 13.8 *Typical Double Check Detector Assembly.*

EXHIBIT 13.9 *Backflow Preventer with Outside Stem and Yoke (OS&Y) Valve with its Position Supervised by a Tamper Switch.*

EXHIBIT 13.10 *Backflow Preventer and Fire Department Connection. (Courtesy of American Fire Sprinkler Association)*

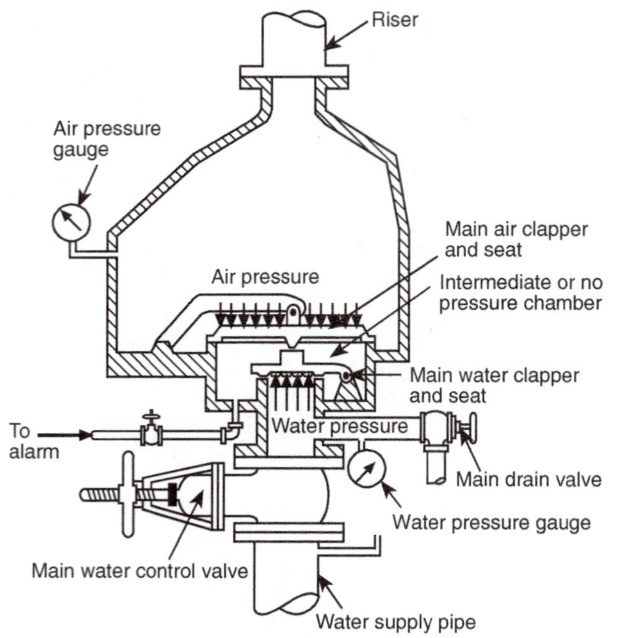

FIGURE A.13.1(e) *Dry Pipe Valve.*

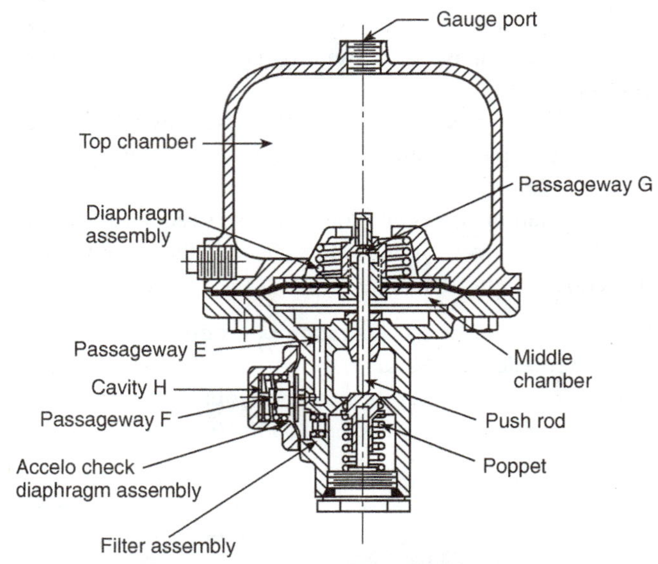

FIGURE A.13.1(f) *Dry Pipe System Accelerator. (Courtesy of The Reliable Automatic Sprinkler Co., Inc.)*

Drip Valves. Drip valves automatically drain condensation or small amounts of water that have leaked into system piping or valves. Drip valves close when exposed to system pressure.

Dry Pipe Valves. Dry pipe valves control the flow of water to areas that could be exposed to freezing conditions. Water is held at the valve by air pressure in the system piping. When the air pressure is reduced, the valve operates and floods the system. [*See Figure A.13.1(e) and Figure A.13.1(f).*]

Indicating Valves. Indicating valves provide a dependable, visible indication of the open position, even at a distance.

Exhibit 13.11 shows two outside screw and yoke (OS&Y) valves: one open and one closed. The stem of the valve on the right is flush with the hand wheel or yoke, so it visually indicates being in the closed position. The open status of the valve on the left is indicated by the protruding stem. Valves controlling connections to water supplies and water supplies to sprinklers must be an indicating type valve per NFPA 13. There is an exception for permitting the use of roadway-box type gate valves under certain conditions and where use is acceptable to the authority having jurisdiction (AHJ). These conditions include a requirement that an on-site T-wrench be accessible to open or close the valve in an emergency.

EXHIBIT 13.11 *OS&Y Indicating Valves Showing Open (left) and Closed (right).*

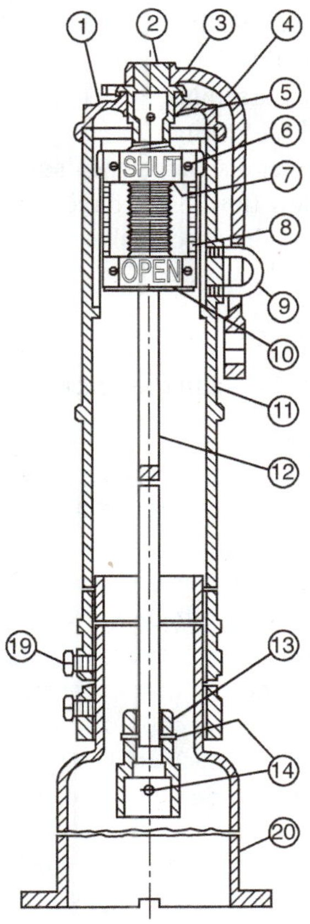

No	Parts list Description
1	Cap
2	Operating stem
3	Operating stem oil hole screw
4	Operating wrench
5	Retaining ring
6	Target plate screw and nut
7	Target plate — shut
8	Target
9	Indicator post staple
10	Target plate — open
11	Indicator post
12	Extension rod — specify length
13	Extension rod coupling
14	Coupling pin
15	Window glass
16	Window frame
17	Window frame screw
18	Cap bolt and nut
19	Set screw
20	Sleeve bonnet

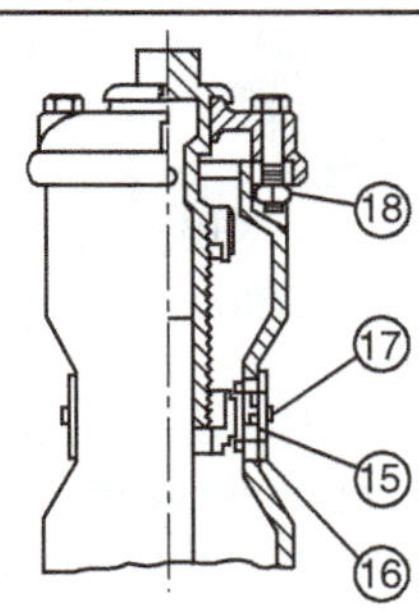

FIGURE A.13.1(g) *Vertical Indicator Post.*

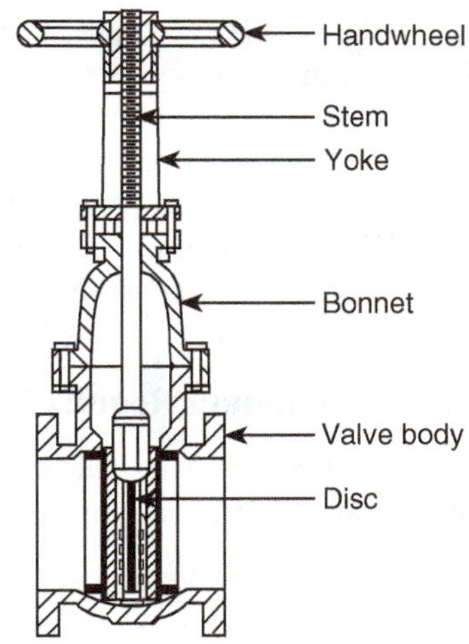

FIGURE A.13.1(h) *OS&Y Gate Valve.*

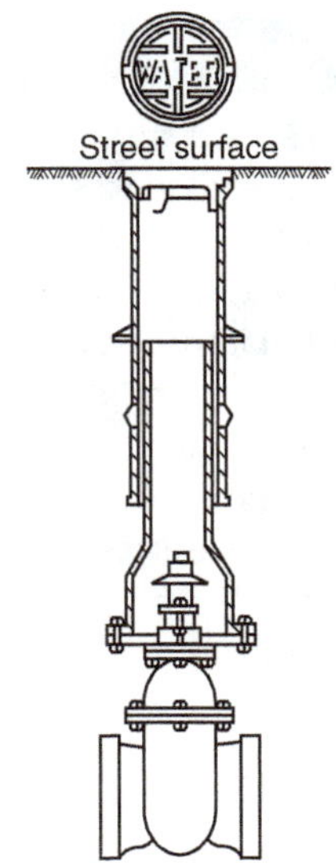

FIGURE A.13.1(i) *Nonindicating-Type Gate Valve.*

Indicator Posts. Indicator posts include wall and underground types and are intended for use in operating inside screwed pattern gate valves and for indicating the position of the gates in the valves. [*See Figure A.13.1(g).*]

NRS Gate Valves, OS&Y Gate Valves. Nonrising stem (NRS) gate valves are used underground with indicator posts attached or as roadway box valves (curb-box installation). Outside screw and yoke (OS&Y) gate valves are used indoors and in pits outdoors. The valve stem moves out when the valve is open and moves in when it is closed. The stem indicates the position of the valve. [*See Figure A.13.1(h) and Figure A.13.1(i).*]

Care should be taken when inspecting indicator posts connected to underground nonrising stem (NRS) valves. It is important to verify that the indicator post is set with the "target" indicating the correct position of the valve ("open" or "shut"). When fully closing and opening the valve during the annual test, it is best practice to count and record the number of revolutions, which will provide an indication of the position of the valve for comparison to future tests. This number should be included in test forms as well. The manufacturer should be contacted for the number of turns needed to fully open or close a particular valve if it is not known and an issue is suspected.

RPA. A reduced-pressure zone principle assembly (RPA) consists of two independently spring-loaded check valves separated by a differential-sensing valve. The differential-sensing valve includes a relief port to atmosphere that discharges excess water resulting from supply system fluctuations. The assembly includes two resilient-seated isolation valves and four test cocks required for testing.

Critical Point

NFPA 13 states that all materials and devices essential to successful system operation shall be listed, but it specifically indicates that components that do not affect system performance are not required to be listed.

RPDA. A reduced-pressure detector assembly (RPDA) is hydraulically balanced to include a metered bypass assembly to detect system leakage. The main valve assembly and bypass assembly afford equal levels of backflow prevention, and each assembly is equipped with two resilient-seated isolation valves and four test cocks required for testing.

Strainers. Strainers are used for protection against clogging of water discharge openings.

Waterflow Detector Check Valves. Detector-type check valves allow flow in one direction only and have provisions for the connection of a bypass meter around the check valve. [*See Figure A.13.1(c).*]

13.1.1 Minimum Requirements.

13.1.1.1 This chapter shall provide the minimum requirements for the routine inspection, testing, and maintenance of valves, valve components, and trim.

The term *valve* as used in this standard includes valves that are used to control water supplies and sprinkler systems as well as trim valves. NFPA 13 requires that listed indicating valves be used where controlling connections *to* water supplies and sprinkler systems. These valves must be listed for fire service. The term *valve* also includes valves considered to be trim used to control waterflow discharge *from* sprinkler systems. These can, but are not required to be, individually listed. Some valves that control water flow *from* fire protection systems or are secondary as trim are specifically excluded in NFPA 13 from the listing requirement, including hose valves, inspector's test valves, drain valves, relief valves, or trim valves for dry pipe, preaction, and deluge valves. This chapter also covers backflow prevention apparatus valves, which are typically standard OS&Y valves that are attached to a backflow device and then included as part of an installed backflow assembly.

NFPA 13 states that all materials and devices essential to successful system operation shall be listed, but it specifically indicates that components that do not affect system performance are not required to be listed. This includes valve trim valve components, which therefore do not need to be individually listed. The Underwriters Laboratories (UL) listing program for valves classifies the various types as follows (shown with the specific UL testing protocol):

- ANSI/UL 193, *Alarm Valves for Fire-Protection Service*
- ANSI/UL 260, *Dry Pipe and Deluge Valves for Fire-Protection Service*
- UL 262, *Gate Valves for Fire-Protection Service*
- ANSI/UL 312, *Check Valves for Fire-Protection Service*
- UL 1091, *Butterfly Valves for Fire-Protection Service*
- Subject 258, *Outline of Investigation for Shutoff Valves for Trim and Drain Purposes*

Exhibit 13.12 illustrates a post indicator valve (PIV) and Exhibit 13.13 shows a wall post indicator valve (WPIV), which are both types of control gate valves.

13.1.1.2 Table 13.1.1.2 shall be used to determine the minimum required frequencies for inspection, testing, and maintenance.

Case In Point

The term *valve components* can refer to the parts that make up an operational assembly or control valve. These components can also be called *trim*. Trim refers to the exterior parts of an operational valve or system component, such as the valve components (piping and smaller ball valves) that make up an alarm check valve or the connection to a water motor gong and retard chamber. A main drain valve connection could also be considered to be trim on a sprinkler riser.

TABLE 13.1.1.2 *Summary of Valves, Valve Components, and Trim Inspection, Testing, and Maintenance*

Item	*Frequency*	*Reference*
Inspection		
Control Valves		
Sealed	Weekly	13.3.2.1
Locked or electrically supervised	Monthly	13.3.2.1.1
Valve Supervisory Signal Initiating Device	Quarterly	13.3.2.1.2
Alarm Valves		
Exterior	Monthly	13.4.1.1
Interior	5 years	13.4.1.2
Strainers, filters, orifices	5 years	13.4.1.2
Check Valves		
Interior	5 years	13.4.2.1
Preaction/Deluge Valves		
Enclosure (during cold weather)	Daily/weekly	13.4.3.1
Exterior	Monthly	13.4.3.1.6
Interior	Annually/5 years	13.4.3.1.7
Strainers, filters, orifices	5 years	13.4.3.1.8
Dry Pipe Valves/Quick-Opening Devices		
Gauges	Weekly/monthly	13.4.4.1.2.4, 13.4.4.1.2.5
Enclosure (during cold weather)	Daily/weekly	13.4.4.1.1
Exterior	Monthly	13.4.4.1.4
Interior	Annually	13.4.4.1.5
Strainers, filters, orifices	5 years	13.4.4.1.6
Pressure-Reducing and Relief Valves		
Sprinkler systems	Quarterly	13.5.1.1
Hose connections	Annually	13.5.2.1
Hose racks	Annually	13.5.3.1
Fire pumps		
Casing relief valves	Weekly	13.5.7.1, 13.5.7.1.1
Pressure-relief valves	Weekly	13.5.7.2, 13.5.7.2.1
Backflow Prevention Assemblies		
Reduced pressure	Weekly/monthly	13.6.1
Reduced-pressure detectors	Weekly/monthly	13.6.1
Fire Department Connections	Quarterly	13.7.1
Testing		
Main Drains	Annually/quarterly	13.2.5, 13.2.5.1, 13.3.3.4
Gauges	5 years	13.2.7.2
Waterflow Alarms	Quarterly/semiannually	13.2.6
Control Valves		
Position	Annually	13.3.3.1
Operation	Annually	13.3.3.1
Supervisory	Semiannually	13.3.3.5

(continues)

TABLE 13.1.1.2 *Continued*

Item	*Frequency*	*Reference*
Preaction/Deluge Valves		
Priming water	Quarterly	13.4.3.2.1
Low air pressure alarms	Quarterly/annually	13.4.3.2.13, 13.4.3.2.14
Full flow	Annually	13.4.3.2.2
Air leakage	3 years	13.4.3.2.6
Dry Pipe Valves/Quick-Opening Devices		
Air leakage	3 years	13.4.4.2.9
Priming water	Quarterly	13.4.4.2.1
Low air pressure alarm	Quarterly	13.4.4.2.6
Quick-opening devices	Quarterly	13.4.4.2.4
Trip test	Annually	13.4.4.2.2
Full flow trip test	3 years	13.4.4.2.2.2
Pressure-Reducing and Relief Valves		
Sprinkler systems	5 years	13.5.1.2
Circulation relief	Annually	13.5.7.1.2
Pressure relief valves	Annually	13.5.7.2.2
Hose connections	5 years	13.5.2.2
Hose racks	5 years	13.5.3.2
Backflow Prevention Assemblies	Annually	13.6.2
Maintenance		
Control Valves	Annually	13.3.4
Preaction/Deluge Valves	Annually	13.4.3.3.2
Dry Pipe Valves/Quick-Opening Devices	Annually	13.4.4.3

EXHIBIT 13.12 *Post Indicator Valve and Fire Department Connection. (Courtesy of Josh Elvove)*

EXHIBIT 13.13 *Wall Post Indicator Valve.*

13.2 General Provisions

13.2.1 The property owner or designated representative shall have manufacturers' literature available to provide specific instructions for inspecting, testing, and maintaining the valves and associated equipment.

The phrase *property owner or designated representative* in 13.2.1 applies to whoever is responsible for the ITM of the equipment (see Section 4.1). Any portion or all of the ITM is permitted to be contracted with an ITM service, but ultimate responsibility falls on the building owner unless these responsibilities have been contractually transferred.

Critical Point

The OEM manuals are an excellent source of information for the inspector and maintenance contractor and should include a system narrative report that describes all aspects of the initial construction project and a complete list of equipment installed, including the contact information for the equipment suppliers.

Case In Point

The manufacturer's literature mentioned in 13.2.1 refers to the operation and equipment maintenance (OEM) manuals as required by the Contractor's Material and Test Certificate at the time of installation.

The OEM manuals are an excellent source of information for the inspector and maintenance contractor and should include a system narrative report that describes all aspects of the initial construction project and a complete list of equipment installed, including the contact information for the equipment suppliers. These documents should also include operation and maintenance instructions for all system components, any previous inspection or test reports, and recommended spare parts. If such a manual is not present, the owner should ask the maintenance contractor to provide one for future reference, complete with a telephone number for emergency service. See NFPA 3, *Recommended Practice for Commissioning and Integrated Testing of Fire Protection and Life Safety Systems*, for more information and guidance on this topic.

13.2.2 All pertinent personnel, departments, authorities having jurisdiction, or agencies shall be notified that testing or maintenance of the valve and associated alarms is to be conducted.

It is not necessary to notify the AHJ when inspection activities will take place, because alarm initiating devices are not necessarily activated. However, when testing is conducted, alarms typically will be activated and the AHJ should be notified. It is important to prepare and retain inspection records, as specified in Section 4.3, for review by the AHJ and/or insurance inspectors to demonstrate compliance with the standard. Such records should include noted deficiencies and corrective action taken to address these issues.

13.2.3* All system valves shall be protected from physical damage and shall be accessible.

A.13.2.3 The valves are not required to be exposed. Doors, removable panels, or valve pits can be permitted to satisfy this requirement. Such equipment should not be obstructed by features such as walls, ducts, columns, direct burial, or stock storage.

NFPA 13 indicates that all control valves must be accessible — and accessible without obstructions that must be moved — but otherwise provides little specific guidance. Most manufacturers, however, will provide specific clearance dimensions for their valves in product data literature. These dimensions should be maintained on all sides of the valve.

For example, operational valves and alarm check, dry pipe, and deluge valves on sprinkler system risers should have adequate clearance on all sides to allow accessibility for normal ITM activities. Exhibit 13.14 shows a typical valve clearance dimension diagram. The dimensions shown should be used as a minimum to maintain access to the valve for operation and maintenance purposes.

Critical Point

NFPA 13 indicates that all control valves must be accessible — and accessible without obstructions that must be moved — but otherwise provides little specific guidance.

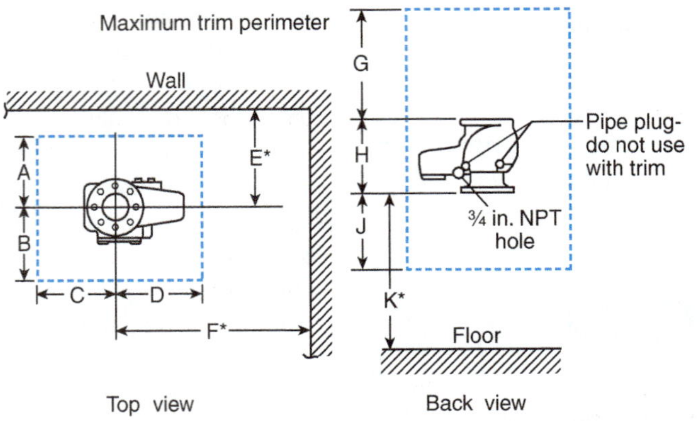

Dimension	2-½ in. (63.5 mm) valve	4 in. (101.6 mm) valve	6 in. (152.4 mm) valve
A	8-¾ in. (222.2 mm)	9-½ in. (241.3 mm)	9 in. (228.6 mm)
B	9 in. (228.6 mm)	10-¼ in. (260.3 mm)	11-¼ in. (285.7 mm)
C	9-¾ in. (247.6 mm)	9-¾ in. (247.6 mm)	11 in. (279.4 mm)
D	10-½ in. (266.7 mm)	15-¾ in. (400.0 mm)	13-½ in. (342.9 mm)
E*	12 in. (304.8 mm)	12-½ in. (317.5 mm)	15 in. (381.0 mm)
F*	24 in. (609.6 mm)	24 in. (609.6 mm)	24 in. (609.6 mm)
G	10-½ in. (266.7 mm)	7 in. (177.8 mm)	5 in. (127.0 mm)
H	9 in. (228.6 mm)	12-¼ in. (311.1 mm)	18 in. (457.2 mm)
J	9-¼ in. (234.9 mm)	10-¼ in. (260.3 mm)	5-¼ in. (133.3 mm)
K*	48 in. (1219.2 mm)	48 in. (1219.2 mm)	48 in. (1219.2 mm)

*These dimensions represent minimum recommended clearances.

EXHIBIT 13.14 *Typical Manufacturer's Valve Clearance Dimensions.*

Critical Point

Water damage that results from opening main drains and other test connections, without first verifying where water will flow to, is one of the most common causes of damage and customer complaints faced by inspectors.

13.2.4 Before opening a test or drain valve, it shall be verified that adequate provisions have been made for drainage.

Water damage that results from opening main drains and other test connections, without first verifying where water will flow to, is one of the most common causes of damage and customer complaints faced by inspectors. Before conducting tests that involve the flowing of water under pressure, the inspector should make sure that the area around the discharge fitting is clear. If the discharge is into a floor drain or sink, the flow should be monitored to make sure that the drain is not clogged or undersized, since either situation could result in water spilling onto the floor and creating a hazard. If the discharge location is to a building exterior, make sure water will not flow into trenches recently dug for construction, into dirt basins that could project rocks or mud at clean cars, onto walkways where pedestrians could slip, or into vegetation that could be destroyed.

Case In Point

In high-rise buildings that have sprinkler risers in the stairway, a separate drain line is often installed that is one diameter larger than the largest main drain valve connected to the drain line. The drain line should not be considered a main drain connection but should be viewed as an auxiliary drain for each floor. Unlike the main drain, the auxiliary drain is not intended to be flow tested. The auxiliary drain line can be used to conduct flow tests on pressure-regulating valves installed in the fire sprinkler system and in the fire department standpipe system and also to test the system flow switch.

13.2.5* Main Drain Test. A main drain test shall be conducted annually for each water supply lead-in to a building water-based fire protection system to determine whether there has been a change in the condition of the water supply.

A.13.2.5 Main drains are installed on system risers for one principal reason: to drain water from the overhead piping after the system is shut off. This allows the contractor or plant maintenance department to perform work on the system or to replace nozzles after a fire or other incident involving system operation.

Testing Procedure Alert

ITM NFPA 25

In order to assess the adequacy of the system's water source, a main drain test must be conducted. While the main drain test is not a quantifiable assessment of the flow and pressure, it will give the owner and inspector a general sense if there has been a reduction in flow when compared to the previous main drain test. A reduction in flow would alert the inspector and owner to a partially closed valve or obstruction, which might lead to an obstruction investigation. For more information on conducting a main drain test, including procedures for systems supplied by fire pumps and those that are not, please refer to the testing procedure at the end of this chapter.

Data collected from the suction gauges during a fire pump flow test that test the water supply would satisfy the requirements for a main drain test.

These drains also are used to determine whether there is a major reduction in waterflow to the system, such as could be caused by a major obstruction, a dropped gate, a valve that is almost fully closed, or a check valve clapper stuck to the valve seat.

A satisfactory main drain test (i.e., one that reflects the results of previous tests) does not necessarily indicate an unobstructed passage, nor does it prove that all valves in the upstream flow of water are fully opened. However, these tests provide a reasonable level of confidence that the water supply has not been compromised.

The main drain test is conducted in the following manner:

(1) Record the pressure indicated by the supply water gauge.
(2) Close the alarm control valve on alarm valves.
(3) Fully open the main drain valve.
(4) After the flow has stabilized, record the residual (flowing) pressure indicated by the water supply gauge.
(5) Close the main drain valve slowly.

FAQ What is the purpose of the main drain test?

The two major reasons to conduct a main drain test are to verify the following:

1. Water supply pressure is similar to the original acceptance test or to other previous tests.
2. A system control valve is not fully closed after having been closed and reopened during annual testing.

Exhibit 13.15 shows the discharge from a 2 in. (50 mm) main drain test. It is critical to understand that the purpose of a main drain test is not to confirm if a control valve is fully open but to confirm that it is not fully closed. A main drain test might not reveal a partially closed valve in a large-diameter supply pipe. In most typical systems, a gate valve closure up to approximately 70 percent to 80 percent would not be detected with a main drain test, since the area still open under the gate is larger than the size of the main drain [typically a 2 in. (50 mm) diameter or about 6 percent of an 8 in. (200 mm) diameter pipe]. The results of the main drain test should be compared to those of reliable previous tests, including the original acceptance test. One good way to identify previous tests is to mark the static and residual marks on the gauge. This method also makes it easier to detect a 10 percent drop, as required by 13.2.5.3. This "alert" point can also be marked on gauges for reference.

13.2.5.1 Where the lead-in to a building supplies a header or manifold serving multiple systems, a single main drain test shall be permitted.

13.2.5.2 In systems where the sole water supply is through a backflow preventer and/or pressure-reducing valves, the main drain test of at least one system downstream of the device shall be conducted on a quarterly basis.

EXHIBIT 13.15 *Discharge from a Main Drain Test.*

FAQ Why do backflow preventers in water supplies change the frequency of the main drain test requirement from annually to quarterly?

The intent of 13.2.5.2 is to exercise the backflow preventers by producing periodic flow through them. Fire protection systems are relatively static, only flowing when a fire occurs. The backflow prevention device was originally designed to be used for domestic water applications that continually cycle water through these devices. In a fire protection system where there is no day-to-day exercising of the equipment, a means is needed to regularly exercise them and determine that the components are still functioning properly.

13.2.5.3 When there is a 10 percent reduction in full flow pressure when compared to the original acceptance test or previously performed tests, the cause of the reduction shall be identified and corrected if necessary.

The requirement in 13.2.5.3 does not necessarily require any corrective action for a water supply that exhibits lower pressure readings during the main drain test. However, it does require an investigation as to the cause of the depleted pressure. While a reduction in the available pressure from the attached water supply is a concern, it is critical that the water supply can still meet or exceed the system demand. The 10 percent (or more) reduction in available pressure based on previous tests is enough to lead to an investigation.

Testing Procedure Alert

A waterflow alarm device, such as a vane-type water flow switch, is the connection between the water-based system and the fire alarm system, ensuring that notification of a fire is not only sent to a notification appliance but also, in many cases, to the local fire department. For procedures of testing both mechanical and electrical type waterflow alarm initiating devices, see the testing procedure at the end of this chapter.

13.2.6 Alarm Devices.

13.2.6.1 Mechanical waterflow alarm devices, including but not limited to water motor gongs, shall be tested quarterly.

As required in 5.3.3.1, water motor gongs must be tested quarterly. According to 5.3.3.2, however, vane-type waterflow alarm devices (called paddle type waterflow alarms in NFPA 13) and pressure switches only need semiannually testing. This difference in requirements is a result of the excellent performance record of vane-type waterflow alarm devices, which, after review, persuaded the NFPA 25 committee to change the testing frequency from quarterly to semi-annually in the 2008 edition. Quarterly visual inspections of waterflow alarm devices are still required per 5.2.5 of NFPA 25 to verify there has not been any physical damage to the device.

Exhibit 13.16 shows two vane-type waterflow alarm devices mounted in a sprinkler riser. See also Exhibit 13.4, which shows a typical riser-installed vane-type waterflow device, and Exhibit 13.2, which shows pressure waterflow alarm devices that are mounted on a water motor gong supply line or alarm bypass line.

See Exhibit 13.3, which illustrates the closed (set) position and the open (operating) position for an alarm valve. Exhibit 13.17 illustrates the closed (set) position and the open (activated) position for a deluge valve. Understanding the differences and installation limitations between the various valve positions is important for individuals conducting visual inspections on the devices. NFPA 13 prohibits paddle (i.e., vane-type) waterflow alarm devices on any systems other than wet pipe systems.

The waterflow alarm device test can have two purposes. One is an NFPA 13 requirement to measure the time to annunciate locally. The other is an *NFPA 72, National Fire Alarm and Signaling Code,* requirement to measure the time to annunciate on an alarm initiating device or at an alarm monitoring company, if provided. Both tests require a simulation of one sprinkler being activated. This must be based on the smallest orifice sprinkler on the system being tested. For example, if a building uses early suppression fast-response (ESFR) K-25 sprinklers in

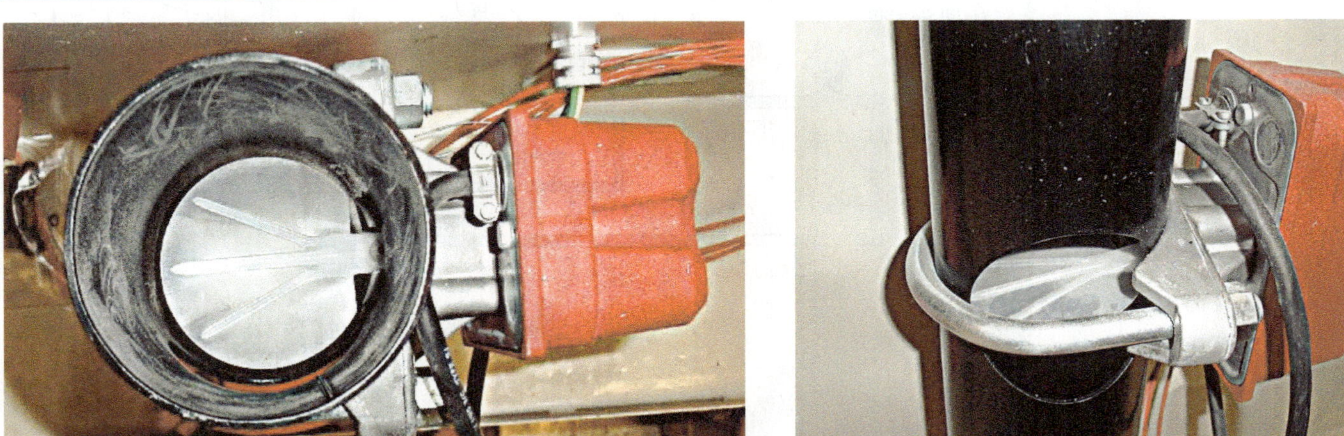

EXHIBIT 13.16 *Vane (or Paddle) Type Waterflow Alarm Devices.*

EXHIBIT 13.17 *Pressure Type Waterflow Alarm on Typical Deluge System: (a) Closed (Set) and (b) Open (Activated).*

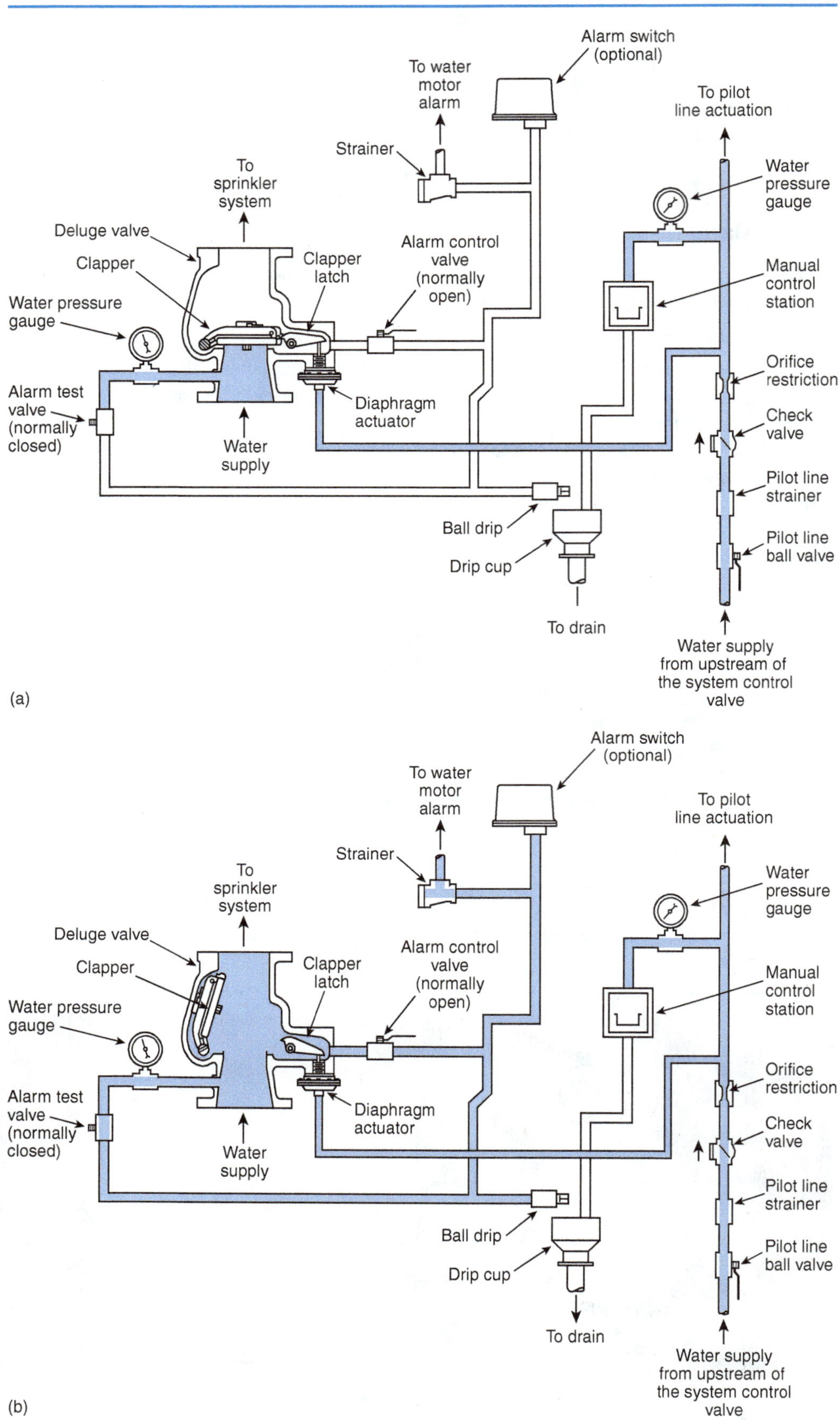

a warehouse area and K-5.6 sprinklers in an office area, and each are fed from separate system risers, the warehouse waterflow device test should simulate flow from an orifice equal in size to one K-25 sprinkler and the office system must be tested based on the flow from a K-5.6 orifice or smaller sprinkler. If systems supply more than one size of sprinklers, the system must be set to an orifice equal to or smaller than the smallest sprinkler. For example, if a building uses ESFR K-25 sprinklers in a warehouse area and K-5.6 sprinklers in an office fed from a common sprinkler riser, the building alarm test should simulate flow from an orifice equal in size to one K-5.6 orifice or smaller sprinkler.

Critical Point

Understanding the differences and installation limitations between the various valve positions is important for individuals conducting visual inspections on the devices.

Per 6.9.1 of NFPA 13, waterflow alarm devices must create an audible alarm on site within 5 minutes of one sprinkler activating. Per 17.12.2 of *NFPA 72,* waterflow alarm devices must create activation of the alarm initiating circuit locally and annunciate at a monitoring company within 90 seconds of one sprinkler head activating. Testing a waterflow alarm device is typically completed by opening an inspector's test valve or alarm bypass valve and recording the time. For locations that have alarm monitoring companies, the operator at this location must also verify that alarm signals are received.

Although NFPA 25 does not specify that the receipt of the alarm signals be confirmed with a central station or monitoring provider, it is good practice to have the receipt of the alarms verified by the owner when the system is reset and removed from the test mode. The exact description on the signal received should also be verified to ensure alarm signals do not annunciate as trouble signals, which could delay a fire department response when required.

Dry pipe and deluge systems equipped with waterflow alarm devices should operate similarly to wet pipe systems. However, these typically annunciate using devices covered in *NFPA 72.*

Most preaction systems, with the exception of double interlock preaction systems, do not have an inspector's test connection that uses water. In these systems, testing the supervisory alarm involves lowering the supervisory air pressure, which can be done by opening the inspector's test connection to bleed off supervisory air. When the air pressure is lowered to the alarm set point, a signal will sound at the preaction control panel. This will activate the preaction valve, allowing the system to fill with water. When the system is filled with water, the waterflow alarm pressure switch or water motor gong will sound, sending a signal to the preaction control panel and sounding the audible alarm signal.

Double interlock preaction systems have an inspector's test connection similar to that of a dry pipe system.

13.2.6.2 Vane-type and pressure switch–type waterflow devices shall be tested semiannually.

System Tagging

The paddle in this control valve is cracked and will make stopping the flow of water difficult. Furthermore, as a result of the crack, a section of the paddle could break off and become lodged somewhere in the system, which could cause reduced flow during a fire event. This condition would be considered a critical deficiency and would need to be corrected quickly.

Noncritical Deficiency

Critical Deficiency

Impairment

(Courtesy of Byron Blake and SimplexGrinnell)

13.2.7 Gauges.

System Tagging

The needle on this gauge has been bent, making it difficult to read the pressure correctly. Because the gauge does not have any effect on system operation, this would be considered a noncritical deficiency.

Noncritical Deficiency

Critical Deficiency

Impairment

(Courtesy of Byron Blake and SimplexGrinnell)

13.2.7.1 Gauges shall be inspected monthly to verify that they are in good condition and that normal pressure is being maintained.

The term *good condition* in 13.2.7.1 means that the gauge does not appear to be physically damaged. Since this is an inspection, actual verification of operation is not expected.

Exhibit 13.18 shows a pressure gauge that is not mechanically damaged but has become difficult to read and has questionable reliability due to its general condition. Thus, this would require replacement.

13.2.7.1.1 Where other sections of this standard have different frequency requirements for specific gauges, those requirements shall be used.

13.2.7.2 Gauges shall be replaced every 5 years or tested every 5 years by comparison with a calibrated gauge.

The requirement in 13.2.7.2 provides the option of replacing a gauge in lieu of testing the gauge. In many cases, replacement might be more economical than testing. If after 5 years an installed gauge is still operating within the original range and with the required accuracy, replacement is not required.

Testing Procedure Alert

ITM NFPA 25

While an inaccurate gauge will not lead to a sprinkler or standpipe system not operating or not receiving enough water, it is important to make sure that the gauges read properly. Gauges on water-based suppression systems allow for an owner, inspector, or AHJ to obtain an instant read on a system's pressure condition and can also provide some insight into closed valves. For more information on conducting the gauge accuracy test required by 13.2.7.2, see the testing procedure at the end of this chapter.

EXHIBIT 13.18 Pressure Gauge that Is Difficult to Read. (Courtesy of Wiginton Fire Systems)

13.2.7.3 Gauges not accurate to within 3 percent of the full scale shall be recalibrated or replaced.

In 13.2.7.3, inaccurate gauges are required to be recalibrated or replaced. The gauge's accuracy should be checked and a comparison pressure table developed to determine the actual readings at various gauge settings. Results from this table should be used as an alternative to replacement to maintain normal pressure. Most gauges used in fire protection systems cannot realistically be recalibrated due to their design. When such a gauge is found to be inaccurate, there is a replacement alternative.

13.2.8 Records. Records shall be maintained in accordance with Section 4.3.

13.3 Control Valves in Water-Based Fire Protection Systems

13.3.1* Each control valve shall be identified and have a sign indicating the system or portion of the system it controls.

A.13.3.1 Signs identifying underground fire service main control valves in roadway boxes should indicate the direction of valve opening, the distance and direction of the valve from the sign location (if the valve is subject to being covered by snow or ice), and the location of the wrench if not located with the sign.

13.3.1.1 Systems that have more than one control valve that must be closed to work on a system shall have a sign on each affected valve referring to the existence and location of other valves.

13.3.1.2* When a normally open valve is closed, the procedures established in Chapter 15 shall be followed.

A.13.3.1.2 Valves that normally are closed during cold weather should be removed and replaced with devices that provide continuous fire protection service.

Paragraph 13.3.1.2 stipulates under what conditions the requirements in Chapter 15 — which refer to fire protection impairments — must be followed.

If the annual testing of a valve requires the closing of that valve, the valve should be returned to its normal open position immediately after testing. The number of turns required to close the valve should be recorded. That information can then be used to make sure the valve is reopened all the way. The manufacturer's literature will also indicate the number of turns necessary to operate the valve through its full range of motion.

Any supervisory device should be returned to its normal position and no trouble signal should be observed at the fire alarm panel.

13.3.1.2.1 When the valve is returned to service, a valve status test (either main or sectional drain, as appropriate) shall be conducted to determine that the valve is not closed.

The intent of a valve status test is consistent with the intent of a main drain test. Previous editions have indicated that a valve must be confirmed open, but this is not possible solely with a drain test as discussed in 13.2.5. Main drain test results should not be solely relied upon to determine whether valves are fully open. A main drain test might not reveal a partially closed or a partially open valve in a large-diameter supply pipe. In most typical systems, a gate valve closure up to approximately 70 percent to 80 percent would not be detected with a main drain test, because the area that is still open under the gate has a combined area at least equal to that of the drain pipe and valve.

Paragraph 13.3.1.2.1 is not typically meant to apply to underground sectional control valves. These valves can be verified by opening a hydrant connected to the underground main system downstream from a sectional valve or, if by no other way, by opening a main drain on a system riser that is supplied by the underground main. Depending on the size of the underground water main system and distance from the valve to the sprinkler riser, a main drain test might have to be flowed for a long time if the interest is to determine if a valve is fully closed. This would be the time to drain an entire water main down to affect pressure, and this test would most likely never accurately indicate a partially closed valve.

13.3.1.3 Each normally open valve shall be secured by means of a seal or a lock or shall be electrically supervised in accordance with the applicable NFPA standards.

13.3.1.4 Normally closed valves shall be secured by means of a seal or shall be electrically supervised in accordance with the applicable NFPA standard.

Electrically supervised valves must send a signal within the first two valve revolutions or one-fifth of the distance from its normal position, per 13.3.3.5.2. Thus, when testing these valves, if a tamper alarm is not transmitted to an alarm panel within this time, the test is considered a failure and maintenance must be performed to adjust the tamper switch. Exhibit 13.19 shows a control valve with an electronic supervisory device that has not been connected.

13.3.1.5 Sealing or electrical supervision shall not be required for hose valves.

Testing Procedure Alert

ITM NFPA 25

The valve status test is new to the 2014 edition of NFPA 25, but its importance cannot be overstated. The most common cause for system ineffectiveness in water-based fire protection systems is a closed valve. A rather simple test allows the owner or inspector to confirm that not only do the valves appear to be open but that water is in fact flowing through the control valve that had been recently closed. A process for conducting this test is provided at the end of this chapter.

EXHIBIT 13.19 Butterfly Valve with Built-In Supervisory Switch (Not Wired).

13.3.2 Inspection.

13.3.2.1 All valves shall be inspected weekly.

13.3.2.1.1 Valves secured with locks or supervised in accordance with applicable NFPA standards shall be permitted to be inspected monthly.

13.3.2.1.2 Control valve supervisory alarm devices shall be inspected quarterly to verify that they are free of physical damage.

13.3.2.1.3 After any alterations or repairs, an inspection shall be made by the property owner or designated representative to ensure that the system is in service and all valves are in the normal position and properly sealed, locked, or electrically supervised.

13.3.2.2* The valve inspection shall verify that the valves are in the following condition:

(1) In the normal open or closed position
(2)* Sealed, locked, or supervised
(3) Accessible
(4) Post indicator valves (PIVs) are provided with correct wrenches
(5) Free from external leaks
(6) Provided with applicable identification

The valve inspection required by 13.3.2 is specifically for system control valves. During inspection of these valves, observations must be made and the following questions answered:

- Is the valve open? (Check the target or the position of the indicating feature.)
- Is the valve leaking water?
- Is the valve accessible?
- Is the valve wrench for a yard PIV available?

Exhibit 13.20 shows a WPIV that has been rendered inoperable by the growth of a nearby tree branch.

EXHIBIT 13.20 *Inoperable WPIV. (Courtesy of John Jensen)*

A.13.3.2.2 Valves should be kept free of snow, ice, storage, or other obstructions so that access is ensured.

A.13.3.2.2(2) The purpose of the valve sealing program is as follows:

(1) The presence of a seal on a control valve is a deterrent to closing a valve indiscriminately without obtaining the proper authority.
(2) A broken or missing seal on a valve is cause for the plant inspector to verify that protection is not impaired and to notify superiors of the fact that a valve could have been closed without following procedures.

ITM Deficiency, Impairment, or Hazard Evaluation?

(Courtesy of John Johnson, Tyco Fire Protection Products)

ANSWER: Impairment

While performing the monthly inspection required by 13.3.2.1.1, this normally open water control valve was found in the closed position. When the indicating flag is parallel to the piping, the valve is open, and when perpendicular, the valve is closed. The flag position will also indicate when a valve is only partially open. The water gauge is showing no water pressure below the seat of this dry pipe valve, which is another indicator of a closed water control valve.

This system will not function at all during a fire event and should be considered impaired. The impairment procedures in Chapter 15 should be implemented immediately.

ITM Deficiency, Impairment, or Hazard Evaluation?

ANSWER: ITM Deficiency or Impairment

At first glance, this post indicator valve (PIV) and post appear to be just fine, with the wrench attached and locked and no physical damage. However, upon closer examination, the target that indicates whether the valve is open or shut is not visible in the window. In 13.3.2.2(1), the inspector is required to verify whether the water control valve is "in the normal open or closed position." Without the visible target as the indicator, the position of the valve cannot be determined, and the assembly is either broken or the target is off track.

As defined in 3.3.23, an inspection is "a visual examination of a system or portion thereof to verify that it appears to be in operating condition and is free of physical damage." In this case, the PIV is obviously not in good operating condition, and as long as the valve can be confirmed to be in its normal open or closed position, this finding should be classified as a deficiency. If a further investigation reveals that the valve is not in its normal open position, then the system should be considered impaired and the impairment procedures in Chapter 15 should be implemented immediately.

(Courtesy of Byron Blake and SimplexGrinnell)

Testing Procedure Alert

Confirmation that a control valve can operate through its full range of motion is important for several reasons. First, it is imperative that the valve can be opened 100 percent so that the full flow of water that is needed can get through the device. It is also important to confirm that a valve can close so that routine testing and maintenance can be done on schedule. When valves do not operate from fully closed to fully open, it typically signifies an obstruction or other impairment. For more information and a procedure for conducting the control valve for several different types of control valves, see the testing procedure at the end of this chapter.

13.3.3 Testing.

13.3.3.1 Each control valve shall be operated annually through its full range and returned to its normal position.

When testing a nonindicating gate valve such as an underground sectional valve as required by 13.3.3.1, the number of turns needed to fully close and open the valve should be noted and compared to the number of turns needed to open and close as noted in the manufacturer's literature. This number should be recorded on the inspection form and retained for future reference.

Valves typically require lubrication annually per manufacturer's specifications. Valves that are difficult to turn might require maintenance. Forcing sticking valves to turn could result in breaking the valve shaft and creating an unplanned system impairment until repairs can be made. Valves should be able to turn smoothly and with minimal effort. If valves require the use of a person's full weight or a cheater bar to turn, the valve should be repaired or replaced. In the case of an emergency, this could lead to breaking the shaft limiting or eliminating the functionality of the valve.

FAQ How often do PIVs need to be cycled?

All control valves are required to be operated annually through their full range of motion per 13.3.3.1. Often, PIVs are not exercised, since they are not considered a control valve because they sit outside of the building envelope.

13.3.3.2* Post indicator valves shall be opened until spring or torsion is felt in the rod, indicating that the rod has not become detached from the valve.

A.13.3.3.2 A proper wrench needs to be used for this test. Using an improper wrench such as a pipe wrench has resulted in damage to the operating nut. The use of break over bars and extensions on the wrench can damage the valve and/or the post. If the valve cannot be closed and reopened using the proper wrench with reasonable force, then some maintenance and/or repairs are necessary so the valve can be operated when needed in a fire event. These "spring tests" are made to verify that a post indicator valve is fully open. If an operator feels the valve is fully open, he or she should push in the "open" direction. The handle usually moves a short distance (approximately a one-quarter turn) and "springs" back toward the operator in a subtle move when released. This spring occurs when the valve gate pulls up tight against the top of its casting and the valve shaft (being fairly long) twists slightly. The spring indicates that the valve is fully opened and that the gate is attached to the handle. If the gate is jammed due to a foreign particle, the handle is not likely to spring back. If the gate is loose from the handle, the handle continues to turn in the "open" direction with little resistance.

13.3.3.2.1 This test shall be conducted every time the valve is closed.

13.3.3.3 Post indicator and outside screw and yoke valves shall be backed a one-quarter turn from the fully open position to prevent jamming.

13.3.3.4 A main drain test shall be conducted any time the control valve is closed and reopened at system riser.

This requirement is intended to verify that the control valve has not been left fully closed. It is not meant to test that a valve has been returned to the fully open position. A main drain test might not reveal a partially closed (or partially open) valve in a large-diameter supply pipe. In most typical systems, a gate valve closure up to approximately 70 percent to 80 percent would not be detected with a main drain test, because the area that is still open under the gate has a combined area at least equal to that of the drain pipe and valve.

13.3.3.5* Supervisory Switches.

A.13.3.3.5 For further information, see *NFPA 72, National Fire Alarm and Signaling Code.*

13.3.3.5.1 Valve supervisory switches shall be tested semiannually.

13.3.3.5.2 A distinctive signal shall indicate movement from the valve's normal position during either the first two revolutions of a hand wheel or when the stem of the valve has moved one-fifth of the distance from its normal position.

This requirement is not meant to imply each valve tamper device must have its own distinctive signal at an annunciator panel. While highly advisable on larger systems with more than one valve, all tamper devices can annunciate as a single signal. Per *NFPA 72,* valve tamper signals initiated per the requirements in 13.3.3.5.2 must only be distinctive from alarm and test signals. These signals should also be distinctive from trouble signals.

13.3.3.5.3 The signal shall not be restored at any valve position except the normal position.

Some approved valve tamper devices are easy to reset. These devices allow the alarm panel to indicate that all valves are open, when in actuality, the valves are closed. When circuit lacing devices are used, special care must be taken in restoration. Exhibit 13.21 shows a valve monitored by lacing a wire around the valve handle.

13.3.4 Maintenance.

13.3.4.1 The operating stems of outside screw and yoke valves shall be lubricated annually.

13.3.4.2 The valve then shall be completely closed and reopened to test its operation and distribute the lubricant.

EXHIBIT 13.21 Valve Monitored by Wire Laced Through Handle.

13.4 System Valves

13.4.1 Inspection of Alarm Valves. Alarm valves shall be inspected as described in 13.4.1.1 and 13.4.1.2.

13.4.1.1* Alarm valves and system riser check valves shall be externally inspected monthly and shall verify the following:

(1) The gauges indicate normal supply water pressure is being maintained.
(2) The valve is free of physical damage.
(3) All valves are in the appropriate open or closed position.
(4) The retarding chamber or alarm drains are not leaking.

A.13.4.1.1 A higher pressure reading on the system gauge is normal in variable pressure water supplies. Pressure over 175 psi (12.1 bar) can be caused by fire pump tests or thermal expansion and should be investigated and corrected.

13.4.1.2* Alarm valves and their associated strainers, filters, and restriction orifices shall be inspected internally every 5 years unless tests indicate a greater frequency is necessary.

A.13.4.1.2 The system should be drained for internal inspection of valve components as follows:

(1) Close the control valve
(2) Open the main drain valve
(3) Open the inspector's test valve
(4) Wait for the sound of draining water to cease and for all gauges to indicate 0 psi (0 bar) before removing the handhole cover or dismantling any component

FAQ Is the requirement for the 5-year internal inspection of an alarm check valve valid if the valve no longer has an active retard chamber?

The internal inspection of an alarm valve is not contingent upon retard chambers or any other trim. The purpose of the internal inspection is to verify the operation of the clapper and the condition of the valve seats, as well as to look for evidence of corrosion or obstructing material (inorganic and organic). This can also meet the requirement of an internal obstruction

Critical Point

The purpose of the internal inspection is to verify the operation of the clapper and the condition of the valve seats, as well as to look for evidence of corrosion or obstructing material (inorganic and organic).

EXHIBIT 13.22 *Alarm Valve Internal Inspection.*

Exhibit 13.23 *Interior Valve Corrosion. (Courtesy of Wayne Automatic Fire Sprinklers, Inc.)*

investigation if required per 14.3.2.2(2). Exhibit 13.22 illustrates the internal components of an alarm check valve that would be seen upon an internal inspection. Exhibit 13.23 shows the interior corrosion of an alarm valve during an internal inspection.

FAQ What test results would indicate that more frequent internal inspections of alarm valves are needed?

Test results that might indicate more frequent inspections are needed include the following:

- Internal biological growth
- Significant corrosion on internal metal parts such as the check valve plate
- Previously found obstructions that, due to water supply sources or other reasons, could be expected to occur again

13.4.1.3 Maintenance.

13.4.1.3.1 Internal components shall be cleaned/repaired as necessary in accordance with the manufacturer's instructions.

13.4.1.3.2 The system shall be returned to service in accordance with the manufacturer's instructions.

13.4.2 Check Valves.

13.4.2.1 Inspection. Valves shall be inspected internally every 5 years to verify that all components operate correctly, move freely, and are in good condition.

13.4.2.2 Maintenance. Internal components shall be cleaned, repaired, or replaced as necessary in accordance with the manufacturer's instructions.

13.4.3 Preaction Valves and Deluge Valves.

FAQ What is the difference between a preaction valve and a deluge valve?

Preaction and deluge valves have identical main valves but with different trim, which is arranged based on the type of operation. For example, a deluge valve activates and discharges water into a piping system of open nozzles or sprinklers, whereas a preaction system trips and floods a

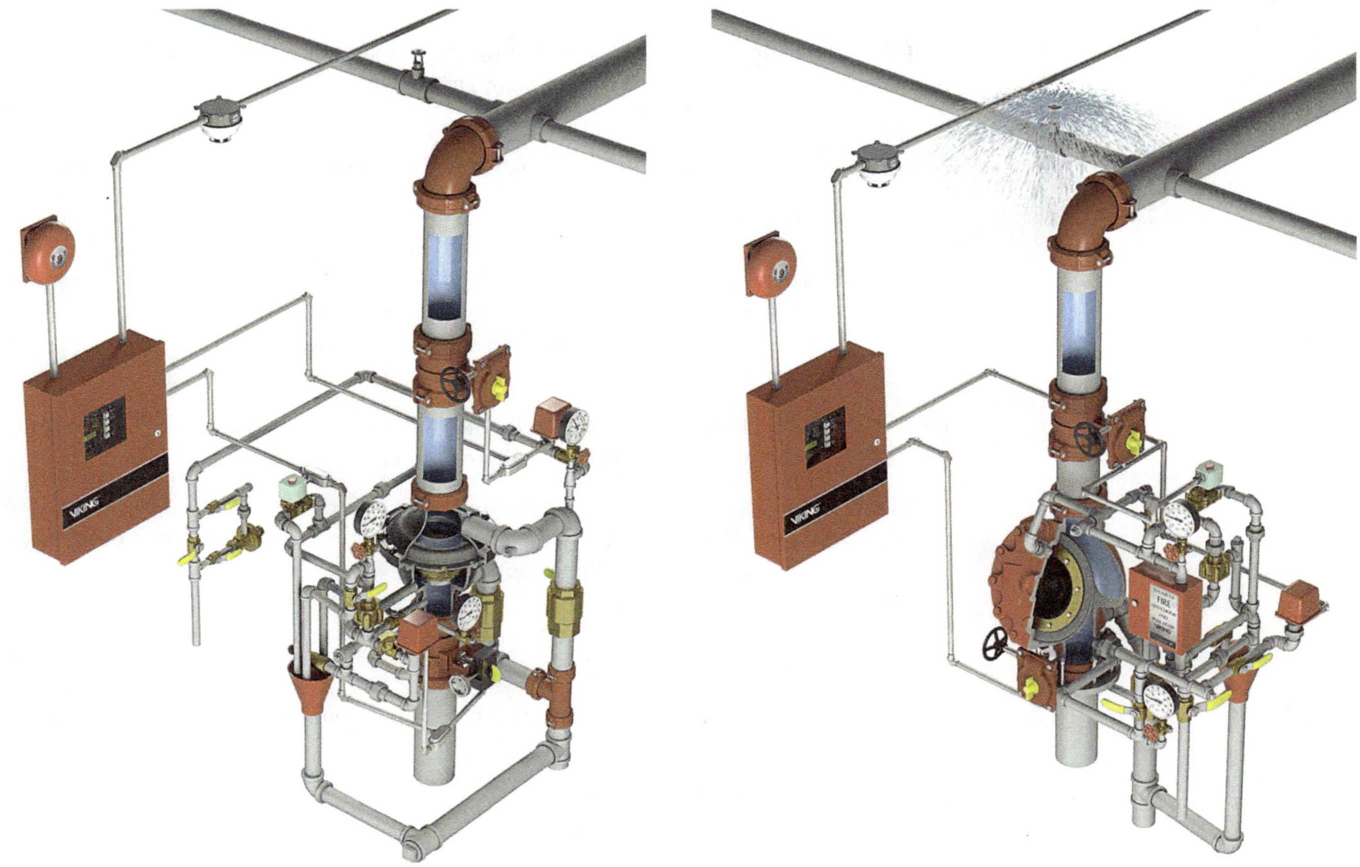

EXHIBIT 13.24 Preaction Valve with Trim Piping and Accessories (left) and Deluge Valve with Trim Piping and Accessories (right). (Courtesy of Jesse Cappon and Viking®)

piping system of closed sprinklers and will not discharge water until a sprinkler activates. Therefore, a preaction system has a check valve installed immediately above the preaction valve to isolate it from the system air pressure. Except for this difference, the two valve bodies (preaction and deluge) are basically the same. See Exhibit 13.24 for a typical preaction valve arrangement and a typical deluge valve arrangement.

Critical Point

Preaction and deluge valves have identical main valves but with different trim, which is arranged based on the type of operation.

13.4.3.1 Inspection.

13.4.3.1.1 Valve enclosures for preaction and deluge valves subject to freezing shall be inspected daily during cold weather to verify a minimum temperature of 40°F (4.0°C).

13.4.3.1.1.1 Valve enclosures equipped with low temperature alarms shall be inspected weekly.

13.4.3.1.2 Low temperature alarms, if installed in valve enclosures, shall be inspected annually at the beginning of the heating season to verify that they are free of physical damage.

Testing Procedure Alert

ITM NFPA 25

The purpose of completing a trip test (partial or full) for a deluge or preaction valve is to ensure the system operates in a manner consistent with the original design application. A comparative assessment of the water delivery time during the full trip test of a double interlock preaction and combined dry pipe/preaction system can reveal significant delays in system operation. For more information on conducting this test, see the testing procedure at the end of this chapter.

13.4.3.1.3 Gauges shall be inspected weekly.

13.4.3.1.3.1 The gauge on the supply side of the preaction or deluge valve shall indicate that the normal supply water pressure is being maintained.

13.4.3.1.4 The gauge monitoring the preaction system supervisory air pressure, if provided, shall be inspected monthly to verify that it indicates that normal pressure is being maintained.

Normally, the supervisory pressure in 13.4.3.1.4 is low [1 psi to 2 psi (0.06 bar to 0.14 bar)] for non-interlocked preaction systems. NFPA 13 requires a minimum air pressure of 7 psi (0.48 bar), but older systems may have pressures as low as 1 psi or 2 psi (0.06 bar or 0.14 bar). The pressure of supervisory air or alarm air used in a double interlock preaction system will be higher, similar to pressures used with a dry sprinkler system. The air pressure should be in accordance with the manufacturer's specifications.

13.4.3.1.5 The gauge monitoring the detection system pressure, if provided, shall be tested monthly to verify that it indicates that normal pressure is being maintained.

The requirement in 13.4.3.1.5 refers to a pilot sprinkler or rate-of-rise thermal detection system used to activate a preaction or deluge system. The air should be dried and kept at the pressure required in the manufacturer's operations and maintenance manual.

An air dryer is filled with a chemical desiccant. The desiccant, which is usually white, turns purple when it is no longer capable of absorbing moisture. When that color change occurs, the desiccant should be replaced.

13.4.3.1.6 The preaction or deluge valve shall be externally inspected monthly to verify the following:

(1) The valve is free from physical damage.
(2) All trim valves are in the appropriate open or closed position.
(3) The valve seat is not leaking.
(4) Electrical components are in service.

The external inspection of the preaction or deluge valve required by 13.4.3.1.6 should also include inspection of pneumatic components.

13.4.3.1.7 The interior of the preaction or deluge valve and the condition of detection devices shall be inspected annually when the trip test is conducted.

13.4.3.1.7.1 Internal inspection of valves that can be reset without removal of a faceplate shall be permitted to be conducted every 5 years.

The requirement in 13.4.3.1.7.1 acknowledges the fact that some listed deluge valves do not have to be opened to be reset. The arrangement of the deluge valve with accessories shown in Exhibit 13.24 illustrates an externally reset deluge valve.

13.4.3.1.8 Strainers, filters, restricted orifices, and diaphragm chambers shall be inspected internally every 5 years unless tests indicate a greater frequency is necessary.

A situation in which a more frequent inspection is required per 13.4.3.1.8 would be a reservoir, where water supply conditions can lead to obstructing material or rapid corrosion.

13.4.3.2 Testing.

13.4.3.2.1* The priming water level in supervised preaction systems shall be tested quarterly for compliance with the manufacturer's instructions.

A.13.4.3.2.1 High priming water levels can adversely affect the operation of supervisory air. Test the water level as follows:

(1) Open the priming level test valve.
(2) If water flows, drain it.
(3) Close the valve when water stops flowing and air discharges.
(4) If air discharges when the valve is opened, the priming water level could be too low. To add priming water, refer to the manufacturer's instructions.

13.4.3.2.2* Each deluge valve shall be trip tested annually at full flow in warm weather and in accordance with the manufacturer's instructions.

A.13.4.3.2.2 Preaction and deluge valves in areas subject to freezing should be trip tested in the spring to allow time before the onset of cold weather for all water that has entered the system or condensation to drain to low points or back to the valve.

13.4.3.2.2.1* Full flow tests shall incorporate full functionality of the system as a unit, including automatic and manual activation.

Low temperature supervisory signals must be tested on a 6-month schedule, in accordance with *NFPA 72*. Each year, one of the low-temp supervisory tests should be performed prior to the start of the heating season.

A.13.4.3.2.2.1 It is necessary that the full flow test incorporate the full functionality of the system, which would include any solenoid valves or other actuation devices. It was a common practice in the past to test the detection system or manual pull station up to the solenoid valve or actuator, and to separately test the deluge valve and system after the solenoid valve or actuator. The detectors on the system can be tested separately as long as the functional test includes activation of the actuator or solenoid when it receives an actual or simulated signal.

13.4.3.2.2.2 Protection shall be provided for any devices or equipment subject to damage by system discharge during tests.

13.4.3.2.2.3 Where the nature of the protected property is such that water cannot be discharged for test purposes, the trip test shall be conducted in a manner that does not necessitate discharge in the protected area.

One method of complying with 13.4.3.2.2.3 is to throttle the water supply by closing the control valve so that it is one or two turns open. When the deluge valve operates, the water supply control valve can be closed quickly before water discharges into the protected area.

A method for conducting a trip test without flowing water entails removing the head of the deluge valve solenoid. When a signal is received at the solenoid, the solenoid becomes magnetized. The magnetization confirms that the alarm system is functioning. This method is not preferred for performing a discharge test of a deluge valve, because it does not verify that the valve is operating properly, only that the system functions.

13.4.3.2.2.4 Where the nature of the protected property is such that water cannot be discharged unless protected equipment is shut down (e.g., energized electrical equipment), a full flow system test shall be conducted at the next scheduled shutdown.

13.4.3.2.2.5 The full flow test frequency shall not exceed 3 years.

13.4.3.2.2.6 The water discharge patterns from all of the open spray nozzles or sprinklers shall be observed to ensure that patterns are not impeded by plugged nozzles, that nozzles are correctly positioned, and that obstructions do not prevent discharge patterns from wetting surfaces to be protected.

(A) Where the nature of the protected property is such that water cannot be discharged, the nozzles or open sprinklers shall be inspected for correct orientation and the system tested with air to ensure that the nozzles are not obstructed.

(B) Where obstructions occur, the piping and sprinklers or nozzles shall be cleaned and the system retested.

13.4.3.2.3 Except for preaction systems covered by 13.4.3.2.5, every 3 years the preaction valve shall be trip tested with the control valve fully open.

13.4.3.2.4 During those years when full flow testing in accordance with 13.4.3.2.3 is not required, the preaction valve shall be trip tested with the control valve partially open.

13.4.3.2.5 Preaction or deluge valves protecting freezers shall be trip tested in a manner that does not introduce moisture into the piping in the freezer.

13.4.3.2.6 Preaction systems shall be tested once every 3 years for air leakage, using one of the following test methods:

(1) A pressure test at 40 psi (3.2 bar) for 2 hours. The system shall be permitted to lose up to 3 psi (0.2 bar) during the duration of the test. Air leaks shall be addressed if the system loses more than 3 psi (0.2 bar) during this test.
(2) With the system at normal system pressure, shut off the air source (compressor or shop air) for 4 hours. If the low air pressure alarm goes off within this period, the air leaks shall be addressed.

Paragraph 13.4.3.2.6 addresses the integrity of preaction systems that have supervisory air pressure. Single interlock preaction systems are not typically equipped with supervisory air pressure and, as a result, would not be subject to the test required in this paragraph.

13.4.3.2.7 Deluge System Pressure Readings.

13.4.3.2.7.1 Pressure readings shall be recorded at the hydraulically most remote nozzle or sprinkler.

Deluge systems typically protect high-hazard or high-challenge risks. It is important to verify the condition of the piping and sprinklers or nozzles. The measurement required in 13.4.3.2.7.1 is intended to confirm that there is adequate pressure at the discharge device to deliver the required density. To make this evaluation, it is necessary to reference the original design criteria and system hydraulic information. If this information is not available, the owner might need to have a hydraulic evaluation of the system completed.

13.4.3.2.7.2 A second pressure reading shall be recorded at the deluge valve.

Most hydraulic calculation programs calculate pressure at the base of the riser. The pressure should be found by referring to the hydraulic design information sign. A reading taken at the deluge valve, as required by 13.4.3.2.7.2, provides an adequate comparison.

13.4.3.2.7.3 These readings shall be compared to the hydraulic design pressures to ensure the original system design requirements are met by the water supply.

13.4.3.2.7.4 Where the hydraulically most remote nozzle or sprinkler is inaccessible, nozzles or sprinklers in other than foam-water systems shall be permitted to be inspected visually without taking a pressure reading on the most remote nozzle or sprinkler.

13.4.3.2.7.5 Where the reading taken at the riser indicates that the water supply has deteriorated, a gauge shall be placed on the hydraulically most remote nozzle or sprinkler and the results compared with the required design pressure.

If the pressure readings are lower than those indicated on the design documents or hydraulic design information signs, the inspector should notify the owner, who should immediately investigate to determine the cause and correct the problem.

13.4.3.2.8 Multiple Systems. The maximum number of systems expected to operate in case of fire shall be tested simultaneously to inspect the adequacy of the water supply.

EXHIBIT 13.25 *Multiple System Manifold.*

Paragraph 13.4.3.2.8 requires the simultaneous testing of multiple systems because the discharge of multiple systems will have an effect on the water supply. It is important to test under conditions that might prevail during a fire. It is not the intent to require testing of multiple systems wherever they exist, but rather only where the discharge of multiple systems is needed, as based on the system design documents. Exhibit 13.25 illustrates several types of systems in a building, not all of which would require simultaneous discharge or testing.

13.4.3.2.9 Manual Operation. Manual actuation devices shall be operated annually.

13.4.3.2.10 Return to Service. After the full flow test, the system shall be returned to service in accordance with the manufacturer's instructions.

13.4.3.2.11 Grease or other sealing materials shall not be applied to the seating surfaces of preaction or deluge valves.

13.4.3.2.12* Records indicating the date the preaction or deluge valve was last tripped and the tripping time, as well as the individual and organization conducting the test, shall be maintained at a location or in a manner readily available for review by the authority having jurisdiction.

A.13.4.3.2.12 Methods of recording maintenance include tags attached at each riser, records retained at each building, and records retained at one building in a complex.

See Supplement 5 for sample report forms. Note that 13.4.3.2.12 does not preclude the use of electronic reporting systems or other methods for documenting the service history of valves and other equipment.

13.4.3.2.13 Low air pressure alarms, if provided, shall be tested quarterly in accordance with the manufacturer's instructions.

13.4.3.2.14 Low temperature alarms, if installed in valve enclosures, shall be tested annually at the beginning of the heating season.

13.4.3.2.15 Automatic air pressure maintenance devices, if provided, shall be tested yearly at the time of the annual preaction or deluge valve trip test, in accordance with the manufacturer's instructions.

13.4.3.3 Maintenance.

13.4.3.3.1 Leaks causing drops in supervisory pressure sufficient to sound warning alarms, and electrical malfunctions causing alarms to sound, shall be located and repaired.

13.4.3.3.2 During the annual trip test, the interior of the preaction or deluge valve shall be cleaned thoroughly and the parts replaced or repaired as necessary.

13.4.3.3.2.1 Interior cleaning and parts replacement or repair shall be permitted every 5 years for valves that can be reset without removal of a faceplate.

13.4.3.3.3* Auxiliary drains in preaction or deluge systems shall be operated after each system operation and before the onset of freezing conditions (and thereafter as needed).

A.13.4.3.3.3 Suitable facilities should be provided to dispose of drained water. Low points equipped with a single valve should be drained as follows:

(1) Open the low-point drain valve slowly
(2) Close the drain valve as soon as water ceases to discharge, and allow time for additional accumulation above the valve
(3) Repeat this procedure until water ceases to discharge
(4) Replace plug or nipple and cap as necessary

Low points equipped with dual valves should be drained as follows:

(1) Close the upper valve
(2) Open the lower valve, and drain the accumulated water
(3) Close the lower valve, open the upper valve, and allow time for additional water accumulation
(4) Repeat this procedure until water ceases to discharge
(5) Replace plug or nipple and cap in lower valve

Removing water from a preaction or deluge system is an essential part of a good maintenance program. Failure to keep these systems free of water can result in damage and expensive repairs to both the system and building. A program for monitoring the condition of the system and the operation of the auxiliary drains should be instituted. Auxiliary drains should be operated on a daily basis after a system operation until several days pass with no discharge of water from the drain valve. Thereafter, it might be possible to decrease the frequency to weekly or longer intervals depending on the volume of water discharged. Likewise, when preparing for cold weather, the auxiliary drains should be operated daily, with the frequency of operation decreasing depending on the discharge of accumulated water. In many cases, the frequency of the operation can decrease significantly if a system is shown to be dry.

It is important that the piping be kept dry, particularly in areas subject to freezing. When a system is tripped during testing or as the result of a fire or other event, it is important to follow a plan to monitor the removal of any water as it accumulates in the system. Systems located in conditioned spaces that have supervised air pressure are particularly susceptible to moisture accumulating during the summer season as a result of warm moist air being introduced into the system.

13.4.3.3.4 Additional maintenance as required by the manufacturer's instructions shall be provided.

13.4.4 Dry Pipe Valves/Quick-Opening Devices.

A dry pipe system valve is a differential-type valve and is more sophisticated than an alarm check valve. Consequently, a dry pipe valve requires more testing and maintenance than an alarm check valve. See Exhibit 13.26 for a typical dry pipe valve.

13.4.4.1 Inspection.

13.4.4.1.1 Valve enclosures for preaction and deluge valves subject to freezing shall be inspected daily during cold weather to verify a minimum temperature of 40°F (4°C).

EXHIBIT 13.26 Typical Dry Pipe Valve.

13.4.4.1.1.1 Valve enclosures equipped with low temperature alarms shall be inspected weekly.

13.4.4.1.1.2 Low temperature alarms, if installed in valve enclosures, shall be inspected annually at the beginning of the heating season to verify that they are free of physical damage.

13.4.4.1.2 Gauges.

13.4.4.1.2.1 The gauge on the supply side of the dry pipe valve shall indicate that the normal supply water pressure is being maintained.

13.4.4.1.2.2 The gauge on the system side of the dry pipe valve shall indicate that the proper ratio of air or nitrogen pressure to water supply pressure is being maintained in accordance with the manufacturer's instructions.

13.4.4.1.2.3* The gauge on the quick-opening device, if provided, shall indicate the same pressure as the gauge on the system side of the dry pipe valve.

A.13.4.4.1.2.3 A conflict in pressure readings could indicate an obstructed orifice or a leak in the isolated chamber of the quick-opening device, either of which could make the quick-opening device inoperative.

13.4.4.1.2.4 Gauges on systems with low air or nitrogen pressure alarms shall be inspected monthly.

13.4.4.1.2.5 Gauges on systems other than those with low air or nitrogen pressure alarms shall be inspected weekly.

13.4.4.1.3 Systems with auxiliary drains shall require a sign at the dry or preaction valve indicating the number of auxiliary drains and location of each individual drain.

The requirement in 13.4.4.1.3 was added in the 2008 edition of the standard and is retroactive, so systems maintained under previous editions of the standard must be retrofitted to comply. Many systems have auxiliary drains that are located above ceilings or in other hard-to-find locations. Service providers should consult with the owner before performing trip tests or maintenance work to verify that all drains have been located and identified.

13.4.4.1.4 The dry pipe valve shall be externally inspected monthly to verify the following:

(1) The valve is free of physical damage.
(2) All trim valves are in the appropriate open or closed position.
(3) The intermediate chamber is not leaking.

13.4.4.1.5 The interior of the dry pipe valve shall be inspected annually when the trip test is conducted.

13.4.4.1.6 Strainers, filters, and restricted orifices shall be inspected internally every 5 years unless tests indicate a greater frequency is necessary.

13.4.4.2 Testing.

13.4.4.2.1* The priming water level shall be tested quarterly.

A.13.4.4.2.1 High priming water levels can affect the operation of supervisory air or nitrogen pressure maintenance devices. Test the water level as follows:

(1) Open the priming level test valve.
(2) If water flows, drain it.
(3) Close the valve when water stops flowing and air discharges.
(4) If air discharges when the valve is opened, the priming water level could be too low. To add priming water, refer to the manufacturer's instructions.

13.4.4.2.2* Each dry pipe valve shall be trip tested annually during warm weather.

A.13.4.4.2.2 Dry pipe valves should be trip tested in the spring to allow time before the onset of cold weather for all water that has entered the system or condensation to drain to low points or back to the valve.

13.4.4.2.2.1 Dry pipe valves protecting freezers shall be trip tested in a manner that does not introduce moisture into the piping in the freezers.

13.4.4.2.2.2* Every 3 years and whenever the system is altered, the dry pipe valve shall be trip tested with the control valve fully open and the quick-opening device, if provided, in service.

Testing Procedure Alert

ITM NFPA 25

The purpose of completing a trip test for a dry pipe valve, be it a partial test or a full test, is to ensure that the system will operate. The full trip test of a dry pipe system allows for a comparative assessment of the water delivery time that can reveal significant delays in system operation. For a series of steps that can be taken to execute this test, refer to the testing procedure at the end of this chapter.

A.13.4.4.2.2.2 A full flow trip test generally requires at least two individuals, one of whom is situated at the dry pipe valve while the other is at the inspector's test. If possible, they should be in communication with each other. A full flow trip test is conducted as follows:

(1) The main drain valve is fully opened to clean any accumulated scale or foreign material from the supply water piping. The main drain valve then is closed.
(2) The system air or nitrogen pressure and the supply water pressure are recorded.
(3) The system air or nitrogen pressure is relieved by opening the inspector's test valve completely. Concurrent with opening the valve, both testers start their stopwatches. If two-way communication is not available, the tester at the dry valve is to react to the start of downward movement on the air pressure gauge.
(4) Testers at the dry pipe valve note the air pressure at which the valve trips and note the tripping time.
(5) Testers at the inspector's test note the time at which water flows steadily from the test connection. This time is noted for comparison purposes to previous tests and is not meant to be a specific pass/fail criterion. Note that NFPA 13, *Standard for the Installation of Sprinkler Systems*, does not require water delivery in 60 seconds for all systems.
(6) When clean water flows, the test is terminated by closing the system control valve.
(7) The air or nitrogen pressure and the time elapsed are to be recorded as follows:
    (a) From the complete opening of the test valve to the tripping of the valve
    (b) From the complete opening of inspector's valve to the start of steady flow from the test connection
(8) All low-point drains are opened and then closed when water ceases to flow.
(9) The dry pipe valve and quick-opening device are reset, if installed, in accordance with the manufacturer's instructions, and the system is returned to service.

For dry pipe systems that were designed and installed using either a manual demonstration or a computer calculation to simulate multiple openings to predict water delivery time, a full flow trip test from a single inspector's test connection should have been conducted during the original system acceptance and a full flow trip test from the single inspector's test should continue to be conducted every 3 years. The system is not required to achieve water delivery to the inspector's test connection in 60 seconds, but comparison to the water delivery time during the original acceptance will determine if there is a problem with the system.

Given the wide range of acceptable water delivery times, it is not the intent of NFPA 25 to reverify any specific time limits with the full flow trip test. Rather, it is the intent of NFPA 25 to reveal substantial delays for the water delivery time of dry pipe systems, since significant differences from one test to another are an indication of possible operational problems with the system. A drastic increase in the water delivery time from one full flow trip test of a dry pipe system to the next is of major concern, because such an increase is most likely due to corrosion of the system piping. This is one condition that would trigger the requirement for an internal obstruction investigation per 14.3.1(14).

An increase in delivery time might also be an indication that physical changes have been made to the system, such as a revision in the piping layout or the orifice size of sprinklers. It might also indicate that the water supply characteristics have changed. Any substantial change in water delivery time from one test to another is cause for investigation into the source of the problem.

Dry systems are limited in size by the time needed for the system to deliver water to the inspector's test connection per NFPA 13. NFPA 13 places no time limit on water delivery for systems with a volume of 500 gal (1893 L) or less, or for systems of 750 gal (2839 L) or less that are equipped with a quick opening device, because systems of these sizes typically deliver the water in a relatively short time. Systems with a volume greater than 750 gal (2839 L) must deliver water in not more than 60 seconds as determined by a listed calculation program, or the water discharge must be verified by an appropriately sized test manifold (see Exhibit 13.27) in accordance with the time limits shown in Commentary Table 13.1.

Critical Point

A drastic increase in the water delivery time from one full flow trip test of a dry pipe system to the next is of major concern, because such an increase is most likely due to corrosion of the system piping.

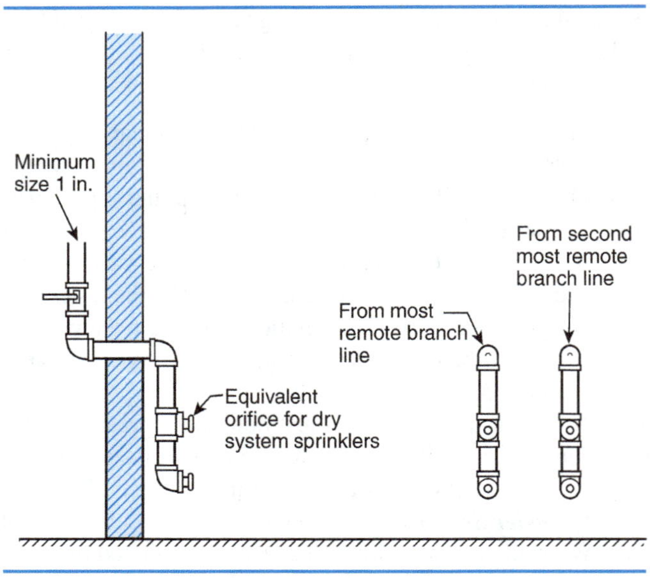

EXHIBIT 13.27 *Example Manifold Arrangement — Four Sprinklers. (Source: NFPA 13, 2013, Figure A.7.2.3.7)*

COMMENTARY TABLE 13.1 *Dry System Water Delivery*

Hazard	Number of Most Remote Sprinklers Initially Open	Maximum Time of Water Delivery (seconds)
Light	1	60
Ordinary I	2	50
Ordinary II	2	50
Extra I	4	45
Extra II	4	45
High piled	4	40

13.4.4.2.2.3* During those years when full flow testing in accordance with 13.4.4.2.2.2 is not required, each dry pipe valve shall be trip tested with the control valve partially open.

A.13.4.4.2.2.3 A partial flow trip test is conducted in the following manner:

(1) Fully open the main drain valve to clean any accumulated scale or foreign material from the supply water piping
(2) Close the control valve to the point where additional closure cannot provide flow through the entire area of the drain outlet
(3) Close the valve controlling flow to the device if a quick-opening device is installed
(4) Record the system air or nitrogen pressure and the supply water pressure
(5) Relieve system air or nitrogen pressure by opening the priming level test valve
(6) Note and record the air or nitrogen pressure and supply water pressure when the dry pipe valve trips
(7) Immediately close the system control valve and open the main drain valve to minimize the amount of water entering the system piping
(8) Trip test the quick-opening device, if installed, in accordance with the manufacturer's instructions

(9) Open all low point drains and close when water ceases to flow
(10) Reset the dry pipe valve and quick-opening device, if installed, in accordance with the manufacturer's instructions and return the system to service

CAUTION: A partial flow trip test does not provide a high enough rate of flow to latch the clappers of some model dry pipe valves in the open position. When resetting such valves, check that the latching equipment is operative.

13.4.4.2.2.4 When refilling a dry system, the air supply shall be capable of restoring normal air pressure in the system within 30 minutes.

13.4.4.2.2.5 The requirements of 13.4.4.2.2.4 shall not apply in refrigerated spaces maintained below 5°F (–15°C), where normal system air pressure shall be permitted to be restored within 60 minutes.

13.4.4.2.3 Grease or other sealing materials shall not be applied to the seating surfaces of dry pipe valves.

13.4.4.2.4* Quick-opening devices, if provided, shall be tested quarterly.

A.13.4.4.2.4 Except when a full flow trip test is conducted in accordance with A.13.4.4.2.2.2, a quick-opening device should be tested in the following manner:

(1) Close the system control valve
(2) Open the main drain valve, and keep it in the open position
(3) Verify that the quick-opening device control valve is open
(4) Open the inspector's test valve. (Note that a burst of air from the device indicates that it has tripped)
(5) Close the device's control valve
(6) Return the device to service in accordance with the manufacturer's instructions and return the system to service

13.4.4.2.5 A tag or card that shows the date on which the dry pipe valve was last tripped, and the name of the person and organization conducting the test, shall be attached to the valve.

13.4.4.2.5.1 Separate records of initial air and water pressure, tripping air pressure, and dry pipe valve operating conditions shall be maintained on the premises for comparison with previous test results.

13.4.4.2.5.2 Records of tripping time shall be maintained for full flow trip tests.

See Supplement 5 for sample trip test report forms.

13.4.4.2.6 Low air pressure alarms, if provided, shall be tested quarterly in accordance with the manufacturer's instructions.

13.4.4.2.7 Low temperature alarms, if installed in valve enclosures, shall be tested annually at the beginning of the heating season.

13.4.4.2.8 Automatic air pressure maintenance devices, if provided, shall be tested annually during the dry pipe valve trip test in accordance with the manufacturer's instructions.

Testing Procedure Alert

ITM NFPA 25

The low air pressure switch is an important piece of equipment in a dry system, because it is the device that signals the dry valve to open at the desired system pressure. Failure of the pressure switches to operate can delay the opening of the dry valve or, in some cases, can cause the valve to open and fill the system when there is no fire event. For more information on conducting this test, see the testing procedure at the end of this chapter.

13.4.4.2.9 Dry pipe systems shall be tested once every 3 years for gas leakage, using one of the following test methods:

(1) A gas (air or nitrogen) pressure test at 40 psi (3.2 bar) shall be performed for 2 hours.
   (a) The system shall be permitted to lose up to 3 psi (0.2 bar) during the duration of the test.
   (b) Gas leaks shall be addressed if the system loses more than 3 psi (0.2 bar) during this test.
(2) With the system at normal system pressure, the gas source (nitrogen supply, compressor, or shop air) shall be shut off for 4 hours. If the low pressure alarm goes off within this period, the leaks shall be addressed.

The test required by 13.4.4.2.9 was introduced in the 2008 edition of the standard. The previous requirement called for a 10 psi (0.7 bar) pressure loss per week, but this loss was difficult to ascertain because the automatic air supply would generally maintain pressure by cycling on and off. The intent of the air pressure test is to determine if a significant problem exists. A dry pipe system must be capable of staying closed without any trips of the dry pipe valve for short periods of time, such as during a power outage, without the air compressor maintaining pressure. Either method listed in 13.4.4.2.9 can be used to demonstrate the integrity of the system.

If a dry pipe valve is designed to trip at 25 psi (1.7 bar), the air supply should be set to provide 20 psi (1.4 bar) in excess of this pressure or 45 psi (3.1 bar). The automatic air compressor could be set to turn on at 35 psi (2.4 bar) and turn off at 45 psi (3.1 bar). The low air pressure switch might be set at 30 psi (2.1 bar) to produce a low air alarm. In this case, assuming the air compressor is about to turn on (a worst case scenario), the system would be permitted to lose only 5 psi (0.3 bar). At the high end of the air compressor cycle, the system would be permitted to lose 15 psi (1 bar).

Testing Procedure Alert

ITM NFPA 25

Dry pipe systems that have excessive leakage can become costly for owners who constantly have to drain dry piping that becomes inadvertently filled. When dry systems leak, the low air pressure switch acknowledges the pressure drop and sends a signal to the valve to open. If this is due to leakage, as opposed to a sprinkler activating, it is important to drain the system so that a freeze situation does not occur. To learn how to conduct a test to make sure dry systems are sufficiently "tight," refer to the testing procedure at the end of this chapter.

13.4.4.3 Maintenance.

13.4.4.3.1 During the annual trip test, the interior of the dry pipe valve shall be cleaned thoroughly, and parts replaced or repaired as necessary.

13.4.4.3.2* Auxiliary drains in dry pipe sprinkler systems shall be drained after each operation of the system, before the onset of freezing weather conditions, and thereafter as needed.

A.13.4.4.3.2 Removing water from a dry system is an essential part of a good maintenance program. Failure to keep the dry system free of water can result in damage and expensive repairs to both the system and building. A program for monitoring the condition of the system and the operation of the auxiliary drains should be instituted. Auxiliary drains should be operated on a daily basis after a dry sprinkler system operation until several days pass with no discharge of water from the drain valve. Thereafter, it might be possible to decrease the frequency to weekly or longer intervals depending on the volume of water discharged. Likewise, when preparing for cold weather, the auxiliary drains should be operated daily with the frequency of operation decreasing depending on the discharge of accumulated water. In many cases, the frequency of the operation can decrease significantly if a system is shown to be dry. A quick-opening device, if installed, should be removed temporarily from service prior to draining low points.

The low point drain in a dry pipe system is commonly referred to as a condensate nipple. The condensate nipple consists of a 1 ft (0.3 m) length of 2 in. (50 mm) pipe with a 1 in. (25 mm) valve on each end. The upper valve is closed to isolate the condensate nipple from the dry pipe system. Next, the lower valve is opened to drain any moisture from within. Once draining is complete, the upper valve is opened to allow moisture to drain into the condensate nipple.

Exhibit 13.28 shows a low point drain on a dry pipe system where any excess moisture is drained following the test. Exhibit 13.29 shows a corroded pipe from a dry pipe system that was not drained properly. See the commentary following A.13.4.3.3.3 for more information on dry piping.

EXHIBIT 13.28 Low Point Drain.

EXHIBIT 13.29 Corroded Pipe.

13.5 Pressure-Reducing Valves and Relief Valves

Pressure-reducing valves, such as the one shown in Exhibit 13.30, and relief valves are installed to control pressures in fire protection systems. They are usually set to reduce pressures on the system side of the valve to pressures not exceeding 175 psi (12.1 bar), the maximum working pressure for most listed sprinkler system components. The maximum pressure allowed for larger standpipe hose connections is 175 psi (12.1 bar). The maximum pressure allowed for connections supplying small hose is 100 psi (6.9 bar). Pressure-reducing and relief valves, as illustrated in Exhibit 13.31, are installed to control pressures in fire protection systems such as sprinkler and standpipe systems. Exhibit 13.32 illustrates a system pressure relief valve located in a supply room.

EXHIBIT 13.30 Master Pressure-Reducing Valve Typically Seen on the Discharge Side of a Fire Pump.

EXHIBIT 13.31 Standpipe with Both a Sprinkler System Pressure-Reducing Valve and a Hose Connection Pressure Reducing Valve.

EXHIBIT 13.32 Pressure-Relief Valve Typically Seen on the Discharge Side of a Fire Pump.

13.5.1 Inspection and Testing of Sprinkler Pressure-Reducing Valves. Sprinkler pressure-reducing valves shall be inspected and tested as described in 13.5.1.1 and 13.5.1.2.

13.5.1.1 All valves shall be inspected quarterly to verify that the valves are in the following condition:

(1) In the open position
(2) Not leaking
(3) Maintaining downstream pressures in accordance with the design criteria
(4) In good condition, with handwheels installed and unbroken

The system pressure-reducing valve is responsible for maintaining downstream system pressures at a level lower than the pressure rating of the system components on the system side of the valve. In 13.5.1.2, NFPA 25 requires a full flow test of this valve to provide the owner and other stakeholders with assurance that the downstream system components will not see pressures in excess of what they are designed to handle. To learn more about this test, including an approach on how to conduct it, please refer to the testing procedures at the end of this chapter.

Testing Procedure Alert

ITM NFPA 25

13.5.1.2* A full flow test shall be conducted on each valve at 5-year intervals and shall be compared to previous test results.

A.13.5.1.2 The sectional drain valve should be opened to compare the results with the original installation or acceptance tests.

As noted in A.13.5.1.2, the sectional drain valve is opened to produce flow for testing purposes. Pressures should be measured on each side of the valve to verify that high pressures on the supply side of the valve are reduced to a pressure that is less than the rating of the system components on the system side of the valve. Sprinkler systems are typically rated for a maximum of 175 psi (12.1 bar). However, there are systems equipped with components — including the sprinklers — that are rated for pressures higher than 175 psi (12.1 bar). Discharge should be to a drain connection or receptacle that can accommodate the flow without spilling. Exhibit 13.33 illustrates apparatus for testing a pressure-reducing valve.

13.5.1.2.1 Adjustments shall be made in accordance with the manufacturer's instructions.

Not all valves are field adjustable. Valves that are adjustable should be adjusted in accordance with the design and installation documents and, as required by 13.5.1.2.1, with the manufacturer's instructions.

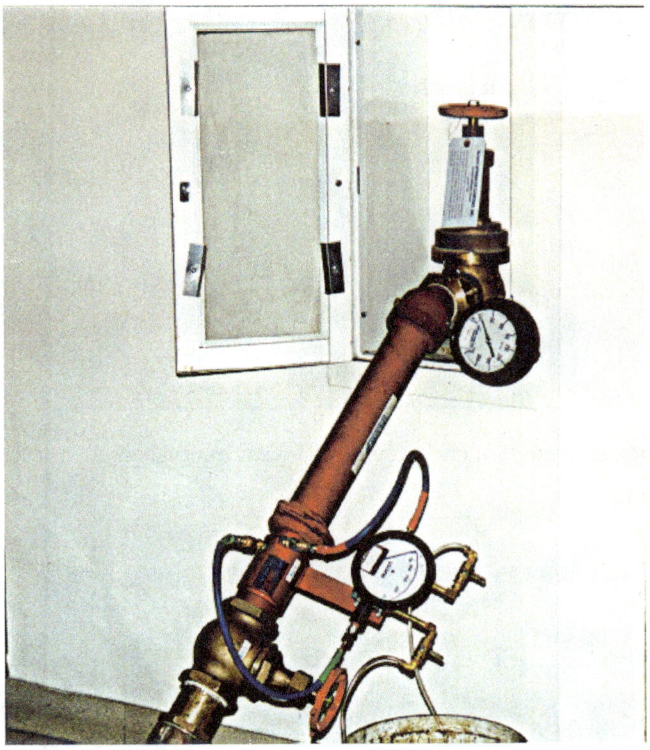

EXHIBIT 13.33 *Pressure-Reducing Valve Test Assembly. (Courtesy of Oliver Sprinkler Co., Inc.)*

13.5.1.3 A partial flow test adequate to move the valve from its seat shall be conducted annually.

The test required in 13.5.1.3 is intended to exercise the valve and to ensure that it is not obstructed and will reseat. A full flow test is not required because a properly functioning valve will move and reseat with a relatively small flow of water.

13.5.2 Hose Connection Pressure-Regulating Devices.

Critical Point

Pressure-reducing valves for hose connections are installed to reduce pressures to a safe level for those individuals that are trained in the use of fire hose.

Pressure-reducing valves for hose connections are installed to reduce pressures to a safe level for those individuals that are trained in the use of fire hose.

Small hose connections are provided in 1½ in. (40 mm) diameters and must be restricted to not more than 100 psi (6.9 bar). Connections for heavy hose [2½ in. (65 mm) diameter] must be restricted to not more than 175 psi (12.1 bar). For more information regarding hose connection pressures, see Chapter 7 of NFPA 14, *Standard for the Installation of Standpipe and Hose Systems.*

13.5.2.1 All devices shall be inspected annually to verify the following:

(1) The handwheel is not broken or missing.
(2) The outlet hose threads are not damaged.
(3) No leaks are present.
(4) The hose adapter and the cap are not missing.

The hose connection cap, which is permitted to be plastic, is intended to protect the hose threads only. However, it also serves the vital function of protecting the integrity of the system by preventing obstructions. If the cap is removed, material such as trash could enter the valve, creating an obstruction and rendering the system ineffective.

System Tagging

In Table A.3.3.7, a distinction is made between dripping and spraying leaks in system piping. This drain valve is dripping and would therefore be considered a noncritical deficiency. The component is part of the drain system, which is not vital to suppressing a fire, and there is only a limited amount of water leaking from the valve.

Noncritical Deficiency

Critical Deficiency

Impairment

(Courtesy of Byron Blake and SimplexGrinnell)

13.5.2.2* A full flow test shall be conducted on each device at 5-year intervals and shall be compared to previous test results.

Testing Procedure Alert

ITM NFPA 25

Hose connection pressure-regulating devices are deigned to maintain downstream pressures to an appropriate level based on the type of hose connection present. This test is designed to confirm that these devices are functioning properly and that there are not excessive pressures at the hose outlet that can damage equipment and potentially endanger users. For information on conducting this test, refer to the testing procedure at the end of this chapter.

A.13.5.2.2 PRV devices can be bench tested in accordance with the manufacturer's instructions or tested in place. To test in place, a gauge is connected on both the inlet side and the outlet side of the device, and flow readings are taken using a Pitot tube or a flowmeter. Water is discharged through a roof manifold, if available, or through hose to the exterior of the building. Another acceptable method for systems having at least two risers is to take one standpipe out of service and use it as a drain by removing PRV devices and attaching hoses at the outlets near the ground floor level. When testing in this manner, a flowmeter should be used and a hose line utilized to connect the riser being tested and the drain riser.

Readings are to be compared to the system's hydraulic demands at the test location. Field-adjustable valves are to be reset in accordance with manufacturer's instructions. Nonadjustable valves should be replaced. Extreme caution should be exercised because of the high pressure involved when testing.

Standpipe systems installed prior to 1993 must be tested as described in A.13.5.2.2. Since 1993, NFPA 14 has required the installation of a 3 in. (75 mm) drain riser where pressure-reducing devices are installed in a standpipe system. The drain riser must provide a 2½ in. (65 mm) swivel outlet on every other floor for connection of hose.

13.5.2.2.1 Adjustments shall be made in accordance with the manufacturer's instructions.

13.5.2.3 A partial flow test adequate to move the device from its seat shall be conducted annually.

As described in 13.5.1.3, the test in 13.5.2.3 is intended to exercise the valve and to ensure that it is not obstructed and will reseat. A partial flow test is sufficient to indicate whether the valve is functioning properly.

13.5.3 Hose Rack Assembly Pressure-Regulating Devices.

Hose rack pressure-reducing valves, covered in 13.5.3, can be either 1½ in. (40 mm) or 2½ in. (65 mm) in size and are pre-connected to a hose rack complete with hose and nozzle. A hose rack can be connected to a sprinkler system or be part of a standpipe system. In either case, the test requirements for the valve are identical to those specified in 13.5.2.

13.5.3.1 All devices shall be inspected annually to verify the following:

(1) The handwheel is not missing or broken.
(2) No leaks are present.

13.5.3.2 A full flow test shall be conducted on each device at 5-year intervals and compared to previous test results.

13.5.3.2.1 Adjustments shall be made in accordance with the manufacturer's instructions.

13.5.3.3 A partial flow test adequate to move the device from its seat shall be conducted annually.

13.5.4 Master Pressure-Regulating Devices.

Beginning with the 2007 edition, NFPA 14 has recognized the use of a master pressure-reducing valve (PRV) in standpipe systems. This device is permitted to be installed downstream of the fire pump discharge piping and can be installed in such a way as to control the pressure to an entire system or an entire zone of a standpipe system. The master PRV must be installed with a bypass arrangement that will permit maintenance of the device without initiating an impairment procedure. Valves must also be installed to allow isolation of the device for maintenance and testing purposes. Some installations might include multiple master PRVs: one for the higher flow of standpipe systems and one for the lower flows of sprinkler systems. Other installations might include two PRVs installed in series to prevent over-pressurization of a standpipe system should one valve fail. See Exhibit 13.34 for typical PRV arrangements, including a dual pressure-regulating device arrangement.

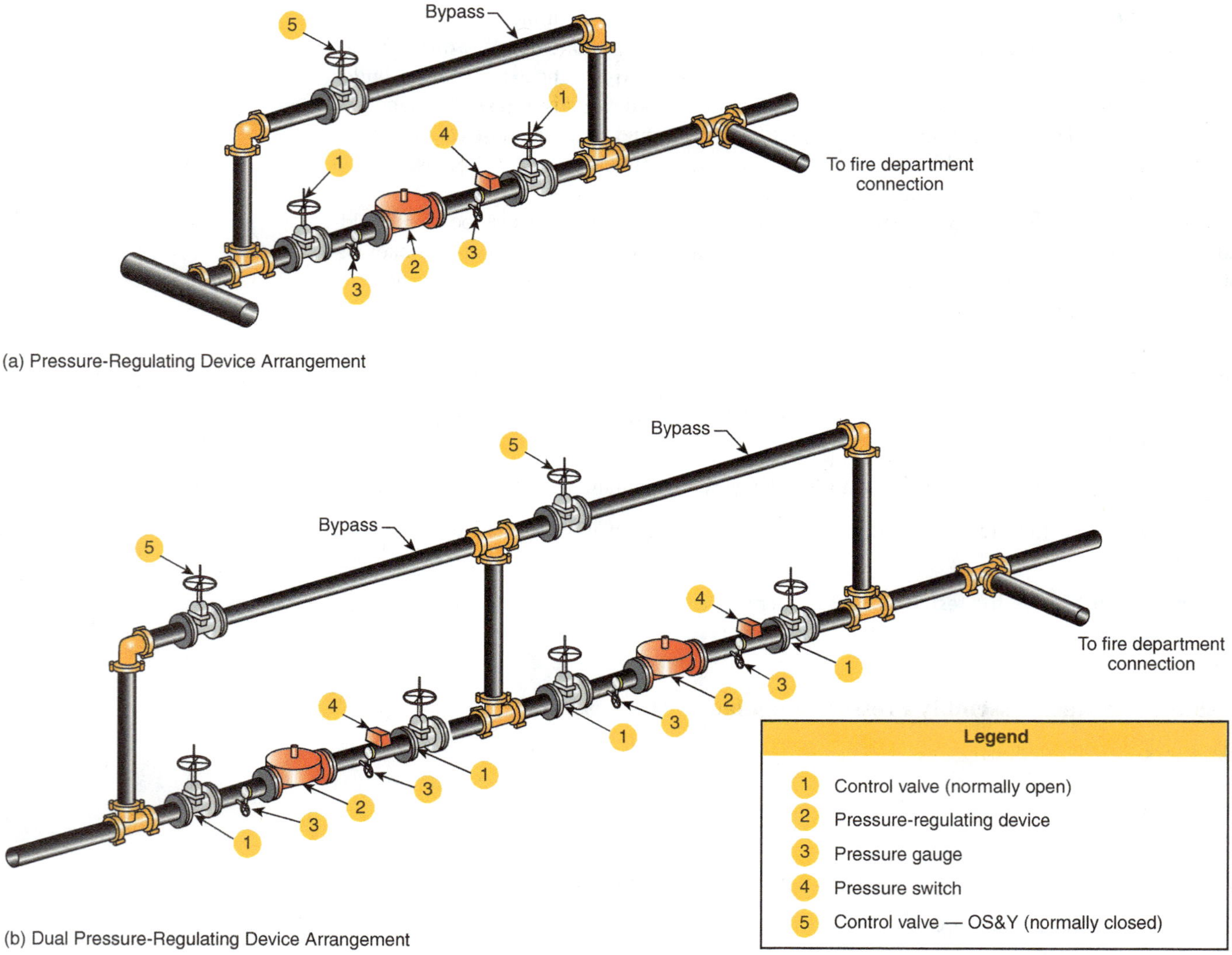

EXHIBIT 13.34 *Pressure-Regulating Device Arrangements. (Courtesy of Stephan Laforest)*

13.5.4.1* Devices shall be inspected weekly to verify that the devices are in the following condition:

(1)* The downstream pressures are maintained in accordance with the design criteria.
(2) The supply pressure is in accordance with the design criteria.
(3) The devices are not leaking.
(4) The devices and trim are in good condition.

A.13.5.4.1 When the PRV is located in or immediately downstream of the fire pump discharge, the weekly inspection of the master PRV can be performed during the weekly fire pump operating test.

A.13.5.4.1(1) Pressures downstream of the master PRV should not exceed the maximum pressure rating of the system components.

13.5.4.2* A partial flow test adequate to move the valve from its seat shall be conducted quarterly.

A.13.5.4.2 The partial flow test of the master PRV can be performed during the quarterly main drain test. (*See 13.2.5.2.*)

13.5.4.3* A full flow test shall be conducted on each valve annually and shall be compared to previous test results.

A.13.5.4.3 When the PRV is located in the fire pump discharge, the full flow test of the master PRV can be performed during the annual fire pump flow test.

13.5.4.4 When valve adjustments are necessary, they shall be made in accordance with the manufacturer's instructions.

13.5.5 Pressure-Reducing Valves.

13.5.5.1 All pressure-reducing valves installed on fire protection systems not covered by 13.5.1, 13.5.2, 13.5.3, or 13.5.4 shall be inspected in accordance with 13.5.1.1.

13.5.5.2 All pressure-reducing valves installed on fire protection systems not covered by 13.5.1, 13.5.2, 13.5.3, or 13.5.4 shall be tested in accordance with 13.5.1.2.

13.5.6 Hose Valves.

13.5.6.1 Inspection.

13.5.6.1.1 Hose valves shall be inspected quarterly to verify that the valves are in the following condition:

(1) Hose caps are in place and not damaged.
(2) Hose threads are not damaged.
(3) Valve handles are present and not damaged.
(4) Gaskets are not damaged or showing signs of deterioration.
(5) No leaks are present.
(6) Valves are not obstructed or otherwise not capable of normal operation.

13.5.6.1.2 Hose valves shall be inspected to ensure that hose caps are in place and not damaged.

13.5.6.1.3 Hose threads shall be inspected for damage.

13.5.6.1.4 Valve handles shall be present and not damaged.

13.5.6.1.5 Gaskets shall be inspected for damage or deterioration.

13.5.6.1.6 Hose valves shall be inspected for leaks.

13.5.6.1.7 Hose valves shall be inspected to ensure no obstructions are present.

13.5.6.1.8 Hose valves shall be inspected to ensure that restricting devices are present.

13.5.6.2 Testing.

The testing and maintenance requirements described in 13.5.6.2 should be performed by a qualified contractor or service provider. A *qualified* individual is defined in 3.3.34 as someone who has met the requirements and training for a given field acceptable to the AHJ. This qualification is not standardized and can vary by jurisdiction.

13.5.6.2.1* Class I and Class III standpipe system hose valves shall be tested annually by fully opening and closing the valves.

System Tagging

Each of these photos shows a broken control valve, which will typically be considered a noncritical deficiency. The ability to turn the control valve does not impact the operation of a system. However, it should be noted that, for control valves that are in the closed position when a handle breaks, this issue can be considered an impairment or critical deficiency if there is no way to get the system back on-line.

(Photos Courtesy of Byron Blake and SimplexGrinnell)

Noncritical Deficiency   Critical Deficiency   Impairment

A.13.5.6.2.1 Hose valves can be tested without a full flow if the cap is left on the hose threads. The purpose of this requirement is to exercise the valve so it can be operated easily.

FAQ Do the valves need to be flowed or just opened and closed to comply with 13.5.6.2.1?

The valves do not need to be flowed for this test. The purpose of this requirement is to exercise the valve so it can be operated easily. This test often will indicate whether the valve is sticking, which allows maintenance personnel to correct the issue so that the valve functions properly when it is needed. This test is not intended to be a prolonged flow test used for the removal of foreign objects in the systems.

13.5.6.2.1.1 Class I and Class III standpipe system hose valves that are difficult to operate or leak shall be repaired or replaced.

13.5.6.2.2* Hose valves on hose stations attached to sprinkler systems and Class II standpipe systems shall be tested every 3 years by opening and closing the valves.

A.13.5.6.2.2 See A.13.5.6.2.1.

13.5.6.2.2.1 Hose valves on hose stations attached to sprinkler systems and Class II standpipe systems that are difficult to operate or leak shall be repaired or replaced.

13.5.6.3 Maintenance. Hose valves that do not operate smoothly or open fully shall be lubricated, repaired, or replaced.

13.5.7 Fire Pump Relief Valves.

13.5.7.1 Circulation Relief Valves.

13.5.7.1.1 Where installed, circulation relief valves shall be inspected on the same frequency as the no-flow (churn) test.

13.5.7.1.2 The inspection shall verify that water flows through the valve when the fire pump is operating at shutoff pressure (i.e., churn) to prevent the pump from overheating.

13.5.7.1.3 On completion of any fire pump test, the closure of the circulation relief valve shall be verified.

13.5.7.2 Main Pressure Relief Valve.

13.5.7.2.1 Where installed, main pressure relief valves shall be inspected during any fire pump test.

13.5.7.2.2 The inspection shall verify that the pressure downstream of the relief valve fittings in the fire pump discharge piping does not exceed the pressure for which the system components are rated.

Paragraph 13.5.7.2.2 applies only to those pump installations equipped with a pressure relief valve. Such installations are those that produce pressures in excess of the listing for system components. An example of this type of system is illustrated in Exhibit 13.35. Other installations will not be so equipped.

EXHIBIT 13.35 Pressure Relief Valve with Sight Glass. (Courtesy of John Jensen)

13.5.7.2.3 During the annual fire pump flow test, the pressure relief valve shall be verified to be correctly adjusted and set to relieve at the correct pressure and to close below that pressure setting.

13.5.8 Maintenance.

All damaged or missing components noted during the inspections specified in 13.5.6.1 through 13.5.6.2.2 shall be repaired or replaced in accordance with the manufacturer's instructions.

13.6 Backflow Prevention Assemblies

13.6.1 Inspection. Inspection of backflow prevention assemblies shall be as described in 13.6.1.1 through 13.6.1.4.

13.6.1.1 The isolation valves on double check assemblies (DCA) and double check detector assemblies (DCDA) shall be inspected weekly to ensure that the valves are in the normal open position.

Critical Point

Backflow assemblies contain OS&Y valves to isolate backflow devices for maintenance and for fire system control.

Backflow assemblies contain OS&Y valves to isolate backflow devices for maintenance and for fire system control. These control valves must be inspected and tested in accordance with NFPA 25. While the backflow itself might require a licensed contractor for testing control valve inspections, it can be accomplished by the property owner or designee subject to AHJ approval.

13.6.1.1.1 Valves secured with locks or electrically supervised in accordance with applicable NFPA standards shall be inspected monthly.

13.6.1.2 The isolation valves on reduced-pressure assemblies (RPA) and reduced-pressure detector assemblies (RPDA) shall be inspected weekly to ensure that the valves are in the normal open position.

13.6.1.2.1 Valves secured with locks or electrically supervised in accordance with applicable NFPA standards shall be inspected monthly.

13.6.1.2.2* RPAs and RPDAs shall be inspected weekly to ensure that the differential-sensing valve relief port is not continuously discharging.

A.13.6.1.2.2 Intermittent discharge from a differential-sensing valve relief port is normal. Continuous discharge is a sign of malfunction of either or both of the check valves, and maintenance is necessary.

13.6.1.3 After any testing or repair, an inspection by the property owner or designated representative shall be made to ensure that the system is in service and all isolation valves are in the normal open position and properly locked or electrically supervised.

13.6.1.4* Backflow prevention assemblies shall be inspected internally every 5 years to verify that all components operate correctly, move freely, and are in good condition.

A.13.6.1.4 Where annual maintenance includes an internal inspection performed by a qualified person, this requirement is satisfied.

13.6.2 Testing.

13.6.2.1* All backflow preventers installed in fire protection system piping shall be exercised annually by conducting a forward flow test at a minimum flow rate of the system demand.

A.13.6.2.1 The full flow test of the backflow prevention valve can be performed with a test header or other connections downstream of the valve. A bypass around the check valve in the fire department connection line with a control valve in the normally closed position can be an acceptable arrangement. When flow to a visible drain cannot be accomplished, closed loop flow can be acceptable if a flowmeter or sight glass is incorporated into the system to ensure flow.

The 2014 edition of this standard requires only a forward flow test of the system through the backflow preventer. The backflow test was deemed by the technical committee to be outside the scope of the standard, because it is a requirement of water quality standards and performed by an individual with specific training and licensing. However, many contractors have obtained the required licenses to perform the backflow test and do so as part of their fire protection system servicing.

•

13.6.2.1.1 Where water rationing is enforced during shortages lasting more than 1 year, an internal inspection of the backflow preventer to ensure the check valves will fully open shall be permitted in lieu of conducting the annual forward flow test.

•

13.6.2.1.2 The forward flow test shall not be required where annual fire pump testing causes the system flow rate to flow through the backflow preventer device.

Many local health authorities require that a backflow preventer be the first device installed on a water supply pipe. In such cases, the annual fire pump test will produce sufficient flow through the backflow preventer to meet the full forward flow test requirement. In some cases, a backflow preventer can be installed on the system side of a fire pump — ordinarily between the fire department connection (FDC) and the potable water supply — so that the annual fire pump test would not produce any flow through the backflow preventer. When this is the case, a separate flow test is necessary.

13.6.2.2 Where hydrants or inside hose stations are located downstream of the backflow preventer, the forward flow test shall include hose stream demand.

On systems installed under the requirements of the 1996 and subsequent editions of NFPA 13, the forward flow test at system demand has been a part of the system acceptance test. However, due to a lack of enforcement, few systems are equipped with a means to conduct the test. As a result, the technical committee has limited this test requirement to only require flow to the amount of water equivalent to the system demand to configurations where some means of testing, such as a fire hydrant or hose connections, are available downstream of the backflow device. Paragraph 13.6.2.2 allows the test to be conducted using the maximum flow rate possible. Flow testing at the highest rate possible does not necessarily mean a simple main drain test. It might necessitate opening inspector's test connections along with the main drain or, where multiple systems are present, opening multiple main drains. For example, a typical big box retail warehouse has multiple systems, with each individual system requiring a flow of 1500 gpm (5678 L/min) or more. The main drain valves on these multiple systems could be opened simultaneously to flow a volume of water that is close to the system demand.

Poorly performing backflow check valve assemblies can result in an impaired fire protection sprinkler system. Backflow assemblies must be listed. Part of this listing for use in a fire protection system includes evaluations for durability, reliability, and flow characteristics. Each assembly has flow characteristics that show pressure drop related to flow. This must be evaluated with system demands, especially when installed on a pre-existing system.

13.6.2.3 Where connections do not permit verification of the forward flow test at the minimum flow rate of system demand, tests shall be conducted at the maximum flow rate possible.

13.6.3 Maintenance. Maintenance of all backflow prevention assemblies shall be conducted by a qualified individual following the manufacturer's instructions in accordance with the procedure and policies of the authority having jurisdiction.

•

13.7 Fire Department Connections

13.7.1 Fire department connections shall be inspected quarterly to verify the following:

(1) The fire department connections are visible and accessible.
(2) Couplings or swivels are not damaged and rotate smoothly.

System Tagging

A failed FDC should be repaired when it is found, because the FDC is an integral part of fire-fighting operations. A leaky FDC caused by a broken gasket will not prevent water from getting to the system, so this is typically considered a noncritical deficiency. However, it is important to check with the fire department, because they might have a different perspective as the users of the device.

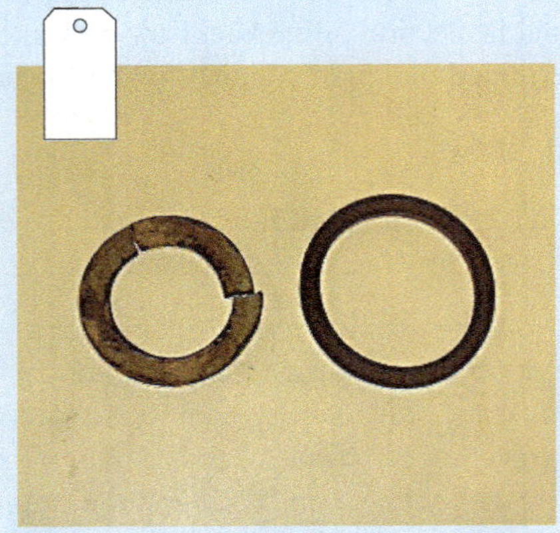

(Courtesy of Byron Blake and SimplexGrinnell)

- Noncritical Deficiency
- Critical Deficiency
- Impairment

System Tagging

The FDC shown in this picture is blocked by shrubbery and would be difficult, if not impossible, for the fire department to see from a roadway or parking lot. Table A.3.3.7 classifies an FDC that is not visible as a critical deficiency. In cases where the fire department cannot access the device, it would be considered an impairment per Table A.3.3.7. In this case, the device is accessible and will become more so when responding personnel simply cut down the shrub to hook up to the FDC during a fire.

(Photos Courtesy of Byron Blake and SimplexGrinnell)

- Noncritical Deficiency
- Critical Deficiency
- Impairment

(3) Plugs or caps are in place and undamaged.
(4) Gaskets are in place and in good condition.
(5) Identification signs are in place.
(6) The check valve is not leaking.
(7) The automatic drain valve is in place and operating properly.
(8) The fire department connection clapper(s) is in place and operating properly.

Exhibit 13.36 illustrates a typical FDC inspection where the caps are removed. The inspector examines inside the FDC for evidence of obstructing material and checks that the couplings are not damaged, the gaskets are in good condition, and the swivels rotate properly. Exhibit 13.37 illustrates an FDC that is obscured from view by vegetation. Visibility, particularly at night, is important so that the fire department can locate the device easily. Exhibit 13.38 illustrates an FDC that has been broken or vandalized. In this case, a threaded outlet and cap are missing, which can allow obstructing material to enter the device, as demonstrated by the trash visible in the valve openings of the FDC shown in Exhibit 13.39. Exhibit 13.40 illustrates an FDC

EXHIBIT 13.36 *FDC Inspection.*

EXHIBIT 13.37 *FDC Obstructed by Vegetation.*

EXHIBIT 13.38 *FDC Missing Threaded Outlet and Cap.*

EXHIBIT 13.39 *FDC Showing Clear Access with Caps Missing.*

EXHIBIT 13.40 *Leaking FDC Indicating Replacement of the Check Valve Is Likely Required.*

EXHIBIT 13.41 *Inaccessible FDC. (Courtesy of Wayne Automatic Fire Sprinklers, Inc.)*

EXHIBIT 13.42 *FDC Heavily Corroded Due to Lack of Maintenance and Inspections. (Courtesy of Wayne Automatic Fire Sprinklers, Inc.)*

that is leaking as evidenced by the staining on the building façade. Exhibit 13.41 shows an FDC that cannot accept a fire department hose due to the FDC's proximity to a large tree. Exhibit 13.42 shows an FDC that has not been maintained and inspected and can no longer accept a threaded hose fitting due to corrosion.

(9)* Interior of the connection is inspected for obstructions.

A.13.7.1(9) It is not the intent of this section for all fire department connection piping to be inspected for obstructions but only the interior of the connection itself.

While 13.7.1(9) does not require all of the FDC piping to be internally inspected, where the connection itself shows signs of obstructions, it should warrant a further investigation to determine the extent of the problem.

System Tagging

Table A.3.3.7 suggests that missing caps on FDCs should be treated as a critical deficiency. The FDC is essential to fire-fighting operations and the equipment is needed to remain in proper working order. In many instances, the fire department will also require a flushing test or an investigation into the FDC piping where it has been recorded that the caps are missing. It is not uncommon for people passing by to put trash such as soda bottles or cans in an FDC, which can render the piping impaired.

(Photos Courtesy of Byron Blake and SimplexGrinnell)

Noncritical Deficiency   Critical Deficiency   Impairment

13.7.2 Components shall be repaired or replaced as necessary in accordance with the manufacturer's instructions.

While it is not the intent of this requirement that the entire run of piping connecting the fire department connection to the check valve be internally inspected, this could be the only way to ensure there are not interior obstructions. A small rock or ball dropped into an FDC could roll all the way down to the check valve. This valve could be underground or many feet from the actual FDC header. The easiest way to do a complete inspection in these cases is through the use of a borescope. If foreign material is seen with this inspection, it must be removed. This might require excavation of the FDC piping up to the check valve.

13.7.3 Any obstructions that are present shall be removed.

13.7.4 The piping from the fire department connection to the fire department check valve shall be hydrostatically tested at 150 psi (10 bar) for 2 hours at least once every 5 years.

13.8 Component Testing Requirements

13.8.1 Whenever a valve, valve component, and/or valve trim is adjusted, repaired, reconditioned, or replaced, the action required in Table 13.8.1 shall be performed.

13.8.2 Where the original installation standard is different from the cited standard, the use of the appropriate installing standard shall be permitted.

•

13.8.3* These actions shall not require a design review.

A.13.8.3 Design review is outside the scope of this standard.

TABLE 13.8.1 Summary of Component Replacement Action Requirements

Component	*Adjust*	*Repair/ Recondition*	*Replace*	*Inspection, Test, and Maintenance Procedures*
Water delivery components				
Post indicator and wall indicator valves	X	X	X	(1) Inspect for leaks at system pressure (2) Perform full operational test conforming to 13.3.3.1 (3) Perform spring torsion inspection conforming to 13.3.3.1 and 13.3.3.2 (4) Verify target visibility at shut and full open position (5) Test supervisory device (6) Main drain test
Control valves other than post indicator and wall indicator valves	X	X	X	(1) Inspect for leaks at system pressure (2) Perform full operational test conforming to 13.3.3.1 (3) Perform spring torsion inspection for OS&Y valves conforming to 13.3.3.2 (4) Verify supervisory device (5) Main drain test
Alarm check valve	X	X	X	(1) Inspect for leaks at system pressure per 13.4.1 (2) Test all alarms and supervisory signals affected by the alarm valve (3) Main drain test
Dry pipe valve	X	X	X	(1) Inspect for leaks at system pressure (2) Trip test per 13.4.4.2 (3) Inspect condition of valve seat (4) Test all dry pipe system alarms and supervisory signals (5) Main drain test
Deluge/preaction valve	X	X	X	(1) Inspect for leaks at system pressure per 13.4.3 (2) Trip test (3) Inspect condition of valve seat (4) Test all deluge/preaction system alarms and supervisory signals (5) Main drain test
Quick-opening device	X	X	X	(1) Inspect for leaks at system pressure per 13.4.4.2.2 (2) Trip test (3) Main drain test
Pressure-regulating device — hose valves	X	X	X	(1) Inspect for leaks at system pressure per 13.5.1 (2) Full flow test (3) Main drain test (Only when a control valve has been closed)
Pressure-regulating devices — other than hose valves	X	X	X	(1) Inspect for leaks at system pressure per Section 13.5 (2) Test pressure setting with full flow and without flow (3) Test supervisory device and alarm (4) Main drain test

TABLE 13.8.1 Continued

Component	*Adjust*	*Repair/ Recondition*	*Replace*	*Inspection, Test, and Maintenance Procedures*
Hose valve	X	X	X	(1) Inspect for leaks at system pressure per 13.5.6 (2) Main drain test
Backflow prevention device	X	X	X	(1) Inspect for leaks at system pressure per Section 13.6 (2) Forward flow test per 13.6.2.1 (3) Test supervisory device and alarm (4) Main drain test
Check valves	X	X	X	(1) Inspect for leaks at system pressure per 13.4.2 (2) Inspect for leaking through check valve (3) Main drain test
Fire department connection	X	X		(1) Inspect for leaks at system pressure per Section 13.7 (2) Main drain test (Only when a control valve has been closed)
Fire department connection — sprinkler system(s)			X	(1) Isolate and hydrostatic test for 2 hours at 150 psi (10 bar) (2) Main drain test (Only when a control valve has been closed)
Fire department connection — other than sprinkler system(s)			X	(1) Isolate and hydrostatic test for 2 hours at 50 psi (3.5 bar) above the normal working pressure [200 psi (14 bar) minimum] (2) Main drain test (Only when a control valve has been closed)
Strainers	X	X	X	Inspect and clean in accordance with manufacturer's instructions
Main drain valves	X	X	X	Main drain test per 13.2.5
Gauges			X	Calibrate per 13.2.7
Alarm and supervisory components				
Alarm device	X	X	X	Test for conformance with NFPA 13 and/or *NFPA 72*
Supervisory device	X	X	X	Test for conformance with NFPA 13 and/or *NFPA 72*
System protection components				
Pressure relief valve — fire pump installation	X	X	X	See 8.3.3.3 and 13.5.7
Pressure relief valve — other than fire pump installation			X	Verify relief valve is listed or approved for the application and set to the correct pressure
Informational components				
Identification signs	X	X	X	Inspect for compliance with NFPA 13 and 13.3.1

References Cited in Commentary

National Fire Protection Association, 1 Batterymarch Park, Quincy, MA 02169-7471.

NFPA 3, *Recommended Practice for Commissioning and Integrated Testing of Fire Protection and Life Safety Systems*, 2012 edition.

NFPA 13, *Standard for the Installation of Sprinkler Systems*, 2013 edition.

NFPA 14, *Standard for the Installation of Standpipe and Hose Systems*, 2013 edition.

NFPA 20, *Standard for the Installation of Stationary Pumps for Fire Protection*, 2013 edition.

NFPA 72®, National Fire Alarm and Signaling Code, 2013 edition.

American Water Works Association, 6666 West Quincy Avenue, Denver, CO 80235.

Manual M14, *Recommended Practice for Backflow Prevention and Cross-Connection Control*, 3rd edition, 2004.

U.S. Government Printing Office, Washington, DC 20402.

Title 42, United States Code, "Safe Drinking Water Act," Public Law 93-523, 1974.

Underwriters Laboratories Inc., 333 Pfingsten Road, Northbrook, IL 60062-2096.

ANSI/UL 193, *Alarm Valves for Fire-Protection Service*, 2004.

ANSI/UL 260, *Dry Pipe and Deluge Valves for Fire-Protection Service*, 2004.

UL 262, *Gate Valves for Fire-Protection Service*, 2004.

ANSI/UL 312, *Check Valves for Fire-Protection Service*, 2010.

UL 1091, *Butterfly Valves for Fire-Protection Service*, 2004.

Subject 258, *Outline of Investigation for Shutoff Valves for Trim and Drain Purposes*, 2004.

Main Drain Test

Purpose

The purpose of conducting a main drain test is to determine the adequacy of the connected water supply and to detect significant reductions that might adversely impact the system operation by comparing the results of the current test to a previous baseline test pressure. The test does not provide absolute assurance of an unobstructed waterflow or fully opened valves, but it provides a reasonable check against gross changes in the water supply conditions. This testing is required for each water supply lead-in to the building. Where a single lead-in supplies multiple risers, the conduct of a single main drain is acceptable, except that additional main drain tests must be conducted so that every system backflow preventer and/or pressure-reducing valve is exercised. These additional tests would be necessary for situations such as those that include individual floor control valves equipped with pressure-reducing valves.

Tools/Equipment

No special tools are required.

Procedure Steps

Systems Supplied by a Fire Pump

1. Prior to any testing, notify the fire department and/or the alarm monitoring company, as well as the facility representatives, that testing is going to be conducted.
2. The pressure gauge readings collected on the suction side of the fire pump during the annual fire pump test can be used for comparison purposes at a common flow rate in lieu of conducting a separate annual main drain test.
   a. During the annual fire pump test, record the static pressure (no flow) reading on the pressure gauge on the suction side of the fire pump.
   b. During the rated flow (or other common flow rate) of the annual fire pump test, allow the pressure to stabilize and then record the residual pressure reading on the pressure gauge on the suction side of the fire pump.
3. Where the sole supply of water to the fire pump is through a backflow preventer and/or pressure-reducing valve requiring a quarterly main drain test, the test should be conducted in accordance with the following procedure with the fire pump in the operating position.
   a. Check the area surrounding the main drain outlet to ensure that there are no obvious conditions that would prevent water from being discharged safely or cause direct damage in the immediate vicinity. Where the discharge of water is to an area subject to potential freezing conditions, the facility representative should be advised of the potential for icing conditions.
   b. On systems where the main drain is connected to an alarm valve or other arrangement where the flow of water from the main drain will result in the activation of an alarm device or water motor gong, the shutoff valve (where equipped) in the connection to the alarm sensing line should be closed to avoid unnecessary sounding of the alarm.
   c. Record the static pressure (no flow) reading on the pressure gauge on the suction side of the fire pump.
   d. Open the main drain valve slowly until it has reached the fully open position initiating the fire pump operation. Allow the pressure to stabilize, then record the residual pressure reading on the pressure gauge on the suction side of the fire pump.
   e. Slowly close the main drain valve to avoid water hammer, record the time for the pressure to stabilize, and record the return static pressure reading on the pressure gauge on the suction side of the fire pump.
   f. Allow the fire pump to operate for 10 minutes for an electric fire pump, or 30 minutes for a diesel fire pump, followed by a shutdown of the fire pump and restoration of automatic operating condition.
   g. Return any shutoff valves in alarm sensing lines that were closed during Step 3b to the normal operating position.
4. Upon completion of all testing, notify the fire department and/or the alarm monitoring company, as well as the facility representatives, that testing is complete. Reset the fire alarm system as necessary.

Systems Not Supplied by a Fire Pump

1. Prior to any testing, notify the fire department and/or the alarm monitoring company, as well as the facility representatives, that testing is going to be conducted.
2. Check the area surrounding the main drain outlet to ensure that there are no obvious conditions that would prevent water from being discharged safely or cause direct damage in the immediate vicinity. Where the discharge of water is to an area subject to potential freezing conditions, the facility representative should be advised of the potential for icing conditions.
3. On systems where the main drain is connected to an alarm valve or other arrangement where the flow of water from the main drain will result in the activation of an alarm device or water motor gong, the shutoff valve (where equipped) in the connection to the alarm sensing line should be closed to avoid unnecessary sounding of the alarm.
4. Record the static pressure reading on the supply pressure gauge. Note that gauge readings are to be taken at a supply water gauge on the supply side of any alarm valves or check valves on the system riser.

5. Open the main drain valve slowly until it has reached the fully open position. Allow the pressure to stabilize, and then record the residual pressure reading on the supply pressure gauge.

6. Slowly close the main drain valve to avoid water hammer, record the time for the pressure to stabilize, and record the return static pressure reading on the supply pressure gauge.

7. Return any alarm valve and shutoff valves in alarm sensing lines that were closed during Step 3 to the normal operating position.

8. Upon completion of all testing, notify the fire department and/or the alarm monitoring company, as well as the facility representatives, that testing is complete. Reset the fire alarm system as necessary.

Evaluation of Results

The recorded results of the residual pressure reading on the supply pressure gauge during the main drain test, or the gauge on the suction side of the fire pump, should be compared to the baseline reading from the original acceptance test of the system or previously preformed tests. Where the main drain residual test pressure is found to be more than 10 percent less than the baseline test pressure, the conditions warrant further investigation and correction as necessary. Additionally, a slow return in static pressure reading after the closing of the main drain valve could also be indicative of a water supply issue such as a partially closed valve. It is important to note the differences between a degraded water supply resulting from a closed or partially closed valve or an obstruction in the piping from that occurring as a result of normal water system fluctuations. For example, a partially closed valve or obstruction will result in an increased drop between the static and residual pressure readings during the main drain test, whereas a water system fluctuation might result in both a lowered static and residual pressure but have a similar drop in pressure under flowing conditions during the main drain test.

The procedure(s) outlined above is not mandated by NFPA 25 and represents only one approach to conducting this test.

Waterflow Alarm Device Test

Purpose

The purpose of conducting a test of the waterflow alarm device is to ensure the operability of the device and initiation of an alarm condition. Waterflow alarm devices include two separate types, a mechanical waterflow device such as a water motor gong that uses a mechanical action created by water flow to create an audible signal and an electronic waterflow alarm device such as a vane-type or pressure switch-type device that transmits an electronic signal upon waterflow. The testing of such devices is performed by simulating the flow that would result from system operation. This can be accomplished by a number of methods. The most common method for a wet pipe system is the opening of the inspector's test connection; however, the use of the alarm test line on an alarm valve, dry pipe valve, preaction valve, or deluge valve is also quite common and advisable for those systems where a trip of the system is not desirable or for systems that are not equipped with an inspector's test connection. Additionally, the alarm test line is required to be used where the discharge of water during freezing weather or under other conditions that prohibit the use of the inspector's test connection.

Tools/Equipment

A timing device is needed to measure the time between flow initiation and establishment of an alarm condition.

Procedure Steps

Wet System: Mechanical Water Motor Gong / Pressure Switch–Type Alarm Device / Vane-Type Alarm Device Using the Inspector's Test Connection

1. Prior to any testing, notify the fire department and/or the alarm monitoring company, as well as the facility representatives, that testing is going to be conducted.
2. Locate the inspector's test connection and check the area surrounding the discharge location for the released water from this operation to ensure that there are no obvious conditions that would prevent water from being discharged safely or cause direct damage in the immediate vicinity. Where the discharge of water is to an area subject to potential freezing conditions, the facility representative should be advised of the potential for icing conditions.
3. Where connected to a fire alarm system, check the fire alarm panel or remote annunciator to determine that the waterflow alarm device is not currently active on the system.
4. Temporarily place any connected fire pumps in the "off" position.
5. Open the inspector's test valve until it has reached the fully open position and note the time of valve operation. Allow the flow to continue until an alarm condition is established, such as the sounding of the water motor gong or alarm bell or the transmission of an alarm signal to the fire alarm control panel or remote annunciator.
6. Record the time between the operation of the inspector's test valve and the establishment of an alarm condition.
7. Close the inspector's test valve.
8. Return any fire pumps that were placed in the "off" position in Step 4 to the automatic position.
9. Upon completion of all testing, notify the fire department and/or the alarm monitoring company, as well as the facility representatives, that testing is complete. Reset the fire alarm system as necessary.

Wet or Dry System: Mechanical Water Motor Gong / Pressure Switch–Type Alarm Device Using Alarm Test Line

1. Prior to any testing, notify the fire department and/or the alarm monitoring company, as well as the facility representatives, that testing is going to be conducted.
2. Locate the alarm test line on the alarm/dry pipe/preaction/deluge valve and check the area surrounding the discharge location for the released water from operation to ensure that there are no obvious conditions that would prevent water from being discharged safely or cause direct damage in the immediate vicinity. Where the discharge of water is to an area subject to potential freezing conditions, the facility representative should be advised of the potential for icing conditions.
3. Where connected to a fire alarm system, check the fire alarm panel or remote annunciator to determine that the waterflow alarm device is not currently active on the system.
4. Temporarily place any connected fire pumps in the "off" position.
5. Open the alarm test valve until it has reached the fully open position and note the time of valve operation. Allow the flow to continue until an alarm condition is established, such as the sounding of the water motor gong or alarm bell or the transmission of an alarm signal to the fire alarm control panel or remote annunciator.
6. Record the time between the operation of the alarm test line and the establishment of an alarm condition.
7. Close the alarm test valve.
8. Return any fire pumps that were placed in the "off" position in Step 4 to the automatic position.
9. Upon completion of all testing, notify the fire department and/or the alarm monitoring company, as well as the facility representatives, that testing is complete. Reset fire alarm system as necessary.

Wet System: Vane-Type Alarm Device Using Re-Circulating Pump

1. Prior to any testing, notify the fire department and/or the alarm monitoring company, as well as the facility representatives, that testing is going to be conducted.

2. Where connected to a fire alarm system, check the fire alarm control panel or remote annunciator to determine that the waterflow alarm device is not currently active on the system.
3. Operate the re-circulating pump, flowing water in a loop past the vane pressure switch, and note the start time of re-circulating pump operation. Allow the re-circulating pump to continue operating until an alarm condition is established, such as the sounding of the alarm bell or transmission of an alarm signal to the fire alarm control panel or remote annunciator.
4. Record the time between the start of operation of the re-circulating pump and the establishment of an alarm condition.
5. Shut off the re-circulating pump operation.
6. Upon completion of all testing, notify the fire department and/or the alarm monitoring company, as well as the facility representatives, that testing is complete. Reset the fire alarm system as necessary. A verification of the receipt of initiated signals with the fire department and/or the alarm monitoring company, where so connected, would be appropriate as part of this effort.

Evaluation of Results

The recorded time for establishment of an alarm condition must be less than 5 minutes for a local alarm per NFPA 13, *Standard for the Installation of Sprinkler Systems*, or 90 seconds for a remotely monitored alarm condition per *NFPA 72®*, *National Fire Alarm and Signaling Code.* Alarm activation times that exceed these values require further investigation, adjustment, and/or corrective action as necessary.

References

National Fire Protection Association, 1 Batterymarch Park, Quincy, MA 02169-7471.

NFPA 13, *Standard for the Installation of Sprinkler Systems*, 2013 edition.

NFPA 72®, *National Fire Alarm and Signaling Code*, 2013 edition.

The procedure(s) outlined above is not mandated by NFPA 25 and represents only one approach to conducting this test.

Testing Procedures for 13.2.7.2

Gauge Accuracy Test

Purpose

The purpose of conducting an accuracy test on the pressure gauges is to ensure a reasonable level of accuracy of the gauge reading. Comparative tests are made against a calibrated pressure gauge to achieve accuracy within a 3 percent range of the full scale of the gauge consistent with the maximum original accuracy tolerances for the manufactured gauge.

Tools/Equipment

- A test pressure gauge[1] is required.
- Wrenches are required for pressure gauge installation or removal.

Procedure Steps

1. Prior to any testing, notify the fire department and/or the alarm monitoring company, as well as the facility representatives, that testing is going to be conducted.
2. For pressure gauge connections equipped with a gauge test valve, close the valve, remove the ¼ in. (6 mm) plug from the test outlet, and install the calibrated test gauge. Reopen the valve, and note the readings on each gauge.
3. Repeat this process for each pressure gauge to be tested, except that where multiple pressure gauges are located at the same elevation and are installed within piping connected to a common water source (i.e., multiple risers connected to a common manifold without intervening check valves), only one pressure gauge needs be tested, where all of the installed pressure gauges have readings within 3 percent of the full gauge scale of one another.
4. Close the valve on the gauge test valve and remove the calibrated test gauge. Reinstall the ¼ in. (6 mm) plug in the test outlet, and reopen the valve.
5. Upon completion of all testing, notify the fire department and/or the alarm monitoring company, as well as the facility representatives, that testing is complete. Reset the fire alarm panel as necessary.

Evaluation of Results

Pressure gauges found to have a reading of more than 3 percent of the full scale of the gauge difference from the calibrated test gauge need to be recalibrated or replaced. For example, a pressure gauge having a 200 psi (13.8 bar) full scale range would be allowed a pressure reading difference of ± 6 psi (0.4 bar) from the calibrated test gauge.

References

American Society of Mechanical Engineers, Three Park Avenue, New York, NY 10016-5990.

ASME B40.1, *Gauges — Pressure Indicating Dial Type — Elastic Element, Grade 1A*, 2005.

1. Test-quality gauges should be used in accordance with ASME B40.1, *Gauges — Pressure Indicating Dial Type — Elastic Element, Grade 1A* having an accuracy of ±1%. The use of quality test gauges produces results that are considered reasonably accurate within the scope of the test procedure. Care should be taken to protect the gauges from rough handling. Gauges should be calibrated at least annually by means of a dead-weight or calibrated tester throughout the range of operation before a test series is begun. Calibration sheets should be kept for each gauge and correction factors affixed to the back of each gauge.

The procedure(s) outlined above is not mandated by NFPA 25 and represents only one approach to conducting this test.

Valve Status Test

Purpose

The purpose of completing a valve status test is to verify that the control valves upstream of the valve status test connection are not closed. This test is conducted by flowing water from a point downstream of the control valves being checked after they have been returned to their normal open position. The test does not provide an absolute assurance of an unobstructed water flow or a fully opened valve, but it provides a reasonable check against a fully closed valve or dropped valve gate within the water supply upstream of the test.

Tools/Equipment

No special tools are required.

Procedure Steps

1. Prior to any testing, notify the fire department and/or the alarm monitoring company, as well as the facility representatives, that testing is going to be conducted.
2. Locate a suitable valve status test connection downstream of the control valve being checked. Such suitable connections might include the main drain, fire pump test header, backflow preventer forward flow test connection, fire hydrant, and other similar locations. Where these devices are not available, an inspector's test connection or other similar, small drain connection can be used, provided the connection discharges to a suitable location. The selection of the larger available valve status test connections listed above is preferable, since they allow for a greater flow rate and are more likely for observation of potential valve closure conditions as opposed to the smaller inspector's test or smaller drain connections. Check the area surrounding the discharge location for the released water from this operation to ensure that there are no obvious conditions that would prevent water from being discharged safely or cause direct damage in the immediate vicinity. Where the discharge of water is to an area subject to potential freezing conditions, the facility representative should be advised of the potential for icing conditions.
3. Open the selected valve status test connection fully, allow the flow to stabilize, and observe the flow conditions noting the character of such. Note that this operation can result in the operation of connected fire pumps. If a main drain connection is used for this purpose, it is suggested that the main drain test procedure be followed to allow for measurable results as to character of the flow conditions.
4. Slowly close the valve status test connection to avoid water hammer.
5. Allow any fire pumps that might have automatically operated to run for 10 minutes for an electric fire pump, or 30 minutes for a diesel fire pump, followed by a shutdown of the fire pump and restoration of automatic operating condition.
6. Upon completion of all testing, notify the fire department and/or the alarm monitoring company, as well as the facility representatives, that testing is complete. Reset the fire alarm system as necessary.

Evaluation of Results

The evaluation of the results, other than a main drain test, is purely subjective, where the flow of water is the confirmation that the valve is not closed. A partially closed valve might not be identifiable using this test method, and it is not intended for such purpose. Observed water flow conditions that raise concern about the position of all control valves should be verified, and the valve status test should be repeated. Conditions that persist should be investigated and corrected as required.

The procedure(s) outlined above is not mandated by NFPA 25 and represents only one approach to conducting this test.

Testing Procedures for 13.3.3

Control Valve Test

Purpose

The purpose of completing a test of the control valves is to exercise valves that control the water supply to fire protection systems through the full range of operation and to ensure the valve is in the open position.

Tools/Equipment

A special wrench is needed for the operation of a post indicator valve or roadway box valve. The wrench for a post indicator valve is commonly found attached to the body of the post.

Procedure Steps

OS&Y Valve and Indicating Butterfly Valve (IBV)

1. Prior to any testing, notify the fire department and/or the alarm monitoring company, as well as the facility representatives, that testing is going to be conducted.
2. Note the position of the valve status indicator (e.g., valve stem, position paddle). The valve position indicated should be fully open for water supply control valves.
3. Turn the valve handle or operating wheel through the full operating range of the valve noting that the position movement of the valve status indicator (e.g., valve stem, position paddle) from open to closed and back. OS&Y valves must be backed off one-quarter turn from the fully open position to prevent jamming. Excessive operating force requiring additional applied mechanical force is indicative of the need for maintenance and/or repairs.
4. A valve status test must be conducted downstream of the closed valve in accordance with 13.3.1.2.1.
5. Upon completion of all testing, notify the fire department and/or the alarm monitoring company, as well as the facility representatives, that testing is complete. Reset the fire alarm system as necessary.

Post Indicator Valve (PIV) and Wall Post Indicator Valve (WPIV)

1. Prior to any testing, notify the fire department and/or the alarm monitoring company, as well as the facility representatives, that testing is going to be conducted.
2. Note the position of the "target" indicator within the view window. The valve position should be indicated as fully open for water supply control valves.
3. Using the appropriate wrench or the operating wheel, turn the valve through the full operating range of the valve, noting the number of turns to fully close and reopen the valve as well as the position movement of the valve position indicator "target" within the view window from open to closed and back. The number of turns to fully close and open a gate valve should be equal to three times the valve diameter plus two or three additional turns. Valves that do not open upon the anticipated number of turns, or that involve excessive operating force requiring additional applied mechanical force, are indicative of needing additional maintenance and/or repairs.
4. Reopening the valve must be continued until a resistance spring or torsion is felt and then must be backed off one-quarter turn from the fully open position to prevent jamming. Valve closure that does not result in a spring back or continues to turn in the open direction needs maintenance and/or repairs.
5. A valve status test must be conducted downstream of the closed valve in accordance with 13.3.1.2.1.
6. Upon completion of all testing, notify the fire department and/or the alarm monitoring company, as well as the facility representatives, that testing is complete. Reset the fire alarm control panel as necessary.

Nonindicating Roadway Box Valve

1. Prior to any testing, notify the fire department and/or the alarm monitoring company, as well as the facility representatives, that testing is going to be conducted.
2. Using the appropriate wrench (T-wrench), turn the valve through the full operating range of the valve, noting the number of turns to fully close and reopen the valve. The number of turns to fully close and open a gate valve should be equal to three times the valve diameter plus two or three additional turns. Valves that do not open upon the anticipated number of turns, or that involve excessive operating force requiring additional applied mechanical force, are indicative of needing maintenance and/or repairs.
3. After reopening the valve, it must then be backed off one-quarter turn from the fully open position to prevent jamming.
4. A valve status test must be conducted downstream of the closed valve in accordance with 13.3.1.2.1.
5. Upon completion of all testing, notify the fire department and/or the alarm monitoring company, as well as the facility representatives, that testing is complete. Reset the fire alarm panel as necessary.

Evaluation of Results

The control valve should cycle through the full range of operation without excessive operating force or disengagement as indicated in the test procedure above. Abnormal conditions require further investigation and corrective action as necessary.

The procedure(s) outlined above is not mandated by NFPA 25 and represents only one approach to conducting this test.

Deluge/Preaction Valve Trip Test

Purpose

The purpose of completing a trip test (partial or full) for a deluge/preaction valve is to ensure the system operates in a manner consistent with the original design application. The operation and testing of the valve must follow the manufacturer's instructions. A full trip test of the deluge valve also verifies that the operating water pressures at the deluge valve and most remote nozzle/sprinkler are consistent with the original design application. As a result, the original design basis must be known for an evaluation of the test results for these types of systems. Additionally, a comparative assessment of the water delivery time during the full trip test of a double interlock preaction and combined dry pipe/preaction system can reveal significant delays in system operation.

The primary automatic releasing means for deluge/preaction valves is generally one of two types that follow, pilot line (wet or dry) or electronic:

- The pilot line contains either pressurized water or gas (air or nitrogen) which, in combination with the valve trim, holds the valve in the closed position with a specific drop in pressure allowing for the release of the valve. This drop in pressure is designed to occur automatically with the fusing of an installed pilot sprinkler head on the pilot line. These systems will typically also contain a means of manually releasing pressure in the pilot line.
- The electronic release using an installed detection system, such as heat, smoke, or ultraviolet/infrared (UV/IR), is connected to a fire alarm/releasing panel, which upon operation will cause the release of the valve, typically through the opening of a solenoid valve in the deluge/preaction valve trim.

A preaction system can generally be designed in one of the three manners that follow:

1. Single/non-interlock preaction system — designed for release of the valve by the operation of the pilot/electronic release mechanism.
2. Double interlock preaction system — designed for release of the valve by a combination of the operation of the pilot/electronic release mechanism along with the release of gas (air or nitrogen) pressure in the preaction system piping.
3. Combined dry pipe/preaction system — designed to release in the same manner as the single/non-interlock preaction, except that the valve will also release upon a loss of gas pressure in the preaction system piping alone, similar to a conventional dry pipe system.

Prior to any testing of these systems, a thorough knowledge of the system type and original design basis should be known. Additionally, since the specific trim for the valve and associated operation/reset procedures vary between manufacturers, the manufacturers' data sheets should be reviewed and followed.

Full trip testing of deluge valves must only be conducted during periods of warm weather. Some system designs include the operation of multiple valves simultaneously requiring the operational testing of deluge valves at the same time.

In addition to the completion of the procedure steps continued herein, for electronic release deluge/preaction systems, complete testing in accordance with the manufacturer's instructions and *NFPA 72®, National Fire Alarm and Signaling Code,* must be completed for the supporting detection system.

Tools/Equipment

1. A timing device is needed to measure the trip and water delivery times for full trip tests for the double interlock preaction and preaction/dry pipe systems.
2. Various wrenches and tools are required to facilitate the reset of the deluge/preaction valve.

Deluge System Trip Test

Procedure Steps

1. Prior to any testing, notify the fire department and/or the alarm monitoring company, as well as the facility representatives, that testing is going to be conducted.
2. Check the area surrounding the discharge points for the main drain and wet pilot line inspector's test (if so installed) to ensure that there are no obvious conditions that would prevent water from being discharged safely or cause direct damage in the immediate vicinity.
3. Temporarily place any connected fire pumps in the "off" position. Note that, for systems where the sole source of water supply is from the fire pump, the pump might be required to remain in the automatic position.
4. Where a partial trip test is to be conducted:
   a. Fully open the main drain valve to clean any accumulated scale or foreign material from the supply water piping. Note that if the fire pump remains in the automatic position, this will result in the operation of connected fire pumps.
   b. Slowly close the system control valve to the point where additional closure cannot provide flow through the entire area of the drain outlet.
5. Where a full trip test is to be conducted:
   a. Verify with the facility representative that the location of system discharge has been properly prepared for the release of water. Check the vicinity surrounding the discharge area to ensure that there are no obvious conditions that would prevent water from being discharged safely or cause direct damage in the immediate vicinity, and take appropriate caution to ensure personnel and property are protected from the resulting discharge, including runoff.
   b. Install, or verify the installation of, a test gauge at or near the hydraulically most remote nozzle/sprinkler. Note that provisions for the installation of a test gauge should have been included as part of the system design or installation.

Deluge/Preaction Valve Trip Test *(Continued)*

   c. Fully open the main drain valve to clean any accumulated scale or foreign material from the supply water piping and slowly reclose the main drain valve. Note that if the fire pump remains in the automatic position, this will result in the operation of connected fire pumps. Where this occurs, allow any fire pumps that might have automatically operated to run for 10 minutes for an electric fire pump, or 30 minutes for a diesel fire pump, followed by a shutdown of the fire pump and restoration of automatic operating condition.

6. Record the water pressure on the supply side of the deluge valve.
7. Where a full trip test is to be conducted, return any fire pumps that might have been shut off in Step 3 to the automatic position.
8. Trip the deluge valve by operating the releasing system. For pilot line operated valves, release gas (air or nitrogen) or water pressure from the pilot line by completely opening the inspector's test connection or other appropriate outlet. For electrically operated valves, simulate an alarm activation at the fire alarm/releasing panel.
9. Where a partial trip test is to be conducted:
   a. Verify the trip of the deluge valve.
   b. Immediately close the system control valve, close the priming water line, and fully open the main drain valve to minimize the amount of water entering the system piping.
10. Where a full trip test is to be conducted:
    a. Verify the trip of the deluge valve.
    b. Ensure all the operating pressures stabilize.
    c. Read and record the operating pressure at the deluge valve gauge.
    d. Read and record the operating pressure at the gauge at or near the hydraulically most remote nozzle/sprinkler.
    e. Observe the discharge patterns from all of the nozzles/sprinklers to verify that the patterns provide complete unobstructed coverage of the protected area and that the nozzles/sprinklers are correctly aimed and are free of any plugging.
    f. Close the system control valve, close the priming water line, and fully open the main drain valve to drain the system.
11. Allow any fire pumps that might have automatically operated to run for 10 minutes for an electric fire pump, or 30 minutes for a diesel fire pump, followed by a shutdown of the fire pump and restoration of automatic operating condition.
12. Close any drain valves, inspector's test connections, or other equipment that was opened during the test.
13. Reset the deluge valve, and return the system to service in accordance with the manufacturer's instructions.
14. Open the system control valve.
15. Repeat Step 6 through Step 14, except:
    a. The trip of the deluge valve in Step 8 is to be initiated by operating the manual release mechanism for the system.
    b. In addition to opening the main drain valve, open all auxiliary/low point drains until the flow of water stops prior to reset of the deluge valve.
    c. Check any drainage weep holes to ensure free drainage.
    d. After the system is reset, recheck for accumulation of water at the auxiliary/low point drains as needed.
16. Return any fire pumps that still remain in the "off" position to the automatic position.
17. A valve status test must be conducted downstream of the closed control valves, in accordance with 13.3.1.2.1, using the main drain valve.
18. Upon completion of all testing, notify the fire department and/or the alarm monitoring company, as well as the facility representatives, that testing is complete. Reset the fire alarm system as necessary.

Single/Non-Interlock Preaction System Trip Test

1. Prior to any testing, notify the fire department and/or the alarm monitoring company, as well as the facility representatives, that testing is going to be conducted.
2. Check the area surrounding the discharge points for the main drain, system inspector's test connection, and wet pilot line inspector's test (if so installed) to ensure that there are no obvious conditions that would prevent water from being discharged safely or cause direct damage in the immediate vicinity.
3. Temporarily place any connected fire pumps in the "off" position. Note that for systems where the sole source of water supply is from the fire pump, the pump might be required to remain in the automatic position.
4. Where a partial trip test is to be conducted:
   a. Fully open the main drain valve to clean any accumulated scale or foreign material from the supply water piping. Note that if the fire pump remains in the automatic position, this will result in the operation of connected fire pumps.
   b. Slowly close the system control valve to the point where additional closure cannot provide flow through the entire area of the drain outlet.
5. Where a full trip test is to be conducted:
   a. Fully open the main drain valve to clean any accumulated scale or foreign material from the supply water

piping, and slowly reclose the main drain valve. Note that if the fire pump remains in the automatic position, this will result in the operation of connected fire pumps. Where this occurs, allow any fire pumps that might have automatically operated to run for 10 minutes for an electric fire pump, or 30 minutes for a diesel fire pump, followed by a shutdown of the fire pump and restoration of automatic operating condition.

6. Record the water pressure on the supply side of the preaction valve, and if supervisory pressure is used on the system, record the gas (air or nitrogen) pressure.
7. Where a full trip test is to be conducted, return any fire pumps that might have been shut off in Step 3 to the automatic position.
8. Trip the preaction valve by operating the releasing system. For pilot line operated valves, release gas (air or nitrogen) or water pressure from the pilot line by completely opening the inspector's test connection or other appropriate outlet. For electrically operated valves, simulate an alarm activation at the fire alarm/releasing panel.
9. Where a partial trip test is to be conducted:
    a. Verify the trip of the preaction valve.
    b. Immediately close the system control valve, close the priming water line, and fully open the main drain valve to minimize the amount of water entering the system piping.
10. Where a full trip test is to be conducted:
    a. Verify the trip of the preaction valve, and allow the system pressure to stabilize.
    b. Close the system control valve, close the priming water line, and fully open the main drain valve to drain the system.
11. Allow any fire pumps that might have automatically operated to run for 10 minutes for an electric fire pump, or 30 minutes for a diesel fire pump, followed by a shutdown of the fire pump and restoration of automatic operating condition.
12. Close any drain valves, inspector's test connections, or other equipment that was opened during the conduct of the test.
13. Reset the preaction valve, and return the system to service in accordance with the manufacturer's instructions.
14. Open the system control valve.
15. Repeat Step 6 through Step 14, except:
    a. The trip of the preaction valve in Step 8 is to be initiated by operation of the manual release mechanism for the system.
    b. In addition to opening the main drain valve, open all auxiliary/low point drains until the flow of water stops prior to reset of the preaction valve.
    c. After the system is reset, recheck for accumulation of water at the auxiliary/low point drains at drain as needed.
16. Return any fire pumps that still remain in the "off" position to the automatic position.
17. A valve status test must be conducted downstream of the closed control valves, in accordance with 13.3.1.2.1, using the main drain valve.
18. Upon completion of all testing, notify the fire department and/or the alarm monitoring company, as well as the facility representatives, that testing is complete. Reset the fire alarm system as necessary.

Double Interlock Preaction System Trip Test

1. Prior to any testing, notify the fire department and/or the alarm monitoring company, as well as the facility representatives, that testing is going to be conducted.
2. Check the area surrounding the discharge points for the main drain, system inspector's test connection, and wet pilot line inspector's test to ensure that there are no obvious conditions that would prevent water from being discharged safely or cause direct damage in the immediate vicinity.
3. Temporarily place any connected fire pumps in the "off" position. Note that, for systems where the sole source of water supply is from the fire pump, the pump might be required to remain in the automatic position.
4. Where a partial trip test is to be conducted:
    a. Fully open the main drain valve to clean any accumulated scale or foreign material from the supply water piping. Note that if the fire pump remains in the automatic position, this will result in the operation of connected fire pumps.
    b. Slowly close the system control valve to the point where additional closure cannot provide flow through the entire area of the drain outlet.
    c. Where installed, close the valve controlling the quick opening device.
5. Where a full trip test is to be conducted:
    a. Fully open the main drain valve to clean any accumulated scale or foreign material from the supply water piping, and slowly reclose the main drain valve. Note that if the fire pump remains in the automatic position, this will result in the operation of connected fire pumps. Where this occurs, allow any fire pumps that might have automatically operated to run for 10 minutes for an electric fire pump, or 30 minutes for a diesel fire pump, followed by a shutdown of the fire pump and restoration of automatic operating condition.
6. Record gas (air or nitrogen) pressure, along with the water pressure on the supply side of the preaction valve.

Deluge/Preaction Valve Trip Test *(Continued)*

7. Where a full trip test is to be conducted, return any fire pumps that might have been shut off in Step 3 to the automatic position.
8. Initiate the operation of the releasing system. For pilot line operated valves, release gas (air or nitrogen) or water pressure from the pilot line by completely opening the inspector's test connection or other appropriate outlet. For electrically operated valves, simulate an alarm activation at the fire alarm/releasing panel.
9. Verify the preaction valve does not trip but that the solenoid valve on the releasing line has engaged. This can be done by listening for an audible engagement of the solenoid or checking for the presence of operating voltage at the solenoid.
10. Where a partial trip test is to be conducted:
    a. Open the condensate drain or priming water test level valve on the preaction system to release the internal gas (air or nitrogen) pressure.
    b. Verify the trip of the preaction valve, noting the gas (air or nitrogen) pressure at which the valve trips.
    c. Immediately close the system control valve, close the priming water line, and fully open the main drain valve to minimize the amount of water entering the system piping.
    d. Where installed, test the quick opening device in accordance with the manufacturer's instructions.
11. Where a full trip test is to be conducted:
    a. Completely open the inspector's test valve of the preaction system to release the internal gas (air or nitrogen) pressure, noting the start time when the valve is fully open.
    b. Verify the trip of the preaction valve, noting the gas (air or nitrogen) pressure at which the valve trips and the time for the valve to trip from the time the inspector's test connection is fully opened.
    c. Allow the system flow to continue, and note the time at which a steady stream of water issues from the inspector's test connection from the time the inspector's test connection is fully opened.
    d. Close the system control valve, close the priming water line, and fully open the main drain valve to drain the system.
12. Allow any fire pumps that might have automatically operated to run for 10 minutes for an electric fire pump, or 30 minutes for a diesel fire pump, followed by a shutdown of the fire pump and restoration of automatic operating condition.
13. Close any drain valves, inspector's test connections, or other equipment that was opened during the conduct of the test.
14. Reset the preaction valve and associated quick opening device, if installed, and return the system to service in accordance with the manufacturer's instructions.
15. Open the system control valve.
16. Repeat Step 6 through Step 15, except:
    a. The trip of the preaction valve in Step 8 is to be initiated by operation of the manual release for the system.
    b. In addition to opening the main drain valve, open all auxiliary/low point drains until the flow of water stops prior to reset of the preaction valve.
    c. After the system is reset, recheck for accumulation of water at the auxiliary/low point drains as needed.
17. Return any fire pumps that still remain in the "off" position to the automatic position.
18. A valve status test must be conducted downstream of the closed control valves, in accordance with 13.3.1.2.1, using the main drain valve.
19. Upon completion of all testing, notify the fire department and/or the alarm monitoring company, as well as the facility representatives, that testing is complete. Reset the fire alarm system as necessary.

Combined Dry Pipe/Preaction System Trip Test

1. Prior to any testing, notify the fire department and/or the alarm monitoring company, as well as the facility representatives, that testing is going to be conducted.
2. Check the area surrounding the discharge points for the main drain, inspector's test connection, and wet pilot line inspector's test to ensure that there are no obvious conditions that would prevent water from being discharged safely or cause direct damage in the immediate vicinity.
3. Temporarily place any connected fire pumps in the "off" position. Note that, for systems where the sole source of water supply is from the fire pump, the pump might be required to remain in the automatic position.
4. Where a partial trip test is to be conducted:
    a. Fully open the main drain valve to clean any accumulated scale or foreign material from the supply water piping. Note that if the fire pump remains in the automatic position, this will result in the operation of connected fire pumps.
    b. Slowly close the system control valve to the point where additional closure cannot provide flow through the entire area of the drain outlet.
5. Where a full trip test is to be conducted, fully open the main drain valve to clean any accumulated scale or foreign material from the supply water piping, and slowly reclose the main drain valve. Note that if the fire pump remains

in the automatic position, this will result in the operation of connected fire pumps. Where this occurs, allow any fire pumps that might have automatically operated to run for 10 minutes for an electric fire pump, or 30 minutes for a diesel fire pump, followed by a shutdown of the fire pump and restoration of automatic operating condition.

6. Record gas (air or nitrogen) pressure along with the water pressure on the supply side of the preaction valve.

7. Where a full trip test is to be conducted, return any fire pumps that might have been shut off in Step 3 to the automatic position.

8. Trip the combined dry pipe/preaction valve by operating the releasing system. For pilot line operated systems, release gas (air or nitrogen) or water pressure from the pilot line by completely opening the inspector's test connection or other appropriate outlet. For electrically operated systems, simulate an alarm activation at the fire alarm/releasing panel.

9. Where a partial trip test is to be conducted:

   a. Verify the trip of the combined dry pipe/preaction valve.

   b. Immediately close the system control valve, close the priming water line, and fully open the main drain valve to minimize the amount of water entering the system piping.

10. Where a full trip test is to be conducted:

    a. Verify the trip of the combined dry pipe/preaction valve and allow pressures to stabilize.

    b. Close the system control valve, close the priming water line, and fully open the main drain valve to drain the system.

11. Allow any fire pumps that might have automatically operated to run for 10 minutes for an electric fire pump, or 30 minutes for a diesel fire pump, followed by a shutdown of the fire pump and restoration of automatic operating condition.

12. Close any drain valves, inspector's test connections, or other equipment that might have opened during the conduct of the test.

13. Reset the combined dry pipe/preaction valve and associated quick opening device, if installed, and return the system to service in accordance with the manufacturer's instructions.

14. Open the system control valve.

15. Repeat Step 6 through Step 14, except the trip of the combined dry pipe/preaction valve in Step 8 is to be initiated by operation of the manual release for the system.

16. Where a partial trip test is to be conducted:

    a. Where installed, close the valve controlling the quick opening device.

    b. Open the condensate drain or priming water test level valve on the combined dry pipe/preaction system to release the internal gas (air or nitrogen) pressure.

    c. Verify the trip of the combined dry pipe/preaction valve, noting the gas (air or nitrogen) pressure at which the valve trips.

    d. Immediately close the system control valve, close the priming water line, and fully open the main drain valve to minimize the amount of water entering the system piping.

    e. Where installed, test the quick opening device in accordance with the manufacturer's instructions.

17. Where a full trip test is to be conducted:

    a. Completely open the inspector's test connection of the preaction/dry pipe system to release the internal gas (air or nitrogen) pressure, noting the start time when the valve is fully open.

    b. Verify the trip of the combined dry pipe/preaction valve, noting the gas (air or nitrogen) pressure at which the valve trips and the time for the valve to trip from the time the inspector's test connection is fully opened. Note that this operation will result in the operation of connected fire pumps.

    c. Allow the system flow to continue, and note the time at which a steady stream of water discharges from the inspector's test connection from the time the inspector's test connection is fully opened.

    d. Close the system control valve, close the priming water line, and fully open the main drain valve and all auxiliary drains for the system until the flow of water stops prior to the reset of the system.

18. Allow any fire pumps that might have automatically operated to run for 10 minutes for an electric fire pump, or 30 minutes for a diesel fire pump, followed by a shutdown of the fire pump and restoration of automatic operating condition.

19. Reset the combined dry pipe/preaction valve and associated quick opening device, if installed, and return the system to service in accordance with the manufacturer's instructions.

20. Recheck for accumulation of water at the auxiliary/low point drains as needed.

21. Open the system control valve.

22. Return any fire pumps that still remain in the "off" position to the automatic position.

23. A valve status test must be conducted downstream of the closed control valves, in accordance with 13.3.1.2.1, using the main drain valve.

24. Upon completion of all testing, notify the fire department and/or the alarm monitoring company, as well as the facility representatives, that testing is complete. Reset the fire alarm system as necessary.

Deluge/Preaction Valve Trip Test *(Continued)*

Evaluation of Results

Instances where the deluge, preaction, or combined dry pipe/preaction valve fails to operate at the specified points in the test procedure or where it cannot be properly reset should be investigated and corrective action taken as necessary.

During a full trip (discharge) of a deluge valve, any nozzles/sprinklers that are found to be misaligned, obstructed, or plugged require adjustment or other corrective action as necessary. The operating pressure readings at the deluge valve and at or near the hydraulically most remote nozzle/sprinkler must be compared to the hydraulic design pressures for the system, to ensure the performance meets or exceeds that required by the original design. Pressures found to be less than the required design pressures should be investigated and corrective action taken as necessary.

During a full trip test of a double interlock preaction valve, a comparative assessment of the water delivery time to the inspector's test connection should be conducted. With the wide range of acceptable water delivery times, it is not the intent of NFPA 25 to reverify any specific time limits with the full flow trip test. Rather, it is the intent of NFPA 25 to reveal substantial delays for the water delivery time of these systems, since significant differences from one test to another are an indication of possible operational problems with the system. A drastic increase in the water delivery time from one full flow trip test to the next is of major concern, because such an increase is most likely due to corrosion of the system piping. An increase in delivery time might also indicate that physical changes have been made to the system, such as a revision in the piping layout. It might also indicate that the water supply characteristics have changed. Any substantial change in water delivery time from one test to another is cause for investigation into the source of the problem.

Double interlock preaction systems and combined dry pipe/preaction systems are limited in size by the time needed for the system to deliver water to the inspector's test connection. For example, NFPA 13, *Standard for the Installation of Sprinkler Systems,* places no time limit on water delivery for double interlock preaction systems with a volume of 500 gal (1893 L) or less, because systems of these sizes typically deliver the water in a relatively short time. Double interlock preaction systems with a volume greater than 500 gal (1893 L) must deliver water in not more than 60 to 70 seconds, be calculated with a listed calculation program, or the water discharge must be verified by an appropriately sized test manifold (see Exhibit 13.25). Combined dry pipe/preaction systems must deliver water in not more than 1 minute for each 400 ft (122 m) of common feed main in a system, with a maximum time not to exceed 3 minutes.

References

National Fire Protection Association, 1 Batterymarch Park, Quincy, MA 02169-7471.

NFPA 13, *Standard for the Installation of Sprinkler Systems*, 2013 edition.

NFPA 72®, National Fire Alarm and Signaling Code, 2013 edition.

The procedure(s) outlined above is not mandated by NFPA 25 and represents only one approach to conducting this test.

Dry Pipe Valve Trip Test

Purpose

The purpose of completing a trip test (partial or full) for a dry pipe valve is to ensure the functionality of the system to operate. The operation and testing of the valve must follow the manufacturer's instructions. The full trip test of a dry pipe system allows for a comparative assessment of the water delivery time that can reveal significant delays in system operation. Full trip testing of dry pipe valves must only be conducted during periods of warm weather.

Tools/Equipment

1. A timing device is needed to measure the trip and water delivery times for full trip tests of the dry pipe valve.
2. Various wrenches and tools are required to facilitate the reset of the dry pipe valve.

Procedure Steps

1. Prior to any testing, notify the fire department and/or the alarm monitoring company, as well as the facility representatives, that testing is going to be conducted.
2. Record gas (air or nitrogen) pressure along with the water pressure on the supply side of the dry pipe valve.
3. Check the area surrounding the discharge points for the main drain and system inspector's test connection to ensure that there are no obvious conditions that would prevent water from being discharged safely or cause direct damage in the immediate vicinity.
4. Temporarily place any connected fire pumps in the "off" position. Note that, for systems where the sole source of water supply is from the fire pump, the pump might be required to remain in the automatic position.
5. Where a partial trip test is to be conducted:
   a. Fully open the main drain valve to clean any accumulated scale or foreign material from the supply water piping. Note that if the fire pump remains in the automatic position, this will result in the operation of connected fire pumps.
   b. Slowly close the system control valve to the point where additional closure cannot provide flow through the entire area of the drain outlet.
   c. Where installed, close the valve controlling the quick opening device.
6. Where a full trip test is to be conducted:
   a. Fully open the main drain valve to clean any accumulated scale or foreign material from the supply water piping, and slowly reclose the main drain valve. Note that if the fire pump remains in the automatic position, this will result in the operation of connected fire pumps. Where this occurs, allow any fire pumps that might have automatically operated to run for 10 minutes for an electric fire pump, or 30 minutes for a diesel fire pump, followed by a shutdown of the fire pump and restoration of automatic operating condition.
7. Where a partial trip test is to be conducted:
   a. Open the priming water test level valve on the dry pipe valve to release the internal gas (air or nitrogen) pressure.
   b. Verify the trip of the dry pipe valve, noting the gas (air or nitrogen) pressure at which the valve trips.
   c. Immediately close the system control valve, close the priming water line, and fully open the main drain valve to minimize the amount of water entering the system piping.
   d. Where installed, test the quick opening device in accordance with the manufacturer's instructions.
8. Where a full trip test is to be conducted:
   a. Return any fire pumps that might have been shut off in Step 4 to the automatic position.
   b. Completely open the inspector's test valve of the dry pipe system to release the internal gas (air or nitrogen) pressure, noting the start time when the valve is fully open.
   c. Verify the trip of the dry pipe valve, noting the gas (air or nitrogen) pressure at which the valve trips and the time for the valve to trip from the time the inspector's test connection is fully opened. Note that this will allow the system flow to continue, and note the time at which a steady stream of water discharges from the inspector's test connection from the time the inspector's test connection is fully opened.
   d. Close the system control valve, and fully open the main drain valve.
9. Allow any fire pumps that might have automatically operated to complete an elapsed run time of 10 minutes for an electric fire pump, or 30 minutes for a diesel fire pump, followed by a shutdown of the fire pump and restoration of automatic operating condition.
10. Fully open all auxiliary/low point drains for the system until the flow of water stops.
11. Close any drain valves, inspector's test connections, or other equipment that might have opened during the conduct of the test.
12. Reset the dry pipe valve and associated quick opening device, if installed, and return the system to service in accordance the manufacturer's instructions.
13. Starting from the initial gas (air or nitrogen) pressure required to set the valve clapper, observe the duration required to return the system pressure to its normal setting. Note that any installed air maintenance devices should have the bypass valves in the open position to

Dry Pipe Valve Trip Test *(Continued)*

allow for the maximum available fill rate from the gas (air or nitrogen) source.

14. Recheck for accumulation of water at the auxiliary/low point drains at drain as needed.
15. Open the system control valve.
16. Return any fire pumps that still remain in the "off" position to the automatic position.
17. A valve status test must be conducted downstream of the closed control valves, in accordance with 13.3.1.2.1, using the main drain valve.
18. Upon completion of all testing, notify the fire department and/or the alarm monitoring company, as well as the facility representatives, that testing is complete. Reset the fire alarm system as necessary.

Evaluation of Results

Instances where the dry pipe valve fails to operate or where it cannot be properly reset should be investigated and corrective action should be taken as necessary. During a full trip test of the dry pipe valve, a comparative assessment of the water delivery time to the inspector's test connection should be conducted. With the wide range of acceptable water delivery times, it is not the intent of NFPA 25 to reverify any specific time limits with the full flow trip test. Rather, it is the intent of NFPA 25 to reveal substantial delays for the water delivery time of these systems, since significant differences from one test to another are an indication of possible operational problems with the system. A drastic increase in the water delivery time from one full flow trip test to the next is of major concern, because such an increase is most likely due to corrosion of the system piping. An increase in delivery time might also indicate that physical changes have been made to the system, such as a revision in the piping layout or the orifice size of sprinklers. It might also indicate that the water supply characteristics have changed. Any substantial change in water delivery time from one test to another is cause for investigation into the source of the problem.

Dry pipe systems are limited in size by the time needed for the system to deliver water to the inspector's test connection. For example, NFPA 13, *Standard for the Installation of Sprinkler Systems,* places no time limit on water delivery for dry pipe systems with a volume of 500 gal (1893 L) or less, or for systems of 750 gal (2839 L) or less that are equipped with a quick opening device, because systems of these sizes typically deliver the water in a relatively short time. Dry pipe systems with a volume greater than 750 gal (2839 L) must deliver water in not more than 60 seconds, be calculated with a listed calculation program, or the water discharge must be verified by an appropriately sized test manifold (see Exhibit 13.25).

The time required to refill the system with normal gas (air or nitrogen) is not to exceed 30 minutes. Longer refill durations require further investigation and corrective action as necessary.

References

National Fire Protection Association, 1 Batterymarch Park, Quincy, MA 02169-7471.

NFPA 13, *Standard for the Installation of Sprinkler Systems*, 2013 edition.

The procedure(s) outlined above is not mandated by NFPA 25 and represents only one approach to conducting this test.

Testing Procedures for 13.4.4.2.6

Dry Pipe System Low Air Pressure Switch Test

Purpose

The purpose of conducting a test of the low air pressure switch is to ensure the operability of the device to signal a low air condition prior to reaching the trip pressure for the dry pipe valve. Some devices might also include a feature to monitor the system for high-pressure conditions. While not specifically mandated by NFPA 25, the testing of this feature is advisable and is included as part of the procedure below should a test of this feature be desired.

Tools/Equipment

No special tools are required.

Procedure Steps

1. Prior to any testing, notify the fire department and/or the alarm monitoring company, as well as the facility representatives, that testing is going to be conducted.
2. Check the fire alarm control panel or remote annunciator to determine that the low air supervisory device is not currently active on the system.
3. Close the control valve for the dry pipe system to ensure that any low gas pressure condition does not result in an unintended trip of the dry pipe valve.
4. Where installed, close the control valve to the quick opening device.
5. Shut off the gas supply (air or nitrogen) to the system.
6. Slowly bleed off pressure to the low air pressure switch until activation of the low air switch. Some dry pipe valves include a trim feature to allow the pressure to be released at the low air pressure switch independently, while others require the entire system pressure to be reduced.
7. Observe and record the pressure at which the low air switch activates and transmits to the fire alarm panel or remote annunciator. Should a pressure drop of more than 10 psi (0.7 bar) occur without an activation of the low air switch, terminate the test to avoid a potential trip of the dry pipe valve.
8. Open the gas supply (air or nitrogen) to the system, and return the system to normal operating pressure.
9. Where high-pressure monitoring is provided:
   a. Continue to slowly increase the system pressure until activation of the switch. Note that this might require an adjustment of pressure settings such as the compressor or regulator on the gas (air or nitrogen) supply or opening of the bypass valve on air maintenance devices to achieve higher pressures, and in some cases it might not be possible due to the limited capacity of the gas (air or nitrogen) supply.
   b. Observe and record the pressure at which the switch activates and transmits to the fire alarm control panel or remote annunciator. Should a pressure rise of more than 10 psi (0.7 bar) occur without an activation of the switch, terminate the test to avoid potentially damaging the dry pipe valve.
   c. Bleed off system pressure to return the system to normal operating pressure.
10. Where installed, open the control valve to the quick opening device.
11. Open the control valve for the dry pipe system.
12. A valve status test must be conducted downstream of the closed control valves in accordance with 13.3.1.2.1.
13. Upon completion of all testing, notify the fire department and/or the alarm monitoring company, as well as the facility representatives, that testing is complete. Reset the fire alarm system as necessary.

Evaluation of Results

The recorded pressure drop/rise for activation of the low air switch from the normal operating pressure is to be not more than 10 psi (0.7 bar). Pressure drops/increases in excess of 10 psi (0.7 bar) without activation of the low air switch require further investigation, adjustment, and/or corrective action as necessary.

The procedure(s) outlined above is not mandated by NFPA 25 and represents only one approach to conducting this test.

Testing Procedures for 13.4.4.2.9

Dry Pipe System Air Leakage Test

Purpose

The purpose of completing an air leakage test is to ensure the dry pipe system is reasonably tight and free of excessive leakage. Two methods are presented for such a determination.

Tools/Equipment

A timing device is needed to measure the specified duration indicated by the procedure.

Procedure Steps

Leakage Rate Test

1. Prior to any testing, notify the fire department and/or the alarm monitoring company, as well as the facility representatives, that testing is going to be conducted.
2. Close the control valve for the dry pipe system to ensure that any low gas pressure condition does not result in an unintended trip of the dry pipe valve. Note that the closure of the system control valve requires the establishment of impairment procedures in accordance with Chapter 15.
3. Adjust the gas pressure (air or nitrogen) on the system to a minimum of 40 psi (2.75 bar).
4. Shut off the gas supply (air or nitrogen) to the system.
5. Record the start time and gas pressure (air or nitrogen) of the test, and allow the system to sit for 2 hours.
6. Record the stop time of the test and the gas (air or nitrogen) pressure.
7. Open the gas supply (air or nitrogen) to the system, and allow the system gas pressure to return to normal.
8. Return the control valve for the dry pipe system to the open position.
9. Conduct a valve status test downstream of the closed control valves in accordance with 13.3.1.2.1.
10. Upon completion of all testing, notify the fire department and/or the alarm monitoring company, as well as the facility representatives, that testing is complete. Reset the fire alarm system as necessary.

Low Air Supervisory Switch Activation Test

1. Prior to any testing, notify the fire department and/or the alarm monitoring company, as well as the facility representatives, that testing is going to be conducted.
2. Close the control valve for the dry pipe system to ensure that any low gas pressure condition does not result in an unintended trip of the dry pipe valve. Note that the closure of the system control valve requires the establishment of impairment procedures in accordance with Chapter 15.
3. Check the gas pressure (air or nitrogen) on the system to ensure that it is at normal system pressure.
4. Check the fire alarm panel or remote annunciator to determine that the low air supervisory switch is not currently active on the system.
5. Shut off the gas supply (air or nitrogen) to the system.
6. Record the start time, and allow the system to sit for 4 hours.
7. Record the stop time of the test.
8. Check the fire alarm panel or remote annunciator to determine if a low air supervisory signal has been received for the tested dry pipe system.
9. Open the gas supply (air or nitrogen) to the system, and allow the system gas pressure to return to normal.
10. Return the control valve for the dry pipe system to the open position.
11. Conduct a valve status test downstream of the closed control valves in accordance with 13.3.1.2.1.
12. Upon completion of all testing, notify the fire department and/or the alarm monitoring company, as well as the facility representatives, that testing is complete. Reset the fire alarm system as necessary.

Evaluation of Results

Testing completed using the air leakage rate method above that results in a loss of more than 3 psi (0.2 bar) over the 2 hour test duration requires further investigation and/or repair. Testing completed using the low air supervisory switch activation method above that results in the activation of the system low air supervisory switch over the 4 hour test duration requires further investigation and/or repair.

The procedure(s) outlined above is not mandated by NFPA 25 and represents only one approach to conducting this test.

System Pressure Reducing Valve Test

Purpose

The purpose of conducting a full flow test of the system pressure-reducing valve is to ensure the operability of the device in maintaining downstream system pressures to less than the pressure rating of the system components on the system side of the valve. Sprinkler systems are typically rated for a maximum of 175 psi (12.1 bar). However, there are systems equipped with components — including the sprinklers — that are rated for pressures higher than 175 psi (12.1 bar). These valves should be set with an outlet pressure of 10 psi (0.7 bar) below the maximum working pressure for the system, to allow a pressure differential between the controlled outlet pressure and that of the installed pressure relief and to allow for reaction time for the valve.

Tools/Equipment

No special tools are required.

Procedure Steps

1. Prior to any testing, notify the fire department and/or the alarm monitoring company, as well as the facility representatives, that testing is going to be conducted.
2. Locate a drain connection or other suitable discharge connection on the outlet side of the pressure-reducing valve. Check the area surrounding the discharge location for the released water from operation thereof to ensure that there are no obvious conditions that would prevent water from being discharged safely or cause direct damage in the immediate vicinity. Where the discharge of water is to an area subject to potential freezing conditions, the facility representative should be advised of the potential for icing conditions.
3. Open the drain or other suitable connection slowly until it has reached the fully open position, allowing for the establishment of automatic operation of any fire pumps. Observe the inlet and outlet pressures on the pressure-reducing valve to ensure the outlet pressures remain less than the pressure rating of the system components throughout the range of flow. Allow the pressure to stabilize, then record the residual pressure on both the inlet and outlet side of the pressure-reducing valve. Note that system pressure-reducing valve installations should include an inlet and outlet pressure gauge; however, where such gauges are not provided, the use of field installed gauges would be required.
4. Slowly close the drain or other suitable discharge connection to avoid water hammer, and record the return pressure on both the inlet and outlet side of the pressure-reducing valve.
5. Allow any fire pumps that might have automatically operated to run for 10 minutes for an electric fire pump, or 30 minutes for a diesel fire pump, followed by a shutdown of the fire pump and restoration of automatic operating condition.
6. Upon completion of all testing, notify the fire department and/or the alarm monitoring company, as well as the facility representatives, that testing is complete. Reset the fire alarm system as necessary.

Evaluation of Results

The pressure-reducing valve outlet pressures should remain at least 10 psi (0.7 bar) less than the pressure rating of the system components through the full ranges of test flows. Higher pressures require further investigation, adjustment, and/or corrective action as necessary.

The procedure(s) outlined above is not mandated by NFPA 25 and represents only one approach to conducting this test.

Hose Connection Pressure-Regulating Device Test

Purpose

The purpose of conducting a test of the hose connection pressure-regulating device is to ensure the operability of the device in maintaining downstream pressures to not more than 100 psi (6.9 bar) for small hose connections [1½ in. (40 mm) diameter] and not more than 175 psi (12.1 bar) for connections for heavy hose [2½ in. (65 mm) diameter]. These devices are separated into the following two types: pressure-restricting and pressure-reducing. The pressure-restricting device is designed to reduce the downstream water pressure under flowing (residual) conditions only; whereas the pressure-reducing device is designed to reduce the downstream pressure under both flowing (residual) pressure and non-flowing (static) pressure conditions.

Tools/Equipment

1. Sufficient hose lengths with the appropriate thread connection are required to allow for the connection of the tested hose connection to a suitable discharge location. Typically 2½ in. (65 mm) hose with a coupling having a thread pattern matching that of the hose outlet are used for this purpose. The number and length(s) of hose will depend on the specific location and the distance from the tested outlet to a suitable location.
2. An in-line flow measurement test device[2] or nozzles to permit flow measurement from the tested hose connection are needed. In-line flow measurement devices are best suited for this test; however, other approved test nozzles can be selected where an exterior discharge location is used. The flow characteristics of the test device must be known.
3. Where an exterior flow with test nozzles is used, a pitot tube together with a test-pressure gauge[1] suitable for the pressures to be expected is needed. [Usually, a 60 psi (4.1 bar) gauge is satisfactory.]
    - A pitot-tube-and-gauge assembly is indispensable in conducting flow tests from nozzles. The small opening at the end of the tube — not more than 1⁄16 in. (1.6 mm) in diameter — is inserted in the center of the stream, in a direct line with the flow, at a distance in front of the opening equal to one-half the opening diameter. Velocity pressure is registered on the gauge attached to the tube.
    - Some available test nozzles include an integral pitot tube or pitotless flow measuring feature with an attached test gauge[1] allowing for direct measurement of flows without the use of a separate pitot tube.
4. Pressure gauges[1] are needed to measure the upstream and downstream pressures on the pressure-regulating device.
5. Various wrenches and tools are needed to facilitate the necessary connections.

Procedure Steps

1. Prior to any testing, notify the fire department and/or the alarm monitoring company, as well as the facility representatives, that testing is going to be conducted.
2. Set up a flow test connection from the hose connection being tested. See Exhibit 13.33 for an example of such a connection. This usually includes the following steps:
    a. Install a pressure gauge at the upstream and downstream side of the pressure-regulating valve. The downstream pressure gauge can be installed using an adapter connected to the hose connection outlet. The installation of an upstream pressure gauge on newer standpipe systems is allowed through the mandated installation of a valved gauge connection outlet; however, older systems might not include such connections, making the installation of an upstream gauge impracticable.

    b. Connect and route a fire hose from the valve connection to a proper discharge location, including the connection to a suitable flow measuring device and control valve to allow for measurement of the flow rate. For convenience, the flow measurements are best done using an in-line flow measurement device at the hose outlet but can also be done with discharge flow measure devices at the point of discharge. Standpipes installed in compliance with the 1993 or newer edition of NFPA 14, *Standard for the Installation of Standpipe and Hose Systems,* will include a 3 in. (75 mm) drain riser specifically for testing of hose connection pressure-regulating devices and should be used for the discharge during the flow test contained herein. Other standpipes require that the discharge be routed to suitable locations. This typically includes routing to the exterior of the building. Alternatively, where a second standpipe is available in the building, it can be isolated and taken out of service to be used as a temporary drain pipe with an extension for drainage to the exterior through one of the ground level hose connections using an additional length of fire hose. This would require that the hose connection pressure-regulating valves in the path of intended drainage within that temporary drain standpipe be removed and replaced with a standard hose connection.
3. Check the area surrounding the discharge point for a drain outlet or the connected discharge devices to ensure that there are no obvious conditions that would prevent water from being discharged safely or cause direct damage in the immediate vicinity. Take appropriate caution to ensure

1. Test-quality gauges should be used in accordance with ASME B40.1, *Gauges — Pressure Indicating Dial Type — Elastic Element, Grade 1A* having an accuracy of ±1%. The use of quality test gauges produces results that are considered reasonably accurate within the scope of the test procedure. Care should be taken to protect the gauges from rough handling.

2. Gauges should be calibrated at least annually by means of a deadweight or calibrated tester throughout the range of operation before a test series is begun. Calibration sheets should be kept for each gauge and correction factors affixed to the back of each gauge.

personnel and property are protected from the resultant high pressure hose stream discharge. The use of a flow diffuser can aid in breaking the directed stream and managing the resultant energy of the discharge. Where the discharge of water is to an area subject to potential freezing conditions, the facility representative should be advised of the potential for icing conditions.

4. Ensure the control valve in the established test connection apparatus installed in Step 2 is in the closed position.
5. Fully open the hose connection valve.
6. Where installed, initiate the operation of the fire pump, and allow pressures to stabilize.
7. Observe and record the reading on the pressure gauge installed on the inlet (if it were possible to install such a gauge) and outlet side of the hose connection.
8. Slowly open the valve in the test connection apparatus installed in Step 2, measuring the flow rate and adjusting until the a flow rate of 100 gpm (379 L/min) is established if testing a 1½ in. (40 mm) connection, or 250 gpm (946 L/min) if testing a 2½ in. (65 mm) hose connection. Record the flow rate.
9. Observe and record the reading on the pressure gauge installed on the inlet (if it were possible to install such a gauge) and outlet side of the hose connection.
10. Slowly close the valve in test connection apparatus installed in Step 2 to avoid water hammer and read the return pressure on both the inlet and outlet side of the pressure-reducing valve.
11. Allow any fire pumps that might have automatically operated to run for 10 minutes for an electric fire pump, or 30 minutes for a diesel fire pump, followed by a shutdown of the fire pump and restoration of automatic operating condition.
12. Close the hose connection valve, remove the test apparatus connection, and replace the cap on the hose connection.
13. Repeat this process for every pressure-regulating valve on the system.
14. Upon completion of all testing, notify the fire department and/or the alarm monitoring company, as well as the facility representatives, that testing is complete. Reset the fire alarm system as necessary.

Evaluation of Results

The pressure-regulating valve performance is based on the class of standpipe system provided and operating conditions as follows:

- For 1½ in. (40 mm) hose connections, the flowing (residual) pressure is not to exceed 100 psi (6.9 bar).
- For 1½ in. (40 mm) hose connections subject to a static pressure upstream of the pressure-regulating device of more than 175 psi (12.1 bar), both the non-flowing (static) and flowing (residual) pressure on the downstream side of the pressure-regulating device are not to exceed 100 psi (6.9 bar). Where a gauge can be installed on the upstream side of the pressure-regulating device, this static pressure reading can be taken directly; however, where the gauge connection is not possible, a determination of static pressure at another location within the system would be required, with necessary elevation adjustments made to determine the static pressure at the test location.
- For 2½ in. (65 mm) hose connections subject to a static pressure upstream of the pressure-regulating device of more than 175 psi (12.1 bar), both the non-flowing (static) and flowing (residual) pressure on the downstream side of the pressure-regulating device are not to exceed 175 psi (12.1 bar). Where a gauge can be installed on the upstream side of the pressure-regulating device, this static pressure reading can be taken directly; however, where the gauge connection is not possible, a determination of static pressure at another location within the system would be required, with necessary elevation adjustments made to determine the static pressure at the test location.

Instances where pressure conditions are found to be in excess of those stated above require further investigation, adjustment, and/or corrective action as necessary.

References

American Society of Mechanical Engineers, Three Park Avenue, New York, NY 10016-5990.

ASME B40.1, *Gauges — Pressure Indicating Dial Type — Elastic Element, Grade 1A*, 2005.

National Fire Protection Association, 1 Batterymarch Park, Quincy, MA 02169-7471.

NFPA 14, *Standard for the Installation of Standpipe and Hose Systems*, 2013 edition.

The procedure(s) outlined above is not mandated by NFPA 25 and represents only one approach to conducting this test.

ITM NFPA 25 — INSPECTION TESTING MAINTENANCE

14 INTERNAL PIPING CONDITION AND OBSTRUCTION INVESTIGATION

Chapter 14 of NFPA 25 addresses the need for continued operability of water-based fire protection systems by providing requirements for monitoring the interior condition of the piping system and conducting obstruction investigations. Metal piping systems can lose structural integrity due to corrosion and can become obstructed due to corrosion by-products such as pipe scale. The requirements for periodic internal inspections of the system piping are intended to address this issue. Any type of system that uses piping can become inoperative when the pipe is filled with obstructing material. Sprinkler systems are no exception. Any maintenance program must include means for revealing potential obstructions and removing any obstructions that are found. Obstructions are addressed in this chapter in Section 14.3, and 14.3.1 provides a list of potential causes of piping obstructions and identifies what type of situation requires an obstruction investigation when any of the listed conditions occur.

The title of this chapter was modified because the 2011 edition title addressed only one of the two major concepts that are tackled in this chapter. Internal assessments of piping, referred to as inspections in the 2011 edition, are different from obstruction investigations. This was not clear in the 2011 chapter title, which was simply called "Obstruction Investigations." In the 2014 edition, both concepts were spelled out in the title to clarify that these are two different activities.

14.1* General

This chapter shall provide the minimum requirements for conducting investigations of fire protection system piping for possible sources of materials that could cause pipe blockage.

A.14.1 For obstruction investigation and prevention, see Annex D.

14.2 Assessment of Internal Condition of Piping

14.2.1* An assessment of the internal condition of piping shall be conducted on a frequency determined by 14.2.1.1 or 14.2.1.2 for the purpose of inspecting for the presence of foreign organic and inorganic material.

FAQ Does 14.2.1 apply to only dry sprinkler systems?

The requirement in 14.2.1 applies to all four types of sprinkler systems — wet, dry, preaction, and deluge — as well as foam-water sprinkler systems.

EXHIBIT 14.1 *Example of Internal Corrosion. (Courtesy of Byron Blake and SimplexGrinnell)*

It is important to note that requirements in NFPA 25 for periodic internal inspection of fire protection piping systems have changed over the years. They were first included in the 2002 edition in an attempt to prevent the destruction of piping systems by microbiologically influenced corrosion (MIC), which is shown in Exhibit 14.1. (See Supplement 1 for more information on MIC.) The 2008 edition of the standard modified this requirement to specifically indicate that 14.2.1 requires an inspection, not a complete (and subsequently, expensive) obstruction investigation. This periodic inspection was intended to reveal the presence of MIC, zebra mussels, or inorganic material such as rust and scale. The 2014 edition made even further changes and no longer specifically requires an inspection but rather an *assessment*, which can be conducted in accordance with either 14.2.1.1 or 14.2.1.2.

A.14.2.1 It is the intent of this requirement to provide a reasonable assurance that corrosion and obstruction issues within fire protection systems are identified. It is not the intent to require verification that every piece of pipe in the system is free from corrosion and obstructions. An assessment of the internal condition of piping can be accomplished by several methods that meet the intent of this section. These methods include the following:

Critical Point

By requiring an *assessment* rather than an *inspection,* as mandated in previous editions of NFPA 25, the owner now has options available to find the most efficient method for determining the condition of the system piping.

By requiring an *assessment* rather than an *inspection,* as mandated in previous editions of NFPA 25, the owner now has options available to find the most efficient method for determining the condition of the system piping. The intent of 14.2.1 is to place the smallest burden possible on the property owner by providing these options. For example, if the inspection option is exercised, it can be coordinated with the internal inspection requirement of system valves, such as is required by 13.4.1.2 for alarm valves, or it can be conducted during system renovations or maintenance.

Assessments can reveal the presence of inorganic types of obstructing material such as work gloves, weld coupons, and other material. When these types of obstructions are found, all feed and cross mains should be inspected. Exhibit 14.2 and Exhibit 14.3 show inorganic obstructions found in system piping.

(1) Opening a flushing connection at the end of one main, and removing the end fitting or piece of branch line or a sprinkler for the purpose of inspecting for the presence of foreign organic and inorganic material.
  (a) In dry pipe systems and preaction systems, the branch line inspected should be the most remote one from the source of water that is not equipped with the inspector's test valve.

Case In Point

Zebra mussels and quagga mussels can enter system piping when the fire protection system is supplied by a natural source of water, such as a river or lake. The photo below shows a typical zebra mussel, which is named for the stripes on the shell. The mussels can enter the piping as larvae or young adults and then can grow into obstructions. The zebra mussel species has become invasive in the last few years. They entered the United States through the Great Lakes, attached to the hulls of foreign ships. The mussels then spread throughout the Great Lakes region and beyond. The map below illustrates the extent of zebra mussel infestation.

Zebra Mussel. (Courtesy of Michigan Sea Grant)

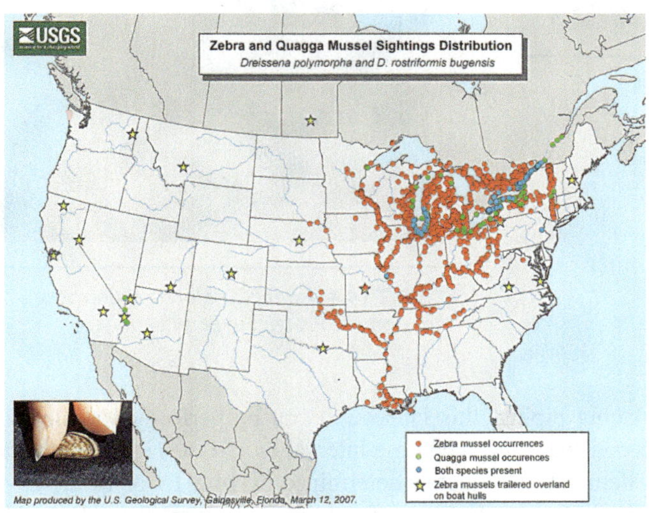

Zebra Mussel Map. (Courtesy of Michigan Sea Grant)

After a dry pipe or preaction system has been tripped, the end of the system has the greatest potential for collecting the most scale and sediment. The branch line with the inspector's test connection, however, is expected to allow the material to flush out of that portion of the system. Examining the branch line adjacent to the branch line containing the inspector's test connection can provide a more realistic portrayal of the amount of material to be expected.

(b) When performing normal maintenance that involves draining down a system to modify a system such as for tenant fit out or building renovations, or when removing

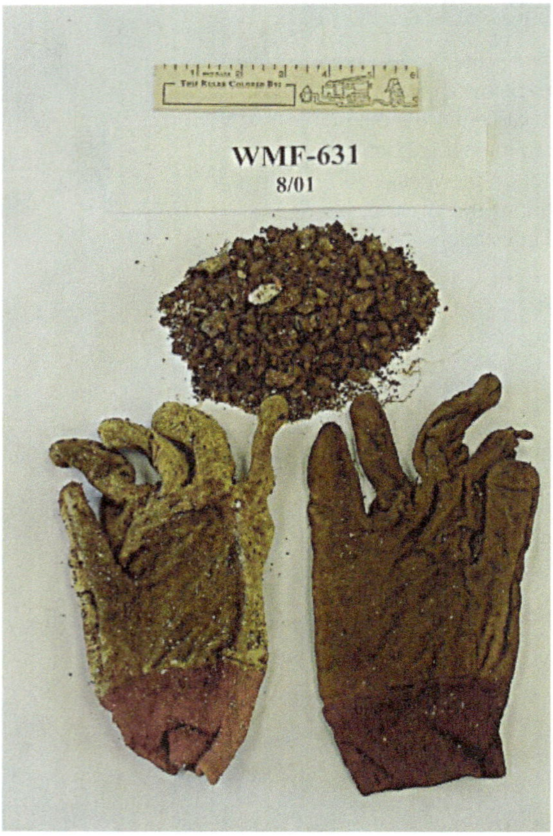

EXHIBIT 14.2 Work Gloves Removed from System Piping. (Courtesy of John Jensen)

EXHIBIT 14.3 Weld Coupons Found in System Piping.

EXHIBIT 14.4 Example of Internal Corrosion. (Courtesy of Wiginton Fire Protection Engineering)

or replacing piping, this inspection can be performed as described and properly recorded at that time. The time interval would then start for the next assessment of that system at the frequency determined by 14.2.1.1 or 14.2.1.2.

(c) If a sprinkler is removed to perform this inspection, 5.4.1.1 requires a new sprinkler matching the characteristics of the replaced sprinkler.

FAQ How can the presence of MIC be detected without an inspection?

In the absence of an inspection and field or laboratory tests, MIC is usually revealed by the presence of a pinhole leak, like the one shown in Exhibit 14.4. Unfortunately, by the time such a leak develops and is noticed, it is usually too late to save the affected pipe. It might even be too late to salvage much of the system.

Yet, even if there is an inspection, MIC might not be detected. For example, an internal inspection that is poorly conducted or conducted too soon following the filling of a system might not reveal the presence of MIC.

FAQ What steps should be taken if MIC is found in the fire protection system?

Test kits and services are available to analyze test samples for the presence of microbes affiliated with the development of MIC. If MIC is found, special equipment and flushing services are available for treatment. For more information on MIC and other causes of obstruction, see Annex D and Supplement 1. See Exhibit 14.5 and Exhibit 14.6 for examples of the effects of MIC in fire protection system piping.

EXHIBIT 14.5 *Pipe Severely Corroded by MIC.*

EXHIBIT 14.6 *Localized Corrosion Caused by MIC.*

(2) Utilizing alternative examination methods such as the following:
   (a) Using video inspection equipment that is inserted into the system at strategic points to observe the internal condition of pipes. This equipment provides a visual exam of the pipes using a camera and lighting system on the end of a push cable. Video inspection equipment can be inserted in alarm, dry, and preaction valves for a look into risers, feed mains, some cross mains, and some branch lines, depending on the system configuration. The push cable can also be inserted in a check valve when performing the five-year internal inspection required by 13.4.2.1 to view additional areas of a system, and in the fire department connection to perform the interior inspection required by 13.7.2.
   (b) Ultrasonic or similar technology that allows the pipe wall to be tested to determine the extent of any deterioration due to microbiologically influenced corrosion (MIC) or other forms of corrosion. This method would not typically be used for the internal inspection of piping required by this section because it might not detect the presence of solid material in the piping, such as wood, plastic, or other foreign obstructions, that are not a by-product of corrosion, because only small representative sections of pipe are examined.
   (c) A laboratory analysis of water samples obtained from the fire protection system, combined with collecting and inspecting solid material from fire protection system water discharged from a main drain, and an inspector's test connection, can provide an indication of the presence of corrosion, MIC, and/or foreign materials. If a high level of MIC is identified, or if a significant amount of foreign materials is found, further investigation might be warranted to verify the extent of corrosion, MIC, or other obstructions in the system. The solid materials should be collected with an

appropriately sized strainer. If inspection of the solid materials identifies excessive rust, black water color, or sulfur (rotten egg) odors, an obstruction investigation as described in Section 14.3 is warranted.

FAQ Does NFPA 25 permit X-ray technology to substitute for the traditional opening of the pipe to inspect for scale and debris within a dry pipe system?

The alternative examination methods discussed in A.14.2.1(2) include the use of X-ray, ultrasound, and remote video techniques for the internal examination of piping systems. These methods are permitted in lieu of impairing the systems to conduct a visual examination. See also 14.3.2.3 and 14.4.1.

Case In Point

The traditional method for conducting the required internal assessment includes draining the system piping, opening up the piping, and looking inside. This tried-and-true method provides the inspector with an opportunity to obtain a visual account of the internal condition of the pipe and take samples of any organic growth found within the piping. This approach does require the system to be taken off line for a period of time, however, which can lead to a level of business interruption that is unacceptable to owners of some constantly operating facilities such as hospitals and power plants. Furthermore, this approach can be costly in high-rise buildings where each floor is considered a separate "system" and must undergo an independent internal evaluation.

As identified in A.14.2.1, draining the system and opening a flushing connection is not the only means for conducting the internal assessment. This section specifically states that nondestructive or noninvasive methods for conducting the assessment are permitted. One such nondestructive approach involves the use of ultrasound, or ultrasonic, technology. Ultrasonic technology allows an assessment to be performed without shutting the system down as the testing is being conducted. By applying ultrasonic transducers to the pipe's exterior surface, as shown in the photo below on the left, internal conditions can be identified quickly and accurately. Pulse echo and guided wave technology are two types of ultrasound that have been used in the sprinkler industry for years with great success. The photo below on the right shows a field technician looking at the test output data on the portable display screen.

Inspector Conducting Ultrasound Inspection. (Courtesy of David Martinez, Sonic Inspection Corporation)

Real-Time Wave Captured on Portable Display Unit. (Courtesy of David Martinez, Sonic Inspection Corporation)

Pulse echo ultrasonic technology works by measuring the amount of time it takes for an ultrasonic straight beam signal to travel through the wall of the pipe and reflect off the back wall, resulting in a "remaining wall thickness" measured down to 1/1000 of an inch. Guided wrap wave (or lamb wave) ultrasonic technology travels around the circumference of the pipe, exposing the internal surface conditions. The concept of a circumferential wave concept is displayed in the diagram below. The wave generated by the pipe being tested, also referred to as the "real time wave," is then compared to a "signature wave," which is a representation of a pristine pipe in the form of "energy." This energy data represents how an ultrasonic surface wave would travel through a piece of pipe of the same material and wall thickness. Discrepancies between the signature wave and the real time wave signify that there has been a change in the internal condition of the piping. The output data are available as the test is being conducted on the portable device. Differences between these two waves could represent degradation of the pipe wall, ice plugs, MIC, material corrosion, scaling, or any other change that could modify the flow characteristics of the piping. The photo below shows an ice plug identified using ultrasonic technology.

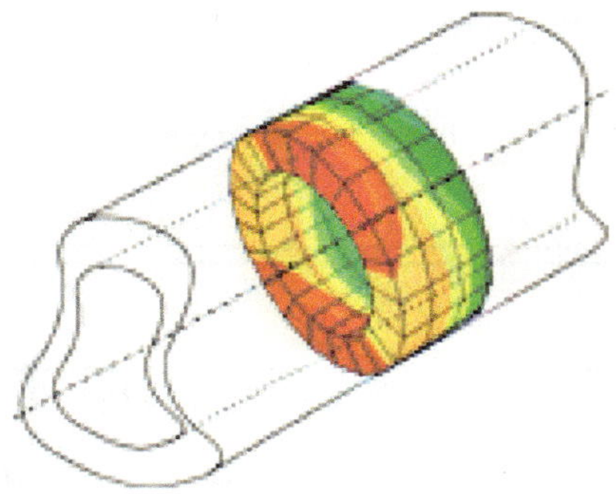

Circumferential Wave. (Courtesy of David Martinez, Sonic Inspection Corporation)

Ice Plug Detected by Ultrasonic Technology. (Courtesy of David Martinez, Sonic Inspection Corporation)

Two sets of output data and corresponding photos of the pipes being inspected are shown on the next page. The top photo (a) shows a piece of severely corroded pipe. Note how the green real-time wave in the diagram does not match the signature wave above it. The vast differences in these two waves would indicate to the person conducting the test that there is a significant problem inside of the pipe. The photo in (b) shows a clean piece of pipe. The real-time wave and signature wave in the diagram look very similar, which tells the inspector that the pipe is in good condition.

While A.14.2.1 states that this technology might not be appropriate for all internal assessments because only "small representative sections of pipe are examined," a larger, more comprehensive study using ultrasonic technology can be conducted as a means for compliance with this section. The next page shows a diagram of a building that has been inspected at dozens of locations. The green, yellow, and red markers identify varying levels of severity in the output data from the test. An assessment of this scope and magnitude provides the owner, inspector, and AHJ with a great deal of knowledge about the internal condition of the piping.

Once the change in the piping is identified by comparing the two waves, the owner and the fire sprinkler contractor can then look at solutions for resolving the conditions identified. The use of this technology as a means for conducting the required internal assessment can be beneficial as a cost-effective, noninvasive means for establishing the internal condition of system piping without causing any business interruption. An example of this situation is presented in the following case study that involves the use of this technology.

(continues)

Case In Point

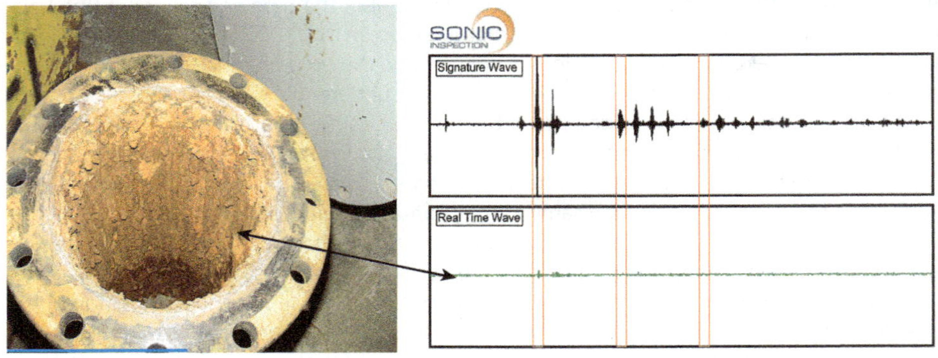

(a)

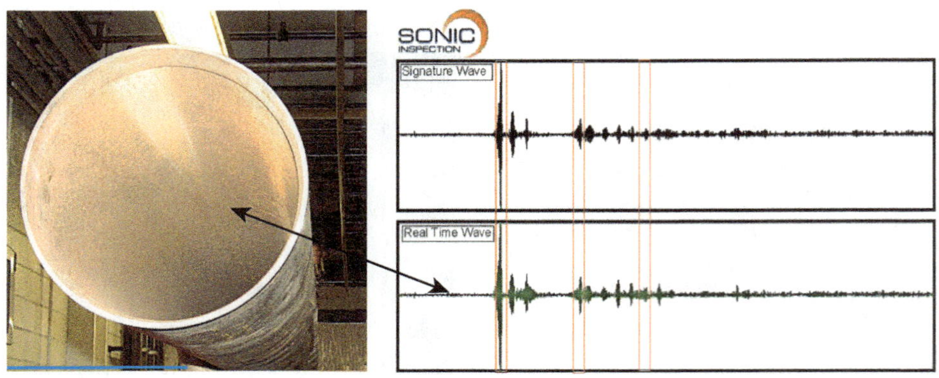

(b)

Output Data from Internal Inspections Using Guided Wrap Wave Technology: (a) Pipe Shows No Lamb Wave Reflections and Should Be Taken Out of Service, and (b) Pipe Shows Excellent Lamb Wave Reflections and Will Be Left in Service. (Courtesy of David Martinez, Sonic Inspection Corporation)

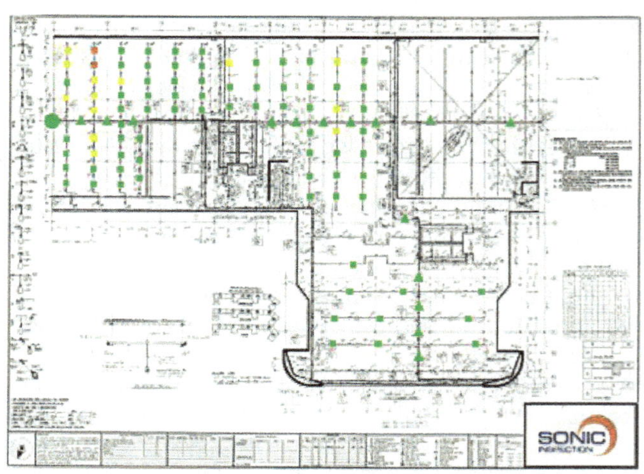

Diagram Identifying Pipe Inspection Locations and Results. (Courtesy of David Martinez, Sonic Inspection Corporation)

CASE STUDY

Problem

The fire sprinkler piping for a dry system in a large data center was not properly pitched. This improper pitch allowed water to be trapped in low areas, causing corrosion, wall thinning (see photo below), and obstructions. Originally the entire piping network was scheduled to be replaced at a substantial cost to the owner.

Solution

An ultrasonic inspection program was implemented using guided wave technology to identify wall thinning, corrosion, obstructions, and water trapped in the low areas. All of this work was conducted while the system was fully operational, which allowed the data center to function as normal. Only the problematic pipe was identified as needing to be replaced, and the pipe that was still in good working order was not touched and required no further internal inspection.

Resolution

The local fire sprinkler contractor was able to remove the bad pipe that was identified and flagged by ultrasonic inspection personnel. The contractor redesigned a more robust system by deploying a large nitrogen dryer, installing Schedule 40 hot dipped galvanized seamless pipe, and laser-pitching lines from the mains to risers with a positive 1 percent fall.

The inspection team retained specimens for the newly installed pipe at its headquarters for future reference. These specimens will now be used as the data center's benchmark for future testing. Specimens are measured to 1/1000 of an inch so future testing can be monitored for wall loss against the benchmark.

Savings to the Data Center

The use of this technology saved the organization $600,000 and reduced the operational impacts of the pipe replacement project by limiting the number of pipes being replaced and the number of impacted zones.

Pipe Wall Thinning Detected by Ultrasound Technology. (Courtesy of David Martinez, Sonic Inspection Corporation)

14.2.1.1 An assessment of the internal condition of piping shall be conducted at a minimum of every 5 years or in accordance with 14.2.1.2 for the purpose of inspecting for the presence of foreign organic and inorganic material.

14.2.1.2* Where an assessment frequency has been established by an approved risk analysis, the assessment shall be performed at a frequency determined by the approved risk analysis.

A.14.2.1.2 The purpose of the assessment of the internal condition of piping is to identify whether foreign organic and inorganic material is present, since the presence of such material can potentially be detrimental to the performance of the sprinkler system. Corrosion, which includes microbiologically influenced corrosion (MIC), can be prevalent in sprinkler piping, but not necessarily in all systems, buildings, or geographic locations. Therefore, the risk analysis should be based upon evidence noted and/or data obtained from proximate sprinkler systems, known problems from the water supply, as well as other factors that could affect the integrity of the system piping. The decision of when, how often, and which method to use to conduct an assessment should also consider the risk of emptying and refilling the sprinkler system with water, which could adversely impact the sprinkler system by introducing foreign or inorganic material into the sprinkler system.

Critical Point

It is important to note that the risk analysis must be *approved,* and any frequencies greater than 5 years should be approved by the authority having jurisdiction.

The requirements in 14.2.1.1 and 14.2.1.2 address the frequency of the assessment required by 14.2.1. The 5-year minimum frequency is intended to limit the burden placed on the owner by aligning the assessment with the 5-year internal inspection requirement for check valves.

When determining an acceptable frequency based on a risk analysis, as permitted by 14.2.1.2, it is important that the analysis be performed by a qualified entity or person, such as a registered engineer, loss control consultant, or others possessing sufficient training and knowledge in system piping obstructions. It is important to note that the risk analysis must be *approved,* and any frequencies greater than 5 years should be approved by the authority having jurisdiction (AHJ). Previous editions of NFPA 25 did not specifically state that an alternative internal inspection or assessment frequency could be developed through a risk assessment. However, the opportunity to develop "performance-based frequencies" was available. As such, this concept is not entirely new, as many building owners had previously revised the internal inspection frequency by using a performance-based approach.

14.2.1.3 Tubercules or slime, if found, shall be tested for indications of microbiologically influenced corrosion (MIC).

14.2.1.4* If the presence of sufficient foreign organic or inorganic material is found to obstruct pipe or sprinklers, an obstruction investigation shall be conducted as described in Section 14.3.

Internal assessments and obstruction investigations are different activities; however, it is not uncommon for the findings of an internal assessment to be the catalyst for an obstruction investigation. Exhibit 14.7 shows a pipe found in a sprinkler system during an internal assessment filled with dirt and sediment that is significant enough to warrant an obstruction investigation.

A.14.2.1.4 Most piping systems contain some foreign material or other evidence of corrosion but not sufficient to trigger an obstruction investigation. Furthermore, an internal inspection is primarily an inspection for determining corrosion of the pipe, but it can result in finding the presence of material that would be an obstruction to piping or sprinklers. If such is found, an obstruction investigation in Section 14.3 would be required.

EXHIBIT 14.7 *Pipe Obstructed With Dirt and Sediment. (Courtesy of Byron Blake and SimplexGrinnell)*

Small amounts of scale or soft sediment are often found within a sprinkler system. The presence of such scale should not necessarily trigger an obstruction investigation; however, significant corrosion, such as the internal corrosion shown in Exhibit 14.8, should trigger an investigation. It is important to remember that where a significant amount of sediment or corrosion is present in the piping, much of the deposits will settle in sprinkler drops and outlets. Whether it is MIC or corrosion due to material degradation, it is not uncommon for sprinklers to become plugged like the sprinkler shown in Exhibit 14.9.

EXHIBIT 14.8 *Example of Interior Corrosion. (Courtesy of Byron Blake and SimplexGrinnell)*

EXHIBIT 14.9 *Example of Plugged Sprinkler. (Courtesy of Wiginton Fire Protection Engineering)*

14.2.1.5 Nonmetallic pipe shall not be required to comply with Section 14.2.

Corrosion is associated with metal pipe. Since nonmetallic piping is corrosion resistant, it is not necessary to perform internal inspections for this type of pipe. However, nonmetallic piping systems are subject to the requirements regarding obstruction investigations as mandated in Section 14.3.

•

14.2.2* In buildings having multiple wet pipe systems, every other system shall have an assessment of the internal condition of piping as described in 14.2.1.

A.14.2.2 In large warehouses, high-rise buildings, and other buildings having multiple systems, it is reasonable to perform the assessment on half of the systems, and conclude that these are representative of all systems in the building. The systems in the building not assessed during one frequency cycle should be assessed during the next one. As long as there is no evidence of any foreign organic and/or inorganic material found in any of the systems being assessed, every other system would be assessed once every frequency cycle. However, if foreign organic and/or inorganic material is found in any system in the building during the frequency cycle, all systems must would then be assessed during that frequency cycle.

14.2.2.1 During the next inspection frequency required by 14.2.1.1 or 14.2.1.2, the alternate systems not assessed during the previous assessment shall be assessed as described in 14.2.1.

Case In Point

During the revision cycle that generated the 2013 edition of NFPA 13, *Standard for the Installation of Sprinkler Systems*, the technical committee responsible for the definition of the term *sprinkler system* modified that definition to essentially define a system as the piping network downstream of a control valve, water source, flow switch, and drain. In the case of high-rise buildings, the system piping that is downstream of a floor control valve assembly (FCVA), is now defined as a *sprinkler system*.

Putting this new concept into practice, consider a 50-story building that is sprinklered throughout. The entire building was once considered to be protected by a single sprinkler system. With the revised definition, the building is now considered to have 50 sprinkler systems, with one on each floor.

The NFPA 13 definition for *sprinkler system* was added to the 2014 edition of NFPA 25. The impact of this change is that wherever the standard requires an inspection, testing, and maintenance (ITM) task to be completed for a "system," it will need to be done for each floor of the building that has an FCVA or each portion of a building that is sectioned off by a control valve, flow switch, and drain. The greatest impact for the application of this definition would be the internal assessment required by 14.2.1 and 14.2.2, where every other floor of a high-rise building would need to have an internal assessment every 5 years. In previous editions of the standard, the common application of this provision was to open the system and conduct an internal inspection in two or three areas of the building. This approach is no longer appropriate due to the modification of the definition, which would now require the system to be opened up on every other floor as described in 14.2.2.

Owners who do not wish to conduct the assessment on every other floor of a building every 5 years would need to illustrate an alternative assessment plan that is appropriate through a risk assessment.

14.2.2.2 If foreign organic and/or inorganic material is found in any system in a building, all systems shall be assessed.

14.3 Obstruction Investigation and Prevention

Obstruction investigations are different from the assessment required by 14.2.1. They are more extensive and do not have a specified frequency. Obstruction investigations are triggered by specific events or conditions that can occur either within the system itself or within the water supply infrastructure for the system. Fifteen events or conditions are listed in 14.3.1 that would warrant an obstruction investigation.

14.3.1* An obstruction investigation shall be conducted for system or yard main piping wherever any of the following conditions exist:

(1) Defective intake for fire pumps taking suction from open bodies of water
(2) The discharge of obstructive material during routine water tests
(3) Foreign materials in fire pumps, in dry pipe valves, or in check valves
(4) Foreign material in water during drain tests or plugging of inspector's test connection(s)
(5) Unknown materials are heard in the system piping during draining, refilling, or otherwise flowing water through the system
(6) Plugged sprinklers
(7) The presence of sufficient foreign organic or inorganic material is found in the pipe
(8) Failure to flush yard piping or surrounding public mains following new installations or repairs

System Tagging

Typically, this level of internal corrosion or organic growth would be discovered during an internal assessment or obstruction investigation conducted in accordance with Chapter 14. Corrosion or organic growth that has developed to this point would impact the flow characteristic and pressure loss in this piping, which makes this a critical deficiency. If an examination of adjacent sections of pipes reveals additional corrosion or growth which is equally widespread or becomes worse, this could be considered an impairment.

(Courtesy of Byron Blake and SimplexGrinnell)

Noncritical Deficiency

Critical Deficiency

Impairment

(9) A record of broken public mains in the vicinity
(10) Abnormally frequent false tripping of a dry pipe valve(s)
(11) A system that is returned to service after an extended shutdown (greater than 1 year)
(12) There is reason to believe that the sprinkler system contains sodium silicate or highly corrosive fluxes in copper systems
(13) A system has been supplied with raw water via the fire department connection
(14) Pinhole leaks
(15) A 50 percent increase in the time it takes water to travel to the inspector's test connection from the time the valve trips during a full flow trip test of a dry pipe sprinkler system when compared to the original system acceptance test

A.14.3.1 Most piping systems contain some foreign material or other evidence of corrosion but not sufficient to trigger an obstruction investigation. Furthermore, an internal inspection is primarily an inspection for determining corrosion of the pipe, but it can result in finding the presence of material that would be an obstruction to piping or sprinklers. If such is found, an obstruction investigation in Section 14.3 would be required. For obstruction investigation procedures, see Section D.3. The type of obstruction investigation should be appropriately selected based on the observed condition. For instance, ordering an internal obstruction investigation would be inappropriate where the observed condition was broken public mains in the vicinity. On the other hand, such an investigation would be appropriate where foreign materials are observed in the dry pipe valve.

14.3.2* Systems shall be examined for internal obstructions where conditions exist that could cause obstructed piping.

A.14.3.2 For obstruction prevention program recommendations, see Section D.4.

14.3.2.1 If the condition has not been corrected or the condition is one that could result in obstruction of the piping despite any previous flushing procedures that have been performed, the system shall be examined for internal obstructions every 5 years.

Obstruction investigations are only required to be repeated on a 5-year basis when the cause of the obstruction cannot be corrected, such as with a zebra mussel infestation. In all other cases, an obstruction investigation is required only when one of the conditions listed in 14.3.1 is found. An obstruction investigation is not intended to be a routinely scheduled activity.

Critical Point

Obstruction investigations are only required to be repeated on a 5-year basis when the cause of the obstruction cannot be corrected, such as with a zebra mussel infestation.

14.3.2.2* Internal examination shall be performed at the following minimum four points:

(1) System valve
(2) Riser
(3) Cross main
(4) Branch line

FAQ How many pipes have to be examined if obstructions in the system are suspected?

The intent of 14.3.2.2 is for each of the main system components to be visually examined. If an obstruction is found in one of these main areas of the system, it would be good practice to open a few other areas of the system to verify whether the obstruction is an isolated incident. For branch lines, the intent is to open a representative number of branch lines rather than open every branch line in the system.

A.14.3.2.2 The indicated locations are presented as the minimum basis for determining the need for the initiation of a more extensive obstruction investigation effort. Additional points of examination can be appropriately dictated to assess the extent and severity of the obstructing material as determined by the observed conditions at these locations. See Section D.3 for further discussion.

14.3.2.3* Alternative nondestructive examination methods shall be permitted.

System Tagging

Internal inspections and obstruction investigations often yield varying levels of corrosion. These images show the inside of an OS&Y and fire department connection with internal corrosion. Table A.3.3.7 considers internal conditions like this to be a critical deficiency. In other instances however, where the corrosion is more widespread, it can be considered an impaired system.

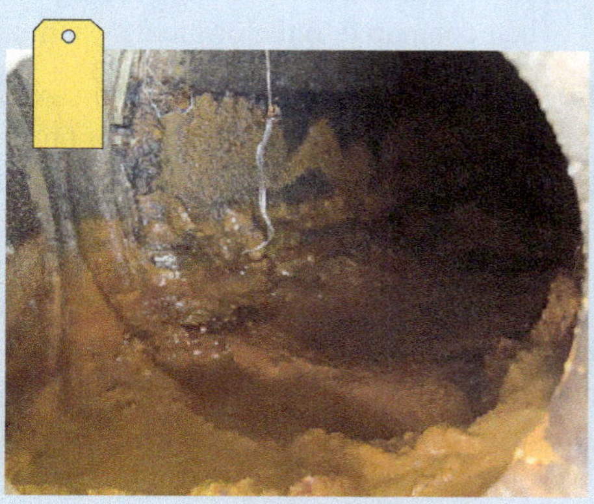

(Photos Courtesy of Byron Blake and SimplexGrinnell)

Noncritical Deficiency   Critical Deficiency   Impairment

A.14.3.2.3 Alternative examination methods can include the following:

(1) Using video inspection equipment that is inserted into the system at strategic points to observe the internal condition of pipes. This equipment provides a visual exam of the pipes using a camera and lighting system on the end of a push cable. Video inspection equipment can be inserted in alarm, dry, and preaction valves, in the risers, and into cross mains and branch lines, depending on the system configuration. This equipment can also be used to determine the results of any flushing that was performed.
(2) Ultrasonic or similar technology that allows the pipe wall to be externally tested to determine the extent of any deterioration due to MIC or other forms of corrosion. This method has advantages including allowing the system to be left in service while performing the examination, and pinpointing where pipes are about to fail before the actual failure occurs, allowing preventive action to be taken. This technology does have its limitations, and to be thorough all pipes would require access, even those in concealed spaces such as those above ceilings. This technology can detect the presence of a buildup of sludge, scale, or other by-products of corrosion, and can detect existing air pockets in wet systems as well as trapped sections of pipe in dry systems that aren't adequately drained. This technology cannot detect the presence of MIC itself, but can detect tubercules caused by MIC. In most cases this technology will not detect the presence of solid material in the piping such as wood, plastic, or other foreign obstructions that are not a by-product of corrosion, because only small representative sections of pipe are examined.

14.3.3* If an obstruction investigation indicates the presence of sufficient material to obstruct pipe or sprinklers, a complete flushing program shall be conducted by qualified personnel.

A.14.3.3 For obstruction investigation flushing procedures, see Section D.5.

The flushing procedures outlined in Section D.5 of Annex D are intended to be carried out with water. However, in cases where the obstruction investigation has identified severe MIC, chemical scouring might be necessary. This process is highly specialized and should be attempted only by professionals who are thoroughly experienced in the procedure. Chemicals should not be added to water used for flushing without a determination of the possible effects on sprinkler and valve seats.

14.3.4 Tubercules or slime, if found during an obstruction investigation, shall be tested for indications of microbiologically influenced corrosion (MIC).

14.4 Ice Obstruction

Dry pipe or preaction sprinkler system piping that protects or passes through freezers or cold storage rooms shall be inspected internally on an annual basis for ice obstructions at the point where the piping enters the refrigerated area.

The inspection described in Section 14.4 should be conducted regularly to ensure that piping is not obstructed with ice and that fittings are not fractured due to ice buildup. Exhibit 14.10 shows piping with ice buildup.

14.4.1 Alternative nondestructive examinations shall be permitted.

The alternative nondestructive examination methods in 14.4.1 include the use of ultrasound, X-ray, and remote video camera technology.

14.4.2 All penetrations into the cold storage areas shall be inspected and, if an ice obstruction is found, additional pipe shall be examined to ensure no ice blockage exists.

EXHIBIT 14.10 Dry System With Ice Buildup. (Courtesy of Byron Blake and SimplexGrinnell)

Reference Cited in Commentary

National Fire Protection Association, 1 Batterymarch Park, Quincy, MA 02169-7471.

NFPA 13, *Standard for the Installation of Sprinkler Systems,* 2013 edition.

15

IMPAIRMENTS

The requirements in Chapter 15 cover all important aspects of a suitable impairment program. An impairment is a condition where a water-based fire protection system or a portion thereof will not function during a fire event. A formal impairment program establishes procedures for ensuring that system control valves remain in their correct position. Furthermore, implementation of an impairment program can help minimize the length of time that a system is out of service. This chapter provides the basic requirements for such an impairment program.

The success of any impairment program is determined by the support it receives from those responsible for implementing it. Often, the best way to ensure that an impairment program is properly set up is to review it with the authority having jurisdiction (AHJ), which can be the fire department, the insurance company, or other officials. If any situations arise that are not covered by the impairment program, it is good practice to seek the advice of the AHJ.

15.1 General

The most common impairments involve closed water-control valves, but other examples can include conditions such as a nonfunctioning fire pump, painted sprinklers, heavily corroded sprinklers, or a broken hose valve. (See A.3.3.7 for more examples of potential impairments.)

This chapter deals primarily with preplanned impairments — those impairments in which the system or a portion thereof is out of service due to work, such as modifications to the water supply or sprinkler system piping, that has been planned in advance.

Rarely is a facility more exposed to a potentially catastrophic fire loss than when a fire protection system or portion of a system is out of service. Many devastating fires could have been mitigated had a proper impairment program been in place and adhered to. Whether an impairment is preplanned or occurs in an emergency situation seems to have little bearing on its tendency to lead to catastrophe. A major factor in previous large-loss fires included the failure to do the following:

- Recognize the risks created by the impairment.
- Establish a fire watch and provide backup fire protection.
- Control ignition sources and/or shut down hazardous processes.
- Expedite repairs.
- Ensure that the fire protection system is properly placed back into service.

Critical Point

Rarely is a facility more exposed to a potentially catastrophic fire loss than when a fire protection system or portion of a system is out of service.

15.1.1 Minimum Requirements.

15.1.1.1 This chapter shall provide the minimum requirements for a water-based fire protection system impairment program.

15.1.1.2 Measures shall be taken during the impairment to ensure that increased risks are minimized and the duration of the impairment is limited.

15.2 Impairment Coordinator

15.2.1 The property owner or designated representative shall assign an impairment coordinator to comply with the requirements of this chapter.

The impairment coordinator is usually someone from the building maintenance department. In larger facilities, the coordinator might be a member of the safety or loss control department.

The impairment coordinator should be someone who is on site and thoroughly familiar with the building layout, operations, and systems. In addition, he or she should have the authority and means to initiate proper notification when a system impairment is planned. This notification includes, but is not limited to, the fire department, insurance carrier, alarm company, property owner, area supervisors, and tenants.

Prior to removing a system from service, the impairment coordinator should ensure that the necessary tools, equipment, and replacement parts are on hand to minimize the duration of the impairment. This typically involves coordination with the contractor or service provider who is scheduled to perform the work that necessitates the shutdown of the system.

> **Critical Point**
>
> Prior to removing a system from service, the impairment coordinator should ensure that the necessary tools, equipment, and replacement parts are on hand to minimize the duration of the impairment.

15.2.2 In the absence of a specific designee, the property owner or designated representative shall be considered the impairment coordinator.

15.2.3 Where the lease, written use agreement, or management contract specifically grants the authority for inspection, testing, and maintenance of the fire protection system(s) to the tenant, management firm, or managing individual, the tenant, management firm, or managing individual shall assign a person as impairment coordinator.

15.3 Tag Impairment System

Many jurisdictions have requirements to identify the system's operational status. These often include tags that are placed at the system riser or other specified locations to identify systems as operational, operational but having deficiencies that might impact the performance of the system, or impaired. The tags are typically color-coded, based on the status of the system. Operational status tagging is similar to the tag impairment system required by Section 15.3

15.3.1* A tag shall be used to indicate that a system, or part thereof, has been removed from service.

A.15.3.1 A clearly visible tag alerts building occupants and the fire department that all or part of the water-based fire protection system is out of service. The tag should be weather resistant, plainly visible, and of sufficient size [typically 4 in. × 6 in. (100 mm × 150 mm)]. The tag should identify which system is impaired, the date and time impairment began, and the person responsible. Figure A.15.3.1 illustrates a typical impairment tag.

Anyone who is shutting down a system should use tagging procedures even if the owner does not. Tags should also be itemized in a list to facilitate proper restoration of the system to working order. Once retrieved, a tag can also be used for verification that a valve or system has been restored to service. Exhibit 15.1 illustrates a valve that has been closed for testing purposes and has an impairment tag attached.

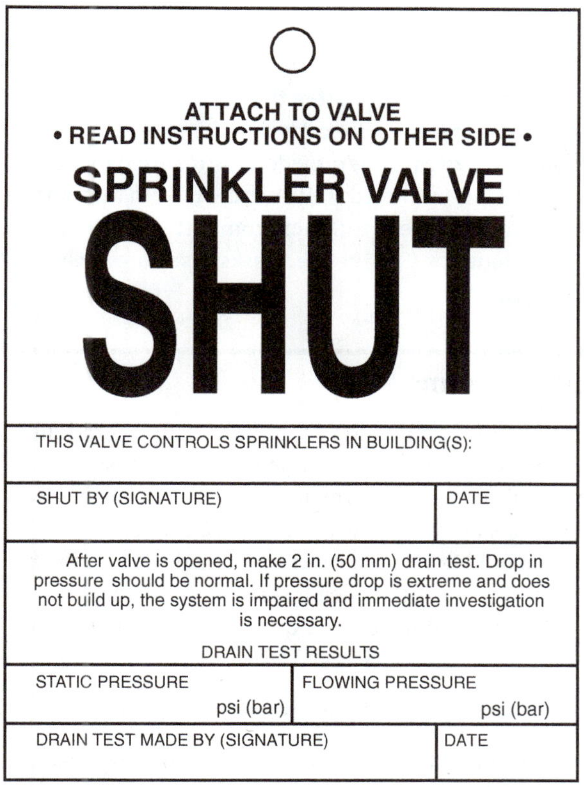

ATTACH TO VALVE
• READ INSTRUCTIONS ON OTHER SIDE •
SPRINKLER VALVE
SHUT

THIS VALVE CONTROLS SPRINKLERS IN BUILDING(S):	
SHUT BY (SIGNATURE)	DATE

After valve is opened, make 2 in. (50 mm) drain test. Drop in pressure should be normal. If pressure drop is extreme and does not build up, the system is impaired and immediate investigation is necessary.

DRAIN TEST RESULTS

STATIC PRESSURE psi (bar)	FLOWING PRESSURE psi (bar)
DRAIN TEST MADE BY (SIGNATURE)	DATE

FIGURE A.15.3.1 Sample Impairment Tag.

EXHIBIT 15.1 *Closed and Tagged Valve.*

15.3.2* The tag shall be posted at each fire department connection and the system control valve, and other locations required by the authority having jurisdiction, indicating which system, or part thereof, has been removed from service.

A.15.3.2 An impairment tag should be placed on the fire department connection to alert responding fire fighters of an abnormal condition. An impairment tag that is located on the system riser only could go unnoticed for an extended period if fire fighters encounter difficulty in gaining access to the building or sprinkler control room.

15.4 Impaired Equipment

15.4.1 The impaired equipment shall be considered to be the water-based fire protection system, or part thereof, that is removed from service.

15.4.2 The impaired equipment shall include, but shall not be limited to, the following:

(1) Sprinkler systems
(2) Standpipe systems
(3) Fire hose systems
(4) Underground fire service mains
(5) Fire pumps
(6) Water storage tanks
(7) Water spray fixed systems
(8) Foam-water systems

(9) Water mist systems
(10) Fire service control valves

Due to the modification to the scope of the technical committee that prepares NFPA 25, water mist systems were added to this list of systems that must be considered impaired. Historically, NFPA 750, *Standard on Water Mist Fire Protection Systems,* has provided the maintenance requirements for water mist systems, and NFPA 25 simply extracted these concepts from that standard. The NFPA Standards Council clarified the scope of these technical committees in 2012, which put all of the inspection, testing, and maintenance (ITM) related tasks associated with water mists systems, including impairments, into NFPA 25.

15.5* Preplanned Impairment Programs

A.15.5 The need for temporary fire protection, termination of all hazardous operations, and frequency of inspections in the areas involved should be determined. All work possible should be done in advance to minimize the length of the impairment. Where possible, temporary feedlines should be used to maintain portions of systems while work is completed.

Water-based fire protection systems should not be removed from service when the building is not in use. Where a system that has been out of service for a prolonged period, such as in the case of idle or vacant properties, is returned to service, qualified personnel should be retained to inspect and test the systems.

15.5.1 All preplanned impairments shall be authorized by the impairment coordinator.

15.5.2 Before authorization is given, the impairment coordinator shall be responsible for verifying that the following procedures have been implemented:

(1) The extent and expected duration of the impairment have been determined.
(2) The areas or buildings involved have been inspected and the increased risks determined.
(3) Recommendations to mitigate any increased risks have been submitted to management or the property owner or designated representative.
(4) Where a fire protection system is out of service for more than 10 hours in a 24-hour period, the impairment coordinator shall arrange for one of the following:
    (a) Evacuation of the building or portion of the building affected by the system out of service
    (b)* An approved fire watch
    (c)* Establishment of a temporary water supply
    (d)* Establishment and implementation of an approved program to eliminate potential ignition sources and limit the amount of fuel available to the fire

An approved fire prevention program, as referred to in 15.5.2(4)(d), should include the following:

1. Good housekeeping
2. On-site security
3. Immediate activation of new or repaired fire protection systems as construction and repairs allow
4. Preservation of existing fire protection systems during building construction or renovation
5. Development of a prefire plan with the local fire department
6. Accurate and rapid communication process
7. Consideration of any special hazards resulting from construction operations
8. Protection of existing structures and equipment from exposure fires resulting from construction, alteration, and demolition operations

Temporary fire extinguishers should be provided in accordance with NFPA 10, *Standard for Portable Fire Extinguishers*. At least one approved fire extinguisher should be provided in plain sight on each floor at each usable stairway, and access to permanent, temporary, or portable fire equipment should be maintained at all times.

(5) The fire department has been notified.

Although many fire departments might not require a permit for maintenance activities, notification to the department that a system or a portion of a system is out of service is still required. This notification of an impaired system serves two purposes, as follows:

1. It assists fire department pre-incident planning should an emergency response to the affected property be necessary.
2. It allows the fire department to monitor progress in repairs and avoid a system impairment that is longer than necessary.

(6) The insurance carrier, the alarm company, property owner or designated representative, and other authorities having jurisdiction have been notified.

Notifying the insurance carrier, the alarm company, the property owner or manager, and other AHJs is important for the following reasons:

1. In addition to ensuring that the alarm company does not inadvertently initiate a fire department response to an alarm signal, alarm companies often require an estimated time period for the impairment and will respond if the system is not placed in service at the end of the specified time period.
2. Insurance companies might have requirements or recommendations for temporary protection during impairments and provide assistance in monitoring the impairment. Many insurance companies have specific tag impairment systems that must be used during impairments.
3. Water or health departments might have special requirements when systems are shut off and brought back on line, such as testing of backflow prevention devices.

(7) The supervisors in the areas to be affected have been notified.
(8) A tag impairment system has been implemented. *(See Section 15.3.)*

Prior to any scheduled inspection or testing that could impair the system, the contractor or service company should consult with the property owner or impairment coordinator. The following issues should be discussed and agreed upon by all parties prior to the start of any work:

1. Any requirements for advance notification to the building occupants
2. Coordination of any business interruption
3. Procedures for the evacuation or for the restoration of protection
4. Any additional related issues

Anyone who might be affected by fire protection system testing in a protected premises should be notified that testing will take place. This notification should include, but might not be limited to, the property owner, property manager, switchboard operator, building engineer, building or floor fire wardens, and building maintenance personnel. In addition, the fire department communication center (if required), the alarm supervising station, and building occupants should also be notified.

Effective methods of notification for the building occupants include emails, bulletin board postings, public address announcements, and lobby signs.

(9) All necessary tools and materials have been assembled on the impairment site.

A.15.5.2(4)(b) A fire watch should consist of trained personnel who continuously patrol the affected area. Ready access to fire extinguishers and the ability to promptly notify the fire department are important items to consider. During the patrol of the area, the person should not only be looking for fire, but making sure that the other fire protection features of the building such as egress routes and alarm systems are available and functioning properly.

FAQ What are the qualifications and responsibilities of the fire watch?

Fire watch personnel should be trained to understand the hazards at the job site and the nature of hot work where hot work operations are used. They should ensure that safe conditions are maintained during all hot work operations and should have the authority to stop work if unsafe conditions develop. Fire-extinguishing equipment should be readily available, and fire watch personnel should be trained in its use. The fire watch personnel should be familiar with the facilities and procedures for sounding an alarm in the event of a fire.

Fire watch personnel should be trained in proper incipient fire response procedures, such as notifying the fire department and persons in the fire area before attempting to extinguish a fire. They should attempt to extinguish a fire only if it is obviously within the capacity of the equipment available. It is important to note that many AHJs have specific requirements that fire watch personnel must meet. Fire watch personnel may be allowed to perform additional tasks, but those tasks should not distract them from their watch responsibilities. For additional guidance on fire watch operations, see NFPA 51B, *Standard for Fire Prevention During Welding, Cutting, and Other Hot Work.*

For torch-applied roofing operations, a fire watch must be posted for the duration of the work and for 60 minutes thereafter.

A.15.5.2(4)(c) Temporary water supplies are possible from a number of sources, including use of a large-diameter hose from a fire hydrant to a fire department connection, use of a portable tank and a portable pump, or use of a standby fire department pumper and/or tanker.

Temporary water supplies can be established through the use of existing systems or portions of systems that are not involved in the maintenance or repair activity. Exhibit 9.9 shows one type of temporary water supply in the form of portable bladder tanks. Exhibit 15.2 shows another type of temporary water supply where a hard-piped connection was made to the fire department connection (FDC) in order to bypass work on the main water supply. Furthermore, the temporary water supply for fire protection should be available as soon as existing fire protection systems are shut down for maintenance or repair.

In situations involving automatic sprinklers, the system should be returned to service as soon as possible. The operation of sprinkler control valves should be limited to properly authorized personnel only. Where the sprinkler protection is regularly turned off and on to facilitate

Critical Point

If at all possible, sprinkler systems should be isolated from standpipes in order to keep the standpipe system in service and readily available for manual fire fighting.

EXHIBIT 15.2 *Temporary Water Supply Connection to FDC.*

connection of newly repaired segments, the sprinkler control valves should be checked at the end of each work shift and the impairment program followed to verify that protection is in service. If at all possible, sprinkler systems should be isolated from standpipes in order to keep the standpipe system in service and readily available for manual fire fighting.

A.15.5.2(4)(d) Depending on the use and occupancy of the building, it could be enough in some circumstances to stop certain processes in the building or to cut off the flow of fuel to some machines. It is also helpful to implement "No Smoking" and "No Hot Work" (cutting, grinding, or welding) policies while the system is out of service because these activities are responsible for many fire ignitions.

Hot work is responsible for 5 percent of all nonresidential fires and accounts for 8 percent of property damage, according to *Introduction to Employee Fire and Life Safety*. However, should welding or cutting be necessary during system repair, these activities should be carried out in accordance with NFPA 51B. OSHA 29 CFR 1910.252, "Welding, Cutting, and Brazing, General," references NFPA 51B and ANSI/AWS Z49.1, *Safety in Welding, Cutting, and Allied Processes*, so the requirements of those two documents must be followed for OSHA compliance.

FAQ Is welding of pipe and pipe supports permitted?

Although shop welding of pipe is preferred, welding of pipe and pipe supports is permitted by NFPA 13, *Standard for the Installation of Sprinkler Systems*, provided that welding operations are conducted in accordance with NFPA 51B. In some areas, a permit issued by the local AHJ might be required.

15.6* Emergency Impairments

A.15.6 Emergency impairments include, but are not limited to, system leakage, interruption of water supply, frozen or ruptured piping, equipment failure, or other impairments found during inspection, testing, or maintenance activities.

15.6.1 Emergency impairments shall include, but are not limited to, interruption of water supply, frozen or ruptured piping, and equipment failure, and includes impairments found during inspection, testing, or maintenance activities.

FAQ What is an emergency impairment?

An emergency impairment is a condition in which the water-based fire protection system or a portion thereof is out of service due to an unexpected occurrence. Such an impairment might occur as the result of a successful system operation during a fire.

Following a fire, sprinklers that have operated must be replaced and the system must be returned to service. The replacement of sprinklers illustrates the importance of the sprinkler cabinet inspections required by Chapter 5, since the impairment time can be substantially reduced if replacement sprinklers are on site.

Critical Point

Following a fire, sprinklers that have operated must be replaced and the system must be returned to service.

15.6.2* The coordinator shall implement the steps outlined in Section 15.5.

The impairment coordinator can use an emergency impairment notice, such as the one shown in Exhibit 15.3, to coordinate emergency repairs of a fire protection system.

A.15.6.2 When one or more impairments are discovered during inspection, testing, and maintenance activities, the owner or owner's authorized representative should be notified in writing. See Figure A.15.6.2 for an example of written notification.

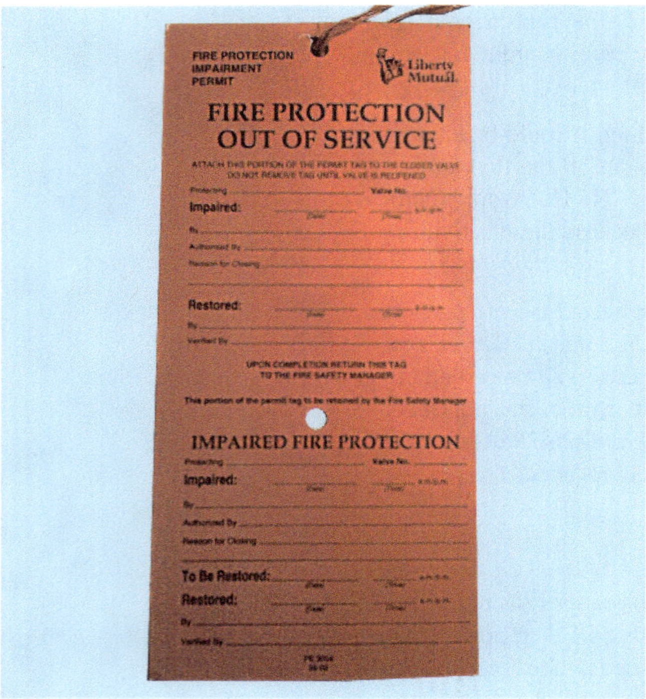

EXHIBIT 15.3 *Impairment Tag. (Courtesy of Liberty Mutual Commercial Markets)*

IMPAIRMENT NOTICE

DURING A RECENT INSPECTION OF YOUR FIRE PROTECTION SYSTEM(S), AN ***EMERGENCY IMPAIRMENT*** WAS DISCOVERED AND INDICATED ON THE INSPECTION REPORT. AS DEFINED BY NFPA 25, AN ***EMERGENCY IMPAIRMENT*** IS "A CONDITION WHERE A WATER-BASED FIRE PROTECTION SYSTEM OR PORTION THEREOF IS OUT OF ORDER DUE TO AN UNEXPECTED OCCURRENCE, SUCH AS A RUPTURED PIPE, OPERATED SPRINKLER, OR AN INTERRUPTION OF WATER SUPPLY TO THE SYSTEM." NFPA 25 FURTHER STATES, "EMERGENCY IMPAIRMENTS INCLUDE BUT ARE NOT LIMITED TO SYSTEM LEAKAGE, INTERRUPTION OF WATER SUPPLY, FROZEN OR RUPTURED PIPING, AND EQUIPMENT FAILURE."

WE RECOMMEND THAT IMMEDIATE STEPS BE TAKEN, AS DESCRIBED IN THE ATTACHED COPY OF CHAPTER 15 OF NFPA 25, TO CORRECT THE FOLLOWING IMPAIRMENT(S) TO YOUR FIRE PROTECTION SYSTEM(S):

- ❑ CONTROL VALVE SHUT. SYSTEM OUT OF SERVICE.
- ❑ LOW WATER PRESSURE DURING FLOW TEST. POSSIBLE OBSTRUCTION IN WATER SUPPLY OR PARTIALLY SHUT VALVE.
- ❑ PIPE(S) FROZEN.
- ❑ PIPE(S) LEAKING.
- ❑ PIPE(S) ARE OBSTRUCTED.
- ❑ SYSTEM PIPING OR PORTIONS OF SYSTEM PIPING ARE DISCONNECTED.
- ❑ FIRE DEPT. CONNECTION MISSING OR DAMAGED OR OBSTRUCTED.
- ❑ DRY PIPE VALVE CANNOT BE RESET.
- ❑ DRY PIPE SYSTEM QUICK-OPENING DEVICE IS OUT OF SERVICE.
- ❑ SPRINKLERS ARE PAINTED, CORRODED, DAMAGED, OR LOADED.
- ❑ FIRE PUMP IS OUT OF SERVICE.
- ❑ DETECTION/ACTUATION SYSTEM IS OUT OF SERVICE.
- ❑ OTHER:______________________________

FIGURE A.15.6.2 *Sample Impairment Notice.*

15.7* Restoring Systems to Service

When all impaired equipment is restored to normal working order, the impairment coordinator shall verify that the following procedures have been implemented:

(1) Any necessary inspections and tests have been conducted to verify that affected systems are operational. The appropriate chapter of this standard shall be consulted for guidance on the type of inspection and test required.

The most common tests required following a system repair include the hydrostatic pressure test to verify that the piping system will not leak and the main drain test to verify that all water supply control valves have been returned to their full open positions. Other tests, such as a running test for fire pumps and flow or alarm tests to check the functioning of replaced components, are required when components are repaired or replaced. The specific actions required for placing a system or system component in service following repair, replacement, or other corrective actions can be found in the appropriate chapter in the standard.

(2) Supervisors have been advised that protection is restored.
(3) The fire department has been advised that protection is restored.
(4) The property owner or designated representative, insurance carrier, alarm company, and other authorities having jurisdiction have been advised that protection is restored.

All parties involved in the original impairment notification should be advised when protection is restored. See the commentary following 15.5.2(5) for further information.

(5) The impairment tag has been removed.

Impairment tags are required to be removed and should be accounted for. As stated in the commentary following A.15.3.1, counting these tags can be a means of verifying that all system control valves have been returned to their proper operating position and the system is fully operational.

A.15.7 Occasionally, fire protection systems in idle or vacant buildings are shut off and drained. When the equipment is eventually restored to service after a long period of not being maintained, it is recommended that qualified personnel or a qualified contractor perform the work. The following is an example of a procedure:

(1) All piping should be traced from the extremities of the system to the main connections with a careful check for blank gaskets in flanges, closed valves, corroded or damaged sprinklers, nozzles, or piping, insecure or missing hangers, and insufficient support. Proper repairs or adjustments should be made, and needed extensions or alterations for the equipment should be completed.
(2) An air test at low pressure (40 psi) should be conducted prior to allowing water to fill the system. When the piping has been proven tight by passing the air test, water can be introduced slowly into the system, with proper precautions against damage by escape of water from previously undiscovered defects. When the system has been filled under normal service pressure, drain valve tests should be made to detect any closed valve that possibly could have been overlooked. All available pipes should be flushed, and an obstruction investigation completed to make sure that the system is clear of debris.
(3) Where the system was known to have been damaged by freezing or where other extensive damage might have occurred, a full hydrostatic test can be performed in accordance with NFPA 13 to determine whether the system integrity has been maintained.
(4) Dry pipe valves, quick-opening devices, alarm valves, and all alarm connections should be examined, put in proper condition, and tested.
(5) Fire pumps, pressure and gravity tanks, reservoirs, and other water supply equipment should receive proper attention before being placed in service. Each supply should be tested separately and then together if they are designed to work together.
(6) All control valves should be operated from the closed to fully open position and should be left sealed, locked, or equipped with a tamper switch.

References Cited in Commentary

National Fire Protection Association, 1 Batterymarch Park, Quincy, MA 02169-7471.

Colonna, G., ed., *Introduction to Employee Fire and Life Safety*, 2001.
NFPA 10, *Standard for Portable Fire Extinguishers*, 2013 edition.
NFPA 13, *Standard for the Installation of Sprinkler Systems*, 2013 edition.
NFPA 51B, *Standard for Fire Prevention During Welding, Cutting, and Other Hot Work*, 2014 edition.
NFPA 750, *Standard on Water Mist Fire Protection Systems*, 2014 edition.

American Welding Society, 550 NW LeJeaune Road, Miami, FL 33126.

ANSI/AWS Z49.1, *Safety in Welding, Cutting, and Allied Processes*, 2005 edition.

U.S. Government Printing Office, Washington, DC 20402.

Title 29, Code of Federal Regulations, Part 1910.252, "Welding, Cutting, and Brazing, General."

16

SPECIAL REQUIREMENTS FROM OTHER NFPA DOCUMENTS

Chapter 16 has been added to the 2014 edition of NFPA 25 to provide guidance for inspection, testing, and maintenance (ITM) of systems installed in buildings that have not previously been addressed by the standard. These include systems in small residential board and care facilities, residential occupancies installed in accordance with NFPA 13D, *Standard for the Installation of Sprinkler Systems in One- and Two-Family Dwellings and Manufactured Homes,* and systems installed per NFPA *101*®, *Life Safety Code*®. The inspection, testing, and maintenance (ITM) of these systems is just as important as for any other system, and perhaps more so, because these systems protect people first, as opposed to buildings and/or their contents. The addition of Chapter 16 formalizes the requirements provided in NFPA *101* and NFPA 13D, and it helps users of NFPA 25 by providing ITM requirements for all water-based systems in a single standard.

16.1 General

16.1.1 Application.

16.1.1.1* This chapter shall include the inspection, testing, and maintenance requirements for water-based fire protection systems found in other NFPA standards that are different from those included in this standard.

A.16.1.1.1 There are numerous NFPA documents that have special requirements for the inspection, testing, and maintenance of water-based fire protection systems that are different from those in NFPA 25. In many cases those documents reference NFPA 25 and could have either more stringent or even less stringent requirements than those in NFPA 25.

16.1.1.2* The requirements of this chapter shall be extracted from the other referenced standards.

The requirements provided throughout Chapter 16 are reprinted from the standard referenced (NFPA *101*) without modification or consideration by the NFPA 25 technical committee. The requirements in each case are dictated by the originating committee and are not addressable by the NFPA 25 technical committee, based on *NFPA Regulations Governing the Development of NFPA Standards*. So, the rules provided in this chapter apply only to the occupancies or building/system types specifically noted. These modified rules cannot be applied to water-based systems that are designed in accordance with the standards that follow, which are included in the scope of NFPA 25 in Chapter 1:

- NFPA 13, *Standard for the Installation of Sprinkler Systems*
- NFPA 13R, *Standard for the Installation of Sprinkler Systems in Low-Rise Residential Occupancies*

Critical Point

Where the basis of design is not known due to a lack of information, it should be assumed that the system design was based on the worst-case scenario until the design basis can be established through researching original installation documents.

- NFPA 14, *Standard for the Installation of Standpipe and Hose Systems*
- NFPA 15, *Standard for Water Spray Fixed Systems for Fire Protection*
- NFPA 16, *Standard for the Installation of Foam-Water Sprinkler and Foam-Water Spray Systems*
- NFPA 20, *Standard for the Installation of Stationary Pumps for Fire Protection*
- NFPA 22, *Standard for Water Tanks for Private Fire Protection*
- NFPA 24, *Standard for the Installation of Private Fire Service Mains and Their Appurtenances*
- NFPA 750, *Standard on Water Mist Fire Protection Systems*

Where the basis of design for a residential occupancy was based on NFPA 13, the ITM requirements provided in Chapter 5 and Chapter 13 should be applied, including all frequencies for ITM activities.

Where the basis of design is not known due to a lack of information, it should be assumed that the system design was based on the worst-case scenario until the design basis can be established through researching original installation documents.

A.16.1.1.2 These requirements are not written by or able to be amended by the technical committee of this standard, and therefore the text is an extract from the document that has the requirements.

16.1.1.3 Where the requirements of the referenced standard differ from the requirements of this standard, the referenced standard shall take precedence.

16.1.2 Definitions. For terms not defined in Chapter 3, the definitions of the reference standard shall apply.

16.2 Small Residential Board and Care Occupancies

Because there are very minimal maintenance requirements for sprinkler systems installed in accordance with NFPA 13D, and because the scope of NFPA 25 excludes NFPA 13D sprinkler systems, 33.2.3.5.8 of NFPA *101* specifies a series of testing and maintenance requirements for such systems where installed in small board and care occupancies. Where NFPA 13D is utilized outside its originally intended scope (one- and two-family dwellings and manufactured homes), NFPA *101* supplements the requirements of NFPA 13D to ensure a high level of reliability, because the system is either a mandatory system or it is being used to modify some other requirement.

The testing and maintenance requirements of 33.2.3.5.8 in NFPA *101* are not onerous. Several of the provisions, such as monthly visual inspection of control valves, can be accomplished by the facility's owner or operator. Such routine visual inspection will help to ensure a control valve is not inadvertently closed, which could lead to the system's failure in the event of a fire. These requirements have been extracted into NFPA 25 so that inspectors who conduct ITM services in accordance with NFPA 25 will have this information at their fingertips. Many inspectors would not know to look in NFPA *101*, because they might not be aware that NFPA *101* has an exception that allows NFPA 13D systems to be installed in these occupancies.

Critical Point

The rules provided in this chapter apply only to the occupancies or building/system types specifically noted.

16.2.1 The requirements in this section shall only apply to residential board and care facilities with sprinkler systems installed in accordance with NFPA 13D, *Standard for the Installation of Sprinkler Systems in One- and Two-Family Dwellings and Manufactured Homes*, as described in NFPA *101*, *Life Safety Code*.

16.2.1.1 Systems installed in accordance with NFPA 13D, *Standard for the Installation of Sprinkler Systems in One- and Two-Family Dwellings and Manufactured Homes*, shall be inspected, tested, and maintained in accordance with 33.2.3.5.8.1 through 33.2.3.5.8.15 of NFPA *101*, which reference specific sections of NFPA 25. The frequency of the inspection, test, or maintenance shall be in accordance with [NFPA *101*], whereas the purpose and procedure shall be from NFPA 25. [***101***:33.2.3.5.8]

16.2.1.1.1 Control valves shall be inspected monthly in accordance with 13.3.2 of NFPA 25. [*101*:33.2.3.5.8.1]

16.2.1.1.2 Gages shall be inspected monthly in accordance with 13.2.7.1 of NFPA 25. [*101*:33.2.3.5.8.2]

16.2.1.1.3 Alarm devices shall be inspected quarterly in accordance with 5.2.5 of NFPA 25. [*101*:33.2.3.5.8.3]

16.2.1.1.4 Alarm devices shall be tested semiannually in accordance with 5.3.3 of NFPA 25. [*101*:33.2.3.5.8.4]

16.2.1.1.5 Valve supervisory switches shall be tested semiannually in accordance with 13.3.3.5 of NFPA 25. [*101*:33.2.3.5.8.5]

16.2.1.1.6 Visible sprinklers shall be inspected annually in accordance with 5.2.1 of NFPA 25. [*101*:33.2.3.5.8.6]

16.2.1.1.7 Visible pipe shall be inspected annually in accordance with 5.2.2 of NFPA 25. [*101*:33.2.3.5.8.7]

16.2.1.1.8 Visible pipe hangers shall be inspected annually in accordance with 5.2.3 of NFPA 25. [*101*:33.2.3.5.8.8]

16.2.1.1.9 Buildings shall be inspected annually prior to the onset of freezing weather to ensure that there is adequate heat wherever water-filled piping is run in accordance with 5.2.5 of NFPA 25. [*101*:33.2.3.5.8.9]

16.2.1.1.10 A representative sample of fast-response sprinklers shall be tested once the sprinklers in the system are 20 years old in accordance with 5.3.1.1.1.3 of NFPA 25. If the sample fails the test, all of the sprinklers represented by that sample shall be replaced. If the sprinklers pass the test, the test shall be repeated every 10 years thereafter. [*101*:33.2.3.5.8.10]

16.2.1.1.11 A representative sample of dry-pendent sprinklers shall be tested once the sprinklers in the system are 10 years old in accordance with 5.3.1.1.1.6 of NFPA 25. If the sample fails the test, all of the sprinklers represented by that sample shall be replaced. If the sprinklers pass the test, the test shall be repeated every 10 years thereafter. [*101*:33.2.3.5.8.11]

16.2.1.1.12 Antifreeze solutions shall be tested annually in accordance with 5.3.4 of NFPA 25. [*101*:33.2.3.5.8.12]

The use of antifreeze solutions, which are not listed for use in a water-based fire protection system, have been severely limited due to the hazards associated with applying combustible liquids to open flames. NFPA Tentative Interim Amendments (TIAs) and advisory bulletins are available at the official NFPA website (www.nfpa.org) and should be reviewed and included in a required evaluation of any antifreeze solution system before a decision is made to allow it to remain in service.

Existing systems incorporating non-listed antifreeze solutions, such as glycerine or propylene glycol, can be dangerous to the occupants of the building. Owners should consider all risks and evaluate the possibility of conditioning the space where piping is installed to protect it from freezing or replace the system with a small dry pipe system.

Critical Point

Owners should consider all risks and evaluate the possibility of conditioning the space where piping is installed to protect it from freezing or replace the system with a small dry pipe system.

16.2.1.1.13 Control valves shall be operated through their full range and returned to normal annually in accordance with 13.3.3.1 of NFPA 25. [*101*:33.2.3.5.8.13]

16.2.1.1.14 Operating stems of OS&Y valves shall be lubricated annually in accordance with 13.3.4 of NFPA 25. [*101*:33.2.3.5.8.14]

16.2.1.1.15 Dry-pipe systems that extend into the unheated portions of the building shall be inspected, tested, and maintained in accordance with 13.4.4 of NFPA 25. [*101*:33.2.3.5.8.15]

References Cited in Commentary

National Fire Protection Association, 1 Batterymarch Park, Quincy, MA 02169-7471.

NFPA 13, *Standard for the Installation of Sprinkler Systems*, 2013 edition.

NFPA 13D, *Standard for the Installation of Sprinkler Systems in One- and Two-Family Dwellings and Manufactured Homes*, 2013 edition.

NFPA 13R, *Standard for the Installation of Sprinkler Systems in Low-Rise Residential Occupancies*, 2013 edition.

NFPA 14, *Standard for the Installation of Standpipe and Hose Systems*, 2013 edition.

NFPA 15, *Standard for Water Spray Fixed Systems for Fire Protection*, 2012 edition.

NFPA 16, *Standard for the Installation of Foam-Water Sprinkler and Foam-Water Spray Systems*, 2011 edition.

NFPA 20, *Standard for the Installation of Stationary Pumps for Fire Protection*, 2013 edition.

NFPA 22, *Standard for Water Tanks for Private Fire Protection*, 2013 edition.

NFPA 24, *Standard for the Installation of Private Fire Service Mains and Their Appurtenances*, 2013 edition.

NFPA *101®*, *Life Safety Code®*, 2012 edition.

NFPA 750, *Standard on Water Mist Fire Protection Systems*, 2014 edition.

ANNEX A

EXPLANATORY MATERIAL

Annex A is not a part of the recommendations of this NFPA document but is included for informational purposes only. This annex contains explanatory material, numbered to correspond with the applicable text paragraphs.

Annex material is useful information that is included in this standard solely to help the user understand the intent of the requirements by providing further information, diagrams, examples, or other details.

The Annex A material is not only explanatory material to specific sections of NFPA 25, but it is voted on by the document's technical committee along with the rest of NFPA 25.

The commentary in this handbook differs from the annex material in that it consists of the opinions of the editors and contributors of this handbook — not the committee — to help further explain provisions in the document.

For the convenience of the readers of this handbook, Annex A text is inserted after the appropriate paragraphs in Chapter 1 through Chapter 16 and, therefore, is not repeated here.

Please note that the deficiency classification table from Annex E of the 2011 edition of NFPA 25 has been incorporated into Annex A of the 2014 edition. This annex table provides examples of how field conditions can be classified per NFPA 25. This table is now provided as part of A.3.3.7 and is linked to the definition of the term *deficiency*.

ANNEX B

FORMS FOR TESTING, INSPECTION, AND MAINTENANCE

This annex is not a part of the requirements of this NFPA document but is included for informational purposes only.

B.1 General

Forms need to be complete with respect to the requirements of NFPA 25 for the system being inspected, tested, or maintained, or any combination thereof. Because water-based fire protection systems are comprised of many components, it could be necessary to complete more than one form for each system.

Authorities having jurisdiction are legitimately concerned that the forms used are comprehensive. Therefore, they could develop their own forms or utilize those already developed and reviewed by their jurisdiction.

At least five formats can be used and are described as follows:

(1) One form in which all requirements for NFPA 25 are specified and large sections of information do not apply to most systems
(2) Individual forms that provide requirements corresponding to each chapter of NFPA 25 and address the following:
    (a) Sprinkler systems
    (b) Standpipe systems
    (c) Private fire service mains
    (d) Fire pumps
    (e) Storage tanks
    (f) Water spray systems
    (g) Foam-water sprinkler systems
(3) Forms that include information from the specific system chapter: Chapter 1, Chapter 13, and Chapter 14
(4) A series of forms similar to option (2) but with a more detailed breakdown of system types. For example, fire sprinkler systems are divided into five separate forms such as:
    (a) Wet pipe fire sprinkler systems
    (b) Dry pipe fire sprinkler systems
    (c) Preaction fire sprinkler systems
    (d) Deluge fire sprinkler systems
    (e) Foam-water sprinkler systems

(5) Separate forms for each individual component of each fire protection system

Any forms that are used for checking a fire protection system must comply with the reporting requirements of NFPA 25. Note that NFPA 25 does not have a specific form or series of forms that must be used when conducting inspection, testing, and maintenance (ITM) tasks. There is a common misconception that the inspector must fill out "the NFPA 25 form," but the standard has no forms that are required to be used. This allows the owner and inspector to develop custom forms for their properties based on the systems that are present.

There are several electronic reporting systems that are also available. Any system that meets the minimum reporting requirements of the standard can be used.

B.2 Sample Forms

Sample forms are available for downloading at www.nfpa.org, www.nfsa.org, and www.sprinklernet.org.

ANNEX C

POSSIBLE CAUSES OF PUMP TROUBLES

This annex is not a part of the requirements of this NFPA document but is included for informational purposes only.

This annex is extracted from NFPA 20, *Standard for the Installation of Stationary Pumps for Fire Protection.*

The information provided in Annex C is intended to assist the user in troubleshooting fire pump problems. Figure C.1 supplements the text of Annex C by identifying potential problems with fire pumps and determining the likely causes of those problems in an easy-to-read, concise matrix format. Figure C.1 illustrates that a given problem can be attributed to a variety of causes in various components and locations in piping associated with the fire pumps.

C.1 Causes of Pump Troubles

This annex contains a partial guide for locating pump troubles and their possible causes *(see Figure C.1)*. It also contains a partial list of suggested remedies. *(For other information on this subject, see Hydraulic Institute Standard for Centrifugal, Rotary and Reciprocating Pumps.)* The causes listed here are in addition to possible mechanical breakage that would be obvious on visual inspection. In case of trouble, it is suggested that those troubles that can be inspected easily should be corrected first or eliminated as possibilities.

C.1.1 Air Drawn into Suction Connection Through Leak(s). Air drawn into suction line through leaks causes a pump to lose suction or fail to maintain its discharge pressure. Uncover suction pipe and locate and repair leak(s).

C.1.2 Suction Connection Obstructed. Examine suction intake, screen, and suction pipe and remove obstruction. Repair or provide screens to prevent recurrence.

Exhibit C.1 shows a playground ball that was drawn in through the suction piping and obstructed the intake into the pump.

C.1.3 Air Pocket in Suction Pipe. Air pockets cause a reduction in delivery and pressure similar to an obstructed pipe. Uncover suction pipe and rearrange to eliminate pocket.

C.1.4 Well Collapsed or Serious Misalignment. Consult a reliable well drilling company and the pump manufacturer regarding recommended repairs.

	Suction				Pump																Driver and/or Pump					Driver						
	Air drawn into suction connection through leak(s)	Suction connection obstructed	Air pocket in suction pipe	Well collapsed or serious misalignment	Stuffing box too tight or packing improperly installed, worn, defective, too tight, or incorrect type	Water seal or pipe to seal obstructed	Air leak into pump through stuffing boxes	Impeller obstructed	Wearing rings worn	Impeller damaged	Wrong diameter impeller	Actual net head lower than rated	Casing gasket defective, permitting internal leakage (single-stage and multistage pumps)	Pressure gauge is on top of pump casing	Incorrect impeller adjustment (vertical shaft turbine-type pump only)	Impellers locked	Pump is frozen	Pump shaft or shaft sleeve scored, bent, or worn	Pump not primed	Seal ring improperly located in stuffing box, preventing water from entering space to form seal	Excess bearing friction due to lack of lubrication, wear, dirt, rusting, failure, or improper installation	Rotating element binds against stationary element	Pump and driver misaligned	Foundation not rigid	Engine-cooling system obstructed	Faulty driver	Lack of lubrication	Speed too low	Wrong direction of rotation	Speed too high	Rated motor voltage different from line voltage	Faulty electrical circuit, obstructed fuel system obstructed steam pipe, or dead battery
Fire pump troubles	1	2	3	4	5	6	7	8	9	10	11	12	13	14	15	16	17	18	19	20	21	22	23	24	25	26	27	28	29	30	31	32
Excessive leakage at stuffing box					X													X					X									
Pump or driver overheats				X	X	X		X			X				X			X	X	X	X	X	X	X	X		X		X	X	X	
Pump unit will not start				X	X										X	X	X				X					X	X					X
No water discharge	X	X	X					X											X													
Pump is noisy or vibrates				X	X			X		X								X			X	X	X	X			X					
Too much power required				X	X			X	X		X		X		X			X			X	X	X	X			X		X	X	X	
Discharge pressure not constant for same gpm	X				X	X	X																									
Pump loses suction after starting	X	X	X			X	X													X												
Insufficient water discharge	X	X	X			X	X	X	X	X	X	X	X		X													X	X		X	
Discharge pressure too low for gpm discharge	X	X	X		X	X	X	X	X	X	X	X	X	X														X	X		X	

FIGURE C.1 *Possible Causes of Fire Pump Troubles.*

C.1.5 Stuffing Box Too Tight or Packing Improperly Installed, Worn, Defective, Too Tight, or of Incorrect Type. Loosen gland swing bolts and remove stuffing box gland halves. Replace packing.

C.1.6 Water Seal or Pipe to Seal Obstructed. Loosen gland swing bolt and remove stuffing box gland halves along with the water seal ring and packing. Clean the water passage to and in the water seal ring. Replace water seal ring, packing gland, and packing in accordance with manufacturer's instructions.

C.1.7 Air Leak into Pump Through Stuffing Boxes. Same as possible cause in C.1.6.

EXHIBIT C.1 *Obstructed Intake. (Courtesy of Damon Pietraz and Underwood Fire Equipment)*

C.1.8 Impeller Obstructed. Does not show on any one instrument, but pressures fall off rapidly when an attempt is made to draw a large amount of water.

For horizontal split-case pumps, remove upper case of pump and remove obstruction from impeller. Repair or provide screens on suction intake to prevent recurrence.

For vertical shaft turbine-type pumps, lift out column pipe and pump bowls from wet pit or well and disassemble pump bowl to remove obstruction from impeller.

For close-coupled, vertical in-line pumps, lift motor on top pull-out design and remove obstruction from impeller.

Exhibit C.2 and Exhibit C.3 show obstructions to suction through the impeller.

C.1.9 Wearing Rings Worn. Remove upper case and insert feeler gauge between case wearing ring and impeller wearing ring. Clearance when new is 0.0075 in. (0.19 mm). Clearances of more than 0.015 in. (0.38 mm) are excessive.

EXHIBIT C.2 *Horizontal Split Case Fire Pump with Large Rocks Obstructing Suction Through the Impeller. (Courtesy of Damon Pietraz and Underwood Fire Equipment)*

EXHIBIT C.3 *Pieces of a Failed Backflow Preventer Obstructing Suction Through the Impeller. (Courtesy of Damon Pietraz and Underwood Fire Equipment)*

C.1.10 Impeller Damaged. Make minor repairs or return to manufacturer for replacement. If defect is not too serious, order new impeller and use damaged one until replacement arrives.

C.1.11 Wrong Diameter Impeller. Replace with impeller of proper diameter.

C.1.12 Actual Net Head Lower Than Rated. Check impeller diameter and number and pump model number to make sure correct head curve is being used.

C.1.13 Casing Gasket Defective, Permitting Internal Leakage (Single-Stage and Multistage Pumps). Replace defective gasket. Check manufacturer's drawing to see whether gasket is required.

C.1.14 Pressure Gauge Is on Top of Pump Casing. Place gauges in correct location.

C.1.15 Incorrect Impeller Adjustment (Vertical Shaft Turbine-Type Pump Only). Adjust impellers according to manufacturer's instructions.

C.1.16 Impellers Locked. For vertical shaft turbine-type pumps, raise and lower impellers by the top shaft adjusting nut. If this adjustment is not successful, follow the manufacturer's instructions.

For horizontal split-case pumps, remove upper case and locate and eliminate obstruction.

C.1.17 Pump Is Frozen. Provide heat in the pump room. Disassemble pump and remove ice as necessary. Examine parts carefully for damage.

C.1.18 Pump Shaft or Shaft Sleeve Scored, Bent, or Worn. Replace shaft or shaft sleeve.

C.1.19 Pump Not Primed. If a pump is operated without water in its casing, the wearing rings are likely to seize. The first warning is a change in pitch of the sound of the driver. Shut down the pump.

For vertical shaft turbine-type pumps, check water level to determine whether pump bowls have proper submergence.

C.1.20 Seal Ring Improperly Located in Stuffing Box, Preventing Water from Entering Space to Form Seal. Loosen gland swing bolt and remove stuffing box gland halves along with the water-seal ring and packing. Replace, putting seal ring in proper location.

C.1.21 Excess Bearing Friction Due to Lack of Lubrication, Wear, Dirt, Rusting, Failure, or Improper Installation. Remove bearings and clean, lubricate, or replace as necessary.

C.1.22 Rotating Element Binds Against Stationary Element. Inspect clearances and lubrication and replace or repair the defective part.

C.1.23 Pump and Driver Misaligned. Shaft running off center because of worn bearings or misalignment. Align pump and driver according to manufacturer's instructions. Replace bearings according to manufacturer's instructions.

C.1.24 Foundation Not Rigid. Tighten foundation bolts or replace foundation if necessary.

In 6.4.1 of NFPA 20, *Standard for the Installation of Stationary Pumps for Fire Protection*, the pumps and driver are required to be mounted on a common grouted base plate. The use of rubber footings or mats, as shown in Exhibit C.4 and Exhibit C.5, does not comply with NFPA 20 and can lead to differential movement between pump components, which can lead to failure.

EXHIBIT C.4 Flexible Mat Installed Under Base Plate — Not in Compliance with NFPA 20. (Courtesy of Indonesian Fire and Rescue Foundation)

EXHIBIT C.5 Rubber Mounting Underneath the Base Plate — Not in Compliance with NFPA 20. (Courtesy of Indonesian Fire and Rescue Foundation)

C.1.25 Engine-Cooling System Obstructed. Heat exchanger or cooling water systems too small. Cooling pump faulty. Remove thermostats. Open bypass around regulator valve and strainer. Inspect regulator valve operation. Inspect strainer. Clean and repair if necessary. Disconnect sections of cooling system to locate and remove possible obstruction. Adjust engine-cooling water-circulating pump belt to obtain proper speed without binding. Lubricate bearings of this pump.

If overheating still occurs at loads up to 150 percent of rated capacity, contact pump or engine manufacturer so that necessary steps can be taken to eliminate overheating.

C.1.26 Faulty Driver. Inspect electric motor, internal combustion engine, or steam turbine, in accordance with manufacturer's instructions, to locate reason for failure to start.

Exhibit C.6 shows a diesel engine that has seized up.

EXHIBIT C.6 A Seized Diesel Engine that Was Found to Have Antifreeze in the Oil. (Courtesy of Damon Pietraz and Underwood Fire Equipment)

C.1.27 Lack of Lubrication. If parts have seized, replace damaged parts and provide proper lubrication. If not, stop pump and provide proper lubrication.

C.1.28 Speed Too Low. For electric motor drive, confirm that rated motor speed corresponds to rated speed of pump, voltage is correct, and starting equipment is operating properly.

Low frequency and low voltage in the electric power supply prevent a motor from running at rated speed. Low voltage can be due to excessive loads and inadequate feeder capacity or (with private generating plants) low generator voltage. The generator voltage of private generating plants can be corrected by changing the field excitation. When low voltage is from the other causes mentioned, it can be necessary to replace transformer taps or increase feeder capacity.

Low frequency usually occurs with a private generating plant and should be corrected at the source. Low speed can result in older type squirrel-cage-type motors if fastenings of copper bars to end rings become loose. The remedy is to weld or braze these joints.

For steam turbine drive, confirm that valves in steam supply pipe are wide open; boiler steam pressure is adequate; steam pressure is adequate at the turbine; strainer in the steam supply pipe is not plugged; steam supply pipe is of adequate size; condensate is removed from steam supply pipe, trap, and turbine; turbine nozzles are not plugged; and setting of speed and emergency governor is correct.

For internal combustion engine drive, confirm that setting of speed governor is correct; hand throttle is opened wide; and there are no mechanical defects such as sticking valves, timing off, or spark plugs fouled, and so forth. The latter can require the services of a trained mechanic.

C.1.29 Wrong Direction of Rotation. Instances of an impeller turning backward are rare but are clearly recognizable by the extreme deficiency of pump delivery. Wrong direction of rotation can be determined by comparing the direction in which the flexible coupling is turning with the directional arrow on the pump casing.

With polyphase electric motor drive, two wires must be reversed; with dc driver, the armature connections must be reversed with respect to the field connections. Where two sources of electrical current are available, the direction of rotation produced by each should be inspected.

C.1.30 Speed Too High. See that pump- and driver-rated speed correspond. Replace electric motor with one of correct rated speed. Set governors of variable-speed drivers for correct speed. Frequency at private generating stations can be too high.

C.1.31 Rated Motor Voltage Different from Line Voltage. For example, a 220 or 440 V motor on 208 or 416 V line. Obtain motor of correct rated voltage or a larger size motor.

C.1.32 Faulty Electric Circuit, Obstructed Fuel System, Obstructed Steam Pipe, or Dead Battery. Inspect for break in wiring open switch, open circuit breaker, or dead battery. If circuit breaker in controller trips for no apparent reason, make sure oil is in dash pots in accordance with manufacturer's specifications. Make sure fuel pipe is clear, strainers are clean, and control valves open in fuel system to internal combustion engine. Make sure all valves are open and strainer is clean in steam line to turbine.

C.2 Warning

Chapters 6 and 7 of NFPA 20, *Standard for the Installation of Stationary Pumps for Fire Protection,*, include electrical requirements that discourage the installation of disconnect means in the power supply to electric motor–driven fire pumps. This requirement is intended to ensure the availability of power to the fire pumps. When equipment connected to those circuits is serviced or maintained, the employee can have unusual exposure to electrical and other hazards. It can be necessary to require special safe work practices and special safeguards, personal protective clothing, or both.

C.3 Maintenance of Fire Pump Controllers After Fault Condition

C.3.1 Introduction. In a fire pump motor circuit that has been properly installed, coordinated, and in service prior to the fault, tripping of the circuit breaker or the isolating switch indicates a fault condition in excess of operating overload.

It is recommended that the following general procedures be observed by qualified personnel in the inspection and repair of the controller involved in the fault. These procedures are not intended to cover other elements of the circuit, such as wiring and motor, which can also require attention.

C.3.2 Caution. All inspections and tests are to be made on controllers that are de-energized at the line terminal, disconnected, locked out, and tagged so that accidental contact cannot be made with live parts and so that all plant safety procedures will be observed.

C.3.2.1 Enclosure. Where substantial damage to the enclosure, such as deformation, displacement of parts, or burning has occurred, replace the entire controller.

C.3.2.2 Circuit Breaker and Isolating Switch. Examine the enclosure interior, circuit breaker, and isolating switch for evidence of possible damage. If evidence of damage is not apparent, the circuit breaker and isolating switch can continue to be used after the door is closed.

If there is any indication that the circuit breaker has opened several short-circuit faults, or if signs of possible deterioration appear within either the enclosure, circuit breaker, or isolating switch (e.g., deposits on surface, surface discoloration, insulation cracking, or unusual toggle operation), replace the components. Verify that the external operating handle is capable of opening and closing the circuit breaker and isolating switch. If the handle fails to operate the device, this would also indicate the need for adjustment or replacement.

C.3.2.3 Terminals and Internal Conductors. Where there are indications of arcing damage, overheating, or both, such as discoloration and melting of insulation, replace the damaged parts.

C.3.2.4 Contactor. Replace contacts showing heat damage, displacement of metal, or loss of adequate wear allowance of the contacts. Replace the contact springs where applicable. If deterioration extends beyond the contacts, such as binding in the guides or evidence of insulation damage, replace the damaged parts or the entire contactor.

C.3.2.5 Return to Service. Before returning the controller to service, inspect for the tightness of electrical connections and for the absence of short circuits, ground faults, and leakage current.

Close and secure the enclosure before the controller circuit breaker and isolating switch are energized. Follow operating procedures on the controller to bring it into standby condition.

References Cited in Commentary

National Fire Protection Association, 1 Batterymarch Park, Quincy, MA 02169-7471.

NFPA 20, *Standard for the Installation of Stationary Pumps for Fire Protection*, 2013 edition.

ITM NFPA 25 INSPECTION TESTING MAINTENANCE

ANNEX D

OBSTRUCTION INVESTIGATION

This annex is not a part of the requirements of this NFPA document but is included for informational purposes only.

Annex D outlines sources of obstruction, treatment strategies, and flushing procedures for piping and discharge outlets in fire protection systems. While Chapter 14 of NFPA 25 provides the enforceable requirements for conducting an obstruction investigation, this annex provides a recommended systematic approach to correcting obstruction problems.

In addition to the methods discussed in Annex D, new technologies employing ultrasonic examination and remote video are now available for interior piping investigations. Methods other than those mentioned here for cleaning obstructed pipe (e.g., flushing with chemical solutions) are also available.

D.1

For effective control and extinguishment of fire, automatic sprinklers should receive an unobstructed flow of water. Although the overall performance record of automatic sprinklers has been very satisfactory, there have been numerous instances of impaired efficiency because sprinkler piping or sprinklers were plugged with pipe scale, corrosion products, including those produced by microbiologically influenced corrosion, mud, stones, or other foreign material. If the first sprinklers to open in a fire are plugged, the fire in that area cannot be extinguished or controlled by prewetting of adjacent combustibles. In such a situation, the fire can grow to an uncontrollable size, resulting in greater fire damage and excessive sprinkler operation and even threatening the structural integrity of the building, depending on the number of plugged sprinklers and fire severity.

Keeping the inside of sprinkler system piping free of scale, silt, or other obstructing material is an integral part of an effective loss prevention program.

D.1.1 While this chapter provides minimum requirements for the investigation and prevention of obstructions, authorities having jurisdiction must also consider regional, local, and project-specific propensities and histories to determine reasonable testing and obstruction mitigation measures.

D.2 Obstruction Sources

D.2.1 Pipe Scale. Loss studies have shown that dry pipe sprinkler systems are involved in the majority of obstructed sprinkler fire losses. Pipe scale was found to be the most frequent obstructing material (it is likely that some of the scale was composed of corrosion products, including those produced by microbiologically influenced corrosion). Dry pipe systems that have been maintained wet and then dry alternately over a period of years are particularly susceptible to the accumulation of scale. Also, in systems that are continuously dry, condensation of moisture in the air supply can result in the formation of a hard scale, microbiological materials, and corrosion products along the bottom of the piping. When sprinklers open, the scale is broken loose and carried along the pipe, plugging some of the sprinklers or forming obstructions at the fittings.

The development of scaling within dry pipe systems highlights the need to properly drain a system after water is introduced into the piping network. Although NFPA 25 does not require a check of the pitch on dry system piping, insufficiently pitched piping can lead to internal corrosion and scaling that can create obstructions. In instances where obstructions due to scaling have been identified, the owner should consider having an assessment of pipe pitch conducted. This assessment would occur outside of the scope of NFPA 25 and as such would not be part of a contract of services for inspectors complying with this standard. Exhibit D.1 shows internal corrosion in a dry system due to water not properly draining from the pipe after it was charged.

D.2.2 Careless Installation or Repair. Many obstructions are caused by careless workers during installation or repair of yard or public mains and sprinkler systems. Wood, paint brushes, buckets, gravel, sand, and gloves have been found as obstructions. In some instances, with welded sprinkler systems and systems with holes for quick-connect fittings, the cutout discs or coupons have been left within the piping, obstructing flow to sprinklers.

Paragraph 6.5.2.4.8(2) of NFPA 13, *Standard for the Installation of Sprinkler Systems*, requires that discs removed prior to attachment of a welded outlet be retrieved. This procedure can be verified by attaching the disc to the system at the point from which it was removed. Exhibit D.2

EXHIBIT D.1 *Corrosion Due to Improperly Pitched Pipe. (Courtesy of M. Steven Welsh, Koffel Associates, Inc.)*

shows discs attached to the system from which they were cut. Exhibit D.3 shows weld coupons that were not properly retrieved and presented an obstruction to the system piping. Exhibit D.4 shows work gloves and other obstructive material removed from an underground pipe in a system water supply.

D.2.3 Raw Water Sources. Materials can be sucked up from the bottoms of rivers, ponds, or open reservoirs by fire pumps with poorly arranged or inadequately screened intakes and then forced into the system. Sometimes floods damage intakes. Obstructions include fine, compacted materials such as rust, mud, and sand. Coarse materials, such as stones, cinders, cast-iron tubercles, chips of wood, and sticks, also are common.

In addition to the solid, inorganic objects mentioned in D.2.3, bodies of raw water are commonly the place where organic materials, such as the ones mentioned in D.2.4, are introduced into a system.

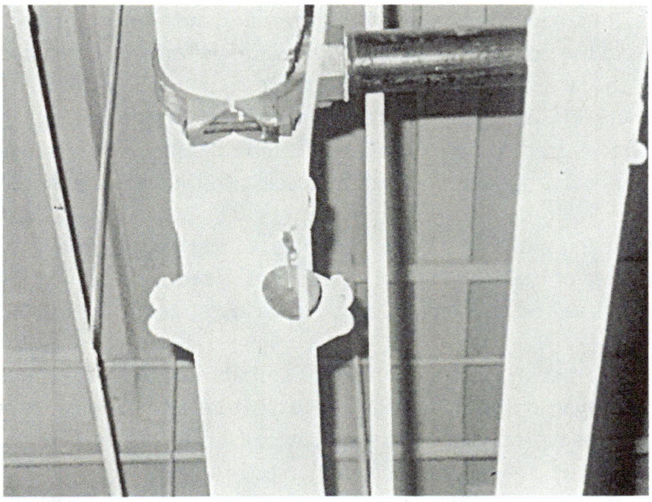

***EXHIBIT D.2** Discs Attached to Piping from Which They Were Cut. (Source: Automatic Sprinkler Systems Handbook, 2010, Exhibit 6.24)*

***EXHIBIT D.3** Weld Coupons Found in Early Suppression Fast-Response (ESFR) Sprinkler Main During Obstruction Investigation.*

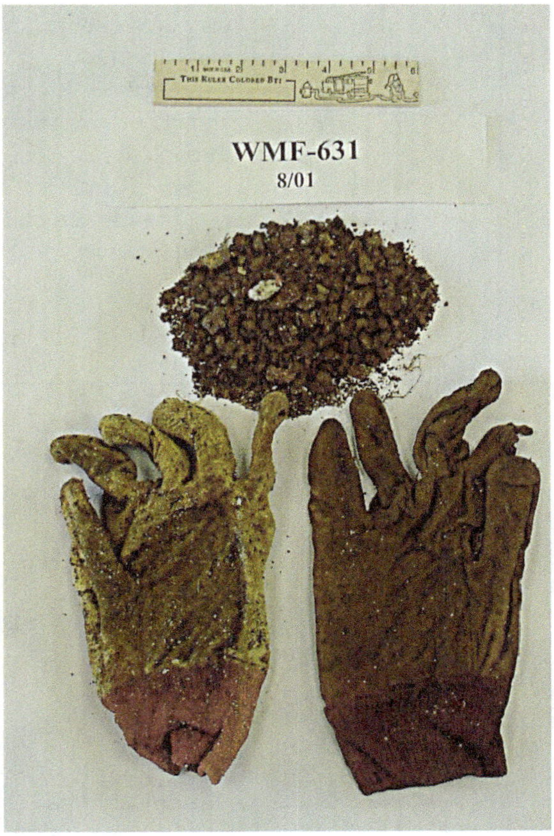

***EXHIBIT D.4** Work Gloves and Other Obstructive Material. (Courtesy of John Jensen)*

D.2.4 Biological Growth. Biological growth has been known to cause obstructions in sprinkler piping. The Asiatic clam has been found in fire protection systems supplied by raw river or lake water. With an available food supply and sunlight, these clams grow to approximately ⅜ in. to 7⁄16 in. (9 mm to 11 mm) across the shell in 1 year and up to 2⅛ in. (54 mm) and larger by the sixth year. However, once in fire mains and sprinkler piping, the growth rate is much slower. The clams get into the fire protection systems in the larval stage or while still small clams. They then attach themselves to the pipe and feed on bacteria or algae that pass through.

Originally brought to Washington state from Asia in the 1930s, the clams have spread throughout at least 33 states and possibly are present in every state. River areas reported to be highly infested include the Ohio River, Tennessee River Valley, Savannah River (South Carolina), Altamaha River (Georgia), Columbia River (Washington), and Delta-Mendota Canal (California).

In addition to the Asian clam, the zebra mussel can create obstructions to fire protection piping systems. Since their introduction to the Great Lakes in the mid-1980s, zebra mussels have spread throughout the Great Lakes into many inland waterways, including the Mississippi Delta to the south and the Hudson River to the northeast. Like the Asian clam, the zebra mussels can enter the fire protection system in the larval stage or as young adults.

The mussel attaches to the inner surface of a pipe and feeds on algae or bacteria. Exhibit D.5 shows a 2 in. (50 mm) diameter pipe infested with zebra mussels. Exhibit D.6 shows the relative size of a zebra mussel. Exhibit D.7 indicates the extent of zebra mussel infestation in the United States and Canada.

The presence of zebra mussels can be revealed by reduced flow or discharge of the mussels or shells from the inspector's test or main drain test connections. If reduced flow or the discharge of obstructive material is observed, a complete obstruction investigation is required per Section 14.3.

A strainer listed for use in a fire protection system can help to combat the problem of clams and mussels obstructing piping. However, in the larval stage, these creatures can certainly pass through the ⅛ in. (3.2 mm) perforations typically found in such devices. If clams or mussels are known to be present, the strainer must be cleaned more often than once a year.

The use of potassium and heat has led to high mortality rates in zebra mussel populations. However, the use of more common chemicals, such as chlorine, should be considered for

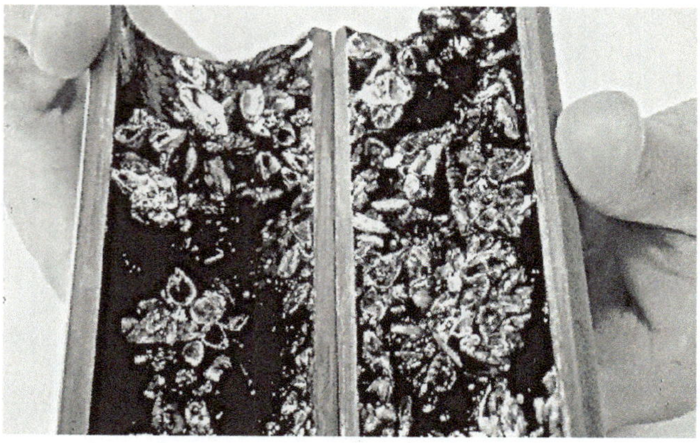

EXHIBIT D.5 *Pipe Infested with Zebra Mussels. (Courtesy of Michigan Sea Grant)*

EXHIBIT D.6 *Relative Size of Adult Zebra Mussel. (Courtesy of Michigan Sea Grant)*

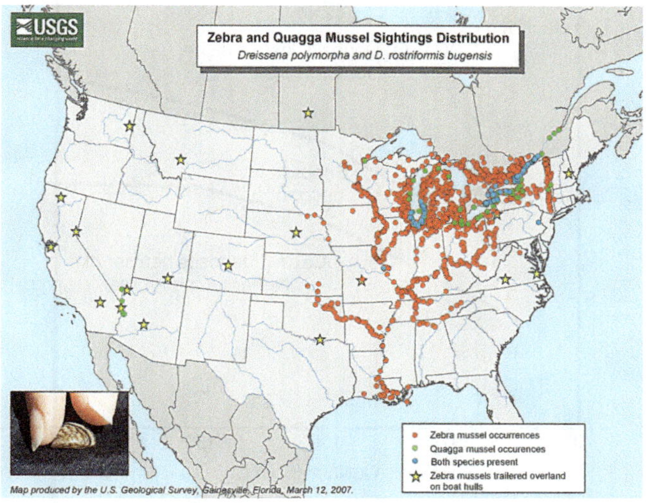

EXHIBIT D.7 Extent of Zebra Mussel Infestation. (Courtesy of the U.S. Geological Survey)

treatment of system water supplies. To prevent repopulation of systems with mussels or clams, flushing and testing in excess of the minimum requirements of the standard should be avoided.

D.2.5 Sprinkler Calcium Carbonate Deposits. Natural freshwaters contain dissolved calcium and magnesium salts in varying concentrations, depending on the sources and location of the water. If the concentration of these salts is high, the water is considered hard. A thin film composed largely of calcium carbonate, $CaCO_3$, affords some protection against corrosion where hard water flows through the pipes. However, hardness is not the only factor to determine whether a film forms. The ability of $CaCO_3$ to precipitate on the metal pipe surface also depends on the water's total acidity or alkalinity, the concentration of dissolved solids in the water, and its pH. In soft water, no such film can form.

In automatic sprinkler systems, the calcium carbonate scale formation tends to occur on the more noble metal in the electrochemical series, which is copper, just as corrosion affects the less noble metal, iron. Consequently, scale formation naturally forms on sprinklers, often plugging the orifice. The piping itself could be relatively clear. This type of sprinkler obstruction cannot be detected or corrected by normal flushing procedures. It can be found only by inspection of sprinklers in suspected areas and then removed.

Most public water utilities in very hard water areas soften their water to reduce consumer complaints of scale buildup in water heaters. Thus, the most likely locations for deposits in sprinkler systems are where sprinklers are not connected to public water but supplied without treatment directly from wells or surface water in areas that have very hard water. These areas generally include the Mississippi basin west of the Mississippi River and north of the Ohio River, the rivers of Texas and the Colorado basin, and other white areas in Figure D.2.5(a). (The water of the Great Lakes is only moderately hard.)

Within individual plants, the sprinklers most likely to have deposits are located as follows:

(1) In wet systems only.
(2) In high temperature areas, except where water has unusually high pH [*see Figure D.2.5(b)*]. High temperature areas include those near dryers, ovens, and skylights or at roof peaks.
(3) In old sprinkler systems that are frequently drained and refilled.
(4) In pendent sprinklers that are located away from air pockets and near convection currents.

Section D.2 describes where the requirements to perform a main drain test (see Chapter 13), flow test (see Chapter 7), and records retention (see Chapter 4) come into play. This annex

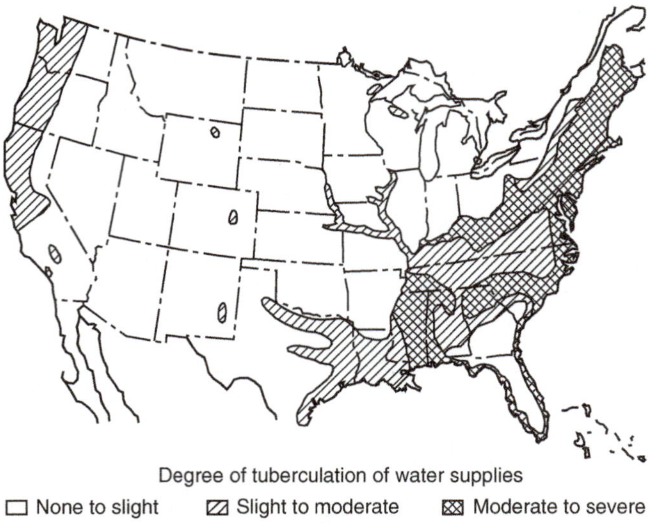

FIGURE D.2.5(a) Map of Hard Water Areas. (Provided by Cast Iron Pipe Research Association. Used with permission.)

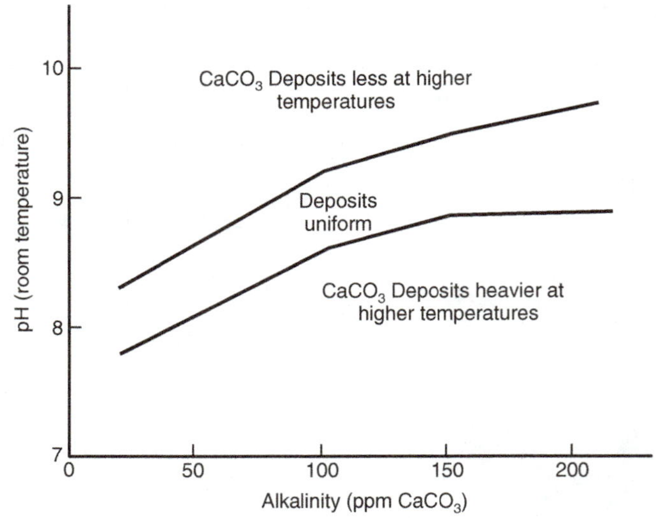

FIGURE D.2.5(b) Scale Deposition as Function of Alkalinity/pH Ratio.

section describes working examples of the requirements from the standard, including the requirement for a main drain test.

Although the main drain test is primarily intended to reveal the presence of a closed or partially closed valve, a comparison of results from one year to the next, or to the results obtained from the acceptance test, can be valuable. A periodic flow test can also be used for comparison purposes. A gradual decrease in available residual pressures from either test might indicate a buildup of tuberculation from calcium carbonate ($CaCO_3$) deposits. Although a slight deterioration can be expected over time, deterioration to the point where the water supply can no longer adequately supply the fire protection system must result in corrective action. Exhibit D.8 illustrates a pipe severely affected by $CaCO_3$ deposits. The effective diameter of this pipe has been substantially reduced to the point where the pipe was removed from service.

D.2.6 Forms of Corrosion. Corrosion is defined as the deterioration of a material, usually a metal, resulting from a chemical or electrochemical reaction. The eight main forms of corrosion include (1) uniform corrosion, (2) pitting, (3) galvanic corrosion, (4) crevice

EXHIBIT D.8 *Pipe Affected by $CaCO_3$ Deposits.*

corrosion, (5) selective leaching (parting), (6) erosion corrosion, (7) environmental cracking, and (8) intergranular corrosion. Microbiologically influenced corrosion (MIC) is included herein as a ninth form of corrosion, although it is usually a secondary factor that accelerates or exacerbates the rate of another form of corrosion. Definitions of the different forms of corrosion are discussed next.

(1) *Uniform (or general) corrosion:* A regular loss of a small quantity of metal over the entire area or over a large section of the total area, which is evenly distributed within a pipe(s).
(2) *Pitting:* A localized form of corrosion that results in holes or cavities in the metal. Pitting is considered to be one of the more destructive forms of corrosion and is often difficult to detect. Pits can be covered or open and normally grow in the direction of gravity — for example, at the bottom of a horizontal surface.

Exhibit D.9 shows the interior of a pipe that has been affected by pitting. Pitting often can go undetected due to the relatively small size of the cavities and the potential for corrosion to obstruct the view of the cavities. Unlike pinhole leaks, which might require non-emergency maintenance or replacement, pitting often leads to catastrophic pipe failure, since the cavities can form in a line and create a perforation in the pipe. This linear perforation can cause the pipe to rupture and could require emergency pipe replacement that could render a system ineffective or impaired.

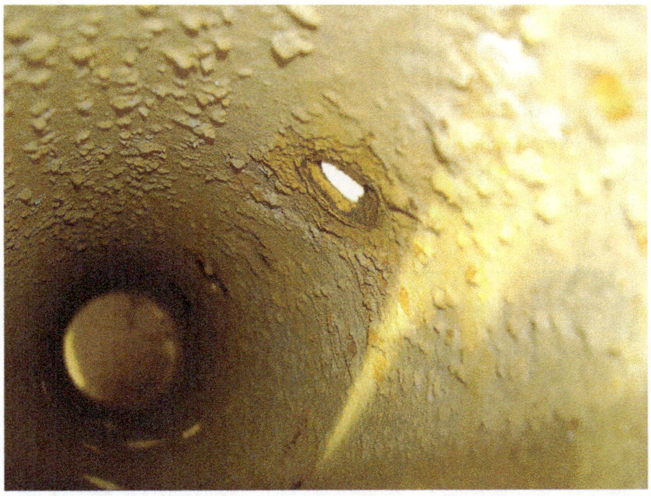

EXHIBIT D.9 *Effects of Pitting on Pipe. (Courtesy of Wiginton Fire Systems)*

(3) *Galvanic corrosion:* An electric potential exists between dissimilar metals in a conductive (corrosive) solution. The contact between the two materials allows electrons to transfer from one metal to the other. One metal acts as a cathode and the other as an anode. Corrosion usually occurs at anodic metal only.
(4) *Crevice corrosion:* A localized form of corrosion that occurs within crevices and other shielded areas on metal surfaces exposed to a stagnant corrosive solution. This form of corrosion usually occurs beneath gaskets, in holes, surface deposits, in thread and groove joints. Crevice corrosion is also referred to as gasket corrosion, deposit corrosion, and under-deposit corrosion.
(5) *Selective leaching:* The selective removal of one element from an alloy by corrosion. A common example is dezincification (selective removal of zinc) of unstabilized brass, resulting in a porous copper structure.

(6) *Erosion corrosion:* Corrosion resulting from the cumulative damage of electrochemical reactions and mechanical effects. Erosion corrosion is the acceleration or increase in the rate of corrosion created by the relative movement of a corrosive fluid and a metal surface. Erosion corrosion is observed as grooves, gullies, waves, rounded holes, or valleys in a metal surface.

(7) *Environmental cracking:* An acute form of localized corrosion caused by mechanical stresses, embrittlement, or fatigue.

(8) *Integranular corrosion:* Corrosion caused by impurities at grain boundaries, enrichment of one alloying element, or depletion of one of the elements in the grain boundary areas.

(9) *Microbiologically influenced corrosion (MIC):* Corrosion initiated or accelerated by the presence and activities of microorganisms, including bacteria and fungi. Colonies (also called bio-films and slimes) are formed in the surface of pipes among a variety of types of microbes. Microbes deposit iron, manganese, and various salts into the pipe surfaces, forming nodules, tubercles, and carbuncles. The formation of these deposits can cause obstruction to flow and dislodge, causing blockage (plugging) of system piping, valves, and sprinklers.

Microbiologically influenced corrosion (MIC) is not limited to water-based fire protection systems but can be found in many types of piping systems. MIC is thought to be responsible for 10 to 30 percent of corrosion in all piping systems in the United States.

Generally, two types of bacteria are responsible for MIC: sulfate-reducing bacteria (SRB) and acid-producing bacteria (APB). SRB are anaerobic (not requiring oxygen) and are typically found in carbon steel systems. APB are also anaerobic but can survive in aerated environments as well as in anaerobic environments. APB produce acids that stimulate the growth of SRB.

MIC can survive in a variety of environments from 0 to 100 percent oxygen-saturated water and in water with a pH from 1 to 10. MIC can be recognized by the presence of a gray or black mud-like slime (typical of anaerobic bacteria) or a brown or rusty color (typical of aerobic bacteria).

Once established, MIC creates nodules on the inner surface of pipe. If left untreated, this can result in pitting of the pipe wall and eventual pinhole leaks (although pinhole leaks can be caused by other factors). Nodules can also increase in size, causing obstruction of water flow. Exhibit D.10 through Exhibit D.12 illustrate pipe with nodules caused by MIC.

EXHIBIT D.10 *Pipe Sample Showing Effect of Nodule Formation. (Courtesy of Bioindustrial Technologies, Inc.)*

EXHIBIT D.11 *Pipe Sample Showing Effect of Severe Nodule Formation and Corrosion. (Courtesy of Bioindustrial Technologies, Inc.)*

EXHIBIT D.12 *Close-Up View of Nodules. (Courtesy of Bioindustrial Technologies, Inc.)*

If the presence of MIC is suspected, water samples should be taken and analyzed before treatment is initiated. In cases of extreme pitting, the affected pipe should be replaced. In addition, the water supplying the system should be treated with a biocide, such as chlorine, to prevent recurrence of the MIC.

Exhibit D.13 and Exhibit D.14 illustrate pipe with moderate to severe pitting, and Exhibit D.15 illustrates copper pipe with pitting. Where pitting is not extensive, flushing the entire system with a chemical cleaning agent to remove nodules is recommended. After flushing, the system should be refilled with treated water to prevent recurrence of MIC.

D.2.7 Microbiologically Influenced Corrosion (MIC). The most common biological growths in sprinkler system piping are those formed by microorganisms, including bacteria and fungi. These microbes produce colonies (also called bio-films, slimes) containing a variety of types of microbes. Colonies form on the surface of wetted pipe in both wet and

EXHIBIT D.13 *Pipe Sample with Pinhole Leak. (Courtesy of Bioindustrial Technologies, Inc.)*

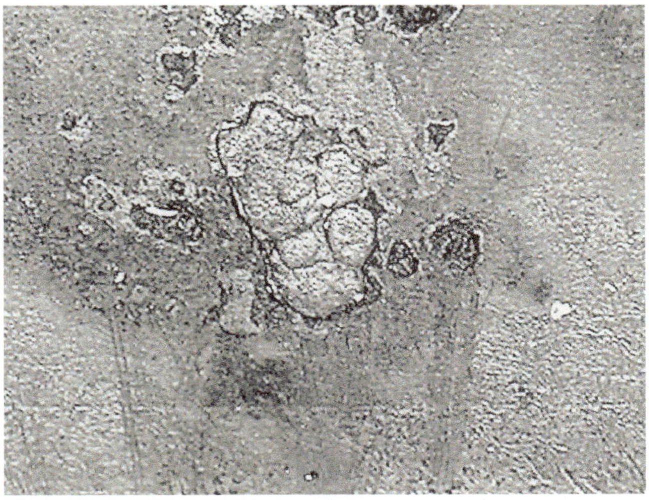

EXHIBIT D.14 *Pipe Sample with Severe Pitting. (Courtesy of Bioindustrial Technologies, Inc.)*

EXHIBIT D.15 Copper Pipe with Localized Pitting. (Courtesy of Bioindustrial Technologies, Inc.)

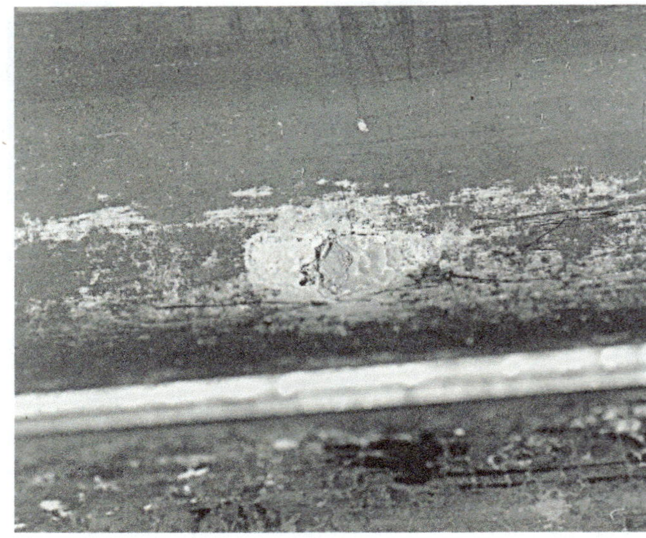

dry systems. Microbes also deposit iron, manganese, and various salts onto the pipe surface, forming discrete deposits (also termed nodules, tubercles, and carbuncles). These deposits can cause obstruction to flow and dislodge, causing plugging of fire sprinkler components. Subsequent under-deposit pitting can also result in pinhole leaks.

Microbiologically influenced corrosion (MIC) is corrosion influenced by the presence and activities of microorganisms. MIC almost always occurs with other forms of corrosion (oxygen corrosion, crevice corrosion, and under-deposit corrosion). MIC starts as microbial communities (also called bio-films, slimes) growing on the interior surface of the wetted sprinkler piping components in both wet and dry systems. The microbial communities contain many types of microbes, including slime formers, acid-producing bacteria, iron-depositing bacteria, and sulfate-reducing bacteria, and are most often introduced into the sprinkler system from the water source. The microbes deposit iron, manganese, and various salts onto the pipe surface, forming discrete deposits (also termed nodules, tubercles, and carbuncles). These deposits can cause obstruction to flow and dislodge, causing plugging of fire sprinkler components. MIC is most often seen as severe pitting corrosion occurring under deposits. Pitting is due to microbial activities such as acid production, oxygen consumption, and accumulation of salts. Oxygen and salts, especially chloride, can greatly increase the severity of MIC and other forms of corrosion.

In steel pipe, MIC is most often seen as deposits on the interior surface of the pipes. The deposits can be orange, red, brown, black, and white (or combinations thereof), depending on local conditions and water chemistry. The brown, orange, and red forms are most common in oxygenated portions of the system and often contain oxidized forms of iron and other materials on the outside, with reduced (blacker) corrosion products on the inside. Black deposits are most often in smaller diameter piping farther from the water source and contain reduced forms (those with less oxygen) of corrosion products. White deposits often contain carbonate scales.

MIC of copper and copper alloys occurs as discrete deposits of smaller size, which are green to blue in color. Blue slimes can also be produced in copper piping or copper components (e.g., brass heads).

MIC is often first noticed as a result of pinhole leaks after only months to a few years of service. Initial tests for the presence of MIC should involve on-site testing for microbes and chemical species (iron, pH, oxygen) important in MIC. This information is also very important in choosing treatment methods. These tests can be done on water samples from source waters and various locations in the sprinkler system (e.g., main drain, inspector's test valve). Confirmation of MIC can be made by examination of interior of pipes for deposits and under-deposit corrosion with pit morphology consistent with MIC (cup-like pits within pits and striations).

The occurrence and severity of MIC is enhanced by the following:

(1) Using untreated water to test and fill sprinkler piping. This is made worse by leaving the water in the system for long periods of time.
(2) Introduction of new and untreated water containing oxygen, microbes, salts, and nutrients into the system on a frequent basis (during repair, renovation, and/or frequent flow tests).
(3) Leaving dirt, debris, and especially oils, pipe joint compound, and so forth in the piping. These provide nutrients and protection for the microbes, often preventing biocides and corrosion inhibitors from reaching the microbes and corrosion sites.

Once the presence of MIC has been confirmed, the system should be assessed to determine the extent and severity of MIC. Severely affected portions should be replaced or cleaned to remove obstructions and pipe not meeting minimal mechanical specifications.

For more information about the causes of MIC, please refer to Supplement 1 of this handbook.

D.3 Investigation Procedures

If unsatisfactory conditions are observed as outlined in Section 14.3, investigations should be made to determine the extent and severity of the obstructing material. From the fire protection system plan, determine the water supply sources, age of underground mains and sprinkler systems, types of systems, and general piping arrangement. Consider the possible sources of obstruction material.

Examine the fire pump suction supply and screening arrangements. If necessary, have the suction cleaned before using the pump in tests and flushing operations. Gravity tanks should be inspected internally, with the exception of steel tanks that have been recently cleaned and painted. If possible, have the tank drained and determine whether loose scale is on the shell or if sludge or other obstructions are on the tank bottom. Cleaning and repainting could be in order, particularly if it has not been done within the previous 5 years.

Investigate yard mains first, then sprinkler systems.

Where fire protection control valves are closed during investigation procedures, the fire protection impairment precautions outlined in Chapter 15 should be followed.

Large quantities of water are needed for investigation and for flushing. It is important to plan the safest means of disposal in advance. Cover stock and machinery susceptible to water damage, and keep equipment on hand for mopping up any accidental discharge of water.

Sections D.3.1 through D.3.5 provide direction on how to conduct various investigation procedures. These procedures should be used as guidelines on how to conduct an investigation, but they might not consider all types of system arrangements. Additional system assessment might be required in addition to what is provided below to properly investigate certain systems or certain portions of systems.

D.3.1 Investigating Yard Mains. Flow water through yard hydrants, preferably near the extremes of selected mains, to determine whether mains contain obstructive material. It is preferable to connect two lengths of 2½ in. (65 mm) hose to the hydrant. Attach burlap bags to the free ends of the hose from which the nozzles have been removed to collect any material flushed out, and flow water long enough to determine the condition of the main being investigated. If there are several water supply sources, investigate each independently, avoiding any unnecessary interruptions to sprinkler protection. In extensive yard layouts, repeat the tests at several locations, if necessary, to determine general conditions.

If obstructive material is found, all mains should be flushed thoroughly before investigating the sprinkler systems. *(See Section D.5.)*

D.3.2 Investigating Sprinkler Systems. Investigate dry systems first. Tests on several carefully selected, representative systems usually are sufficient to indicate general conditions throughout the plant. If, however, preliminary investigations indicate the presence of obstructing material, this justifies investigating all systems (both wet and dry) before outlining needed flushing operations. Generally, the system can be considered reasonably free of obstructing material, provided the following conditions apply:

(1) Less than ½ cup of scale is washed from the cross mains.
(2) Scale fragments are not large enough to plug a sprinkler orifice.
(3) A full, unobstructed flow is obtained from each branch line inspected.

Where other types of foreign material are found, judgment should be used before considering the system unobstructed. Obstruction potential is based on the physical characteristics and source of the foreign material.

In selecting specific systems or branch lines for investigation, the following should be considered:

(1) Lines found obstructed during a fire or during maintenance work
(2) Systems adjacent to points of recent repair to yard mains, particularly if hydrant flow shows material in the main

Tests should include flows through 2½ in. (65 mm) fire hose directly from cross mains *[see Figure D.3.2(a) and Figure D.3.2(b)]* and flows through 1½ in. (40 mm) hose from representative branch lines. Two or three branch lines per system is a representative number of branch lines where investigating for scale accumulation. If significant scale is found, investigation of additional branch lines is warranted. Where investigating for foreign material (other than scale), the number of branch lines needed for representative sampling is dependent on the source and characteristic of the foreign material.

If provided, fire pumps should be operated for the large line flows, since maximum flow is desirable. Burlap bags should be used to collect dislodged material as is done in the investigation of yard mains. Each flow should be continued until the water clears (i.e., a minimum of 2 to 3 minutes at full flow for sprinkler mains). This is likely to be sufficient to indicate the condition of the piping interior.

D.3.3 Investigating Dry Pipe Systems. Flood dry pipe systems one or two days before obstruction investigations to soften pipe scale and deposits. After selecting the test points of a

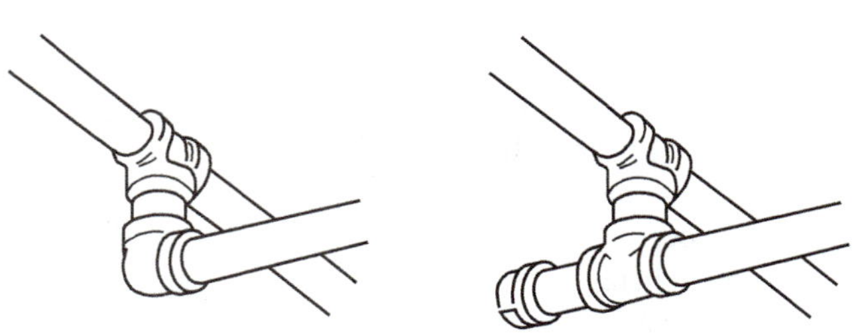

FIGURE D.3.2(a) *Replacement of Elbow at End of Cross Main with Flushing Connection Consisting of 50 mm (2 in.) Nipple and Cap.*

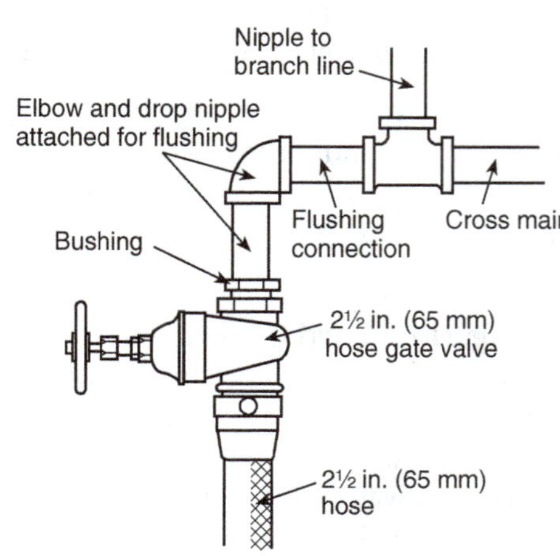

FIGURE D.3.2(b) *Connection of 65 mm (2½ in.) Hose Gate Valve with 50 mm (2 in.) Bushing and Nipple and Elbow to 50 mm (2 in.) Cross Main.*

dry pipe system, close the main control valve and drain the system. Check the piping visually with a flashlight while it is being dismantled. Attach hose valves and 1½ in. (40 mm) hose to the ends of the lines to be tested, shut the valves, have air pressure restored on the system, and reopen the control valve. Open the hose valve on the end branch line, allowing the system to trip in simulation of normal action. Any obstructions should be cleared from the branch line before proceeding with further tests.

After flowing the small end line, shut its hose valve and test the feed or cross main by discharging water through a 2½ in. (65 mm) fire hose, collecting any foreign material in a burlap bag.

After the test, the dry pipe valve should be cleaned internally and reset. Its control valve should be locked open and a drain test performed.

D.3.4 Investigating Wet Pipe Systems. Testing of wet systems is similar to that of dry systems, except that the system should be drained after closing the control valve to permit installation of hose valves for the test. Slowly reopen the control valve and make a small hose flow as specified for the branch line, followed by the 2½ in. (65 mm) hose flow for the cross main.

In any case, if lines become plugged during the tests, piping should be dismantled and cleaned, the extent of plugging noted, and a clear flow obtained from the branch line before proceeding further.

Perform similar tests on representative systems to indicate the general condition of the wet systems throughout the plant, keeping a detailed record of the procedures performed.

D.3.5 Other Obstruction Investigation Methods. Other obstruction investigation methods, such as technically proven ultrasonic and X-ray examination, have been evaluated and if applied correctly, are successful at detecting obstructions.

The sources of the obstructing material should be determined and steps taken to prevent further entrance of such material. This entails work such as inspection and cleaning of pump suction screening facilities or cleaning of private reservoirs. If recently laid public mains appear to be the source of the obstructing material, waterworks authorities should be requested to flush their system.

D.4 Obstruction Prevention Program

D.4.1 Dry Pipe and Preaction Systems — Scale.

(1) Dry pipe and preaction systems using noncoated ferrous piping should be thoroughly investigated for obstruction from corrosion after they have been in service for 15 years, for 25 years, and every 5 years thereafter.
(2) Dry pipe systems with noncoated ferrous piping should be kept on air year-round, rather than on air and water alternately, to inhibit formation of rust and scale.

D.4.2 Flushing Connections. Sprinkler systems installed in accordance with recent editions of NFPA 13, *Standard for the Installation of Sprinkler Systems*, should have provisions for flushing each cross main. Similarly, branch lines on gridded systems should be capable of being readily "broken" at a simple union or flexible joint. Property owners of systems installed without these provisions should be encouraged to provide them when replacement or repair work is being done.

D.4.3 Suction Supplies.

(1) Screen pump suction supplies and screens should be maintained. Connections from penstocks should be equipped with strainers or grids, unless the penstock inlets themselves are so equipped. Pump suction screens of copper or brass wire tend to promote less aquatic growth.

(2) Extreme care should be used to prevent material from entering the suction piping when cleaning tanks and open reservoirs. Materials removed from the interior of gravity tanks during cleaning should not be allowed to enter the discharge pipe.
(3) Small mill ponds could need periodic dredging where weeds and other aquatic growth are inherent.

D.4.4 Asian Clams. Effective screening of larvae and small-size, juvenile Asian clams from fire protection systems is very difficult. To date, no effective method of total control has been found. Such controls can be difficult to achieve in fire protection systems.

D.4.5 Calcium Carbonate. For localities suspected of having hard water, sample sprinklers should be removed and inspected yearly. Paragraph D.2.5 outlines sprinkler locations prone to the accumulation of deposits where hard water is a problem. Sprinklers found with deposits should be replaced, and adjacent sprinklers should be inspected.

D.4.6 Zebra Mussels. Several means of controlling the zebra mussel are being studied, including molluscicides, chlorines, ozone, shell strainers, manual removal, robotic cleaning, water jetting, line pigging, sonic pulses, high-voltage electrical fields, and thermal backwashing. It is believed that these controls might need to be applied only during spawning periods when water temperatures are 57°F to 61°F (14°C to 16°C) and veligers are present. Several silicon grease-based coatings also are being investigated for use within piping systems.

While it appears that the use of molluscicides could provide the most effective means of controlling the mussel, these chemicals are costly. It is believed that chlorination is the best available short-term treatment, but there are problems associated with the use of chlorine, including strict Environmental Protection Agency regulations on the release of chlorine into lakes and streams. The use of nonselective poison, such as chlorine, in the amounts necessary to kill the mussels in large bodies of water could be devastating to entire ecosystems.

To provide an effective means of control against zebra mussels in fire protection systems, control measures should be applied at the water source, instead of within the piping system. Effective controls for growth of the zebra mussel within fire protection systems include the following:

(1) Selecting a water source that is not subject to infestation. This could include well water or potable or pretreated water.
(2) Implementing a water treatment program that includes biocides or elevated pH, or both.
(3) Implementing a water treatment program to remove oxygen, to ensure control of biological growth within piping.
(4) Relying on a tight system approach to deny oxygen and nutrients that are necessary to support growth.

D.5 Flushing Procedures

D.5.1 Yard Mains. Yard mains should be flushed thoroughly before flushing any interior piping. Flush yard piping through hydrants at dead ends of the system or through blow-off valves, allowing the water to run until clear. If the water is supplied from more than one direction or from a looped system, close divisional valves to produce a high-velocity flow through each single line. A velocity of at least 10 ft/sec (3 m/sec) is necessary for scouring the pipe and for lifting foreign material to an aboveground flushing outlet. Use the flow specified in Table D.5.1 or the maximum flow available for the size of the yard main being flushed.

Connections from the yard piping to the sprinkler riser should be flushed. These are usually 6 in. (150 mm) mains. Although flow through a short, open-ended 2 in. (50 mm) drain can create sufficient velocity in a 6 in. (150 mm) main to move small obstructing material, the restricted waterway of the globe valve usually found on a sprinkler drain might not allow stones and other large objects to pass. If the presence of large size material is suspected, a larger outlet is needed to pass such material and to create the flow necessary to move it. Fire department connections on sprinkler risers can be used as flushing outlets by removing the clappers. Yard mains also can be flushed through a temporary Siamese fitting attached to the riser connection before the sprinkler system is installed. *[See Figure D.5.1.]*

TABLE D.5.1 *Flushing Rates to Accomplish Flow of 10 ft/sec (3 m/sec)*

	Steel			*Copper*				*Polybutylene*	
Pipe Size	*SCH 10 (gpm)*	*SCH 40 (gpm)*	*XL (gpm)*	*K (gpm)*	*L (gpm)*	*M (gpm)*	*CPVC (gpm)*	*CTS (gpm)*	*IPS (gpm)*
¾	—	—	—	14	15	16	19	12	17
1	29	24	30	24	26	27	30	20	27
1¼	51	47	52	38	39	41	48	30	43
1½	69	63	70	54	55	57	63	42	57
2	114	105	114	94	96	99	98	72	90
2½	170	149	163	145	149	152	144	—	—
3	260	230	251	207	212	217	213	—	—
4	449	396	—	364	373	379	—	—	—
5	686	623	—	565	582	589	—	—	—
6	989	880	—	807	836	846	—	—	—
8	1665	1560	—	1407	1460	1483	—	—	—
10	2632	2440	—	2185	2267	2303	—	—	—
12	—	3520	—	—	—	—	—	—	—

For SI units, 1 gpm = 3.785 L/min.

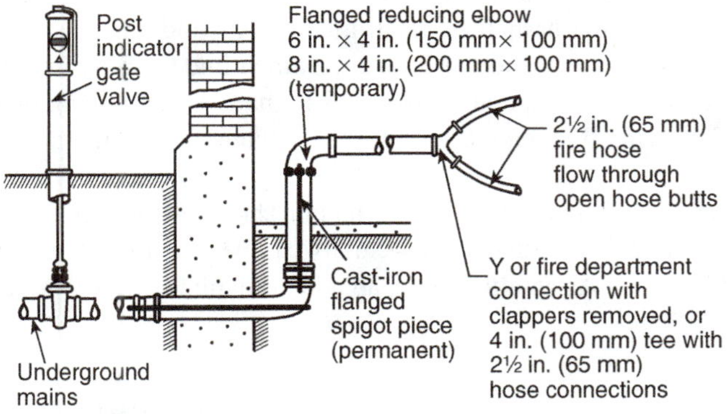

FIGURE D.5.1 *Arrangement for Flushing Branches from Underground Mains to Sprinkler Risers.*

D.5.2 Sprinkler Piping. Two methods commonly are used for flushing sprinkler piping:

(1) The hydraulic method
(2) The hydropneumatic method

The hydraulic method consists of flowing water from the yard mains, sprinkler risers, feed mains, cross mains, and branch lines, respectively, in the same direction in which water would flow during a fire.

The hydropneumatic method uses special equipment and compressed air to blow a charge of about 30 gal (114 dm^3) of water from the ends of branch lines back into feed mains and down the riser, washing the foreign material out of an opening at the base of the riser.

The choice of method depends on conditions at the individual plant and the type of material installed. If examination indicates the presence of loose sand, mud, or moderate amounts of pipe scale, the piping generally can be flushed satisfactorily by the hydraulic method. Where the material is more difficult to remove and available water pressures are too low for effective scouring action, the hydropneumatic method generally is more satisfactory. The hydropneumatic method should not be used with listed CPVC sprinkler piping.

In some cases, where obstructive material is solidly packed or adheres tightly to the walls of the piping, the pipe needs to be dismantled and cleaned by rodding or other means.

Dry pipe systems should be flooded one or two days before flushing to soften pipe scale and deposits.

Successful flushing by either the hydraulic or hydropneumatic method is dependent on establishing sufficient velocity of flow in the pipes to remove silt, scale, and other obstructive material. With the hydraulic method, water should be moved through the pipe at least at the rate of flow indicated in Table D.5.1.

Where flushing a branch line through the end pipe, sufficient water should be discharged to scour the largest pipe in the branch line. Lower rates of flow can reduce the efficiency of the flushing operation. To establish the recommended flow, remove the small end piping and connect the hose to a larger section, if necessary.

Where pipe conditions indicate internal or external corrosion, a section of the pipe affected should be cleaned thoroughly to determine whether the walls of the pipe have seriously weakened. Hydrostatic testing should be performed as outlined in NFPA 13, *Standard for the Installation of Sprinkler Systems*.

Pendent sprinklers should be removed and inspected until it is reasonably certain that all are free of obstruction material.

Painting the ends of branch lines and cross mains is a convenient method for keeping a record of those pipes that have been flushed.

D.5.3 Hydraulic Method. After the yard mains have been thoroughly cleaned, flush risers, feed mains, cross mains, and finally the branch lines. In multistory buildings, systems should be flushed by starting at the lowest story and working up. Branch line flushing in any story can immediately follow the flushing of feed and cross mains in that story, allowing one story to be completed at a time. Following this sequence prevents drawing obstructing material into the interior piping.

To flush risers, feed mains, and cross mains, attach 2½ in. (65 mm) hose gate valves to the extreme ends of these lines *[see Figure D.5.3]*. Such valves usually can be procured from the manifold of fire pumps or hose standpipes. As an alternative, an adapter with 2½ in. (65 mm)

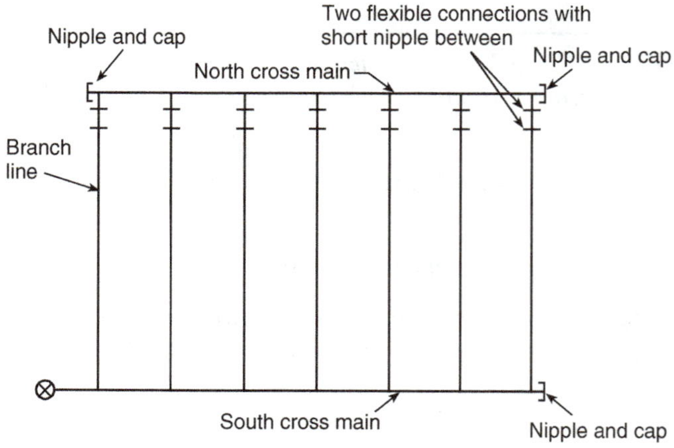

FIGURE D.5.3 *Gridded Sprinkler System Piping.*

hose thread and standard pipe thread can be used with a regular gate valve. A length of fire hose without a nozzle should be attached to the flushing connection. To prevent kinking of the hose and to obtain maximum flow, an elbow usually should be installed between the end of the sprinkler pipe and the hose gate valve. Attach the valve and hose so that no excessive strain will be placed on the threaded pipe and fittings. Support hose lines properly.

Where feed and cross mains and risers contain pipe 4 in., 5 in., and 6 in. (100 mm, 125 mm, and 150 mm) in diameter, it could be necessary to use a Siamese with two hose connections to obtain sufficient flow to scour this larger pipe.

Flush branch lines after feed and cross mains have been thoroughly cleared. Equip the ends of several branch lines with gate valves, and flush individual lines of the group consecutively. This eliminates the need for shutting off and draining the sprinkler system to replace a single hose line. The hose should be 1½ in. (40 mm) in diameter and as short as practicable. Branch lines can be permitted to be flushed in any order that expedites the work.

Branch lines also can be permitted to be flushed through pipe 1½ in. (40 mm) in diameter or larger while extended through a convenient window. If pipe is used, 45 degree fittings should be provided at the ends of branch lines. Where flushing branch lines, hammering the pipes is an effective method of moving obstructions.

Figure D.5.3 shows a typical gridded piping arrangement prior to flushing. The flushing procedure is as follows:

(1) Disconnect all branch lines and cap all open ends.
(2) Remove the cap from the east end of the south cross main, flush the main, and replace the cap.
(3) Remove the cap from branch line 1, flush the line, and replace the cap.
(4) Repeat step (3) for the remaining branch lines.
(5) Reconnect enough branch lines at the west end of the system so that the aggregate cross-sectional area of the branch lines approximately equals the area of the north cross main. For example, three 1¼ in. (32 mm) branch lines approximately equal a 2½ in. (65 mm) cross main. Remove the cap from the east end of the north cross main, flush the main, and replace the cap.
(6) Disconnect and recap the branch lines. Repeat step (5), but reconnect branch lines at the east end of the system and flush the north cross main through its west end.
(7) Reconnect all branch lines and recap the cross main. Verify that the sprinkler control valve is left in the open and locked position.

D.5.4 Hydropneumatic Method. The apparatus used for hydropneumatic flushing consists of a hydropneumatic machine, a source of water, a source of compressed air, 1 in. (25 mm) rubber hose for connecting to branch lines, and 2½ in. (65 mm) hose for connecting to cross mains.

The hydropneumatic machine *[see Figure D.5.4(a)]* consists of a 30 gal (114 dm^3) (4 ft^3) water tank mounted over a 185 gal (700 dm^3) (4 ft^3) compressed air tank. The compressed air tank is connected to the top of the water tank through a 2 in. (50 mm) lubricated plug cock. The bottom of the water tank is connected through hose to a suitable water supply. The compressed air tank is connected through suitable air hose to either the plant air system or a separate air compressor.

To flush the sprinkler piping, the water tank is filled with water, the pressure is raised to 100 psi (6.9 bar) in the compressed air tank, and the plug cock between tanks is opened to put air pressure on the water. The water tank is connected by hose to the sprinkler pipe to be flushed. The lubricated plug cock on the discharge outlet at the bottom of the water tank then is snapped open, allowing the water to be "blown" through the hose and sprinkler pipe by the compressed air. The water tank and air tank should be recharged after each blow.

Outlets for discharging water and obstructing material from the sprinkler system should be arranged. With the clappers of dry pipe valves and alarm check valves on their seats and cover plates removed, sheet metal fittings can be used for connection to 2½ in. (65 mm)

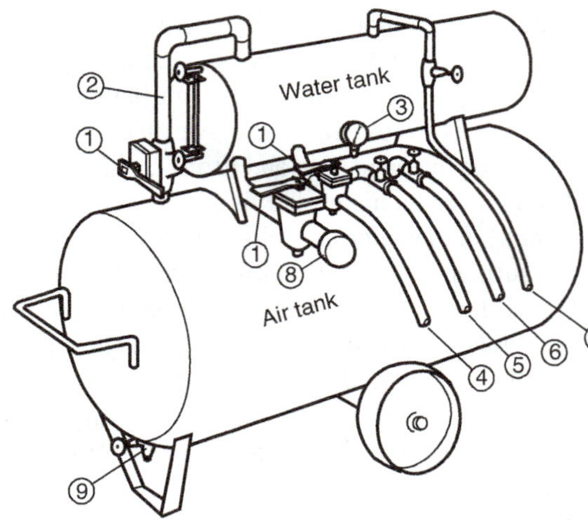

1 Lubricated plug cocks
2 Pipe connection between air and water tanks (This connection is open when flushing sprinkler system.)
3 Air pressure gauge
4 1 in. (25 mm) rubber hose (air type) (Used to flush sprinkler branch lines.)
5 Hose connected to source of water (Used to fill water tank.)
6 Hose connected to ample source of compressed air (Used to supply air tank.)
7 Water tank overflow hose
8 2½ in. (65 mm) pipe connection [Where flushing large interior piping, connect woven jacket fire hose here and close 1 in. (25 mm) plug cock hose connection (4) used for flushing sprinkler branch lines.]
9 Air tank drain valve

FIGURE D.5.4(a) *Hydropneumatic Machine.*

hose lines or for discharge into a drum [maximum capacity per blow is approximately 30 gal (114 dm^3)]. If the 2 in. (50 mm) riser drain is to be used, the drain valve should be removed and a direct hose connection made. For wet pipe systems with no alarm check valves, the riser should be taken apart just below the drain opening and a plate inserted to prevent foreign material from dropping to the base of the riser. Where dismantling of a section of the riser for this purpose is impractical, the hydropneumatic method should not be used.

Before starting a flushing job, each sprinkler system to be cleaned should be studied and a schematic plan prepared showing the order of the blows.

To determine that the piping is clear after it has been flushed, representative branch lines and cross mains should be investigated, using both visual examination and sample flushings.

(1) *Branch Lines.* With the yard mains already flushed or known to be clear, the sprinkler branch lines should be flushed next. The order of cleaning individual branch lines should be laid out carefully if an effective job is to be done. In general, the branch lines should be flushed, starting with the branch closest to the riser and working toward the dead end of the cross main. *[See Figure D.5.4(b).]* The order for flushing the branch lines is shown by the circled numerals. In this example, the southeast quadrant is flushed first, then the southwest, followed by the northeast, and, finally, the northwest. Air hose 1 in. (25 mm) in diameter is used to connect the machine with the end of the branch line being flushed. This hose air pressure should be allowed to drop to 85 psi (5.9 bar) before the valve is

closed. The resulting short slug of water experiences less friction loss and a higher velocity and, therefore, cleans more effectively than if the full 30 gal (114 dm^3) of water were to be used. One blow is made for each branch line.

(2) *Large Piping.* Where flushing cross mains, fill the water tank completely and raise the pressure in the air receiver to 100 psi (6.9 bar) (690 kPa). Connect the machine to the end of the cross main to be flushed with no more than 50 ft (15.2 m) of 2½ in. (65 mm) hose. After opening the valve, allow air pressure in the machine to drop to zero (0). Two to six blows are necessary at each location, depending on the size and length of the main. In Figure D.5.4(b), the numerals in squares indicate the location and order of the cross main blows. Because the last branch line blows performed were located west of the riser, clean the cross main located east of the riser first. Where large cross mains are to be cleaned, it is best, if practical, to make one blow at 38, one at 39, the next again at 38, then again at 39, alternating in this manner until the required number of blows has been made at each location.

(3) Where flushing cross mains and feed mains, arrange the work so that the water passes through a minimum of right-angle bends. In Figure D.5.4(b), blows at 38 should be adequate to flush the cross mains back to the riser. Do not attempt to clean the cross main from location A to the riser by backing out branch line 16 and connecting the hose to the open side of the tee. If this were to be done, a considerable portion of the blow would pass northward up the 3 in. (76 mm) line supplying branches 34 to 37, and the portion passing eastward to the riser could be ineffective. Where the size, length, and condition of cross mains necessitate blowing from a location corresponding to location A, the connection should be made directly to the cross main corresponding to the 3½ in. (90 mm) pipe so that the entire flow travels to the riser. Where flushing through a tee, always flush

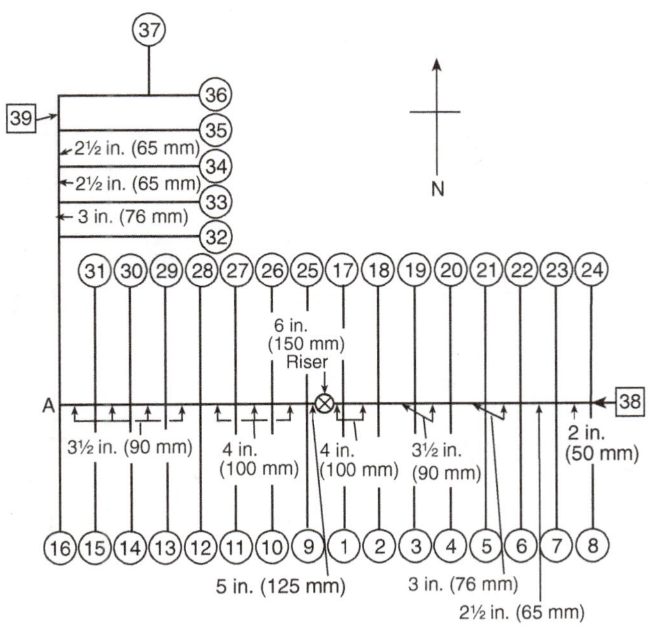

FIGURE D.5.4(b) *Schematic Diagram of Sprinkler System Showing Sequence to Be Followed Where Hydropneumatic Method Is to Be Utilized.*

the run of the tee after flushing the branch. Note the location of blows 35, 36, and 37 in Figure D.5.4(b). Gridded systems can be flushed in a similar fashion. With branch lines disconnected and capped, begin flushing the branch line closest to the riser (branch line 1 in Figure D.5.3), working toward the most remote line. Then flush the south cross main in Figure D.5.3 by connecting the hose to the east end. Flushing the north cross main involves connecting the hose to one end while discharging to a safe location from the other end.

REFERENCES CITED IN COMMENTARY

National Fire Protection Association, 1 Batterymarch Park, Quincy, MA 02169-7471.

Lake, J., ed., *Automatic Sprinkler Systems Handbook*, 2010 edition.

NFPA 13, *Standard for the Installation of Sprinkler Systems*, 2013 edition.

ANNEX E

HAZARD EVALUATION FORM

This annex is not a part of the requirements of this NFPA document but is included for informational purposes only.

E.1

This annex provides one example of a hazard evaluation form. A hazard evaluation is not part of a system inspection. *(See Figure E.1.)*

The property owner or the designated representative must not make changes in the occupancy, the use or process, or the materials used or stored in the building without evaluation of the fire protection systems for their capability to protect the new occupancy, use, or materials. While this evaluation is required by 4.1.6, it is not considered a part of the normal inspection, testing, and maintenance (ITM) required by the standard.

Prior editions of NFPA 25 have included this requirement in 4.1.6 and 4.1.7 for the need to address changes in hazard, but there was no specific trigger for the need to conduct a hazard evaluation. As a result, such evaluations were rare even though they are extremely important, and there was a common misunderstanding that these evaluations took place during the course of a normal system inspection. In response, a new hazard evaluation form was added for the 2011 edition of the standard as a means of assisting in the differentiation between a system inspection and a system hazard evaluation. When such an evaluation is considered appropriate, as part of the evaluation, the owner or authority having jurisdiction (AHJ) can ask an individual qualified to address system design issues to use this form.

This form provides a focus and is intended to assist the user in considering factors that that are not directly related to the system but may impact system performance, such as occupancy changes; process or material changes such as metal stamping to molded plastics; building revisions such as relocated walls, added mezzanines, and ceilings added below sprinklers; and removal of heating systems in spaces with piping subject to freezing.

This form is intended to be used as a tool for meeting the requirements of 4.1.6 and 4.1.7.

FIRE SPRINKLER SYSTEM HAZARD EVALUATION

Changes in building occupancy, use, or process, or material used or stored, create the need for evaluation of the installed fire protection systems. This form is intended to identify and evaluate such changes and should be completed only by an individual properly qualified in the area of system design.

Owner: ______________________ Owner's address: ______________________

Property being evaluated: ______________________

Property address: ______________________

Date of work: ______________________

(All responses refer to the current hazard evaluation performed on this date.)

Section 1. Identification of Sprinklered Occupancy and Storage Hazards

(Use additional pages as needed.)

Area of Property (List nonsprinklered areas separately in Section 3.)	Type of System and Sprinklers	Design Capability of System	Hazard Protected (Uses or storage arrangements, including commodity)	Improvements Needed to Address Hazard
1.				
2.				
3.				
4.				
5.				

 NFPA 25 (p. 1 of 4)

FIGURE E.1 *Fire Sprinkler Hazard Evaluation.*

FIRE SPRINKLER SYSTEM HAZARD EVALUATION *(continued)*

Section 2. Evaluation of Protection

For each area of the property evaluated in Section 1, please answer the following questions with a "yes," "no," "N/A," or "?" and explain all "no" and "?" responses below by row and column identification:

Answer the following for each identified property area:	1	2	3	4	5
a. Are all sprinklers the correct type for their application?					
b. Are the obstructions to sprinklers in all areas within acceptable limits for the specific types of sprinklers used?					
c. Are hazards associated with all occupancy areas consistent with hazards typical for that occupancy hazard classification?					
d. Are stockpiles of combustibles located within occupancy areas limited to appropriate heights?					
e. Are miscellaneous and dedicated storage areas properly identified and managed?					
f. Are all dedicated storage areas protected in accordance with the proper storage configuration and commodity classification?					
g. Is the storage or use of flammable liquids, combustible liquids, or aerosol products in any area properly addressed?					
h. Is all idle pallet storage properly protected?					
i. Is there any presence of nitrate film, pyroxylin plastic, compressed or liquefied gas cylinders, liquid or solid oxidizers, or organic peroxide formulations except where specifically addressed by appropriate protection measures?					
j. Are all sprinklers spaced appropriately for the hazard and the type of sprinkler?					
k. Do the available sources of heat and cooling appear adequate for the type of system and temperature rating of sprinklers?					

Explanation of "no" and "?" answers:

Examples:
b2 – no – Obstructions to ESFR sprinklers exceed currently accepted standards
e3 – ? – Owner must provide information on type of plastic involved in product before evaluation can be finalized.

 NFPA 25 (p. 2 of 4)

FIGURE E.1 *Continued*

FIRE SPRINKLER SYSTEM HAZARD EVALUATION *(continued)*

Section 3. Evaluation of Unsprinklered Areas

Area of Property for Which Protection Is Not Provided	Basis of Lack of Protection (if known)	Basis for Omission Under Current Codes/Standards
1.		
2.		
3.		
4.		
5.		

Section 4. Water Supply Evaluation

If this hazard evaluation is the result of a reduction in the residual pressure during routine inspections, explain the results of the investigation made to determine the reasons for this change:

Explain the basis of continued acceptability of the water supply or proposed improvements:

 NFPA 25 (p. 3 of 4)

FIGURE E.1 Continued

FIRE SPRINKLER SYSTEM HAZARD EVALUATION *(continued)*

Section 5. Hazard Evaluator's Information and Certification

Evaluator: ____________________

Company: ____________________

Company address: ____________________

I state that the information on this form is correct at the time and place of my review of my evaluation.

Is this hazard evaluation completed? (Note: All "?" must be resolved.) ❑ Yes ❑ No

Explain if answer is not "yes":

____

____

Have deficiencies in protection been identified that should be improved or corrected? ❑ Yes ❑ No

Summarize improvements of corrections needed:

____

____

____

____

Signature of Evaluator: ____________________ Date: ____________________

License or Certification Number (if applicable): ____________________

 NFPA 25 (p. 4 of 4)

FIGURE E.1 *Continued*

ANNEX F

INFORMATIONAL REFERENCES

F.1 Referenced Publications

The documents or portions thereof listed in this annex are referenced within the informational sections of this standard and are not part of the requirements of this document unless also listed in Chapter 2 for other reasons.

F.1.1 NFPA Publications. National Fire Protection Association, 1 Batterymarch Park, Quincy, MA 02169-7471.

NFPA 13, *Standard for the Installation of Sprinkler Systems*, 2013 edition.

NFPA 13R, *Standard for the Installation of Sprinkler Systems in Low-Rise Residential Occupancies*, 2013 edition.

NFPA 14, *Standard for the Installation of Standpipe and Hose Systems*, 2013 edition.

NFPA 15, *Standard for Water Spray Fixed Systems for Fire Protection*, 2012 edition.

NFPA 16, *Standard for the Installation of Foam-Water Sprinkler and Foam-Water Spray Systems*, 2011 edition.

NFPA 20, *Standard for the Installation of Stationary Pumps for Fire Protection*, 2013 edition.

NFPA 22, *Standard for Water Tanks for Private Fire Protection*, 2013 edition.

NFPA 24, *Standard for the Installation of Private Fire Service Mains and Their Appurtenances*, 2013 edition.

NFPA 70E®, Standard for Electrical Safety in the Workplace®, 2012 edition.

NFPA 72®, National Fire Alarm and Signaling Code, 2013 edition.

NFPA 291, *Recommended Practice for Fire Flow Testing and Marking of Hydrants*, 2013 edition.

NFPA 551, *Guide for the Evaluation of Fire Risk Assessments*, 2013 edition.

NFPA 750, *Standard on Water Mist Fire Protection Systems*, 2010 edition.

NFPA 780, *Standard for the Installation of Lightning Protection Systems*, 2014 edition.

NFPA 1962, *Standard for the Care, Use, Inspection, Service Testing, and Replacement of Fire Hose, Couplings, Nozzles, and Fire Hose Appliances*, 2013 edition.

NFPA's Future in Performance Based Codes and Standards, July 1995.

NFPA Performance Based Codes and Standards Primer, December 1999.

F.1.2 Other Publications.

F.1.2.1 ASTM Publications. ASTM International, 100 Barr Harbor Drive, P.O. Box C700, West Conshohocken, PA 19428-2959.

IEEE/ASTM-SI-10, *American National Standard for Use of the International System of Units (SI): The Modern Metric System*, 2002.

F.1.2.2 AWWA Publications. American Water Works Association, 6666 West Quincy Avenue, Denver, CO 80235.

AWWA, *Manual of Water Supply Practices — M42 Steel Water-Storage Tanks*, 1998.

F.1.2.3 Hydraulic Institute Publications. Hydraulic Institute, 9 Sylvan Way, Parsippany, NJ 07054.

Hydraulic Institute Standards for Centrifugal, Rotary and Reciprocating Pumps, 14th edition, 1983.

F.1.2.4 SFPE Publications. Society of Fire Protection Engineers, 7315 Wisconsin Avenue, Suite 1225 W, Bethesda, MD, 20814.

SFPE *Engineering Guide: Fire Risk Assessment,* 2006.

F.1.2.5 Other Publications.

Edward K. Budnick, P.E., "Automatic Sprinkler System Reliability," *Fire Protection Engineering*, Society of Fire Protection Engineers, Winter 2001.

Fire Protection Equipment Surveillance Optimization and Maintenance Guide, Electric Power Research Institute, July 2003.

William E. Koffel, P.E., *Reliability of Automatic Sprinkler Systems*, Alliance for Fire Safety.

F.2 Informational References

The following documents or portions thereof are listed here as informational resources only. They are not a part of the requirements of this document.

F.2.1 NFPA Publications. National Fire Protection Association, 1 Batterymarch Park, Quincy, MA 02169-7471.

NFPA 1, *Fire Code*, 2012 edition.

F.2.2 Other Publications.

F.2.2.1 AWWA Publications. American Water Works Association, 6666 West Quincy Avenue, Denver, CO 80235.

AWWA D101, *Inspecting and Repairing Steel Water Tanks, Standpipes, Reservoirs, and Elevated Tanks, for Water Storage,* 1986.

F.2.2.2 SSPC Publications. Society of Protective Coatings, 40 24th Street, 6th Floor, Pittsburgh, PA 15222.

SSPC Chapter 3, "Special Pre-Paint Treatments," 1993.

SSPC-PA 1, *Shop, Field, and Maintenance Painting,* 1991.

SSPC Paint 8, *Aluminum Vinyl Paint,* 1991.

SSPC Paint 9, *White (or Colored) Vinyl Paint,* 1995.

SSPC-SP 6, *Commercial Blast Cleaning,* 1994.

SSPC-SP 8, *Pickling,* 1991.

SSPC-SP 10, *Near-White Blast Cleaning,* 1994.

F.2.2.3 U.S. Government Publications. U.S. Government Printing Office, Washington, DC 20402.

Bureau of Reclamation Specification VR-3.

Federal Specification TT- P-86, *Specifications for Vinyl Resin Paint,* M-54, 1995.

F.3 References for Extracts in Informational Sections

NFPA 13, *Standard for the Installation of Sprinkler Systems,* 2013 edition.

NFPA 14, *Standard for the Installation of Standpipe and Hose Systems*, 2013 edition.

NFPA 15, *Standard for Water Spray Fixed Systems for Fire Protection*, 2012 edition.

NFPA 20, *Standard for the Installation of Stationary Pumps for Fire Protection*, 2013 edition.

NFPA 24, *Standard for the Installation of Private Fire Service Mains and Their Appurtenances*, 2013 edition.

PART 2

Supplements

In addition to the 2014 edition of NFPA 25 and the commentary presented in Part 1, the *Water-Based Fire Protection Systems Handbook* includes eight supplements. The supplements included in Part 2 of this handbook are not part of NFPA 25; they are included as additional information for handbook users. The supplements in Part 2 are as follows:

1. Microbiologically Influenced Corrosion in Fire Sprinkler Systems
2. Conducting Fire Pump Flow Tests
3. ITM Roles and Responsibilities
4. Role of the Inspector
5. Role of the Owner or Designated Representative
6. Role of the Authority Having Jurisdiction
7. Fire Pump Field Data Collection and Analysis
8. Technical/Substantive Changes from the 2011 Edition to the 2014 Edition of NFPA 25

SUPPLEMENT

1

MICROBIOLOGICALLY INFLUENCED CORROSION IN FIRE SPRINKLER SYSTEMS

By Anthony M. Aguilera and Bruce H. Clarke

Editor's Note: *Supplement 1 has been included to provide the user with background information on microbiologically influenced corrosion (MIC) and its impact on fire protection systems. While MIC has been in existence for some time, only recently have owners and operators of fire protection systems recognized the source of corrosion and begun to address MIC as a maintenance issue. This supplement discusses the microbial corrosion process and potential treatment of MIC in fire protection systems.*

Beginning in the early 1990s, concerns began to increase about microbiologically influenced corrosion (MIC) affecting fire sprinkler systems as documented cases began to arise involving the abnormally rapid development of pinhole-sized leaks and highly obstructive interior biological growth in water-based wet and dry fire protection systems. While most of these failures occurred in systems with 5 to 20 years of service, cases have been documented where leaks related to MIC occurred in systems after as little as 18 months of service.

While there is not a standard industry answer to exactly how long a fire sprinkler system should be expected to last due to the various installation and environmental factors that can affect longevity, history tells us that building owners should expect such an expensive investment to last for many years, especially when made with steel pipe manufactured to ASTM standards as required by NFPA 13, *Standard for the Installation of Sprinkler Systems* [1]. Yet some cases that have been documented show system failure well before the end of these systems' life expectancies.

Research into the cause of these leaks, which have greatly increased in the last two decades, has led to a growing awareness of the problem. In the past 20 years, MIC has grown from a relatively unknown topic of regional discussions to one now generating widespread concern throughout several countries. MIC in fire protection systems also has become the subject of a wide array of speculation, debate, and, in some cases, gross inaccuracies. But despite this progress, no standard investigation and treatment regimen exists that can be universally implemented with every pinhole leak and every obstructed pipe that is discovered.

Today there are many options for both detection and treatment of MIC, but there are few that have been correlated from detection through time-proven successful treatment and acknowledged as a universally accepted "best practice" in the fire protection industry. There are several detection and treatment systems being marketed that appear to be effective, but long-term, large-scale data to support overall effectiveness claims are still limited. While the control of MIC and biological growth in other mechanical systems such as cooling towers and gas pipelines has been extensively researched by allied engineering fields for decades, detection and treatment in fire sprinkler systems is still relatively new. This supplement will provide an overview of the issues related to MIC in the fire protection industry.

For related topics, see the following chapters in Section 16 of the *Fire Protection Handbook* [2]: Chapter 1, "Principles of Automatic Sprinkler System Performance;" Chapter 2, "Automatic Sprinklers;" Chapter 3, "Automatic Sprinkler Systems;" and Chapter 11, "Care and Maintenance of Water-Based Extinguishing Systems."

Defining Corrosion

General Corrosion

Generating a universal "working definition" for MIC in fire protection is complex. In order to understand the relationship of MIC and sprinkler components, general corrosion and its various causes must first be understood. One edition of *Webster's* defines *corrosion* as "the wearing away of materials by chemical action(s)." Another edition simply defines it as "the wearing away of material gradually." The online encyclopedia *Encarta* describes corrosion as "specifically being related to the gradual action of natural agents such as air or salt water on metals." All these definitions are true generally but are also confusing to those in the fire protection industry who are looking for answers. Even the more detailed definitions presented by the corrosion engineering communities can at times appear contradictory and confusing.

Corrosion can occur from many biophysical reactions and be described from a multitude of scientific viewpoints. The term *general corrosion* typically refers to uniform corrosion that occurs on most unprotected metallic systems. This can be associated with the uniform "rust" layer seen on many steel structures. Besides rust from oxidation, several other types of general corrosion include stray current, uniform biological, galvanic, molten salt, dealloying, chemical, high-temperature, and general carburization. Nonbiological corrosion processes having adverse effects on fire protection sprinkler systems must always be considered with MIC. Some specific examples include oxygen, acid, and oxygen-acid corrosion. [3]

Microbiologically Influenced Corrosion

In contrast to general, or uniform, corrosion, MIC is a form of localized corrosion. Material is lost at discrete points instead of universally across an entire surface. There are several types of localized corrosion, including pitting, crevices, cratering, and filiform. (See Exhibit S1.1.)

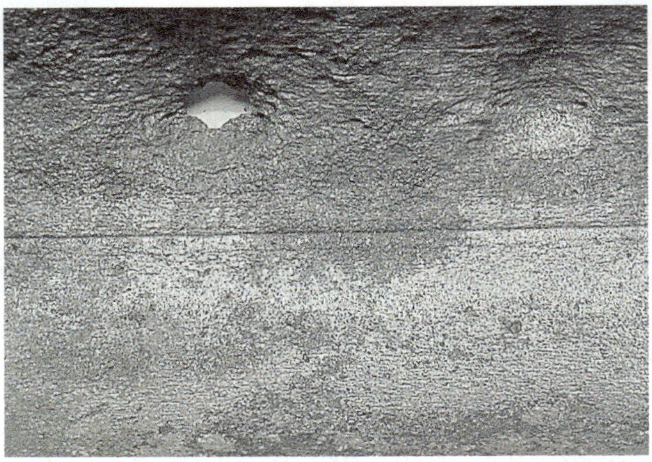

EXHIBIT S1.1 Interior Pinhole Leak.

As with the various other types of corrosion, MIC can take many forms and affect different systems in unique ways. For the fire protection profession, however, an industry-specific definition can be developed. This definition captures both the causes and effects of this problem with several continuous generalities to avoid some of the current inaccuracies. Thus, microbiologically influenced corrosion in fire protection systems can be described as "an electrochemical corrosion process that is concentrated and accelerated by the activity of specific bacteria within a fire sprinkler system resulting in the premature failure of metallic system components." [4]

The Corrosion Problem

Concentrated and Accelerated Corrosion

All metallic systems normally begin to corrode to various degrees from the instant that moisture contacts metal. Typically, this appears as general corrosion or "rust," which in itself is not a major concern due to a self-induced slow rate of decay. By definition, and in comparison, the MIC process is both *concentrated* and *accelerated* from that of typical "rust" corrosion. With general (or typical) corrosion, when moisture is introduced into a system, a thin layer of oxidation occurs relatively evenly throughout the pipe wall surface. This type of corrosion is typically not a significant concern in fire sprinkler systems and does not require specialized treatment as it does not drastically change a pipe's interior surface roughness (i.e., C-factor) unevenly or at a radically increased rate. While the rate of decay is typically slow with normal corrosion, with MIC, the corrosion rate is abnormally accelerated and no longer evenly dispersed across the pipe wall surface. The result changes a relatively smooth pipe to one with pits and valleys.

The Hazen-Williams formula, while empirical, is universally accepted for calculating friction loss in sprinkler system pipe. As the formula indicates, pressure drop in pipe is relative to the amount of water flow, physical properties of the pipe, and the pressure drop caused by friction:

For U.S./Imperial Units:

$$P_d = \frac{4.52Q^{1.85}}{C^{1.85}d^{4.87}}$$

where:

P_d = pressure drop (psi/ft)

Q = flow (gpm)

C = frictional loss coefficient (roughness based on material)

d = interior pipe diameter (in.)

For International/SI Units:

$$S = \frac{1067Q^{1.85}}{C^{1.85}d^{4.87}}$$

where:

S = Head loss (m of water per m)

Q = flow (m^3/s)

C = frictional loss coefficient (roughness based on material)

d = interior pipe diameter (m)

While there has not been a detailed study on exactly what amount of pipe change (or changes in pipe wall roughness) is unacceptable, the Hazen-Williams formula makes the following points clear:

1. Having an interior pipe surface free from irregularities, roughness, or projections is critical in sprinkler system effectiveness. The C-factor numerically relates to pipe wall smoothness; increases in interior pipe roughness subsequently decrease the value of C and thereby exponentially increase the pressure loss in each foot of sprinkler pipe. Even a small amount of internal corrosion, especially in main feed areas, could potentially make a system ineffective in fire control due to decreased pressure delivery to operating sprinklers.
2. Pipe diameter is also a significant contributor to friction loss. A pipe's C-factor affects friction loss, and so the smaller the interior pipe size, the more significant is the exponential decrease in water flow. In a small pipe, a decrease in interior pipe size even more dramatically affects pipe friction loss and ultimately overall system performance.

The Hazen-Williams formula is not intended to be used for non-uniform pipe surfaces where C changes within pipe lengths. Random pockets of biological growth have the potential to affect performance. This is especially true if a sprinkler system has a minimum design buffer, which is the case with many installations today. (See Exhibit S1.2 and Exhibit S1.3.)

MIC is the result of specific bacteria (see the section titled "Defining MIC"). A multitude of bacteria are always present in all ecosystems, including the interior of sprinkler systems. Just as only a small number of bacteria on earth have the potential to cause human sickness, only a relatively small number of bacteria have the potential to cause the rapid system destruction currently linked to MIC. Thus, as defined, only a few specific bacteria concentrate and accelerate the general or uniform corrosion process.

EXHIBIT S1.2 *Exterior View of Pipe with No Signs of Internal Corrosion.*

EXHIBIT S1.3 *Interior View of Pipe Shown in Exhibit S1.2 with Approximate 65 Percent Obstruction to 6 In. Tee Connection.*

Premature Failure

The ultimate effect of MIC in fire protection systems is the premature failure of metallic components.

Premature Failure Defined. It is clear that MIC and other related forms of corrosion cause sprinkler systems to fail prematurely. What constitutes "premature" with regard to the integrity of specific system components has not officially been defined in the industry. Long-term warranties addressing system premature failure are not typical for metallic components developed by the industry to last more than 50 years before major repairs are required. In most cases of MIC, it appears that after treatment of the affected areas, healthy systems can be expected to have a normal life expectancy.

However, as discussed, failure is a function of both integrity and design. And a system without leaks is never considered acceptable if it fails functionally (i.e., in fire control). Any time a system is in service and fails to operate as designed, it has experienced an unacceptable premature failure.

Forms of Premature Failure. Premature failure can take two forms, both requiring individual consideration. First is the failure of a system to hold water — in other words, the presence of pinhole leaks that require component replacement (see Exhibit S1.4). Leaks can cause significant damage and require immediate action. Leaks can also lead to excessive direct and indirect costs, as well as inherent risks from repeated system impairments.

In one known case, MIC-related leaks resulted in the shutdown of an aerospace manufacturer's global computing center. [5] A pinhole leak developed in a wet pipe sprinkler branch line located over the mainframe equipment of the computing center. Water from the leak not only damaged computing equipment but also resulted in more than 5000 hours of lost operations time. Property damage from similar leaks at other sites has also been documented in the millions of dollars. [5,6]

The second and more worrisome effect is the failure of a system to operate as designed to achieve fire control. This failure not only affects property loss but also could threaten lives. Several systems with MIC have been found with obstructed sprinkler drops, the result of debris generated as a by-product of microbial activity (called biofilm or biosludge). Many in the industry have found systems with feed mains over 60 percent obstructed from biological growth. [6,7] An analysis of corrosion buildup indicates that thousands of pounds of debris can accumulate in medium-sized piping. [8] This buildup of debris presents an obvious hydraulic concern, as many affected sprinkler systems today will not provide fire control because the required discharge criteria, in terms of flow and pressure, are not available due to obstruction and associated frictional pressure loss.

EXHIBIT S1.4 Exterior Pinhole.

Extent of MIC in the Sprinkler Industry

Currently only a limited number of credible national studies on the extent of the MIC problem in the fire protection industry exist. Most data are primarily anecdotal. No detailed, comparative engineering-based study on the effectiveness of the various treatment options currently being marketed has been completed. Complicating this lack of data, reported MIC cases rarely have secondary analysis to differentiate MIC as a primary contributing cause over other types of corrosion involving similar symptoms. Equally important, follow-up engineering data on treatment are rarely published.

The results of MIC in other industries are well documented. The Energy and Power Research Institute has estimated that corrosion in the U.S. electric power industry costs 5 to 10 billion dollars each year. Corrosion is said to be the culprit in half the forced outages each year in steam-generating plants. [9] And data collected from before the 2001 "U.S. power shortage" showed corrosion to be the primary factor in more than 10 percent of U.S. power generation costs. [9] In the gas and nuclear industries, MIC is specifically said to account for 15 to 30 percent of corrosion-related failures. [10] And, reports indicate that to prevent such problems, North American companies spend in excess of $1.5 billion per year (over $7 billion globally) on treatment chemicals to prevent microbial corrosion and fouling. Similar information, as mentioned, is not available for the fire protection industry. Although several industry groups have been attempting to compile data on this problem for several years, only two national studies have been published for the fire protection industry. [11,12] Thus, the true extent and cost of the problem are still not fully known.

In 1996, the National Fire Sprinkler Association sent out a questionnaire in its quarterly membership magazine requesting member information on sprinkler system failures that have been experienced. Results yielded approximately 40 responses from across the United States and Canada, which appeared to indicate that MIC was a "widespread problem." However, due to the reasons previously stated, the quality of analytical techniques used by respondents to confirm MIC as a cause, and the lack of investigation after detection and treatment bring the results of this study into question.

The only other related studies or data compilations completed have been conducted by FM Global. In their studies of losses involving sprinkler systems, corrosion is the fifth leading cause of system failures. This is based on a review of data from 1988 to 1997. In another study by the FM Metallurgical Laboratory between 1994 and 2000 reviewing 155 cases of sprinkler system leaks, MIC was found to be present in approximately 40 percent of cases. Details on whether the presence of MIC directly caused leaks were not indicated in all these cases. [12] And, in another FM study of piping field samples between 1991 and 2002, over 60 percent of these failures are said to be attributed to MIC. [13]

Pinning the Blame

Part of the problem in obtaining detailed conclusive evidence is the complexity and cost of such investigations. This is also a major point of frustration with owners experiencing these types of problems. Due to the nature of electrochemical corrosion, most corrosion engineers agree that the degree to which MIC specifically increases or contributes to general corrosion can never be conclusively determined. In fact, the complex biological interactions between bacteria and host materials are still not fully understood in many cases. There are simply too many variables and uncertainties that affect all corrosion reactions, especially those involving bacterial interactions. Thus, a percentage of blame or rate of corrosion from MIC likely cannot be numerically defined in any study. And, any database generated at this point will likely list only reported cases, not confirmed cases. Thus, the best that can be achieved in the future is a regularly updated database indicating where reported cases are occurring, method of failure detection, interior condition of pipe, methods of testing, and type of treatment with future corrosion monitoring.

Materials Affected by MIC

Although steel sprinkler pipe is the typically observed first point of failure, sprinkler orifice caps, control valves, fittings, and supply tanks can also be subject to damage from MIC. Numerous cases of obstructive growth and pinhole leaks associated with MIC have been found within 20 feet of the discharge side of site fire pumps — an interesting phenomenon due to the velocity of water flow at this point in the system piping. While evidence suggests that only metallic components are susceptible to MIC, it is clear that some grades and alloys of steel are definitely more susceptible than others. As an example, certain grades of stainless steel are more susceptible than regular steel pipe, while others appear to show signs of resistivity. And although plastic components, such as underground water mains, are not subject to direct MIC, they are subject to the effects of biofouling or bacterial debris blockage from upstream corrosion activity.

Defining MIC

Oxygen Tolerance

MIC-related bacteria are classified primarily by oxygen tolerance; that is, they are *aerobic* or *anaerobic*. Aerobic bacteria require oxygen to flourish and reproduce. Anaerobic bacteria do not require oxygen to flourish and reproduce. [11] The most damaging MIC appears to take place within a highly complex community with multiple species of bacteria. This community includes not only aerobic and anaerobic bacteria, but also facultative bacteria — those MIC bacteria that function in both aerobic and anaerobic environments. All three types of bacteria can play a role in the somewhat random interactions that can occur in microbiologically influenced corrosion. [14]

Metabolism

In further defining MIC bacteria, classification is not absolute and can become confusing. The most commonly used method of categorizing bacteria associated with MIC is by metabolism. These categories are basically definitions of what each bacteria type eats (or metabolizes) and excretes as a by-product. As these terms imply, where plants use photosynthesis (i.e., light) to develop energy, bacteria use chemosynthesis (i.e., eating/breathing various chemicals or minerals) to sustain life.

Metabolic classifications are not universally replicated among scientists and can be somewhat confusing. A single bacteria type can fall under more than one metabolic definition. Some of the commonly referenced categories include sulfur-reducing bacteria, metal-reducing bacteria, acid-producing bacteria, iron-depositing bacteria, low-nutrient bacteria, iron-related bacteria, iron-reducing bacteria, iron-oxidizing bacteria, sulfate-oxidizing bacteria, slime-forming bacteria, and iron bacteria. [11,14,15,16]

Scientific Nomenclature

Finally, all bacteria (i.e., all plants and animals) can be classified by their scientific name under phylum, class, order, family, genus, and species. For example, one type of sulfate-reducing bacteria is anaerobic and metabolizes sulfate to sulfide. The sulfate-reducing bacteria group includes the genera *Desulfovibrio*, *Desulfobacter*, and *Desulfomaculum*. [4] All are of the phylum Thiopneutes, which interestingly translates from Greek to "sulfur-breathers."

Sources of MIC Infection

Although there are no conclusive relational studies in the fire protection industry, as noted there are growing beliefs that a sprinkler system's water supply is not the only (and possibly primary) source of bacterial infection. Bacteria capable of causing MIC are potentially present in soil, air, and cutting oils, as well as in water. Thus, the manufacture, shipping, storage, and flushing of system materials should be addressed in all MIC investigations. MIC does not occur only in water-filled systems. Dry pipe systems are also susceptible. Dry systems might even be more susceptible to damage than are wet systems, due to the humidified atmosphere that is created after trip testing. A trip test and subsequent drain can create the right atmospheric moisture content for some bacterial colonies to thrive. Complete drainage and the subsequent use of truly dried air or nitrogen gas appear to mitigate this problem. With regard to tubercle growth alone, dissolved oxygen content, not bacteria, might be the only significant factor in prevention.

MIC Corrosion Process

As with other forms of corrosion, MIC removes material through a series of electrochemical interactions. As such, both an electrical and a chemical component occur with MIC. The electrical component occurs through electron transfer. Electron transfer removes pipe wall material one electron at a time. With MIC, this exact interaction is highly dependent on the specific bacteria involved. Within a sprinkler system, metallic parts become anodic in relation to the cathodic corrosion cell and surrounding water. Basic cathodic depolarization occurs as electrons are stripped away through various forms of oxidation and are pulled to an atom with another electrical potential. Although many of the complex cellular interactions of bacteria are still unclear and can vary by system, there appear to be several somewhat universal steps in the MIC process. They are as follows: [4]

1. Bacteria enter the system, attach to metallic components, and begin to rapidly colonize and reproduce.
2. Aerobic colonies metabolize nutrients from the water and/or the metal surfaces they are attached to and subsequently excrete a polymer film by-product that bonds together to form crustaceous nodules called tubercles.
3. Tubercles and associated biofilms create microenvironments on the metallic material surface (under the tubercles). Tubercles are hard protective shells formed by biological activity. Tubercles typically have an open interior fluid cavity over the corrosion floor area with an approximate pH of 3 to 4. (See Exhibit S1.5.)
4. The underdeposit area (i.e., under the tubercles) becomes oxygen depleted (i.e., anaerobic and anodic) in relation to the surrounding system water or air, which remains aerobic and cathodic. Thus, electrons in the anodic metal flow to the cathode through a reduction reaction.
5. Underdeposit anaerobic bacteria metabolize pipe wall materials and excrete acids (such as acetic acid) as by-products, which are aggressive to the carbon steels used in sprinkler piping. Relative acidity and alkalinity levels within the tubercle shells are reduced to an approximate pH of 2 to 4, which chemically attacks the metallic component surface.

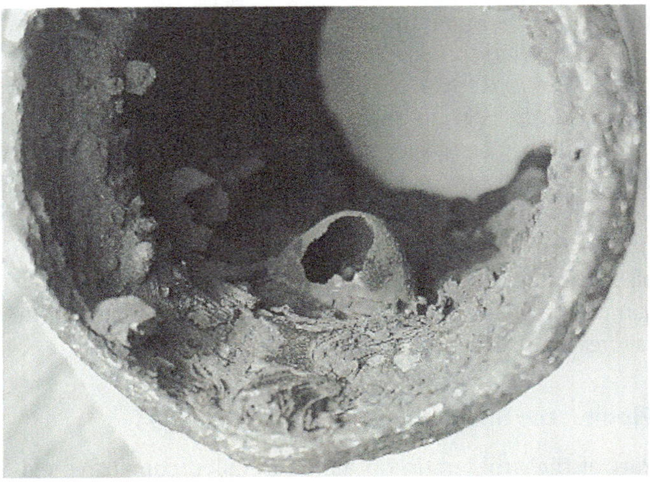

EXHIBIT S1.5 *Interior Tubercle.*

On painted sprinkler piping, it is common to observe blisters where through-wall penetration has occurred. Testing of the fluid within these blisters has shown a pH of less than 3.

The described corrosion process can continue indefinitely until the aerobic and anaerobic bacteria in the system are killed. The tubercles created from bacterial colonization must also be broken down to destroy the underdeposit microenvironment. Even without bacteria in the underdeposit of a corrosion cell, the process can continue indefinitely, as the corrosion chain in its final phases is no longer reliant on their activity.

Prevention and Treatment

References from Allied Fields

Currently, the fire protection industry has a limited number of directly usable references supported by scientific data. However, there is excellent information on data from other industries. The National Association of Corrosion Engineers (NACE) has published multiple studies about MIC detection and treatment for many years. The American Society for Testing and Materials (ASTM) offers several publications on proper bacterial testing practices.

Also, the American Water Works Association (AWWA) offers standards describing the proper management of the somewhat hazardous chemicals typically used in injection devices attached to sprinkler systems for microbial control.

Fire Protection Codes and Standards

Chapter 14 of NFPA 25 addresses the need for an inspection of piping and branch line conditions for the presence of foreign organic and inorganic material and for conducting an obstruction investigation if necessary. However, this section is not meant to imply MIC is the only cause of interior pipe obstruction that is a concern. An individual conducting an internal inspection or obstruction investigation must keep in mind that MIC is not the only and possibly not even the major cause of abnormal pipe corrosion and obstructions. Additionally, while internal inspections and obstruction investigations related to MIC and other obstructions are discussed, the scope of required testing is still debatable when related to the definition of how a "system" is defined.

The first time MIC was discussed in NFPA 25 was in the 2002 edition. A new chapter dedicated to obstruction investigations was added after the majority of the technical committee brought in outside experts and toured an affected site to learn more about this issue. This was the first attempt that an NFPA document made at addressing this issue. And this chapter was significantly modified in both the 2008 and 2011 editions.

The current edition of NFPA 25 now requires two potential inspection regimes: internal obstruction inspections and internal obstruction investigations.

Internal obstruction inspections per Section 14.2 are required every 5 years by opening a flushing connection at the end of one main and by removing a sprinkler toward the end of one branch line. Any tubercules or slime found must be tested for MIC potential and any foreign material (organic and inorganic) should be noted and hydraulically analyzed for the potential to impede system flow. If the presence of sufficient foreign material is found and has the potential to obstruct pipe or sprinklers, then a secondary *obstruction investigation* must be conducted per Section 14.3.

In buildings having multiple wet pipe systems, every other system must have an internal 5-year inspection such that all systems or risers are inspected every 10 years.

If at any time foreign material is found in any system part that could result in system obstruction, all systems must then be internally inspected per 14.2.2.2. It should be noted that since MIC can affect one system or sprinkler line while leaving and adjacent system clean and undamaged, more extensive inspections are recommended if issues are expected.

An internal obstruction investigation is required whenever any of the 14 conditions specified per 14.3.1 are found to exist. This list includes examining the discharge material capable of obstructing sprinklers during water testing, as is routinely done for plant hydrants, fire pumps, dry pipe valves, check valves, and drain tests, and flowing inspector's tests connections during fire alarm testing. Other triggers, which might be more difficult to detect and interpret, include knowing there was a failure to properly flush facility yard piping or surrounding public mains following a new installation or repairs; recently having local broken public water mains without flushing; and recording over a 50 percent increase in the time it takes water to travel to an inspector's test connection during a dry pipe full flow trip test. This last item would be based on a required comparison to the original system acceptance test performance. Obstruction investigation programs must also be implemented if pinhole leaks develop in overhead sites, which are typically the first signs that building owners visibly see when MIC activity has been affecting internal piping.

While there are several actions that are included in a complete obstruction investigation, NFPA 25 dictates an internal four-point inspection. These four points are comprised of each system control valve, sprinkler riser, one cross main, and at least one branch line. If problems are detected during this internal inspection, a complete flushing program is required by qualified personnel. It should be noted that in many cases where advanced MIC has occurred, standard flushing will not acceptably remove biological growth, and replacement of affected pipe sections might be the only acceptable option to successfully return a system to reliable service. The requirement in 14.4.1 also allows alternative nondestructive investigation methods, but these should be reviewed closely by someone experienced with MIC in fire systems. The requirements of Section 14.3 also apply to underground pipe systems with the same 14 action triggers.

Previous editions of NFPA 13 contained little information regarding MIC. But with the recently released 2013 edition comes several changes and additions that could significantly affect owners and contractors. Section 4.3, which covers the owner's certificate, has a new provision requiring that any special knowledge of the water supply that might be responsible for corrosion, specifically including MIC, be noted on the form. And 7.11.1 now requires that all additives to sprinkler system water supplies intended for control of microbiological (or other corrosion) be listed. The ramifications of this requirement will be interesting, as there are many chemicals being sold for this purpose but few have listings. This requirement is also extended in 7.11.2 to include the listing of internal pipe coatings intended for control of microbiological or other corrosion.

Chapter 24 of NFPA 13, which covers fire protection water supplies, now requires in 24.1.5.1 that water supplies be evaluated for the existence of microbes and conditions that contribute to MIC. Compliance with this requirement is hard to interpret and harder to enforce. In basic terms, testing should be done by a qualified laboratory if MIC could be an issue due to known local cases or other issues. The term *local* would be defined as the entire community served by the same municipal water supply or supplies if more than one. Also, per Chapter 24, where conditions are found from testing or other activities, the owner(s) is required to notify the sprinkler system installer. And more importantly, a plan must be developed to treat the system using MIC-resistant pipe, water treatment (using an approved biocide), or through implementing an approved plan for monitoring the interior conditions of the pipe. This program must include established time intervals and specific test locations. This monitoring can consist of one of several corrosion monitoring stations on the market but must also include established intervals. While 24.1.5.1 is not new to NFPA 13, 24.1.5.3 is a new provision requiring that water supply treatment biocides and/or corrosion inhibitors are not only confirmed to be compatible with system components, but that they also be compatible with each other. Many comprehensive treatment programs include both chemicals: one to kill the microbes and one as a protective interior pipe coating. Companies providing these chemicals should issue a letter to the property owner confirming compliance with 24.1.5.3 before any contracts are signed.

New requirements were added in Chapter 25 of NFPA 13 that also cover system signage. In cases where injection systems are installed to treat MIC (or other corrosion), 25.6.2(17) now requires that the type of chemical, concentration of the chemical, and where information can be found as to the proper disposal of the chemical be listed on a placard on each fire sprinkler system riser being treated.

Although these requirements have generated more industry discussion based on scientific methods and follow-up, questions about effective treatment still remain. These include the following:

1. Who is best qualified to make the determination of whether a sprinkler pipe failure is the result of MIC or some other cause and whether a biocidal treatment program will return the system to new condition and prevent future failures?
2. What is the most cost-effective and accurate means of testing to confirm MIC?
3. What follow-up is recommended to document success or failure in available treatment options?
4. Why does MIC affect one pipe or system and leave another directly adjacent to it unharmed?

Almost anything requiring laboratory analysis can be taken to excessive levels; however, under-testing can lead to a false sense of security. The cost to achieve a firm conclusion in many cases may be cost prohibitive or even impossible. Discussions and solid answers to these questions are still evolving.

Operational Considerations

As the industry continues to develop methods to treat and prevent MIC, building owners are faced with the challenge of managing associated risks to their assets and business operations. At-risk businesses should review the interdependence of various operations to identify critical locations.

Proper Diagnosis

The analysis required to properly select a course of action to address MIC is typically outside the scope of most sprinkler contractors and engineers. Thus, until treatment methods become universally proven and standardized, the most critical step in proper mitigation begins with the selection of a qualified corrosion control consultant.

With the wrong choice, a large amount of money can be spent on a problem that might not be correctable. A poor treatment choice could actually accelerate the corrosion rate and increase the affected area beyond that experienced before treatment. The company chosen to determine treatment must have detailed knowledge not only of microbial corrosion control but also of metallurgy and sprinkler system dynamics. Fire sprinkler systems have flow characteristics and concerns that are much different from other industrial processes and systems where MIC is typically addressed.

Specific Considerations for Fire Sprinkler Systems

Other industries deal with MIC in systems containing fluids that are either always static or always flowing, such as in cooling towers. Unlike sprinkler systems, dynamic systems have flow rates that are relatively constant, requiring that prescribed chemical dose rates be constant. A constant flow rate does not occur in sprinkler systems. Variable differences are seen with system drains and refills, inspector testing, and main drain tests. The dose rate for each of these flows must be considered to ensure that the chemical injection rate is always effective. Other industrial systems also have multiple points where biocidal chemicals can be injected. In contrast, sprinkler system water can realistically only be treated at system risers, backflow apparatus, or suction tanks.

Finally, it must again be stated that it is critical to understand that premature system failure can be a function of both bacterial infection and water quality that is incompatible with components. In the majority of premature system failure cases, water chemistry or poor design might be the only likely factors requiring consideration. A high bacteria count does not always indicate that MIC will occur, and conversely, a low bacterial count does not discount that MIC has occurred in the past, in a given system, and will not occur again in the near future.

Detection in Existing Systems

In existing systems suspected of being infected, the first step is to have all possible water supply sources (e.g., tank, city mains, ponds, rivers, etc.) and the interior of each system tested for bacterial levels and activity. Although current technology makes this detection easy, analysis of the results is somewhat complex. And, as previously stated, in determining treatment, bacterial detection is worthless without factoring in water quality. The laboratory used for analysis should be capable of giving conclusive details of water supply mineral and chemical levels, pipe wall deposit compositions, and type-specific bacterial counts. Multiple tests are used in these analyses, from simple bacterial incubation with visual inspection to sulfur print or DNA testing. Obviously, not all tests are required or are necessarily needed. Current preferred analysis methods run the spectrum, depending on the consultant chosen. Costs for such testing can also vary widely.

Mitigation in Affected Systems

When MIC is confirmed in operational systems, the building owner is first faced with a fundamental question. Can the system be salvaged (i.e., cleaned), or does it have to be replaced? Currently, this cost/benefit decision requires further study and is not supported one way or the other by documented best practices in the fire protection industry.

Pipe cleaning is typically an option when corrosion (i.e., pitting) is not excessive. However, "excessive" is a relative term. The after-cleaning quality of the pipe must be considered, for both future longevity and system hydraulics. The resulting frictional loss from numerous pits after cleaning could affect system performance. This, of course, is typically outside the scope of work of most corrosion control consultants. When replacement materials are chosen that are different from those of the original system, they must also be accounted for in hydraulic analysis of the post-treated system. Case studies suggest that pipe cleaning may remove corrosion by-products that are, in effect, stopping the flow of water through existing pipe penetrations, and subsequently leaks develop. [14] Prudence dictates that complete mitigation must include some form of treatment once pipes are cleaned.

Prevention in New Systems

In new systems, it is critical that susceptibility be determined before any systems are filled or tested with any water. If water tests are positive for MIC-related bacteria, a chemical injection system must be installed and used from the first fill. This includes hydrostatic testing and preliminary fills. The treatment methodology must also consider that water delivery of treatment chemicals might not effectively treat high spots, remote areas, or areas that trap air within the system.

If MIC is anticipated, one form of risk reduction might be to specify thicker piping in the design. Although this option only tends to buy time, it might serve as a prudent measure until universally accepted treatment methods are developed. Thicker walls buy time, not because they are less susceptible to MIC — a point on which evidence to date is inconclusive — but simply because a thicker wall has more material to corrode before a through-wall leak develops.

Another possibility in design is to remove the risk of leaks before MIC treatment is even evaluated. Areas such as those occupied by critical energized equipment can often be addressed by removing overhead sprinkler piping and limiting leak exposure through use of sidewall sprinkler coverage. Where it is not feasible to relocate large feed mains, piping can be sleeved so as not to permit potential leaks to contact energized equipment. Although these efforts provide some measure of immediate relief, they do not address the root cause or present a long-term solution to the problem.

Finally, the frequency of sprinkler and waterflow device testing must be addressed. Most agree that repeated draining and refilling of sprinkler systems can increase both biological and nonbiological forms of corrosion. Draining and refilling provide nutrients and oxygen to bacterial colonies and oxygen for general tuberculation. Some facilities have noted substantial reductions in the frequency of leaks by reducing the frequency of drain and alarm testing.

It is critical to note that once a system is filled with infected water, treatment can become exponentially more complex, because any future treatment from a chemical injection system must now be effective in remote and stagnant system legs. In bacteria-positive areas, several additional water quality tests should be completed throughout the first year of service to ensure that contamination has not occurred from any other sources.

Chemical Injection

Once system components have been cleaned and sterilized or replaced, a chemical injection system must be installed to prevent recurrence. Once installed, this system will be required to be operational continuously. As with any other mechanical system, this will require continuous system preventive maintenance.

Several commercially available chemical injection systems have been specifically designed for installation on fire protection systems. Some simply use existing hardware and chemicals modified from MIC treatment in other utilities, such as cooling towers. At the time of this writing, none of the systems currently available was believed to be specifically listed or approved for use as a sprinkler system component. Although most systems appear to be effective when properly installed and maintained, reliability and effectiveness have not been time-proven when compared with other industrial system benchmarks. Past references should always be investigated with any choice.

Most injection systems currently available are designed to work with specific chemicals. These selected chemicals and dose rates are critical. Some bacteria can develop chemical resistance over time if doses are not strong enough and bacteria are not quickly killed. A small number of bacteria believed to be related to MIC, such as the genera *Bacillus* and *Clostridium*, are known to have the ability to convert to a spore state when they encounter adverse conditions that are not lethal. [4, 17] Spores are impervious to penetration by most chemicals and can thus survive biocide treatments indefinitely. Although subsequent treatments might slow or stop their activity, spores will reappear and resume colonization if and when treatments are stopped. With a weak chemical attack, bacteria might also become resistant to the chemicals chosen.

As with other factors involved in treatment, the choice of chemical depends on the consultant. These chemicals generally include penetrants and biodispersants to break up the tubercles that protect underdeposit colonies, a biocide to kill the bacteria in the colonies, and a corrosion inhibitor to protect the interior surfaces of the system.

When such a system is chosen, the applicable authority having jurisdiction (AHJ) should be consulted. In addition to frictional loss concerns mentioned from changes in pipe surface roughness, increased backflow prevention hardware might be required. This could mean a pressure drop of 10 psi (0.7 bar) or more to sprinkler systems in addition to that created by pitting if cleaning is chosen. In new system designs, added alarm system contacts should also be planned for in order to monitor injection system chemical levels, operational status, and trouble conditions such as loss of power. Many preengineered systems provide contact points for these signals. As with fire detection, the perceived "best choice" is at the discretion of the person choosing, and opinions on this subject are highly variable.

Unlike other industrial systems treated for MIC, several unique interactions must be considered. First, sprinkler systems are typically located directly over people. The chemicals used must therefore be nontoxic in contemplation of accidental discharge or exposure to fire fighters under fire conditions. The effect of chosen chemicals on fire fighting (i.e., heat absorption) and chemical reaction with fire (i.e., heat) also needs consideration. Second, system designs typically place water discharge (such as from an inspector's test ports) into foliage or biologically sensitive drains and dry wells. Most municipal wastewater treatment plants (to which typical drains ultimately flow) require bacterial activity to decompose waste. Too large a quantity of biocides in municipal drains could cause a problem.

In conclusion, a complete toxicity review with the highest possible biocidal chemical concentration must be completed. As much as possible, these chemicals should be noncombustible, colorless, odorless, and nontoxic. These chemicals must also be nondeteriorating to rubbers and polymers such as those used on pipe couplings, sprinkler O-rings, and valves. Chemical storage should also be reviewed, as several chemicals currently in use degrade rapidly with heat and might create relatively toxic vapors.

Opinions of which chemicals are believed to be "most effective" in control vary. Choices are available from those currently used for treatment in cooling towers and boilers to specifically patented compounds for fire protection systems. Some of the more common chemicals currently in use specifically for microbial control in sprinkler systems include quaternary ammonium compounds, organosulfur compounds, bromines, carbamates, isothiazalone, phosphates, and chlorines. Sodium silicate is effectively used in bulk quantities by several municipalities as an inhibitor, but this should be avoided for individual systems due to the potential for sprinkler plugging that overdosing can cause. At the writing of this article, no chemical can conclusively be said to be proven as the "most effective." The chemical choice can greatly depend on the bacteria present and system water quality.

Summary

Current testing and treatment options can be confusing. Treatment is slowly evolving and research is continuing. Several industry groups, allied groups, and insurance companies are looking at the problem and applicable solutions. Many universities, governments, and private industry groups also continue to research microbial control in general industry, as they have for the past several decades. These efforts will continue to provide improved treatment options in our industry.

As the need to address this problem draws on the industry's creativity for resolution, prudent thinking requires us to evaluate each solution's impact on overall sprinkler system integrity and performance. In light of this problem, the overwhelming value of sprinkler systems should not be regarded as tarnished or tainted. Although some with repeated MIC problems might view sprinklers as a risk to property, the reduction of risk to life and property from fire that these systems provide should never be overshadowed.

References Cited

1. NFPA 13, *Standard for the Installation of Sprinkler Systems*, National Fire Protection Association, Quincy, MA, 2013 edition.
2. Cote, Arthur E., ed., Fire Protection Handbook®, 20th edition, National Fire Protection Association, Quincy, MA, 2008.
3. Christ, Bruce W., Ph.D., "Corrosion Process Inside Steel Fire Sprinkler Piping," Fire Protection Engineering, Summer 2005.

4. Clarke, B., "Microbiologically Influenced Corrosion in Fire Protection Systems," *Fire Protection Engineering*, Society of Fire Protection Engineers, No. 9, Winter 2001, pp. 4–16.
5. Cappers, M. A., "Investigation of Microbiological Influenced Corrosion in Sprinkler Systems," *Proceedings of Fire Suppression and Detection Research Application Symposium, Research and Practice: Bridging the Gap*, February 12–14, 1997, Orlando, FL, National Fire Protection Research Foundation, Quincy, MA, 1997, pp. 69–81.
6. Kammen, J., "Bacteria Spell Doom for Fire Sprinklers," *The Arizona Republic*, October 24, 1999.
7. Shenkiryk, M., "Pipe-Klean Project—McCarran Airport," *HERC Products*, July 25, 1996.
8. Duncan, Bill, "Pipe Corrosion And Its Growing Threat to Office Building and Plant Operations," *Corrosioneering Journal*, July 2002.
9. Hoffman, S., *Bugging Water Systems for Corrosion Control*, Hoffman Publications, Inc., 1999.
10. "The Mitigation and Detection of Microbial Corrosion," Argonne National Laboratory, Programs and Capabilities Database No. 526-002, 1999.
11. Bsharat, T. K., "Detection, Treatment, and Prevention of Microbiologically Influenced Corrosion in Water-Based Fire Protection Systems," National Fire Sprinkler Association, Inc., June 1998.
12. *Data Sheet 2-1: Internal Corrosion in Automatic Sprinkler Systems*, FM Global Property Loss Prevention Data Sheet, 2001.
13. Yee, Geary G., Ph.D., "Detection and Diagnostic Studies of MIC in Fire Protection Systems." Presented at the 2005 NFPA World Safety Conference and Exposition, June 9, 2005.
14. Little, B. J., Ray, R. I., and Wagner, P. A., "Tame Microbiologically Influenced Corrosion," *Chemical Engineering Progress*, September 1998.
15. Borenstein, S.W., *Microbiologically Influenced Corrosion Handbook*, Woodhead Publications Ltd., 1994.
16. Pope, D. H., Duquette, D. J., Johannes, A. H., and Wayner, P. C., "Microbiologically Influenced Corrosion of Industrial Alloys," *Materials Performance*, July 1984.
17. Hero, H. M., *The Nalco Guide to Cooling Water System Failure Analysis*, McGraw-Hill, Inc., 1993.

SUPPLEMENT 2

CONDUCTING FIRE PUMP FLOW TESTS

By Bruce H. Clarke

Editor's Note: *In many buildings, the fire pump is a critical element for delivering a sufficient amount of water to a sprinkler or standpipe system. It is imperative that, where a pump is installed, it is properly maintained and that the functional performance of the pump is confirmed through the required flow tests in Chapter 8. This supplement looks at the process for conducting these tests, and more importantly, how the information ascertained during the testing should be interpreted to confirm pump functionality.*

NFPA 25 does not provide a specific procedure or set of steps that must be followed to conduct a flow test in accordance with Chapter 8. Chapter 8 simply provides a requirement that the test must be conducted. The following sections provide an approach for conducting these flow tests. It is important to keep in mind that the approaches shown below are not mandated by NFPA 25 and might not be appropriate for all installations. The types and models of system components installed vary greatly from installation to installation; therefore, specific procedures outlined in the manufacturer's literature should also be reviewed.

Sample Testing Procedure

Purpose

The following is a description of the testing procedures necessary to conduct the annual test required in 8.3.3.1.2. This test allows the owner/inspector to determine the fire pump's capability based on its current operating condition. The owner/inspector can then compare the current data to the certified shop test curve to quantify the pump degradation since its installation.

Tools/Equipment

The test equipment required should be in good working condition and of such quality as to not be questioned should a flow test fail. All gauges used in the test (i.e., those for the suction and discharge sides of the pump as well as the pitot tube) should have recent calibration within the past 6 to 12 months. Gauges should be calibrated at least annually by means of a dead-weight or calibrated tester throughout the range of operation. Calibration sheets should be kept for each gauge and correction factors affixed to the back of each gauge before a test series is begun. The flow characteristics of the test nozzles should be known as well. While listed play pipes have standard flow performance curves, some proprietary flow devices, such as hose monster test nozzles, have equipment specific flow charts. If flow meters are used, they require pre-test calibration. Typical tools equipment used in annual pump flow testing are as follows:

1. A handheld tachometer[1] is required to measure the operating speed of the fire pump and verify that on the engine.
2. A pitot tube together with a test-pressure gauge[2],suitable for the pressures to be expected, is required. [Usually, a 60 psi (4 bar) gauge is satisfactory.] Test-quality gauges should be used in accordance with ASME B40.1, *Gauges — Pressure Indicating Dial Type — Elastic Element, Grade 1A,* having an accuracy of ±1 percent. The use of quality test gauges produces results that are considered reasonably accurate within the scope of the testing procedure. Care should be taken to protect the gauges from

rough handling. While pitot tubes for fire protection flow testing are not listed, the gauges used should be listed and should be acceptable for water or water/air systems and not only air systems. A pitot-tube-and-gauge assembly, such as the one shown in Exhibit 7.1, is indispensable in conducting flow tests from hydrants and nozzles. The small opening at the end of the tube — not over 1/16 in. (1.6 mm) in diameter — is inserted in the center of the stream, in a direct line with the flow, at a distance in front of the opening equal to one-half the opening diameter. Velocity pressure is registered on the gauge attached to the tube. Some available test nozzles, including an integral pitot tube, are pitotless flow measuring features with an attached test gauge allowing for direct measurement of flows without the use of a separate pitot tube.

3. Test-pressure gauges suitable for measurement of expected the suction and discharge pressures on the fire pump are required. Note that the gauge used on the suction side of the fire pump should be a compound pressure and vacuum gauge having a pressure range of at least twice the static pressure available from the suction supply. The gauge used on the discharge side of the fire pump should have a range of at least twice the rated working pressure of the fire pump.
4. A sufficient number of test nozzles are required, including hoses for connection to test headers where needed [typically 50 ft (15.2 m) lengths of lined 2½ in. (65 mm) hose might be used for this purpose], to permit the measurement of a flow rate up to 150 percent of the rated capacity of the fire pump. Test nozzles can include UL play pipes or other approved test nozzles. The flow characteristics of the test nozzles must be known.
5. The various wrenches and other specific tools necessary to allow for required adjustments during the test and for the installation of required test equipment are required.
6. For electric pump tests, a multimeter and/or clamp-on ammeter is required for taking voltage and amperage readings.
7. A timing device is required to measure the duration of the fire pump run (on acceptance tests).
8. A form for recording test data is required.
9. While not required for the actual pump test data collections, approved hearing protection and personal protective equipment (PPE) are typically required. If PPE is not handy, then the start of pump testing can often be delayed while the PPE is being located.

Positive Displacement Pump Flow Tests. The pump flow for positive displacement pumps is completed to meet the specified rated performance criteria. Typically, one performance point is required to establish positive displacement pump performance. Many positive displacement pumps also have special applications, such as a component in foam or water mist systems. For these types of pumps, manufacturers' data should be referenced. The pump flow test for foam positive displacement pumps typically is completed using a flowmeter or orifice plate installed in a test loop back to the foam concentrate tank or the inlet side of a water pump. The flowmeter reading or discharge pressure should be recorded and should be in accordance with the pump manufacturer's flow performance data. If orifice plates are used, the orifice size and corresponding discharge pressure to be maintained on the upstream side of the orifice plate should be made available to the authority having jurisdiction (AHJ). Flow rates should be as specified while operating at the system design pressure.

In addition to manufacturers' recommendations, systems with positive displacement pumps should reference ANSI/Hydraulic Institute 3.6, *Rotary Pump Tests,* and Chapter 12, Chapter 13, and Annex C of NFPA 750, *Standard on Water Mist Fire Protection Systems.* ANSI 3.6 applies to industrial rotary positive displacement pumps, and it includes detailed procedures on the setup and methods for conducting hydrostatic test and performance tests of such pumps.

Steam-Driven Turbine Fire Pumps. While Chapter 13 of NFPA 20, *Standard for the Installation of Stationary Pumps for Fire Protection*, is dedicated to these types of pumps, they are rarely installed today due to maintenance and reliability concerns both on the driver and steam source. A steam turbine driver for a fire pump should always be kept "warmed up" to permit instant operation at full rated speed. The automatic starting of the turbine should not be dependent on any manual valve operation or period of low-speed operation, and this should be verified as part of the test. Steam turbines are provided with governors to maintain a predetermined speed, with some adjustment for higher or lower speeds. Desired speeds below this range can be obtained by adjusting the main throttle valve.

Vertical Turbine and Horizontal Split Case Pump Flow Tests. As required by 8.3.3.1, pump flow and pressure, as well as drive shaft speed (rpm), is measured at a minimum of three points: churn (no flow), rated (100 percent flow), and peak/overload (150 percent flow).

Procedure Steps

Pre-Flow Activities/Planning.

1. If the site has a fire alarm system, notify the alarm monitoring company and place the fire alarm system on test.
2. Notify all facility representatives that testing is going to be conducted. In some cases, the amount of water flowing on the ground might require extra maintenance personnel. Or, suction pressure requirements could affect

site production equipment (i.e., site cooling towers could trip off line due to low supply pressure, and so forth.)

3. Notify the local fire department and/or any other AHJ that might want to attend, such as property insurance representatives.
4. Complete the required activities in Section 8.2 to ensure pump system readiness for full flow testing. This maintenance should include checking battery electrolyte levels and specific gravity, inspecting cable conditions, and checking for any corrosion on pipes and other system equipment.
5. Review the fire pump assembly nameplates, noting the following:
   a. Voltage rating (electric pumps)
   b. Rated speed
   c. Rated capacity (flow)
   d. Churn pressure
   e. Rated pressure
   f. Overload pressure
6. Ensure all valves are in the proper position. Normally, pump discharge valves that supply site fire systems or fire system loops should be kept fully open throughout both weekly testing and the annual pump test. These are closed sometimes due to concern that the "high pump pressure" or "pressure fluctuations" might damage the fire system where pipe integrity is a concern. However, if fire mains are subject to damage due to pump pressures, it is best to discover this during an annual test rather than an actual fire. However, where the pump churn pressure will significantly exceed the rated pressure for the system, the fire pump discharge control valve might be closed prior to the operation of the fire pump to limit the exposure of the connected systems. Consideration might be given to closure of the fire pump discharge control valve prior to conducting the test, to limit the exposure of the connected systems to the resultant pressure surge during the start-up of the fire pump. Another method to avoid potential water hammer affects is to partially open a test header outlet at start-up, and then close it for the churn pressure reading.
7. Replace suction and discharge pressure gauges with the appropriate calibrated test gauges (for a vertical turbine fire pump there will only be a discharge gauge).
8. Install flow test equipment on the test header as required, to allow for the maximum flow rate. This might include the use of hoses connected to the test header and extending them to flow measurement devices, such as listed play pipes or other test nozzles.
9. Confirm that all test equipment, such as play pipes, hoses, play pipe test stands, and so forth, are adequately secured or protected against movement. Manually securing play pipes (i.e., by hand holding) is not acceptable and could cause serious personal damage. Swivel elbows, if installed on test headers, must be locked in place with set screws to avoid spinning during testing, which could cause serious personal damage. In some cases, the connection of flow test equipment such as play pipes can be made directly to the test header. The use of a flow diffuser can aid in preventing movement as well as breaking the directed stream.
10. For testing using hose-connected flow-measuring devices or UL play pipes directly connected to the test header, ensure that the individual hose connection valves on the test header are in the closed position, and slowly open the main test header control valve pressurizing the test header. For other flow measuring devices directly connected to the test header, ensure that the main test header control valve is closed. Then fully open a single hose connection valve, readying the arrangement for the first flow point, with the flow being controlled by the main test header control valve. Any time the use of an additional hose connection valve is required with this latter arrangement the additional hose connection valves are to be fully opened and the flow again controlled by the position of the main test header control valve rather than the individual hose connection valves.
11. Check the area surrounding the test per 8.3.3.1.2.1(B) for potential uncontrolled water flow that could cause damage — plugged drain pipes, local construction with open pits, berms subject to hydraulic erosion, and so forth. Check any relief valve or cooling water discharge outlets as well as the discharge points for the fire pump flow to ensure that there are no obvious conditions that would prevent water from being discharged safely or cause direct damage in the immediate vicinity. Where the discharge of water is to an area subject to potential freezing conditions, the facility representative should be advised of the potential for icing conditions.
12. Check the fire pump packing glands for a slight discharge (slow drip) of water, adjusting the packing gland nuts as needed to achieve approximately 1 drip per second. If the packing is completely dry before start-up, it could overheat and fail once the pump is turned for the pump test, The overheating of bearings could also quickly occur. For safety purposes, the adjustment of the packing gland should be made while the pump is not running. Care should be exercised to ensure the glands are not tightened to the point of breaking.
13. For vertical turbine pumps, make sure the pump is completely primed.
14. In cases where a handheld tachometer will be used, verify where the point readings will be taken from and that reflective tape is installed as required. Some handheld tachometers require advance preparation prior to the test for proper measurement, such as the application of a reflective tape on the shaft. Care should be taken

not to apply reflective tape on a shaft that already has it applied, because false "double readings" might occur. Since a reading will be required at each point during the actual test, care should be taken of where the location tape is applied, considering the proximity to casing spray, hot engine exhaust, accessibility to the sides of a rotating shaft, and so forth.

Flow Test Activities/Planning.

1. For pressure-actuated fire pump controllers, simulate an automatic start for the fire pump by creating a pressure drop in the sensing line to the fire pump controller. This is typically accomplished by slowly opening the drain valve on the sensing line located near the fire pump controller until the fire pump starts automatically. The use of the "start" button on the fire pump controller is not acceptable for the purpose of simulating an automatic start. Note the start time of the fire pump, and record the starting pressure.

    For pressure-actuated fire pump controllers that use an automatic timer, an automatic opening of a solenoid valve in the sensing line to the fire pump controller might be used to simulate the automatic start of the fire pump. These systems must include a record of the pressure drop on the pressure recorder for the controller.

    For non-pressure-actuated fire pump controllers, the automatic start can be simulated by other means.
2. Check the fire pump packing gland for a slight discharge (slow drip) of water, adjusting the packing gland nuts as needed to achieve approximately 1 drip per second. For safety purposes, the adjustment of the packing gland should be made while the pump is not running. Care should be exercised to ensure the glands are not tightened to the point of breaking.
3. Monitor the fire pump operation for any unusual noise, vibration, or other signs of malfunction.
4. For electric-driven pumps, verify that the operation of the circulation (casing) relief valve has a steady stream of water to ensure proper cooling of the pump case.
5. Check the packing gland box, shaft bearings, and pump casing for overheating approximately every 5 minutes during the test. The packing gland box and shaft bearings might be warm to the touch, but the pump casing should remain cool throughout the test.
6. If the fire pump is equipped with a main pressure relief valve, verify the operation of the pressure relief valve such that outlet pressures do not exceed the pressure rating of the piping downstream of the fire pump. Usually, this rating is 175 psi (12.1 bar); however, some systems might be designed for higher pressures.
7. For electric-driven pumps, motor voltage and current on all phases (lines) should be recorded. This can be done with a multimeter and clamp-on ammeter; however, these readings should only be taken by individuals trained and qualified in electrical hazards and equipped with the needed safety equipment as outlined in *NFPA 70E®, Standard for Electrical Safety in the Workplace.* Also see 8.3.3.6 and Section 4.8, along with the associated commentary, for additional guidance.
8. Record the suction and discharge pressures. Note that only the discharge pressure is recorded for vertical turbine pumps.
9. Record the operating speed (rpm) of the motor.
10. The data recorded in Step 8 and Step 9 should be noted as the churn test point.
11. Initiate the flow of the fire pump by slowly opening either hose control valves on the test header or by slowly opening the main test header control valve, depending on the arrangement selected. As the valve is opened, take flow measurements to determine the flow rate, adjusting each valve position until a combined flow rate equal to the rated capacity from the nameplate on the fire pump is achieved. Where the flow rate with the valve fully open is inadequate to meet this rated capacity, additional hose valves must be opened and adjustments made until the total measured flow rate equals the rated capacity. For systems equipped with a main pressure relief valve, this should now be verified as closed and not flowing water. If it is flowing water, adjust the pressure relief valve to temporarily close.
12. Repeat Step 8 and Step 9.
13. The data recorded in Step 11 and Step 12 should be noted as the rated (or 100 percent flow) test point.
14. Continue to open hose valves and/or the main test header control valve, as in Step 11, and take flow measurements to determine the flow rate, adjusting the valve(s) position until a flow rate equal to the 150 percent of the rated capacity from the nameplate on the fire pump is achieved, or until the maximum available flow rate is achieved, whichever is lower. Care must be taken to ensure that the suction pressure (at the point of the city connection) does not fall below an acceptable level. NFPA 20 allows a suction pressure of 0 psi (0 bar) for fire pumps connected to municipal supplies and -3 psi (-0.2 bar) for fire pumps taking suction from a grade level tank. Some local water authorities might require a higher maintained pressure [typically 20 psi (1.4 bar) where this is a regulation], as the flow rate is increased to 150 percent.
15. Repeat Step 8 and Step 9.
16. The data recorded in Step 14 and Step 15 should be noted as the overload (or 150 percent flow) test point.
17. Shut down the flow by slowly closing the main test header control valve to avoid water hammer.
18. If the run time for the fire pump has not reached 10 minutes, then allow the fire pump to continue operating

until a total run time of 10 minutes is reached, rechecking for overheating periodically during the test, and then shut down manually upon completion of the test.

19. Plot the test data against the manufacturer's specifications *before* removing test equipment, to verify a retest is not required due to bad data recording and other reasons.

20. Remove all attached test equipment from the test header, restore the closed position of all hose connection valves, and reinstall hose caps. In locations subject to freezing, check that the test header properly drains or pumps out as necessary.

21. Remove the calibrated gauges, and reinstall the original gauges on the suction and discharge side of the fire pump. Any gauges found out of calibration should be replaced.

22. Inspect and clean any installed intake screens.

23. For electric-driven pumps, verify a fire pump phase reversal supervisory signal by jumping across the monitored points in the fire pump controller (typically labeled on a wiring diagram posted on the inside the fire pump controller) to simulate an activation.

24. Verify that all of the fire pump supervisory signals (e.g., fire pump running, loss of power, phase reversal) required by NFPA 20 and NFPA 25 (see 8.3.3.5) have been indicated at the fire pump controller as well as being transmitted to any connected fire alarm panels.

25. Verify the main pressure relief valve is properly set and adjusted.

26. Verify fire pump start-stop settings (see below).

27. If closed, reopen the fire pump discharge control valve, and conduct a valve status test downstream of the closed valve in accordance with 13.3.1.2.1.

28. Restore the fire pump to the automatic operating position.

29. Complete a coupling alignment check.

30. Upon completion of all testing, notify the fire department and/or the alarm monitoring company (as needed), as well as the facility representatives, that testing is complete. Reset the fire alarm system as necessary.

31. Complete the pump test data analysis to determine compliance with 8.3.3.1 and 8.3.6.1.

At each flow test point, the quantity of water discharging from the fire pump test assembly should be stabilized as best as possible before taking pitot readings. When using multiple outlets for pitot readings, final readings on all outlets cannot accurately be made until all adjustments are completed. Due to hydraulic principles, each time one test header outlet is adjusted, the flow on others will be affected. That is, opening an additional test header will cause the flow (and related pitot readings) on others flowing to drop to a different number than the reading taken before the additional valve was opened.

Fire Pump Operational Settings. The fire pump system, when started by pressure drop, should be arranged as follows:

1. The jockey pump stop point should equal the pump churn pressure plus the minimum static supply pressure.
2. The jockey pump start point should be at least 10 psi (0.7 bar) less than the jockey pump stop point.
3. The fire pump start point should be 5 psi (0. 4 bar) less than the jockey pump start point. Use 10 psi (0.7 bar) increments for each additional pump.
4. Where minimum run timers are provided, the pump will continue to operate after attaining these pressures. The final pressures should not exceed the pressure rating of the system.
5. Where the operating differential of pressure switches does not permit these settings, the settings should be as close as equipment will permit. The settings should be established by pressures observed on test gauges.

Evaluation of Results

The text following this Testing Procedure section provides an in-depth review of interpreting the results of the annual flow test and determining if the pump has seen more than a 5 percent degradation when compared to the certified shop test curve.

References

1. American Society of Mechanical Engineers, Three Park Avenue, New York, NY 10016-5990.
2. ASME B40.1, *Gauges — Pressure Indicating Dial Type — Elastic Element, Grade 1A,* 2005.
3. National Fire Protection Association, 1 Batterymarch Park, Quincy, MA 02169-7471.
4. NFPA 20, *Standard for the Installation of Stationary Pumps for Fire Protection,* 2013 edition.
5. *NFPA 70E®, Standard for Electrical Safety in the Workplace®,* 2012 edition.
6. NFPA 750, *Standard on Water Mist Fire Protection Systems,* 2010 edition.
7. Hydraulic Institute, 6 Campus Drive First Floor North, Parsippany, NJ 07054-4406.
8. ANSI/Hydraulic Institute 3.6, *Rotary Pump Tests,* 2010 edition.

Endnotes

1. All test equipment should be calibrated by an approved testing laboratory within 12 months prior to the test.
2. Test-quality gauges should be used in accordance with ASME B40.1, *Gauges — Pressure Indicating Dial Type — Elastic Element, Grade 1A* having an accuracy of ±1%.

The use of quality test gauges produces results that are considered reasonably accurate within the scope of the testing procedure. Care should be taken to protect the gauges from rough handling.

Gauges should be calibrated at least annually by means of a dead-weight or calibrated tester throughout the range of operation before a test series is begun. Calibration sheets should be kept for each gauge and correction factors affixed to the back of each gauge.

The procedure(s) outlined above is not mandated by NFPA 25 and represents only one approach to conducting this test.

Fire Pump Flow Test Analysis

The inspection, testing, and maintenance (ITM) tasks outlined in Chapter 8 of NFPA 25 are intended to ensure that a fire pump, when installed, will function at or near (within 5 percent of the original specification documentation) the performance criteria the pump was installed for. Confirming functionality is achieved by properly maintaining the equipment and conducting the necessary tests to perform a quantifiable analysis of the pump's capability based on the data gathered during the test. NFPA 25 has two specific data analysis requirements for pump flow performance after conducting an annual fire pump test to determine pump acceptance. These are as follows:

1. Determine if a fire pump net performance, relative to the original design/installation, is acceptable.

   Per 8.3.7.3, pump *net performance* adjusted for rated speed must be graphically plotted and evaluated with a comparison to the net pressure curve from owner documents, copies of original manufacturers pump curves, pump nameplate data, or pump retrofit/rebuild documents. And, per 8.3.7.3, when graphically plotted the annual net pump performance flow should be no less than 95 percent of the original pump performance across the complete flow performance curve (i.e., churn through 150 percent of rated flow).

2. Determine if a fire pump gross performance, relative to connected fire protection system demands, is acceptable.

   Per 8.3.7.7, pump *gross performance* not adjusted for speed must be graphically plotted and evaluated with a comparison to owner-provided fire system demands. And when graphically plotted, the annual gross pump performance flow should not be less than the system requirement demands of all water-based fire protection supplied by the fire pump.

There are many more actions for ITM that are required as part of an acceptable annual fire pump test, but these two items listed are the requirements for engineering analysis of the actual pump data. And both are required to determine a "pass/fail" for the test. Other requirements include inspecting general pump room conditions, including heating and cooling systems; inspecting and testing control valves and controller alarm signals; testing controller start-stop pressures; and inspecting and testing numerous mechanical performance features such as master relief valves, circulating relief valves, and heat exchanger solenoid valves. These requirements are covered throughout NFPA 25, primarily in Chapter 8.

Whenever deficiencies are found during ITM procedures, the owner (or designee) is responsible for repairs per 4.1.5. And supplemental action after repair might be required per summary tables in each chapter, including Table 8.6.1 for fire pump systems. To help ensure quick repair, it is also required that any deficiencies and abnormalities observed during inspection or testing must be reported promptly to the property owner or designated representative by the test contractor per 8.4.1 and any documented performance issues be formally recorded and retained for future reference per Section 4.3. Thus deficiencies must be formally reported and presented to building owners, or those designated as responsible for initiating repairs, and not simply verbally communicated.

Data analysis is also required for electric-driven fire pumps regarding recorded current and voltage readings per 8.3.7.5 for pump tests to be considered acceptable, but this is not part of the flow test analysis.

For data analysis requirement 1, per 8.3.7.2.1 a test must be compared to the original pump performance. The original pump performance data is available typically from a manufacturer's pump curve. It can also be recorded from a fire pump design placard or nameplate. Data on these should be the same, as the nameplate data is usually stamped directly from the pump curve data.

Previous editions of NFPA 25 required that acceptance of the annual fire pump performance be determined by comparing the speed-adjusted test flow data to the "original certified pump curve." The 2014 edition of NFPA 25 has been changed to require in 8.3.7.2.1 that the speed-adjusted data be compared to the original pump performance. Net speed-adjusted data can be compared to owner document pressure curves, copies of the original manufacturer's pump curves, pump nameplate data, or pump retrofit/rebuild documents per 8.3.7.3.

This change was made to acknowledge that the original factory-new performance data, to use in comparison, can come from (or be generated from) numerous sources, as now listed. All four acceptable sources now listed should contain the exact same data, because all should be generated from the original manufacturer's pump curve. The term *certified* was removed from the requirement because this was never an NFPA 25–defined term. In relation to technical documents, the term typically implies that something is official, is recognized by an authority having jurisdiction, and is based upon a review relative to a specific recognized quality standard. While it could be said that pumps are designed to be listed and meet the requirements of NFPA 20, *Standard for the Installation of Stationary Pumps for Fire Protection,* the certified pump curves have always only been self-certified by manufacturers, representing the performance of the pump at the factory that should be seen once installed. Since these curves are not reviewed and approved by an AHJ or outside authority at the factory, the term could have been misleading when used with pump curve documents. A certified manufacturer's curve is the same as an original manufacturer's pump curve.

On pumps where this information is missing, original manufacturers can be contacted for this information. Manufacturers typically keep original pump curve data stored by pump serial numbers. If pump serial numbers cannot be located, many times information can be located by manufacturers based on the address of the pump installation. Exhibit S2.1 shows the relationship between a manufacturer's curve and placard.

Before an annual test can be completed, as implied by the two primary data analysis requirements, some design information is required. This information should be collected, reviewed, and analyzed before the annual fire pump test is scheduled and made available to those responsible for the analysis. This

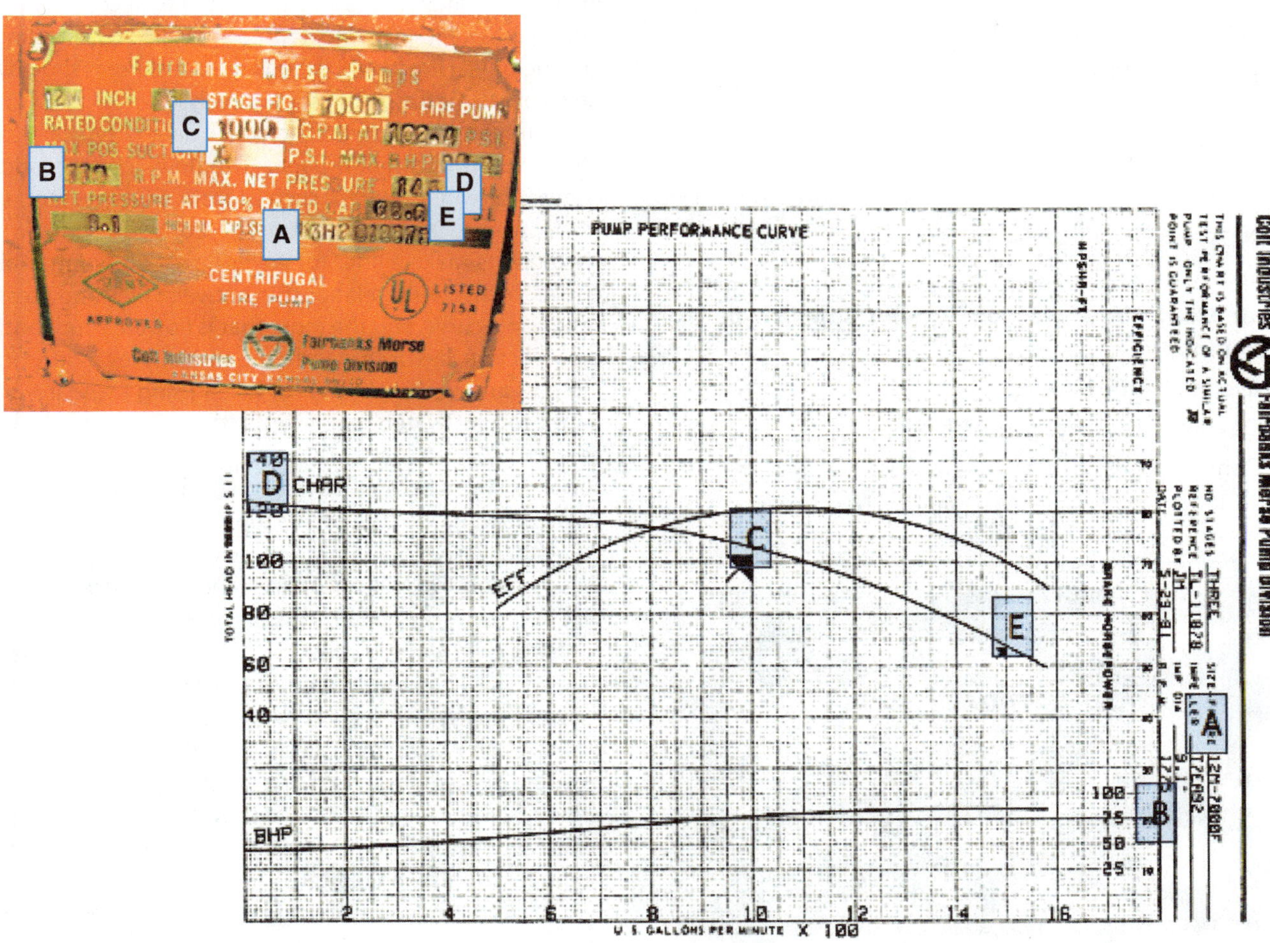

A: Serial number: Unique to each pump. Curve and placard must match.
B: RPM: Pump and curve must match.
C: Rated point:
D: Churn pressure
E: 150% rated capacity

EXHIBIT S2.1 *Relationship Between Pump Placard and Manufacturer's Certified Pump Curve. (Courtesy of Liberty Mutual Insurance Fire Lab)*

evaluation includes original pump performance data system requirement demands of all water-based fire protection systems supplied by the fire pump. It is also helpful to have recent and historical copies of past water supply test results.

Fire Pump Testing Analysis Walkthrough

The following sections of this supplement use a real-world scenario to walk through a fire pump testing analysis.

Site Information

For this example, assume the following data have been collected, reviewed for accuracy, and are available before the test (see Exhibit S2.2):

Original Pump Performance Data (Determined from One of the Means in 8.3.7.3).

Driver type: Diesel

Rated flow: 1000 gpm

Rated pressure: 85 psi

Rated rpm: 1760 rpm

Churn pressures (i.e., maximum net pressure): 97.0 psi

Net pressure at 150 percent: 56.2 psi

Building Sprinkler System Placard Design Information (Water-Based Fire Protection Systems Supplied by the Fire Pump).

System 1, Office Area:

Design: 0.10 gpm/ft² over 1500 ft²

Demand for design (including hoses): 268 gpm at 78 psi

EXHIBIT S2.2 Manufacturer's Original Pump Performance Data Placard. (Courtesy of Liberty Mutual Insurance Fire Lab)

System 2, Manufacturing Area:

Design: 0.28 gpm/ft² over 2500 ft²

Demand for design (including hoses): 1254 gpm at 102 psi

System 3, Warehouse Area:

Design: ESFR sprinkler system

Demand for design (including hoses): 1440 gpm at 63 psi

Pump Suction Water Supply Test Data (from City Fire Hydrant Test or a Hydrant on the Fire Pump Suction Supply).

Static pressure: 45 psi

Residual pressure at flow: 35 psi

Flow: 1250 gpm

2014 Annual Fire Pump Recoded Raw Test Data. The information shown in Table S2.1 should be recorded directly during the pump test. And as noted in 8.3.3.2(2), if testing an electric-driven fire pump, the voltage and current should also be recorded for the power supply analysis.

The plotting of hydraulic water data is typically completed using $N^{1.85}$ graph paper. The flow characteristics of water flow when plotted, such as from a fire hydrant, visually appear as a curve. Contrary to common sense where "two points make a line," for water flow, two points make a curve. The hydraulic relationship between flow and pressure results in a curve, and the more that is flowed, exponentially more friction loss occurs and exponentially less corresponding pressure is available. Thus, when using standard graph paper to plot water supply tests, the results appear more as a "curve" than a line. This can be complicated for water supply analysis because several points must be plotted, and those drawing the plot points must be competent in using French curves.

By using the $N^{1.85}$ semi-exponential log paper, the lines on the vertical axis remain constant, but the lines on the horizontal axis increase inwards as the grid moves from left to right, with $N^{1.85}$ taking the curve out of the line. This allows accurate analysis to be achieved easily and with only two points. Exhibit S2.3 and Exhibit S2.4 illustrate water supply test data plotted on straight-line graph paper and $N^{1.85}$ paper. Exhibit S2.5 and Exhibit S2.6 also show the straight-line graph paper and $N^{1.85}$ paper, this time with both water supply and pump test data being plotted.

While the benefits of using $N^{1.85}$ graph paper for water supply analysis does not transfer to fire pump net performance, it is recommended for fire pump gross performance analysis and is especially useful if questionable pump test results require a suction supply review. And if used for fire pump net analysis, secondary plots are not required for fire pump gross analysis.

TABLE S2.1 *Pressure and Flow Data for Fire Pump Testing*

Test	Test Flow Info	Pitot Reading	S (rpm)	Q (gpm)	P_s (psi)	P_d (psi)	P_v (psi)	P_n (psi)	Q_N (gpm)	P_N (gpm)
1-Churn	No flow	n/a	1777	0	45	140				
2-Rated	2-1.75 in. play pipes	32, 32	1763	1028	38	121				
3-150% Rated	3-1.75 in. play pipes	32, 32, 32	1731	1542	29	84				

S = pump speed — typically recorded at pump shaft with tachometer
Q = pump discharge flow — combined pitot tube pressure readings converted to flow for given orifice
P_s = pump suction pressure — from reading suction flange gauge at flow
P_d = pump discharge pressure — from reading discharge flange gauge at flow
P_v = pump velocity head — correction calculated based on suction and discharge pipe sizes
P_n = pump net discharge — total pump discharge performance accounting for P_v
Q_N = pump flow — gross pump flow adjusted for speed
P_N = pump pressure — gross pump pressure adjusted for speed

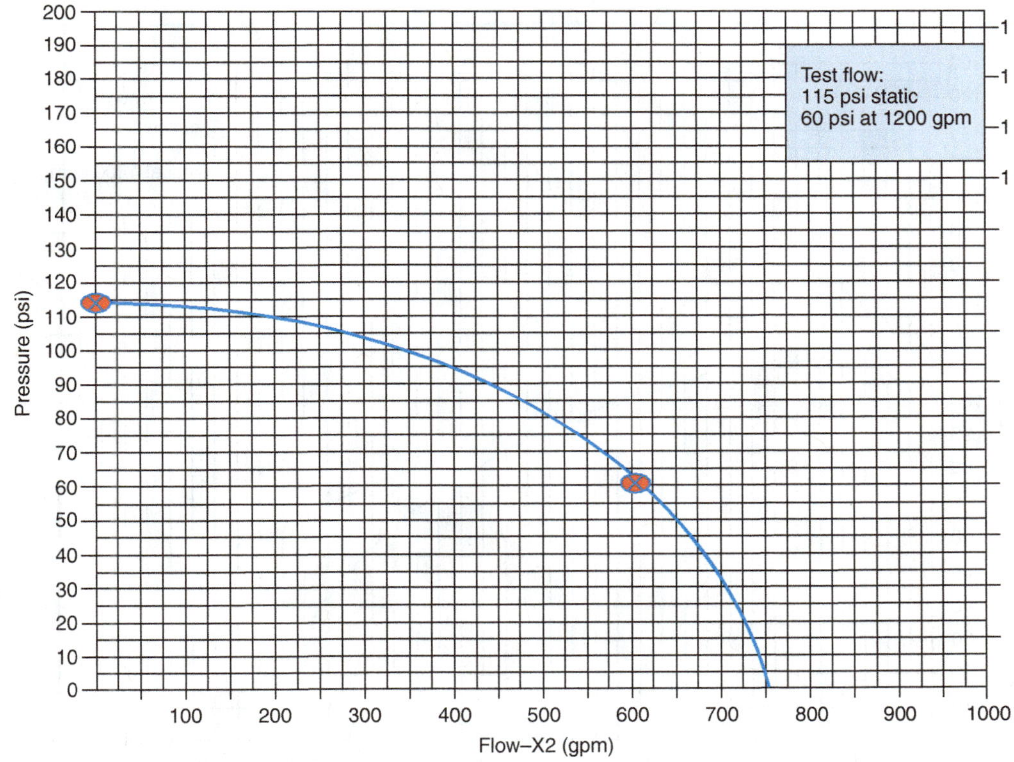

EXHIBIT S2.3 *Water Supply Test Plotted on Straight-Line Graph Paper.*

EXHIBIT S2.4 *Water Supply Test Plotted on $N^{1.85}$ Graph Paper.*

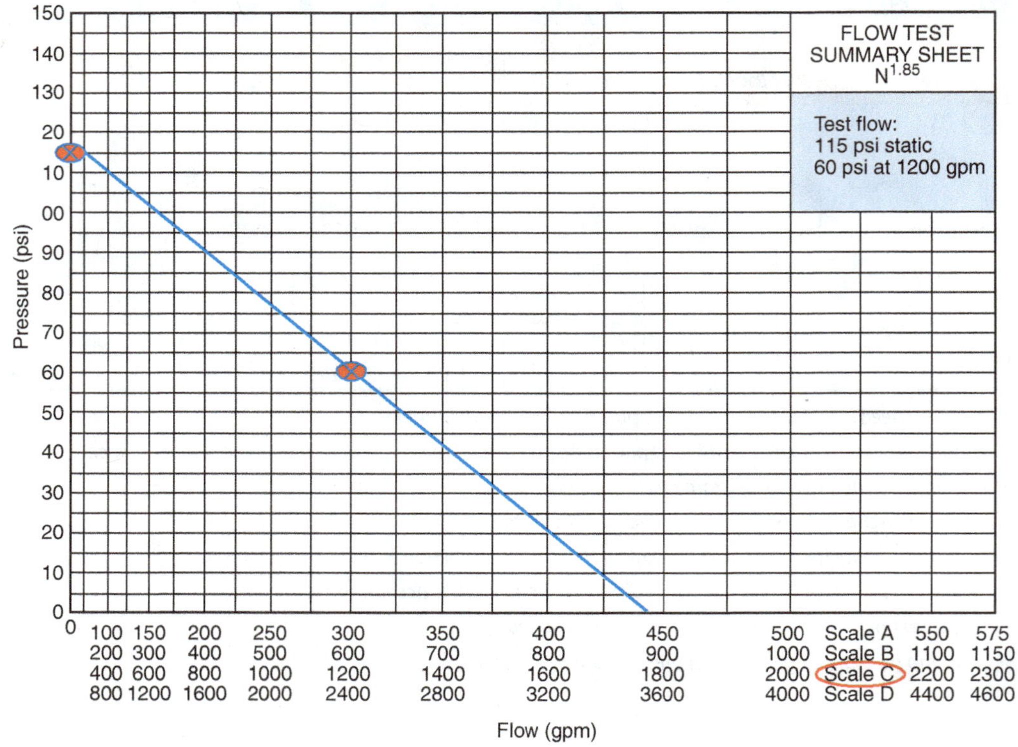

EXHIBIT S2.5 *Water Supply and Pump Test Plotted on Straight-Line Graph Paper.*

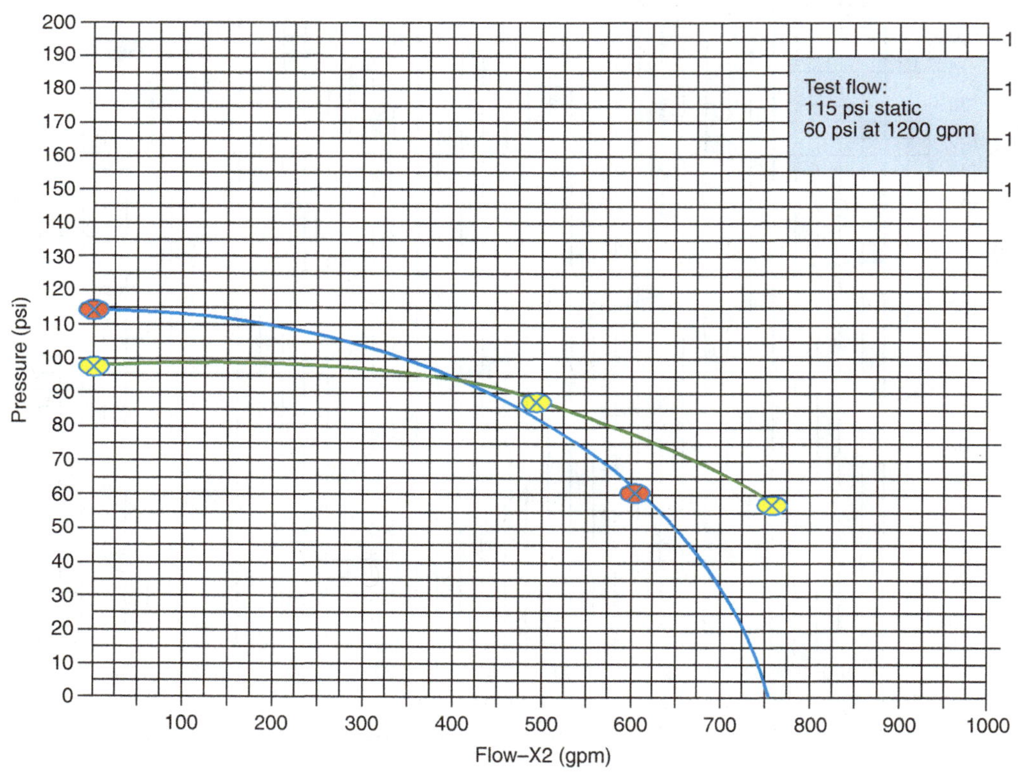

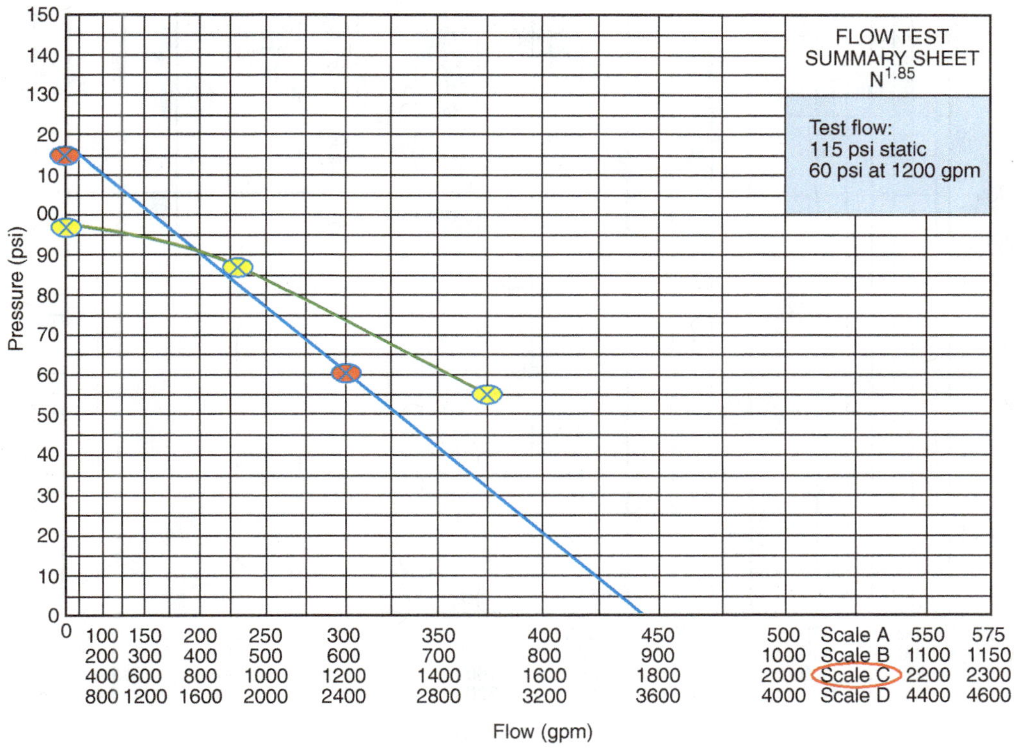

EXHIBIT S2.6 Water Supply and Pump Test Plotted on $N^{1.85}$ Graph Paper.

There are drawbacks to using the $N^{1.85}$ graph paper too. A water supply might not exactly follow the set exponential relationship, such as in cities where municipal pumps are phased on to meet added demands. In these types of cases, no matter which graph paper is used for analysis, systematic and detailed analysis is required. Secondly, $N^{1.85}$ paper can make reading flow points on the curve somewhat difficult, but this would only typically account for a 1 psi (0.07 bar) to 3 psi (0.2 bar) difference at most.

Analysis 1

The purpose of the first analysis is to look at the overall fire pump output to confirm that it will support the necessary demand.

Determine if a fire pump gross performance, relative to connected fire protection system demands, is acceptable.

Step 1: Plot gross fire pump performance, which is "*Q*" at the associated "P_d". (See Exhibit S2.7.)

Step 2: Plot all provided fire system demands. (See Exhibit S2.8.)

Step 3: Make a determination for final report if a fire pump gross (unadjusted) performance, relative to the original design/installation, is acceptable.

This analysis is completed by reviewing closely the data in the gross pump test with system demands as shown in Exhibit S2.8. If the gross pump discharge has a flow and pressure that exceeds what each fire system demand needs, the test is considered acceptable. From the example plot shown in Exhibit S2.8, all fire system demands fall under the gross test data and thus appear acceptably supplied by the fire pump.

Figure A.8.3.7.3(1)(b) shows another example of a pump test result plotted and compared with system demands. That figure shows that the actual pump discharge can meet system demands. Some things to consider in this analysis are as follows:

1. Sprinkler system demands are not always calculated to the pump. If the pump is significantly detached from the sprinkler system, this could be an issue. Many are only calculated to the base of the riser. Thus, this information must be known.

   For example, if the System 2 design as listed on the placard was only calculated to the base of the riser, there could easily be another 5 psi (0.34 bar) to 20 psi (1.38 bar) pressure loss between what was plotted with the gross pump discharge and what is available for the fire protection system. In the plot example shown, there is only an approximate buffer of 3 psi (0.2 bar). After analysis, this

EXHIBIT S2.7 Pump Gross Test Data Plotted on $N^{1.85}$ Graph Paper.

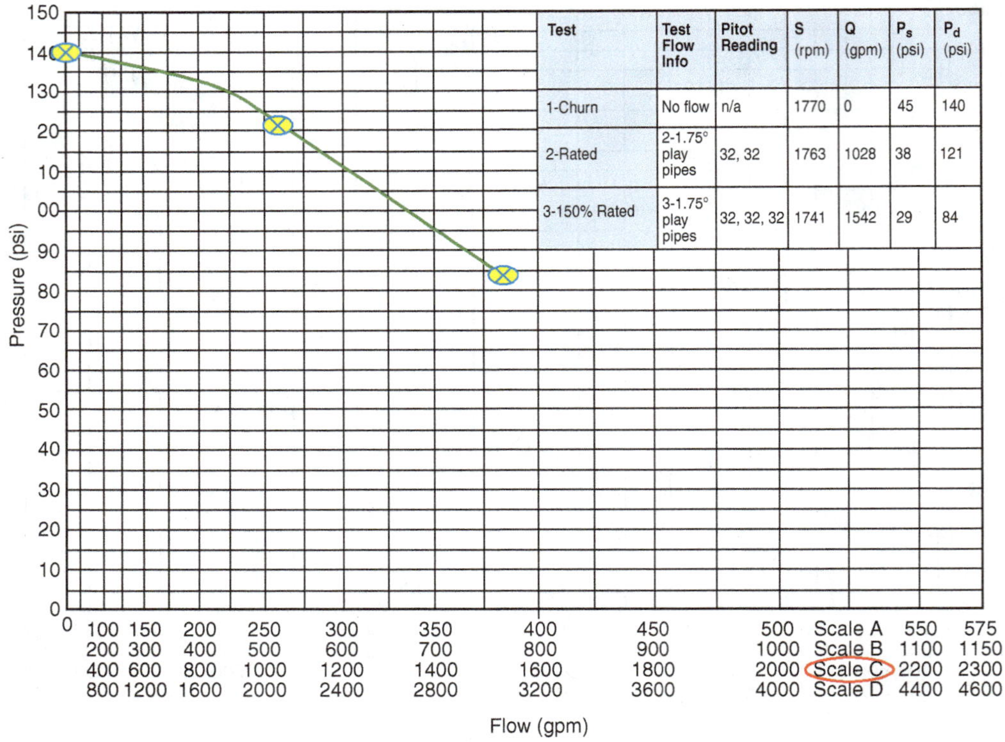

Test	Test Flow Info	Pitot Reading	S (rpm)	Q (gpm)	P_s (psi)	P_d (psi)
1-Churn	No flow	n/a	1770	0	45	140
2-Rated	2-1.75° play pipes	32, 32	1763	1028	38	121
3-150% Rated	3-1.75° play pipes	32, 32, 32	1741	1542	29	84

EXHIBIT S2.8 Pump Gross Test Data with System Demand Plotted on $N^{1.85}$ Graph Paper.

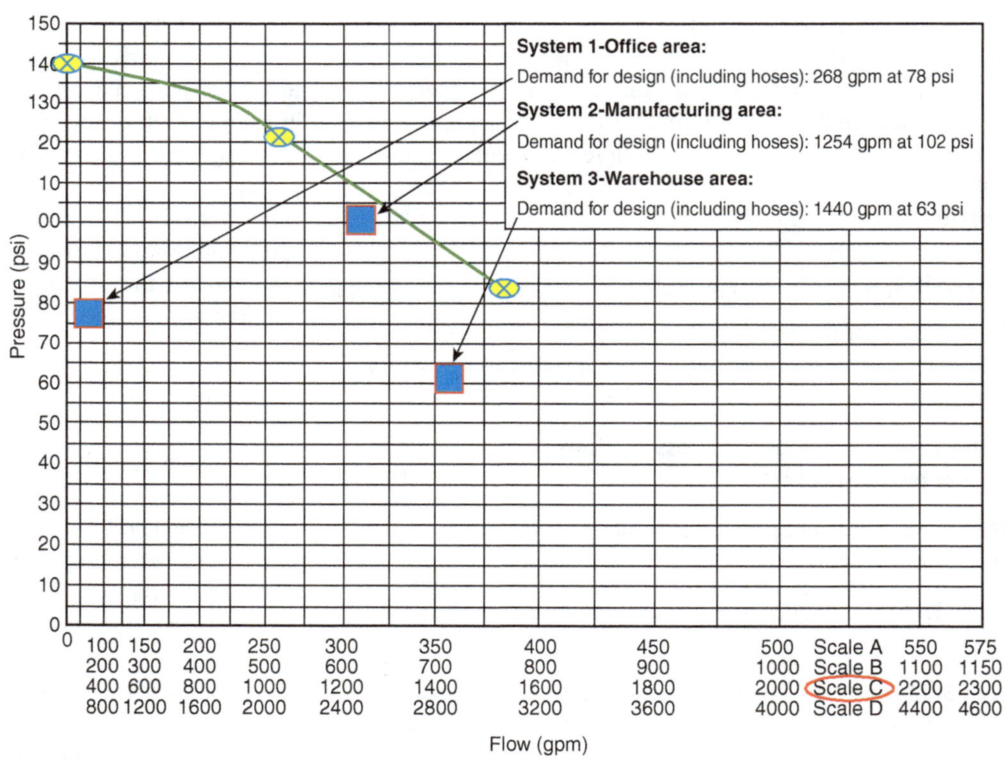

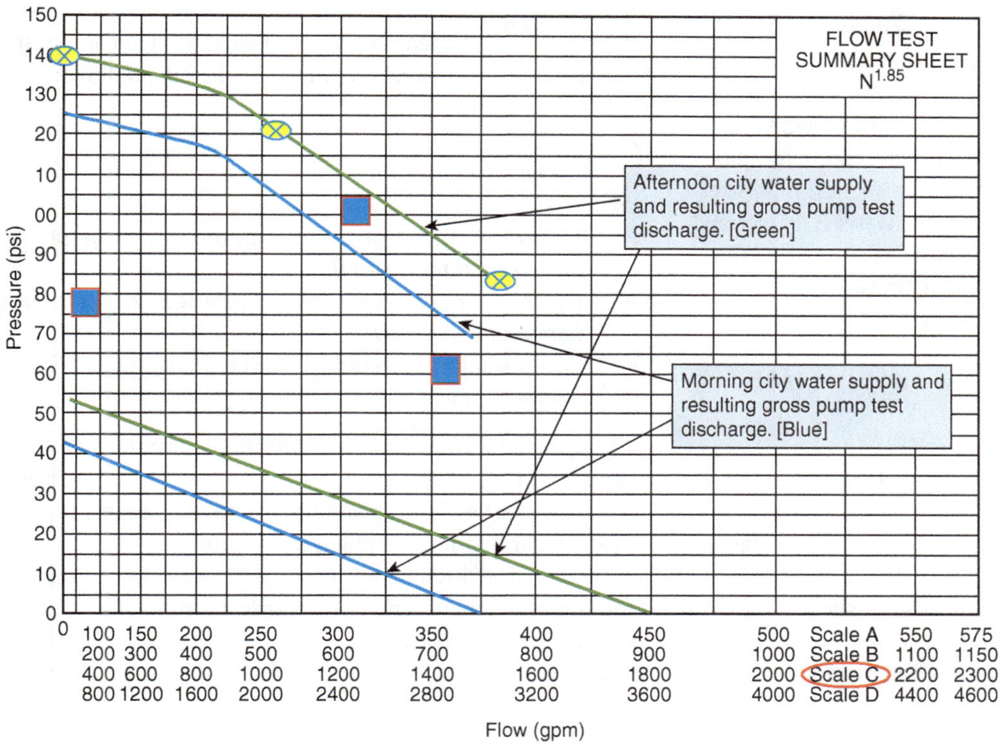

EXHIBIT S2.9 *Pump Gross Test Data Plotted on $N^{1.85}$ Graph Paper.*

could mean the pump test is not acceptable and that fire system reinforcement or a pump replacement is warranted.

2. Sprinkler system demands are not always calculated to include hose allowances per NFPA 13, *Standard for the Installation of Sprinkler Systems*. Also, data shown on riser placards do not always include the sprinkler system demand with hose allowances included. Thus, this information must be known.

   For example, if the System 2 design listed on the riser placard was calculated without hose allowance requirements, there could be another 100 gpm (379 L/min) to 750 gpm (2835 L/min) required to be added to the system demand. And in the example shown, there is only an approximate 45 gpm (170 L/min) buffer. After analysis, this could mean the pump test is not acceptable and that fire system reinforcement or a pump replacement is warranted.

3. Fire pump water supply fluctuations (P_s) directly affect fire pump discharge (P_d). As a municipal water supply pressure drops, so does the fire pump gross performance. Municipal supply pressures can fluctuate for various reasons. Some cities now cycle or adjust municipal pump pressures down to force water conservation in areas that experience drought. Less pressure relates to less flow, and in turn, this creates less household and industry consumption. Pressure fluctuations can also occur on shared water mains where heavy equipment or landscape irrigation drops the total system pressure.

   For this example, if the pump test were completed in the late afternoon when the water supply from a city was the highest, or if the same test was completed in the morning or during the highest demand hours, the results could have indicated that System 2 only had an acceptable water supply from the installed fire pump at certain times of the day.

Any significant drop in gross pump output, as shown in Exhibit S2.9, should always be a concern and should be investigated, since this can also be an indication of a hidden closed valve somewhere on the suction supply leading to the fire pump. This is why it is recommended that historical water supply data, such as from past city fire hydrant tests or hydrant flow tests on the fire pump suction supply, be maintained and available during the pump test for comparison.

Analysis 2

This analysis of the pump's capabilities will examine the functionality of the pump when compared to the performance of the pump during acceptance testing conducted in accordance with NFPA 20.

Determine if a fire pump net performance, relative to the original design/installation, is acceptable.

Step 1: Determine velocity head adjustment relative to net pressure (P_n).

The pump test data analysis should include velocity head correction. Velocity head is created when the discharge pipe size leading to a fire pump differs from the suction pipe size. Changes in pipe size change velocity. Since water is an incompressible fluid, when the pipe size increases or decreases without compression as an option, the velocity must change.

For fire pump analysis purposes, the following calculation can be applied:

$$P_{n:}\quad (P_d - P_s) + (P_v)$$

Where:

P_n = pump net discharge — total pump discharge performance accounting for P_v

P_d = pump discharge pressure — from reading discharge flange gauge at flow

P_s = pump suction pressure — from reading suction flange gauge at flow

P_v = pump velocity head — correction calculated based on suction and discharge pipe sizes

To mathematically calculate velocity head, the following calculation can be applied:

$$P_v = 0.001123\ Q^2(1/D_d^{\,4} - 1/D_s^{\,4})$$

Where:

P_v = velocity pressure (psi)

Q = measured flow (gpm)

D_d = suction pipe diameter measured at gauge attachment flange (inches)

D_s = discharge pipe diameter measured at gauge attachment flange (inches)

However, while velocity head can be worth noting in some cases, such as for very small pumps with small pipes, this adjustment is negligible and can be ignored for almost all standard fire protection pump installations. And for this reason, most contractor pump test sheets do not even have a column to record this calculation.

For example, for installations where both suction and discharge pipes or flanges (measured where gauges are mounted) are the same, this can always be disregarded, since there is no velocity head potential. And most fire pumps have between 6 in. (150 mm) and 10 in. (255 mm) pipes where, even if sizes change, the P_v created is typically negligible — less than 2 psi (0.14 bar). For example, a pump with an 8 in. (200 mm) suction pipe and 10 in. (255 mm) discharge flowing 2000 gpm (7570 L/min) and a pump flowing 500 gpm (1890 L/min) with a 4 in. (100 mm) suction and a 6 in. (150 mm) discharge would both result in less than a 1 psi (0.07 bar) adjustment. For this reason, P_v should be considered to be 0 psi, as is the typical industry practice. And therefore, P_n is simply calculated as follows:

$$P_n = (P_d - P_s)$$

Where:

P_n = pump net discharge — total pump discharge performance accounting for P_v

P_d = pump discharge pressure — from reading discharge flange gauge at flow

P_s = pump suction pressure — from reading suction flange gauge at flow

Calculating Vertical Turbine Pump Adjustments. One other adjustment must also be accounted for on vertical turbine pumps. Since vertical turbine pumps have impellers that are typically submerged in a static water source and must lift the water up to the discharge piping, there are no suction gauges, and therefore, there is no P_s. For vertical turbine pumps, pressure loss due to the required lift, or P_e, must also be included, and P_n is as follows:

$$P_n = P_d + P_e$$

Where:

$P_e = E \times 0.433$

P_n = pump net discharge — total pump discharge performance accounting for P_v

P_d = pump discharge pressure — from reading discharge flange gauge at flow

P_e = pump negative pressure (loss) due to elevation head

E = distance (ft) between pump discharge gauge and suction water level

In the rare instances where a vertical turbine pump is connected to a pressurized water main, the adjustment methodology discussed for horizontal split case pumps must be used with the required elevation adjustments. See Table S2.2.

Step 2: Determine speed-adjusted net flow and pressure (Q_N and P_N).

Per 8.3.7.2.1, mathematical adjustment must be made for correction of recorded test data to the original pump rated speed (may also include a velocity head adjustment), and it must be applied when determining flow test performance relative to the original pump performance. These corrections are shown in Table S2.2.

The three data points on a manufacturer's pump curve and placard are generated as part of the manufacturing quality assurance process, such that each pump is tested for flow and pressure at the speed listed on the manufacturer's pump

TABLE S2.2 *Fire Pump Output Data with Velocity Head and Speed Corrections*

Test	Test Flow Info	Pitot Reading	S (rpm)	Q (gpm)	P_s (psi)	P_d (psi)	P_v (psi)	P_n (psi)	Q_N (gpm)	P_N (gpm)
1-Churn	No flow	n/a	1777	0	45	140	0	95		
2-Rated	2-1.75 in. play pipes	32, 32	1763	1028	38	121	0	83		
3-150% Rated	3-1.75 in. play pipes	32, 32, 32	1731	1542	29	84	0	55		

S = pump speed — typically recorded at pump shaft with tachometer
Q = pump discharge flow — combined pitot tube pressure readings converted to flow for given orifice
P_s = pump suction pressure — from reading suction flange gauge at flow
P_d = pump discharge pressure — from reading discharge flange gauge at flow
P_v = pump velocity head — correction calculated based on suction and discharge pipe sizes
P_n = pump net discharge — total pump discharge performance accounting for P_v
Q_N = pump flow — gross pump flow adjusted for speed
P_N = pump pressure — gross pump pressure adjusted for speed

curve or placard. At each pump manufacturing site, when a manufacturer's curve is generated, the pump is connected to a test driver and the driver is manually adjusted to the same speed at each test point before flow is recorded. So, on the manufacturer's curve the churn, rated, and 150 percent flow points are all taken at the exact same speed, which is listed as the pump speed (rpm) at rated flow and pressure. In the same way a car does not maintain the same speed along each hill and valley, a pump and driver that are installed in the field do not really maintain the same speed across all flow points.

In a real installation, for most diesel drivers (and to a smaller degree, for some electric drivers), the more work the pump does (i.e., more flow), the more work or load is required of the driver. And as with most drivers, increased work load results in decreased speed. This is similar to what is seen in automobiles. A vehicle towing a heavy load has a slower speed than the same vehicle towing nothing unless more fuel is supplied via the gas pedal.

For this reason, to make an equal comparison of a fire pump installed in the field to a test flow to that on a manufacturer's pump curve, the test data (both flow and pressure) must be mathematically adjusted to the pump's rated speed (or rpm).

Adjusting flow and pressure mathematically for speed related to fire pump analysis is typically completed using the Affinity Laws as follows:

1. For flow: $Q_N = Q$ [rated speed (rpm)/measured speed (S)]
2. For pressure: $P_N = P_n$ [rated speed (rpm)/measured speed (S)]^2

For our case, placard data indicates the pump speed rating is 1760 rpm. Therefore, the following applies:

Churn:

$Q_N = 0$ gpm

$P_N = 95 \text{ psi} \times (1760/1777)^2 = 93$ psi

Rated:

$Q_N = 1028 \text{ gpm} \times (1760/1763) = 1028$ gpm

$P_N = 83 \times (1760/1763)^2 = 83$ psi

150 percent:

$Q_N = 1542 \text{ gpm} \times (1760/1731) = 1569$ gpm

$P_N = 55 \times (1760/1731)^2 = 57$ psi

This data is summarized in Table S2.3.

Step 3: Make determination for final report if a fire pump net (adjusted) performance, relative to the original design/installation, is acceptable.

This calculation is completed by comparing the manufacturer's performance curve plot with the adjusted flow and pressure plot (three Q_N and P_N points), with judgment relative to meeting the requirements of 8.3.6.2. If both curves run within +/− 5 percent of each other, the test is acceptable. This is shown in our final plot of Exhibit S2.10.

Figure A.8.3.7.3(1)(a) shows another example of a net pump test plotted and compared to an original pump manufacturer's test curve. Adjusted results of this test closely

TABLE S2.3 *Gross Pump Output Data Following Adjustments for Speed*

Test	Test Flow Info	Pitot Reading	S (rpm)	Q (gpm)	P_s (psi)	P_d (psi)	P_v (psi)	P_n (psi)	Q_N (gpm)	P_N (gpm)
1-Churn	No flow	n/a	1777	0	45	140	0	95	0	93
2-Rated	2-1.75 in. play pipes	32, 32	1763	1028	38	121	0	83	1028	83
3-150% Rated	3-1.75 in. play pipes	32, 32, 32	1731	1542	29	84	0	55	1569	57

S = Pump speed — typically recorded at pump shaft with tachometer

Q = pump discharge flow — combined pitot tube pressure readings converted to flow for given orifice

P_s = pump suction pressure — from reading suction flange gauge at flow

P_d = pump discharge pressure — from reading discharge flange gauge at flow

P_v = pump velocity head — correction calculated based on suction and discharge pipe sizes

P_n = pump net discharge — total pump discharge performance accounting for P_v

Q_N = pump flow — gross pump flow adjusted for speed

P_N = pump pressure — gross pump pressure adjusted for speed

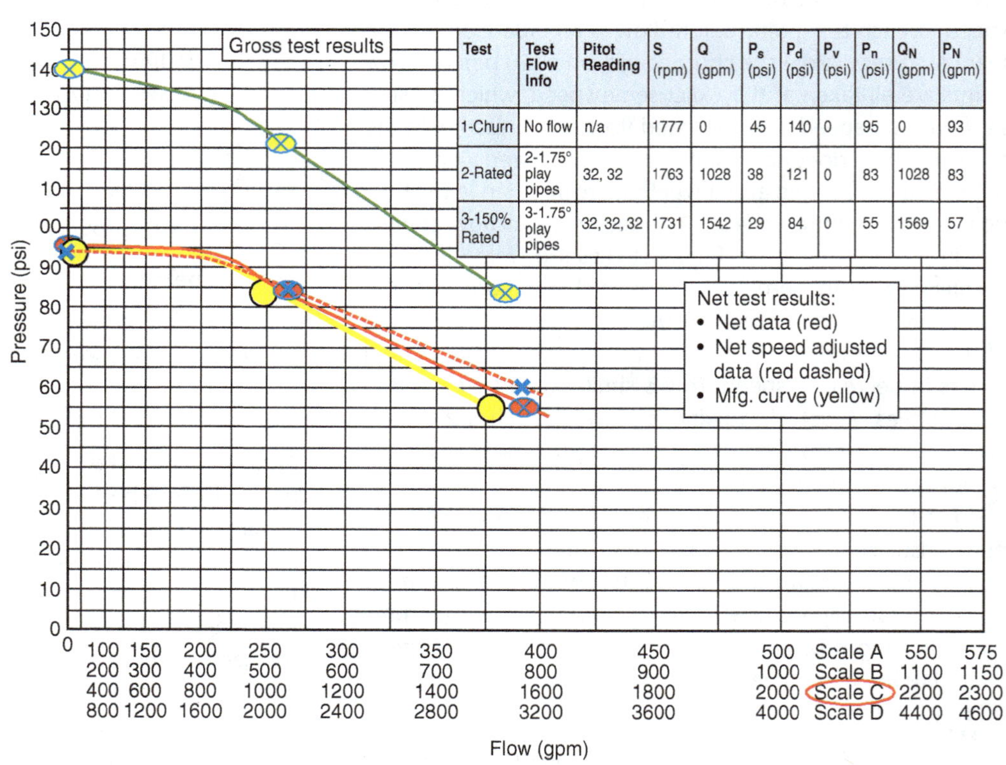

EXHIBIT S2.10 *Pump Plot: Gross vs. Net Data Plotted on $N^{1.85}$ Graph Paper.*

overlap (i.e., the pump is performing at or above 95 percent of the original design specifications per the manufacturer's performance curve), which indicate the pump is performing close to the way it did when it was factory new. And, this plot is a good indication that the internal parts of the pump are functioning well and the driver is operating as required for the attached pump.

A final step in any pump test is a written evaluation of the conclusions. Per 8.3.7.1.2.1, 8.4.1, and 8.4.2, the property owner or designated representative must be given a written copy of the data analysis (i.e., interpretation of results) with the test results as part of an annual pump test that includes any noted abnormalities. And any documented performance issues must be recorded and retained for future comparison purposes.

This analysis leads to one final question: What must be done if a fire pump does not pass an annual test? If the failure is in the associated Analysis 1, this must be addressed immediately. Without further engineering review, a failed test would indicate that in the event of a fire, there might not be an adequate water supply to acceptably supply the fire systems connected to the fire pump. This translates to the potential of an uncontrolled fire and, potentially, extended loss of life and property.

The solutions are varied and a complete engineering analysis must be completed, but a logical first step is to verify that all gauges and flow measuring devices are properly calibrated and working. This might also involve a pump retest to verify the results.

If results still show a failed pump test, a possible solution would be replacement of the fire pump with a larger unit. Since the driver size is based on pump size, this will likely mean a new, more powerful driver as well. And since controller components, power supply transformers, and wiring for electric drivers are sized based on motor size, this could also mean that a new transformer, a new controller, and rewiring are required.

The flow also must be evaluated to see if the issue is with a water supply source. As previously discussed, a water supplier could be purposely decreasing a municipal water supply for reasons related to fire protection, such as forced water conservation. In these cases, a request can be made to increase the pump suction water supply. Reduced flow could also be the result of a valve partially closed downstream of the pump. Thus, there are the reasons a full engineering analysis must be completed.

In some cases, the solution is connecting the pump to another water supply source or parallel source to counter frictional loss, such as from a recent backflow apparatus installation. In extreme cases, this could also mean adding a fire pump suction tank as a new source of site water. And in this case, an entirely new fire pump and driver would likely also be required.

In completing the evaluation for Analysis 2, a failed test per 8.3.7.4 that required further investigation is pump flow performance degradation in excess of 5 percent. While the causes for failure here could be different than for Analysis 1, they could also be the same or related. Thus, a full engineering analysis must be completed for this, along with following the same logic and process of issue investigations. A logical first step would also be to verify that all gauges and flow measuring devices were properly calibrated and working. A pump retest to verify the results would also be recommended. If results are confirmed, maintenance involving taking apart the pump for visual inspection is then typically required. Poor performance indicated here is usually the result of a damaged impeller or blocked impeller vane.

The following are some examples of poor net pump test curves and possible problems.

Exhibit S2.11 shows a failed test because there is more than a 5 percent performance degradation. The test shown in Exhibit S2.12 is not a failed test per NFPA 25; however, an investigation is still warranted. This test, taken as plotted, would mean the pump and driver actually improved performance by almost 10 percent from when it was factory new. This is not likely. This would most likely be the result of test equipment not calibrated correctly or poor test data recording. In either case, the test has to be questioned as "too good to be true."

There are numerous reasons for a test as shown in Exhibit S2.11. Exhibit S2.9 shows a gross test plot with the water supply. Any significant drop in gross pump output per Exhibit S2.9 should always be a concern and should be investigated, since this can also be an indication of a hidden closed valve somewhere on the suction supply leading to the fire pump. This is why it is recommended that historical water supply data such as from past city fire hydrant tests or hydrant flow tests on a fire pump suction supply be maintained and available during the pump test for comparison. Of course, since Exhibit S2.11 only looks at net performance, this would not be detected if only one of the two analyses were completed. This could be the result of impeller wear, uncalibrated gauges reading low, or using the wrong flow device pitot reading chart. It could also be from a pressure relief valve stuck open in a certain position.

Exhibit S2.13 is a failure for the same reason as Exhibit S2.11. However, this might not be as big of a concern, because performance is within the requirements at the rated and 150 percent point. This result could simply be a pressure relief valve closing as flow increases, as should be the case. For testing, it should be ensured that this valve closes as soon as water flow through the header is started, so as not to affect the performance curve data.

Exhibit S2.14 is a failure for the same reason as Exhibit S2.11 and Exhibit S2.13. And this is much more critical, as the flow region with the poor performance is typically

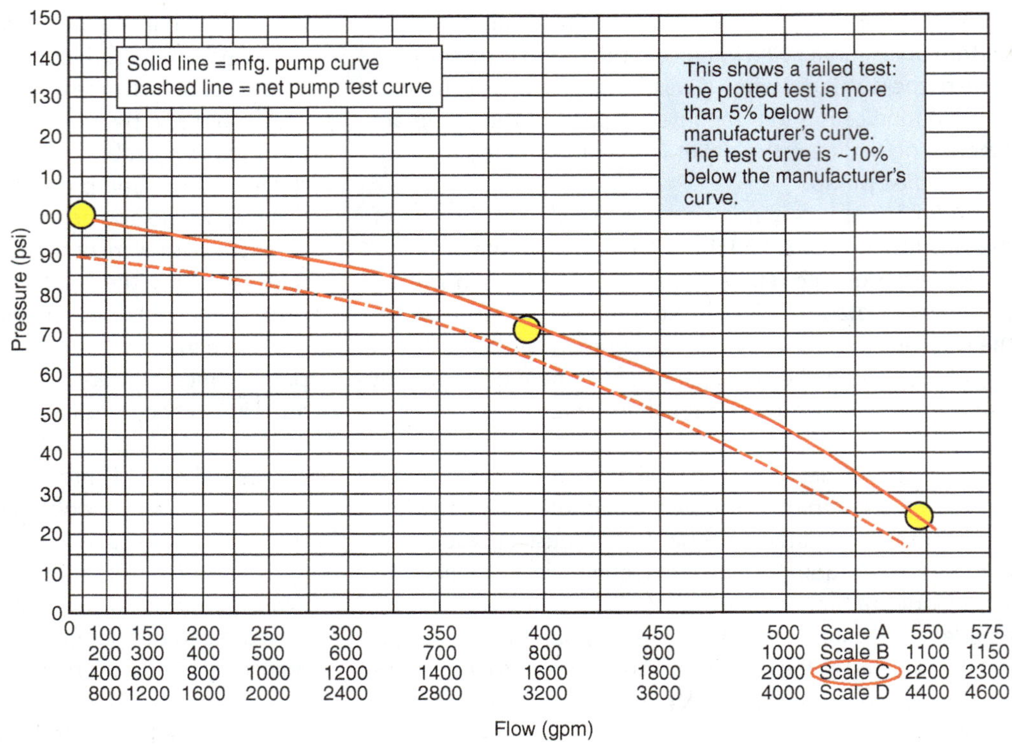

EXHIBIT S2.11 *Example of Failed Test.*

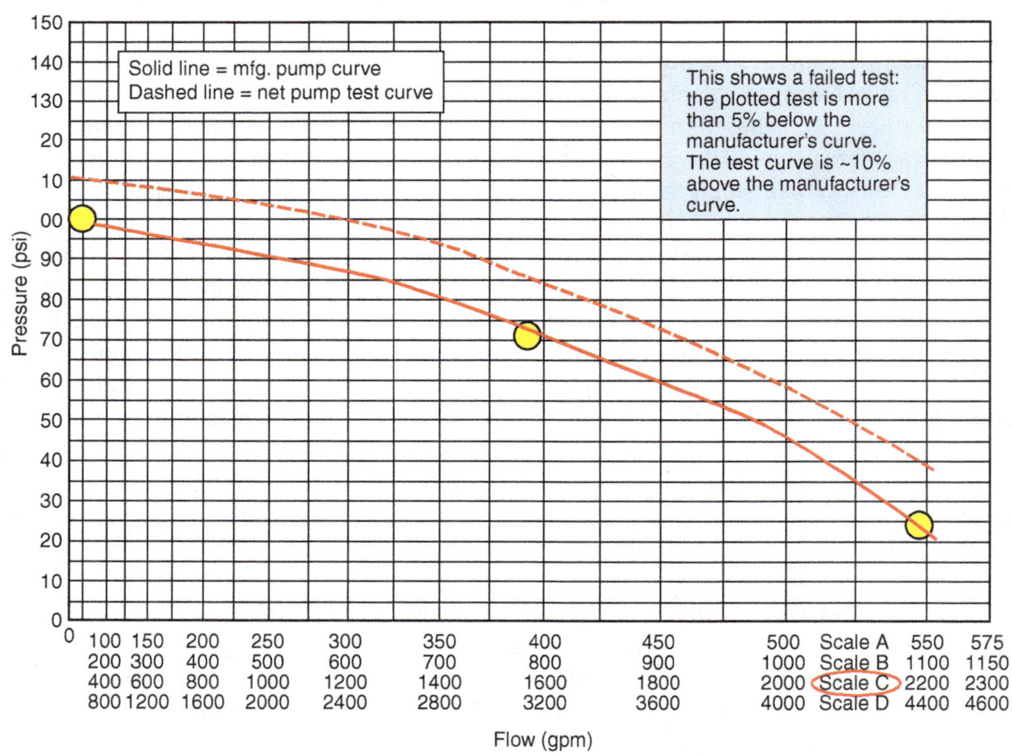

EXHIBIT S2.12 *Example of Test Results That Would Trigger an Investigation.*

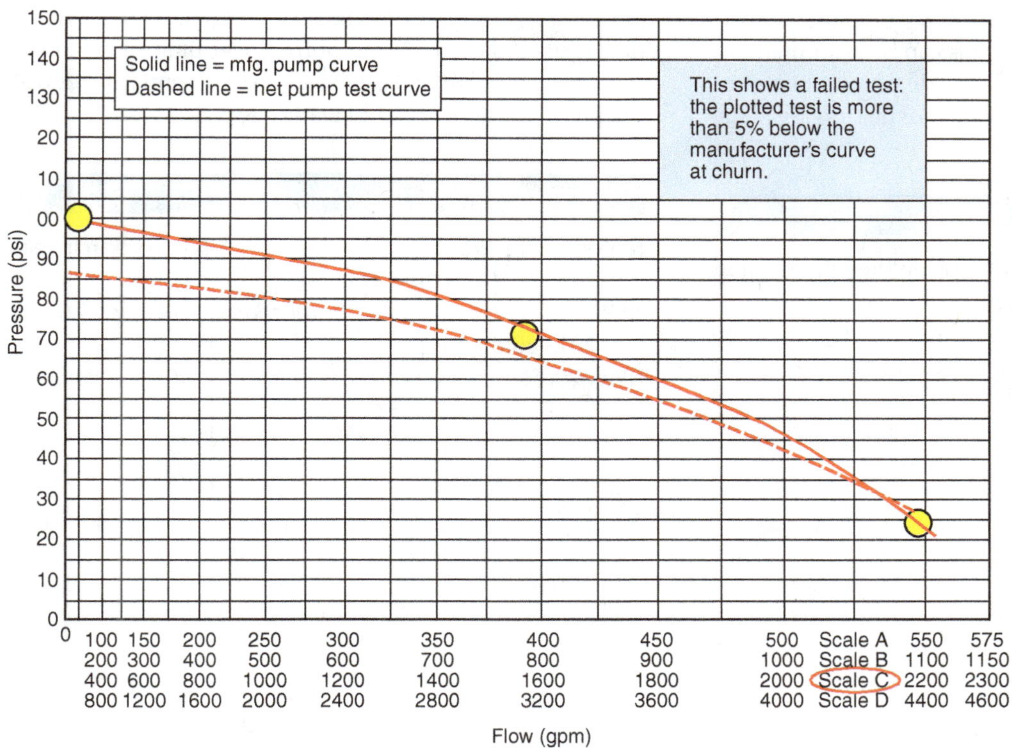

EXHIBIT S2.13 *Example of Failed Test Data.*

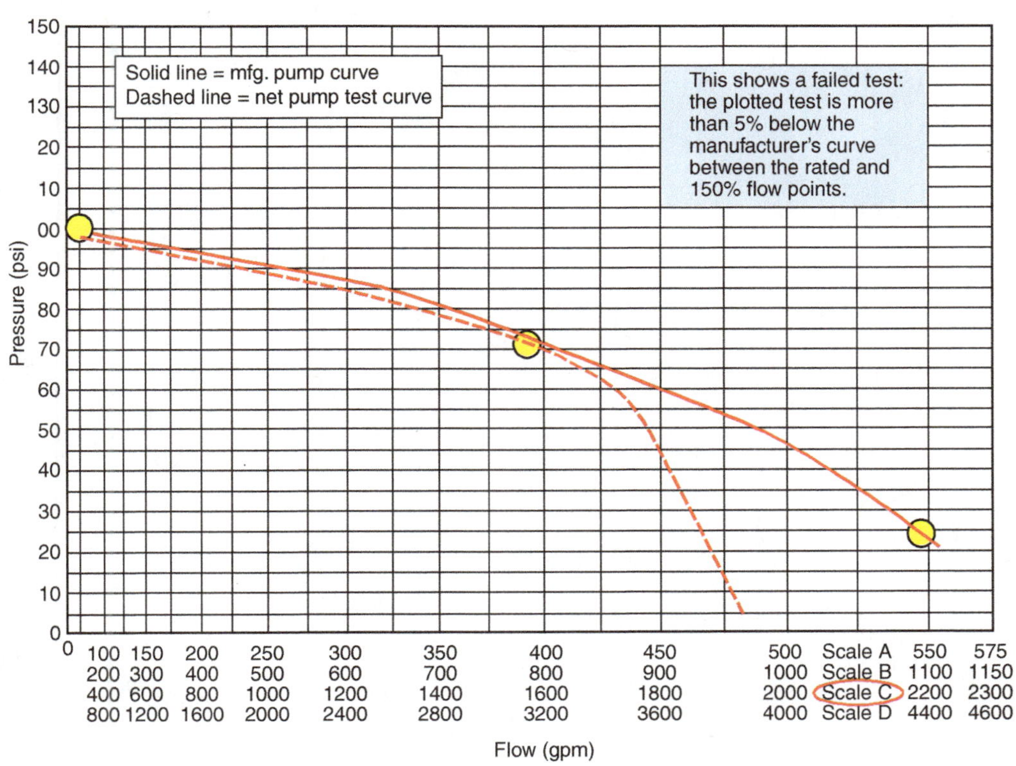

EXHIBIT S2.14 *Example of Failed Test Data.*

where most sprinkler system demands would fall on the graph. Reviewing the pump affinity laws, pressure is exponentially affected relative to flow. This can be attributed to a partially closed valve on the suction side of a fire pump or, more likely, an impeller that is damaged or has foreign material lodged in it.

Summary

Confirming a fire pump's ability to provide the required flow and pressure based on system demand is imperative to maintaining the level of life safety anticipated for a given property. Properly conducting fire pump flow tests to quantify a pump's functionality is a critical step to making this confirmation. In addition to confirming that the pump will supply the necessary pressure and flow, it is important to assess the performance of the pump against its original performance data. Degradation in pump output over time can lead to a system impairment and can be a sign of potential issues with the pump, suction piping, or water supply.

References Cited

National Fire Protection Association, 1 Batterymarch Park, Quincy, MA 02169-7471.

NFPA 13, *Standard for the Installation of Sprinkler Systems*, 2013 edition.

NFPA 20, *Standard for the Installation of Stationary Pumps for Fire Protection*, 2013 edition.

SUPPLEMENT

3

ITM ROLES AND RESPONSIBILITIES

By Rich Ray, Russell Leavitt, and John A.M. Lake

Editor's Note: *This supplement is included in the handbook to provide the reader with a snapshot of some of the roles and responsibilities associated with several key players in the water-based system inspection, testing, and maintenance (ITM) process.*

The inspector, owner, and authority having jurisdiction (AHJ) are key cogs in the ITM machine. In order for these individuals to be effective in executing ITM activities, it is important for them to understand not only what they need to be doing, but also why they are doing it. This table summarizes many of their responsibilities for a series of tasks that are associated with the ITM process. This table is not intended to be all inclusive, but rather to hit the high points and basic principles for each of the tasks outlined. For a more detailed look at the role of the inspector, owner, and AHJ, refer to the full-length narrative supplements that are provided as follows: Supplement 4, Role of the Inspector; Supplement 5, Role of the Owner or Designated Respresentative; and Supplement 6, Role of the Authority Having Jurisdiction.

Roles and Responsibilities of Key Stakeholders

Responsibility	*Role*
Overall ITM Program	**Inspector** ▪ Provide information to all interested AHJs and the property owner regarding the requirements of NFPA 25. ▪ Provide competitive pricing to the building owner for the services necessary to comply with NFPA 25. ▪ Conduct the tasks outlined in NFPA 25 in accordance with the scope of the document, which addresses wear and tear issues.
	Owner ▪ Ensure that the systems are inspected, tested, and maintained. ▪ Provide access to system components as needed, have needed repairs and corrections completed, and keep records. ▪ Evaluate the adequacy of existing systems if changes are made to the building, its use, or other conditions that could impact system performance. ▪ Understand and comply with any local requirements regarding qualifications for performing inspections, tests, and maintenance along with reporting and record keeping. ▪ Put procedures in place to address any impairments that occur, whether planned or unplanned.

Responsibility	Role
Overall ITM Program (continued)	**AHJ** ▪ Determine if an existing situation involves a distinct hazard to life or property that requires inspection, testing, and maintenance methods in excess of those required by the standard. ▪ Approve alternative programs. Enforce the requirements of a code or standard or approve equipment, materials, an installation, or a procedure. ▪ Educate the property owner of the requirements of the various codes and standards that affect the facility. ▪ Review the fire protection system as it compares to the original installation by requesting copies of the original design documentation from the owner to help identify any changes that might inhibit the effectiveness of the system. ▪ Know the requirements of NFPA 25, recognize its limited scope for AHJs, and know where to obtain the information necessary to ensure the protection system will operate correctly.
Qualification/ Competency of ITM Personnel	**Inspector** ▪ Become completely and exhaustively familiar with the scope of NFPA 25 and what NFPA does *not* require of an inspecting company. ▪ Obtain credentials to be deemed "qualified" in the jurisdiction where work is being performed, and become familiar with all types of system components from fire pumps to pressure switches. Some states require specific credentials, such as NICET Certification or Academy Exam Certification through the Academy of Fire Sprinkler Technology. ▪ Be aware if there are other agencies that impose regulations regarding qualifications and competency.
	Owner ▪ Understand that NFPA 25 is, by design, quite generic regarding qualifications for those that will perform inspections, testing, and maintenance. ▪ Know that the owner (or owner's personnel) can perform many of the NFPA 25 tasks, particularly inspections. However, many inspections and tests require a progressively higher level of technical competence. ▪ Exercise sound judgment when determining whether inspections, tests, and maintenance are to be performed by in-house personnel. ▪ Be aware of any state and local jurisdictions that have strict qualification requirements for conducting any inspection, test, or maintenance of a water-based suppression system.

	AHJ
	▪ Approve the right personnel for the task, such as those who have the following qualifications:
	  • Factory trained and certified for the specific type and brand of system being inspected
	  • Certified by a nationally recognized fire protection certification organizations
	  • Registered, licensed, or certified by a state or local authority
	  • Employed and qualified by an organization listed by a nationally recognized testing laboratory for the servicing of fire protection systems
	▪ Determine specific qualifications and competencies found throughout the standard. The requirements are often found in the text, such as in the specification of "qualified" for performing particular functions.
	▪ Know the areas in the standard where the need for qualified personnel becomes an issue for the AHJ.

Equipment Accessibility	**Inspector** ▪ Inform the property owner of the items to which access must be made available, such as pump and valve rooms, test valves, and tenant spaces. ▪ Notify the owner of the scheduled inspection in a timely fashion to allow the owner ample time to prepare for the inspection and testing activities.
	Owner ▪ Provide access to all components requiring ITM. While NFPA 25 places this responsibility with the owner, it can be problematic for properties such as apartments, hotels, and multi-unit condominiums. ▪ Coordinate access with tenants and occupants, and know that the standard gives no exceptions to the accessibility requirement except in some instances where access might be unsafe. ▪ Understand that not performing inspections, tests, or maintenance in an area because it might be inconvenient is not a valid reason for the owner to defer required tasks.

Responsibility	Role
Equipment Accessibility (continued)	**AHJ** ▪ Understand that in NFPA 25, the AHJ has a limited role in relation to accessibility to equipment. ▪ Know that the AHJ has a more significant role in the installation phase where decisions on the fire protection equipment locations are discussed and approved. When initial plans are submitted for review, the AHJ should review the design for operational aspects as well as the need to gain access to the equipment for inspection, repair, and replacement. ▪ Remember that, when looking at the installation, there are standards other than NFPA 25 that apply to accessibility. One example would be the requirement for a controller for a fire pump in 10.2.4, where it states that "Working clearances around controllers shall comply with *NFPA 70, National Electrical Code*, Article 110."
Notification of Impairment	**Inspector** ▪ Identify impaired conditions found while conducting an inspection and document the impairment. Where possible, the inspector should identify impairments to the owner during or shortly after the inspection, rather than waiting to file a written report days later. ▪ Be ready and willing to assist the owner by helping to identify the approximate duration of the shutdown and impairment until repairs can be made. ▪ Know that the notification of impairment may apply to the inspector provided that individual has the competency to make such a determination. If not, this might become the duty of another segment of the inspecting contractor's firm.
	Owner ▪ Notify all interested stakeholders when an impairment occurs, including the fire department, insurance carrier, alarm company, and other AHJs. ▪ Coordinate notifications with the service provider when a provider such as a contractor is employed by the owner. ▪ Verify that all required notifications were completed in cases where the service provider makes the notifications.
	AHJ ▪ Understand that NFPA 25 requires the AHJ to be notified of impairments to the fire protection system. ▪ Request specific information to determine what the nature of the impairment is and whether or not there are additional notifications or investigations that need to take place. The level of response from the AHJ will directly relate to the type of impairment and the length of time it will take to correct. ▪ Require standby fire personnel or an approved fire watch when potentially hazardous conditions or a reduction in a life safety feature exist due to impairment to the fire protection system. ▪ Know that when a fire protection system is out of service for more than 4 hours in a 24-hour period, the AHJ has the ability to require the building to be evacuated or an approved fire watch to be provided for all portions left unprotected by the fire protection system shutdown until the fire protection system has been returned to service (NFPA 1: 13.1.9).
Notification of System Shutdown	**Inspector** ▪ Understand that notifying interested AHJs of system shutdowns is the responsibility of the property owner. ▪ Assist the owner with tasks, as specifically requested and directed by the owner, such as providing fire watches, assisting in tagging equipment affected by the shutdown as "out of service," and providing temporary water supplies.
	Owner ▪ Understand that a majority of all system failures are the result of a shut water control valve. ▪ Notify the fire department and alarm receiving company when a system is to be shut down, and know that NFPA 25 provides strict and detailed instructions for this notification. When a service provider is employed by the owner, often the service provider will assist with the notifications. ▪ Verify that notifications were performed as needed, since the responsibility of the notification rests with the owner.

Responsibility	*Role*
Notification of System Shutdown (continued)	**AHJ** ▪ Develop a facility-specific plan with the AHJ and stakeholders to address emergency response prior to responding to the need to shut down a fire protection system. These plans should include response with a partially or fully impaired fire protection system. ▪ Know that the role and responsibility of the AHJ is the same for a system shutdown as it is for impairment. ▪ Ask simple questions such as: • Why is the system being shut down? • How long will it be shut down? ▪ Determine and establish procedures for the type and level of response that can be initiated after the AHJ receives the needed information.
Changes in Occupancy, Use, Process, or Material	**Inspector** ▪ Understand that these changes cannot be handled by the everyday inspector — a higher level of education and training is necessary, such as that of a licensed professional fire protection engineer. ▪ Know that the "more qualified" person could assist the property owner in determining the effect that such changes will have in regards to the fire sprinkler system. ▪ Understand that, in instances where the inspector becomes aware of a change in occupancy or hazard during an inspection and chooses to notify the owner, it should be done on a separate report with a disclaimer that this information is beyond the scope of an NFPA 25 inspection and does not establish the fact that a full design evaluation has been conducted.
	Owner ▪ Conduct an evaluation of existing systems whenever changes to a building or its use are contemplated. ▪ Understand that the evaluation is not a part of the normal ITM required by the standard and must be performed by someone who is qualified. ▪ Make whatever corrections are needed to ensure that the system will perform as needed if the evaluation finds that the existing system is inadequate to protect the building or the hazard. ▪ Request that a service provider perform an evaluation of the fire protection systems if many existing buildings have had numerous changes to the original layout or use. ▪ Know that many contractors have personnel that are qualified to do this evaluation or can assist the owner in finding a qualified consultant or engineer.
	AHJ ▪ Evaluate the system using NFPA 25 and the installation standard for the year the system was installed. By using both standards, the AHJ can reasonably be assured that the fire protection system will function correctly. ▪ Determine if any changes have been made to the fire protection system. The standard assumes that the original system design and installation have not changed and the water supply or building use is the same as originally constructed. This is an incorrect assumption that should not be perpetuated. ▪ Conduct an evaluation of any proposed building changes before any change is incorporated. Permits must be applied for and the required plans submitted for approval prior to making any changes. ▪ Be aware the standard provides that if a property owner is contemplating changing a process or material, then it is the owner's responsibility to arrange for the evaluation of the fire protection system. However, there might be changes that the property owner is not aware of, including the following: • Moving stock from one area to another (change of commodity in a given area) • Increase in storage height (change in volume of combustibles) • Component material changes not by the property owner (commodity class change) • Material stored (material owner is not the same as the property owner) ▪ Communicate with the property owner on a regular basis and make visits to the facility, such as with the responding fire station that performs the preplanning for the facility. The engine crews can provide valuable information to the AHJ, gather important information for response, and participate in preparation of response plans (in case of impairment or system shutdown), as well as help the owner with questions that might arise. ▪ Look at the fire protection system as it compares to the original installation, and wherever any device, equipment, system, condition, arrangement, level of protection, fire-resistive construction, or any other feature is changed, notify the property owner of the need to correct the deficiency.

Responsibility	Role
Records Retention	**Inspector** ▪ Assist the owner by explaining the requirements of NFPA 25 as they pertain to record keeping. ▪ Suggest methods to organize and store records, and ensure the owner is aware that he or she is responsible for maintaining records. ▪ The inspector might choose to keep a set of records to make future inspection and tests easier, but the inspector is not responsible for retaining the documents for other parties.
	Owner ▪ Maintain records of inspection, testing, and maintenance of water-based systems. ▪ Keep records of the original acceptance tests, as-built drawings, and other close-out documents. The standard allows the keeping of these records electronically, and many service providers maintain records for owners that can be accessed at any time using the Internet. ▪ Verify that the owner retains ownership of the records should the owner choose to change service providers.
	AHJ ▪ Retain acceptance test records for the life of the system, and refer to these records when performing a facility inspection. Use the records to compare the existing conditions with the original installation to determine if any changes have been made that will affect the fire protection system. ▪ Require the ITM service to provide copies of their reports, and incorporate these reports into the AHJ's property file for the facility. It is recommended that the AHJ request copies of the inspection reports from all the ITM service providers in their jurisdiction. ▪ Give a specific timeline to the ITM service for providing the reports so that the report can be evaluated and deficiencies can be corrected in a timely fashion. ▪ Require that a log book be kept at the facility containing the original plans, acceptance test documents, and any inspection or repair documents.
Inspection Frequency	**Inspector** ▪ Strictly follow the requirements of NFPA 25 as they pertain to the frequency of inspections. ▪ Educate the AHJs and property owner with what NFPA 25 requires in regard to inspection frequency. ▪ Note that other agencies might impose regulations regarding inspection frequencies that differ from NFPA 25. These should be considered only when they are more stringent than the frequencies in NFPA 25.
	Owner ▪ Follow the inspection frequencies provided in NFPA 25, and understand that there is nothing to restrict more frequent inspections. ▪ Know that if the owner desires to perform inspections less frequently, a performance-based program can be instituted that involves the owner, demonstrating that less frequent inspections will not decrease the reliability of the system below acceptable standards. ▪ Have all performance-based programs be approved. ▪ Use a performance-based program only after following a methodology such as that outlined in A.4.7.
	AHJ ▪ Evaluate the inspection reports to determine if the inspections are being performed in accordance with the standard and, if not, formulate an action plan to address the deficiency. ▪ Note that jurisdictions might allow the inspection frequency to be modified to less than those listed in the standard. These modifications are based on a well-documented performance-based inspection schedule. Copies of this documentation should be maintained in the facility log book as well as in the AHJ's property record. ▪ Document any undocumented variance to the standard's inspection frequencies as a failure of the system. ▪ Evaluate what has been presented and determine if the schedule is acceptable. However, if the AHJ feels uncomfortable approving the performance-based schedule, he or she can always require a third-party expert to evaluate the schedule, the costs of which are borne by the proponent of the design.

Responsibility	*Role*
Tools and Safety Equipment	**Inspector** ▪ Keep tools and testing equipment — hoses, gauges, amp/volt meters, diffusers, and so forth — clean, calibrated, and in good working condition. ▪ Wear the proper personal protective equipment (PPE) — safety glasses, hard topped boots, and gloves — as part of a daily routine. ▪ Comply with the requirements of *NFPA 70E, Standard for Electrical Safety in the Workplace,* for fire pump testing, including arc-protection PPE such as insulated gloves, rubber boots, face shields, and body aprons, because many older controllers require the inspector to directly measure amp and volt readings while the controller is fully energized.
	Owner ▪ Maintain strict safety protocols that must be followed by employees and vendors. ▪ Ask for confirmation or evidence that service providers comply with state and federal OSHA requirements along with site-specific safety needs. ▪ Inform the service providers of any hazard materials that might be encountered along with the current Material Safety Data Sheets (MSDS).
	AHJ ▪ Understand that the typical municipal AHJ working for the building or fire department has a limited role in the preparation of tools and safety equipment for the ITM inspector. ▪ Be aware that there might be several different AHJs for an activity, since the numbers of regulations for the various activities have different enforcers. ▪ Determine the level of involvement needed. The local AHJ might be involved in permitting and inspection of work sites, such as confined space entry, or in an inspection that involves entry into an area that requires use of respiratory protection.
System ITM Activities	**Inspector** ▪ Always be prompt, professional, and proficient in his/her job (the three *P*s). ▪ Explain what you are doing, and always ask the owner if an escort is required. ▪ Clearly explain the results and findings of inspections and tests. ▪ Think ahead of time where the water will travel to prior to performing any given test that involves the discharge of water. ▪ Be respectful of the owner's property and inform all occupants and tenants of any alarms that will be sounding and any other ancillary building systems that might be affected during the testing.
	Owner ▪ Determine the level of involvement. When the owner contracts ITM tasks to a service provider, the level of involvement for many items is subject to the discretion of the owner, such as whether or not to accompany those performing ITM. ▪ Require direct involvement by the owner for the items that follow: • Providing access to all system components that require ITM • Notifying all persons occupying the property of any alarms, flowing water, or other potential disruptions associated with ITM • Ensuring that all required impairment procedures are performed
	AHJ ▪ For the most part, the AHJ doesn't attend or conduct the ITM activities outlined in NFPA 25. These activities are performed by employees of fire protection contractors, who are contracted by the property owner to comply with the requirements of specific law. ▪ The reports generated by these inspections are useful tools to identify deficiencies in the fire protection equipment.

Responsibility	*Role*
ITM Reports/Forms	**Inspector** ▪ Make sure all reports, forms, and any written correspondence is clear, legible, and well written. ▪ Submit all reports in a timely manner. ▪ Note all deficiencies and refer to the appropriate section(s) of NFPA 25.
	Owner ▪ Make sure that all required records are complete and legible when using a service provider. ▪ Verify that all reported tasks were completed. ▪ Understand all reports or forms before signing, as many terms, descriptions, or deficiencies might be unfamiliar to the owner. ▪ Ask for and refer to a copy of NFPA 25 to be certain that all required work was performed.
	AHJ ▪ Require a copy of the ITM reports. ▪ Evaluate the report to determine if any corrective actions are needed to ensure that the protection system is operational. ▪ Request copies of all inspection reports, with specific timelines for providing the reports, to ensure that the deficiencies are corrected as quickly as possible. ▪ Review the inspection report to make sure it includes a detailed explanation of every deficiency and an indication of whether the inspection frequency is a weekly, monthly, quarterly, or annual inspection.
Corrective Actions for Deficiencies and Impairments	**Inspector** ▪ Provide a proposal for making any necessary corrections. ▪ Make sure these proposals are competitive, clear in scope, and provided in a timely manner. ▪ Clearly identify any long lead time items. ▪ Note what equipment or building areas will need to be accessed to perform the work.
	Owner ▪ Correct any deficiencies identified during ITM. ▪ Verify that the person or entity making corrections or repairs is qualified to do so, and ensure that any jurisdictional requirements for licensing or the certification of those making corrections are followed. ▪ Understand that while simple corrections can be made by the owner's personnel, such as locking a control valve in the appropriate position, more complex corrections might require a contractor. ▪ Note that there is no requirement for the inspecting entity to make corrections.
	AHJ ▪ Contact all of the ITM service providers in the jurisdiction, and request copies of the inspection reports. ▪ Provide a specific timeline to the ITM service provider for completing the reports, so there is time for the report to be evaluated and deficiencies to be corrected in accordance with the policies of the jurisdiction. ▪ Investigate the cause of any impairments to ensure that other systems in the area are not affected by the same issues. ▪ Take the necessary action to ensure a level of response commensurate with the decrease in protection provided by the impaired system. ▪ Determine the level of protection for a facility and the level of protection required in the event a system is impaired.
System Tagging	**Inspector** ▪ Assist the owner with tagging procedures and in complying with Chapter 15. ▪ Note that the inspecting firm can also provide assistance with the tags and help establish the most effective wording.
	Owner ▪ Understand that NFPA 25 has limited tagging requirements. ▪ Attach tags to system components, showing the date of inspections, tests, or maintenance along with other information. ▪ Ensure that tags have been properly completed and placed in cases where a jurisdiction requires the use of tags.

Responsibility	*Role*
System Tagging (continued)	**AHJ** ▪ Use fire protection system tags as an initial indication of a system's condition, but do not rely solely on the information. ▪ Review the inspection report to identify the information needed to determine system readiness. ▪ Remember that the inspection report and tagging of the fire protection system relates strictly to the NFPA 25 inspection and does not look at all of the conditions that could cause a fire protection system to fail. ▪ Remind the property owner of the scope of the NFPA 25 inspection and educate them on the other conditions that could cause failure.
Follow-Up on Corrective Actions	**Inspector** ▪ Re-perform the inspection and testing activities as outlined in NFPA 25. ▪ For examples, see Table 8.6.1 for fire pumps and Table 5.5.1 for fire sprinkler systems.
	Owner ▪ Make all corrective actions and repairs as directed. ▪ Verify that all corrections have been completed and documentation has been provided to the appropriate authorities when using a service provider.
	AHJ ▪ Enforce the provisions of NFPA 25 if the AHJ has adopted the appropriate enabling document, such as NFPA 1, *Fire Code*. ▪ Contact the ITM service provider when a report is received that lists deficiencies, and determine if the repairs have been made. ▪ Determine the level of deficiency and severity of hazard if the repairs have not been made. ▪ Contact the facility owner/agent if the deficiency is a relatively low hazard, and request a timeline for repair or a fire safety improvement plan.
Planned Impairment Programs	**Inspector** ▪ Assist the owner by providing fire watches, establishing temporary water supplies, and determining the duration of impairment. ▪ Assemble and procure the needed materials, tools, manpower, and equipment to perform any needed repairs *prior* to commencing the planned impairment.
	Owner ▪ Ensure that all steps listed in Chapter 15 are completed, as these are the minimum requirements for impairment programs. ▪ Obtain assistance from service providers in implementing an impairment program. ▪ Ensure that a system is in compliance with the insurer's requirements, as many insurance companies have strict impairment procedures. ▪ Notify the fire department that a system is impaired, and obtain approval from the fire department when using alternative water supplies, fire watches, and other impairment activities that could impact fire department response and fire fighting tactics.
	AHJ ▪ Note that when the ITM service provider has determined that a fire protection system must be taken out of service, the provisions of the facility emergency plan and policies of the AHJ should be discussed, and those portions of the plans should be implemented prior to taking the system out of service. The planned impairment might also require plans to be submitted and a permit obtained. ▪ Where a fire protection system is out of service for more than 10 hours in a 24-hour period, the impairment coordinator must arrange for one of the following: • Evacuation of the building or portion of the building affected by the system out of service • An approved fire watch • Establishment of a temporary water supply • Establishment and implementation of an approved program to eliminate potential ignition sources and limit the amount of fuel available to the fire

Responsibility	Role
Emergency Impairments	**Inspector** ▪ Assist the owner, at the owner's specific direction, by providing fire watches, establishing temporary water supplies, and determining the duration of impairment. ▪ Assemble and procure the needed materials, tools, manpower, and equipment to perform any needed repairs once authorized to do so.
	Owner ▪ Know the importance of having a plan to follow when an unplanned impairment occurs. ▪ Have someone on site at all times who is knowledgeable about the system(s), including the location and operation of water supply control valves and drain valves. ▪ Make sure the water supply is shut off promptly and drains are opened to significantly reduce water damage when an unplanned water discharge occurs, such as a broken sprinkler. ▪ Maintain spare sprinklers, because acquiring specialized or hard-to-obtain critical components can lengthen the time a system is out of service and increase the associated costs.
	AHJ ▪ Preplanned actions need to be discussed with the ITM service provider in all cases involving impairments to ensure a consistent level of protection for the property. ▪ Discuss policies involving the disabling of any fire protection system prior to that system being taken out of service. ▪ Provide contact information that will allow the ITM service provider to notify the appropriate authorities any time a system is taken out of service. ▪ Determine what level of alternative protection must be provided once the AHJ or designee is notified that a system is out of service. Part of that evaluation might include the following: • Level of impairment • Amount of time the system might be out of service • Type of hazard without protection • Potential for an uncontrolled fire • Risk to the community
Restoring System to Service	**Inspector** ▪ Re-perform the inspection and testing activities as delineated in NFPA 25. ▪ Assist the owner with removing all tags and informing interested AHJs of system status.
	Owner ▪ Ensure that systems are back in service. ▪ Ask for documentation that all required tests were performed if a service provider is used. ▪ Ensure tags are removed and that the fire department, other AHJs, and the alarm receiving station are notified. ▪ Ensure that all required tasks are completed either by the service provider or owner's representative.
	AHJ ▪ Consider requiring that the fire protection systems be evaluated to ensure that the design addresses the hazards. ▪ Consider having the work performed by a qualified contractor. ▪ Inspect the fire protection system after being notified that a repair to the system is complete to ensure it is operational. ▪ Keep in mind that NFPA 25 provides for a certain level of inspection to be performed. However, depending on the level of repair, the AHJ might require additional inspection/testing to be conducted.
Obstruction Investigations	**Inspector** ▪ Clearly explain the importance of obstruction investigations and the procedures that will be followed in performing the investigation, such as what system components will need to be accessed, which areas of the facility will be disrupted, and so forth. ▪ Clearly identify and take precautionary efforts regarding where the water will travel. ▪ Identify and communicate to the owners and AHJ the duration of all system shutdowns, and assist the owner in complying with NFPA 25 regarding the shutdown. ▪ Explain all findings of the investigation, including submitting a detailed report complete with pictures as necessary. ▪ Note that some of what is outlined here might be the duty of the actual inspector, while some might be that of another entity of the inspecting firm.

<table>
<tr><th>Responsibility</th><th>Role</th></tr>
<tr><td rowspan="2">Obstruction Investigations (continued)</td><td>Owner
▪ Follow a prescribed frequency, or determine the frequency of internal pipe inspections, by completing a risk assessment. The owner should not take this assessment lightly and should have a qualified individual or entity conduct it. A 5-year inspection frequency for internal pipe inspections is designed to keep the cost at a minimum by having the inspection correspond to the internal inspection of check valves, which are also on a 5-year frequency.
▪ Conduct an obstruction investigation when an inspection or other issues reveal the potential for obstructions. Section D.3 contains procedures for conducting an obstruction investigation.</td></tr>
<tr><td>AHJ
▪ Note that an obstruction investigation should be viewed similar to an impairment. The AHJ should be notified of obstruction investigations and to what degree the fire protection system will be impaired.
▪ Request specific information to determine what the nature of the investigation is and whether or not there are additional notifications or investigations that need to take place.
▪ Determine a level of response, but note that it will directly relate to the type of obstruction found.
▪ Consider impairment procedures if the fire protection system is going to be out of service. The fire department must be able to respond accordingly to a particular incident.</td></tr>
<tr><td rowspan="3">Performance-Based ITM Programs</td><td>Inspector
▪ Inform and educate owners and AHJs of alternative means and methods of complying with NFPA 25.
▪ Provide variances on costs, activities to be performed, equipment to be accessed, and so forth.
▪ Make sure the owner and AHJ understand clearly the advantages and disadvantages associated with differing means of complying with NFPA 25.</td></tr>
<tr><td>Owner
▪ Understand that although NFPA 25 gives the owner the option to use a performance-based ITM program, great care should be exercised in doing so.
▪ Include the fire department and other interested AHJs when developing a performance-based program.
▪ Determine a failure rate that is acceptable to all stakeholders.
▪ Note that performance-based programs are generally suitable to owners with large numbers of properties and/or components in order to develop a database large enough to make a realistic assessment of potential frequencies and failure rates.</td></tr>
<tr><td>AHJ
▪ Evaluate performance-based options as a means to provide a greater degree of safety and, potentially, a lower cost to the facility owner.
▪ Look to provide the highest level of protection with the lowest degree of risk when developing the program to provide an alternative means of compliance.
▪ Evaluate testing programs using a method that is acceptable to the AHJ and appropriate for the conditions. Each design should be realistic and apply to conditions found in the facility.</td></tr>
</table>

References Cited

National Fire Protection Association, 1 Batterymarch Park, Quincy, MA 02169-7471.

NFPA 1, *Fire Code*, 2012 edition.

NFPA 70®, *National Electrical Code®*, 2014 edition.

NFPA 70E®, *Standard for Electrical Safety in the Workplace®*, 2012 edition.

SUPPLEMENT 4

ROLE OF THE INSPECTOR

By Terry L. Victor

Editor's Note: *In an effort to assist those involved in training inspectors and those who are being trained, this supplement has been included in the handbook to outline the training needs, safety equipment, and basic duties and approach to the performance of work for the professional inspector. It includes information for the training and education of the professional water-based fire protection systems inspector. This supplement can be used for formal training classes or for individual study. It is not part of NFPA 25 and is not intended to be an enforceable part of the standard.*

The inspection and testing of a fire protection system requires specialized knowledge of system components, including valves, piping, hangers, test connections, and sprinklers. This specialized knowledge includes recognition of different types of components, their intended use, manufacturer's testing requirements, and Underwriters Laboratories listing criteria. The inspection and testing of a water-based fire protection system requires expertise that is not usually associated with building engineers or building maintenance personnel, and 4.1.1.2 states that "Inspection, testing, and maintenance shall be performed by qualified personnel."

Therefore, many property owners and occupants rely on and contract with fire protection contractors who offer inspection and testing services. Although NFPA 25 requires some training and experience for personnel involved in performing the tasks required by the standard, it cannot dictate the exact training and experience for every situation in which the standard might be adopted and used. For example, some jurisdictions might require licensing for individuals providing professional inspection, testing, and maintenance (ITM) services, whereas other jurisdictions do not.

This supplement is intended to offer examples illustrating the role and duties of the inspector. It is not intended to be an all-inclusive list of responsibilities, but rather an outline of the background and qualifications necessary to perform the work. The discussion of preplanning, tools and equipment, and the approach to an inspection is intended to provide the inspector and property owner with a basic plan for inspecting any building or system.

Many activities in NFPA 25 require only a visual inspection to determine the status and/or the condition of a component or system. These activities can be taught to on-site personnel and require a minimal amount of training. Most of the weekly and monthly inspection requirements fall into this category, and training on-site personnel might be the most cost-effective way to comply with the standard. This supplement deals primarily with the activities of the inspectors employed by fire protection contractors and the methods used to conduct the inspection and/or tests required by NFPA 25. Methods used to compile and record results are also addressed. For those who feel they are qualified to perform this work, proceed with caution and have the telephone number of a reputable contractor in your area available, in case you are unable to silence the fire alarms, reset the dry pipe valve, or restore the fire pump to normal service.

Inspector Responsibilities

The purpose of periodic inspection and testing of fire protection systems is to provide a reasonable amount of assurance that, at that moment in time, the system will operate as intended during a fire emergency. Note the following in A.1.2 of NFPA 25:

> History has shown that the performance reliability of a water-based fire protection system under fire-related conditions increases where comprehensive inspection, testing, and maintenance procedures are enforced.

The contract with a fire protection contractor to perform the ITM of a water-based fire protection system is not an insurance policy or an engineering evaluation. If the property owner is ever uncertain whether the fire protection system as installed will provide the minimum protection required by the applicable codes, the owner should contact a fire protection engineer or qualified contractor to perform an analysis of the building, the building occupancy, the commodity classification of any storage present, the system design criteria, and the water supply.

The inspection and testing of a fire protection system in accordance with NFPA 25 is *not* equivalent to any of the following:

1. An engineering or authority having jurisdiction (AHJ) review or evaluation of the adequacy of the system to control or extinguish a fire in the protected occupancy
2. An engineering or AHJ review or evaluation of the adequacy of the water supply, including water tanks and fire pumps, to provide the necessary water flow and pressure needed to meet the fire protection system demand as designed
3. An engineering or AHJ review or evaluation of the hazards present in the facility to determine the required system criteria for minimum protection levels
4. An engineering or AHJ review, evaluation, acceptance, inspection, or test of the fire protection system to determine whether the system was designed and installed in accordance with the applicable installation standard, including NFPA 13, *Standard for the Installation of Sprinkler Systems*; NFPA 13R, *Standard for the Installation of Sprinkler Systems in Low-Rise Residential Occupancies*; NFPA 14, *Standard for the Installation of Standpipe and Hose Systems*; NFPA 15, *Standard for Water Spray Fixed Systems for Fire Protection*; NFPA 16, *Standard for the Installation of Foam-Water Sprinkler and Foam-Water Spray Systems*; NFPA 20, *Standard for the Installation of Stationary Pumps for Fire Protection*; NFPA 22, *Standard for Water Tanks for Private Fire Protection*; and NFPA 214, *Standard on Water-Cooling Towers*
5. An engineering or AHJ review or evaluation of all areas of the facilities to determine whether all areas of the facility are protected in accordance with the original applicable installation standard

The NFPA Technical Committee on Inspection, Testing, and Maintenance of Water-Based Systems has made it clear that the role of the inspector performing inspection and testing activities in accordance with NFPA 25 is not that of a fire protection engineer or an AHJ. The inspector is not required to be a fire protection engineer, nor is the inspector required to have detailed knowledge of the fire protection system installation requirements in the applicable NFPA standards. Therefore, when property owners contract with a fire protection contractor, they should not expect the inspector to provide an engineering analysis detailing the adequacy of the system design when compared to the hazard, or evaluate the accuracy and completeness of the installation.

The inspector is not required to perform the following functions:

- Review the design criteria for the system to determine whether it is adequate for the commodity or hazard
- Review the installation shop drawing to make sure all pipes are correctly sized and the sprinklers are the correct type, orifice size, and temperature rating
- Perform a water supply analysis to determine whether the water supply is sufficient to provide the required gallons per minute at the required pounds per square inch for the system
- Perform a complete building inspection to determine whether closets, blind spaces, attics, crawl spaces, and all other special building features are adequately protected in accordance with the version of the installation standard in force at the time of the installation
- Recognize that storage commodities and/or arrangements are different from those anticipated when the system was designed and installed
- Research the installation contract files to determine whether special requirements were required by the AHJ

Inspector Qualifications and Training

The inspector of a water-based fire protection system, whether an employee of the property owner, the owner's authorized representative, or an employee of a fire protection contractor, has a tremendous responsibility. Water-based fire protection systems protect both life and property. These systems have a history of reliability that exceeds most other mechanical systems. In those instances when a water-based fire protection system fails to extinguish or control a fire, poor maintenance and shut water control valves are the most frequent reasons for the failures. Although a properly conducted ITM program cannot guarantee the performance of a water-based fire protection system, the chances of timely system operation and fire control during a fire emergency are increased dramatically. The inspector is responsible for conducting inspections and tests in accordance with NFPA 25, industry-accepted practices, and manufacturer's recommendations. Given the importance of these systems, the choice of whom to hire and train as an inspector should be given careful consideration.

Inspector Qualifications

A person hired as an inspector who has little or no knowledge of fire protection systems will obviously have a more difficult time during training than someone who has worked in or around the fire protection sprinkler industry and has knowledge of these systems. The inspector-in-training with little or no knowledge of fire protection systems will have to learn the systems and components, as well as the requirements of NFPA 25, and the proper ways to conduct an inspection and/or test.

Whatever the background of the individual hired as a water-based fire protection systems inspector or trainee, several personal traits must be considered as well. The role of the inspector is to properly perform the activities required by NFPA 25, which are meant to provide the highest level of assurance that systems will respond as expected during a fire emergency. Without a certain work ethic and sense of responsibility, the performance of these activities will not be carried out as accurately or thoroughly as expected or, in some cases, not carried out at all. The following sections describe the desired personal traits of the person hired to perform inspection and testing activities and allow the reader to fully understand the total scope of competency and training required of the NFPA 25 inspector.

Basic Traits. There are some basic traits that an inspector should possess to do the job completely, accurately, efficiently, and professionally. Aside from related work experience, the person being hired to be an inspector should meet the following minimum qualifications:

Conscientiousness and Motivation. An inspector must be conscientious and aware of the importance of the job being performed. The need to perform every inspection completely and thoroughly must be the number one priority of the inspector. An inspector must possess a good work ethic, be self-motivated, and have the ability to self-supervise. Some contractors schedule their inspectors and even set up appointments to have systems inspected, whereas others rely on the inspectors to establish their own schedule and make their own appointments. Whichever the case, the inspector needs to be at the property when scheduled. The owner, the owner's authorized representative, and/or the occupants expect the inspection to take place as scheduled. The monitoring station and the local authorities might expect alarms to be tested at certain times. An inspector who is aware of schedules and appointments shows regard for the other individuals involved in supervising and maintaining water-based fire protection systems.

Dexterity. An inspector should be mechanically inclined. Some of the activities that must be performed by an inspector include resetting or restoring to service dry valves, preaction valves, deluge valves, accelerators, exhausters, air compressors, fire pumps, controllers, alarm switches, and numerous other system components. The inspector must also be able to interpret test results, perform basic troubleshooting, and make recommendations to correct deficiencies or impairments found.

Education. An inspector must be educated and have legible writing. Local codes as well as NFPA 25 must be read, understood, and applied as inspections are performed. The inspector must have the ability to read the manufacturer's maintenance information and follow the instructions. Inspection forms must be filled out so that those responsible for system maintenance can read them and correct any deficiencies found. In recent years, computer programs have been developed to aid in the inspection process. These programs include question sets that correspond to the activities required by NFPA 25. In addition, the programs provide standardized descriptions of, and describe corrective actions for, impairments and deficiencies. If computers are used as part of the inspection process, then computer skills are also necessary.

Communication Skills. An inspector must be able to communicate well, both orally and in writing. Once an inspection is complete, the inspector must be able to translate the results of the inspection and/or test onto the inspection report. These results must then be explained to the owner or the owner's authorized representative. For most fire protection contractors, the inspector works face-to-face with the customer more than anyone else in the company; therefore, the inspector's relationship with the customer reflects the attitude and professionalism of the contractor.

Organization. An inspector must be organized and have an organized office behind the scenes. Keeping track of the names and addresses of the properties to be inspected, the testing frequencies required for the types of systems and components in each building or structure, and the previous inspection reports requires organized tracking methods. Some contractors rely entirely on the inspector to keep these records organized, whereas others provide office staff to assist.

Appearance and Hygiene. An inspector should be neat in appearance and have good personal hygiene. The inspector has to interact with owners, owners' authorized representatives, building occupants, building engineers, approving authorities, and other inspectors. Having a neat appearance is especially important for the inspector who works for a fire protection contractor, as he or she represents the professionalism of the contractor. Having good personal hygiene is a common courtesy to those who have to work with the inspector. The pride that an inspector has in himself or herself is usually reflected in the quality of the work they do.

Character and Background. Any inspector, whether working for a fire protection contractor or directly employed by the property owner, must also possess the personal characteristics necessary to perform work without being influenced by those who may not want the inspections or tests done properly. Every inspector should approach the inspection and testing of a water-based fire protection system as the most important job he or she has ever had. Lives and property are at stake. Therefore, the inspector should be public safety–oriented, self-motivated, responsible, a good communicator, thorough, organized, ethical, and honest (i.e., would not falsify a report for any reason). Many facilities being inspected require security clearance and background checks of their own employees as well as contractor employees entering their sites. Most fire protection contractors in turn perform background checks on candidates before hiring an inspector or inspector trainee.

The candidate who does not have the complete set of minimum qualifications and basic traits described here will struggle in the performance of the responsibilities of an inspector. Some of the traits described, such as communication skills, appearance, and education level, can be learned. The others, such as organizational ability or character, are inherent for the most part, and either cannot be taught or are taught with much difficulty and effort.

Experience. Many inspectors working for fire protection contractors have previous experience with water-based fire protection systems either as a sprinkler fitter, an engineering technician, a fire service inspector, or an insurance underwriter. Inspectors with previous experience in these fields have a varied range of knowledge of water-based fire protection systems and how these systems should operate during a fire emergency.

The sprinkler fitter who installed fire protection systems knows how to read installation shop drawings, knows different system configurations, and knows all of the standard system components. An inspector with this background also knows the methods of concealing piping systems, which is useful in locating pipes when inspecting a system for the first time. Many valves and system components that need to be inspected or tested are concealed behind access panels in walls or above ceilings. The sprinkler fitter who serviced and repaired water-based fire protection systems has a more thorough knowledge of how these systems work and what the possible reasons are for bad test results or failures. The service sprinkler fitter has replaced the gaskets in dry pipe valves, rebuilt exhausters and accelerators, and cleared obstructions from all parts of a system, from mains to branch lines to trim piping.

The engineering technician has a good theoretical knowledge of systems and their components. The layout of a fire protection system involves the ability to visualize how the pipe, fittings, and hangers fit in the building or structure. All of the various valves and system components are detailed on the shop drawings. With a background in performing and translating hydraulic calculations for various types of systems, the engineering technician should be able to better interpret main drain test and fire pump test results.

The fire service inspector has experience and skills that closely parallel those needed by the inspector of water-based fire protection systems. Knowledge of applicable codes and the ability to interpret them is required in both positions. The fire service inspector has performed a visual walkthrough of entire systems to ensure that they were installed in accordance with the approved shop drawings and has witnessed the testing of systems conducted by the installing contractor.

The insurance underwriter has performed the same inspections and tests that the fire service inspector has, and has seen firsthand how the systems perform. The insurance underwriter has been on the scene after a fire emergency and has documented how problems in a water-based fire protection system, such as a painted sprinkler or a clogged branch line, have impaired the system's ability to control fires. The insurance underwriter has also been involved in the rating of properties with and without fire systems and therefore knows the financial benefit of properly maintaining these systems.

Training the Inspector

Inspector training is a process of teaching skills to individuals over a period of time until a minimum competency level is obtained that will allow the inspector to perform basic inspection and test activities. However, training does not stop at this point, unless the inspector is limited to performing basic activities such as inspecting and testing wet and dry systems for the remainder of his or her career. In reality, the training process never stops. As the requirements of NFPA 25 change, as new types of systems and components are developed for the industry, and as more sophisticated inspection and testing equipment are introduced, additional training will be necessary.

A program developed for the training of inspectors should address the various levels of experience of the individuals being trained. If the individual is new to the fire protection industry, training must start at a rudimentary level. Terminology within the fire protection industry is somewhat specialized and the knowledge of the components that make up a fire protection system must first be taught. Training seminars are available from several organizations, such as NFPA, National Fire Sprinkler Association (NFSA), and American Fire Sprinkler Association (AFSA). Internet-based training is also available. Some colleges and universities have developed curriculums specific to training inspectors of water-based fire protection systems.

Certifications

The National Institute for Certification in Engineering Technologies (NICET) sponsors certification programs for inspectors of water-based fire protection systems, in addition to other fire protection subfields. NICET certification can be used to determine the competency of inspectors in the application of NFPA 25 and other standards and practices used in the field. The following three levels of certification are available, based on written verification of the individual's experience and ability to pass a written exam:

- Level I — less than 2 years of experience
- Level II — 2 to 5 years of experience
- Level III — over 5 years of experience

Once certified, inspectors must maintain their certification by accumulating professional development points from on-the-job experience, by attending training seminars, or by participating in specific industry-related activities. Inspectors must accumulate 90 professional development points every three years to maintain an active certification. For more information regarding NICET certification, see www.nicet.org. Many states and local jurisdictions now require NICET certification for the licensing of fire protection contractors and/or inspectors. Contact the state or local jurisdiction for their licensing requirements.

Training should be documented, not only for the advancement of personnel but also to protect against possible lawsuits. If the individual has a background in fire protection, knows the industry terminology, and can identify system types and components, training can concentrate on the actual inspection and test activities. In today's world, any incident that involves a loss of life or property becomes a target for lawsuits. One way to defend against lawsuits that attempt to prove negligence is to submit training records and personnel records that prove the competency of both the contractor and the individual inspectors. Maintaining high standards in hiring practices and providing continuous training is necessary if legal problems are to be avoided.

Before Conducting an Inspection and/or Test

Conducting an inspection and/or test for the first time in a building will require more time than will subsequent inspections and/or tests. The initial inspection will require that the inspector meet the owner or the owner's authorized representative for the first time. The inspector must become familiar with the building, the systems installed in the building, the system components, and the records of the previous inspections and tests performed, then determine the necessary inspection and test tools, equipment, and instruments needed. The inspector must also plan the inspection and test sequence and prepare the inspection reports. Subsequent inspections and tests take less time because the building and systems are known and the inspector can focus on inspecting and testing the systems and components.

Written Agreements

An inspection agreement and its specifications should be read carefully by both the owner and the occupant. Paragraph 4.1.1.3 of NFPA 25 states the following:

> Where the property owner or designated representative is not the occupant, the property owner or designated representative shall be permitted to delegate the authority for inspecting, testing, maintenance, and the managing of impairments of the fire protection system to a designated representative.

As with any contract for services, a properly executed document removes any ambiguities, misunderstandings, or confusion. More importantly, the inspection agreement defines the financial responsibilities and liabilities of the owner and the fire protection contractor.

An inspection/test agreement describes the activities provided by a fire protection contractor for the owner, occupant, management firm, or managing individual. These agreements normally include the following information:

Building address
Description and quantity of the system(s) being inspected/tested
Frequency of the inspection/test activities
Reports
Annual cost of the agreement
Payment terms
Renewal terms
Limits of liability
Signature space
Other specifications, including:
  Criteria
  Notification
  Working Hours
  Cancellation
  Access

Building Address. Referencing the street address of the building being inspected is necessary to avoid any misunderstandings by the property owner, the fire protection contractor, and the inspector(s). Also, a building's name, owner, or management company might change, making it difficult to maintain a historical record of inspections and tests of the fire protection systems. The building's street address generally remains a consistent point of reference.

Description and Quantity of the System(s) Being Inspected/Tested. As in any contract, the services being provided for the price indicated must be clearly defined. Many buildings have several fire protection systems that are required to be inspected and tested by NFPA 25. For instance, a typical high-rise office building will have a backflow prevention device, one or more fire pumps, a standpipe system, wet pipe sprinkler systems on each floor, and dry pipe systems in the parking garage and mechanical penthouse. There might even be one or more preaction systems in computer rooms and a fixed water spray system in a cooling tower. Does the contract include all water-based fire protection systems in the building? Even if the intent of the contract is to include all systems in the building at the time the contract is executed, it is still advisable to include a list of all systems in the building. Systems can be added or changed, which will affect the contract price.

NFPA 25 contains a series of tables that describe what action must be performed on systems and/or components when corrective action is taken to address a deficiency or impairment. Each table is referenced in the subsection in which it appears, for example: "Whenever a component in a [type of] system is adjusted, repaired, reconditioned, or replaced, the action required in Table 5.5.1 shall be performed." Many of the actions required include reinspecting and/or retesting of systems or components after they are adjusted, repaired, reconditioned, or replaced. The fire protection contractor might be contacted to reinspect and/or retest the system or component, and the inspector might be called to return to the facility to perform the inspection or test. These inspections and tests are performed separately from those conducted on a periodic basis per NFPA 25, and the agreement should specify if these additional inspections and/or tests are included in the contract price. Because the types of reinspections or retests required or the frequency of these inspections and tests cannot accurately be determined ahead of time, most contracts will not include them. The fire protection contractor can price and provide these inspections and tests on an as-needed basis.

Frequency of the Inspection/Test Activities. NFPA 25 requires that certain activities be performed at prescribed frequencies. However, rarely does an inspection and test agreement assign all of these activities and frequencies to a fire protection contractor. To comply fully, an implied partnership is formed between the owner or the owner's representative and the fire protection contractor. All of the requirements of NFPA 25 are the responsibility of the owner, and the fire protection contractor relies on the owner to perform all inspections and tests not specified in the agreement.

A common agreement for a wet sprinkler system will specify that the fire protection contractor enter the building on a quarterly or semiannual basis to inspect the gauges, control valves, alarm devices, hydraulic nameplate, and the fire department connection, as well as to test some of the alarm devices. In addition to these activities, flow switches and valve tamper switches will be tested semiannually; the building, pipe hangers, pipes, fittings, and spare sprinklers will be inspected annually; and the main drain will be tested annually. Such an agreement would specify a quarterly or semiannual inspection and test in accordance with NFPA 25. All other weekly and monthly activities are expected to be performed by the owner, occupant, management firm, or managing individual.

Unless specified in the agreement, all other less-frequent activities are not included, such as testing gauges every 5 years; testing sprinklers every 5, 10, 20, or 50 years; and internally inspecting pipes, alarm valve, check valves, strainers, filters, and orifices every 5 years. These activities are normally performed by the fire protection contractor but are invoiced as additional periodic services.

Reports. Records of ITM activities are to be maintained by the property owner. The fire protection contractor generally provides reports of the inspection and test activities performed at each visit. Many local jurisdictions require that such reports be kept at the site. Local ordinances might also require that a copy of any reports be submitted directly to the AHJ.

Reports come in as many versions as there are contractors providing services. Many forms are still filled out by hand, although computerized programs with printed reports are becoming more and more popular. The electronic reports are more professional and legible. When preprinted report forms are filled out by hand, multiple copies can be provided on the spot using multi-sheet forms or the occupant's copy machine. Some companies provide printers for technicians using computer-based inspection programs, but the more common practice is to input the test results off-site, create the report in the office, and e-mail it to the customer.

The inspector or the office administrator should be able to reference the agreement to determine who has been authorized by the owner or the owner's representative to receive the inspection report forms.

Annual Cost of the Agreement. Of course, the most important part of any agreement is the price. A typical agreement is priced for a one-year term. The price can be broken down into quarterly payments and in some cases per system. However, most agreements state a lump sum that covers all described activities and frequencies.

Payment Terms. Payment terms are extremely important to both the fire protection contractor and the owner. The owner needs to know when to expect billing for budgeting purposes, and the fire protection contractor needs to meet payroll and expense needs. Typically, one of two types of payment terms is used in the majority of agreements.

The first type stipulates that the full annual amount is invoiced up front and paid before any inspection or test activities begin. This ensures that the fire protection contractor will have the necessary funds available to meet payroll and expense needs throughout the term of the agreement and creates less paperwork for both parties. There is one invoice, one accounts payable transaction, one check posting, and no collection effort.

The second type stipulates that invoicing is done after each inspection in an amount equal to the work being performed. Many agreements with quarterly inspection frequencies invoice one-fourth of the annual amount after each inspection/test. This payment arrangement usually forces the fire protection contractor to wait until the last inspection of the agreement term to perform any time-intensive activities, such as tripping dry pipe valves and flow testing fire pumps. The owner might be able to budget more favorably with this payment schedule, but it requires four times the office work: four invoices, four accounts payable transactions, and four check postings. Another downside to invoicing quarterly after each inspection is that any delay in the payment of one of the invoices can affect the schedule for the subsequent inspections. Good business practices dictate that a fire protection contractor does not perform additional work if payment is late for the previous work performed.

Many other payment term options, such as one-twelfth invoiced per month, are commonly used by local governments on large agreements, but this option is the exception to the norm.

Renewal Terms. Every agreement should spell out the renewal terms. Typically, an annual agreement renews automatically each year with a cost of living increase tied to the federal index or to pre-described increase percentages per year. Most agreements can be cancelled by either party with 30 days' written notice, provided all payments have been paid for services rendered.

Some agreements require special action on the part of the owner and/or fire protection contractor. The owner might require a new agreement each year, including any revised pricing, or a separate purchase order might have to be issued each year. These provisions should be spelled out in the agreement.

Limits of Liability. Every agreement should state the limits of liability for the fire protection contractor. This legal terminology protects the fire protection contractor for losses that are beyond the fire protection contractor's control and responsibility. No one can guarantee that a water-based fire protection system will work as designed and installed at any given time. As previously stated, fire protection systems are more reliable than just about any other mechanical system, and the fire protection industry has an excellent track record. However, unexpected and improbable occurrences do happen. A clogged sprinkler cannot be detected during an inspection or test, nor can an inspector predict when a leak in a pipe is going to develop. Despite the difficulty of predicting every potential problem or failure of a system, litigation is very common for a loss event that is associated with a fire protection system, in an attempt to assign blame to everyone from the architect, to the engineer, to the general contractor, to the installing sprinkler contractor, to the fire protection contractor performing inspections and tests. The unanticipated loss could cost any of these parties hundreds of thousands of dollars to defend, unless contractual language exists in the agreement limiting this exposure.

Signature Space. The agreement is a contractual document and therefore must be signed to be legally binding. The agreement should include designated lines at the end of the document for the fire protection contractor and the owner or the owner's authorized representative to sign, print their name and title, and insert the date. An unsigned agreement is not legally considered an agreement.

Agreement Specifications

Specifications are a binding part of an agreement when referenced as an attachment in the body of the agreement. Specifications normally include the more detailed requirements of a contract. Some specifications include the following:

- *Criteria* indicate which specific NFPA 25 requirements are included.
- *Notification* describes who is responsible for notifying occupants of an inspection or test.
- *Working hours* are the hours during which the inspection/test activities can be performed.
- *Cancellation* refers to a cancellation clause in the agreement.
- *Access* refers to the buyer's responsibility to provide access to the inspector.

Communication with the Property Owner

The fire protection contractor is required by NFPA 25 to communicate with the property owner or owner's authorized representative, or the occupant, management firm, or managing individual, as described in the agreement, to obtain certain information concerning changes in occupancy, use, or layout of the building.

Communication with the owner or owner's authorized representative, or the occupant, management firm, or managing individual, is also necessary to schedule an inspection or test. Throughout the inspection and testing process, communication will be both verbal and written. Once an executed contract exists, communication then involves the actual performance of the inspection and/or test.

In some cases, arrangements are made so the inspector can arrive at the property without an appointment and perform the inspection and/or test. In such a case, however, the owner or the owner's authorized representative, or the occupant, management firm, or managing individual, must be available either in person or by phone to answer questions on the inspection report, and the inspector must be provided with the necessary keys and access codes to inspect throughout the protected areas of the property.

Safety Laws and Regulations

Generally, it is the responsibility of the property owner to advise anyone working in the building of specific hazards present. The inspector should, however, request copies of such items as a material safety data sheet (MSDS) and should determine whether any confined spaces or other hazardous conditions exist.

Tools and Equipment

Essentially three categories of tools and equipment are necessary to properly inspect and/or test a fire protection system: safety equipment, standard inspection tools and equipment, and standard test equipment.

Safety Equipment. Safety must be a matter of primary importance and should be second nature to the inspector. Federal and state regulations require compliance with safety laws. Common sense dictates that care and caution must be taken to ensure the safety of the inspector, and those working around the inspector, from certain hazards inherent in inspection and test procedures. As a minimum, federal Occupational Safety and Health Administration (OSHA) regulations must be adhered to. If not, employee safety is at risk. OSHA levies stiff fines on companies whose employees do not follow these national regulations. If violations are repeated, the fines increase exponentially.

Table S4.1 describes examples of the safety equipment needed and some general uses for each.

Inspection Tools and Equipment. Table S4.2 lists some of the inspection tools and equipment commonly used by inspectors.

Test Equipment. Table S4.3 lists some of the standard test equipment for the inspection and testing of water-based fire protection equipment.

TABLE S4.1 *Safety Equipment*

Safety Equipment	General Use for Equipment
Hard hat	Hard hats are required when head injury is a possibility from climbing, crawling through confined spaces, or falling objects.
Safety shoes	Safety shoes are required in most industrial and manufacturing plants and at any location where foot injury is possible from stepping on sharp objects or from heavy objects falling on the feet.
Proper clothing	Loose clothing is not to be worn around machinery. In some plants, fire and/or chemical protective clothing must be worn.
Eye protection	Safety glasses should always be worn when inspecting or testing fire protection systems.
Hearing protection	Hearing protection should be a part of the standard safety equipment worn by inspectors. Many areas in industrial occupancies post signs where hearing protection must be worn.
Safety harness	See OSHA regulations for the requirements for safety harnesses. Safety harnesses are required when inspectors must be on ladders or lifts above a certain height.
Confined space equipment	Confined space equipment requirements are not limited to pits or cellars but can also extend to more common spaces, such as unventilated attics, crawl spaces, concealed spaces, and vaults. See OSHA regulations for the requirements for confined space entry.

Inspection Procedure

NFPA 25 does not prescribe how to perform an inspection or test. However, the standard does list the systems and components that are to be inspected and tested, spell out the acceptance criteria, and identify the corrective action requirements.

The actual inspection process is unique to each facility, building, and system. The NFPA 25 report forms available from NFPA's website, if followed line by line, lead the inspector through the process in one way. However, if these report forms or any other report forms are followed line by line, duplication of effort and doubling back over the same parts of the same systems will occur.

This section describes one method of performing an inspection. However, as each system is unique, so is each inspector. What works for one will be inefficient for another. Inspectors must develop their own processes. However, certain bare essentials are universal. Before the experienced inspector can put together a game plan to perform the activities in the most efficient and comprehensive way possible, certain information must be reviewed. This information can be divided into three basic areas: preparation for the test, performance of the test, and post-test processing of information.

TABLE S4.2 *Standard Inspection Tools and Equipment*

Tool	General Use and Notes
Ladders	The inspector should use the right height ladder at all times, to avoid overextending his or her reach. When using extension ladders, always follow the "1 to 4 rule," setting the base of the ladder 1 ft (0.3 m) away from the wall for every 4 ft (1.2 m) in height. The inspector must follow all other safety rules concerning ladders as detailed in OSHA and local requirements, including safety harnesses and fall protection.
Flashlight	A flashlight is particularly useful when working in confined spaces.
Binoculars	Binoculars are typically used for examination of sprinklers from the floor level. The inspector is not expected to climb a ladder to inspect each sprinkler, but binoculars can be helpful when inspecting sprinklers for leakage, corrosion, foreign materials, loading, paint, physical damage, or improper orientation.
Tape measure or folding rule	The size of some pipes or components might not be evident, and the inspector might need to measure the size to provide an accurate report.
Assorted hand tools	To perform certain internal inspections, the inspector will have to remove face plates from valves or remove sprinklers or flushing connections. Screwdrivers, a socket set, crescent wrench, hammer, and a set of open-end wrenches should be included in the inspector's toolkit.

TABLE S4.3 *Standard Test Equipment*

Equipment	General Notes on Use
12 in. pipe wrench	
Crescent wrench	
Calibrated test gauges	
Volt meter	Primarily used during the annual fire pump test. Proper safety equipment should be worn while using this equipment.
Amp meter tachometer	
Hydrometer or refractometer	Used to measure the concentration of antifreeze.
Pitot tube	Used to measure velocity pressure during flow and fire pump tests. (See Chapters 6 and 8.)
Stopwatch	A watch that reads in minutes and seconds. Used to measure trip times for dry pipe systems.
Heat lamp	Used to activate heat detection systems on water spray fixed systems.
Digital gauges	
Water analysis kit	Used when performing the five-year obstruction inspection if tubercules or slime are found in the piping.

Preparing for the Test

Familiarity with the Contract. The inspector needs to know the entire scope of the agreement and the types of systems being inspected. He or she must become familiar with the layout of the building and identify where the system controls and valves are.

Scope. The inspector must know how many systems are to be inspected and tested, what type of systems they are, and the frequency of the inspections to be performed. With this information the inspector can plan the time it will take to perform the inspection or test and determine the equipment needed to perform these activities. The terms and conditions of the contract must be reviewed at this time to determine whether the inspection and test activities are to be performed during off-hours, whether there are special payment terms, and whether there are any other special requirements that will affect the inspection and test process.

Property to Be Inspected. The inspector should review any as-built drawings available or should, as a minimum, review previous inspection reports or sketches that are retained in the owner's inspection file.

Contact Information. Usually, the inspection contract will have contact information. A scheduled appointment might be necessary prior to performing the inspection; contact with the property owner or owner's representative, or the occupant, management firm, or managing individual, should be made at this time.

Planning. Based on the inspection contract and type of systems involved, the NFPA 25 inspection and test requirements should be reviewed prior to entering the building.

Preplanning will save time and can reveal components or systems that require special attention.

Preplanning will reveal the specific inspection and test tasks needed to be performed, which in turn will dictate the types of tools and equipment needed to perform them. Preplanning will also highlight the needed report forms.

Before the Inspection/Test

The following procedures represent normal practices used when performing a sprinkler inspection and/or test.

Walk the Perimeter. Upon arrival at the location, the inspector should drive or walk the exterior perimeter of the building to become familiar with the building layout. This helps to determine some of the features of the fire protection systems and the location of many of the components being inspected and tested. The perimeter walk enables the inspector to locate the fire department connection, motor water gong, main valve room, inspector's test valves, main drains, any outside valves [i.e., post indicator valve (PIV), wall post indicator valve (WPIV)], fire pump header — which also indicates the approximate size of the pump — and any specific conditions that the inspector needs to be aware of before flowing any water.

Meet the Owner or the Owner's Authorized Representative. Upon entering the building, the inspector should report to the owner or owner's representative, or the occupant, management firm, or managing individual; provide company identification, which includes the inspector's name; and explain the purpose of the visit, including the steps involved in performing the inspection. The inspector should then ask a series of questions that deal with occupancy changes, content changes, building changes, and system(s) status, filling out the inspection report forms as responses are provided. The inspector should also ask to be directed to the room housing the interior valves controlling the fire protection system(s). Generally, a conscientious inspector will observe the areas being passed through on the way to the valve room and make note of anything that is pertinent to the inspection.

Locate Systems and Equipment. At the system valve room, the following steps are to be taken before starting any actual testing:

- Note the type, manufacturer, size, and number of systems to be inspected and/or tested.
- Note the location, type, manufacturer, size, and quantity of any flow switches or tamper switches.
- Alert the building occupants. Determine whether the alarms are monitored through the building fire alarm system and, if so, warn the building occupants of a possible alarm.
- Locate the fire alarm panel. Verify accessibility to the fire alarm equipment and determine whether the system is monitored by a central station.
- Notify the monitoring company. If it is monitored, call the central station and the appropriate local fire response authorities.
- Verify that an authorized building engineer or in-house user is on site to reset any alarm systems.
- Notify the AHJ/local fire department. Every inspector's worst nightmare is hearing sirens wailing as the fire trucks pull up in front of the building while conducting a flow test of a fire protection system. Not only is this embarrassing, but a false alarm can be costly to all parties involved. A false alarm could potentially divert a fire department from a real emergency. The owner, or the fire protection contractor, or both, can be fined by the fire department for causing a false alarm. The business at the property loses productivity because of the disruption. The inspector loses valuable time while meeting with the fire officials to assure them it was a mistake and not intentional. Finally, after the fire engines leave, the building occupants return to work, and the alarm company is properly notified, the test can continue. Do not depend on the central station, the property owner, or anyone else to notify local fire response authorities. Write down the names of individuals that were notified and the times they were notified.

Conducting the Inspection or Test

Inspect/Test. Once the preparatory steps have been taken, actual inspection and testing can begin. The activities described here are for an annual inspection and test (with the exception of full flow tripping dry, preaction, or deluge systems). The requirements in NFPA 25 should be followed for these specific activities and for inspections and tests during other frequencies.

Perform a Visual Inspection. The purpose of a walkthrough is to visually inspect all areas in the building visible from the floor level. It is the owner's responsibility to declare any such added areas or rooms when answering the general questions at the beginning of the inspection. However, during the walkthrough the inspector should look for areas of the building that are not protected by the fire protection system, such as new additions, new rooms, or new mezzanines. These should be brought to the owner's attention and recorded, as a courtesy to the owner, not as an obligation of the agreement or as a requirement of NFPA 25.

Additional observations and testing that must be conducted include the following:

- Perform a visual inspection of the alarm, dry, preaction and/or deluge system valve, and all associated trim. Look for any signs of leakage, rust, and any unusual problems or conditions.

- Determine where test water will discharge. Verify that the water from the main drain will discharge in a safe location.
- Conduct a main drain test, which includes the following steps:
  1. Note and record the static pressure on the supply side of the system valve.
  2. Conduct the main drain test by opening the main drain valve fully.
  3. Note and record the residual pressure with the valve wide open.
  4. Observe the water discharging from the main drain test fitting and note any debris or restrictions of any kind.
  5. Flow water through the main drain until the water stream is clear.
  6. Close the main drain valve and record the gauge readings.
  7. Note and record the static pressure after the valve is fully closed.
- Verify the operation of the water motor alarm (if applicable). On a wet system it will ring during the main drain test. On dry, preaction, and deluge systems, it will ring when the test valve in the system valve trim is opened.
- Close and fully open all control valves to verify proper operation. Test the tamper devices if applicable.
- Locate and flow all inspector's test valves for the required 90 seconds and verify alarm activation. Reset any alarms and continue with walkthrough.
- Look for proper coverage and orientation of sprinklers. Also observe sprinklers for evidence of loading, paint, corrosion, damage, and any physical obstructions to discharge patterns.
- Observe any exposed system piping and hangers for corrosion, leakage, misalignment, and physical damage.
- Record all conditions observed during the walkthrough. When this has been done, the inspection is complete.

Note that the inspector is not required by NFPA 25 to determine whether the fire protection system as installed provides adequate coverage in the building or whether all areas of the building are protected in accordance with the original applicable installation standard. The installation codes change every three to five years, and what is required in the most recent code might not have been required in earlier versions, and vice versa. The inspector is not required to examine concealed spaces, closets, attic spaces, or any other building feature to determine whether the fire protection system extends into those areas. If the property owner is ever uncertain whether the fire protection system as installed will provide the minimum protection required by the applicable codes, the owner should contact a fire protection engineer or qualified contractor to perform an analysis of the building, the building occupancy, the commodity classification of any storage present, the system design criteria, and the water supply.

Compile Results. Many inspectors fill out inspection and test reports as they perform the activities. Others compile results, observations, and deficiencies on a separate piece of paper and transfer the information to the actual report at the end of the inspection. Either method is acceptable. Using the inspection form as the inspection and test is performed helps to eliminate missed activities and aids in establishing a systematic process.

Complete Reports. All pressures, trip times, flows, and other recorded results are to be written in the report. All deficiencies are to be clearly detailed, and corrective action recommended.

Many inspectors employed by fire protection contractors started their fire protection careers in other areas of the industry, so they have specific knowledge of codes and standards other than NFPA 25. Because of this specific knowledge, these inspectors might observe potential problems with the water-based fire protection systems, to other fire protection systems, or to general safety requirements that are unrelated to the NFPA 25 inspection being performed. Examples of some problems typically observed include the following:

- Sprinklers appear to be too far from walls or appear inadequately spaced.
- Sprinklers appear too close to building construction features that could obstruct spray patterns.
- Emergency exits are chained and locked.
- Flammable liquids are present and not stored properly.
- New kitchen cooking equipment has been added and is not protected by the hood suppression system.
- Fire extinguishers are discharged or are missing.
- Fire alarm zones are out of service.
- Fire doors are propped open.

Inspectors are often conscientious, so they are compelled to alert the owner or owner's representative, or the occupant, management firm, or managing individual, of their observations. The inspector should not note these observations on the NFPA 25 inspection form. Instead, the inspector should provide a separate attachment describing the observations and advise the owner or owner's representative to have them checked out by the appropriate fire protection system or building code professional.

Completing the Inspection and/or Test

To complete the inspection or test, many of the processes required to start these activities must be repeated in reverse.

It is just as important to go through the following steps upon the completion of the inspection or test as it was before starting.

- All systems are to be returned to normal service.
- Test and drain valves are to be closed tightly until no water flow or dripping is observed. All flow and tamper switches are to be returned to normal settings.
- All alarm panels are to be returned to their normal condition. To do this, confirm that all alarm, trouble, or supervisor signals have been verified at the panel, and then reset the panel.
- Fill out any tags provided and/or required.

Some jurisdictions have a visual tag procedure that the inspector must follow. A green tag indicates the system did not have any deficiencies or impairments. A yellow tag indicates one or more deficiencies were found and therefore the application of one or more components is not within its designed limits or specifications, but there were no impairments found. A red tag indicates one or more impairments were found and the system might not function as intended during a fire emergency. If a red tag is issued, the owner or owner's representative, or the occupant, management firm, or managing individual, must be notified immediately so emergency impairment procedures can be initiated.

Advise Occupants. If occupants were notified of the test, notify them that the test is complete.

Notify the AHJ/Local Fire Department. Notify the proper authorities that the test is complete and the system has been returned to normal service. Write down the name of the individual who received the notification and the time of notification.

Notify the Monitoring Company. Notify the monitoring company, if applicable, that the test is complete. Have the individual perform a system check to verify they are receiving the proper signals. Write down the name of the individual who performed the system check and the time of notification.

Review Reports. After all systems are back in service and all notifications have been made, meet with the owner or owner's authorized representative, or the occupant, management firm, or managing individual, and review the results of the inspection and test. The report should be explained line by line, and any unusual readings or findings discussed. Deficiencies should be clearly explained, and if necessary show them the deficiencies. Explain the requirements of NFPA 25 pertaining to acceptable results and corrective action that must be taken to address deficiencies.

Follow Impairment Procedures, if Applicable. If impairments are found, impairment procedures must be implemented by the owner. It is important for the owner to tag systems and/or components that are not working properly, especially those impaired. The inspector should advise the owner of the impairments and the steps required by NFPA 25 to correct those impairments.

Sign Reports. When the inspector has completely explained the inspection and test results with the owner or the owner's authorized representative and is satisfied that the results and deficiencies have been clearly explained, the report is to be signed and dated by both parties. The owner or the owner's representative should be given a copy of the signed report.

After the Inspection

Inspection Follow-Up. An inspection and/or test performed without proper follow-up is useless. Protection of lives and property is dependent on properly operating water-based fire protection systems. By contracting with a reputable fire protection contractor, the owner has made a commitment to the employees, the insurance company, stockholders, and the community at large that the system(s) is in proper working order. The fire protection contractor has an obligation to follow up with the inspection in a timely fashion.

Notification Process. Although not required by NFPA 25, it is good practice to directly notify the property owner upon completion of an inspection and/or test. In most cases, the owner is not the individual who meets with the inspector on site and reviews the report after the inspection. NFPA 25 does require that the owner maintain records of all inspections and tests for a period of time. Therefore, the fire protection contractor should send a copy of each report to the owner. In addition, it is a good practice, as well as a preferred legal procedure, to attach a cover letter to any report that contains deficiencies or impairments. The cover letter should clearly explain all deficiencies and/or impairments and should detail the requirements of NFPA 25 to have them corrected. If the fire protection contractor is also in the repair and installation business, an offer to make the corrections and the price for those corrections can be included in the letter. For liability reasons, it is a common practice to send cover letters that describe impairments by certified mail and to call the owner to notify them of the impairment. Should a fire emergency occur between the time the impairment is discovered and the owner is notified of the impairment, lawsuits might be filed as insurance companies assess blame for the system failure.

Corrective Action. When an owner performs repairs to the system or contracts the repair to a fire protection contractor, the date the repairs were made and the corrective action taken should be recorded. This information is then passed on to the inspector at the next inspection and noted on the new inspection report.

Component Replacement Action Requirements. NFPA 25 includes a series of tables describing what action must be performed on systems and/or components when corrective action is taken to address a deficiency or impairment. Each table is referenced in the subsection in which it appears, for example: "Whenever a component in a [type of] system is adjusted, repaired, reconditioned, or replaced, the action required in Table 5.5.1 shall be performed." Many of the actions required include reinspecting and/or retesting of systems or components after they are adjusted, repaired, reconditioned, or replaced. The fire protection contractor might be contacted to reinspect and/or retest the system or component, and the inspector might be called to return to the facility to perform the inspection or test. These inspections and tests are performed separately from those conducted on a periodic basis per NFPA 25, and the results should be recorded on a separate inspection/test report. These additional reports should be retained by both the owner and the fire protection contractor for future reference.

Maintaining Inspection Files. Proper filing of reports by the fire protection contractor is necessary to provide ongoing inspection and test services to the customer. The reports must be accessible for reference should the property owner call with questions about the reports or if an additional copy is needed for the AHJ or insurance underwriter. The inspector will also need copies of previous reports when the next inspection is performed. Comparisons between current test results and previous test results must be made to determine whether degradation of water supplies, system piping, fire pump performance, and so forth is taking place.

Code Enforcement

Laws enacted by local, city, and state legislatures allow for the enforcement of building codes as well as NFPA codes, including NFPA 25. Many of these codes are modified by local ordinances, with some mandating more stringent requirements and some relaxing requirements. Each jurisdiction adopts those sections of a code that meet its needs and the needs of the community. Noncompliance with these laws can result in fines, loss of business license, and in some cases, businesses being shut down. In the case of NFPA 25, the inspection and testing requirements might be adopted without exception, but some frequencies might be amended to require fewer inspections and/or tests per year. If requirements are relaxed by local ordinance, the fire protection contractor might decide to still comply with NFPA 25 frequencies. NFPA 25 is a minimum standard, and therefore, many fire protection contractors make the business decision to fully comply with NFPA 25 requirements even though they are not required to do so by local ordinances.

References Cited

National Fire Protection Association, 1 Batterymarch Park, Quincy, MA 02169-7471.

NFPA 13, *Standard for the Installation of Sprinkler Systems,* 2013 edition.

NFPA 13R, *Standard for the Installation of Sprinkler Systems in Low-Rise Occupancies,* 2013 edition.

NFPA 14, *Standard for the Installation of Standpipe and Hose Systems,* 2013 edition.

NFPA 15, *Standard for Water Spray Fixed Systems for Fire Protection,* 2012 edition.

NFPA 16, *Standard for the Installation of Foam-Water Sprinkler and Foam-Water Spray Systems,* 2011 edition.

NFPA 20, *Standard for the Installation of Stationary Pumps for Fire Protection,* 2013 edition.

NFPA 22, *Standard for Water Tanks for Private Fire Protection,* 2013 edition.

NFPA 214, *Standard on Water-Cooling Towers,* 2011 edition.

SUPPLEMENT

5

ROLE OF THE OWNER OR DESIGNATED REPRESENTATIVE

By Russell Leavitt and J. William Sheppard

Editor's Note: *In an effort to assist building owners in selecting the right inspector or in training their employees in inspection procedures, this supplement has been included in the handbook to outline the training needs, safety equipment, and basic duties and approach to the performance of an inspection or test of a water-based fire protection system. This supplement is intended to illustrate the methods for in-house inspection, testing, and maintenance of water-based extinguishing systems and the use of reporting forms to document the required activities in NFPA 25. A sample dry pipe sprinkler system complete with a fire pump is used as an example of the systems being maintained.*

Owner's Responsibility

Choosing Personnel for Training

There are many instances where an owner or designated representative has qualified personnel and equipment to adequately perform inspection and testing of water-based extinguishing systems. Further, the owner or designated representative might have the capability to perform the necessary maintenance required to correct any deficiencies noted as a result of the inspection or test.

The qualifications for this in-house performance can be as a result of "skilled trades" training in mechanical and electrical issues, training of employees by the installing contractor using the manufacturers' documentation of the equipment, and other outside training materials and methods that are available.

In-house expertise is not always available. However, where the talent is available, the primary consideration is the cost to perform tasks effectively and to assure that satisfactory inspection, testing, and maintenance (ITM) of the systems and components are accomplished.

The personnel responsible for the inspections, testing, and maintenance should be mechanically and electrically trained. Some of the activities that must be performed include resetting or restoring to service dry valves, preaction valves, deluge valves, accelerators, exhausters, air compressors, fire pumps, controllers, alarm switches, and numerous other system components.

Some of the basic skills that the owner or designated representative should look for in a candidate for this training include critical thinking ability, familiarity with codes and standards, communication skills, and organizational ability.

Critical Thinking. In-house personnel must have the ability to read the manufacturer's maintenance information and follow the instructions. They must also be able to interpret test results, perform basic troubleshooting, and make recommendations to correct deficiencies or impairments found.

Familiarity with Codes and Standards. Inspection personnel must understand and apply local codes, as well as NFPA 25, as inspections are performed. Inspection forms must be filled out completely so that those responsible for system maintenance can read them and correct any deficiencies noted. In recent years, computer programs have been developed to aid in the inspection process. These programs include question sets that correspond to the activities required by NFPA 25. In addition, the programs provide standardized descriptions and corrective actions for impairments and deficiencies. If computers are used as part of the inspection process, then computer skills are also necessary.

Communication Skills. An inspector must be able to communicate well, both orally and in writing. Once an inspection is complete, the inspector must be able to translate the results of the inspection and/or test onto the inspection report. These results must then be explained to the facility or maintenance staff responsible for corrective action.

Organizational Ability. An inspector must be organized. Keeping track of the testing frequencies and the previous inspection reports required for all of the types of systems and components in each building or structure requires organized tracking methods. Some owners or designated representatives rely entirely on the inspector to keep these records organized, whereas others provide office staff for this purpose.

The candidate who does not have the minimum qualifications as described will struggle in the performance of the responsibilities of an inspector. Communication skills and education level can be improved. The other characteristics, however, are for the most part inherent, and either cannot be taught or are taught with much difficulty and effort.

Training and Certification of Personnel

There are many activities stated in NFPA 25 that require only a visual inspection to determine the condition of the equipment. Training of in-house personnel by the installing contractor is often the most cost-efficient method to assure that the program is effective. In addition to the "skilled trades" and contractor training that might be available, there are various programs such as those provided by the National Fire Protection Association (NFPA), the National Fire Sprinkler Association (NFSA), and the American Fire Sprinkler Association (ASFA). Further, Internet-based training is also available. Some colleges and universities have also developed curriculums specific to training personnel to inspect and test water-based fire protection systems.

Finally a certification program is offered by the National Institute of Certification in Engineering Technologies (NICET) for Inspection and Testing of Water-Based Systems. This three-level certification program is designed for engineering technicians in the automatic fire sprinkler industry who are engaged in the physical and mechanical aspects of the ITM of water-based systems.

More information on training programs is available at the websites for each of these organizations.

If the individual has a background in fire protection, knows the industry terminology, and can identify system types and components, training can concentrate on the actual inspection and test activities.

Conducting an Inspection and/or Test

Conducting an inspection and/or test for the first time in a building will require more time than will subsequent inspections and/or tests. The inspector must become familiar with the building, the systems installed in the building, the system components, and the records of the previous inspections and tests that have been performed. The inspector must also determine the necessary inspection and test tools, equipment, and instruments needed, plan the inspection and test sequence, and prepare the inspection reports. Subsequent inspections and tests will take less time to complete because the building and systems are known, and the inspector can focus on inspecting and testing the systems and components.

Many buildings have several fire protection systems that are required to be inspected and tested by NFPA 25. For instance, a typical large facility might have a backflow prevention device, one or more fire pumps, yard main systems including hydrants and valves, a standpipe system, wet pipe sprinkler systems and control valves on each floor, and dry pipe systems or antifreeze systems for some processes or areas of the facility. There might even be one or more preaction systems in computer rooms or electrical service rooms, and fixed water spray systems for cooling towers and other processes. Further, depending upon the facility and occupancy, there might be standpipe systems employed for hose stream application.

NFPA 25 contains a series of tables that describe an action that must be performed on systems and/or components when corrective action is taken to address a deficiency or impairment. Each table is referenced in the subsection in which it appears, for example: "Whenever a component in a [type of] system is adjusted, repaired, reconditioned, or replaced, the action required in Table 5.5.1 shall be performed." Many of the actions required include reinspecting and/or retesting of systems or components after they are adjusted, repaired, reconditioned, or replaced. These inspections and tests are performed separately from those conducted on a periodic basis per NFPA 25.

Frequency for Common Inspection and Testing Activities

An example of a common approach for a wet sprinkler system is to arrange for the following steps as required by NFPA 25:

Quarterly. Inspect the gauges, control valves, alarm devices, hydraulic nameplate, and the fire department connection, and test some of the alarm devices.

Semiannually. Test flow switches and valve tamper switches.

Annually. Manually inspect the building, pipe hangers, pipes, fittings, and spare sprinklers, and test the main drain.

All other less-frequent activities are not normally included in the inspection program, including testing gauges every 5 years; testing sprinklers every 5, 10, 20, or 50 years;

and internally inspecting piping, alarm valves, check valves, strainers, filters, and orifices every 5 years.

These activities are normally performed by a fire protection contractor. However, depending upon the experience level, these activities may be performed by the in-house inspector.

Reports

Records of ITM activities are to be maintained by the building owner or designated representative. Many local jurisdictions require that such reports be kept at the site. Local ordinances might also require that a copy of any reports be submitted directly to the authority having jurisdiction (AHJ).

Reports are not standardized and come in many versions. Many forms are still filled out by hand, although computerized programs with printed reports are becoming more and more popular. The electronic reports tend to be more professional and legible.

Tools and Equipment

Essentially three categories of tools and equipment are necessary to properly inspect and/or test a fire protection system: safety equipment, standard inspection tools and equipment, and standard test equipment.

Safety Equipment. Safety must be a matter of primary importance and be second nature to the inspector. Federal and state regulations require compliance with safety laws. Common sense dictates that care and caution must be taken to ensure the safety of the inspector, and those working around the inspector, from certain hazards inherent with performing inspection and test procedures. As a minimum, Federal Occupational Safety and Health Administration (OSHA) regulations must be adhered to. If not, employee safety is at risk.

Table S5.1 describes examples of the safety equipment needed and some general uses for each.

TABLE S5.1 *Safety Equipment*

Safety Equipment	General Use for Equipment
Hard hat	Hard hats are required when head injury is a possibility from climbing, crawling through confined spaces, or falling objects.
Safety shoes	Safety shoes are required in most industrial and manufacturing plants and at any location where foot injury is possible from stepping on sharp objects, or from heavy objects falling on the feet.
Proper clothing	Loose clothing is not to be worn around machinery. In some plants, fire and/or chemical protective clothing must be worn.
Eye protection	Safety glasses should always be worn when inspecting or testing fire protection systems.
Hearing protection	Hearing protection should be a part of the standard safety equipment worn by inspectors. Many areas in industrial occupancies post signs where hearing protection must be worn.
Safety harness	See OSHA regulations for the requirements for safety harnesses. Safety harnesses are required when inspectors must be on ladders or lifts above a certain height.
Confined space equipment	Confined space equipment requirements are not limited to pits or cellars but can also extend to more common spaces, such as unventilated attics, crawl spaces, concealed spaces, and vaults. See OSHA regulations for the requirements for confined space entry.

Inspection Tools and Equipment. Table S5.2 lists some of the inspection tools and equipment commonly used by inspectors.

Test Equipment. Table S5.3 lists some of the standard test equipment for the inspection and testing of water-based fire protection equipment.

Inspection Procedure

NFPA 25 does not prescribe how to perform an inspection or test. The systems and components that are to be inspected and tested are listed, the acceptance criteria are spelled out, and the corrective action requirements are clear.

The owner or designated representative must realize that the actual inspection process is unique to each facility, building, and system. The NFPA 25 report forms available from NFPA's website, if followed line by line, lead the inspector through the process.

As each system is unique, so is each inspector. What works for one will be inefficient for another. Inspectors must develop their own processes. However, certain essentials are universal. Before the experienced inspector can put together a game plan to perform the activities in the most efficient and comprehensive way possible, certain information must be reviewed. The inspector needs to know the types of systems being inspected. He or she must become familiar with the layout of the building and identify where the system controls and valves are.

The following describes one method of performing an inspection.

TABLE S5.2 *Standard Inspection Tools and Equipment*

Tool	General Use and Notes
Ladders	The inspector should use the right height ladder at all times, to avoid overextending his or her reach. When using extension ladders, always follow the "1 to 4 rule," setting the base of the ladder 1 ft (0.3 m) away from the wall for every 4 ft (1.2 m) in height. The inspector must follow all other safety rules concerning ladders as detailed in OSHA and local requirements, including safety harnesses and fall protection.
Flashlight	A flashlight is particularly useful when working in confined spaces.
Binoculars	Binoculars are typically used for examination of sprinklers from the floor level. The inspector is not expected to climb a ladder to inspect each sprinkler, but binoculars can be helpful when inspecting sprinklers for leakage, corrosion, foreign materials, loading, paint, physical damage, or improper orientation.
Tape measure or folding rule	The size of some pipes or components might not be evident, and the inspector might need to measure the size to provide an accurate report.
Assorted hand tools	To perform certain internal inspections, the inspector will have to remove face plates from valves or remove sprinklers or flushing connections. Screwdrivers, a socket set, crescent wrench, hammer, and a set of open-end wrenches should be included in the inspector's tool kit.

TABLE S5.3 *Standard Test Equipment*

Equipment	General Notes on Use
12 in. pipe wrench	
Crescent wrench	
Calibrated test gauges	
Volt meter	Primarily used during the annual fire pump test. Proper safety equipment should be worn while using this equipment.
Amp meter tachometer	
Hydrometer or refractometer	Used to measure the concentration of antifreeze.
Pitot tube	Used to measure velocity pressure during flow and fire pump tests. (See Chapters 6 and 8.)
Stopwatch	A watch that reads in minutes and seconds. Used to measure trip times for dry pipe systems.
Heat lamp	Used to activate heat detection systems on water spray fixed systems.
Digital gauges	
Water analysis kit	Used when performing the five-year obstruction inspection if tubercules or slime are found in the piping.

Locate Systems and Equipment

At the system valve room, the following steps are taken before starting any actual inspection or testing:

- Note the type, manufacturer, size, and number of systems to be inspected and/or tested.
- Note the location, type, manufacturer, size, and quantity of any flow switches or tamper switches.
- Alert the building occupants. Determine whether the alarms are monitored through the building fire alarm system and, if so, warn the building occupants of a possible alarm.
- Locate the fire alarm panel. Verify accessibility to the fire alarm equipment and determine whether the system is monitored by a proprietary or central station.
- Notify the monitoring company or, if the systems are monitored within the premises, the personnel monitoring the proprietary alarm system.
- If the local procedures at the facility require it, notify the AHJ or local fire department.

Performing the Inspection and/or Test

Once the preparatory steps are taken, the actual inspection and testing can begin. The activities described in this section are for an annual inspection and test, with the exception of full flow tripping dry, preaction, or deluge systems. The requirements in NFPA 25 should be followed for the specific activities that follow and for inspections and tests during other frequencies.

- Perform a visual inspection of the alarm, dry, preaction and/or deluge system valve, and all associated trim. Look for any signs of leakage, rust, and any unusual problems or conditions.

- Determine where test water will discharge. Verify that the water from the main drain will discharge in a safe location.
- Conduct a main drain test, including the following steps:
  1. Note and record the static pressure on the supply side of the system valve.
  2. At least annually, conduct the main drain test by opening the main drain valve fully.
  3. Note and record the residual pressure with the valve wide open.
  4. Observe the water discharging from the main drain test fitting and note any debris or restrictions of any kind.
  5. Flow water through the main drain until the water stream is clear.
  6. Close the main drain valve and record the gauge readings.
  7. Note and record the static pressure after the valve is fully closed.
- Verify the operation of the water motor alarm (if applicable). On a wet system it will ring during the main drain test. On dry, preaction, and deluge systems, it will ring when the test valve in the system valve trim is opened.
- Close and fully open all control valves to verify proper operation. Test the tamper devices if applicable.
- Locate and flow all inspector's test valves for the required 90 seconds and verify alarm activation. Reset any alarms and continue with the walkthrough.
- Look for proper coverage and orientation of sprinklers. Also observe sprinklers for evidence of loading, paint, corrosion, damage, and any physical obstructions to discharge patterns.
- Observe any exposed system piping and hangers for corrosion, leakage, misalignment, and physical damage.
- Record all conditions observed during the walkthrough. When this has been done, the inspection is complete.

Compile Results. Many inspectors fill out inspection and test reports as they perform the activities. Others compile results, observations, and deficiencies on a separate piece of paper and transfer the information to the actual report at the end of the inspection. Either method is acceptable. Using the inspection form as the inspection and test is performed helps to eliminate missed activities and aids in establishing a systematic process.

Complete Reports. All pressures, trip times, flows, and other recorded results are written in the report. All deficiencies are to be clearly detailed, and corrective action recommended.

Completing the Inspection and/or Test

To complete the inspection or test, many of the processes required to start these activities must be repeated in reverse. It is just as important to go through these steps upon the completion of the inspection or test as it was before starting. Some of the steps that should be undertaken in completing the inspection or test are as follows:

- All systems are to be returned to normal service.
- Test and drain valves are to be closed tightly until no water flow or dripping is observed. All flow and tamper switches are to be returned to normal settings.
- All alarm panels are to be returned to their normal condition. To do this, confirm that all alarm, trouble, or supervisor signals have been verified at the panel, and then reset the panel.
- Fill out any tags provided and/or required.

Tags. Some jurisdictions have a visual tag procedure that the inspector must follow. A green tag indicates the system did not have any deficiencies or impairments. A yellow tag indicates one or more deficiencies were found and therefore the application of one or more components is not within its designed limits or specifications, but there were no impairments found. A red tag indicates one or more impairments were found and the system might not function as intended during a fire emergency. If one or more impairments are found, impairment procedures must be initiated as necessary, until the impaired condition is properly corrected. Table A.3.3.7 of NFPA 25 provides guidance on determining the severity of the deficiency for the purposes of providing the correct tag.

Advise Occupants. If occupants were notified of the test, notify them that the test is complete.

Notify the AHJ/Local Fire Department. Notify the proper authorities that the test is complete and the system has been returned to normal service. Write down the name of the individual who received the notification and the time of notification.

Notify the Monitoring Company. Notify the monitoring company, if applicable, that the test is complete. Have the individual perform a system check to verify they are receiving the proper signals. Write down the name of the individual who performed the system check and the time of notification.

Review Reports. After all systems are back in service and all notifications have been made, review the results of the inspection and test. The report should be explained line by line, and any unusual readings or findings discussed with the appropriate management representative. Deficiencies should be clearly explained. Review the requirements of

NFPA 25 pertaining to acceptable results and the corrective action that must be taken to address deficiencies.

Follow Impairment Procedures, if Applicable. As previously stated, if impairments are found, impairment procedures must be implemented as required by NFPA 25 for prompt corrective action. It is important to tag systems and/or components that are not working properly, especially those that are impaired.

After the Inspection

Inspection Follow-Up. An inspection and/or test performed without proper follow-up is useless. Protection of lives and property is dependent on properly operating water-based fire protection systems. The owner or designated representative has made a commitment to the employees, the insurance company, stockholders, and the community at large to have the system(s) in proper working order.

Corrective Action. When an owner or designated representative performs repairs to the system, or contracts the repair to a fire protection contractor, the date the repairs were made and the corrective action taken should be recorded. This information is then processed for the subsequent inspection.

Component Replacement Action Requirements. NFPA 25 includes a series of tables that describe what action must be performed on systems and/or components when corrective action is taken to address a deficiency or impairment. Each table is referenced in the subsection in which it appears, for example: "Whenever a component in a [type of] system is adjusted, repaired, reconditioned, or replaced, the action required in Table 5.5.1 shall be performed." Many of the actions required include reinspecting and/or retesting of systems or components after they are adjusted, repaired, reconditioned, or replaced. These inspections and tests are performed separately from those conducted on a periodic basis per NFPA 25, and the results should be recorded on a separate inspection/test report. These additional reports should be retained by the owner or designated representative for future reference.

Maintaining Inspection Files. Proper filing of reports is necessary to provide ongoing inspection and test services. The reports must be accessible for possible distribution to the AHJ or insurance underwriter. The inspector will also need copies of previous reports when the next inspection is performed. Comparisons between current test results and previous test results must be made to determine whether degradation of water supplies, system piping, fire pump performance, and so forth is taking place.

Sample Documentation

The following is an example of documentation for the practices described in this supplement. Both the contractor's material and test certificate for aboveground piping and the contractor's material and test certificate for underground piping must be kept on file by the property owner or designated representative for the life of the system.

The material and test certificates provide a basis for many of the subsequent inspections and tests required for their respective systems. Subsequent ITM forms must be kept on file by the property owner or designated representative for a period of one year after the activity is repeated.

This supplement provides three sample forms that can be used by the owner. Exhibit S5.1 illustrates the reporting requirements necessary to verify compliance with the ITM requirements for sprinkler systems found in Chapter 5 of NFPA 25. Exhibit S5.2 and Exhibit S5.3 illustrate the reporting requirements necessary to demonstrate compliance with Chapter 8 for fire pumps.

The annual activities or activities that require specialized training and equipment might be conducted by a qualified fire protection contractor. Some jurisdictions require licensing of inspectors, and such licensing laws are commonly addressed to those involved in the business of providing such services. In most cases, licensing of an employee of the property owner or designated representative is not required for "in-house" work. This should be verified with the local jurisdiction if an employee of the property owner or designated representative is qualified to perform this type of work.

A file of all reports must be kept on the premises as stated in Chapter 4 of NFPA 25. This file provides direct evidence of compliance with NFPA 25, and a review of this file can be requested by the AHJ during an inspection. An accurate and complete file will lead to a brief and successful review.

Inspection, Testing, and Maintenance of Dry Pipe Sprinkler Systems

Property Name: XYZ Corporation **Inspector:** John Doe

Property Address: 100 Main St., Taylorsville, USA **Contract No.:**

Phone Number: (000) 555-0000 **Date:** 4-15-14

This Report Covers: ☐ Monthly ☐ Quarterly ☑ Annual ☐ Three-Year ☐ Five-Year

Inspections

Monthly

☐ Yes ☐ No ☐ N/A Gauges—normal air and water pressure maintained

Control Valves

☐ Yes ☐ No ☐ N/A In the correct (open or closed) position

☐ Yes ☐ No ☐ N/A Sealed, locked, or supervised

☐ Yes ☐ No ☐ N/A Accessible

☐ Yes ☐ No ☐ N/A Free from damage or leaks

☐ Yes ☐ No ☐ N/A Proper signage

Dry Pipe Valves

☐ Yes ☐ No ☐ N/A Exterior is free of damage, trim valves are in correct open or closed position, and intermediate chamber is not leaking

Quarterly

☐ Yes ☐ No ☐ N/A Alarm devices—free of damage

☐ Yes ☐ No ☐ N/A Hydraulic data nameplate—securely attached to riser/legible

Fire Department Connections

☐ Yes ☐ No ☐ N/A Visible and accessible

☐ Yes ☐ No ☐ N/A Coupling/swivels operate correctly

☐ Yes ☐ No ☐ N/A Plugs/caps are in place

☐ Yes ☐ No ☐ N/A Gaskets are not damaged

☐ Yes ☐ No ☐ N/A Identification signs are in place

☐ Yes ☐ No ☐ N/A Ball drip is functional

☐ Yes ☐ No ☐ N/A FDC clapper is functional

Pressure Reducing Valve

☐ Yes ☐ No ☐ N/A In the open position/not leaking

☐ Yes ☐ No ☐ N/A Maintaining downstream pressure

☐ Yes ☐ No ☐ N/A In good condition

1 of 3

EXHIBIT S5.1 *Sample Inspection, Testing, and Maintenance of Dry Pipe Sprinkler Systems.*

Annual — **Sprinklers**

☑ Yes ☐ No ☐ N/A No damage or leaks
☑ Yes ☐ No ☐ N/A Free of corrosion, foreign material, or paint
☑ Yes ☐ No ☐ N/A Installed in proper orientation
☑ Yes ☐ No ☐ N/A Fluid in glass bulbs
☑ Yes ☐ No ☐ N/A Spare sprinklers—proper number and type. Complete with wrench?
☑ Yes ☐ No ☐ N/A Hangers and seismic bracing—not damaged or loose

Pipes and Fittings

☑ Yes ☐ No ☐ N/A In good condition/no external corrosion
☑ Yes ☐ No ☐ N/A No leaks or mechanical damage
☑ Yes ☐ No ☐ N/A Correct alignment—no external loads
☑ Yes ☐ No ☐ N/A Dry pipe valve interior—following trip test
☑ Yes ☐ No ☐ N/A Building—prior to onset of freezing weather—all openings are closed, no water-filled pipe is exposed to freezing temps

Five-Year

☐ Yes ☐ No ☐ N/A Obstruction inspection—no foreign or obstructing material found
☐ Yes ☐ No ☐ N/A Check valve—internal moves freely, in good condition
☐ Yes ☐ No ☐ N/A Dry pipe valve strainers, filters, and orifices internal inspection

Test

Quarterly

☐ Yes ☐ No ☐ N/A Alarm devices—water motor gong
☐ Yes ☐ No ☐ N/A Main drain test—if the sole supply is through a backflow preventer or pressure reducing valve

Static psi ________ Residual psi ________

☐ Yes ☐ No ☐ N/A Do results differ by more than 10% from previous test?
☐ Yes ☐ No ☐ N/A Priming water—test level
☐ Yes ☐ No ☐ N/A Low air alarm—test per manufacturer's instructions
☐ Yes ☐ No ☐ N/A Quick opening device tested

Semi-annual

☐ Yes ☐ No ☐ N/A Supervisory switch functions
☐ Yes ☐ No ☐ N/A Alarm devices—inspectors test or bypass opened/obstructed waterflow

Annual

☑ Yes ☐ No ☐ N/A Main drain test Static psi: 59 Residual psi: 50
☐ Yes ☑ No ☐ N/A Do results differ by more than 10% from previous test?
☑ Yes ☐ No ☐ N/A All control valves operated through full range of motion and returned to normal position.

2 of 3

EXHIBIT S5.1 *Continued.*

☐ Yes ☐ No ☐ N/A Dry pipe valve—partial flow trip test

Water pressure 59 psi Air pressure 45 psi

Tripping air pressure 30 psi Trip time 25 (sec)

☑ Yes ☐ No ☐ N/A Results comparable to previous tests

☑ Yes ☐ No ☐ N/A Backflow preventer—backflow test

☐ Yes ☑ No ☐ N/A Backflow preventer—flow test

Three-Year

☐ Yes ☐ No ☐ N/A Dry pipe valve—full flow trip test

Water pressure ______ Air pressure ______

Tripping air pressure ______ Trip time ______ (sec)

Water delivery time ______ (min.) ______ (sec)

☐ Yes ☐ No ☐ N/A Results comparable to previous years

Five-Year

☐ Yes ☐ No ☐ N/A Gauges tested or replaced

☐ Yes ☐ No ☐ N/A Pressure reducing valve—flow test and comparable to previous results

Routine Maintenance

☐ Yes ☐ No ☑ N/A Sprinklers tested or replaced per appropriate testing schedule

Comments: ______

Signature: *John Doe* Date: 4-15-14

License/Certification No.: Spr100500

3 of 3

EXHIBIT S5.1 *Continued.*

Annual Performance Tests for Fire Pumps

Date: 6-11-14 **Inspector:** Jack Smith **System:** Main Fire Pump

Location: Pump house #1

Y = Satisfactory **N** = Unsatisfactory (explain below)

Pump manufacturer and model: ______

Type: ☑ Centrifugal ☐ Turbine

Controller manufacturer and model: ACME Model #100

Rated capacity: 1000 gpm (L/min)

Water supply source: City Water Supply

Rated pressure: 100 psi (bar) Rated speed: 1770 rpm

Power: ☑ Electric ☐ Diesel ☐ Steam

Automatic starts performed 6 times	Y	Timer indicates total run time: 10 min.	Y
Automatic start functions properly	Y	Timer reset and graph paper changed?	Y
Automatic stop functions properly	Y	Test data and flow charts completed (attach all waterflow charts, electrical power charts, performance curves, etc.)	Y
Automatic start: 150 psi (bar)			
Automatic stop: 160 psi (bar)		Fire pump electrical readings recorded at each flow condition?	Y
Manual starts performed 6 times	Y	Fire pump motor speed: 1770 rpm	Y
Manual start functions properly	Y	Fire pump discharge flow: 1000 gpm (L/min)	Y
Manual start: 150 psi (bar)		Jockey pump operational	Y
Manual stop: 160 psi (bar)		Jockey pump appears properly aligned	Y
Remote start functions properly	Y	Jockey pump valves open	Y
Remote stop functions properly	Y	Jockey pump "turn-on": 150 psi (bar)	Y
Remote start: 150 psi (bar)			
Remote stop: 160 psi (bar)		Jockey pump "turn-off": 165 psi (bar)	Y

Notes: ______

1 of 1

EXHIBIT S5.2 Sample Annual Performance Tests for Fire Pumps.

Annual Test Summary Page for Fire Pumps

Date: 6-11-14 **Inspector:** Jack Smith **System:** Main Fire Pump

Location: Pump house #1

	Test 1	Test 2	Test 3
Approximate percent of rated pump discharge (gpm) (L/min)	0	100%	150%
Nozzle size in inches (mm)	No flow	1 1/8" (4)	1 3/4" (4)
Pitot pressure in psi (bar)	None	45/47 44/48	18/17 19/18
Flow in gpm (L/min)	None	246/251 243/254	376/365 383/376
Pump suction in psi (bar)	8	7	5
Pump discharge in psi (bar)	141	107	70
Net pump head (discharge pressure minus suction pressure)	133	100	65
Pump speed (rpm)	1775	1770	1770
Operate electric circuit breaker	Y	Y	Y
Test emergency power supply	Y	Y	Y
Check for excessive back pressure in exhaust system	NA	NA	NA

Notes:

1 of 1

EXHIBIT S5.3 Sample Annual Test Summary Page for Fire Pumps.

SUPPLEMENT

6

ROLE OF THE AUTHORITY HAVING JURISDICTION

By John A.M. Lake

Editor's Note: *In an effort to assist AHJs in identifying their roles and responsibilities in the ITM process, this supplement has been included in the handbook to provide a quick reference on several different topics. This supplement is intended to describe some best practices that AHJs can follow to assure that the buildings in their jurisdiction are safe not only for the building occupants but also for first responders. This supplement describes some of the interactions that the AHJ may have with other people in the ITM process and also how to address impaired systems.*

The role of the authority having jurisdiction (AHJ) in the inspection, testing, and maintenance (ITM) process is largely misunderstood. In most instances, when an AHJ is responsible for enforcing a standard, the AHJ is required to be present at system testing or actively reviewing and approving system documentation. In the NFPA 25 ITM realm, the AHJ is not asked to attend tests or inspections and is not generating test reports or retaining documents. However, the role of the AHJ in the process is critical, and therefore, it is entirely appropriate for an AHJ to make sure the property owners in the jurisdiction are actively taking care of their systems. Some of the most common reasons for water-based system failure are closed valves and lack of maintenance. Deficiencies and impairments that arise due to these conditions can be dangerous not only for building occupants, but also for responding personnel. The following sections of this supplement elaborate on the topics addressed in the Roles and Responsibilities of Key Stakeholders table provided in Supplement 3.

Overview of the ITM Program

The role of the AHJ in the ITM inspection and the use of NFPA 25 is unique, because it depends on whom the AHJ represents. Because of the broad definition of an AHJ, the view and scope of interest will be as varied as the AHJ's motivation. The motivation for most fire officials is the protection of life and property, whereas the motivation for an insurance company might be to reduce the maximum probable loss. The following information is presented from the perspective of a fire official. In many jurisdictions, the AHJ does not use NFPA 25 as the primary manual for the ITM of water-based fire protection systems. The AHJ usually references a fire or building code that gives the appropriate authority to enforce the provisions of the standard. In looking at those provisions, the AHJ usually has a different view on how a fire protection system should be maintained.

The NFPA Technical Committee on Inspection, Testing, and Maintenance of Water-Based Systems has made it clear that the role of the inspector performing inspection and testing activities in accordance with NFPA 25 is not all-inclusive. The ITM inspector is not required to have detailed knowledge of the fire protection system installation requirements in the applicable NFPA standards. However, this is not the case when the inspector works for the AHJ. In many cases, the AHJ has a different role; the AHJ looks at the facility from the perspective of being the responding agency in the event of an emergency, as well as being the enforcement authority. When a facility and the protection features required are considered, the fire protection system must meet the requirements to protect the hazard present. Upon review of the other applicable NFPA standards, it can be determined that compliance with NFPA 25 might not provide assurance that the fire protection system will work appropriately. This is further complicated by the fact that the installation documents provide minimal guidance beyond requiring compliance with NFPA 25.

It is the responsibility of the inspectors to identify any issues that might cause a fire suppression system to not function properly. This includes many issues that are not covered by NFPA 25. In trying to identify the role of the AHJ in administering the provisions of NFPA 25, the standard does not require the AHJ to have detailed knowledge of the fire protection system installation requirements. In fact, NFPA 25 does not provide any language requiring the enforcement authority to have the requisite knowledge. However, NFPA 25 does provide requirements that give the AHJ the responsibility to do the following:

- Determine that an existing situation involves a distinct hazard to life or property and permitting him/her to require ITM methods in excess of those required by the standard (1.2.2).
- Approve alternative programs (1.3.2).
- Enforce the requirements of a code or standard or for approving equipment, materials, installation, or procedure (3.2.2).
- Determine if the person or company is competent and capable and that they have met the requirements and training for a given field to perform the inspection (3.3.34).
- Approve alternative means of compliance for components and systems inspection, testing, and maintenance (4.7).
- Approve the recognized testing laboratory where sample sprinklers are submitted for field service testing (5.3.1.1).
- Approve an appropriate location for the standpipe flow test when the hydraulically most remote outlet(s) is not practical (6.3.1.1.1).
- Determine where the additional flow can be taken when the 250 gpm (946 L/min) cannot be flowed from each additional standpipe (6.3.1.2.2).
- Determine the standpipe system demand where the standpipe system demand cannot be determined by the inspector (6.3.1.3.1).
- Approve a location or manner where records indicating the date the preaction or deluge valve was last tripped and the tripping time, as well as the individual and organization conducting the test and providing information on procedures and policies for maintenance of all backflow prevention assemblies (13.4.3.2.12).

NFPA 25 does *not* require that the AHJ do the following:

- Review or evaluate the adequacy of the system to control or extinguish a fire in the protected occupancy.
- Review or evaluate the adequacy of the water supply, including water tanks and fire pumps, to provide the necessary water flow and pressure needed to meet the fire protection system demand as designed.
- Review or evaluate of the hazards present in the facility to determine the required system criteria for minimum protection levels.
- Conduct a review, evaluation, acceptance testing, inspection, or other test of the fire protection system to determine whether the system was designed and installed in accordance with the applicable installation standard, including NFPA 13, *Standard for the Installation of Sprinkler Systems*; NFPA 13R, *Standard for the Installation of Sprinkler Systems in Low-Rise Residential Occupancies*; NFPA 14, *Standard for the Installation of Standpipe and Hose Systems*; NFPA 15, *Standard for Water Spray Fixed Systems for Fire Protection*; NFPA 16, *Standard for the Installation of Foam-Water Sprinkler and Foam-Water Spray Systems*; NFPA 20, *Standard for the Installation of Stationary Pumps for Fire Protection*; NFPA 22, *Standard for Water Tanks for Private Fire Protection*; NFPA 24, *Standard for the Installation of Private Fire Service Mains and Their Appurtenances*; and NFPA 750, *Standard on Water Mist Fire Protection Systems.*
- Review or evaluate all areas of the facility to determine whether those areas are protected in accordance with the original applicable installation standard.
- Review the design criteria for the system to determine whether it is adequate for the commodity or hazard.
- Review the installation shop drawing to make sure all pipes are correctly sized and the sprinklers are the correct type, orifice size, and temperature rating.
- Perform a water supply analysis to determine whether the water supply is sufficient to provide the required gallons per minute at the required pounds per square inch for the system.
- Perform a complete building inspection to determine whether closets, blind spaces, attics, crawl spaces, and all other special building features are adequately protected in accordance with the version of the installation standard in force at the time of the installation.
- Recognize that storage commodities and/or arrangements are different from those anticipated when the system was designed and installed.
- Research the installation contract files to determine whether special requirements were required.

This lack of specific requirements for enforcing the provisions of NFPA 25 requires that the AHJ go to other codes and standards to find that authority. Some of the installation standards provide information that requires the system be maintained at least to the same level of performance and protection as designed. The standard then refers the AHJ back to NFPA 25, which does not require the AHJ to perform an analysis of the system to assure the system will function as required to protect the hazard that is present. For those facilities that are inspected using only NFPA 25, the facility and its occupants are at an increased risk, due to the focus on maintenance issues and the failure to address problems that can significantly impact these systems. (See the NFPA report, "U.S. Experience with Sprinklers," for more information.)

The standard does require that the property owner address changes to use, process, or materials by mandating an evaluation of the fire protection system for its capability to protect the new occupancy, use, or material. In the case of a facility that has had changes over a significant period of time, the provision does little to address the potential issues that can arise from a change that is not recognized as such by the property owner. A classic example of this would be a warehouse that has stored materials that are classified as noncombustible, and over a period of decades, the warehouse now includes combustible materials. The owner might not recognize that the change in product could have an impact on their fire protection system. Without expert intervention, the property owner might never know something is wrong until there is a fire.

An issue can also arise when a space is reconfigured resulting in improper sprinkler spacing and obstructions to discharge. The use, process, or materials have not changed, so the owner does not know that an evaluation needs to be done. Again, without expert intervention there is a potential for a significant loss in a property that was considered "protected."

In the case of fire sprinkler systems, it is widely held that the majority of issues related to the failure of these systems can be resolved with an ITM program that deals with shut valves and maintenance. This is an incorrect supposition, as the "U.S. Experience with Sprinklers" report shows in Table 3-A that shut valves account for 42 percent and maintenance accounts for only 10 percent of sprinkler system failures. While it is agreed that the majority of failures relate to shut valves, when it comes to the other categories, there are significantly more failures in systems where the water supply is inadequate (12 percent). When coupled with obstructions to distribution (10 percent) and inappropriate systems (9 percent), the grouping almost equals the same percentage as shut valves. All of these items can be identified by a properly trained inspector. The inspector working for the AHJ must identify these issues and require modification to the system to correct the deficiencies and protect the hazards present.

The installation standards provide some guidance to the AHJ, but they ultimately refer back to NFPA 25.The provisions for ITM in the related standards are as follows:

- NFPA 13, Section 27.1, General.
  A sprinkler system installed in accordance with this standard shall be properly inspected, tested, and maintained by the property owner or their authorized representative in accordance with NFPA 25 to provide at least the same level of performance and protection as designed.
- NFPA 13R, Section 11.3, Inspection, Testing, and Maintenance.
  Sprinkler systems shall be inspected, tested, and maintained in accordance with NFPA 25.
- NFPA 14, Section 13.1, General.
  A standpipe system installed in accordance with this standard shall be properly inspected, tested, and maintained by the property owner or an authorized representative in accordance with NFPA 25, *Standard for the Inspection, Testing, and Maintenance of Water-Based Fire Protection Systems*, to provide at least the same level of performance and protection as designed.
- NFPA 15, 11.1.1.
  A water spray system installed in accordance with this standard shall be properly maintained in accordance with NFPA 25, *Standard for the Inspection, Testing, and Maintenance of Water-Based Fire Protection Systems*, and *NFPA 72®, National Fire Alarm and Signaling Code*, to provide at least the same level of performance and protection as designed.
- NFPA 16, Section 9.3, Inspection, Testing, and Maintenance Frequency.
  Foam-water sprinkler systems shall be tested and inspected in accordance with NFPA 25, *Standard for the Inspection, Testing, and Maintenance of Water-Based Fire Protection Systems*.
- NFPA 20, Section 14.4, Periodic Inspection, Testing, and Maintenance.
  Fire pumps shall be inspected, tested, and maintained in accordance with NFPA 25, *Standard for the Inspection, Testing, and Maintenance of Water-Based Fire Protection Systems*.
- NFPA 22, Section 18.1, General.
  Tanks shall be periodically inspected, tested, and maintained in accordance with NFPA 25.
- NFPA 24, Section 14.1, General.
  A private fire service main and its appurtenances installed in accordance with this standard shall be properly inspected, tested, and maintained in accordance with NFPA 25 to provide at least the same level of performance and protection as designed.
- NFPA 750, 13.2.2, Frequencies.
  The frequency of inspections of components of water mist systems shall be in conformance with the manufacturer's listing requirement and NFPA 25, *Standard for the Inspection, Testing, and Maintenance of Water-Based Fire Protection Systems*.

In each of these cases, the installation standard does not require the system to be maintained in accordance with the installation standard; instead it refers back to NFPA 25, so as the saying goes, "you can't get there from here."

From the perspective of the AHJ, the facility and the protection system must be maintained to address any changes in the protected property that might affect the operation of the protection system. A fire in a protected property that is not controlled reflects poorly on the industry and creates barriers to the installation of protection systems. To minimize the potential for failure of fire protection systems, the AHJ must take on the role of educator, engineer, and enforcer.

First and foremost, the AHJ must educate the property owner/manager. Far too often, the property owner has not been properly informed of the requirements of the various codes and standards that affect his/her facility. This is sometimes due to changes in ownership or management where the information is not been passed on. The AHJ has a unique opportunity to help the owner understand the myriad of regulations and provide guidance on how best to protect the owner's property. This can be accomplished at any point from the preconstruction meetings through permitting, to the change of occupancy discussions, to referrals from the fire department through preplans. By defining the regulations and roles of various inspectors in advance, the AHJ can mitigate some of the common confusion issues.

In the role of engineer, the AHJ must look at the fire protection system as it compares to the original installation. Requiring the property owner to have copies of the original design documentation will help to identify any changes that might inhibit the effectiveness of the system. A review of this documentation and of items such as preplans will help when reviewing process and/or commodity changes. Additional guidance in this area is provided in NFPA 1, *Fire Code*:

> **1.3.6.3** Repairs, renovations, alterations, reconstruction, change of occupancy, and additions to buildings shall conform to this *Code*, NFPA *101*®, and the building code.
>
> **1.3.6.4** Newly introduced equipment, materials, and operations regulated by this *Code* shall comply with the requirements for new construction or processes.

NFPA 1 gives the AHJ the ability to look at all issues that might affect the operation of the fire protection system and require modification of that system to address those changes. This concept is further discussed in NFPA *101*:

> **4.5.8.1** Whenever or wherever any device, equipment, system, condition, arrangement, level of protection, fire-resistive construction, or any other feature is required for compliance with the provisions of this *Code*, such device, equipment, system, condition, arrangement, level of protection, fire-resistive construction, or other feature shall thereafter be continuously maintained. Maintenance shall be provided in accordance with applicable NFPA requirements or requirements developed as part of a performance-based design, or as directed by the AHJ. [***101:***4.6.12.1]
>
> **4.5.8.6** Any device, equipment, system, condition, arrangement, level of protection, fire-resistive construction, or any other feature requiring periodic testing, inspection, or operation to ensure its maintenance shall be tested, inspected, or operated as specified elsewhere in this *Code* or as directed by the AHJ. [***101:***4.6.12.4]

As the enforcer, the AHJ must have the ability to assure that a fire protection system will function correctly. Yet, NFPA 25 does not provide the authority or guidance (see A.4.1.6) on how to accomplish this. As an example, in 1.2.2, NFPA 25 states the following:

> In those cases where it is determined that an existing situation involves a distinct hazard to life or property, the authority having jurisdiction shall be permitted to require inspection, testing, and maintenance methods in excess of those required by the standard.

This leads to such questions as: which methods, or which standards? Without a clear tie to some code or standard, this section can be politically sensitive and difficult to enforce.

The AHJ must go to other codes or standards to address the issues of function and operation of the systems. The *International Fire Code* provides clear guidance to the AHJ in the provisions of 901.4, which state the following:

> Fire protection systems shall be maintained in accordance with the original installation standards for that system. Required systems shall be extended, altered, or augmented as necessary to maintain and continue protection whenever the building is altered, remodeled, or added to. Alterations to fire protection systems shall be done in accordance with applicable standards.

NFPA 1 also provides the following clear direction to the AHJ on maintenance of fire protection systems:

> **1.6 Enforcement.** This *Code* shall be administered and enforced by the AHJ designated by the governing authority.

In addition, NFPA 1 provides guidance in dealing with changes to the building and/or processes and who enforces the code:

> **1.3.6.3** Repairs, renovations, alterations, reconstruction, change of occupancy, and additions to

buildings shall conform to this *Code*, NFPA *101*, and the building code.

1.3.6.4 Newly introduced equipment, materials, and operations regulated by this *Code* shall comply with the requirements for new construction or processes.

This requirement gives the AHJ the ability to look at all issues that might affect the operation of the fire protection system and require modification of that system to address those changes. There are additional provisions found in NFPA 1, such as the following, which address maintenance, inspection, and testing and provide enforceable language:

10.4.1 Whenever or wherever any device, equipment, system, condition, arrangement, level of protection, fire-resistive construction, or any other feature is required for compliance with the provisions of this *Code*, such device, equipment, system, condition, arrangement, level of protection, fire-resistive construction, or other feature shall thereafter be continuously maintained. Maintenance shall be provided in accordance with applicable NFPA requirements or requirements developed as part of a performance-based design, or as directed by the AHJ. [***101*:**4.6.12.1]

NFPA 25 is a valuable tool in the maintenance of water-based fire protection system; however, it does not provide the necessary language for the AHJ to assure the installation will function correctly and provide the level of protection required. The AHJ must look at the requirements of this document and recognize its limited scope and know where to obtain the information necessary to assure the protection system will operate correctly.

Determining Qualification or Competency of ITM Personnel

In determining the qualifications or competency as outlined in 3.3.34, the AHJ should look to personnel who are factory-trained and certified for the specific type and brand of system being inspected, personnel who are certified by nationally recognized fire protection certification organizations such as the National Institute for Certification in Engineering Technologies (NICET), personnel who are registered, licensed, or certified by a state or local authority, and personnel who are employed and qualified by an organization listed by a nationally recognized testing laboratory for the servicing of fire protection systems.

The AHJ has responsibilities for determining qualifications or competencies found throughout NFPA 25. More often than not, the requirements are found in the text, such as in the specification of "qualified" for performing particular functions. It must be noted that the AHJ for one portion of a system might not be the AHJ for another portion of the system. An example might involve 13.6.2, where the AHJ for the sprinkler system is the fire marshal and the AHJ for the backflow preventer is the local utility or health department. Both have a distinct interest in the water supply but for very different reasons. An example of language defining *qualified* is as follows in the Florida State Statute:

The State Fire Marshal shall require the National Institute of Certification in Engineering Technologies (NICET), Sub-Field of Inspection and Testing of Fire Protection Systems Level II or equivalent training and education as determined by the division as proof that the permit holders are knowledgeable about nationally accepted standards for the inspection of fire protection systems.

The areas in NFPA 25 where the term *qualified* must be considered by the AHJ are as follows:

3.3.34 Qualified. A competent and capable person who has met the requirements and training for a given field acceptable to the AHJ. [**96,** 2014]

A.1.1.1.2 There are times when a single inspection or test can meet the requirements of both NFPA 25 and *NFPA 72* (e.g., operation of a tamper switch). This standard does not necessarily require that two separate inspections or tests be conducted on the same component, provided the inspection or test meets the requirements of both standards and the individual performing the inspection or test is ***qualified*** to perform the inspection or test required by both standards.

A.3.3.21 Impairment. The use of the phrase *fire protection system or unit* is a broad reference to those terms used in titles of Chapters 5 through 12. Some fire protection features are referred to as systems in the installation standards (e.g., sprinkler, standpipe, water spray, foam-water, and water mist), or are referred to as units (e.g., fire pumps), and others use neither term (e.g., private service fire mains and water tanks). For the purpose of this standard, the term *unit* refers to a fire pump and its connections required by NFPA 20, or a water storage tank and its connections required by NFPA 22, or a private service fire main and its connections required by NFPA 24. The use of the term *unit* in the definitions of impairment, deficiency, critical deficiency, and noncritical deficiency is not referring to an individual component such as a sprinkler, valve, fitting, switch, piece of pipe, and so forth.

Temporarily shutting down a system as part of performing the routine inspection, testing, and maintenance on that system while under constant attendance by ***qualified*** personnel, and where the system can be restored to service quickly, should not be considered an impairment. Good judgment should be considered for the hazards presented.

3.3.24 Inspection, Testing, and Maintenance Service. A service program provided by a ***qualified*** contractor or qualified property owner's representative in which all components unique to the property's systems are inspected and tested at the required times and necessary maintenance is provided.

4.1.1.2 Inspection, testing, and maintenance shall be performed by ***qualified*** personnel.

A.4.1.5.1 System deficiencies not explained by normal wear and tear, such as hydraulic shock, can often be indicators of system problems and should be investigated and evaluated by a ***qualified*** person or engineer.

4.1.5.2 Corrections and repairs shall be performed by **qualified** maintenance personnel or a **qualified** contractor.

A.4.1.6 The inspections and tests specified in this standard do not address the adequacy of design criteria or the capability of the fire protection system to protect the building or its contents. It is assumed that the original system design and installation were appropriate for the occupancy and use of the building and were approved by all applicable authorities having jurisdiction. If no changes to the water supply or to the building or its use have transpired since it was originally occupied, no evaluation is required. If changes are contemplated, it is the owner's responsibility to arrange for the evaluation of the fire protection system(s). Where the inspections and tests specified in the standard have been contracted to a ***qualified*** inspection provider or contractor, it is not the role of the inspector or contractor to determine if any changes have been made or the subsequent evaluation of the fire protection system. The evaluation of any building changes should be conducted before any proposed change is incorporated and should utilize the appropriate installation standard and input from applicable authorities having jurisdiction.

4.4 Water Supply Status. During inspection, testing, and maintenance, water supplies, including fire pumps, shall remain in service unless under constant attendance by ***qualified*** personnel or unless impairment procedures in Chapter 15 are followed.

5.3.3.4 Fire pumps shall not be taken out of service during testing unless constantly attended by ***qualified*** personnel or all impairment procedures contained in Chapter 15 are followed.

5.3.4.2.1(3) Antifreeze systems with concentrations in excess of 30 percent propylene glycol and 38 percent glycerin shall be permitted based upon an approved deterministic risk assessment prepared by a ***qualified*** person approved by the authority having jurisdiction.

A.8.3.1.1.2 The risk analysis should be prepared and reviewed by ***qualified*** people. Increased test frequencies might be desirable when high impact losses could result from an uncontrolled fire. Examples where increased fire pump test frequencies can be considered could include high piled storage facilities and buildings where the predominant occupancy is protected by an extra hazard density sprinkler system.

A.8.3.1.2.4 The risk analysis should be prepared and reviewed by ***qualified*** people. Increased test frequencies can be desirable when high impact losses could result from an uncontrolled fire. Examples where increased fire pump test frequencies can be considered could include high piled storage facilities and buildings where the predominant occupancy is protected by an extra hazard density sprinkler system.

8.3.2.7 ***Qualified*** personnel shall be in attendance whenever the pump is in operation.

8.3.2.7.1 The use of the automatic timer allowed in 8.3.2.6 shall not eliminate the requirement of 8.3.2.7 to have ***qualified*** personnel present during test.

8.3.3.1 An annual test of each pump assembly shall be conducted by ***qualified*** personnel under no-flow (churn), rated flow, and 150 percent of the pump rated capacity flow of the fire pump by controlling the quantity of water discharged through approved test devices.

8.3.7.1.2 ***Qualified*** individuals shall interpret the test results.

9.5.2.1 Maintenance of all automatic tank fill valves shall be conducted by a ***qualified*** person following the manufacturer's instructions in accordance with the procedure and policies of the authority having jurisdiction.

13.6.3 Maintenance. Maintenance of all backflow prevention assemblies shall be conducted by a ***qualified*** individual following the manufacturer's instructions in accordance with the procedure and policies of the authority having jurisdiction.

14.3.3 If an obstruction investigation indicates the presence of sufficient material to obstruct pipe or sprinklers, a complete flushing program shall be conducted by ***qualified*** personnel.

A.15.5 The need for temporary fire protection, termination of all hazardous operations, and frequency of inspections in the areas involved should be determined. All work possible should be done in advance to minimize the length of the impairment. Where possible, temporary feedlines should be used to maintain portions of systems while work is completed.

Water-based fire protection systems should not be removed from service when the building is not in use. Where a system that has been out of service for a prolonged period, such as in the case of idle or vacant properties, is returned to service, ***qualified*** personnel should be retained to inspect and test the systems.

A.15.7 Occasionally, fire protection systems in idle or vacant buildings are shut off and drained. When the equipment is eventually restored to service after a long period of not being maintained, it is recommended that ***qualified*** personnel or a ***qualified*** contractor perform the work.

C.3.1 Introduction. In a fire pump motor circuit that has been properly installed, coordinated, and in service prior to the fault, tripping of the circuit breaker or the isolating switch indicates a fault condition in excess of operating overload.

It is recommended that the following general procedures be observed by ***qualified*** personnel in the inspection and repair of the controller involved in the fault. These procedures are not intended to cover other elements of the circuit, such as wiring and motor, which can also require attention.

Accessibility to Equipment

In NFPA 25, the AHJ has a limited role in relation to accessibility to equipment. The AHJ has a more significant role in the installation phase where decisions on the fire protection equipment locations are discussed and approved. When initial plans are submitted for review, the AHJ should review the design for operational aspects as well as the need to gain access to the equipment for inspection, repair, and replacement. When considering the installation of equipment, the AHJ needs to remember that there are standards other than water-based fire protection that apply to accessibility, such as a controller for a fire pump, as shown in the following requirement from NFPA 20:

10.2.4 Working clearances around controllers shall comply with *NFPA 70®*, *National Electrical Code®*, Article 110.

The following provisions found in NFPA 1, *Fire Code*, provide for accessibility to the fire protection system:

13.1.3 Obstructions shall not be placed or kept near fire hydrants, fire department inlet connections, or fire protection system control valves in a manner that would prevent such equipment or fire hydrants from being immediately visible and accessible.

13.1.4 A minimum 36 in. (91 mm) of clear space shall be maintained to permit access to and operation of fire protection equipment, fire department inlet connections, or fire protection system control valves. The fire department shall not be deterred or hindered from gaining immediate access to fire protection equipment.

Notification of Impairment

The specific notification requirements found in NFPA 25 require the AHJ to be notified of impairments to the fire protection system. The AHJ should request specific information to determine what the nature of the impairment is and whether or not there are additional notifications or investigations that need to take place. The level of response from the AHJ will directly relate to the type of impairment and the length of time it will take to correct the impairment. The fire department must be able to respond accordingly to a particular incident. For occupancies of a hazardous nature or where special hazards exist, the AHJ should consider requiring additional safeguards such as additional fire safety equipment (including fully staffed apparatus), more than one type of fire safety equipment, or special systems suitable for the protection of the hazard involved.

If the impairment is due to debris in the system, it might be due to a past break in the integrity of the supply system. This could mean other fire protection systems on the same service could also be affected. The AHJ should contact the utility or service provider to determine what had occurred, where it occurred, and what other systems are on that supply and request those systems to be evaluated.

The AHJ has the ability to require standby fire personnel or an approved fire watch when potentially hazardous conditions or a reduction in a life safety feature exist due to impairment to the fire protection system. As per 13.1.9 of NFPA 1, when a fire protection system is out of service for more than

4 hours in a 24-hour period, the AHJ has the authority to require that the building be evacuated or that an approved fire watch be provided for all portions left unprotected by the fire protection system shutdown until the fire protection system has been returned to service.

Notification of System Shutdown

Prior to responding to the need to shut down a fire protection system, the AHJ and stakeholders should develop a facility-specific plan to address emergency response. These plans should include response with a partially or fully impaired fire protection system.

The role and responsibility of the AHJ is the same for a system shutdown as it is for impairment. Simple questions need to be asked, such as: Why is the system being shut down? How long will it be shut down? What is the hazard associated with the system or portion of the facility being shut down? Once the AHJ receives the information, procedures for the type and level of response can be initiated.

Standby fire personnel or an approved fire watch might be part of that response. In some cases, if the hazard is determined to be high due to the potential for loss of life or significant property loss, the AHJ might require that the building be evacuated and all processes be shut down until the fire protection system is restored.

Changes in Occupancy, Use, Process, or Material

The inspections and tests specified in NFPA 25 do not address the adequacy of design criteria or the capability of the fire protection system to protect the building or its contents. The standard assumes that the original system design and installation have not changed and the water supply or building use is the same as originally constructed. This is an incorrect assumption that should not be perpetuated. The AHJ should evaluate the system using NFPA 25 and the installation standard for the year the system was installed. By using both standards, the AHJ can reasonably be assured that the fire protection system will function correctly.

It is the role of the inspector who works for the local AHJ to determine if any changes have been made to the fire protection system. The evaluation of any proposed building changes should be conducted before any change is incorporated. Permits must be applied for and the required plans must be submitted for approval prior to making any changes.

The standard requires that if a property owner is contemplating changing a process or material, then it is the owner's responsibility to arrange for the evaluation of the fire protection system. The concern is to identify any changes that the property owner might not be aware of. Some of these changes can include moving stock from one area to another (change of commodity in a given area), increase in storage height (volume of combustibles), component material changes not by the property owner (commodity class change), or material stored (material owner is not the same as the building owner). All of these items can negatively affect the function of the fire protection system. The standard requires that an expert evaluate the conditions and advise the property owner if there is a need to either modify the conditions or return the facility to the original permitted condition. In most cases, the property owner is not that expert and should not be relied upon to make those judgments.

The AHJ can help in this process by communicating with the property owner on a regular basis. This communication can be accomplished either by regular site visits or by using surrogates, such as the responding fire station performing preplans, to visit the facility. The engine crews can provide valuable information to the AHJ, gather important information for response, and participate in preparation of response plans (in case of impairment or system shutdown), as well as help the facility owner with questions that might arise.

The AHJ must look at the fire protection system as it compares to the original installation, and whenever or wherever any device, equipment, system, condition, arrangement, level of protection, fire-resistive construction, or any other feature is changed, the AHJ must notify the property owner of the need to correct the deficiency.

NFPA 25 does provide some guidance in 4.1.6, but states in 4.1.6.1 that such an evaluation is not part of the "normal" ITM required by the standard. The owner must know that there is a change to cause the evaluation. In most cases, the owners look to the experts to identify those changes and provide them with guidance to protect their property.

Records Retention

Acceptance test records should be retained for the life of the system. The AHJ will frequently refer to these records when performing a facility inspection. The AHJ will use these records to compare the existing conditions with the original installation to determine if any changes have been made that will affect the fire protection system.

The standard requires that all ITM records must be provided upon request. In many cases, the AHJ will require the ITM contractor to provide copies of their reports. The reports can be incorporated into the AHJ's property file for the facility. It is recommended that the AHJ contact all of the ITM contractors providing services in their jurisdiction and request copies of the inspection reports. It is also recommended that a specific timeframe be given to the ITM contractor for providing the reports so that there is adequate time for the report to be evaluated and for any deficiencies to be corrected. If the jurisdiction has adopted NFPA 1, then the property owner is also required to maintain detailed records documenting all systems and equipment testing and maintenance. If the AHJ requires documentation to be maintained for review during inspections, the property owner should be notified as part of the original approval process. Such notification ensures that the property owner is aware of these requirements and will have the documentation when it is needed.

The AHJ might also require that a log book be kept at a facility. The log book would contain the original plans, acceptance test documents, and any inspection or repair documents. The AHJ will review the log book to determine if there are any outstanding deficiencies. If the ITM report contains severe violations, then the AHJ can take action to assure the deficiency is corrected. Subsequent test records should be retained for a period of 1 year after the next test. The comparison of these records and the log book can help determine any deterioration of the system's performance or condition and the need for further testing or maintenance.

The 2014 edition of the standard now allows records to be maintained electronically. The AHJ should provide an e-mail address or some other means to allow the ITM contractor to send reports electronically. This change makes it easier to provide the records required by the standard.

Records retention might also be governed by a different AHJ. In many cases, state and local governments require different record retention schedules than those provided in the standard. It is recommended that the AHJ become aware of these requirements and retain the records in accordance with those policies. These requirements might also affect the ITM contractor if the facility being inspected is covered under this records retention schedule.

Determining Inspection Frequency

When addressing the testing frequencies, reliance should be placed on the provisions of NFPA 25. The AHJ should evaluate the inspection reports to determine if the inspections are being performed in accordance with the standard, and if not, the AHJ should formulate an action plan to address the deficiency. Many inspectors will review the inspection reports to verify that the inspection frequencies are being met, and if not, they will discuss the excessive intervals between ITM tasks with the property owner.

Jurisdictions might allow the inspection frequency to be modified to fewer than those listed in the standard. These modifications are based on a well-documented performance-based inspection schedule. Copies of this documentation should be maintained in the facility log book as well as in the AHJ's property record. Any variance to the standard's inspection frequencies should be well-documented as a failure of the system, because a variance in the inspection schedule can lead to litigation. If the AHJ feels uncomfortable approving the performance-based schedule, a third-party expert can be required by the AHJ to evaluate the schedule. The costs for the third-party evaluator are borne by the proponent of the design. Ultimately, the AHJ must evaluate what has been presented and determine if the schedule is acceptable.

Preparing and Providing Tools and Safety Equipment

The typical AHJ working as a building official or fire marshal has a limited role in the preparation of tools and safety equipment for the ITM inspector. There might be several different AHJs for this activity since the numbers of regulations for the various activities have different enforcers. Many of the issues related to the tools and safety equipment are under the jurisdiction of agencies such as OSHA and state occupational safety and labor departments. These agencies regulate many of the activities that relate to worker safety.

In some cases, the local AHJ might be involved in permitting and inspection of work sites such as confined space entry. The AHJ might require a site visit and, in some cases, a standby rescue crew. The local AHJ might also be involved in an inspection that involves entry into an area that requires use of respiratory protection. This is to ensure that the necessary emergency response personnel are available in case of an incident. If the AHJ determines the hazard to be high enough, standby personnel might be required.

Conducting and Attending System ITM Activities

For the most part, the AHJ does not attend or conduct the ITM activities outlined in NFPA 25. These activities are performed by employees of fire protection contractors, who are contracted by the property owner to comply with the requirements of specific law or rule. The reports generated by these inspections are a useful tool to identify deficiencies in the fire protection equipment.

Preparing ITM Reports

The AHJ role in preparing the ITM report is to simply require a copy of the report. The AHJ should evaluate the report to determine if any corrective actions are needed to ensure the protection system is operational. The AHJ should request copies of all inspection reports along with specific deadlines for providing the reports. This will ensure that the deficiencies are corrected as quickly as possible.

The AHJ should review the inspection report to make sure it includes a detailed explanation of every deficiency and an indication of the inspection frequency, such as whether it is a weekly, monthly, quarterly, or annual inspection. The inspection report should include the name of the permitted water-based fire protection inspector, the inspector permit number, the inspector's signature, the date and time of the inspection, and the signature of the owner or the owner's representative.

Some jurisdictions have provided regulatory language to require the reports, such as in Florida Statute 633.082 or 69A-46.041(7) Florida Administrative Code that follows, which covers inspection of fire control systems, fire hydrants, and fire protection systems:

> **69A-46.041(7) Florida Administrative Code:** The contractor or his or her permitted Water-Based Fire Protection Inspector shall complete in detail the inspection reports as required in NFPA 25, as adopted in Rule Chapter 69A-3, F.A.C. that outlines all points of the inspection, test, and maintenance as required by the applicable NFPA standards. A copy of the inspection report shall be provided to the owner at the completion of each inspection performed.

> **Reporting FS 633.082.** The inspecting contractor shall provide to the building owner or hydrant owner and the local authority having jurisdiction a copy of the applicable inspection report established under this chapter. The maintenance of fire hydrant and fire protection systems as well as corrective actions on deficient systems is the responsibility of the owner of the system or hydrant. Equipment requiring periodic testing or operation to ensure its maintenance shall be tested or operated as specified in the *Fire Prevention Code*, *Life Safety Code*, National Fire Protection Association standards, or as directed by the agency having jurisdiction, provided that such agency shall not require a sprinkler system not required by the *Fire Prevention Code*, *Life Safety Code*, or National Fire Protection Association standards to be removed regardless of its condition. This section does not prohibit governmental entities from inspecting and enforcing fire safety codes.

Corrective Actions for Deficiencies and Impairments

This area is where the AHJ can have the most impact. As previously stated in the reporting section, it is recommended that the AHJ contact all of the ITM contractors providing services in the jurisdiction and request copies of the inspection reports. It is also recommended that a specific time frame be given to the ITM contractor for providing the reports, so that there is adequate time for the report to be evaluated and for any deficiencies to be corrected in accordance with the policies of the jurisdiction. In the case of impairments, the AHJ should investigate the cause of the impairment to ensure that other systems in the area are not affected by the same issues. The AHJ should also take the necessary action to ensure a level of response commensurate with the decrease in protection resulting from the impaired system.

An evaluation of the current operating condition of the fire protection system should be weighed against the potential for loss in the facility. In some high-risk cases, the AHJ might require a full complement of fire equipment and manpower to protect a facility for as long as the fire protection system is not fully operational. This is where communication with the ITM inspection contractors is essential. The ITM contractors need to be aware that notification of system impairment or deficiencies that will affect the system's protection of the facility need to be addressed in the time frame set by the AHJ. It is the responsibility of the AHJ to determine the level of protection for a facility and the level of protection required in the event that a system is impaired.

System Tagging

Many jurisdictions require a fire protection system to be tagged to indicate the readiness condition of the system. The AHJ can use the tag as an initial indication of the system's condition but should not rely solely on this information. A review of the inspection report will reveal the information needed to determine system readiness. The AHJ should remember that the report and tagging of the fire protection system relates strictly to the NFPA 25 inspection and does not consider all of the conditions that might cause a fire protection system to fail.

The AHJ should review the inspection reports to determine the condition of the system as it relates to NFPA 25. In some cases, the tag will relate information on system deficiencies. NFPA 25 provides guidance for the actions the inspector should take for particular types of deficiencies. The AHJ can review the information provided in Annex A, but since this information is annex material and therefore not mandatory but explanatory in nature, the jurisdiction would need to adopt the information.

When many inspectors see a red tag, it might mean the system is not operational and the appropriate emergency actions must be put into effect. This might not be the case, and a review of the inspection report might provide the information necessary for the AHJ to determine what actions, if any, need to be performed. Conversely, an inspector might see a green tag and think the system is operational and that no action is necessary. Again, this might be an inappropriate action because the system might not have an adequate water supply; for example, sprinkler heads might be obstructed or there could be other deficiencies.

The AHJ should remember that the property owner might see a green tag on a riser and receive a report from the ITM inspector that indicates the system is compliant with NFPA 25, but the system might not function correctly to control a fire in the facility. The AHJ must remind the facility owner of the scope of the NFPA 25 inspection and educate them on the other conditions that could cause failure. Then the inspector for the AHJ must inspect the facility, evaluate all of the conditions that might keep the system from performing correctly, and note any deficiencies.

In many cases, the AHJ will provide regulations on the tagging system. This can include everything from the size and color, to the information contained on the tag, to the size of the lettering. The following provisions of 69A-46.041(2), (3), and (4) Florida Administrative Code are an example of these requirements:

Inspection Requirements for Fire Protection Systems

Each system that has been inspected, tested, or maintained by a fire protection contractor, or his or her permitted Water-Based Fire Protection Inspector, shall have a record tag of durable and weather resistant material placed on the riser.

Inspection Tags

After inspection and testing, an inspection tag shall be completed indicating all work that has been done, and then attached to the system riser in such a position as to permit convenient inspection and not hamper activation or operation. A new inspection tag shall be attached to the system riser each time an inspection and test service is performed.

Inspection tags must be a minimum dimension of 5¼ inches (133 mm) in height and 3½ inches (89 mm) in width, and a maximum dimension of 7 inches (178 mm) in height and 4 inches (102 mm) in width.

Inspection tags shall bear the following information in an easily read format:

'DO NOT REMOVE BY ORDER OF THE STATE FIRE MARSHAL.'

- This particular information shall be in a minimum of 10 pt type and in all capital letters
- The licensed Fire Protection System Contractor I or II's name and licensed physical address
- The license number of the Fire Protection System Contractor I or II
- The permit number of the Water-Based Fire Protection Inspector
- The permitted Water-Based Fire Protection Inspector's signature
- The day, month and year
- The facility name and address

The reverse of the noncompliant tag shall include at least four separate boxes for the listing of repair work as follows:

a. Date of Repair
b. Repaired by (Signature)
c. Print Name
d. Type of Repair
e. Permit Number (If repair is made by a Water-Based Fire Protection Inspector, Permit Number must be noted.)

Compliance and Noncompliance Tag

If a fire protection system is found to be in compliance with the applicable NFPA standards, a GREEN Compliance Tag shall be attached to either the main riser or main control valve in such a manner as to be plainly visible and not more than 60 inches above the finished floor.

If a fire protection system is found to have deficiencies and is not in compliance with the applicable NFPA standards, a completed RED Noncompliance Tag shall be attached to the main control valve of the system to indicate that corrective action is necessary. If the system is operational but not in compliance with the applicable NFPA standards, the building owner or authorized representative and occupant shall be notified by copy of the NFPA 25 inspection report within 30 days of the inspection. If the system is not operational, the contractor or inspector shall notify the building owner or authorized representative, the occupant, and the authority having jurisdiction within 24 hours of the time of the inspection.

Pursuant to the provisions of [*Local Jurisdiction's regulation*], it is the owner's responsibility to maintain the fire protection system. Affixing an inspection tag as required herein does not invalidate responsibility nor shall a transfer of risk be construed.

Follow-Up on Corrective Actions

If the AHJ has adopted the appropriate enabling document, such as NFPA 1, then the AHJ has the ability to enforce the provisions of NFPA 25. The AHJ and ITM inspector should be communicating on a regular basis concerning inspection reports. When a report is received that lists deficiencies, the AHJ can contact the ITM contractor and determine if the repairs have been made. If the repairs have not been made, then the AHJ should determine the level of deficiency and severity of the hazard. If the deficiency is a relatively low hazard, then the facility owner or owner's representative should be requested to provide a time frame for repair or for a fire safety improvement plan. If the hazard is such that immediate action in necessary, the AHJ should take the steps necessary to mitigate the hazard. These steps might include requiring a fire watch, shutting down hazardous operations, or evacuating the building.

Planned Impairment Programs

In a situation where the ITM contractor has determined that the fire protection system must be taken out of service, the provisions of the facility emergency plan and policies of the AHJ should be discussed and those portions of the plans should be implemented prior to taking the system out of service. The planned impairment might also require plans to be submitted and a permit obtained. The provisions of 15.5.2(4) in NFPA 25 provide guidance to the ITM inspector as follows:

> Where a fire protection system is out of service for more than 10 hours in a 24-hour period, the impairment coordinator shall arrange for one of the following:
>
> **(a)** Evacuation of the building or portion of the building affected by the system out of service
> **(b)** An approved fire watch
> **(c)** Establishment of a temporary water supply
> **(d)** Establishment and implementation of an approved program to eliminate potential ignition sources and limit the amount of fuel available to the fire

However, if the jurisdiction is governed by the provisions of NFPA 1 or NFPA *101, Life Safety Code,* then the following provisions apply:

> When a fire protection system is out of service for more than 4 hours in a 24-hour period, the AHJ shall be permitted to require the building to be evacuated or an approved fire watch to be provided for all portions left unprotected by the fire protection system shutdown until the fire protection system has been returned to service. [**1:** 13.1.9]

The decisions to evacuate a building, require a fire watch, or establish any alternative programs must be discussed and approved by the AHJ, which in consultation with the fire protection agency, will determine the level of response for a given facility based on the hazard analysis and response plans.

The AHJ should verify that the appropriate actions have been taken and that the system has been tagged at all of the required points. The AHJ should also make sure that the response personnel have updated their response plans to include the potential lack of fire protection at the site of the impairment.

Upon notification that the system has been restored to operational, the AHJ needs to inform the appropriate response personnel that the system is back in service. The AHJ should also verify that the ITM contractor has removed all the warning tags and that the system has been restored to operational.

Emergency Impairments

In all cases involving impairments, preplanned actions need to be discussed with the ITM contractor to ensure there is a consistent level of protection for the property. The AHJ should meet and discuss policies involving the disabling of any fire protection system prior to that system being taken out of service. The AHJ should make available contact information that will allow the ITM contractor to notify the appropriate authorities any time a system is taken out of service.

Once the AHJ is notified that a system is out of service, the AHJ should determine what level of alternative protection must be provided. Part of that evaluation might include the level of impairment, the amount of time the system might be out of service, the type of hazard without protection, the potential for an uncontrolled fire, and the risk to the community. After evaluating the hazard, the AHJ might require actions that include everything from establishing a fire watch to requesting standby fire fighters with apparatus. When a fire protection system is out of service for more than 4 hours in a 24-hour period, the AHJ can require that the building be evacuated or an approved fire watch be provided for all portions left unprotected by the fire protection system shutdown until the fire protection system has been returned to service.

Restoring Systems to Service

When fire protection systems in idle or vacant buildings are being restored to service, it is usually due to the building being reoccupied. In this case, most jurisdictions require permitting and a re-evaluation of the fire protection features to ensure that the installed systems will protect the hazards being introduced into the building. During the inspection process, the AHJ should require that the fire protection systems be evaluated to make certain that the design addresses the hazards. It is recommended that a qualified contractor perform the work.

A sample review note that can be used in a jurisdiction is as follows:

> Interior renovations can affect the sprinkler protection provided by any existing fire sprinkler system(s). A licensed fire sprinkler contractor should be contacted to determine if alterations to any existing fire sprinkler system will be required to comply with the coverage requirements of NFPA 13. Any existing sprinkler system shall cover all areas required. Alteration of any existing sprinkler system (if required) will require a separate permit by a licensed fire sprinkler contractor.

When the AHJ is notified that a repair to the fire protection system is complete, the system should be inspected to ensure it is operational. NFPA 25 provides for a certain level of inspection to be performed; however, depending on the level of repair, the AHJ might want to take the inspection to a higher level. The AHJ should look over the facility to verify that the area protected has not been modified and that the system design is appropriate for the hazard present. The authority to perform this type of inspection is not found in NFPA 25 but is covered in NFPA 1. This will assure that when the AHJ leaves the facility, the fire protection system will function correctly and protect the occupants and property in accordance with the design.

The AHJ should also make sure that all of the systems associated with the fire protection system are operational. In many cases, when a fire sprinkler system is taken out of service the fire alarm is also removed from service to avoid potential false alarms. Thus verification of the alarm service should be performed, including contact through the dispatch service.

Obstruction Investigations

An obstruction investigation should be handled similarly to an impairment. The AHJ should be notified of obstruction investigations and to what degree the fire protection system will be impaired. The AHJ should request specific information to determine what the nature of the investigation is and whether or not there are additional notifications or investigations that need to take place. The level of response from the AHJ will directly relate to the type of obstruction found. The AHJ should request that the ITM inspector provide information on what was found and, if possible, the circumstances surrounding the obstruction. This information will help the AHJ determine if other systems might be affected and what steps to take to correct any potential issues. If the fire protection system is going to be out of service, then impairment procedures should be considered. The fire department must be able to respond accordingly to a particular incident. For occupancies of a hazardous nature or where special hazards exist, the AHJ should consider requiring additional safeguards consisting of additional fire safety equipment (including fully staffed apparatus), more than one type of fire safety equipment, or special systems suitable for the protection of the hazard involved.

Performance-Based ITM Programs

Performance-based test and inspection methods proposed as an alternative means of compliance for testing frequencies require significant involvement of the AHJ. The performance-based options should be evaluated as a means to provide a greater degree of safety and potentially a lower cost to the facility owner. In developing the program to provide the alternative means of compliance, the AHJ should always look to provide the highest level of protection with the lowest degree of risk.

The verification process starts with the submittal to the AHJ. If the AHJ does not consider itself qualified to perform an adequate review of the performance-based design, the AHJ might specify a qualified third-party reviewer. A detailed process for evaluation is outlined in the *SFPE Engineering Guide to Performance-Based Fire Protection Analysis and Design of Buildings*. Additional information on reviewing performance-based design can be found in the *SFPE Enforcer's Guide to Performance-Based Design Review*.

In reviewing the verification process, the AHJ needs to identify a clear set of goals and objectives. Attention must be paid to both those objectives that apply to the facility as a whole and those that apply only to limited aspects of the facility. The purpose of this review is to determine whether the objectives of the designer and the owner are commensurate with the community's objectives.

The AHJ should review the objectives for consistency and judiciousness. The objectives must form a realistic and comprehensive package. Assumptions need to be reasonable, consistent, comprehensive, cohesive, and supported by adequate references.

Ultimately, the interest is in determining whether the designer did a credible job and if there is a sufficient margin of safety acceptable to the AHJ. Because of the complexity involved in this process, considerable discussions should be planned between the AHJ, the designer, and, if used, the third-party reviewer.

Testing programs must be evaluated using a method acceptable to the AHJ and appropriate for the conditions. Each design should be realistic and apply to conditions found in the facility. System components must be evaluated using products that are similar if not the same. If the testing is too general, the results might be inaccurate, since the quality for one manufacturer's product might be different from another. Failure rates for products and components typically follow specific production lots and manufacturers.

All aspects of the design must be documented. The format and content of the documentation must be acceptable to the AHJ. Sufficient documentation must be supplied to support the validity, accuracy, relevance, and precision of the proposed methods. The engineering standards, calculation methods, and other forms of scientific information provided must be appropriate for the application and methodologies used. It is imperative that the AHJ be involved with the development of the performance-based program from its inception. The development of the program, including all of the required documents, requires the AHJ's approval.

Frequencies must not be arbitrarily extended or reduced without a suitable basis and rationale. Failure to properly evaluate, document, and obtain approval for modifications of the testing requirements can result in enforcement actions being taken. Depending on circumstances, the enforcement actions might be taken against the owner of the facility or the ITM provider. It is recommended that the inspector review the local jurisdiction's laws, rules, and codes for requirements on servicing fire protection equipment.

A performance-based program provides that a maximum allowable failure rate be established and approved by the AHJ in advance of implementation. The use of historical system and component fire system inspection records can be used to determine failure rates. The design team and the AHJ should develop a process for reviewing the failure rates and increasing or decreasing the frequency of testing.

Ultimately, if the AHJ has been involved in the entire development process, then approval of the program should be relatively simple. Approval of the process used to determine test frequencies and the final design must be obtained in advance of any alterations to the test program.

References Cited

National Fire Protection Association, 1 Batterymarch Park, Quincy, MA 02169-7471.

NFPA 1, *Fire Code*, 2012 edition.

NFPA 13, *Standard for the Installation of Sprinkler Systems*, 2013 edition.

NFPA 13R, *Standard for the Installation of Sprinkler Systems in Low-Rise Residential Occupancies*, 2013 edition.

NFPA 14, *Standard for the Installation of Standpipe and Hose Systems*, 2013 edition.

NFPA 15, *Standard for Water Spray Fixed Systems for Fire Protection*, 2012 edition.

NFPA 16, *Standard for the Installation of Foam-Water Sprinkler and Foam-Water Spray Systems*, 2011 edition.

NFPA 20, *Standard for the Installation of Stationary Pumps for Fire Protection*, 2013 edition.

NFPA 22, *Standard for Water Tanks for Private Fire Protection*, 2013 edition.

NFPA 24, *Standard for the Installation of Private Fire Service Mains and Their Appurtenances*, 2013 edition.

NFPA 70®, *National Electrical Code®*, 2014 edition.

NFPA 72®, *National Fire Alarm and Signaling Code*, 2013 edition.

NFPA *101®*, *Life Safety Code®*, 2012 edition.

NFPA 750, *Standard on Water Mist Fire Protection Systems*, 2014 edition.

The SFPE Engineering Guide to Performance-Based Fire Protection Analysis and Design of Buildings, 2000 edition.

"U.S. Experience with Sprinklers," John R. Hall, Jr., March 2012.

International Code Council, 500 New Jersey Avenue, NW, 6th Floor, Washington, DC 20001.

International Fire Code, 2012 edition.

Society of Fire Protection Engineers, 7315 Wisconsin Ave., Bethesda, MD, 20814.

SFPE Enforcer's Guide to Performance-Based Design Review, 2003 edition.

Florida State Statute 633.082, "Inspection of Fire Control Systems, Fire Hydrants, and Fire Protection Systems," Title XXXVII, Chapter 633, 2012.

Florida Administrative Code, 69A-46.041(2), (3), (4), and (7), "Inspection Requirements for Fire Protection Systems."

SUPPLEMENT

7

FIRE PUMP FIELD DATA COLLECTION AND ANALYSIS

By T. Gayle Pennel

Editor's Note: *Supplement 7 provides the reader with extracted sections, including the executive summary, directly from the 2012 Fire Protection Research Foundation Report, "Fire Pump Field Data Collection and Analysis." The full text of the report is available from the Fire Protection Research Foundation in the Research section of the NFPA website, www.nfpa.org.*

Abstract

The focus of this report is review the testing frequency of fire pumps and to describe data collection needs, techniques, and analysis to provide a more scientific basis for the required testing frequency of fire pumps, with a longer term view of extending the techniques to provide data for other items that need a more scientific basis for establishing performance based code requirements. The data collected and analyzed for this project provides a basis for reviewing the non-flow testing frequency requirements for fire pumps, but data collection and analysis need to be further refined and continued to remove some anomalies that were found within the data collected.

Introduction

One of the significant hindrances in developing performance based Codes and Standards is the lack of reliable performance information. Many code requirements are ultimately based on verbal antidotal reports instead of actual knowledge of the reliability and performance of system components.

With recent technology changes it is now possible to collect and correlate performance data during normal inspection and testing of systems. This provides a tool that can improve fire protection, both through improved Codes and Standards, and also by potentially identifying problems in new technologies earlier.

The NFPA Research Foundation established this project as an initial step to address this issue.

Fire pumps were selected as the starting point to develop the techniques, protocols, and tools for analyzing inspection and test data. Fire pumps are complicated systems that are critical for fire protection systems. Fire pumps have code required test frequencies and performance evaluation. The evaluation of test data is sufficiently complicated that field input of the test data is the most efficient form of analysis and report generation. In addition the required type and frequency of these tests is the subject of ongoing debate that extends to the earliest days of the NFPA Technical Committee on Fire Pumps established in 1899.

This project was contracted to Aon Fire Protection who worked in concert with a Technical Panel who oversaw the project.

Executive Summary

Moving towards performance based codes and standards requires better quantification so that performance can be measured with a reasonable level of confidence. This project was conceptualized to utilize current technology including the internet to identify, collect and analyze available data to validate testing frequency for fire pumps. In addition this project was to develop protocols and tools that could be extended to address issues that require longer term data collection for validation.

Fire Pump Testing

NFPA 25 currently requires annual performance testing of fire pumps. In a performance test water is discharged and the net performance of the fire pump is analyzed through a minimum of three flow points. NFPA 25 also requires non-flow testing of fire pumps, where the fire pump is started and run for a time interval under churn conditions. Currently weekly non-flow tests are required for diesel fire pumps and monthly non-flow tests are required for electric fire pumps. The non-flow test frequency remains an item that is debated during every NFPA 25 code cycle.

Project Questions

The following questions based on the original project scope and Technical Panel guidance were formulated to provide additional focus to the project.

A. What is the appropriate test frequency for non-flow testing of electric and diesel fire pumps?
B. How well and how frequently do the tests provide evidence of existing or developing problems?
C. What unknown or inadequately defined issues can be improved with systematic data collection? Issues identified in the RFP included (but were not limited to):
   1. Limited service controller reliability;
   2. Variable speed pressure limiting pumps reliability (Pressure relief valves on variable speed fire pumps because of a lack of experience); and,
   3. Unforeseen failure modes that may occur with new technology, product redesign, and manufacturing changes.

Data Collection

Collecting, correlating, and entering existing paper test records into a standardized electronic format for analysis is extremely time consuming, so the project was focused on available data that was already in electronic format. Performance fire pump testing requires analysis that makes entering the original test data into a spreadsheet or data base more efficient than manual analysis, and a significant part of the industry now collects this information electronically. Non-flow fire pump testing only requires a record that the test was conducted and could be a checkmark(s) on a paper form or checklist. While some large firms with multiple locations and some pump test companies maintain electronic records, a significant portion of these records are maintained on paper.

Non-Flow Data Collection. To collect non-flow test data for this project, a standardized Non-Flow Pump Test Data Collection Form was developed and provided to the Technical Panel members who solicited input from their various plant locations. Standardized failure modes were included on the data collection form. The initial form which provided only a Pass/Fail option was later modified to provide a Pass/Repair/Fail Option. Five companies provided a total of 32 data sets with 3396 non-flow tests on 79 fire pumps using the Pass/Fail only forms. A total of 112 failures were reported in these tests. One company provided a total of 6 data sets with 749 non-flow tests on 17 fire pumps using the Pass/Repair/Fail forms. A total of 20 pumps needing repairs and 29 failures were reported in these tests. Comments regarding maintenance issues that do not directly affect the fire pump starting and operating were recoded on 122 tests.

One company provided electronic records in a different format on 983 tests on 41 fire pumps. A total of 103 tests identified repairs were needed, and 11 failures were reported.

Conclusions

Data Significance

The sample size for non-flow weekly testing was adequate to provide a meaningful fire pump failure rate range with 95% confidence. For electric motor pumps there were 19 failures in 1547 tests, or a failure rate of 1.23%, a standard deviation of 0.28%, and a 95% confidence in a failure range between 0.67% and 1.79%. For diesel engine pumps there were 98 failures in 4855 tests, or a failure rate of 2.02%, a standard deviation of 0.20%, and a 95% confidence in a failure range between 1.61% and 2.42%. When electric motor and diesel engine pump data are combined, there are 117 failures in 6402 tests, a failure rate of 1.83%, a standard deviation of 0.17%, and a 95% confidence in a failure range between 1.49% and 2.16%.

The sample size for non-flow monthly testing provided a 95% confidence failure range of between 0% and 7.9% for monthly testing which does not allow meaningful analysis of the monthly test data.

Data Collection and Analysis Issues

The following classifications were used for non-flow data collection:

Y Indicates the fire pump started automatically and was capable of delivering water at the needed pressure for the duration of the test.

F Indicates the pump did not start automatically or appeared incapable of delivering water at the needed pressure for the duration of the test.

R Indicates the fire pump is in need of repairs, but started automatically and was capable of delivering water at the needed pressure for the duration of the test.

Although the classifications appear straightforward, test comments associated with reported failures (see Table 3) show some reported failures that do not appear to classify as failures. As an example data set 17 reported two failures because the test header gauge was not working properly. The test header gauge does not affect the actual fire pump performance. Additional examples include data set 29 where a failure was reported because "out of charts" and data set 30 where a failure was reported because "Disk full (installed new disk)." A second issue is recurring sequential failures because repairs were not completed. Data set 3 reports 8 sequential failures because of a 10 year inspection and repair of a water tank. Other sequential failures are reported in data sets 33, 34, 35, 36, 38, 39, and 40. While these may be actual failures, it may be inappropriate to include them in analyzing the test frequency because after the first failure, the sequential failures are already known and not revealed by the pump test. To avoid injecting any researcher bias into the project, the Pass/Repair/Fail assignment made on the submitted information was maintained for this analysis. For the 7 day test frequency analysis, all data, including sequential failures, was included in the analysis in an attempt to avoid any further skewing of the data by the researcher.

Failure Modes

Ten failure modes were included in the standardized non-flow weekly test form. The modes and reported failures were: 1) Water supply inadequate (12 occurrences); 2) Electric power supply failure (1 occurrence); 3) Pipe, fitting, and/or valve failure (2 occurrences); 4) Controller failure (2 occurrences); 5) Pump running condition failure (0 occurrences); 6) Pump running operation failure (3 occurrences); 7) Relief valve failure (10 occurrences); 8) Diesel engine failure (6 occurrences); 9) Diesel engine system failure (21 occurrences); 10) Other (provide comment), (59 occurrences).

Based on a review of the comments, reports of relief valve failures may have included the casing relief valve as well as the main pressure relief valve.

Appropriate Test Frequency for Non-Flow Testing of Electric and Diesel Fire Pumps

Based on all the data received, there were 19 reported failures in 1547 weekly tests of electric fire pumps. This is a failure rate of 1.23%. Assuming a normal distribution, the sample has standard deviation of 0.28%. A 95% confidence level around the expected failure rate will be ± 45.6% of the estimated failure rate, which is within a failure range between 0.67% and 1.79%. A sample size of 131,734 is required to reduce the 95% confidence around the expected failure rate to 5% of the estimated failure rate. Although achieving 95% confidence requires using a failure range of 0.67% and 1.79%, the calculated failure rate of 1.23% was used to simplify calculations and presentation. Assuming the average failure occurs at the midpoint of the test period, a failure rate of 1.23% equates to an expected yearly down time of 53.8 hours before identification of impairment, and provides a 99.4% yearly reliability.

There were 98 reported failures in 4855 weekly tests of diesel fire pumps. This is a failure rate of 2.02%. With a normal distribution, the sample has a standard deviation of 0.20%. A 95% confidence level around the expected failure rate will be ± 20.0% of the estimated failure rate, which is within a failure range of 1.61% and 2.42%. A sample size of 78,400 is required to reduce the 95% confidence around the expected failure rate to 5% of the estimated failure rate. Although achieving 95% confidence requires using a failure range of between 1.61% and 2.42%, the actual failure rate of 2.02% was used to simplify calculations and presentation Assuming the average failure occurs at the midpoint of the test period, this equates to an expected yearly down time of 88.4 hours before identification of impairment, and provides a 99.0% yearly reliability.

The sample size for a 7 day test interval for both electric fire pumps are adequate to provide a meaningful failure range with 95% confidence.

Data received for a 30 day test frequency for electric fire pumps showed 13 failures in 91 tests covering 7 fire pumps. However, sequential failures with the same reported issue accounts for 10 of these failures. After eliminating sequential failures, this leaves 3 (13-10) failures in 81 (91-10) tests or a failure rate of 3.7%. With a normal distribution, the sample has a standard deviation of 2.09%. A 95% confidence level around the expected failure rate will be ± of 113.3% of the estimated failure rate, or within a failure range between 0% and 7.90%. The 95% confidence failure range of between 0% and 7.9% does not allow meaningful analysis of the monthly test data. Assuming a failure rate of approximately 3.7%, a sample size of approximately 41,067 is required to limit the 95% confidence around the expected failure rate to 5% of the estimated failure rate.

The sample size of 5 tests for 90 day tests interval was insufficient to provide reliable analysis.

Some insight into the appropriate test period can be gained by ignoring the unknown impact of the test frequency on the failure rate, and extrapolating downtime and reliability from weekly test results to monthly rates. For electric fire pumps, extrapolating from weekly to monthly provides an expected yearly down time of 233.7 hours before identification of impairment, and a 97.3% yearly reliability. For diesel fire pumps, extrapolating from weekly to monthly, provides an expected yearly down time of 384.2 hours before identification of impairment, and a 95.6% yearly reliability.

The decrease in reliability going from weekly tests to monthly tests is significant. For electric fire pumps a 99.4% reliability decreased to 97.3%. For diesel fire pumps a 99.0% reliability decreased to 95.6%.

Additional data is needed to establish a relationship between a 30 day test frequency and reliability.

Effectiveness of Tests (Identifying Existing or Developing Problems)

Non-flow weekly test data showed a failure rate of 1.23% for electric fire pumps, and 2.02% for diesel fire pumps and 1.83% for combined electric and diesel fire pumps. Needed repairs were identified in 1.5% of the non-flow tests for electric fire pumps, and 5.2% for diesel fire pumps.

NFPA 20 requires an investigation whenever the performance is more than 5% below the original pump curve. In performance tests, without RPM adjustments, the fire pumps underperformed in 47.7% of the tests analyzed. With RPM adjustments, the fire pumps underperformed in 51.3% of the tests.

Systematic Data Collection for Unknown or Inadequately Defined Issues

In addition to providing a more scientific base for performance based codes, data collection and analysis that identifies issues early in a product cycle life can minimize the cost and negative images associated with major product recalls. These issues are more likely to occur with new technology. Fire protection issues that go undetected may cause a reduction of fire protection and life safety. Products that could be tracked using techniques identified in this research project that would benefit include:

- Backflow prevention devices
- Pressure transducers used in fire pumps
- Electronic accelerators
- Pressure reducing valves used in fire protection applications
- Dry pipe valves
- Preaction and deluge valves

Issues with pressure transducers used in fire pumps should be revealed in long term data collection as outlined in this research project. Extending this research project to cover other devices would require a standardized electronic data collection form and data base.

Recommendations

1. Compare the reliability of weekly tested fire pumps (99.4% for electric and 99.0% for diesel) to the reliability of monthly tested fire pumps (97.3% for electric and 95.6% for diesel) to determine an appropriate test frequency. The decrease in reliability associated with monthly testing appears to justify reestablishing a weekly test requirement for electric fire pumps in NFPA 25.
2. Complete the web site.
3. Develop a solicitation program for users. This should include direct solicitation by NFPA and announcement of the data collection process in fire protection related publications.
4. Make the data collection and analysis permanent.
5. Develop data collection and data base formats for pressure regulating devices, backflow prevention devices, dry pipe valves, and quick opening devices.

SUPPLEMENT 8

TECHNICAL/SUBSTANTIVE CHANGES FROM THE 2011 EDITION TO THE 2014 EDITION OF NFPA 25

Editor's Note: *Supplement 8 contains a table highlighting the significant technical changes to NFPA 25 for the 2014 edition, along with a brief comment regarding the reason for the change. For a complete record of all changes, along with the full committee statements for both editorial and technical changes, consult the NFPA 25 Report on Proposals and Report on Comments for the Annual 2013 Revision Cycle. These reports can be downloaded from the NFPA 25 document page at www.nfpa.org.*

2014 Text	*Notes*
1.1.3* This standard addresses the operating condition of fire protection systems as well as impairment handling and reporting and applies to fire protection systems that have been properly installed in accordance with generally accepted practice. **A.1.1.3** Generally accepted NFPA installation practices for water-based fire protection systems relevant to this standard are found in the following: NFPA 13, *Standard for the Installation of Sprinkler Systems* NFPA 13R, *Standard for the Installation of Sprinkler Systems in Low-Rise Residential Occupancies* NFPA 14, *Standard for the Installation of Standpipe and Hose Systems* NFPA 15, *Standard for Water Spray Fixed Systems for Fire Protection* NFPA 16, *Standard for the Installation of Foam-Water Sprinkler and Foam-Water Spray Systems* NFPA 20, *Standard for the Installation of Stationary Pumps for Fire Protection* NFPA 22, *Standard for Water Tanks for Private Fire Protection* NFPA 24, *Standard for the Installation of Private Fire Service Mains and Their Appurtenances* NFPA 750, *Standard on Water Mist Fire Protection Systems*	This revision clarifies the scope of the standard. Sections 4.1.5 and 4.1.6 discuss this comment; however, the committee decided to add language to Chapter 1 for clarity.
1.1.4* Corrective action needed to ensure that a system operates in a satisfactory manner shall be in accordance with this standard unless this standard specifically refers to an appropriate installation standard. **A.1.1.4** For systems originally installed in accordance with one of these standards, the repair, replacement, alteration, or extension of such systems should also be performed in accordance with that same standard. When original installations are based on other applicable codes or standards, repair, replacement, alteration, or extension practices should be conducted in accordance with those other applicable codes or standards.	This text has been revised because the 2014 edition of NFPA 25 includes Summary of Component Replacement Action Requirements tables in each chapter. Therefore, the user does not have to perform corrective actions per the installation standard unless specifically referred to in the tables.

2014 Text	*Notes*
3.3.1 Adjust. To maintain or regulate, within prescribed limits, by setting the operating characteristics to specified parameters. [**1911,** 2012]	A definition of the term *adjust*, which is used frequently throughout the maintenance tables, has been added.
3.3.5 Automatic Transfer Switch (ATS). Self-acting equipment for transferring the connected load from one power source to another power source. [**110,** 2013]	An extracted definition from NFPA 110 has been added to address the system component associated with electric-driven pumps. In 8.3.3.4 there are testing requirements for automatic transfer switches used with fire pumps. Therefore, the standard should have a definition.
3.3.6 Clean. To remove dirt, scale, and debris.	The definition of this term has been added because it is used throughout NFPA 25 in the maintenance tables.
3.3.7* Deficiency. For the purposes of inspection, testing, and maintenance of water-based fire protection systems, a condition that will or has the potential to adversely impact the performance of a system or portion thereof but does not rise to the level of an impairment. **A.3.3.7 Deficiency.** Depending on the nature and significance of the deficiency, it can result in a system impairment. Critical deficiencies will adversely impact performance but without the need for the implementing impairment procedures. Noncritical deficiencies have the potential to impact performance. Table A.3.3.7 provides examples for classifying conditions needing repair or correction that are identified during the inspection, testing, and maintenance of water-based suppression systems. The conditions are classified as an impairment, critical deficiency, or noncritical deficiency. The table is not all-inclusive but is included to provide guidance in responding to these conditions. For example, an impairment should be addressed promptly by either immediately correcting the condition or implementing the impairment procedures found in Chapter 15. Critical and noncritical deficiencies should be corrected as soon as practical after considering the nature and severity of the risk. It should be noted that many jurisdictions have requirements for the timely correction of impairments and/or deficiencies. The table does not take into account every variation of the conditions needing repair or correction. For example, a single lightly painted sprinkler in a large warehouse might be noncritical in its risk while a single painted sprinkler in a battery charging station might be considered a critical deficiency or perhaps an impairment. In addition, the nature of the hazard or the life safety exposure of the occupancy should be considered when assigning a classification. The table should be used with good judgment and could require input from the authority having jurisdiction.	The phrase "that will or has the potential to adversely impact the performance of a system" has been added to the definition to address the potential for a field condition to lead to an impaired system.
3.3.7.1 Critical Deficiency. A deficiency that, if not corrected, can have a material effect on the ability of the fire protection system or unit to function as intended in a fire event. **3.3.7.2 Noncritical Deficiency.** A deficiency that does not have a material effect on the ability of the fire protection system or unit to function in a fire event, but correction is needed to meet the requirements of this standard or for the proper inspection, testing, and maintenance of the system or unit.	The addition of the term *material* is intended to allow for minor deficiencies that do not impact the functionality of the systems in question.
Table A.3.3.7 — See table for text.	The former Annex E table has been relocated into Annex A and attached to the definition of the term *deficiency*. This table has been significantly modified since the 2011 edition.

2014 Text	*Notes*
3.3.10.1 Main Drain. The primary drain connection located on the system riser.	The phrase "and also utilized as a flow test connection" has been removed from the end of the definition.
3.3.20 Hydrostatic Test. A test of a closed piping system and its attached appurtenances consisting of subjecting the piping to an increased internal pressure for a specified period of duration to verify system integrity and leak rates. [**24,** 2013]	This definition has been added and was extracted from NFPA 24. NFPA 25 contains requirements for performing hydrostatic test(s), and no definition was previously included in the standard.
3.3.21.1* Emergency Impairment. A condition where a water-based fire protection system or portion thereof is out of order due to an unplanned occurrence, or the impairment is found while performing inspection testing or maintenance activities. **A.3.3.21.1 Emergency Impairment.** Examples of emergency impairments might include a ruptured pipe, an operated sprinkler, or an interruption of the water supply to the system.	Impairments can be found while performing inspection, testing, or maintenance (ITM) activities, not just when a problem arises in the building. Examples are provided in the annex.
3.3.22 Inspect. See 3.3.23, Inspection.	The term *inspect* has replaced the term *check* throughout the standard when referring to a visual act that must be conducted by the inspector.
3.3.24* Inspection, Testing, and Maintenance Service. A service program provided by a qualified contractor or qualified property owner's representative in which all components unique to the property's systems are inspected and tested at the required times and necessary maintenance is provided. **A.3.3.24 Inspection, Testing, and Maintenance Service.** This program includes logging and retention of relevant records.	This definition has been split into two parts to comply with the *NFPA Manual of Style* (MOS). The second half was moved to the annex, since definitions should be limited to a single sentence and should not contain requirements.
3.3.29 Performance-Based Program. Methods and frequencies that have been demonstrated to deliver equivalent or superior levels of performance through quantitative performance-based analysis.	This definition has been added because the standard allows an alternative for compliance using performance-based testing but did not have a definition for the term in previous editions.
3.3.35 Rebuild. To restore working condition by replacement or repair of worn or damaged parts.	The term *rebuild* is used throughout the maintenance tables for fire pumps and was previously undefined.
3.3.36 Reduced-Pressure Principle Backflow Prevention Assembly (RPBA). Two independently acting check valves together with a hydraulically operating, mechanically independent pressure differential relief valve located between the check valves, along with two resilient-seated shutoff valves, all as an assembly, and equipped with properly located test cocks.	The term *resilient-seated* is not appropriate for the test cocks as this is not part of the listing process for this device. The revisions to address "upstream" do not provide any additional clarification. The modifications are in line with the submittal and create a single sentence to comply with the MOS.
3.3.37 Remove. To physically take away or eliminate. **3.3.38 Repair.** Restore to sound working condition or to fix damage. **3.3.39 Replace.** To remove a component and install a new or equivalent component.	These terms have been added because they are used throughout the maintenance table for fire pumps and were previously undefined.

2014 Text	*Notes*
3.3.40.1 Installation Orientation. **3.3.40.1.1 Concealed Sprinkler.** A recessed sprinkler with cover plate. [**13,** 2013] **3.3.40.1.2 Flush Sprinkler.** A sprinkler in which all or part of the body, including the shank thread, is mounted above the lower plane of the ceiling. [**13,** 2013] **3.3.40.1.3 Pendent Sprinkler.** A sprinkler designed to be installed in such a way that the water stream is directed downward against the deflector. [**13,** 2013] **3.3.40.1.4 Recessed Sprinkler.** A sprinkler in which all or part of the body, other than the shank thread, is mounted within a recessed housing. [**13,** 2013] **3.3.40.1.5 Sidewall Sprinkler.** A sprinkler having special deflectors that are designed to discharge most of the water away from the nearby wall in a pattern resembling one-quarter of a sphere, with a small portion of the discharge directed at the wall behind the sprinkler. [**13,** 2013] **3.3.40.1.6 Upright Sprinkler.** A sprinkler designed to be installed in such a way that the water spray is directed upwards against the deflector. [**13,** 2013]	It is not assumed that an NFPA 25 inspector is also familiar with the design terminology from NFPA 13. Therefore, the types of sprinklers defined by NFPA 13 have been extracted into NFPA 25 to help inspectors identify the types of devices that might be found in the field.
3.3.42 Semiautomatic Dry Standpipe System. A standpipe system permanently attached to a water supply that is capable of supplying the system demand at all times arranged through the use of a device such as a deluge valve and that requires activation of a remote control device to provide water at hose connections. [**14,** 2013]	This definition has been added because testing of semiautomatic dry standpipe systems are referred to in 6.2.3.1 and a definition was not previously included in the standard.
3.3.43 Standpipe System Classes.	NFPA 14 refers to "classes" of standpipes, and not "types" of standpipes. This modification has been made to synchronize NFPA 14 and NFPA 25 on this subject.
3.3.46 Test. The operation of a device to verify that it is functioning correctly, or the measurement of a system characteristic to determine if it meets requirements.	A test is a functional verification of system readiness, and several types of tests are required by NFPA 25. This definition has been added to clearly differentiate a test from an inspection.
3.3.48 Valve Status Test. Flowing water to verify that valves for a portion of the system are not closed. **3.3.49* Valve Status Test Connection.** A point in the system where water is discharged for purposes of performing a valve status test. **A.3.3.49 Valve Status Test Connection.** These connections can include the main drain, fire pump test header, backflow preventer forward flow test connection, fire hydrant, and other similar locations. In the absence of the aforementioned devices, an inspector's test connection might be used.	These definitions have been added to make a distinction between the main drain test and a valve status test. The valve status test is much different in scope from the main drain test, which measures flow to determine if there was a change in the condition of water supply. The valve status test does not require flow measurements and is intended to confirm whether or not valves are closed.
3.3.52 Waterflow Alarm Device. An attachment to a water-based fire protection system that detects and signals a predetermined waterflow.	This definition has been revised to clearly define the function of the device, without referring to examples of what the device initiates.
3.5.6 [Pressure] Relief Valve. A device that allows the diversion of liquid to limit excess pressure in a system. [**20,** 2013]	This definition has been added because relief valves are referred to in Section 13.5, 13.5.7.2, 13.5.7.2.1, and 13.5.7.2.2. The definition was extracted from NFPA 20.
3.5.6.1 Circulation Relief Valve. A valve used to cool a pump by discharging a small quantity of water. This valve is separate from and independent of the main relief valve. [**20,** 2013]	This definition has been added because circulation relief valves are referenced in 13.5.7.1, 13.5.7.1.1, and 13.5.7.1.3.

2014 Text	Notes
3.6.4* Sprinkler System. A system that consists of an integrated network of piping designed in accordance with fire protection engineering standards that includes a water supply source, a water control valve, a waterflow alarm, and a drain and is commonly activated by heat from a fire, discharging water over the fire area. The portion of the sprinkler system above ground is a network of specifically sized or hydraulically designed piping installed in a building, structure, or area, generally overhead, and to which sprinklers are attached in a systematic pattern. The system is commonly activated by heat from a fire and discharges water over the fire area. [**13,** 2013]	This revised definition, which has been extracted from NFPA 13, includes a major revision, which states that the piping network that exists downstream of a floor control valve assembly (FCVA) is a separate sprinkler system.
3.6.4.1.1 Premixed Antifreeze Solution. A mixture of an antifreeze material with water that is prepared and factory-mixed by the manufacturer with a quality control procedure in place that ensures that the antifreeze solution remains homogeneous and that the concentration is as specified. [**13,** 2013]	Antifreeze solutions that are introduced into an existing sprinkler system must be factory premixed to ensure that sufficient energy has been imparted on the solution and that it will remain a single homogeneous solution in the system.
3.6.4.4 Marine System. A sprinkler system installed on a ship, boat, or other floating structure that takes its supply from the water on which the vessel floats.	NFPA 13 mandates in Chapter 25 that marine systems are maintained in accordance with NFPA 25, and 5.4.3 of NFPA 25 requires certain maintenance of marine systems. Thus, this definition has been included in this edition of the standard.
3.6.5 Water Mist System. A distribution system connected to a water supply or water and atomizing media supplies that is equipped with one or more nozzles capable of delivering water mist intended to control, suppress, or extinguish fires and that has been demonstrated to meet the performance requirements of its listing and [the applicable] standard. [**750,** 2010]	Because NFPA 25 now owns the scope for ITM of water mist systems, the definition of a water mist system has been included in this edition of the standard.
3.6.6.1 Ultra High-Speed Water Spray System. A type of automatic water spray system where water spray is rapidly applied to protect specific hazards where deflagrations are anticipated. [**15,** 2012]	Ultra high-speed water spray systems are covered in Section 10.4. A definition of the system has been added because it should be included in the standard.
3.7 ITM Task Frequencies. **3.7.1* Frequency.** Minimum and maximum time between events. **3.7.1.1 Daily Frequency.** Occurring every day. **3.7.1.2 Weekly Frequency.** Occurring once per calendar week. **3.7.1.3 Monthly Frequency.** Occurring once per calendar month. **3.7.1.4 Quarterly Frequency.** Occurring four times per year with a minimum of 2 months and a maximum of 4 months. **3.7.1.5 Semiannual Frequency.** Occurring twice per year with a minimum of 4 months and a maximum of 8 months. **3.7.1.6 Annual Frequency.** Occurring once per year with a minimum of 9 months and a maximum of 15 months. **3.7.1.7 Three Years Frequency.** Occurring once every 36 months with a minimum of 30 months and a maximum of 40 months. **3.7.1.8 Five Years Frequency.** Occurring once every 60 months with a minimum of 54 months and a maximum of 66 months.	Historically, NFPA 25 has not provided a range for the frequencies outlined in the standard. These definitions have been added to provide guidance to AHJs and owners on what the acceptable ranges should be.

2014 Text	*Notes*
A.3.7.1 Frequency. The frequencies in NFPA 25 are intended to establish an optimal time between tasks that are required by this document. When scheduling conflicts or other conditions do not allow the tasks to be performed on a strict calendar schedule, it is important that the required task frequencies be identified and complied with according to the variances described in the frequency definitions. When the required task frequencies have not been followed, it should be noted on the inspection report, the task should be performed, and the task frequencies should be followed for all future tasks. The variances should not be used to "skip" tasks or to perform fewer tasks than called for in this document.	
4.1.1* Responsibility for Inspection, Testing, Maintenance, and Impairment. The property owner or designated representative shall be responsible for properly maintaining a water-based fire protection system. **4.1.1.1*** Inspection, testing, maintenance, and impairment procedures shall be implemented in accordance with those established in this document and in accordance with the manufacturer's instructions. **4.1.1.2** Inspection, testing, and maintenance shall be performed by qualified personnel. **4.1.1.3*** Where the property owner or designated representative is not the occupant, the property owner or designated representative shall be permitted to delegate the authority for inspecting, testing, maintenance, and the managing of impairments of the fire protection system to a designated representative. **A.4.1.1** Any portion or all of the inspection, testing, and maintenance can be permitted to be contracted with an inspection, testing, and maintenance service. **A.4.1.1.1** In order to ensure compliance, the owner should verify that windows, skylights, doors, ventilators, other openings and closures, concealed spaces, unused attics, stair towers, roof houses, and low spaces under buildings do not expose water-filled piping to freezing. This should occur prior to the onset of cold weather and periodically thereafter.	Subsection 4.1.1 has been rewritten to clarify the roles and responsibilities of the owner in the ITM process. This subection formerly had a requirement that was titled "Buildings," which stated that the building temperature responsibility was assigned to the owner. The language was ambiguous, so it was relocated to a new section and was consolidated with other related requirements.
4.1.1.4 Where a designated representative has received the authority for inspecting, testing, maintenance, and the managing of impairments, the designated representative shall comply with the requirements identified for the property owner or designated representative throughout this standard.	This requirement has been revised to confirm the responsibility of the designated representative upon taking over the responsibilities of the owner. Typically this will be a contractual relationship; however, this is not mandated by NFPA 25.
4.1.2* Freeze Protection. The property owner or designated representative shall ensure that water-filled piping is maintained at a minimum temperature of 40°F (4°C) unless an approved antifreeze solution is utilized. **A.4.1.2** Other means of freeze protection for water-filled piping including heated valve enclosures, heat tracing, insulation, antifreeze solutions, or other methods are allowed by the applicable installation standard. Installation standards require heat tracing protecting fire protection piping against freezing to be supervised. **4.1.2.1** All areas of the building containing water-filled piping that does not have another means of freeze protection shall be maintained at a minimum temperature of 40°F (4°C).	This revision clarifies that the owner is responsible for making sure that the building is at an adequate temperature. It also clarifies which areas must be heated and also specifically addresses other freeze protection options.

2014 Text	Notes
4.1.2.2 Aboveground water-filled pipes that pass through open areas, cold rooms, passageways, or other areas exposed to temperatures below 40°F (4°C), protected against freezing by insulating coverings, frostproof casings, listed heat tracing systems, or other reliable means, shall be maintained at temperatures between 40°F (4°C) and 120°F (48.9°C). **4.1.2.3** Where other approved means of freeze protection for water-filled piping as described in 4.1.2.2 are utilized, they shall be inspected, tested, and maintained in accordance with this standard.	
4.1.4 Notification of System Shutdown or Testing. The property owner or designated representative shall notify the authority having jurisdiction, the fire department, if required, and the alarm-receiving facility before testing or shutting down a system or its supply.	The 2011 title was limiting and was not consistent with the requirements within the subsection. The title has been refined to include "testing."
4.1.4.2 The authority having jurisdiction, the fire department, and the alarm-receiving facility shall be notified when the system, supply, or component is returned to service or when the test is complete.	The technical committee has decided to remove notification requirements in each chapter by stating that this requirement is covered in this section of Chapter 4 and by changing the title to include testing as a reason for notification. The change to the subsequent subsections adds testing to them as well. Part of the notification should include the duration of the shutdown or test, which was added to 4.1.4.2.
4.1.5.1* The property owner or designated representative shall correct or repair deficiencies or impairments that are found during the inspection, test, and maintenance required by this standard. **4.1.5.2** Corrections and repairs shall be performed by qualified maintenance personnel or a qualified contractor. **A.4.1.5.1** System deficiencies not explained by normal wear and tear, such as hydraulic shock, can often be indicators of system problems and should be investigated and evaluated by a qualified person or engineer. Failure to address these issues could lead to catastrophic failure. Examples of deficiencies that can be caused by issues beyond normal wear and tear are as follows: (1) Pressure gauge deficiencies as follows: (a) Gauge not returning to zero (b) Gauge off scale (c) Gauge with bent needle (2) Support devices deficiencies as follows: (a) Bent hangers and/or rods (b) Hangers pulled out/off structure (c) Indication of pipe or hanger movement such as the following: i. Hanger scrape marks on pipe, exposed pipe surface where pipe and hangers are painted ii. Firestop material damaged at pipe penetration of fire-rated assembly (3) Unexplained system damage as follows: (a) Unexplained system damage beyond normal wear and tear (b) Bent or broken shafts on valves (c) Bent or broken valve clappers (d) Unexplained leakage at branch lines, cross main, or feed main piping	The annex material has been relocated to the more appropriate section.

2014 Text	*Notes*
(e) Unexplained leakage at closed nipples (f) Loose bolts on flanges and couplings (4) Fire pump deficiencies as follows: (a) Fire pump driver out of alignment (b) Vibration of fire pump and/or driver (c) Unusual sprinkler system piping noises (sharp report, loud bang)	
4.1.8 Valve Location. The location of shutoff valves shall be identified at the system riser or other approved locations.	This requirement has been moved to 4.1.8 to comply with the MOS.
4.1.10.1 Where an impairment to a water-based fire protection system occurs or is identified during inspection, testing, or maintenance activities, the procedures outlined in Chapter 15 shall be followed, including the attachment of a tag to the impaired system. **4.1.10.2** Where a water-based fire protection system is returned to service following an impairment, the system shall be verified to be working properly by means of an appropriate inspection or test as described in the table "Summary of Component Replacement [Action] Requirements" in the applicable chapters of this document.	This revision clarifies that many impairments are discovered while performing ITM on the system, and the owner or owner's representative needs to know to follow the procedures in Chapter 15 once an impairment is discovered.
4.2 Manufacturer's Corrective Action. Manufacturers shall be permitted to make modifications to their own listed product in the field with listed devices that restore the original performance as intended by the listing, where acceptable to the authority having jurisdiction.	The change to this section better describes the content of this section. This change also helps the user to avoid confusion with 4.1.4, which covers corrections and repairs.
4.3.1* Records shall be made for all inspections, tests, and maintenance of the system and its components and shall be made available to the authority having jurisdiction upon request. **4.3.1.1*** Records shall be permitted to be stored and accessed electronically. **4.3.2** Records shall indicate the following: (1) The procedure/activity performed (e.g., inspection, test, or maintenance) (2) The organization that performed the activity (3) The required frequency of the activity (4) The results and date of the activity (5) The name and contact information of the qualified contractor or owner, including lead person for activity **A.4.3.1** Inspection reports used for system inspections should contain an "Owner's Section" as shown in Figure A.4.3.1 that the property owner or designated representative should complete. Typical records include, but are not limited to, valve inspections; flow, drain, and pump tests; and trip tests of dry pipe, deluge, and preaction valves. Acceptance test records should be retained for the life of the system or its special components. Subsequent test records should be retained for a period of 1 year after the next test. The comparison determines deterioration of system performance or condition and the need for further testing or maintenance. **A.4.3.1.1** Computer programs that file inspection and test results should provide a means of comparing current and past results and should indicate the need for corrective maintenance or further testing.	A number of inspection programs that are on the market today provide for electronic records. These records are still accessible to AHJs upon request and are permitted by NFPA as a means of record retention.

2014 Text	Notes
4.4 Water Supply Status. During inspection, testing, and maintenance, water supplies, including fire pumps, shall remain in service unless under constant attendance by qualified personnel or unless impairment procedures in Chapter 15 are followed.	This section has been revised to address all three phases of the ITM process.
4.6.5* When a component or subsystem is adjusted, repaired, reconditioned, or replaced, it shall be tested in accordance with the original acceptance test required for that subsystem or the requirements where specified by the standard. **A.4.6.5** Examples of subsystems or components include fire pumps, drivers or controllers, pressure-regulating devices, detection systems and controls, alarm check, and dry pipe, deluge, and preaction valves. The required tests for components are contained in the corresponding chapter in tables titled *Summary of [Component] Inspection, Testing, and Maintenance.*	Subsection 4.5.6 has been revised because it was not correlated with the summary component action tables found in each chapter. The revised language is consistent with the requirements as found in the standard and better clarifies the intent of the section.
4.7* Performanced-Based Programs. As an alternative means of compliance and where approved by the authority having jurisdiction, components and systems shall be permitted to be inspected, tested, and maintained under a performance-based program.	This definition has been revised because the wording had inadvertently left out the word "approved."
5.2.1.1.2 Any sprinkler that shows signs of any of the following shall be replaced: (1) Leakage (2)* Corrosion (3) Physical damage (4) Loss of fluid in the glass bulb heat-responsive element (5)* Loading (6) Painting unless painted by the sprinkler manufacturer **A.5.2.1.1.2(2)** Corrosion found on the seat, or built up on the deflector that could affect the spray pattern, or a buildup on the operating elements that could affect the operation can have a detrimental effect on the performance of the sprinkler. Light surface corrosion on the boss, frame arms, and/or the deflector, and/or surface discoloration, not impacting the operation of the sprinkler should not warrant replacement. A degree of judgment should be exercised in the determination of the extent, location, and character of corrosion that would necessitate replacement.	The listed conditions can have a detrimental effect on the performance of the sprinkler through affect of the water distribution patterns, insulation of thermal elements delaying operation, or other conditions rendering the sprinkler inoperable or ineffectual. The annex material has been added to provide sorely needed guidance for the practical application of this intent.
5.2.1.1.3* Any sprinkler that has been installed in the incorrect orientation shall be corrected by repositioning the branchline, drop, or sprig, or shall be replaced.	This requirement has been revised to clarify it is possible to correct a problem with a sprinkler's orientation without having to replace the sprinkler. A qualified person should be able to make the determination on the most economical remedy for an improperly installed sprinkler while still observing all installation standards.
5.2.1.1.4* Sprinklers installed in concealed spaces such as above suspended ceilings shall not require inspection. **A.5.2.1.1.4** Examples include spaces above ceilings, whether the ceilings are lay-in tile or gypsum board, areas under theater stages, pipe chases, and other inaccessible areas, even if access panels or hatches are provided into the areas.	It is often misunderstood that any entry point through an access panel or hatch will automatically make the space accessible, thus eliminating it from being categorized as a concealed space.

2014 Text	*Notes*
5.2.1.1.6 Escutcheons and coverplates for recessed, flush, and concealed sprinklers shall be replaced if found missing during the inspection. **5.2.1.1.7** Escutcheons for pendent sprinklers that are not recessed, flush, or concealed shall not be required to be replaced if found missing during the inspection.	These requirements have been added because the standard has never addressed the issue of how to deal with missing escutcheons and cover plates. Some escutcheons and cover plates are merely decorative while others serve a function in the operation of the sprinkler.
5.2.1.2* The minimum clearance to storage as described in 5.2.1.2.1 through 5.2.1.2.6 shall be maintained below all sprinkler deflectors. **5.2.1.2.1*** Unless greater distances are required by 5.2.1.2.2, 5.2.1.2.3, or 5.2.1.2.4, or lesser distances are permitted by 5.2.1.2.6, clearance between the deflector and the top of storage shall be 18 in. (457 mm) or greater. **5.2.1.2.2** Where standards other than NFPA 13, *Standard for the Installation of Sprinkler Systems*, specify greater clearance to storage minimums, they shall be followed. **5.2.1.2.3*** Clearance between the deflector and the top of storage shall be 36 in. (914 mm) or greater for special sprinklers. **5.2.1.2.4** Clearance from the top of storage to sprinkler deflectors shall be 36 in. (914 mm) or greater where rubber tires are stored. **5.2.1.2.5** In-rack sprinklers shall not be required to meet the obstruction criteria and clearance from storage requirements. **5.2.1.2.6*** Clearance between the deflector and the top of storage shall be permitted to be less than 18 in. (457 mm) where shown to be permitted by the installation standard. **5.2.1.3*** Storage closer to the sprinkler deflector than permitted by the clearance rules of the installation standard described in 5.2.1.2.1 through 5.2.1.2.4 shall be corrected. **A.5.2.1.2** NFPA 13, *Standard for the Installation of Sprinkler Systems*, in the storage definitions defines clearance as the distance from the top of storage to the ceiling sprinkler deflectors. Other obstruction rules are impractical to enforce under this standard. However, if obstructions that might cause a concern are present, the owner is advised to have an engineering evaluation performed. **A.5.2.1.2.1** The 18 in. (457 mm) clearance rule generally applies to standard pendent, upright and sidewall spray sprinklers, extended coverage upright and pendent sprinklers, and residential sprinklers. **A.5.2.1.2.3** The special sprinklers that the minimum 36 in. (915 mm) clearance rule generally applies to includes large drop sprinklers, CMSA sprinklers, and early suppression fast-response (ESFR) sprinklers. **A.5.2.1.2.6** The purpose of maintaining a minimum clearance is to ensure water discharge is not obstructed. There are certain installations where this can be achieved by other means. Examples include library stacks, record storage, and where sprinklers are installed in aisles in between storage shelving. Clearance is also not needed for shelving along perimeter walls since this does not cause an obstruction. NFPA 13 allows a clearance less than 18 in. (457 mm) where full-scale fire tests demonstrate an acceptable sprinkler discharge pattern. Also, where sufficient shielding of the sprinkler spray pattern has resulted in an increase in the hazard classification to Extra Hazard Group 2, a clearance less than 18 in. (457 mm) might be acceptable.	Inspectors should not have to know the minimum clearances required by the installation standard because those rules change over time and it is unreasonable to ask the inspector to know which ones applied when. There are some basic clearance rules in NFPA 13 and the ones that can be inspected should be in NFPA 25, so they have been added. However, there might be situations where the sprinklers will not be obstructed by storage located within 18 in. of the deflector. An example is library stacks and record storage, where sprinklers are installed in the aisle between storage racks that are within 18 in. of the sprinkler deflector and along perimeter walls. Therefore, another provision is needed to permit these situations.

2014 Text	Notes
5.2.1.4 The supply of spare sprinklers shall be inspected annually for the following: (1) The correct number and type of sprinklers as required by 5.4.1.5 (2) A sprinkler wrench for each type of sprinkler as required by 5.4.1.5.5 (3) The list of spare sprinklers as required by 5.4.1.5.6	This language has been revised to be consistent with the requirements of 6.2.9 in NFPA 13. This language was created by an intercommittee task group consisting of members of the RSS, SSI, and NFPA 25 technical committees. This task group was created at the request of the NFPA 13 technical correlating committee.
A.5.2.2.1 Surface corrosion not impacting the integrity of the piping strength or raising concern of potential leakage should not warrant the replacement of piping. A degree of judgment should be exercised in the determination of the extent of corrosion that would necessitate replacement.	Black steel pipe starts to rust shortly after it is first manufactured. This light surface corrosion is always to be expected. Similarly, cast iron fittings often develop a bit of surface corrosion between the time of manufacture and fabrication/installation. However, as it was previously written in the standard, it was implied that when piping starts to lose its luster, it needs to be replaced. This was never the intent, so annex text has been added to provide the necessary clarification.
5.2.3.1 Hangers and seismic braces shall not be damaged, loose, or unattached. **5.2.3.2** Hangers and seismic braces that are damaged, loose, or unattached shall be replaced or refastened.	These added conditions are deficiencies as well and have been included for this edition. Although most inspectors probably noted hangers or seismic braces that were unattached, this standard did not require them to do so.
5.2.4.1* Gauges on wet pipe and deluge sprinkler systems shall be inspected quarterly to ensure that they are in good condition and that normal water supply pressure is being maintained. **5.2.4.2** Gauges on dry and preaction systems shall be inspected weekly to ensure that normal air or nitrogen and water pressures are being maintained.	These requirements have been revised to be more specific. Most but not all sprinkler systems are under contract for the inspection requirements of this standard to be performed. For those that are, a quarterly inspection should suffice. For those that are not, they are most likely not being performed by anyone at any period as specified by this standard. As this is a minimum standard, for those properties that are having inspections performed by their personnel, they might still elect to perform a monthly inspection.
5.2.5 Waterflow Alarm and Supervisory Signal Initiating Device. Waterflow alarm and supervisory signal initiating devices shall be inspected quarterly to verify that they are free of physical damage.	This revision is a global change to the standard to replace each instance of the phrase "supervisory device" with "supervisory signal initiating device."
5.2.6* Hydraulic Design Information Sign. The hydraulic design information sign shall be inspected quarterly to verify that it is provided, attached securely to the sprinkler riser, and is legible. **5.2.6.1** A hydraulic design information sign that is missing or illegible shall be replaced. **5.2.6.2** A pipe schedule system shall have a hydraulic design information sign that reads "Pipe Schedule System."	There is always a question about the need for a hydraulic design information sign when none is present on the system riser. The changes have been made to clarify that if a sign is not present, one needs to be provided, either to replace the one that is missing, or to retrofit a sign if the system is a pipe schedule. When a sign needs to be replaced or added, the owner is to supply the information for the sign based on the records from the original installation or from the most recent system evaluation.

2014 Text	*Notes*
5.2.7 Heat Tracing. Heat tracing shall be inspected and maintained in accordance with manufacturer's requirements.	This section was modified to clarify that the industry associates more with the term *heat tracing* rather than *heat tape*, which was previously used in the standard. Inspecting the heat tracing does not do much unless maintenance is performed if needed.
5.2.8 Information Sign. The information sign required by 4.1.9 shall be inspected annually to verify that it is provided, securely attached, and legible.	The text that has been added clarifies that this sign is to be present on each system control valve, antifreeze loop, and auxiliary system control valve.
5.3.4* Antifreeze Systems. Annually, before the onset of freezing weather, the antifreeze solution shall be tested using the following procedure: (1) Using installation records, maintenance records, information from the owner, chemical tests, or other reliable sources of information, the type of antifreeze in the system shall be determined. (a) If the type of antifreeze is found to be a type that is no longer permitted, the system shall be drained completely and replaced with an acceptable solution. (b) If the type of antifreeze cannot be reliably determined, the system shall be drained completely and replaced with an acceptable solution. (2) If the antifreeze is not replaced in accordance with step 1, test samples shall be taken at the top of each system and at the bottom of each system. (a) If the most remote portion of the system is not near the top or the bottom of the system, an additional sample shall be taken at the most remote portion. (b) If the connection to the water supply piping is not near the top or the bottom of the system, an additional sample shall be taken at the connection to the water supply. (3) The specific gravity of each solution shall be checked using a hydrometer with a suitable scale or a refractometer having a scale calibrated for the antifreeze solution. (4) If any of the samples exhibits a concentration in excess of what is permitted by NFPA 25, the system shall be emptied and refilled with a new acceptable solution. If a concentration greater than what is currently permitted by NFPA 25 was necessary to keep the fluid from freezing, alternate methods of preventing the pipe from freezing shall be employed. (5) If any of the samples exhibits a concentration lower than what is necessary to keep the fluid from freezing, the system shall be emptied and refilled with a new acceptable solution.	Due to the combustibility concerns surrounding traditional antifreeze solutions, testing the solutions becomes critical. Testing the solutions at multiple points allows the inspector to confirm that the solution has not separated and that the concentrations existing in the pipe never exceed the allowable limits. This requirement has been revised to clarify these points.
5.3.4.1 The use of antifreeze solutions shall be in conformity with state and local health regulations. **5.3.4.1.1*** Listed CPVC sprinkler pipe and fittings shall be protected from freezing with glycerin only. **5.3.4.1.1.1** The use of diethylene, ethylene, or propylene glycols shall be specifically prohibited.	This revision clarifies that where a deterministic risk assessment is approved by an AHJ and permits the use of traditional solutions in nonmetallic piping, the solution must be glycerin only.

2014 Text	Notes
5.3.4.2 Except as permitted by 5.3.4.2.1 and 5.3.4.2.2, all antifreeze systems shall utilize listed antifreeze solutions. **5.3.4.2.1*** For systems installed prior to September 30, 2012, listed antifreeze solutions shall not be required until September 30, 2022, where all of the following conditions are met: **A.5.3.4.2.1** All antifreeze systems installed after September 30, 2012, are assumed to meet the minimum requirements of NFPA 13, 2013 edition. For systems installed after September 30, 2012, that do not meet the requirements of the 2013 edition of NFPA 13, consideration should be given to applying 5.3.4.2.1. (1)* The concentration of the antifreeze solution shall be limited to 50 percent glycerin by volume or 40 percent propylene glycol by volume. **A.5.3.4.2.1(1)** The use of factory premixed solutions is required because solutions that are not mixed properly have a possibility of separating from the water, allowing the pure concentrate (which is heavier than water) to drop out of solution and collect in drops or low points of the system. Such concentrations are combustible and could present problems during fires. The properties of glycerine are shown in Table A.5.3.4.2.1(1). (2) Newly introduced solutions shall be factory premixed antifreeze solutions (chemically pure or United States Pharmacopeia 96.5 percent). (3)* Antifreeze systems with concentrations in excess of 30 percent propylene glycol and 38 percent glycerine shall be permitted based upon an approved deterministic risk assessment prepared by a qualified person approved by the authority having jurisdiction.	This revision requires that all traditional, non-listed antifreeze solutions must be replaced by September 30, 2022. This requirement gives the industry 10 years to find solutions and design alternatives to combustible antifreeze solutions. This section establishes the allowable concentrations for antifreeze.
A.5.3.4.2.1(3) Antifreeze solutions with a maximum concentration of 38 percent glycerine or 30 percent propylene glycol do not require a deterministic hazard analysis. The risk assessment should be prepared by individual(s) who can demonstrate an ability to prepare a risk assessment by education and experience and who can demonstrate an understanding of the issues associated with antifreeze sprinkler systems, including the available related fire tests. For additional information regarding the risk assessment process, documentation to be submitted, and the AHJ's role, refer to NFPA 551, *Guide for the Evaluation of Fire Risk Assessments*, and the SFPE *Engineering Guide: Fire Risk Assessment*. Propylene glycol and glycerine antifreeze solutions discharged from sprinklers have the potential to ignite under certain conditions. Research testing has indicated that several variables may influence the potential for large-scale ignition of the antifreeze solution discharged from a sprinkler. These variables include, but are not limited to, the concentration of antifreeze solution, sprinkler discharge characteristics, inlet pressure at the sprinkler, ceiling height, and size of fire at the time of sprinkler discharge. All relevant data and information should be carefully reviewed and considered in the deterministic risk assessment. As appropriate, the risk assessment should consider factors such as the following: (1) Occupancy use group per NFPA 13 (2) Ceiling height (3) Antifreeze solution concentration and type (4) Maximum system pressure (normal static pressures) (5) Sprinkler type, including K-factor (6) Potential and actual fuel load (Christmas trees) (7) Type of structure (construction types) (8) Size of structure	Where antifreeze is used in existing systems, a deterministic risk assessment must be conducted to validate the use of a traditional antifreeze solution for a certain configuration. This annex section has been added to provide a list of items that should be considered when developing the risk assessment. This section also provides references to the various reports on antifreeze solution testing, which are useful in the development of a deterministic risk assessment.

2014 Text	*Notes*
(9) Ability of the sprinkler system to control the fire (10) Occupied spaces versus unoccupied spaces such as trash enclosures and dust collectors as follows: (a) Adjacent occupancies (spaces adjacent to the area protected by antifreeze systems) (b) Separation between areas protected with an antifreeze system and other areas (c) Ventilation of areas protected with an antifreeze system to prevent damage to adjacent areas (d) Duration of antifreeze discharge Tests summarized in Table A.5.3.4.2.1(3) show that large-scale ignition of the sprinkler spray did not occur in tests with 50 percent glycerine and 40 percent propylene glycol antifreeze solutions discharging onto a fire having a nominal heat release rate (HRR) of 1.4 MW. A deterministic risk assessment that demonstrates that the heat release rate for reasonably credible fire scenarios will be less than 1.4 MW at the time of sprinkler activation should be acceptable. The risk assessment should also address issues associated with management of change, such as change in occupancy and temporary fuel loads. A natural Christmas tree can result in an HRR well above 1.4 MW at the time of sprinkler activation. In addition to the variables identified previously, the deterministic risk assessment should include occupancy, quantity of solution, impact on life safety, and potential increase in heat release rate. The following is a list of research reports that have been issued by the Fire Protection Research Foundation (FPRF) related to the use of antifreeze in sprinkler systems that should be considered in the development of the deterministic risk assessment: (1) *Antifreeze Systems in Home Fire Sprinkler Systems — Literature Review and Research Plan*, Fire Protection Research Foundation, June 2010. (2) *Antifreeze Systems in Home Fire Sprinkler Systems — Phase II Final Report*, Fire Protection Research Foundation, December 2010. (3) *Antifreeze Solutions Supplied through Spray Sprinklers — Interim Report*, Fire Protection Research Foundation, February 2012. Table A.5.3.4.2.1(3) provides an overview of the testing conducted by the FPRF.	
5.3.4.2.2 Premixed antifreeze solutions of propylene glycol exceeding 30 percent concentration by volume shall be permitted for use with ESFR sprinklers where the ESFR sprinklers are listed for such use in a specific application.	Prior to the recent fire events where full scale ignition of antifreeze solutions occurred, full-scale testing of any ESFR sprinkler with concentrations of propylene glycol in excess of 30 percent yielded a listed combination of the sprinkler and solution. This combination of listing and sprinkler is still permitted to be used in spite of the language in 5.3.4.2.1.
5.3.4.3 The antifreeze solution shall be tested at its most remote portion and where it interfaces with the wet pipe system.	This requirement has been revised to clarify that the antifreeze solution must be tested at the interface with the wet pipe system to determine if the solution has "slipped" and that there is not the appropriate mixture of water and antifreeze solution.

2014 Text	*Notes*
5.3.4.4 When antifreeze systems have a capacity larger than 150 gal (568 L), tests at one additional point for every 100 gal (379 L) shall be made. **5.3.4.4.1** If the results indicate an incorrect freeze point at any point in the system, the system shall be drained and refilled with new premixed antifreeze. **5.3.4.4.2** For premixed solutions, the manufacturer's instructions shall be permitted to be used with regard to the number of test points and refill procedure.	Where the mixture of antifreeze and water yields a combined solution that does not provide sufficient freeze protection, the solution must be drained and refilled with a solution that meets the requirements of 5.3.4.2.1 or an alternative freeze protection approach must be considered.
5.4.1.1 Where a sprinkler has been removed for any reason, it shall not be reinstalled.	This requirement has been revised to be consistent with the modification to the 2013 edition of NFPA 13, where identical language was included.
5.4.1.5* A supply of at least six spare sprinklers shall be maintained on the premises so that any sprinklers that have operated or been damaged in any way can be promptly replaced. **5.4.1.5.5*** One sprinkler wrench as specified by the sprinkler manufacturer shall be provided in the cabinet for each type of sprinkler installed to be used for the removal and installation of sprinklers in the system. **5.4.1.5.6** A list of the sprinklers installed in the property shall be posted in the sprinkler cabinet. **5.4.1.5.6.1*** The list shall include the following: (1) Sprinkler identification number (SIN) if equipped; or the manufacturer, model, orifice, deflector type, thermal sensitivity, and pressure rating (2) General description (3) Quantity of each type to be contained in the cabinet (4) Issue or revision date of the list	These requirements have been revised to clarify that one sprinkler wrench design can be appropriate for many types of sprinklers and that multiple wrenches of the same design are not required. The minimum information in the list contained in the spare sprinkler cabinet should be marked with the following: the quantity of sprinklers to be maintained in the spare sprinkler cabinet and a general description of the sprinkler, including upright, pendent, residential, ESFR, and so forth.
5.4.1.6* Sprinklers shall not be altered in any respect or have any type of ornamentation, paint, or coatings applied after shipment from the place of manufacture. **A.5.4.1.6** Corrosion-resistant or specially coated sprinklers should be installed in locations where chemicals, moisture, or other corrosive vapors exist.	These changes clarify the entire application of protective coverings by adding a separate section title and using most of the wording from NFPA 13. The use of protective coverings is very limited in NFPA 13 and the current text in NFPA 25 seems to imply that these coverings can be retrofitted in other applications.
5.4.1.7 Sprinklers and automatic spray nozzles used for protecting commercial-type cooking equipment and ventilating systems shall be replaced annually. **5.4.1.7.1** Where automatic bulb-type sprinklers or spray nozzles are used and annual examination shows no buildup of grease or other material on the sprinklers or spray nozzles, such sprinklers and spray nozzles shall not be required to be replaced.	This language has been revised to be consistent with the requirements of 6.2.9 in NFPA 13. This language was created by an intercommittee task group consisting of members of the RSS, SSI and NFPA 25 technical committees.
5.4.1.8 Protective Coverings. **5.4.1.8.1*** Sprinklers protecting spray areas and mixing rooms in resin application areas installed with protective coverings shall continue to be protected against overspray residue so that they will operate in the event of fire. **5.4.1.8.2** Sprinklers installed as described in 5.4.1.8.1 shall be protected using cellophane bags having a thickness of 0.003 in. (0.076 mm) or less or thin paper bags. **5.4.1.8.3** Coverings shall be replaced periodically so that heavy deposits of residue do not accumulate.	These requirements have been revised to clarify that typical sandwich bags purchased in a grocery store are generally plastic, not cellophane. Plastic bags have a tendency to shrink and adhere to the sprinkler prior to sprinkler activation, creating the potential for disruption of sprinkler spray patterns. Bags that are permitted to be placed over sprinklers need to be true cellophane or paper.

2014 Text	Notes
6.2.2.1 Gauges on automatic wet and semiautomatic dry standpipe systems shall be inspected quarterly to ensure that they are in good condition and that normal water supply pressure is being maintained. **6.2.2.2** Gauges on automatic dry standpipe systems shall be inspected weekly to ensure that normal air or nitrogen and water pressure are being maintained.	The language has been revised to more appropriately match the specific types of standpipe systems to which the inspection of gauges apply. Dry systems can include the use of nitrogen as well as air and should be recognized.
6.2.3* Hydraulic Design Information Sign. The hydraulic design information sign for standpipe systems shall be inspected annually to verify that it is provided, attached securely, and legible. **6.2.3.1** A hydraulic design information sign that is missing or illegible shall be replaced. **6.2.3.2** A standpipe system that was not sized by hydraulic design shall have a hydraulic design information sign that reads "Pipe Schedule System."	These changes have been made to clarify that if a sign is not present, one needs to be provided by either replacing the one that is missing or retrofitting a sign if the standpipe system is a pipe schedule. When a sign needs to be replaced or added, the owner is to supply the information for the sign based on the records from the original installation or from the most recent system evaluation.
6.3.1.1* A flow test shall be conducted every 5 years on all Class I and Class III standpipe systems to verify that the required flow and pressure are available at the hydraulically most remote hose valve outlet(s) while flowing the standpipe system demand. **6.3.1.1.1** Where a flow test of the hydraulically most remote outlet(s) is not practical, the authority having jurisdiction shall be consulted for the appropriate location for the test. **6.3.1.2*** The standpipe system demand shall include 500 gpm (1892 L/min) for the first standpipe and 250 gpm (946 L/min) for each additional standpipe until the total system demand is simultaneously flowing. **6.3.1.2.1*** The 250 gpm (946 L/min) required from each additional standpipe shall be allowed to be flowed from the most convenient hose valve on that standpipe. **6.3.1.2.2*** Where the 250 gpm (946 L/min) cannot be flowed from each additional standpipe, the authority having jurisdiction shall determine where the additional flow can be taken. **A.6.3.1.2** When the standpipe system was accepted, NFPA 14 required that each additional standpipe be flowed to simulate the hydraulic calculations. Typically, the lowest hose valve was used to create this simultaneous flow so hoses wouldn't have to be run all the way down each standpipe. **A.6.3.1.2.1** Since the pressures at each standpipe aren't required to be balanced by NFPA 14 or this standard, any hose valve on the standpipe can be flowed to achieve the additional 250 gpm (950 L/min) needed. It might be more convenient to use a hose valve on an upper level rather than the lowest one on the standpipe. **A.6.3.1.2.2** In some instances it isn't reasonable to attach a hose to a standpipe to provide this additional flow point. The authority having jurisdiction can allow the additional flow be made at other outlets on the standpipe system, such as from another standpipe, or from the fire pump test header. Although the results of having the flow points somewhere else in the standpipe system won't match the hydraulic calculations, the test will still prove that the most remote standpipe can provide the necessary flow and pressure required for fire department use while simultaneously flowing the full system demand.	When the standpipe system was accepted, NFPA 14 required that each additional standpipe be flowed to simulate the hydraulic calculations. Typically the lowest hose valve was used to create this simultaneous flow so that hoses wouldn't have to be run all the way down each standpipe.
6.3.1.3 The standpipe system demand shall be based on the design criteria in effect at the time of the installation. **6.3.1.3.1** Where the standpipe system demand cannot be determined, the authority having jurisdiction shall determine the standpipe system demand.	Due to the fact that the demand for the systems would be different depending on which edition of NFPA 14 was used in the design, it can be difficult to pinpoint the correct demand. In these instances, the AHJ dictates the demand.

2014 Text	Notes
6.3.2.1* Hydrostatic tests of not less than 200 psi (13.8 bar) pressure for 2 hours, or at 50 psi (3.4 bar) in excess of the maximum pressure, where maximum pressure is in excess of 150 psi (10.3 bar), shall be conducted every 5 years on manual standpipe systems and semiautomatic dry standpipe systems, including piping in the fire department connection. **6.3.2.1.1** Manual wet standpipes that are part of a combined sprinkler/standpipe system shall not be required to be tested in accordance with 6.3.2.1. **A.6.3.2.1** The intent of 6.3.2.1 is to ascertain whether the system retains its integrity under fire conditions. Minimum leakage existing only under test pressure is not cause for repair.	The intent of 6.3.2.1 is to ascertain whether the system retains its integrity under fire conditions. Minimum leakage existing only under test pressure is not cause for repair.
8.2.2* The pertinent visual observations specified in the following checklists shall be performed weekly: (1) Pump house conditions as follows: (a) Heat is adequate, not less than 40°F (4°C) for pump room with diesel engine–driven pumps with engine heaters. (b) Heat is adequate, not less than 70°F (21°C) for pump room with diesel engine–driven pumps without engine heaters. (c) Ventilating louvers are free to operate.	Many older engines were not furnished with engine (block) heaters and require higher room temperatures.
8.3.1.1* A non-flow test shall be conducted for diesel engine–driven fire pumps without recirculating water back to the pump suction on a test frequency in accordance with 8.3.1.1.1 or 8.3.1.1.2. **8.3.1.1.1** Except as permitted in 8.3.1.1.2, a weekly test frequency shall be required. **8.3.1.1.2*** The test frequency shall be permitted to be established by an approved risk analysis. **8.3.1.2*** A non-flow test shall be conducted for electric motor–driven fire pumps without recirculating water back to the pump suction on a test frequency in accordance with 8.3.1.2.1, 8.3.1.2.2, 8.3.1.2.3, or 8.3.1.2.4. **8.3.1.2.1** Except as permitted in 8.3.1.2.2 and 8.3.1.2.3, a weekly test frequency shall be required for the following electric fire pumps: (1) Fire pumps that serve fire protection systems in high rise buildings that are beyond the pumping capacity of the fire department (2) Fire pumps with limited service controllers (3) Vertical turbine fire pumps (4) Fire pumps taking suction from ground level tanks or a water source that does not provide sufficient pressure to be of material value without the pump **8.3.1.2.2** A monthly test frequency shall be permitted for electric fire pumps not identified in 8.3.1.2.1. **8.3.1.2.3*** A monthly test frequency shall be permitted for electric fire pump systems having a redundant fire pump. **8.3.1.2.4*** The test frequency shall be permitted to be established by an approved risk analysis.	These requirements have been revised to clarify that a risk analysis should be prepared and reviewed by qualified people. Increased test frequencies might be desirable when high impact losses could result from an uncontrolled fire. Examples where increased fire pump test frequencies might be considered could include high piled storage facilities and buildings where the predominant occupancy is protected by an extra hazard density sprinkler system. Increased test frequencies might be desirable when high impact losses could result from an uncontrolled fire. Examples where increased fire pump test frequencies might be considered could include high piled storage facilities and buildings where the predominant occupancy is protected by an extra hazard density sprinkler system.

2014 Text	Notes
A.8.3.1.1 Fire pump systems should be designed so that the pressure relief valve has a minimum flow (to verify pressure relief valve is properly set and operating) at churn, and only allows a larger flow under abnormal conditions (i.e., engine overspeed or failure of a variable speed pressure limiting control). The fire pump imparts more energy into the water when recirculating the water through the pump than when the pump is operating at churn (no flow). NFPA 20 requires a circulation relief valve downstream of the pressure relief valve whenever the pressure relief valve is piped back to the pump suction. Improperly installed and/or operating circulation relief valves can result in unacceptably high water temperature, especially when recirculating the water to the pump suction. High water temperatures can affect the operation of a diesel engine drive. Modern engines, due to EPA requirements, are more sensitive to cooling water temperatures. **A.8.3.1.1.2** The risk analysis should be prepared and reviewed by qualified people. Increased test frequencies might be desirable when high impact losses could result from an uncontrolled fire. Examples where increased fire pump test frequencies can be considered could include high piled storage facilities and buildings where the predominant occupancy is protected by an extra hazard density sprinkler system. Test frequency has been a heavily discussed and researched topic for several years, and is still continuing to be researched. A set of data was submitted in 2008 by a group of owners and maintainers of large numbers of fire pumps. This data was presented to the committee as indicating a decreased test frequency on electric fire pumps did not "significantly" impact "reliability"; however, "reliability" as used in the discussion of the data presentation was actually the failure rate, and did not take into account the effect of test frequency on the fire pump reliability (i.e., the time between failure and discovery of the failure affects reliability). Subsequently, the NFPA Research Council commissioned research, and the resultant "Fire Pump Field Data Collection and Analysis Report" in 2011 (available for download at www.nfpa.org\Foundation) reported that electric fire pumps tested weekly had a failure rate of approximately 0.64 per year. Assuming a failure rate independent of the test frequency, and assuming that on the average the impairment occurs at the midpoint of the test interval, this failure rate provides approximately 99.4 percent reliability with weekly testing and approximately 97.3 percent reliability with monthly testing. Diesel engine fire pumps tested weekly had a failure rate of approximately 1.02 per year. Assuming a failure rate independent of the test frequency and assuming that on the average the impairment occurs at the midpoint of the test interval, this failure rate provides approximately 99.1 percent reliability with weekly testing and approximately 96.0 percent reliability with monthly testing. Based on this data, the lower reliability has not been determined to be acceptable for all facilities. Decisions to decrease test frequency must be based on more than cost savings. A reliability/risk analysis to decrease test frequency should take into account the risk associated with life safety, property values, hazards, and business interruption at the protected property. Fire pump redundancy can impact overall fire system reliability and be used in a reliability/risk analysis. **A.8.3.1.2** For pressure relief valve operation, see 8.3.2.4. **A.8.3.1.2.3** For systems where multiple fire pumps are required to meet the system demand, a one-for-one redundancy is not necessary (i.e., one backup pump for two or more primary pumps meets the intent of this section). **A.8.3.1.2.4** The risk analysis should be prepared and reviewed by qualified people. Increased test frequencies can be desirable when high impact losses could result from an uncontrolled fire. Examples where increased fire pump test frequencies can be considered could include high piled storage facilities and buildings where the predominant occupancy is protected by an extra hazard density sprinkler system.	

2014 Text	*Notes*
8.3.2.1* A test of fire pump assemblies shall be conducted without recirculating water back to the pump suction.	A system must have flow in order to keep the engine and pump cool. The intent is to conduct a churn test. The new annex text has been added to help explain the committee's intent.
8.3.2.6 An automatic timer that meets 8.3.2.6.1 through 8.3.2.6.3 shall be permitted to be substituted for the starting procedure. **8.3.2.6.1** A solenoid valve drain on the pressure control line shall be the initiating means for a pressure-actuated controller. **8.3.2.6.2** In a pressure-actuated controller, performance of this program timer shall be recorded as a pressure drop indication on the pressure recorder. **8.3.2.6.3** In a non-pressure-actuated controller, the test shall be permitted to be initiated by means other than a solenoid valve.	The changes that have been made to these requirements recognize the need for qualified personnel to witness the pump operation but maintain some flexibility in streamlining the fire pump test. The technical committee recognizes the difficulty in getting personnel into these rooms during these tests.
8.3.2.7 Qualified personnel shall be in attendance whenever the pump is in operation. **8.3.2.7.1*** The use of the automatic timer allowed in 8.3.2.6 shall not eliminate the requirement of 8.3.2.7 to have qualified personnel present during test.	These requirements have been revised to address the issue of owners/operators improperly using the timer initiated test to run the test without the presence of a qualified operator.
8.3.2.8 The pertinent visual observations or adjustments specified in the following checklists shall be conducted while the pump is idle: (1) Record the system suction and discharge pressure gauge readings (2) For pumps that use electronic pressure sensors to control the fire pump operation, record the current pressure and the highest and the lowest pressure shown on the fire pump controller event log (3) If the highest or lowest pressure is outside of the expected range, record all information from the event log that helps identify the abnormality **8.3.2.9*** The pertinent visual observations or adjustments specified in the following checklists shall be conducted while the pump is running: (1) Pump system procedure as follows: (a) Record the pump starting pressure from the pressure switch or pressure transducer (b) Record the system suction and discharge pressure gauge readings (c) Inspect the pump packing glands for slight discharge (d) Adjust gland nuts if necessary (e) Inspect for unusual noise or vibration (f) Inspect packing boxes, bearings, or pump casing for overheating (g) Record pressure switch or pressure transducer reading and compare to the pump discharge gauge (h) For pumps that use electronic pressure sensors to control the fire pump operation, record the current pressure and the highest and the lowest pressure shown on the fire pump controller event log (i) For electric motor and radiator cooled diesel pumps, check the circulation relief valve for operation to discharge water (2) Electrical system procedure as follows: (a) Observe the time for motor to accelerate to full speed (b) Record the time controller is on first step (for reduced voltage or reduced current starting) (c) Record the time pump runs after starting (for automatic stop controllers)	Many of the pressure transducers covered by a recent recall were used on fire pumps. The transducers might stick or fail and prevent the fire pump from responding when needed. This language has been revised to help identify pressure transducers that have developed abnormalities. Because NFPA 25 addresses existing installed equipment, controller modifications that could identify many failures when they occur were deemed to be impractical. The changes will not identify transducer failures when they occur but will make it more likely to identify abnormal transducers during fire pump testing.

2014 Text	*Notes*
(3) Diesel engine system procedure as follows: (a) Observe the time for engine to crank (b) Observe the time for engine to reach running speed (c) Observe the engine oil pressure gauge, speed indicator, water, and oil temperature indicators periodically while engine is running (d) Record any abnormalities (e) Inspect the heat exchanger for cooling waterflow (4) Steam system procedure as follows: (a) Record the steam pressure gauge reading (b) Observe the time for turbine to reach running speed	
8.3.3.1* An annual test of each pump assembly shall be conducted by qualified personnel under no-flow (churn), rated flow, and 150 percent of the pump rated capacity flow of the fire pump by controlling the quantity of water discharged through approved test devices. **A.8.3.3.1** Minimum flow for a pump is the churn pressure.	The clarification of 150 percent capacity has been made to give the user a defined meaning to the term that was previously used, *peak*.
8.3.3.1.2.3 Use of Pump Discharge via Bypass Flowmeter to Pump Suction (Closed-Loop Metering). **(A)** Pump suction and discharge pressures and the flowmeter measurements shall determine the total pump output. **(B)** When testing includes recirculating water back to the fire pump suction, the temperature of the recirculating water shall be monitored to verify that it remains below temperatures that could result in equipment damage as defined by the pump and engine manufacturers.	The revisions to this requirement have been added because several fire pump systems have been damaged by using this form of testing.
8.3.3.2 The pertinent visual observations, measurements, and adjustments specified in the following checklists shall be conducted annually while the pump is running and flowing water under the specified output condition: (1) At no-flow condition (churn) as follows: (a) Inspect the circulation relief valve for operation to discharge water (b) Inspect the pressure relief valve (if installed) for proper operation (2) At each flow condition as follows: (a) Record the electric motor voltage and current (all lines) (b) Record the pump speed in rpm (c) Record the simultaneous (approximately) readings of pump suction and discharge pressures and pump discharge flow (3)* For electric motor–driven pumps, do not shut down the pump until it has run for 10 minutes (4) For diesel motor–driven pumps, do not shut down the pump until it has run for 30 minutes	The clarifications that have been added in this requirement allow for not discharging water during this time, but would allow churn for time stated in items 3 and 4. In areas with severe water restrictions, this would define the intent of the standard more clearly. Modifications were made to the originally proposed language for correlation with NFPA 20.
8.3.3.3.2.1 When it is necessary to close the relief valve to achieve minimum rated characteristics for the pump, the pump discharge control valve shall be closed if the pump churn pressure exceeds the system rated pressure.	This revision allows for closing the pump discharge valve as to not permit over pressurization of the building's sprinkler system(s).

2014 Text	*Notes*
8.3.3.3.3 When pressure relief valves are piped back to the fire pump suction, the temperature of the recirculating water shall be monitored to verify that it remains below temperatures that could result in equipment damage as defined by the pump and engine manufacturers.	This requirement has been revised to provide enforceable language to address overheating that can occur if inadequate water for cooling is discharged during any part of the test.
8.3.3.4 For installations having an automatic transfer switch, the following test shall be performed to ensure that the overcurrent protective devices (i.e., fuses or circuit breakers) do not open: (1) Simulate a power failure condition while the pump is operating at peak load (2) Verify that the transfer switch transfers power to the alternate power source (3) Verify that the pump continues to perform at peak horsepower load on the alternate power source for 10 minutes for an alternate utility or 30 minutes if the alternate power source is a standby generator set (4) Remove the power failure condition and verify that, after a time delay, the pump is reconnected to the normal power source	These revisions have been made to clarify that, during annual tests, the alternate power source must also be tested to ensure that circuits and generators are tested to confirm that they perform under peak load. The revision also clarifies that a standby generator requires a 30-minute test while carrying peak electric motor fire pump horsepower load.
8.3.4 Diesel Fuel Testing and Maintenance. **8.3.4.1** Diesel fuel shall be tested for degradation no less than annually. **8.3.4.1.1*** Fuel degradation testing shall comply with ASTM D 975-11b, *Standard Specification for Diesel Fuel Oils,* or ASTM D 6751-11b, *Standard Specification for Biodiesel Fuel Blend Stock (B100) for Middle Distillate Fuels,* as approved by the engine manufacturer, using ASTM D 7462-11, *Standard Test Method for Oxidation Stability of Biodiesel (B100) and Blends of Biodiesel with Middle Distillate Petroleum Fuel (Accelerated Method).* **8.3.4.2*** If diesel fuel is found to be deficient in the testing required in 8.3.4.1.1, the fuel shall be reconditioned or replaced, the supply tank shall be cleaned internally, and the engine fuel filter(s) shall be changed. **8.3.4.2.1** After the restoration of the fuel and tank in 8.3.4.2, the fuel shall be retested every 6 months until experience indicates the fuel can be stored for a minimum of 1 year without degradation beyond that allowed in 8.3.4.1.1. **8.3.4.3** When provided, active fuel maintenance systems shall be listed for fire pump service and installed in accordance with the requirements in NFPA 20 for a new installation. **8.3.4.3.1** Maintenance of active fuel maintenance systems shall be in accordance with the manufacturer's recommendations. **8.3.4.3.2** Maintenance of active fuel maintenance systems shall be performed at a minimum annual frequency for any portion of the system that the manufacturer does not provide a recommended maintenance frequency.	Commercial distillate fuel oils used in modern diesel engines are subject to various detrimental effects from storage. The origin of the crude oil, refinement processing techniques, time of year, and geographical consumption location all influence the determination of fuel blend formulas. Naturally occurring gums, waxes, soluble metallic soaps, water, dirt, blends, and temperature all contribute to the degradation of the fuel as it is handled and stored. These effects begin at the time of fuel refinement and continue until consumption. Proper maintenance of stored distillate fuel is critical for engine operation, efficiency, and longevity. Storage tanks should be kept water-free. Water contributes to steel tank corrosion and the development of microbiological growth where fuel and water interface. This and the metals of the system provide elements that react with fuel to form certain gels or organic acids, resulting in clogging of filters and system corrosion. Scheduled fuel maintenance helps to reduce fuel degradation. Fuel maintenance filtration can remove contaminants and water and maintain fuel conditions to provide reliability and efficiency for standby fire pump engines. Fuel maintenance and testing should begin the day of installation and first fill.

2014 Text	*Notes*
8.3.5 Positive Displacement Pumps. [20:14.2.6.4.3] **8.3.5.1** Except as provided in 8.3.5.1 through 8.3.5.7, positive displacement pumps shall be tested in accordance with 8.3.1 through 8.3.3. **8.3.5.2** The pump flow for positive displacement pumps shall be tested and determined to meet the specified rated performance criteria where only one performance point is required to establish positive displacement pump acceptability. [**20:**14.2.6.4.3.1] **8.3.5.3** The pump flow test for positive displacement pumps shall be accomplished using a flowmeter or orifice plate installed in a test loop back to the supply tank, to the inlet side of a positive displacement water pump, or to drain. [**20:**14.2.6.4.3.2] **8.3.5.4** The flowmeter reading or discharge pressure shall be recorded and shall be in accordance with the pump manufacturer's flow performance data. [**20:**14.2.6.4.3.3] **8.3.5.5** If orifice plates are used, the orifice size and corresponding discharge pressure to be maintained on the upstream side of the orifice plate shall be made available to the authority having jurisdiction. [**20:**14.2.6.4.3.4] **8.3.5.6** Flow rates shall be as specified while operating at the system design pressure. Tests shall be performed in accordance with HI 3.6, *Rotary Pump Tests*. [**20:**14.2.6.4.3.5] **8.3.5.7** Positive displacement pumps intended to pump liquids other than water shall be permitted to be tested with water; however, the pump performance will be affected, and manufacturer's calculations shall be provided showing the difference in viscosity between water and the system liquid. [**20:**14.2.6.4.3.6]	Positive displacement pumps are routinely utilized to supply all types of water mist systems — wet pipe, dry pipe, deluge, and preaction. Since many of these systems are being installed in lieu of standard sprinkler systems for the same application, it is necessary to ensure they are inspected, tested, and maintained to achieve an equivalent level of dependability. The existing annual flow test requirements of Chapter 8 are unique to centrifugal pumps (e.g., the test to ensure 150 percent rated capacity at 65 percent rated head), which are characteristics not mutually inherent to positive displacement pumps. A unique feature of positive displacement pumps is the fact that the flow they supply is directly proportional to driver speed (rpm), and that pressure is typically controlled via a pressure sustaining valve or other regulating bypass device installed downstream of the pump. As a result, they exhibit a fairly flat pump curve, which ends abruptly once the maximum capacity of the pump is reached. In addition, they do not "churn" in the same manner as a standard fire pump. Therefore, an annual flow test program specific to the key operating characteristics of positive displacement pumps is required.
8.3.6.4* Parallel and angular alignment of the pump and driver shall be inspected during the annual test, and any misalignment shall be corrected.	If pumps and drivers were shipped from the factory with both machines mounted on a common base plate, they were accurately aligned before shipment. All base plates are flexible to some extent and, therefore, must not be relied on to maintain the factory alignment. Realignment is necessary after the complete unit has been leveled on the foundation and again after the grout has set and foundation bolts have been tightened.
8.3.7.1* Data Interpretation. **8.3.7.1.1** The interpretation of the flow test results and performance relative to system demand requirements shall be the basis for determining acceptable performance of the pump assembly. **8.3.7.1.2** Qualified individuals shall interpret the test results. **8.3.7.1.2.1** Interpretation of results shall include review of pump test data and written evaluation of conclusions.	Mathematical adjustment is typically completed using Affinity Law calculations based on original and current test speed differences at each test flow point. The original factory pump curves are almost always available from the manufacturer by contacting them with the pump serial number. Manufacturers typically keep this product pump data for perpetuity. The version of the performance curve from the acceptance test that is most useful is the version with the pump running at rated speed. The version of the acceptance test with the pump running at the speed of the manufacturer's shop test might not be as valuable since it might not be at the rated speed of the pump and driver on this particular installation.

2014 Text	Notes
8.3.7.2 Engine Speed Adjustments. **8.3.7.2.1*** Mathematical adjustment shall be made for correction of recorded test data to the original pump rated speed and velocity head, and shall be applied when determining flow test performance relative to the original pump performance. **8.3.7.2.2** Modifying engine speed to affect flow during testing shall not be permitted as a method for meeting acceptable pump performance. **8.3.7.2.3*** Net performance at pump rated speed shall be graphically plotted and evaluated with a comparison to the net pressure curve from owner documents, copies of original manufacturers' pump curves, pump nameplate data, or pump retrofit/rebuild documents. **A.8.3.7.2.1** Mathematical adjustment is typically completed using Affinity Law calculations based on original and current test speed differences at each test flow point. The original factory pump curves are almost always available from manufacturers by contacting them with the pump serial number. Manufacturers typically keep this product pump data for perpetuity. The version of the performance curve from the acceptance test that is most useful is the version with the pump running at rated speed. The version of the acceptance test with the pump running at the speed of the manufacturer's shop test might not be as valuable since it might not be at the rated speed of the pump and driver on this particular installation. **A.8.3.7.2.3** There are rare cases where original fire pump performance data is not available due to lost data, pump/driver replacement, or pump modifications that change discharge. In such cases 8.3.7.3(1) cannot realistically be completed, and a flow test should be conducted using previous flow data for comparison. The performance per 8.3.7.3(2) should still be documented. **8.3.7.3** A fire pump performance flow test shall be considered acceptable when both of the following conditions are determined from test results: (1)* Pump flow performance adjusted for speed per 8.3.7.2.1 is no less than 95 percent of the original specification documentation across the complete flow performance curve. (2) Pump performance unadjusted for speed meets or exceeds all requirements for supplied system demands based on owner-supplied system requirements. **8.3.7.4*** In evaluating adjusted pump flow, performance degradation in excess of 5 percent shall require an investigation to reveal the cause of degraded performance. **8.3.7.4.1** Investigation findings shall be documented through written evaluation as part of the fire pump test documents.	The revisions that have been made to these requirements aim to clarify the rules with respect to when the data get adjusted for rated speed and when they do not.
8.6.2 NFPA 20 shall be consulted for the minimum requirements for design, installation, and acceptance testing. **8.6.3** Replacement parts shall be provided that will maintain the listing for the fire pump component assembly whenever possible. **8.6.3.1** If the part is no longer available from the original equipment manufacturer, then an approved like part shall be permitted to be used.	The intent of the revisions that have been made to these requirements is to reference NFPA 20 for design and installation issues and whenever component replacement warrants a full acceptance test.

2014 Text	*Notes*
9.2.1.1* The water level in tanks equipped with supervised water level alarms that are connected to a constantly attended location shall be inspected quarterly. **9.2.1.2** The water level in tanks not equipped with supervised water level alarms connected to a constantly attended location shall be inspected monthly. **9.2.2.1** The air pressure in pressure tanks that have their air pressure source supervised in accordance with *NFPA 72, National Fire Alarm and Signaling Code*, shall be inspected quarterly.	These requirements have been clarified to state that it is the water level that needs to be inspected, not the tank.
9.2.4.1 The temperature of water in tanks shall not be less than 40°F (4.0°C).	The term *in* has been added to 9.2.4.1 to clarify that the temperature threshold applies to the water in the tank and not the tank itself. This correlates with the other subsections of 9.2.4.
9.2.6.1.2* The interior of all other types of tanks shall be inspected every 5 years. **A.9.2.6.1.2** If written verification of interior corrosion protection for a tank per NFPA 22, *Standard for Water Tanks for Private Fire Protection*, cannot be provided by the building owner, the interior of the tank should be inspected every 3 years.	These clarifications were made because without written verification of corrosion protection, the inspector would not know if the tank required a 5-year inspection or a 3-year inspection.
9.3.3 Low water temperature signals, where provided, shall be tested monthly (cold weather only). **9.3.4*** High water temperature limit switches on tank heating systems, where provided, shall be tested monthly whenever the heating system is in service.	Using the terminology of NFPA 72, an *alarm* is an indication of a condition where the only correct action is to call the fire department. For other indications of problems in a system, the correct term is a *signal*. The correct action when a low temperature or low water condition occurs is not to call the fire department. Therefore, the term *alarm* has been changed to *signal*.
9.5.1.1 Automatic tank fill valves shall be inspected in accordance with Table 9.5.1.1. **9.5.1.1.1** OS&Y isolation valves that are a part of the automatic fill valves shall be inspected in accordance with Chapter 13.	The weekly requirement for the OS&Y valves to be inspected has been determined to be inappropriate. These revisions clarify that OS&Y valves with electronic supervision should be allowed to be inspected monthly as permitted by Chapter 13.
Chapter 12 — Water Mist Systems	The ITM requirements for water mist systems have been removed from NFPA 750 and included in NFPA 25 only. Similar to sprinklers, water mist nozzles have the potential to degrade when exposed to installation environments. Previous editions of NFPA 25 and NFPA 750 do not include ITM requirements for water mist nozzles. Since water mist technology utilizes fast-response heat responsive elements, 20 years has been referenced as a reasonable time period for initial replacement or representative sample testing of these devices, which is consistent with the approach that was applied to sprinklers.

2014 Text	*Notes*
13.2.5* Main Drain Test. A main drain test shall be conducted annually for each water supply lead-in to a building water-based fire protection system to determine whether there has been a change in the condition of the water supply. **13.2.5.1** Where the lead-in to a building supplies a header or manifold serving multiple systems, a single main drain test shall be permitted. **A.13.2.5** Main drains are installed on system risers for one principal reason: to drain water from the overhead piping after the system is shut off. This allows the contractor or plant maintenance department to perform work on the system or to replace nozzles after a fire or other incident involving system operation. Data collected from the suction gauges during a fire pump flow test that test the water supply would satisfy the requirements for a main drain test. These drains also are used to determine whether there is a major reduction in waterflow to the system, such as could be caused by a major obstruction, a dropped gate, a valve that is almost fully closed, or a check valve clapper stuck to the valve seat. A satisfactory main drain test (i.e., one that reflects the results of previous tests) does not necessarily indicate an unobstructed passage, nor does it prove that all valves in the upstream flow of water are fully opened. However, these tests provide a reasonable level of confidence that the water supply has not been compromised. The main drain test is conducted in the following manner: (1) Record the pressure indicated by the supply water gauge. (2) Close the alarm control valve on alarm valves. (3) Fully open the main drain valve. (4) After the flow has stabilized, record the residual (flowing) pressure indicated by the water supply gauge. (5) Close the main drain valve slowly.	This revision adds new definitions for *valve status test* and *valve status test connection*. These definitions were added to make a distinction between the main drain test and a valve status test. The valve status test is much different in scope than the main drain test, which measures flow to determine if there was a change in the condition of water supply. The valve status test does not require flow measurements and is intended to confirm whether or not valves are closed. The revisions also confirm that the intent of the main drain test is for each lead-in into the building, not each system within the building.
13.2.6.1 Mechanical waterflow alarm devices, including but not limited to water motor gongs, shall be tested quarterly.	Revisions have been made to change the phrase "waterflow devices" to "waterflow alarm devices," which is a more appropriate and technically accurate term.
13.3.2.1.2 Control valve supervisory alarm devices shall be inspected quarterly to verify that they are free of physical damage.	This requirement exists in Chapter 5, but also applies to control valves in other chapters of this standard, and therefore has been included in Chapter 13 for continuity.
13.3.2.2* The valve inspection shall verify that the valves are in the following condition: (1) In the normal open or closed position (2)* Sealed, locked, or supervised (3) Accessible (4) Post indicator valves (PIVs) are provided with correct wrenches (5) Free from external leaks (6) Provided with applicable identification	This requirement has been revised to clarify that wrenches are only needed for PIVs and that a proper wrench needs to be used for this test. Using an improper wrench, such as a pipe wrench, has resulted in damage to the operating nut. The use of break over bars and extensions on the wrench can damage the valve and/or the post. If the valve cannot be closed and reopened using the proper wrench with reasonable force, then some maintenance and/or repairs are necessary so the valve can be operated when needed in a fire event.

2014 Text	*Notes*
13.4.3.1.1 Valve enclosures for preaction and deluge valves subject to freezing shall be inspected daily during cold weather to verify a minimum temperature of 40°F (4.0°C). **13.4.3.1.2** Low temperature alarms, if installed in valve enclosures, shall be inspected annually at the beginning of the heating season to verify that they are free of physical damage. **13.4.4.1.1** Valve enclosures for preaction and deluge valves subject to freezing shall be inspected daily during cold weather to verify a minimum temperature of 40°F (4°C). **13.4.4.1.1.2** Low temperature alarms, if installed in valve enclosures, shall be inspected annually at the beginning of the heating season to verify that they are free of physical damage.	These revisions have been made to address that there is no way for the inspector to know whether the heating equipment is working. If the inspector goes into the enclosure on a day where the temperature is over 40 degrees, there is no way to determine if the heating equipment is operational. There is no way to simulate a cold condition to see if the heating comes on. The building owner is already required under 4.1.1.1 to make sure that adequate heat is provided in areas with water-filled piping. This is a more appropriate way to address this issue as an ongoing maintenance requirement rather than a periodic inspection.
13.4.4.2.9 Dry pipe systems shall be tested once every 3 years for gas leakage, using one of the following test methods: (1) A gas (air or nitrogen) pressure test at 40 psi (3.2 bar) shall be performed for 2 hours. (a) The system shall be permitted to lose up to 3 psi (0.2 bar) during the duration of the test. (b) Gas leaks shall be addressed if the system loses more than 3 psi (0.2 bar) during this test. (2) With the system at normal system pressure, the gas source (nitrogen supply, compressor, or shop air) shall be shut off for 4 hours. If the low pressure alarm goes off within this period, the leaks shall be addressed.	Nitrogen is recognized as a legitimate gas to use in dry pipe systems and is gaining popularity due to its ability to prevent corrosion within the piping. These revisions have been made to clarify that, where nitrogen is used, the system integrity needs to be maintained, just as with air.
13.5.2 Hose Connection Pressure-Regulating Devices. **13.5.2.1** All devices shall be inspected annually to verify the following: (1) The handwheel is not broken or missing. (2) The outlet hose threads are not damaged. (3) No leaks are present. (4) The hose adapter and the cap are not missing. **13.5.2.2*** A full flow test shall be conducted on each device at 5-year intervals and shall be compared to previous test results. **13.5.2.2.1** Adjustments shall be made in accordance with the manufacturer's instructions. **13.5.2.3** A partial flow test adequate to move the device from its seat shall be conducted annually.	NFPA 14 requires a permanently installed drain riser to be provided adjacent to each standpipe equipped with pressure-regulating devices to facilitate tests of each device. The drain riser is required to be sized large enough to handle the full flow required from the largest pressure-regulating device.
13.5.7 Fire Pump Relief Valves. **13.5.7.1 Circulation Relief Valves.** **13.5.7.1.1** Where installed, circulation relief valves shall be inspected on the same frequency as the no-flow (churn) test. **13.5.7.1.2** The inspection shall verify that water flows through the valve when the fire pump is operating at shutoff pressure (i.e., churn) to prevent the pump from overheating.	Section titles have been added to clarify which specific relief valves are being described.

2014 Text	*Notes*
13.5.7.1.3 On completion of any fire pump test, the closure of the circulation relief valve shall be verified. **13.5.7.2 Main Pressure Relief Valve.** **13.5.7.2.1** Where installed, main pressure relief valves shall be inspected during any fire pump test. **13.5.7.2.2** The inspection shall verify that the pressure downstream of the relief valve fittings in the fire pump discharge piping does not exceed the pressure for which the system components are rated.	
13.6 Backflow Prevention Assemblies. **13.6.1** Inspection. Inspection of backflow prevention assemblies shall be as described in 13.6.1.1 through 13.6.1.4. **13.6.1.1** The isolation valves on double check assemblies (DCA) and double check detector assemblies (DCDA) shall be inspected weekly to ensure that the valves are in the normal open position. **13.6.1.1.1** Valves secured with locks or electrically supervised in accordance with applicable NFPA standards shall be inspected monthly. **13.6.1.2** The isolation valves on reduced-pressure assemblies (RPA) and reduced-pressure detector assemblies (RPDA) shall be inspected weekly to ensure that the valves are in the normal open position. **13.6.1.2.1** Valves secured with locks or electrically supervised in accordance with applicable NFPA standards shall be inspected monthly. **13.6.1.2.2*** RPAs and RPDAs shall be inspected weekly to ensure that the differential-sensing valve relief port is not continuously discharging. **13.6.1.3** After any testing or repair, an inspection by the property owner or designated representative shall be made to ensure that the system is in service and all isolation valves are in the normal open position and properly locked or electrically supervised.	The original intent of NFPA 25 was to match the inspection rules for backflow devices with the inspection rules for control valves (since there are two control valves as a part of each backflow assembly). But the rules have never quite matched up. Although the inspection of the valves is okay, the inspection of the relief port is required to be weekly, regardless of the supervision on the valve.
13.6.2.1* All backflow preventers installed in fire protection system piping shall be exercised annually by conducting a forward flow test at a minimum flow rate of the system demand. **13.6.2.1.1** Where water rationing is enforced during shortages lasting more than 1 year, an internal inspection of the backflow preventer to ensure the check valves will fully open shall be permitted in lieu of conducting the annual forward flow test. **13.6.2.1.2** The forward flow test shall not be required where annual fire pump testing causes the system flow rate to flow through the backflow preventer device. **13.6.2.2** Where hydrants or inside hose stations are located downstream of the backflow preventer, the forward flow test shall include hose stream demand.	The changes to these requirements were needed to better reflect that the backflow preventer is not a precise test whereby the flow through it must be measured but effort to exercise the device at flows as near as possible to the system demand.
13.6.3 Maintenance. Maintenance of all backflow prevention assemblies shall be conducted by a qualified individual following the manufacturer's instructions in accordance with the procedure and policies of the authority having jurisdiction.	The term *qualified* has been defined by the standard and is the appropriate term for use in this section.

2014 Text	*Notes*
13.7.1 Fire department connections shall be inspected quarterly to verify the following: (1) The fire department connections are visible and accessible. (2) Couplings or swivels are not damaged and rotate smoothly. (3) Plugs or caps are in place and undamaged. (4) Gaskets are in place and in good condition. (5) Identification signs are in place. (6) The check valve is not leaking. (7) The automatic drain valve is in place and operating properly. (8) The fire department connection clapper(s) is in place and operating properly. (9)* Interior of the connection is inspected for obstructions. **A.13.7.1(9)** It is not the intent of this section for all fire department connection piping to be inspected for obstructions but only the interior of the connection itself.	A new item (9) has been added to this requirement, which makes the inspector remove the cap (especially locking) to ensure it can be removed and has not been damaged or oxidized to the connection. It also addresses the issue that no one is sure when the cap was placed on the FDC. It could have been off for weeks and placed on before the inspection. The inspector would then never perform 13.7.2 because a cap was in place. The revision requires the interior to be inspected quarterly to see if debris has been introduced into the connection, thus making 13.7.2 no longer needed.
13.7.4 The piping from the fire department connection to the fire department check valve shall be hydrostatically tested at 150 psi (10 bar) for 2 hours at least once every 5 years.	The piping from the fire department connection to the fire protection system is dry most of the time and subject to corrosion due to the moist atmosphere. Failures of this piping have occurred when fire departments pump into the connections. Revisions have been made to select the 150 psi pressure, since this is the pressure most frequently used in the standard operating procedure of fire departments when supporting fire protection systems.
14.2 Assessment of Internal Condition of Piping. **14.2.1*** An assessment of the internal condition of piping shall be conducted on a frequency determined by 14.2.1.1 or 14.2.1.2 for the purpose of inspecting for the presence of foreign organic and inorganic material. **14.2.1.1** An assessment of the internal condition of piping shall be conducted at a minimum of every 5 years or in accordance with 14.2.1.2 for the purpose of inspecting for the presence of foreign organic and inorganic material. **14.2.1.2*** Where an assessment frequency has been established by an approved risk analysis, the assessment shall be performed at a frequency determined by the approved risk analysis. **A.14.2.1.2** The purpose of the assessment of the internal condition of piping is to identify whether foreign organic and inorganic material is present, since the presence of such material can potentially be detrimental to the performance of the sprinkler system. Corrosion, which includes microbiologically influenced corrosion (MIC), can be prevalent in sprinkler piping, but not necessarily in all systems, buildings, or geographic locations. Therefore, the risk analysis should be based upon evidence noted and/or data obtained from proximate sprinkler systems, known problems from the water supply, as well as other factors that could affect the integrity of the system piping. The decision of when, how often, and which method to use to conduct an assessment should also consider the risk of emptying and refilling the sprinkler system with water, which could adversely impact the sprinkler system by introducing foreign or inorganic material into the sprinkler system.	This section has been changed and reorganized to allow for options when trying to determine the integrity of the fire protection system piping. The title has been changed to more accurately describe that an assessment is now being required, and not just an internal pipe inspection. That assessment could include an internal pipe inspection, but it could also include alternative methods for checking for the presence of obstructing material that could affect the performance of the system in a fire event, as well as for corrosion that could affect the life span of the system. The frequency of the assessment can now be determined through a risk analysis, which could determine that a longer frequency is appropriate, or that a shorter frequency is necessary. However, if an approved risk analysis is not performed, the assessment must be performed every 5 years, with exceptions as provided.

2014 Text	Notes
14.2.1.3 Tubercules or slime, if found, shall be tested for indications of microbiologically influenced corrosion (MIC). **14.2.1.4*** If the presence of sufficient foreign organic or inorganic material is found to obstruct pipe or sprinklers, an obstruction investigation shall be conducted as described in Section 14.3. **14.2.1.5** Nonmetallic pipe shall not be required to comply with Section 14.2. **14.2.2*** In buildings having multiple wet pipe systems, every other system shall have an assessment of the internal condition of piping as described in 14.2.1. **14.2.2.1** During the next inspection frequency required by 14.2.1.1 or 14.2.1.2, the alternate systems not assessed during the previous assessment shall be assessed as described in 14.2.1. **14.2.2.2** If foreign organic and/or inorganic material is found in any system in a building, all systems shall be assessed. **A.14.2.1** It is the intent of this requirement to provide a reasonable assurance that corrosion and obstruction issues within fire protection systems are identified. It is not the intent to require verification that every piece of pipe in the system is free from corrosion and obstructions. An assessment of the internal condition of piping can be accomplished by several methods that meet the intent of this section. These methods include the following: (1) Opening a flushing connection at the end of one main, and removing the end fitting or piece of branch line or a sprinkler for the purpose of inspecting for the presence of foreign organic and inorganic material. (a) In dry pipe systems and preaction systems, the branch line inspected should be the most remote one from the source of water that is not equipped with the inspector's test valve. (b) When performing normal maintenance that involves draining down a system to modify a system such as for tenant fit out or building renovations, or when removing or replacing piping, this inspection can be performed as described and properly recorded at that time. The time interval would then start for the next assessment of that system at the frequency determined by 14.2.1.1 or 14.2.1.2. (c) If a sprinkler is removed to perform this inspection, 5.4.1.1 requires a new sprinkler matching the characteristics of the replaced sprinkler. (2) Utilizing alternative examination methods such as the following: (a) Using video inspection equipment that is inserted into the system at strategic points to observe the internal condition of pipes. This equipment provides a visual exam of the pipes using a camera and lighting system on the end of a push cable. Video inspection equipment can be inserted in alarm, dry, and preaction valves for a look into risers, feed mains, some cross mains, and some branch lines, depending on the system configuration. The push cable can also be inserted in a check valve when performing the five-year internal inspection required by 13.4.2.1 to view additional areas of a system, and in the fire department connection to perform the interior inspection required by 13.7.2. (b) Ultrasonic or similar technology that allows the pipe wall to be tested to determine the extent of any deterioration due to microbiologically influenced corrosion (MIC) or other forms of corrosion. This method would not typically be used for the internal inspection of piping required by this section because it might not detect the presence of solid material in the piping, such as wood, plastic, or other foreign obstructions, that are not a by-product of corrosion, because only small representative sections of pipe are examined.	

2014 Text	*Notes*
(c) A laboratory analysis of water samples obtained from the fire protection system, combined with collecting and inspecting solid material from fire protection system water discharged from a main drain, and an inspector's test connection, can provide an indication of the presence of corrosion, MIC, and/or foreign materials. If a high level of MIC is identified, or if a significant amount of foreign materials is found, further investigation might be warranted to verify the extent of corrosion, MIC, or other obstructions in the system. The solid materials should be collected with an appropriately sized strainer. If inspection of the solid materials identifies excessive rust, black water color, or sulfur (rotten egg) odors, an obstruction investigation as described in Section 14.3 is warranted.	
14.3.2.2* Internal examination shall be performed at the following minimum four points: (1) System valve (2) Riser (3) Cross main (4) Branch line	The word *minimum* has been added to allow for the examination of as many points in the system as is warranted, based on the type of obstructing material found and the configuration of the system. The annex text explains that examining at additional points might be necessary to determine the extent of the obstruction and the corrective action needed.
15.4.2 The impaired equipment shall include, but shall not be limited to, the following: (1) Sprinkler systems (2) Standpipe systems (3) Fire hose systems (4) Underground fire service mains (5) Fire pumps (6) Water storage tanks (7) Water spray fixed systems (8) Foam-water systems (9) Water mist systems (10) Fire service control valves	Water mist systems have been added to the list of potentially impaired systems.
15.6* Emergency Impairments. **A.15.6** Emergency impairments include, but are not limited to, system leakage, interruption of water supply, frozen or ruptured piping, equipment failure, or other impairments found during inspection, testing, or maintenance activities. **15.6.1** Emergency impairments shall include, but are not limited to, interruption of water supply, frozen or ruptured piping, and equipment failure, and includes impairments found during inspection, testing, or maintenance activities. **15.6.2*** The coordinator shall implement the steps outlined in Section 15.5. **A.15.6.2** When one or more impairments are discovered during inspection, testing, and maintenance activities, the owner or owner's authorized representative should be notified in writing. See Figure A.15.6.2 for an example of written notification.	Most impairments are discovered while performing inspection, testing, and/or maintenance on the system, and yet this standard had not clearly stated that this condition is considered an emergency impairment once it is discovered. This has been clarified in this edition.

2014 Text	*Notes*
15.7* Restoring Systems to Service. When all impaired equipment is restored to normal working order, the impairment coordinator shall verify that the following procedures have been implemented: (1) Any necessary inspections and tests have been conducted to verify that affected systems are operational. The appropriate chapter of this standard shall be consulted for guidance on the type of inspection and test required. (2) Supervisors have been advised that protection is restored. (3) The fire department has been advised that protection is restored. (4) The property owner or designated representative, insurance carrier, alarm company, and other authorities having jurisdiction have been advised that protection is restored. (5) The impairment tag has been removed. **A.15.7** Occasionally, fire protection systems in idle or vacant buildings are shut off and drained. When the equipment is eventually restored to service after a long period of not being maintained, it is recommended that qualified personnel or a qualified contractor perform the work. The following is an example of a procedure: (1) All piping should be traced from the extremities of the system to the main connections with a careful check for blank gaskets in flanges, closed valves, corroded or damaged sprinklers, nozzles, or piping, insecure or missing hangers, and insufficient support. Proper repairs or adjustments should be made, and needed extensions or alterations for the equipment should be completed. (2) An air test at low pressure (40 psi) should be conducted prior to allowing water to fill the system. When the piping has been proven tight by passing the air test, water can be introduced slowly into the system, with proper precautions against damage by escape of water from previously undiscovered defects. When the system has been filled under normal service pressure, drain valve tests should be made to detect any closed valve that possibly could have been overlooked. All available pipes should be flushed, and an obstruction investigation completed to make sure that the system is clear of debris. (3) Where the system was known to have been damaged by freezing or where other extensive damage might have occurred, a full hydrostatic test can be performed in accordance with NFPA 13 to determine whether the system integrity has been maintained. (4) Dry pipe valves, quick-opening devices, alarm valves, and all alarm connections should be examined, put in proper condition, and tested. (5) Fire pumps, pressure and gravity tanks, reservoirs, and other water supply equipment should receive proper attention before being placed in service. Each supply should be tested separately and then together if they are designed to work together. (6) All control valves should be operated from the closed to fully open position and should be left sealed, locked, or equipped with a tamper switch.	The annex language has been revised to clarify that there is more than one procedure. The terminology has also been updated to reflect that the commonly used term is *non-use*, not *disuse*. Also, since the standard already has a definition for a qualified person, *knowledgeable* and *responsible* would not provide any further direction for the type of person doing this work, so they have been removed.
Chapter 16 — Special Requirements from Other NFPA Documents	This new chapter has been added to address NFPA *101*® references to ITM work required for NFPA 13D systems installed in small board and care facilities.

INDEX

IMPORTANT NOTICES AND DISCLAIMERS CONCERNING NFPA® STANDARDS

NOTICE AND DISCLAIMER OF LIABILITY CONCERNING THE USE OF NFPA STANDARDS

NFPA® codes, standards, recommended practices, and guides ("NFPA Standards"), of which the document contained herein is one, are developed through a consensus standards development process approved by the American National Standards Institute. This process brings together volunteers representing varied viewpoints and interests to achieve consensus on fire and other safety issues. While the NFPA administers the process and establishes rules to promote fairness in the development of consensus, it does not independently test, evaluate, or verify the accuracy of any information or the soundness of any judgments contained in NFPA Standards.

The NFPA disclaims liability for any personal injury, property or other damages of any nature whatsoever, whether special, indirect, consequential or compensatory, directly or indirectly resulting from the publication, use of, or reliance on NFPA Standards. The NFPA also makes no guaranty or warranty as to the accuracy or completeness of any information published herein.

In issuing and making NFPA Standards available, the NFPA is not undertaking to render professional or other services for or on behalf of any person or entity. Nor is the NFPA undertaking to perform any duty owed by any person or entity to someone else. Anyone using this document should rely on his or her own independent judgment or, as appropriate, seek the advice of a competent professional in determining the exercise of reasonable care in any given circumstances.

The NFPA has no power, nor does it undertake, to police or enforce compliance with the contents of NFPA Standards. Nor does the NFPA list, certify, test, or inspect products, designs, or installations for compliance with this document. Any certification or other statement of compliance with the requirements of this document shall not be attributable to the NFPA and is solely the responsibility of the certifier or maker of the statement.

ADDITIONAL NOTICES AND DISCLAIMERS

Updating of NFPA Standards

Users of NFPA codes, standards, recommended practices, and guides ("NFPA Standards") should be aware that these documents may be superseded at any time by the issuance of new editions or may be amended from time to time through the issuance of Tentative Interim Amendments or corrected by Errata. An official NFPA Standard at any point in time consists of the current edition of the document together with any Tentative Interim Amendments and any Errata then in effect. In order to determine whether a given document is the current edition and whether it has been amended through the issuance of Tentative Interim Amendments or corrected through the issuance of Errata, consult appropriate NFPA publications such as the National Fire Codes® Subscription Service, visit the NFPA website at www.nfpa.org, or contact the NFPA at the address listed below.

Interpretations of NFPA Standards

A statement, written or oral, that is not processed in accordance with Section 6 of the Regulations Governing the Development of NFPA Standards shall not be considered the official position of NFPA or any of its Committees and shall not be considered to be, nor be relied upon as, a Formal Interpretation.

Patents

The NFPA does not take any position with respect to the validity of any patent rights referenced in, related to, or asserted in connection with an NFPA Standard. The users of NFPA Standards bear the sole responsibility for determining the validity of any such patent rights, as well as the risk of infringement of such rights, and the NFPA disclaims liability for the infringement of any patent resulting from the use of or reliance on NFPA Standards.

NFPA adheres to the policy of the American National Standards Institute (ANSI) regarding the inclusion of patents in American National Standards ("the ANSI Patent Policy"), and hereby gives the following notice pursuant to that policy:

> **NOTICE:** The user's attention is called to the possibility that compliance with an NFPA Standard may require use of an invention covered by patent rights. NFPA takes no position as to the validity of any such patent rights or as to whether such patent rights constitute or include essential patent claims under the ANSI Patent Policy. If, in connection with the ANSI Patent Policy, a patent holder has filed a statement of willingness to grant licenses under these rights on reasonable and nondiscriminatory terms and conditions to applicants desiring to obtain such a license, copies of such filed statements can be obtained, on request, from NFPA. For further information, contact the NFPA at the address listed below.

Law and Regulations

Users of NFPA Standards should consult applicable federal, state, and local laws and regulations. NFPA does not, by the publication of its codes, standards, recommended practices, and guides, intend to urge action that is not in compliance with applicable laws, and these documents may not be construed as doing so.

Copyrights

NFPA Standards are copyrighted. They are made available for a wide variety of both public and private uses. These include both use, by reference, in laws and regulations, and use in private self-regulation, standardization, and the promotion of safe practices and methods. By making these documents available for use and adoption by public authorities and private users, the NFPA does not waive any rights in copyright to these documents.

Use of NFPA Standards for regulatory purposes should be accomplished through adoption by reference. The term "adoption by reference" means the citing of title, edition, and publishing information only. Any deletions, additions, and changes desired by the adopting authority should be noted separately in the adopting instrument. In order to assist NFPA in following the uses made of its documents, adopting authorities are requested to notify the NFPA (Attention: Secretary, Standards Council) in writing of such use. For technical assistance and questions concerning adoption of NFPA Standards, contact NFPA at the address below.

For Further Information

All questions or other communications relating to NFPA Standards and all requests for information on NFPA procedures governing its codes and standards development process, including information on the procedures for requesting Formal Interpretations, for proposing Tentative Interim Amendments, and for proposing revisions to NFPA standards during regular revision cycles, should be sent to NFPA headquarters, addressed to the attention of the Secretary, Standards Council, NFPA, 1 Batterymarch Park, P.O. Box 9101, Quincy, MA 02269-9101; email: stds_admin@nfpa.org

For more information about NFPA, visit the NFPA website at www.nfpa.org. All NFPA codes and standards can be viewed at no cost at www.nfpa.org/docinfo.

FEATURES IN THIS EDITION

This fourth edition of the *Water-Based Fire Protection Systems Handbook* contains the complete text of the 2013 edition of NFPA® 25, *Standard for the Inspection, Testing, and Maintenance of Water-Based Fire Protection Systems.* The full text of Chapter 1 through Chapter 16 and Annexes A through F are provided in this handbook. Additional expert commentary is provided to explain the reasoning behind the document's recommendations and clarify the committee's intent.

Part 1 / Chapter 5: Sprinkler Systems 157

Case In Point

After a series of fires where the presence of antifreeze in the system negatively contributed to the fire, the NFPA sprinkler committees responsible for NFPA 13 and NFPA 25 prepared a series of Tentative Interim Amendments (TIAs) to the standards to reflect the most up-to-date information on antifreeze.

This included a series of TIAs that were approved by the NFPA Standards Council following the printing of the 2011 edition of NFPA 25. These TIAs were written to the past edition of the standard (2011 edition), but the content provided in them was also picked up by the NFPA 25 technical committee in the 2014 revision cycle.

The revisions to the testing section for antifreeze systems require that a sample of the antifreeze solution be tested every year before the onset of freezing weather. The first photo shows an antifreeze solution test outlet and a test sample being collected and analyzed. Another photo shows an antifreeze loop with a test outlet and an expansion tank. With the use of a refractometer, such as the one shown in the final photo, the percentage of water and antifreeze in a given solution can be determined. The importance of this test cannot be understated, because it provides the owner and inspector with the concentration of the solution in the system. Certain concentrations of antifreeze have been proven in full-scale testing to significantly increase the heat release rate of a fire.

Antifreeze Loop. (Courtesy of Josh Elvove)

Antifreeze Testing. Obtaining an Antifreeze Sample by Draining a Small Quantity of Antifreeze from the Antifreeze System (top). Applying the Antifreeze Test Sample to a Refractometer (middle). Reading the Refractometric Index of the Antifreeze Sample by Holding the Refractometer Up to a Natural Light Source (bottom).

Digital Refractometer. (Courtesy of MISCO Refractometer, Cleveland, OH)

Water-Based Fire Protection Systems Handbook 2014

Several Case In Point boxes are included throughout the handbook to expand further on material in the commentary.

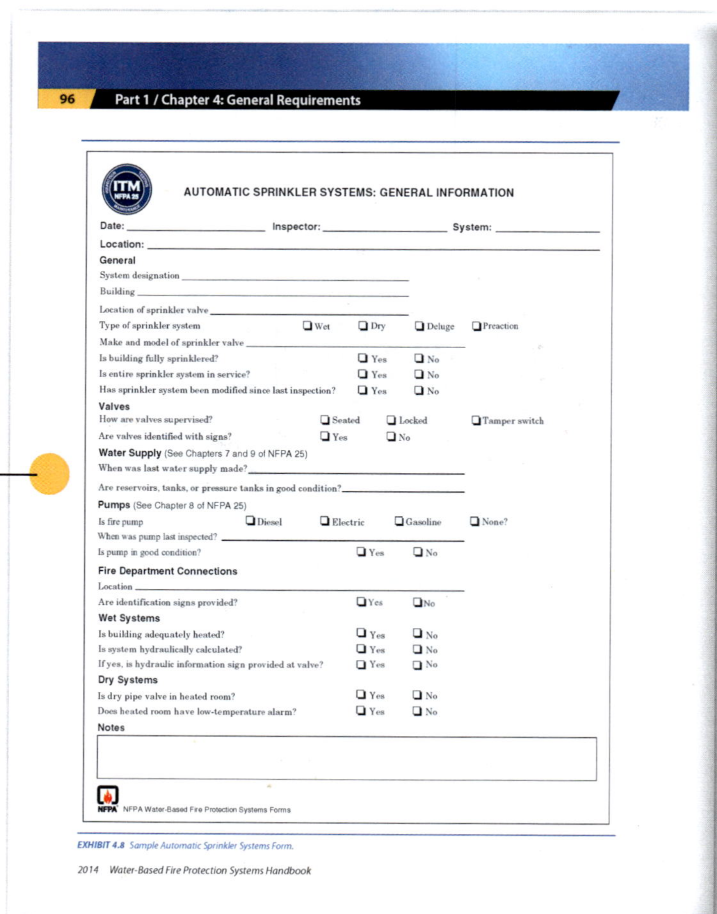

96 Part 1 / Chapter 4: General Requirements

AUTOMATIC SPRINKLER SYSTEMS: GENERAL INFORMATION

Date: ______ Inspector: ______ System: ______

Location: ______

General

System designation ______

Building ______

Location of sprinkler valve ______

Type of sprinkler system ❑ Wet ❑ Dry ❑ Deluge ❑ Preaction

Make and model of sprinkler valve ______

Is building fully sprinklered? ❑ Yes ❑ No

Is entire sprinkler system in service? ❑ Yes ❑ No

Has sprinkler system been modified since last inspection? ❑ Yes ❑ No

Valves

How are valves supervised? ❑ Seated ❑ Locked ❑ Tamper switch

Are valves identified with signs? ❑ Yes ❑ No

Water Supply (See Chapters 7 and 9 of NFPA 25)

When was last water supply made? ______

Are reservoirs, tanks, or pressure tanks in good condition? ______

Pumps (See Chapter 8 of NFPA 25)

Is fire pump ❑ Diesel ❑ Electric ❑ Gasoline ❑ None?

When was pump last inspected? ______

Is pump in good condition? ❑ Yes ❑ No

Fire Department Connections

Location ______

Are identification signs provided? ❑ Yes ❑ No

Wet Systems

Is building adequately heated? ❑ Yes ❑ No

Is system hydraulically calculated? ❑ Yes ❑ No

If yes, is hydraulic information sign provided at valve? ❑ Yes ❑ No

Dry Systems

Is dry pipe valve in heated room? ❑ Yes ❑ No

Does heated room have low-temperature alarm? ❑ Yes ❑ No

Notes

NFPA NFPA Water-Based Fire Protection Systems Forms

EXHIBIT 4.8 Sample Automatic Sprinkler Systems Form.

2014 Water-Based Fire Protection Systems Handbook

Several sample forms are included to show the reader examples of possible ways to record system information or to document an ITM task that has been completed.